Lecture Notes in Computer Science 8687

Preface

The International Colloquium on Theoretical Aspects of Computing (ICTAC) was founded in 2004. In 2014, the 11th edition of ICTAC was organized for the first time in Europe, namely in Bucharest, Romania. Since its early days, the study of computer science in Romania has had a strong theoretical component, owing in part to the mathematical orientation of some of its most notable pioneers, such as Professor Grigore Moisil. In modern times, this legacy is proudly carried on by several prestigious research institutions, such as the University of Bucharest and the Romanian Academy.

A burgeoning metropolis in recent years, Bucharest is a lively, cosmopolitan city. Featuring an eclectic mix of architectural styles and traditions, cultural and artistic life, Bucharest was an ideal setting for the 11th International Colloquium on Theoretical Aspects of Computing. The Colloquium aims to bring together practitioners and researchers from academia, industry, and government, with the purpose of encouraging the presentation of novel research directions, through the exchange of both ideas and experience, related to current theoretical challenges in computing, as well as practical applications of existing theoretical results. An additional goal is that of promoting wide-reaching forms of cooperation in research and education between participants and their institutions, from developing and industrial countries.

We were honored to have three distinguished guests as invited speakers: Cristian Calude (University of Auckland, New Zealand), Jin-Song Dong (National University of Singapore, Singapore), and Razvan Diaconescu (Institute of Mathematics of the Romanian Academy, Bucharest). Professor Solomon Marcus was a special guest, providing a talk on important differences and similarities between Theoretical Computer Science project and the Bourbaki project in Mathematics.

Professor Cristian Calude presented a talk entitled "Probabilistic solutions to undecidable problems" which addressed the (in)famous algorithmically undecidable halting problem. Undecidability is everywhere: in logic, mathematics, computer science, engineering, physics, ...; the talk revealed how a probabilistic method can help deal with undecidable problems.

Professor Jin-Son Dong presented a talk entitled "Event Analytic" in which he described the Process Analysis Toolkit and his vision of event analytics, which goes beyond data analytics. The event analytics research is based on applying model checking to event planning, scheduling, prediction, strategy analysis, and decision making.

Professor Razvan Diaconescu presented a talk entitled "From Universal Logic to Computer Science, and Back," focusing on universal logic. He recalled some important ideas that have shaped the success of institutional model theory in computer science. Moreover, he showed how insights from computer science have led (through institutional model theory) to a (sometimes drastic) reformulation and novel understanding of several important concepts in logic, resulting in a set of new and important results.

ICTAC 2014 welcomed submissions pertaining (but not limited) to the following areas:

- Automata theory and formal languages;
- Principles and semantics of programing languages;
- Theories of concurrency, mobility and reconfiguration;
- Logics and their applications;
- Software architectures and their models, refinement, and verification;
- Relationship between software requirements, models, and code;
- Static and dynamic program analysis and verification;
- Software specification, refinement, verification, and testing;
- Model checking and theorem proving;
- Models of object and component systems;
- Coordination and feature interaction;
- Integration of theories, formal methods and tools for engineering computing systems;
- Service-oriented architectures: models and development methods;
- Models of concurrency, security, and mobility;
- Theories of distributed, grid and cloud computing;
- Real-time, embedded, hybrid and cyber-physical systems;
- Type and category theory in computer science;
- Models for e-learning and education;
- Case studies, theories, tools, and experiments of verified systems;
- Domain-specific modeling and technology: examples, frameworks and practical experience.
- Challenges and foundations in environmental modeling and monitoring, healthcare, and disaster management.

We received submissions from 36 countries around the world; initially we got 93 abstracts, followed by 74 full submissions. The selection process was rigorous, each paper receiving at least four reviews. We got external reviews for papers that lacked expertise within the Program Committee. The Program Committee, after long and careful discussions on the submitted papers, decided to accept only 25 papers. This corresponds to an overall acceptance rate of approximately 33.7%. The accepted papers cover several aspects of theoretical aspects of computing. They provided a scientifically exciting program for the ICTAC meeting, triggering interesting discussions and exchanges of ideas among the participants.

We would like to thank all the authors who submitted their work to ICTAC 2014. We are grateful to the Program Committee members and external reviewers for their high-quality reviews and discussions. Finally, we wish to thank the Steering Committee and Organizing Committee members for their excellent support.

July 2014

Gabriel Ciobanu
Dominique Méry

Colloquium Organization

We are grateful to the people who contributed to this conference.

General Chairs

Gabriel Ciobanu	Romanian Academy, Romania
Florentin Ipate	University of Bucharest, Romania

PC Chairs

Gabriel Ciobanu	Romanian Academy, Romania
Dominique Méry	LORIA, Université de Lorraine, France

Steering Committee

Ana Cavalcanti	University of York, UK
John Fitzgerald	Newcastle University, UK
Martin Leucker	University of Luebeck, Germany
Zhiming Liu	UNU-IIST, China
Tobias Nipkow	Technical University Munich, Germany
Augusto Sampaio	Federal University of Pernambuco, Brazil
Natarajan Shankar	SRI International, USA

Organizing Committee

Radu Gramatovici	University of Bucharest, Romania (Co-chair)
Alin Stefanescu	University of Bucharest, Romania (Co-chair)
Camil-Alexandru Parvu	Alumni Association, University of Bucharest, Romania

Program Committee

Yamine Ait Ameur	IRIT/ENSEIHT, France
Farhad Arbab	CWI and Leiden University, The Netherlands
Michael Butler	University of Southampton, UK
Ana Cavalcanti	University of York, UK
Jérémie Chalopin	LIF, CNRS, Aix Marseille Université, France
Zhenbang Chen	National University of Defense Technology, China

Gabriel Ciobanu	Romanian Academy, Romania
Maximiliano Cristia	CIFASIS and UNR, Argentina
Rocco De nNicola	IMT Lucca, Italy
David Deharbe	University Federal do Rio Grande do Norte, Brasil
José Luiz Fiadeiro	Royal Holloway University of London, UK
John Fitzgerald	Newcastle University, UK
Simon Gay	University of Glasgow, UK
Dimitra Giannakopoulou	NASA Ames Research Center, USA
Stefania Gnesi	ISTI-CNR, Italy
Lindsay Groves	Victoria University of Wellington, New Zealand
Kim Guldstrand Larsen	Aalborg University, Denmark
Geoff Hamilton	Dublin City University, Ireland
Ian J. Hayes	University of Queensland, Australia
Rob Hierons	Brunel University, UK
Jane Hillston	University of Edinburgh, UK
Ross Horne	Romanian Academy, Romania
Zhenjiang Hu	National Institute of Informatics, Japan
Florentin Ipate	University of Bucharest, Romania
Tudor Jebelean	RISC, Johannes Kepler University, Austria
Cliff Jones	Newcastle University, UK
Lila Kari	University of Western Ontario, Canada
Jetty Kleijn	Leiden University, The Netherlands
Maciej Koutny	Newcastle University, UK
Yassine Lakhnech	VERIMAG, France
Insup Lee	University of Pennsylvania, USA
Axel Legay	IRISA/Inria, France
Martin Leucker	University of Lübeck, Germany
Zhiming Liu	Birmingham City University, UK
Dominique Mery	Université de Lorraine, LORIA, France
Marius Minea	Politehnica University of Timisoara, Romania
Victor Mitrana	Polytechnic University of Madrid, Spain
Rosemary Monahan	NUI Maynooth, Ireland
Mohamed Mosbah	LaBRI - University of Bordeaux, France
Tobias Nipkow	Technische Universität München, Germany
Manuel Nuñez	Universidad Complutense de Madrid, Spain
Paritosh Pandya	Tata Institute of Fundamental Research, India
Jun Pang	University of Luxembourg, Luxembourg
Corina Pasareanu	CMU/NASA Ames Research Center, USA
Shengchao Qin	Teesside University, UK
António Ravara	Universidade Nova de Lisboa, Portugal
Grigore Rosu	University of Illinois at Urbana-Champaign, USA

Augusto Sampaio	Federal University of Pernambuco, Brasil
Pierre-Yves Schobbens	University of Namur, Belgium
Emil Sekerinski	McMaster University, Canada
Natarajan Shankar	SRI International, USA
Viorica Sofronie-Stokkermans	University Koblenz-Landau, Germany
Gheorghe Stefanescu	University of Bucharest, Romania
Andrzej Tarlecki	Warsaw University, Poland
Elena Troubitsyna	Abo Akademi, Finland
Emilio Tuosto	University of Leicester, UK
Martin Wirsing	Ludwig-Maximilians-University, Germany
Burkhart Wolff	LRI, University Paris-Sud, France
Jim Woodcock	University of York, UK
Fatiha Zaidi	LRI, University Paris-Sud, France
Naijun Zhan	Chinese Academy of Sciences, China
Jianjun Zhao	Shanghai Jiao Tong University, China
Huibiao Zhu	East China Normal University, China

Additional Reviewers

Abdellatif, Takoua
Aman, Bogdan
Bacci, Giovanni
Bandur, Victor
Belzner, Lenz
Biondi, Fabrizio
Bodeveix, Jean-Paul
Boudjadar, A. Jalil
Bracciali, Andrea
Bucur, Stefan
Busch, Marianne
Carvalho, Gustavo
Ciancia, Vincenzo
Craciun, Florin
Decker, Normann
Edmunds, Andrew
Fahrenberg, Uli
Fratani, Séverine
Gehrke, Mai
Gherghina, Cristian
Griffin, Lewis
Grigorieff, Serge
Hainry, Emmanuel
Hennicker, Rolf
Horbach, Matthias
Huang, Yanhong
Ivanov, Radoslav
Karamichalis, Rallis
Knapp, Alexander
Kopecki, Steffen
Krishna, S.N.
Laibinis, Linas
Li, Qin
Lluch Lafuente, Alberto
Loreti, Michele
Maletti, Andreas
Mardare, Radu
Melissen, Matthijs
Meng, Wenrui
Miyazawa, Alvaro
Mizera, Andrzej
Monteiro, Luis
Moore, Brandon
Paolieri, Marco
Park, Junkil
Pattinson, Dirk
Pereverzeva, Inna
Peters, Kirstin

Piazza, Carla
Popescu, Marius
Poulsen, Danny Bøgsted
Prokhorova, Yuliya
Pérez, Jorge A.
Qamar, Nafees
Reynier, Pierre-Alain
Rivas, Exequiel
Salehi Fathabadi, Asieh
Salomaa, Kai
Scateni, Riccardo
Scheffel, Torben
Shamoi, Pakita
Simjour, Amirhossein
Solin, Kim
Srba, Jiri
Stefanescu, Alin
Stefanescu, Andrei
Struth, Georg
Stümpel, Annette
Su, Wen
Sulzmann, Martin
Sun, Meng
Suzuki, Tomoyuki
Talbot, Jean-Marc
Thoma, Daniel
Tiezzi, Francesco
Tounsi, Mohamed
Tribastone, Mirco
Wilczyński, Bartek
Wilkinson, Toby
Xue, Bingtian
Yan, Dacong
Yang, Xiaoxiao
Yu, Nengkun
Zhang, Sai
Zhao, Yongxin
Zhu, Jiaqi
Zhu, Longfei
Zou, Liang

Sponsors

We are thankful for the support received from the University of Bucharest.

Extended Abstract

Probabilistic Solutions for Undecidable Problems

Cristian S. Calude

Department of Computer Science
University of Auckland
Private Bag 92019, Auckland, New Zealand
`www.cs.auckland.ac.nz/~cristian`

There is no algorithm which can be applied to any arbitrary program and input to decide whether the program stops when run with that input [8, 9]. This is the most (in)famous (algorithmically) undecidable problem: the *halting problem.*

Undecidability is everywhere: in logic, mathematics, computer science, engineering, physics, etc. For example, software testing is undecidable: we can never completely identify all the defects within a given code, hence we cannot establish that a code functions properly under *all* conditions.

Instead, various methods have been developed to establish that a code functions properly under some *specific* conditions. Can we estimate the probability that a code—which has been successfully tested under *specific* conditions—actually functions correctly under *all* conditions?

Many practical undecidable problems can be reduced to the halting problem, that is, they would be solved provided one could solve the halting problem. Hypercomputation (see [7]) explores various nature-inspired proposals—physical, biological—to solve the halting problem. Currently, there is no clear success, but also no impossibility proof.

Here we take a different approach: we try to solve approximately the halting problem and evaluate probabilistically the error.

In this talk we present and critically discuss two methods for approximately solving the halting problem.

1. A probabilistic "anytime algorithm"[1] based on the the fact that a program which is executed up to a certain computable runtime without stoping, cannot stop at any later *algorithmically incompressible* time (see [1, 4, 6, 3]).
2. A probabilistic method of approximating the (undecidable) set of programs which eventually stop (Halt) with a decidable set (see [5, 2]). To this aim we construct a decidable set R of probability $r \in (0, 1]$ such that Halt $\cap R$ is decidable. The closer r gets to 1, the better the approximation is.

Developing mathematical models along the ideas above is interesting, but not the whole story: showing that the models are adequate for different concrete problems (for example, by choosing the "right" probability) is essential and more challenging. Much more work has to be done in both directions.

---

[1] That is, an algorithm that can return a valid solution even if it is interrupted any time after a given time threshold, possibly before the computation ends.

References

1. Calude, C.S.: Information and Randomness. An Algorithmic Perspective, 2nd edn. Revised and Extended. Springer, Berlin (2002)
2. Calude, C.S., Desfontaines, D.: Universality and Almost Decidability, CDMTCS Research Report 462 (2014)
3. Calude, C.S., Desfontaines, D.: Anytime Algorithms for Non-Ending Computations. CDMTCS Research Report 463 (2014)
4. Calude, C.S., Stay, M.A.: Most programs stop quickly or never halt. Advances in Applied Mathematics 40, 295–308 (2008)
5. Hamkins, J.D., Miasnikov, A.: The halting problem is decidable on a set of asymptotic probability one. Notre Dame J. Formal Logic 47(4), 515–524 (2006)
6. Manin, Y.I.: Renormalisation and computation II: Time cut-off and the Halting Problem. Math. Struct. in Comp. Science 22, 729–751 (2012)
7. Ord, T.: Hypercomputation: computing more than the Turing machine, arXiv:math/020933 (2002)
8. Turing, A.: On computable numbers, with an application to the Entscheidungsproblem. Proceedings of the London Mathematical Society 42(2), 230–265 (1936)
9. Turing, A.: On computable numbers, with an application to the Entscheidungsproblem: A correction. Proceedings of the London Mathematical Society 2, 43, 544–546 (1937)

Table of Contents

From Universal Logic to Computer Science, and Back

Răzvan Diaconescu

Simion Stoilow Institute of Mathematics of the Romanian Academy, Romania
Razvan.Diaconescu@imar.ro

Abstract. Computer Science has been long viewed as a consumer of mathematics in general, and of logic in particular, with few and minor contributions back. In this article we are challenging this view with the case of the relationship between specification theory and the universal trend in logic.

1 From Universal Logic...

Although universal logic has been clearly recognised as a trend in mathematical logic since about one decade only, mainly due to the efforts of Jean-Yves Béziau and his colleagues, it had a presence here and there since much longer. For example the anthology [9] traces universal logic ideas back to the work of Paul Herz in 1922. In fact there is a whole string of famous names in logic that have been involved with universal logic in the last century, including Paul Bernays, Kurt Gödel, Alfred Tarski, Haskell Curry, Jerzy Łoś, Roman Suszko, Saul Kripke, Dana Scott, Dov Gabbay, etc.

Universal logic is not a new super-logic, but it is rather a body of general theories of logical structures, in the same way universal algebra is a general theory of algebraic structures (see [8] for a discussion about what universal logic is and is not). Within the last century mathematical logic has witnessed the birth of a multitude of unconventional logical systems, such as intuitionistic, modal, multiple valued, paraconsistent, non-monotonic logics, etc. Moreover, a big number of new logical systems have appeared in computer science, especially in the area of formal methods. The universal logic trend constitutes a response to this new multiplicity by the development of general concepts and methods applicable to a great variety of logical systems. One of the aims of universal logic is to determine the scope of important results (such as completeness, interpolation, etc.) and to produce general formulations and proofs of such results. This is very useful in the applications and helps with the distinction between what is and what is not really essential for a certain particular logical system, thus leading to a better understanding of that logical system. Universal logic may also be regarded as a toolkit for defining specific logics for specific situations; for example, temporal deontic paraconsistent logic. It also helps with the clarification of basic concepts by explaining what is an extension, or what is a deviation, of a given logic, what it means for a logic to be equivalent or translatable to another logic, etc.

G. Ciobanu and D. Méry (Eds.): ICTAC 2014, LNCS 8687, pp. 1–16, 2014.

Paramount researchers in mathematical logic consider universal logic as a true renaissance of the study of logic, that is based on very modern principles and methodologies and that responds to the new mathematical logic perspectives. The dynamism of this area, its clear identity, and its high potential have been materialized through a dedicated new book series (*Studies in Universal Logic*, Springer Basel), a dedicated new journal (*Logica Universalis*, Springer Basel), a dedicated corner of *Journal of Logic and Computation* (Oxford Univ. Press), and through a dedicated series of world congresses and schools (UNILOG: Switzerland 2005, China 2007, Portugal 2010, Brazil 2013; see `www.uni-log.org`).

The analogy between universal algebra and universal logic however fails in the area of the supporting mathematical structures. While the former is in fact a mathematical theory based upon a relatively small set of core mathematical definitions, this is not the case with the latter. There is not a single commonly accepted mathematical base for universal logic. Instead the universal trend in logic includes several theories each of them supported by adequate mathematical structures that share a non-substantialist view on logic phenomena, free of commitment to particular logical systems, and consequently a top-down development methodology. One of the most famous such theories is Tarski's general approach to logical consequence via closure operators [60]. And perhaps now the single most developed mathematical theory in universal logic is the *institution theory* of Goguen and Burstall [39,40].

2 ...to Computer Science,...

2.1 Origins of Institution Theory

Around 1980's there was already a population explosion of logical systems in use computer science, especially in the logic-based areas such as specification theory and practice. People felt that many of the theoretical developments (concepts, results, etc.), and even aspects of implementations, are in fact independent of the details of the actual logical systems. Especially in the area of structuring of specifications or programs, it would be possible to develop the things in a completely generic way. The benefit would be not only in uniformity, but also in clarity since for many aspects of specification theory the concrete details of actual logical systems may appear as irrelevant, with the only role being to suffocate the understanding. The first step to achieve this was to come up with a very general model oriented formal definition for the informal concept of logical system. The model theoretic orientation is dictated by formal specification in which semantics plays a primary role. Due to their generality, category theoretic concepts appeared as ideal tools. However there is something else which makes category theory so important for this aim: its deeply embedded non-substantialist thinking which gives prominence to the relationships (morphisms) between objects in the detriment of their internal structure. Moreover, category theory was at that time, and continues even now to be so, the mathematical field of the upmost importance for computer science. In fact, it was computer science that recovered the status of category theory, at the time much diminished in conventional mathematical areas. The article [37] that Joseph Goguen wrote remains one of the most relevant and beautiful essays on the significance of category theory for computer science and not only.

The categorical model theories existing at the time, although quite deep theoretically, were however unsatisfactory from the perspective of a universal logic approach to specification. Sketches of [35,44,62] had just developed another language for expressing (possibly infinitary) first-order logic realities. The satisfaction as cone injectivity [1,2,3,49,47,46], whilst considering models as objects of abstract categories, lacks a multi-signature aspect given by the signature morphisms and the model reducts, which leads to severe methodological limitations. Moreover in both these categorical model theory frameworks, the satisfaction of sentences by the models is defined rather than being axiomatized, which give them a strong taste of concreteness in contradiction with universal logic aims and ideals. On the other hand, the model theory trend known as 'abstract model theory' [4,5] had an axiomatic approach to the satisfaction relation, it also had a multi-signature aspect, but it was still only concerned with extensions of conventional logic in that the signatures and the models are concrete, hence it lacked a fully universal aspect.

2.2 The Concept of Institution

The definition of institution [14,40] can be seen as representing a full generalisation of 'abstract model theory' of [4,5] in a true universal logic spirit by also considering the signatures and models as abstract objects in categories.

Definition 1 (Institutions). *An* institution $\mathcal{I} = (Sig^{\mathcal{I}}, Sen^{\mathcal{I}}, Mod^{\mathcal{I}}, \models^{\mathcal{I}})$ *consists of*

1. *a category* $Sig^{\mathcal{I}}$*, whose objects are called* signatures,
2. *a functor* $Sen^{\mathcal{I}} : Sig^{\mathcal{I}} \to \mathbf{Set}$ *(to the category of sets), giving for each signature a set whose elements are called* sentences *over that signature,*
3. *a functor* $Mod^{\mathcal{I}} : (Sig^{\mathcal{I}})^{\mathrm{op}} \to \mathbf{CAT}$ *(from the opposite of* $Sig^{\mathcal{I}}$ *to the category of categories) giving for each signature* Σ *a category whose objects are called* Σ-models, *and whose arrows are called* Σ-*(*model*)* homomorphisms, *and*
4. *a relation* $\models^{\mathcal{I}}_{\Sigma} \subseteq |Mod^{\mathcal{I}}(\Sigma)| \times Sen^{\mathcal{I}}(\Sigma)$ *for each* $\Sigma \in |Sig^{\mathcal{I}}|$, *called* Σ-satisfaction,

such that for each morphism $\varphi\colon \Sigma \to \Sigma'$ *in* $Sig^{\mathcal{I}}$, *the* satisfaction condition

$$M' \models^{\mathcal{I}}_{\Sigma'} Sen^{\mathcal{I}}(\varphi)(\rho) \quad \text{if and only if} \quad Mod^{\mathcal{I}}(\varphi)(M') \models^{\mathcal{I}}_{\Sigma} \rho$$

holds for each $M' \in |Mod^{\mathcal{I}}(\Sigma')|$ *and* $\rho \in Sen^{\mathcal{I}}(\Sigma)$.

The functions $Sen^{\mathcal{I}}(\varphi)$ *are called* sentence translation functions *and the functors* $Mod^{\mathcal{I}}(\varphi)$ *are called* model reduct functors.

The literature (e.g. [22,57]) shows myriads of logical systems from computing or from mathematical logic captured as institutions. In fact, an informal thesis underlying institution theory is that any logic may be captured by the above definition. While this should be taken with a grain of salt, it certainly applies to any logical system based on satisfaction between sentences and models of any kind. However the very process of formalising logical systems as institutions is not a trivial one as it has to provide precise and consistent mathematical definitions for basic concepts that are commonly considered in a rather naive style. Moreover these definitions have to obey the axioms of institution. For example we will see below how the template given by Def. 1 shapes a drastically reformed understanding of logical languages (signatures) and variables.

The following example may convey an understanding about the process of capturing of a logical system as institution.

Example 1 (Many sorted algebra as institution). This is a very common logical system in computer science, and constitutes the logical basis of traditional algebraic specification. It is also used frequently in the literature as an example of the definition of institution; however there are some slight differences between various formalisations of many sorted algebra as institution. Here we sketch this institution in accordance with [22] and other papers of the author in the recent years.

The signatures (S, F) consist of a set of sorts (types) S and a family F of functions typed by arities (finite strings of sorts) and sorts, i.e. $F = (F_{w \to s})_{w \in S^*, s \in s}$. Signature morphisms map symbols such that arities are preserved; they can be presented as families of functions between corresponding sets of function symbols. Given a signature, its models M are many-sorted algebras interpreting sorts s as sets M_s, and function symbols $\sigma \in F_{s1 \ldots sn \to s}$ as functions $M_\sigma : M_{s1} \times \cdots \times M_{sn} \to M_s$. Model homomorphisms are many-sorted algebra homomorphisms. Model reduct means reassembling the models components according to the signature morphism, i.e. for any signature morphism $\varphi : (S, F) \to (S', F')$ and any (S', F')-model M' we have $Mod(\varphi)(M')_x = M'_{\varphi(x)}$ for each x sort in S or function symbol in F. The sentences are first-order formulæ formed from atomic equations (i.e. equalities between well formed terms having the same sort) by iteration of logical connectives $(\wedge, \neg)$ and (first-order) quantifiers $\forall X$ (where X is a finite block of S-sorted variables). Sentence translation means replacement of the translated symbols, for example for variables the sort is changed accordingly.

Satisfaction is the usual Tarskian satisfaction of a first-order sentence in a many-sorted algebra that is defined by induction on the structure of the sentences.

When working out the details of this definition, the *Sig*, *Mod* and $\models$ components are straightforward. Less so is *Sen* that requires a careful management of the concept of variable, an issue that will be discussed below in some detail. The proof of the Satisfaction Condition is done by induction on the structure of the sentences, the only non-trivial step corresponding to the quantifications. This involves some mild form of model amalgamation (see [22]).

2.3 The Expanse of Institution Theory

Def. 1 constitutes the starting concept of institution theory. Institution theory currently comprises a rather wide (both in terms of internal developments and applications) and constitutes a dynamic research area. The relationship between the concept of institution and institution theory is somehow similar to that between the concept of group and group theory in algebra. The definition of group is very simple and abstract and it does not convey the depth and expanse of group theory; the same holds for institutions and institution theory. The theory of institutions has gradually emerged as the most fundamental mathematical tool underlying algebraic specification theory (in its wider meaning) [57], also being increasingly used in other areas of computer science. And a lot of model theory has been gradually developed at the level of abstract institutions (see [22]), with manyfold consequences including a systematic supply of model theories to (sometimes sophisticated) non-conventional logical systems, but also new deep results in conventional model theory.

We refrain here from discussing in some details the rather long list of achievements of institution theory, instead we refer to the survey [26] that gives a brief account of the development of institution theory both in computer science and in mathematical logic (model theory).

3 ...and Back

The wide body of abstract model theory results developed within institution theory (many of them collected in [22]) can be regarded as an important contribution of computer science to logic and model theory in general, and to universal logic in particular. However here we will set this aside and instead will focus on something else, which is more basic and subtle in the same time, namely on the reformed understanding of some important basic concepts in logic. Through our analysis we will see that this has been made possible not only because of the universal logic aspect of institution theory, but especially because of its computer science origins. Computer science in general, and formal methods in particular, cannot afford a naive informal treatment of logical entities for the simple reason that often these have to be realised directly in implementations. It is thus no surprise that in many situation issues arising from implementation of formal specification languages can be very consonant with issues regarding the mathematical rigor imposed by the definition of institution and the corresponding solutions are highly convergent.

In this section we will discuss the new understanding of the concepts of logical language, variables, quantifiers, interpolation brought in by institution theory. We conclude with a brief discussion challenging the common view on many sorted logics.

3.1 On Logical Languages

Logical languages are the primary syntactic concept in mathematical logic. Informally they represent structured collections of symbols that, on the one hand are used as extra-logical symbols[1] in the composition of the sentences or formulæ, and on the other hand are interpreted, often in set theory, in order to get semantics. In institution theory the logical languages correspond to the objects of *Sig* and are called *signatures*, a terminology that owes to computer science. Institution theory leads to a more refined understanding of two aspects of logical languages, namely mappings between languages and variables.

Signature Morphisms and Language Extensions. In Def. 1, *Sig* is a category rather than a class; this means that morphisms between signatures play a primary role. In fact the category theoretic thinking leans towards morphisms rather than towards objects, objects are somehow secondary to morphisms. Some early and courageous presentations of category theory [34] even do it without the concept of object since objects can be assimilated to identity morphisms. In concrete situations the fact that *Sig* is only required to be category gives a lot of freedom with respect to the choice for an actual concept of signature morphism. One extreme choice is not to have proper signature morphisms at all or even that *Sig* has only one object. The latter situation is common to logical studies in which no variation in the language is necessary. A less extreme choice is made in the traditional model theory practice, namely to have only language (signature) extensions as morphisms. However, mathematically this may be quite an unconventional choice since usually, in concrete situations, morphisms are structure preserving mappings between objects and from this perspective signature extensions represent a rather strong restriction.

With respect to signature morphisms the practice of formal specification is quite different than that of mathematical logic in that it considers more sophisticated concepts of mappings between languages. The example of the many-sorted algebra institution given above is quite illustrative in this respect. The practice of algebraic specification (especially in the area of parameterised specifications) requires much more than signature extensions, it requires at least the fully general structure-preserving morphisms as in the aforementioned example. Moreover the literature (e.g. [57]) considers also an even more complex concept of signature morphism, the so-called *derived signature morphisms* that are in fact second-order substitutions replacing function symbols by terms. These of course are also immediately accommodated by the *Sig* part of Def. 1. This widening of the concept of language extension to various forms of signature morphisms has manyfold implications in all areas that involve the use of language extensions. For example paramount logical concepts such as interpolation and definability get a much more general formulation (see [22,59,20,51] etc.) with important consequences in the applications.

The case of the derived signature morphisms shows that in some situations the simple criterion of preserving the mathematical structure is not enough for defining a fully usable concept of signature morphism. There is also another famous case that comes from the behavioural specification trend [52,53,41,42,10,45,29,54]. When defining the corresponding institution(s), the use of the mere structure-preserving

[1] Logical symbols are connectors such as $\wedge$, $\neg$,..., or quantifiers $\forall, \exists$, or modalities $\Box, \Diamond$, etc. Sorts (types), function, relations symbols, etc. are extra-logical symbols.

mappings for the signature morphisms leads to the failure of the Satisfaction Condition of Def. 1. In order to get that holding, an additional condition has to be imposed on the signature morphisms known in the literature as the 'encapsulation condition' and which in the concrete applications corresponds clearly to an object-orientation aspect. In both [38] and [41] the authors remark that the derivation of the encapsulation condition on the morphisms of signatures from the meta-principle of invariance of truth under change of notation (the Satisfaction Condition of institutions) seems to confirm the naturalness of each of the principles. We may add here that this shows an inter-dependency between the abstract logic level and pragmatical computer science aspects.

Variables and Quantifiers. The concept of variable is primary when having to deal with quantifications. Mathematical logic has a common way to treat variables which has a *global* aspect to it. A typical example is the following quotation from [15] that refers to the language of first-order logic.

> "To formalize a language $\mathcal{L}$, we need the following *logical symbols*
>
> parentheses), (;
>
> variables $v_0, v_1, \ldots, v_n, \ldots$;
>
> connectives $\wedge$ (and), $\neg$ (not);
>
> quantifier $\forall$ (for all);
>
> and one binary relation symbol $\equiv$ (identity). We assume, of course, that no symbol in $\mathcal{L}$ occurs in the above list."

Upon analysis of this text we can easily understand that variables are considered as logical rather than extra-logical symbols which also implies that, as a collection, they are invariant with respect to the change of the signature. Moreover they have to be disjoint from the signatures. And of course, this collection of variables ought to be infinite.

While such treatment of variables may work well when having to deal only with ad-hoc signature extensions, as it is the case with conventional model theory. However it rises a series of technical difficulties with the institution theoretic approach.

1. Having a set of variables χ as logical symbols means that the respective institution has χ as a parameter. Therefore, strictly speaking, it is improper to talk, for example, about *the* institution of first-order logic.
2. In the concrete situations the category *Sig* is usually defined in the style of Ex. 1, which means that the individual signatures are set theoretic structures that are not restricted in any way on the basis of the fixed set of variables. This of course cannot guarantee the principle of disjointness between the signatures and the variables. For example it is possible that some signatures may contain some of the variables as constants.
3. Moreover, the institution-theoretic approach to quantifiers [22,59,19] etc. abstracts blocks of variables just to signature morphisms $\varphi : \Sigma \to \Sigma'$, where in the concrete situations φ stands for the extension of Σ with a respective block of variables. This means that while the variables have to be disjoint from the signature Σ, *they are actually part of* Σ'.

Unfortunately much of the institution theory literature is quite sloppy about these issues and adopts the common logic view of variables. However starting with [32] a series of works in institution theory adopts a view on variables that responds adequately to the aforementioned issues and therefore is mathematically rigorous. This is based on a *local* rather than the common global view of variable, drawing inspiration from the actual implementations of specification languages. For many sorted algebra (Ex. 1) it goes like this. Given a signature (S, F), a block of variables for (S, F) consists of a finite set of triples $(x, s, (S, F))$ where x is the name of the variable and $s \in S$ is its sort. It is also required that in any block of variables different variables have different names. Because of the qualification by the signature (the third component), by axiomatic set theory arguments we get that a variable for a signature is disjoint from the respective signature. On the other hand, they can be adjoined to the signature. So, given a block X of variables for a many sorted signature (S, F) let $(S, F + X)$ denote the new signature obtained by adding the variables of sort s as new constants of sorts s. Then for any $(S, F+X)$-sentence ρ we have that $(\forall X)\rho$ is a (S, F)-sentence. In this way satisfaction of quantified sentences can be defined only in terms of model reducts, without having to resort to traditional concepts such as valuations of variables that have a strong concrete aspect. An (S, F)-model M satisfies $(\forall X)\rho$ if and only if for each $(S, F+X)$-model M' such that $Mod(\varphi)(M') = M$ we have that M' satisfies ρ, where φ denotes the signature expansion $(S, F) \to (S, F + X)$. Note that this definition is institution theoretic since it does not depend on the many sorted algebra case, it can be formulated in exactly the same way in abstract institutions. Moreover, our local concept of variable also behaves well with respect to the sentence translations induced by signature morphisms. Given a signature morphism $\chi : (S, F) \to (S', F')$, any block of variables X for (S, F) translates to a block of variables X' for (S', F') by mapping each variable $(x, s, (S, F))$ to $(x, \chi(s), (S', F'))$. "Behaves well" here means two things: (1) that we get a block of variables as required, and (2) that the translation is functorial. Then latter aspect is crucial for the functor axioms for *Sen*.

It is very interesting to note that this local view on variables, necessary to meet the mathematical rigor of the definition of institution, fits the way logical variables are treated in actual implementations of specification languages (e.g. CafeOBJ [28], etc.). There variables are declared explicitly and their scope is restricted to the module in which they are declared. The way this fits exactly the aforementioned approach to logical variables is explained by the fact that, according to works such as [24] the institutions underlying specification languages have structured specifications or modules as their signatures, so in this case the qualification by the signature of the institution means qualification by a corresponding module.

The mathematical properties underlying our local approach to logical variables are axiomatised by the following abstract notion which has been used in a series of works (e.g. [23,48,27,31], etc.) for building explicit quantifications in abstract institutions in a way that it yields another sentence functor (and consequently another institution that shares the signatures and the models with the original institution).

Definition 2 (Quantification Space). *For any category Sig a subclass of arrows $\mathcal{D} \subseteq Sig$ is called a* quantification space *if, for any $(\chi : \Sigma \to \Sigma') \in \mathcal{D}$ and $\varphi : \Sigma \to \Sigma_1$, there is a designated pushout*

$$\begin{array}{ccc} \Sigma & \xrightarrow{\varphi} & \Sigma_1 \\ {\scriptstyle \chi}\downarrow & & \downarrow{\scriptstyle \chi(\varphi)} \\ \Sigma' & \xrightarrow[\varphi[\chi]]{} & \Sigma'_1 \end{array}$$

with $\chi(\varphi) \in \mathcal{D}$ and such that the 'horizontal' composition of such designated pushouts is again a designated pushout, i.e. for the pushouts in the following diagram

$$\begin{array}{ccccc} \Sigma & \xrightarrow{\varphi} & \Sigma_1 & \xrightarrow{\theta} & \Sigma_2 \\ {\scriptstyle \chi}\downarrow & & \downarrow{\scriptstyle \chi(\varphi)} & & \downarrow{\scriptstyle \chi(\varphi)(\theta)} \\ \Sigma' & \xrightarrow[\varphi[\chi]]{} & \Sigma'_1 & \xrightarrow[\theta[\chi(\varphi)]]{} & \Sigma'_2 \end{array}$$

$\varphi[\chi]; \theta[\chi(\varphi)] = (\varphi;\theta)[\chi]$ and $\chi(\varphi)(\theta) = \chi(\varphi;\theta)$, and such that $\chi(1_\Sigma) = \chi$ and $1_\Sigma[\chi] = 1_{\Sigma'}$.

The use of designated pushouts is required by the fact that quantified sentences ought to have a unique translation along a given signature morphism. The coherence property of the composition is required by the functoriality of the translations. For example, in the aforementioned concrete case of many sorted algebra, $\mathcal{D}$ consists of the signature extensions $\varphi : (S,F) \to (S, F+X)$ where X is a finite block of variables for (S,F). For any signature morphism $\chi : (S,F) \to (S',F')$ we define $X' = \{(x, \chi(s), (S',F')) \mid (x, s, (S,F)) \in X\}$, $\varphi[\chi]$ to be signature extension $(S',F') \to (S', F'+X')$ and $\chi(\varphi) : (S, F+X) \to (S', F'+X')$ to be the canonical extension of χ that maps each variable $(x, x, (S,F))$ to $(x, \chi(s), (S',F'))$.

3.2 On Interpolation

Because of its many applications in logic and computer science, interpolation is one of the most desired and studied properties of logical systems. Although it has a strikingly simple and elementary formulation as follows,

> given sentences ρ_1 and ρ_2, if ρ_2 is a consequence of ρ_1 (written $\rho_1 \vdash \rho_2$) then there exists a sentence ρ (called *interpolant*) in the common language of ρ_1 and ρ_2 such that $\rho_1 \vdash \rho$ and $\rho \vdash \rho_2$,

in general it is very difficult to establish. The famous result of Craig [16] marks perhaps the birth of the study of interpolation, proving it for (single sorted) first-order logic. The actual scope of Craig's result has been gradually extended to many other logical systems (for example in the world of modal logics, see

[36]), a situation that meets the universal character of interpolation that can be easily detected from its formulation that does not seem to commit inherently to any particular logical system.

The institution theoretic approach to interpolation has lead to a multi dimensional reformation of this important concept that will be discussed below. However, before that, we note that within institution theory the consequence relation $\vdash$ from the above formulation of interpolation is interpreted as the semantic consequence $\models$, i.e. for a given signature Σ and sets E, Γ of Σ-sentences, $E \models \Gamma$ when for each Σ-model M if M satisfies each sentence in E then it satisfies each sentence in Γ too.

From Single Sentences to Sets of Sentences. It has been widely believed that equational logic, the logical system underlying traditional algebraic specification, lacks interpolation; likewise for Horn-clause logic and other such fragments of first-order logic. As far as we know, Piet Rodenburg [55,56] was the first to point out that this is a misconception due to a basic misunderstanding of interpolation, rooted in the heavy dependency of logic culture on classical first-order logic with all its distinctive features taken for granted. Then it follows the grave general fault of exporting a coarse understanding of concepts dependent on details of a particular logical system to other logical systems of a possibly very different nature, where some detailed features may not be available. In the case of interpolation, the gross confusion has to do with looking for an interpolant as a single sentence. In first-order logic, which has conjunction, looking for interpolants as finite sets of sentences ($\{\rho_1, \ldots, \rho_n\}$) is just the same as looking for interpolants as single sentences ($\rho_1 \wedge \cdots \wedge \rho_n$). Hence, the common formulation of interpolation requires single sentences as interpolants. However, this is not an adequate formulation for equational logic which lacks conjunction, i.e., conjunction $\rho_1 \wedge \rho_2$ of universally quantified equations ρ_1 and ρ_2 cannot be captured as a universally quantified equation in general. Rodenburg [55,56] proved that equational logic has interpolation with the interpolant being a finite set of sentences, and this apparently weaker interpolation property is quite sufficient in both computer science and logic applications.

From Language Extensions to Signature Morphisms. The relationship between signatures Σ_1 (of ρ_1), Σ_2 (of ρ_2) and their union $\Sigma_1 \cup \Sigma_2$ (where the consequence $\rho_1 \vdash \rho_2$ happens) and intersection $\Sigma_1 \cap \Sigma_2$ (the signature of the interpolant), is depicted by the following diagram where arrows indicate the obvious inclusions:

$$\begin{array}{ccc} \Sigma_1 \cap \Sigma_2 & \xrightarrow{\subseteq} & \Sigma_1 \\ {\scriptstyle\subseteq}\downarrow & & \downarrow{\scriptstyle\subseteq} \\ \Sigma_2 & \xrightarrow[\subseteq]{} & \Sigma_1 \cup \Sigma_2 \end{array}$$

While intersections $\cap$ and unions $\cup$ are more or less obvious for signatures as used in first-order logic and in many other standard logics, they are not so in some other logical systems, and certainly not at the level of abstract institutions where signatures are just objects of an arbitrary category. One immediate response to this problem would be to add an infrastructure to the abstract category of signatures that would support concepts of $\cap$ and $\cup$; in fact this is already available in the institution theoretic literature and is called *inclusion system* [30,22]. Another solution would be, at the abstract level to use arbitrary signature morphisms and in the applications to restrict the signature morphisms only to those that are required to be, for example, inclusions (i.e. language extensions). Due to the abstraction involved, this means a lot of flexibility. For instance, in many computer science applications it is very meaningful to consider non-inclusive signature morphisms in the role of inclusions in the square above. An example comes from the practice of parameterised specifications (e.g. [57]) where instantiation of the parameters may involve signature morphisms that collapse syntactic entities. A generalised form of interpolation involving such non-injective signature morphisms is needed in order to get the completeness of formal verification for structured specifications (e.g. [12,11]). This generalisation of interpolation that relaxes language extensions to arbitrary signature morphisms has been introduced in [59]. The category-theoretic property of the above intersection-union square that makes things work is that it is a *pushout.* These considerations lead to the following abstract formulation of the interpolation property [22].

Definition 3 (Institution-theoretic Craig Interpolation [22]). [2] *Given* $\mathcal{L}$, $\mathcal{R} \subseteq$ *Sig, the institution has* Craig $(\mathcal{L},\mathcal{R})$-interpolation *when for each pushout square of signatures*

$$\begin{array}{ccc} \Sigma & \xrightarrow{\varphi_1 \in \mathcal{L}} & \Sigma_1 \\ {\scriptstyle \varphi_2 \in \mathcal{R}}\downarrow & & \downarrow{\scriptstyle \theta_1} \\ \Sigma_2 & \xrightarrow[\theta_2]{} & \Sigma' \end{array}$$

and any finite sets of sentences $E_1 \subseteq Sen(\Sigma_1)$ *and* $E_2 \subseteq Sen(\Sigma_2)$*, if* $\theta_1(E_1) \models \theta_2(E_2)$ *then there exists a finite set* E *of* Σ*-sentences such that* $E_1 \models \varphi_1(E)$ *and* $\varphi_2(E) \models E_2$.

The (abstract) restriction to pre-defined classes of signature morphisms, $\mathcal{L}$ for φ_1 and $\mathcal{R}$ for φ_2, constitutes an essential parameter in the above definition. In its absence the interpolation concept would be unrealistically too rigid and strong (for example many-sorted first-order logic would not support it [43,13,22]).

A couple of typical examples of institution-theoretic Craig $(\mathcal{L},\mathcal{R})$-interpolation are as follows:

- many-sorted first-order logic for either $\mathcal{L}$ or $\mathcal{R}$ consisting of the signature morphisms that are injective on the sorts [43,22]; and
- many-sorted Horn clause logic for $\mathcal{R}$ consisting of the signature morphisms that are injective [20,22].

[2] Given a signature morphism $\varphi : \Sigma \to \Sigma'$, we abbreviate $Sen(\varphi)$ as φ, and so for a set of sentences $E \subseteq Sen(\Sigma)$, $\varphi(E)$ is the image of E under $Sen(\varphi)$.

From Craig to Craig-Robinson Interpolation. There is a variety of situations in model theory (e.g. Beth definability [7,15]) and in computer science (e.g. complete calculi for structured specifications [12]) when Craig interpolation is used together with implication. The latter property is so obvious in some logics – such as first-order logic – that it is hardly ever mentioned explicitly in concrete contexts. Its definition at the level of abstract institutions is straightforward [59]: an institution *has implication* when for every signature Σ and Σ-sentences ρ_1, ρ_2, there exists a Σ-sentence ρ such that for each Σ-model M,

$$M \models \rho \text{ if and only if } M \models \rho_2 \text{ whenever } M \models \rho_1.$$

However, in many contexts we may render implication unnecessary by reformulating the interpolation property. Important applications are definability [51] in model theory and the completeness of calculus for structured specifications [22] in computer science. The trick is to 'parameterise' each instance of interpolation by a set of 'secondary' premises. In [33,58,61] this is called *Craig-Robinson interpolation*; it also plays an important role in specification theory, e.g. [6,30,33,22]. Let us recall here explicitly its institution-theoretic formulation.

Definition 4 (Institution-Theoretic Craig-Robinson Interpolation). *An institution has* Craig-Robinson $(\mathcal{L}, \mathcal{R})$-interpolation *when for each pushout square of signatures with $\varphi_1 \in \mathcal{L}$ and $\varphi_2 \in \mathcal{R}$*

$$\begin{array}{ccc} \Sigma & \xrightarrow{\varphi_1} & \Sigma_1 \\ {\scriptstyle \varphi_2}\downarrow & & \downarrow{\scriptstyle \theta_1} \\ \Sigma_2 & \xrightarrow[\theta_2]{} & \Sigma' \end{array}$$

and finite sets of sentences $E_1 \subseteq Sen(\Sigma_1)$ and $E_2, \Gamma_2 \subseteq Sen(\Sigma_2)$, if $\theta_1(E_1) \cup \theta_2(\Gamma_2) \models \theta_2(E_2)$ then there exists a finite set E of Σ-sentences such that $E_1 \models \varphi_1(E)$ and $\varphi_2(E) \cup \Gamma_2 \models E_2$.

Clearly, Craig-Robinson interpolation implies Craig interpolation. In any compact institution with implication, Craig-Robinson interpolation and Craig interpolation are equivalent [25,22] (so for instance within first-order logic, the two properties coincide). This means that Craig-Robinson interpolation alone in principle is weaker than Craig interpolation and implication. But is it properly so? Is there a significant example of an institution lacking implication but having Craig-Robinson interpolation? Through a rather sophisticated technique of so-called Grothendieck institutions [18,21], a result in [22] gives a general method to lift Craig interpolation to Craig-Robinson interpolation in institutions that may not have implication but are embedded in a certain way into institutions having implication. A concrete consequence of this result based on the Craig interpolation property of many-sorted first-order logic that was mentioned above, is as follows.

Corollary 1 (Craig-Robinson Interpolation in Many-Sorted Horn-clause Logic). *Many-sorted Horn-clause logic (with equality) has* $(\mathcal{L}, \mathcal{R})$*-Craig-Robinson interpolation when* $\mathcal{L}$ *consists only of signature morphisms* φ *that are injective on sorts and 'encapsulate' the operations.*[3]

One of the important significance of this result can be seen in conjunction with the upgrade in [22] of the completeness result for structured specifications of [12], that replaces Craig interpolation and implication by Craig-Robinson interpolation as one of the conditions. In the light of [12], the lack of implication has been used in the formal specification community as an argument against the adequacy of equational logic as a specification formalism. However we can see that this was only due to a couple of misunderstandings (1) that implication is not really needed for obtaining the completeness result of [12] and (2) that equational logic does satisfy the kind of interpolation that is really needed there and in a form that meets the requirements of the applications. In practice the only restriction involved by the conditions of Cor. 1 is that all information hidings have to be done with morphisms from $\mathcal{L}$, something that seem to accord well with practical intuitions underlying the concept of information hiding.

3.3 A Short Word on Many-Sortedness

Another significance of the aforementioned Craig-Robinson interpolation property of many-sorted Horn-clause logic is that, if we reduce the context to conditional equational logic by not considering predicate symbols – which is the logic underlying the equational logic programming paradigm (e.g. [17]) – it makes sense only in the many-sorted context. In a single sorted context it is clear that $\mathcal{L}$ collapses to nothing. This is just one of the examples that sharply refutes an idea that, in my opinion, is common among mathematical logicians, namely that many-sorted logics are "inessential" variations of their single-sorted versions (e.g. [50]). Another example is of course the case of generalised Craig interpolation in first-order logic; while the single-sorted variant supports it for all pushout squares of signature morphisms, we have seen that it is not so for the many-sorted variant.

References

1. Andréka, H., Németi, I.: Łoś lemma holds in every category. Studia Scientiarum Mathematicarum Hungarica 13, 361–376 (1978)
2. Andréka, H., Németi, I.: A general axiomatizability theorem formulated in terms of cone-injective subcategories. In: Csakany, B., Fried, E., Schmidt, E.T. (eds.) Universal Algebra, pp. 13–35. North-Holland (1981); Colloquia Mathematics Societas János Bolyai, 29
3. Andréka, H., Németi, I.: Generalization of the concept of variety and quasivariety to partial algebras through category theory. Dissertationes Mathematicae, vol. 204. Państwowe Wydawnictwo Naukowe (1983)

[3] In the sense that no operation symbol outside the image of φ is allowed to have a sort in the image of φ.

4. Barwise, J.: Axioms for abstract model theory. Annals of Mathematical Logic 7, 221–265 (1974)
5. Barwise, J., Feferman, S.: Model-Theoretic Logics. Springer (1985)
6. Bergstra, J., Heering, J., Klint, P.: Module algebra. Journal of the Association for Computing Machinery 37(2), 335–372 (1990)
7. Beth, E.W.: On Padoa's method in the theory of definition. Indagationes Mathematicæ 15, 330–339 (1953)
8. Béziau, J.-Y.: 13 questions about universal logic. Bulletin of the Section of Logic 35(2/3), 133–150 (2006)
9. Béziau, J.-Y. (ed.): Universal Logic: an Anthology. Studies in Universal Logic. Springer Basel (2012)
10. Bidoit, M., Hennicker, R., Wirsing, M.: Behavioural and abstractor specifications. Sci. Comput. Program. 25(2-3), 149–186 (1995)
11. Borzyszkowski, T.: Generalized interpolation in CASL. Information Processing Letters 76, 19–24 (2001)
12. Borzyszkowski, T.: Logical systems for structured specifications. Theoretical Computer Science 286(2), 197–245 (2002)
13. Borzyszkowski, T.: Generalized interpolation in first-order logic. Fundamenta Informaticæ 66(3), 199–219 (2005)
14. Burstall, R., Goguen, J.: The semantics of Clear, a specification language. In: Bjorner, D. (ed.) Abstract Software Specifications. LNCS, vol. 86, pp. 292–332. Springer, Heidelberg (1980)
15. Chang, C.-C., Keisler, H.J.: Model Theory. North Holland, Amsterdam (1990)
16. Craig, W.: Three uses of the Herbrand-Gentzen theorem in relating model theory and proof theory. Journal of Symbolic Logic 22, 269–285 (1957)
17. Diaconescu, R.: Category-based semantics for equational and constraint logic programming. DPhil thesis, University of Oxford (1994)
18. Diaconescu, R.: Grothendieck institutions. Applied Categorical Structures 10(4), 383–402 (2002); Preliminary version appeared as IMAR Preprint 2-2000, ISSN 250-3638 (February 2000)
19. Diaconescu, R.: Institution-independent ultraproducts. Fundamenta Informaticæ 55(3-4), 321–348 (2003)
20. Diaconescu, R.: An institution-independent proof of Craig Interpolation Theorem. Studia Logica 77(1), 59–79 (2004)
21. Diaconescu, R.: Interpolation in Grothendieck institutions. Theoretical Computer Science 311, 439–461 (2004)
22. Diaconescu, R.: Institution-independent Model Theory. Birkhäuser (2008)
23. Diaconescu, R.: Quasi-boolean encodings and conditionals in algebraic specification. Journal of Logic and Algebraic Programming 79(2), 174–188 (2010)
24. Diaconescu, R.: An axiomatic approach to structuring specifications. Theoretical Computer Science 433, 20–42 (2012)
25. Diaconescu, R.: Borrowing interpolation. Journal of Logic and Computation 22(3), 561–586 (2012)
26. Diaconescu, R.: Three decades of institution theory. In: Béziau, J.-Y. (ed.) Universal Logic: an Anthology, pp. 309–322. Springer Basel (2012)
27. Diaconescu, R.: Quasi-varieties and initial semantics in hybridized institutions. Journal of Logic and Computation, doi:10.1093/logcom/ext016
28. Diaconescu, R., Futatsugi, K.: CafeOBJ Report: The Language, Proof Techniques, and Methodologies for Object-Oriented Algebraic Specification. AMAST Series in Computing, vol. 6. World Scientific (1998)

29. Diaconescu, R., Futatsugi, K.: Behavioural coherence in object-oriented algebraic specification. Universal Computer Science 6(1), 74–96 (1998); First version appeared as JAIST Technical Report IS-RR-98-0017F (June 1998)
30. Diaconescu, R., Goguen, J., Stefaneas, P.: Logical support for modularisation. In: Huet, G., Plotkin, G. (eds.) Logical Environments, Cambridge, pp. 83–130 (1993); Proceedings of a Workshop held in Edinburgh, Scotland (May 1991)
31. Diaconescu, R., Madeira, A.: Encoding hybridized institutions into first order logic. Mathematical Structures in Computer Science (to appear)
32. Diaconescu, R., Petria, M.: Saturated models in institutions. Archive for Mathematical Logic 49(6), 693–723 (2010)
33. Dimitrakos, T., Maibaum, T.: On a generalized modularization theorem. Information Processing Letters 74, 65–71 (2000)
34. Ehresmann, C.: Catégories et strcutures, Dunod Paris (1965)
35. Ehresmann, C.: Esquisses et types des structures algébriques. Buletinul Institutului Politehnic Iaşi 14(18), 1–14 (1968)
36. Gabbay, D.M., Maksimova, L.: Interpolation and Definability: modal and intuitionistic logics. Oxford University Press (2005)
37. Goguen, J.: A categorical manifesto. Mathematical Structures in Computer Science 1(1), 49–67 (1991); Also, Programming Research Group Technical Monograph PRG–72, Oxford University (March 1989)
38. Goguen, J.: Types as theories. In: Reed, G.M., Roscoe, A.W., Wachter, R.F. (eds.) Topology and Category Theory in Computer Science, Oxford, pp. 357–390 (1991); Proceedings of a Conference held at Oxford (June 1989)
39. Goguen, J., Burstall, R.: Introducing institutions. In: Clarke, E., Kozen, D. (eds.) Logic of Programs 1983. LNCS, vol. 164, pp. 221–256. Springer, Heidelberg (1984)
40. Goguen, J., Burstall, R.: Institutions: Abstract model theory for specification and programming. Journal of the Association for Computing Machinery 39(1), 95–146 (1992)
41. Goguen, J., Diaconescu, R.: Towards an algebraic semantics for the object paradigm. In: Ehrig, H., Orejas, F. (eds.) Abstract Data Types 1992 and COMPASS 1992. LNCS, vol. 785, pp. 1–34. Springer, Heidelberg (1994)
42. Goguen, J., Malcolm, G.: A hidden agenda. Theoretical Computer Science 245(1), 55–101 (2000)
43. Găină, D., Popescu, A.: An institution-independent proof of Robinson consistency theorem. Studia Logica 85(1), 41–73 (2007)
44. Guitart, R., Lair, C.: Calcul syntaxique des modèles et calcul des formules internes. Diagramme 4 (1980)
45. Hennicker, R., Bidoit, M.: Observational logic. In: Haeberer, A.M. (ed.) AMAST 1998. LNCS, vol. 1548, pp. 263–277. Springer, Heidelberg (1998)
46. Makkai, M.: Ultraproducts and categorical logic. In: DiPrisco, C.A. (ed.) Methods in Mathematical Logic. Lecture Notes in Mathematics, vol. 1130, pp. 222–309. Springer (1985)
47. Makkai, M., Reyes, G.: First order categorical logic: Model-theoretical methods in the theory of topoi and related categories. Lecture Notes in Mathematics, vol. 611. Springer (1977)
48. Martins, M.A., Madeira, A., Diaconescu, R., Barbosa, L.S.: Hybridization of institutions. In: Corradini, A., Klin, B., Cîrstea, C. (eds.) CALCO 2011. LNCS, vol. 6859, pp. 283–297. Springer, Heidelberg (2011)
49. Matthiessen, G.: Regular and strongly finitary structures over strongly algebroidal categories. Canadian Journal of Mathematics 30, 250–261 (1978)

50. Donald Monk, J.: Mathematical Logic. Springer (1976)
51. Petria, M., Diaconescu, R.: Abstract Beth definability in institutions. Journal of Symbolic Logic 71(3), 1002–1028 (2006)
52. Reichel, H.: Behavioural equivalence – a unifying concept for initial and final specifications. In: Proceedings, Third Hungarian Computer Science Conference, Budapest. Akademiai Kiado (1981)
53. Reichel, H.: Initial Computability, Algebraic Specifications, and Partial Algebras, Clarendon (1987)
54. Roşu, G.: Hidden Logic. PhD thesis, University of California at San Diego (2000)
55. Rodenburg, P.-H.: Interpolation in conditional equational logic. Preprint from Programming Research Group at the University of Amsterdam (1989)
56. Rodenburg, P.-H.: A simple algebraic proof of the equational interpolation theorem. Algebra Universalis 28, 48–51 (1991)
57. Sannella, D., Tarlecki, A.: Foundations of Algebraic Specifications and Formal Software Development. Springer (2012)
58. Shoenfield, J.: Mathematical Logic. Addison-Wesley (1967)
59. Tarlecki, A.: Bits and pieces of the theory of institutions. In: Pitt, D., Abramsky, S., Poigné, A., Rydeheard, D. (eds.) Category Theory and Computer Programming. LNCS, vol. 240, pp. 334–360. Springer, Heidelberg (1986)
60. Tarski, A.: On some fundamental concepts of metamathematics. In: Logic, Semantics, Metamathematics, pp. 30–37. Oxford University Press (1956)
61. Veloso, P.: On pushout consistency, modularity and interpolation for logical specifications. Information Processing Letters 60(2), 59–66 (1996)
62. Wells, C.F.: Sketches: Outline with references (unpublished draft)

Event Analytics

Jin Song Dong[1], Jun Sun[2], Yang Liu[3], and Yuan-Fang Li[4]

[1] National University of Singapore, Singapore
[2] Singapore University of Technology and Design, Singapore
[3] Nanyang Technology University, Singapore
[4] Monash University, Australia

Abstract. The process analysis toolkit (PAT) integrates the expressiveness of state, event, time, and probability-based languages with the power of model checking. PAT is a self-contained reasoning system for system specification, simulation, and verification. PAT currently supports a wide range of 12 different expressive modeling languages with many application domains and has attracted thousands of registered users from hundreds of organizations. In this invited talk, we will present the PAT system and its vision on "Event Analytics" (EA) which is beyond "Data Analytics". The EA research is based on applying model checking to event planning, scheduling, prediction, strategy analysis and decision making. Various new EA research directions will be discussed.

1 Introduction

Large complex systems that generate intricate patterns of streaming events arise in many domains. These event streams arise from composite system states and control flows across many interacting components. Concurrency, asynchrony, uncertain environments - leading to probabilistic behaviours- and real-time coordination are key features of such systems. Many of the functionalities are realized in these systems by embedded software (and hardware) that must interact with the physical agents and processes. The proper functioning of such systems depends crucially on whether the software-mediated event patterns that are generated fulfill the required criteria. For example in a public transport system such as the Metro railway systems in large cities the control software must ensure that the distances between two trains sharing a track must never fall below a certain threshold and at the same time must optimize the number of trains deployed on track and their speeds to cater for increased demand for service during peak hours.

The key barriers to designing and deploying software-controlled complex systems are capturing system requirements and parameters and verifying important reliability, security and mission critical properties. There are well known methods (the so-called formal methods) for tackling this problem that are based on mathematical modelling and logic. The history of formal methods can be traced back to an early paper "Checking Large Routine" presented by Alan Turing at a conference on High Speed Automatic Calculating Machines at Cambridge University in 1949. More recently one particularly successful technique called Model Checking [CE81] was recognized through the Turing award being awarded to its creators Clarke, Emerson and Sifakis. At a Computer Aided Verification (CAV) conference, a research group from the Intel Corporation reported that the entire Intel i7 CPU core execution cluster was verified

G. Ciobanu and D. Méry (Eds.): ICTAC 2014, pp. 17–24, 2014.

using model checking without a single test case [KGN+09]. The Static Driver Verifier (SDV) [BCLR04] from Microsoft, a model checking system, has been deployed to verify Windows driver software automatically. Bill Gates in 2002 stated:

> "For things like software verification, this has been the Holy Grail of computer science for many decades. But now, in some very key areas for example driver verification we're building tools that can do actual proofs of the software and how it works in order to guarantee the reliability. "

These two industrial case studies are particularly exciting because they show that the exhaustive search techniques that the model checking method is based on can handle systems of large sizes: Intel i7 has eight cores and millions of registers, whereas driver software typically has thousands of lines of code. For a restricted version of model checking called bounded model checking, which often suffices in many practical settings, the scope of applicability is further enhanced by the so called SMT solvers [DeB08]. As a result many industries have started to actively invest in model checking technology [MWC10].

2 PAT Model Checking Systems

Many model checkers have been developed and successfully applied to practical systems, among which the most noticeable ones include SPIN, NuSMV, FDR, UPPAAL, PRISM and the Java Pathfinder. However those tools are designed for specialized domains and are based on restrictive modeling languages. Users of such systems usually need to manually translate models from the user's domain to the target language. In contrast, the Process Analysis Toolkit (PAT) [SLDP09] support modelling languages that combine the expressiveness of event, state, time and probability based modeling techniques to which model checking can be directly applied. PAT currently supports 12 different formalisms and languages ranging from graphical Timed Automata to programming languages for sensor networks. PAT is a self-contained system for system specification, simulation, and verification. Its core language is called CSP# which is based on Hoare's event based formalism CSP (Communicating Sequential Processes) [Hoare85] and the design of the CSP# is influenced by the integrated specification techniques (e.g. [MD00, TDC04]). The formal semantics of CSP# [SZL+13] is defined in Unified Theory of Programming [HH98]. The key idea is to treat sequential terminating programs, which may be as complex as C# programs, as events. The resulting modeling language is highly expressive and can cover many application domains such as concurrent data structures [LCL+13], web services [TAS+13], sensor networks [ZSS+13], multi-agent systems [HSL+12], mobile systems [CZ13] and cyber security systems [BLM+13]. The PAT system is designed to facilitate the development of customized model checkers and analysis tools. It has an extensible and modularized architecture to support new languages, reduction, abstraction and new model checking algorithms [DSL13]. PAT has attracted more than 3,000 registered users from 800+ organizations in 71 countries, including companies such as Microsoft, HP, Sony, Hitachi, Canon and many others. Many universities use PAT for teaching advanced courses. The following diagram illustrates the PAT architecture.

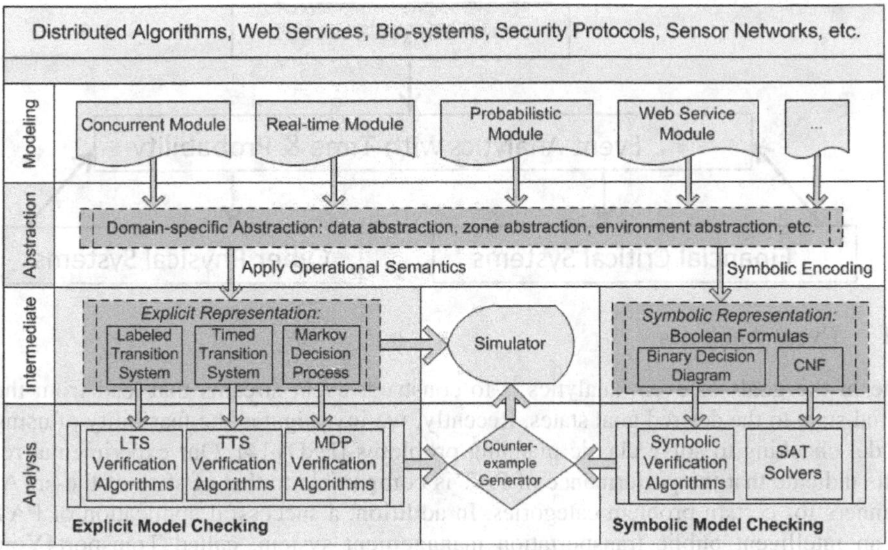

Recently, we have successfully applied PAT to event planning and scheduling problems [LSD+14] and have developed verification modules for real-time and probabilistic systems [SSLD12]. These research results provide a solid foundation for a new future research direction, namely, to develop model checking based approaches to event analytics that can serve the needs of planning and prediction as well as decision making. We note that while "data" is typically static "event" are dynamic and involve causality, communication, timing and probability. We believe event analytics (EA) driven technologies can offer significant advantages that are orthogonal to those based on "data analytics".

3 Event Analytics: A New Proposal

The development of novel parameterised real-time and parameterised probabilistic model checking algorithms and systems can support complex event analytics, i.e., to automatically answer the questions like "what is the maximum time delay of a critical event beyond which the overall system reliability will be compromised" and "what is the minimum probability shift (delta) of a specific event that will tip the balance of the winning strategy". While we believe there are wide applications for event analytics, we have started to work on the mission critical aspects of two major application domains: financial transactions systems and cyber-physical systems. The EA methodology and the accompanying tools can be validated in (applied to and get feedback from) those two major application domains:

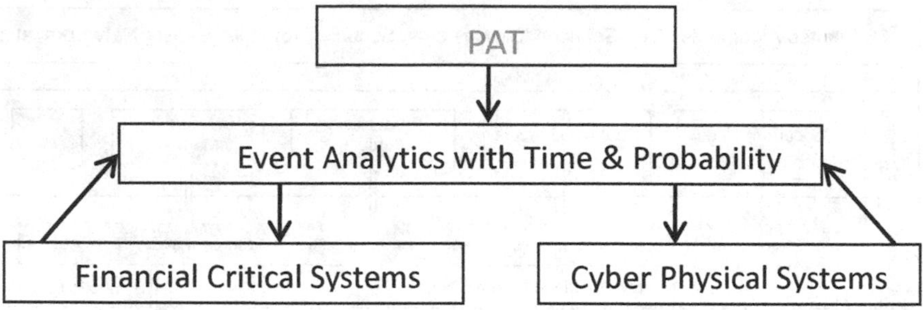

3.1 Event Analytics

One of the goals of event analytics is to construct event streams that lead from the initial state to the desired goal states. Recently, we investigated the feasibility of using model checking to solve classic planning problems [LSD+14]. Our experimental results indicate that the performance of PAT is comparable to that of state-of-the-art AI planners for certain problem categories. In addition, a successful application of PAT to an intelligent public transportation management system, called Transport4You, won the ICSE 2011 SCORE Competition [LYW11]. In the Transport4You project, PAT model checker is used not only as a verification tool but also as a service that computes optimal travel plans. PAT's new real-time and probabilistic verification modules can reason about real-time properties and calculate min/max probabilistic values for a particular events or states (the methodology and some preliminary results are reported in [SSLD12] with fixed value for timing and probability parameters). This sets a solid foundation for the proposed EA research. For EA systems to work with timed and probabilistic events that can evolve dynamically, one must develop sophisticated algorithms that can synthesizes timing and probabilistic parameter variables for real-time and probabilistic concurrent systems. It is important to conduct research to make the techniques scalable by developing new abstraction and reduction techniques, and apply multi-core and many-core verification to improve the performance. EA systems can then be deployed to assist the decision making and risk analysis in financial systems, and they can also provide context based activity/service planning for cyber-physical systems. It will be interesting to investigate the potential integration of optimisation techniques from Operations Research into EA systems.

3.2 Financial Critical Systems Verification

Financial software systems are critical and thus subject to strict requirements on functional correctness, security and reliability. With the recent development of online banking through mobile devices and computerized stock trading, the potential damage which could be the result of software vulnerability (e.g., credit card information leakage, financial loss due to high frequency trading) is high. This is why often strong regulations, e.g., Internet Banking and Technology Risk Management Guidelines from the Monetary Authority of Singapore, are imposed on the financial software design/development/testing process. While the regulations provide a checklist that could contribute to the quality of the software system, there are hardly any formal guarantees. The Mondex project on smart cards has been developed by NatWest bank in collaboration with Oxford

University using formal methods. As a result, Mondex became the first financial product to achieve certification at the highest level, namely, ITSEC level E6 [WSC+08]. While the Mondex project is a success story, demonstrating that applying formal methods does provide a guarantee in terms of system correctness/security, the techniques used in the project are rather limited, i.e., primarily formal modeling and manual theorem proving, which requires a considerable amount of expertise and time.

Recently, we have developed a method that combines hypothesis testing and probabilistic reachability analysis to study real-time trading systems. We identified the weak components inside the system so that the system designer can improve these components to improve the reliability of the whole system (some initial results were reported in [GSL+13]). It is important to investigate this further with EA based techniques, along with reliability predictions of other financial critical systems. It will be also interesting to investigate event based risk analytics for financial decision making which can have potential benefits for e.g., Monetary Authority of Singapore (MAS).

3.3 Cyber-Physical Systems Verification

Cyber-physical systems will play important roles in building smart cities. Such systems are fully automatic and they are 'aware' of their environment and self-adaptive to the environment changes. Many successes have been achieved in research laboratories especially for activity monitoring. However, such systems are not widely deployed due to, not only scalability and a lack of guarantees for correctness and reliability, but also the fact that those system are designed for demonstration purpose with well controlled scenarios in a lab environment. It is important to apply EA technology to analysis real environment and deliver highly reliable systems and reduce the prototyping time and cost. We also plan to apply event analytics to provide automatic intelligent assistive services in the smart city context (initial ideas have been applied to smart transportation systems [LYW11]). We have gained substantial experience will be also based on a recent successful application of the PAT system to "Activity Monitoring and UI Plasticity for supporting Ageing with Mild Dementia at Home" (AMUPADH 2010-12) [BMD10], a joint project with Alexandra Hospital in Singapore. After the Ethics Approval granted in 2011, a smart reminding system with cyber connected sensors has been successfully deployed in Peacehaven nursing home in 2012. The experience and techniques gained in AMUPADH project can certainly be generalized and applied in a wider context. It will be interesting to develop domain specific EA techniques that can automatically analysis the probabilistic and real-time services rules for smart city systems. Furthermore, parameterized probabilistic based verification techniques can be applied to estimate overall reliability of the smart city systems based on component subsystem reliabilities. EA techniques can also be developed to facilitate service compositions.

4 Event Analytics vs Data Analytics

Big Data and Data Analytics have received much hype in recent years. However, we surpassed capacity to store all the data we produce while growth in data creation has continued at an exponential rate [BSH13]. A recent study suggests we are able to analyze only 0.5% of all the data. Another significant limitation of current data analysis technique is the use of black box techniques (which is what Machine Learning techniques are) to generate results that cannot be explained. The ability to extract critical events from Big Data and to synthesize high-level models from such events can allow us to gain insights that are previously unattainable. For instance, better control on analysis that offer guarantees in believability or trust, combined with explanation can allow more confident decision making that rely on Big Data analysis. Furthermore, reducing the reliance on prior training data as is the case with majority of current approaches, possibly substituted or complemented by use of prior domain knowledge, would make Big Data analysis more scalable and robust. One future research can aim to combine model checking, machine learning and knowledge representation techniques and create an event analytics-based decision making engine for Big Data.

4.1 Event Extraction from Big Data

Model checking has been applied to large systems with more than 10^20 states [BCM+92]. However Big Data may still pose challenges to state-of-the-art model checking techniques. It is important to investigate techniques that further improve the scalability of model checking-based Event Analytics techniques. Large amounts of data streams can be generated from different sources such as social media and sensors. The granularity of such data may be too fine, and the quantity may still be too large for model checking techniques even with various reduction techniques. The data generated from these sources are not random: there is often (implicit or explicit) structure and semantics behind it. In other words, knowledge can be extracted from such data. It is important to investigate the integration of data mining techniques to continually extract patterns, continually from raw data. Such patterns, higher-level summaries, will then be turned into event traces which can then be more effectively utilized as inputs to model checking.

Ontologies have been widely applied as a median for sharing data and knowledge within and across domains. Long-standing research on ontologies and knowledge representation has developed ontology-based data integration techniques [Len02], which are especially suitable for this purpose. It will be interesting to investigate the problems of knowledge representation and learning to automatically induce event ontologies from raw data. Event ontologies can be supplied to model checking techniques to alleviate the data heterogeneity and scalability problems. Further research on ontology-based data/event integration, optimisation of ontology reasoning and more accurate ontology learning that involves more expressive ontology languages is thus required.

4.2 Model Synthesis from Events

Events extracted from Big Data are temporal in nature: they occur sequentially or concurrently, and form concurrent event traces that are interact in complex ways. An expressive mathematically based model that represents an entire system using states and events will enable deep analyses of interacting event traces on a globally level. For example, the L_ algorithm [Ang87] is proposed to learn deterministic finite automata (DFA) from a set of events. It will be interesting to investigate the problem of synthesizing, or generating appropriate models from event traces which may base on our early synthesis and verification work [SD06, LAL+14]. Model checking techniques have traditionally been applied to the analysis and verification of software and hardware systems, where complete knowledge of the system and its environment is usually assumed. However, such an assumption is often too strong for open scenarios such as emergency response and infectious disease management. It is important to investigate novel model checking techniques that are capable of handling such organic systems.

Acknowledgement. We would like to thank E. Clarke, D. Rosenblum, A.P. Sheth P.S. Thiagarajan and many others for insightful discussions on these topics.

References

[Ang87] Angluin, D.: Learning Regular Sets from Queries and Counterexamples. Information and Computation 75(2), 87–106 (1987)

[BCLR04] Ball, T., Cook, B., Levin, V., Rajamani, S.K.: SLAM and Static Driver Verifier: Technology Transfer of Formal Methods inside Microsoft. In: Boiten, E.A., Derrick, J., Smith, G.P. (eds.) IFM 2004. LNCS, vol. 2999, pp. 1–20. Springer, Heidelberg (2004)

[BLM+13] Bai, G., Lei, J., Meng, G., Venkatraman, S., Saxena, P., Sun, J., Liu, Y., Dong, J.S.: AUTHSCAN: Automatic Extraction of Web Authentication Protocols from Implementations. In: Proceedings of the Network and Distributed System Security Symposium, NDSS 2013, San Diego, USA (2013)

[BMD10] Biswas, J., Mokhtari, M., Dong, J.S., Yap, P.: Mild dementia care at home – integrating activity monitoring, user interface plasticity and scenario verification. In: Lee, Y., Bien, Z.Z., Mokhtari, M., Kim, J.T., Park, M., Kim, J., Lee, H., Khalil, I. (eds.) ICOST 2010. LNCS, vol. 6159, pp. 160–170. Springer, Heidelberg (2010)

[BCM+92] Burch, J.R., Clarke, E.M., McMillan, K.L., Dill, D.L., Hwang, L.J.: Symbolic model checking: 1020 states and beyond. Information and computation 98(2), 142–170 (1992)

[BSH13] Barnaghi, P., Sheth, A., Henson, C.: From Data to Actionable Knowledge: Big Data Challenges in the Web of Things. IEEE Intelligent Systems (November/December 2013)

[CZ13] Ciobanu, G., Zheng, M.: Automatic Analysis of TiMo Systems in PAT. In: Engineering of Complex Computer Systems, ICECCS (2013)

[CE81] Clarke, E.M., Emerson, E.A.: Design and Synthesis of Synchronization Skeletons Using Branching-Time Temporal Logic. Logic of Programs, 52–71 (1981)

[DeB] de Moura, L., Bjørner, N.S.: Z3: An efficient SMT solver. In: Ramakrishnan, C.R., Rehof, J. (eds.) TACAS 2008. LNCS, vol. 4963, pp. 337–340. Springer, Heidelberg (2008)

[DSL13] Dong, J.S., Sun, J., Liu, Y.: Build Your Own Model Checker in One Month. In: 35th International Conference on Software Engineering (ICSE 2013, tutorial), San Francisco, CA, USA, May18-26 (2013)

[GSL+13] Gui, L., Sun, J., Liu, Y., Si, Y., Dong, J.S., Wang, X.: Combining Model Checking and Testing with an Application to Reliability Prediction and Distribution. In: ISSTA 2013, Lugano, Switzerland, July 15-20 (2013)

[HSL+12] Hao, J., Song, S., Liu, Y., Sun, J., Gui, L., Dong, J.S., Leung, H.-F.: Probabilistic Model Checking Multi-agent Behaviors in Dispersion Games Using Counter Abstraction. In: Rahwan, I., Wobcke, W., Sen, S., Sugawara, T. (eds.) PRIMA 2012. LNCS, vol. 7455, pp. 16–30. Springer, Heidelberg (2012)

[Hoare85] Hoare, C.A.R.: Communicating Sequential Process. Prentice Hall (1985)

[HH98] Hoare, C.A.R., He, J.: Unifying Theory of Programming. Pretice Hall (1998)

[KGN+09] Kaivola, R., Ghughal, R., Narasimhan, N., Telfer, A., Whittemore, J., Pandav, S., Slobodová, A., Taylor, C., Frolov, V., Reeber, E., Naik, A.: Replacing Testing with Formal Verification in Intel® Core™ i7 Processor Execution Engine Validation. In: Bouajjani, A., Maler, O. (eds.) CAV 2009. LNCS, vol. 5643, pp. 414–429. Springer, Heidelberg (2009)

[LAL+14] Lin, S.-W., André, É., Liu, Y., Sun, J., Dong, J.S.: Learning Assumptions for Compositional Verification of Timed Systems. IEEE Trans. Software Eng. 40(2), 137–153 (2014)

[LCL+13] Liu, Y., Chen, W., Liu, Y.A., Sun, J., Zhang, S., Dong, J.S.: Verifying Linearizability via Optimized Refinement Checking. IEEE Transactions on Software Engineering 39(7), 1018–1039 (2013)

[LLX02] Lee, V.Y., Liu, Y., Zhang, X., Phua, C., Sim, K., Zhu, J., Biswas, J., Dong, J.S., Mokhtari, M.: ACARP: Auto correct activity recognition rules using process analysis toolkit (PAT). In: Donnelly, M., Paggetti, C., Nugent, C., Mokhtari, M. (eds.) ICOST 2012. LNCS, vol. 7251, pp. 182–189. Springer, Heidelberg (2012)

[LDS+14] Li, Y., Dong, J.S., Sun, J., Liu, Y., Sun, J.: Model Checking Approach to Automated Planning. Formal Methods in System Design 44(2), 176–202 (2014)

[Len02] Lenzerini, M.: Data integration: A theoretical perspective. In: Proceedings of the Twenty-first ACM SIGMOD-SIGACT-SIGART Symposium on Principles of Database Systems (PODS). ACM (2002)

[LYW11] Li, Y., Yang, H., Wu, H.: PAT Approach to Transport4You, an intelligent public transportation manager (Tutors: J. S. Dong, Y. Liu, J. Sun), ICSE 2011 SCORE winner. 94 teams from 22 countries registered the competition where 56 teams completed the projects; two teams were selected as winners.

[MD00] Mahony, B., Dong, J.S.: Timed Communicating Object Z. IEEE Transactions on Software Engineering 26(2), 150–177 (2000)

[MWC10] Miller, S.P., Whalen, M.W., Cofer, D.D.: Software model checking takes off. Commun. ACM 53(2), 58–64 (2010)

[SZL+13] Shi, L., Zhao, Y., Liu, Y., Sun, J., Dong, J.S., Qin, S.: A UTP Semantics for Communicating Processes with Shared Variables. In: Groves, L., Sun, J. (eds.) ICFEM 2013. LNCS, vol. 8144, pp. 215–230. Springer, Heidelberg (2013)

[SD06] Sun, J., Dong, J.S.: Design Synthesis from Interaction and State-based Specifications. IEEE Transactions on Software Engineering (TSE) 32(6), 349–364 (2006)

[SLDP09] Sun, J., Liu, Y., Dong, J.S., Pang, J.: PAT: Towards flexible verification under fairness. In: Bouajjani, A., Maler, O. (eds.) CAV 2009. LNCS, vol. 5643, pp. 709–714. Springer, Heidelberg (2009)

[SSLD12] Song, S., Sun, J., Liu, Y., Dong, J.S.: A Model Checker for Hierarchical Probabilistic Real-Time Systems. In: Madhusudan, P., Seshia, S.A. (eds.) CAV 2012. LNCS, vol. 7358, pp. 705–711. Springer, Heidelberg (2012)

[SYD+13] Sun, J., Liu, Y., Dong, J.S., Liu, Y., Shi, L., André, É.: Modeling and Verifying Hierarchical Real-time Systems using Stateful Timed CSP. The ACM Transactions on Software Engineering and Methodology (TOSEM) 22(1), 3:1-3:29 (2013)

[TDC04] Taguchi, K., Dong, J.S., Ciobanu, G.: Relating Pi-calculus to Object-Z. In: The 9th IEEE International Conference on Engineering Complex Computer Systems (ICECCS 2004), Florence, Italy, pp. 97–106. IEEE Press (2004)

[TAS+13] Tan, T.H., Andre, E., Sun, J., Liu, Y., Dong, J.S., Chen, M.: Dynamic Synthesis of Local Time Requirement for Service Composition. In: ICSE 2013 (2013)

[WD96] Woodcock, J., Davies, J.: Using Z: specification, refinement, and proof. Prentice-Hall (1996)

[WSC+08] Woodcock, J., Stepney, S., Cooper, D., Clark, J.A., Jacob, J.: The certification of the Mondex electronic purse to ITSEC Level E6. Formal Asp. Comput. 20(1), 5–19 (2008)

[ZSS+13] Zheng, M., Sanán, D., Sun, J., Liu, Y., Dong, J.S., Gu, Y.: State Space Reduction for Sensor Networks using Two-level Partial Order Reduction. In: The 14th International Conference on Verification, Model Checking, and Abstract Interpretation (VMCAI 2013), Rome, Italy, pp. 515–535 (January 2013)

A Logical Descriptor for Regular Languages via Stone Duality

Stefano Aguzzoli[1], Denisa Diaconescu[2], and Tommaso Flaminio[3]

[1] Department of Computer Science, University of Milan, Italy
aguzzoli@di.unimi.it
[2] Department of Computer Science, Faculty of Mathematics and Computer Science, University of Bucharest, Romania
ddiaconescu@fmi.unibuc.ro
[3] DiSTA - Department of Theoretical and Applied Science, University of Insubria, Italy
tommaso.flaminio@uninsubria.it

Abstract. In this paper we introduce a class of descriptors for regular languages arising from an application of the Stone duality between finite Boolean algebras and finite sets. These descriptors, called *classical fortresses*, are object specified in *classical propositional logic* and capable to accept exactly regular languages. To prove this, we show that the languages accepted by classical fortresses and deterministic finite automata coincide. Classical fortresses, besides being propositional descriptors for regular languages, also turn out to be an efficient tool for providing alternative and intuitive proofs for the closure properties of regular languages.

Keywords: regular languages, finite automata, propositional logic, Stone duality.

1 Motivations

Regular languages are those formal languages that can be expressed by Kleene's regular expressions [12] and that correspond to Type-3 grammars in Chomsky hierarchy [13]. As is well-known, there are several ways to recognize if a formal language is either regular or not: regular expressions, regular grammars, deterministic and non-deterministic finite automata.

The aim of this paper is to present an approach to the problem of recognizing regular languages introducing a dictionary for translating deterministic finite automata (DFA) in the language of classical propositional logic. The main idea underlying our investigation is to regard each DFA as a finite set-theoretical object and then applying the finite slice of the Stone duality to move from DFA to algebra and, finally, to logic. The logical objects which arise by this "translation" are called *classical fortresses* (for FORmula, TheoRy, SubstitutionS) and the main result in the paper shows that a language is regular if and only if there exists a fortress that accepts it.

It is known that if one tries to describe the behavior of DFA using a logical language, by Büchi-Elgot-Trakhtenbrot Theorem [3,6,14], one comes up with a

G. Ciobanu and D. Méry (Eds.): ICTAC 2014, LNCS 8687, pp. 25–42, 2014.

formalization in the monadic fragment of classical second-order logic. Hence, it is worth to point out that in this paper we address a different problem: we do not aim at *describing* DFA using logic, but at introducing logico-mathematical objects – classical fortresses – capable to *mimic* them through the mirror of the Stone duality.

Dualities have already been used to study regular languages. For instance, the authors of [9,8] extend Stone duality to Boolean algebras with additional operators to get Stone spaces equipped with Kripke's style relations, shading thus new light on the connection between regular languages and syntactic monoids. In [7], dualities are used to give a nice proof of Brzozowski's minimisation algorithm [2]. In this contribution we use the bare minimum needed of Stone duality theory to point out how a deterministic finite-state automaton can be fully described in classical propositional logic. The logical description we obtain, the classical fortress, could, by virtue of its simplicity, prove itself useful in several directions.

In this paper we show that classical fortresses are an efficient and robust formalism for providing alternative and intuitive proofs for the closure properties of regular languages. In this setting, we shall provide alternative and easy proofs of the classical results stating that the class of regular languages is closed under the usual set theoretical operations of union, intersection and complementation.

Moreover, classical fortresses offer a privileged position to generalize DFA to a non-classical logical setting, thus providing a uniform and reasonably defensible way to define models of computation in these non-classical logics. In fact, if L is any non-classical algebraizable logic for which a Stone-like duality holds between the finite slice of its algebraic semantics and a target category – which plays the rôle of finite sets in the classical case –, then, once fortresses have been defined in L, we can reverse the construction which brings to DFA from classical fortresses and obtain a notion of "automaton" for L. Although a detailed description of this generalization is beyond the scope of the present paper, we shall discuss it in slightly more details in the last section. In the same section we shall also discuss a further generalization of DFA's which is obtained by lifting the definition of fortress by using the full Stone duality between the variety of Boolean algebras and totally disconnected compact Hausdorff topological spaces.

The paper is structured as follows: Section 2 recalls the necessary preliminary notions and results about regular languages, deterministic and non-deterministic finite automata, Boolean algebras and Stone duality, and classical propositional logic. In Section 3 classical fortresses are introduced. In the same section we shall also show the main result of this paper, namely that a language is regular if and only if it is accepted by a classical fortress. The effective procedure to define a classical fortress from a DFA is presented in Section 4, while in Section 4.1 we show that the runs of an automaton over words can be mimicked in the corresponding fortress. Section 5 is dedicated to the problem of reducing the number of variables used by fortresses. In Section 6 we provide alternative proofs for the closure properties of regular languages through classical fortresses. We end this paper with Section 7 in which we discuss some possible generalizations of finite automata taking into account the viewpoints that fortresses offer.

2 Preliminaries

2.1 Deterministic and Nondeterministic Finite Automata

We refer to [10] for all unexplained notions on the theory of finite automata.

Let Σ be a finite alphabet. A *deterministic finite automata* (*DFA* henceforth) over Σ is a tuple

$$A = (S, I, \delta, F),$$

consisting in a finite set S of *states*, an *initial state* $I \in S$, a *transition relation* $\delta \subseteq S \times \Sigma \times S$ such that $|\delta(s,a)| \leq 1$, where $\delta(s,a) = \{s' \in S \mid (s,a,s') \in \delta\}$, for any $s \in S$, $a \in \Sigma$, and a set $F \subseteq S$ of *final states.* A DFA is *complete* if $|\delta(s,a)| = 1$, for any $s \in S$, $a \in \Sigma$.

For a finite word $w = a_1 a_2 \ldots a_n \in \Sigma^*$, a *run* of a DFA A over w is a finite sequence of states $s_1, \ldots, s_{n+1}$ such that $s_1 = I$ and $s_{i+1} \in \delta(s_i, a)$, any $1 \leq i \leq n$. A run is *accepting* if $s_{n+1} \in F$. The *language* accepted by a DFA A, denoted by $\mathcal{L}(A)$, is the set of all words accepted by A and it is a regular language. Every incomplete DFA can be transformed into a complete one, while preserving its language.

2.2 Classical Propositional Logic

We shall work with a finite set of *propositional variables* $V = \{v_1, \ldots, v_n\}$. *Formulas*, denoted by lower case greek letters $\varphi, \psi, \ldots$, are built in the signature $\{\wedge, \vee, \neg, \top\}$ of classical propositional logic as usual:

$$\varphi ::= \top \mid v_i \mid \neg\varphi \mid \varphi \wedge \varphi \mid \varphi \vee \varphi.$$

Denote by $Form(V)$ the set of all formulas defined from V. A *substitution* is a map $\sigma\colon V \to Form(V)$. Given a formula φ in the variables $v_1, \ldots, v_n$, and given a substitution σ, the formula $\varphi[\sigma]$ is obtained by replacing in φ each occurrence of a variable v_i by the formula $\sigma(v_i)$. Substitutions can be composed: if σ_1 and σ_2 are two substitutions on the same set of variables, then $\varphi[\sigma_1 \circ \sigma_2] = (\varphi[\sigma_2])[\sigma_1]$, for any formula φ.

A *valuation* is any map ρ from V to $\{0,1\}$, which uniquely extends to formulas by the usual inductive clauses: $\rho(\top) = 1$; $\rho(\varphi \wedge \psi) = \min\{\rho(\varphi), \rho(\psi)\}$; $\rho(\varphi \vee \psi) = \max\{\rho(\varphi), \rho(\psi)\}$; $\rho(\neg\varphi) = 1 - \rho(\varphi)$. A valuation ρ is a *model* for a formula φ iff $\rho(\varphi) = 1$.

The *deductive closure* $\langle\Gamma\rangle$ of a set Γ of formulas is the set of all formulas φ such that each model of all formulas in Γ is a model of φ, too. We shall write $\langle\gamma\rangle$ instead of $\langle\{\gamma\}\rangle$. In the following we shall use capital greek letters $\Theta, \Gamma, \ldots$ to denote *theories*, that is, deductively closed set of formulas, and hence we shall write $\Theta \models \varphi$ iff every valuation ρ which is a model for every $\vartheta \in \Theta$, is a model of φ, too. Equivalently, $\Theta \models \varphi$ iff $\varphi \in \Theta$. Two formulas φ and ψ are *logically equivalent*, and we write $\varphi \equiv \psi$, if and only if $\langle\varphi\rangle \models \psi$ and $\langle\psi\rangle \models \varphi$. Equivalently, $\varphi \equiv \psi$ if $\rho(\varphi) = \rho(\psi)$ for each valuation ρ. A theory Θ is *prime over* V if for every pair of formulas $\varphi, \psi \in Form(V)$, if $\varphi \vee \psi \in \Theta$, then either

$\varphi \in \Theta$ or $\psi \in \Theta$. In classical propositional logic, prime and maximal theories coincide, whence, Θ is prime iff for every formula φ, either $\varphi \in \Theta$ or $\neg\varphi \in \Theta$.

A *min-term* is a maximally consistent elementary conjunction of literals from V, that is, a formula α is a min-term, if

$$\alpha = \bigwedge_{i=1}^{n} (v_i)^{*(i)},$$

where $*(\cdot) : \{1, \ldots, n\} \to \{0, 1\}$ and, for each variable v_i,

$$(v_i)^{*(i)} = \begin{cases} v_i, & \text{if } *(i) = 1, \\ \neg v_i, & \text{if } *(i) = 0. \end{cases}$$

For each min-term α, the theory $\langle\alpha\rangle$ is prime over V, and every prime theory is the deductive closure of a min-term.

2.3 Finite Boolean Algebras: Duality and Propositional Logic

We refer to [4] for all unexplained notions on the theory of Boolean algebras and to [11] for Stone duality.

Let $B = (B, \wedge, \vee, \neg, 1)$ be a Boolean algebra. A *filter* of B is an upward closed (w.r.t. the lattice order of B) subset $\emptyset \neq \mathfrak{f} \subseteq B$ such that $x, y \in \mathfrak{f}$ implies $x \wedge y \in \mathfrak{f}$. Filters are in bijection with congruences via the maps $\mathfrak{f} \mapsto \bar{\Theta}_{\mathfrak{f}} = \{(a, b) \mid (\neg a \vee b) \wedge (\neg b \vee a) \in \mathfrak{f}\}$, and $\bar{\Theta} \mapsto \mathfrak{f}_{\bar{\Theta}} = \{a \mid (a, 1) \in \bar{\Theta}\}$. Given a Boolean algebra A and a congruence $\bar{\Theta} \subseteq A^2$, consider the system $A/\bar{\Theta} = (\{a/\bar{\Theta} \mid a \in A\}, \wedge_{\bar{\Theta}}, \vee_{\bar{\Theta}}, \neg_{\bar{\Theta}}, \top/\bar{\Theta})$, where $a/\bar{\Theta}$ is the $\bar{\Theta}$-equivalence class of a, and $a/\bar{\Theta} \wedge_{\bar{\Theta}} b/\bar{\Theta} = (a \wedge b)/\bar{\Theta}$, $a/\bar{\Theta} \vee_{\bar{\Theta}} b/\bar{\Theta} = (a \vee b)/\bar{\Theta}$, $\neg a/\bar{\Theta} = (\neg a)/\bar{\Theta}$. Then $A/\bar{\Theta}$ is a Boolean algebra, called the *quotient* of A *modulo* $\bar{\Theta}$.

A *prime filter* of B is a proper filter $\mathfrak{p}$ such that $x \vee y \in \mathfrak{p}$ implies $x \in \mathfrak{p}$ or $y \in \mathfrak{p}$. A *maximal filter* of B is a proper filter which is not included in any other proper filter. Notice that, for every Boolean algebra B, prime filters and maximal filters coincide. The set of all prime filters of B is called the *prime spectrum* of B and it is denoted by Spec B.

An *atom* of B is an element $a \in B$ such that $a \neq 0$ and if $0 \neq c \leq a$ (w.r.t. the lattice order of B), then $a = c$. Notice that each prime filter $\mathfrak{p}$ of B is the upward closure of a unique atom a of B. We say that $\mathfrak{p}$ is *generated* by a, and write $\mathfrak{p} = \uparrow a$.

The *free n-generated Boolean algebra* is denoted by $F_n(\mathcal{B})$ and it is, up to isomorphism, the unique Boolean algebra A such that: (i) A is generated by a set X of n of its elements, (ii) for any Boolean algebra B, each set function $X \to B$ extends uniquely to a homomorphism $A \to B$. The structure $F_n(\mathcal{B})$ is uniquely determined, up to isomorphisms, as the only Boolean algebra having 2^n atoms and 2^{2^n} elements. $F_n(\mathcal{B})$ is isomorphic to the direct product of 2^n copies of the two-element Boolean algebra $\{0, 1\}$, or, equivalently is the algebra of all functions $f: \{0, 1\}^n \to \{0, 1\}$ endowed with pointwise defined operations. The *(free) generators* of $F_n(\mathcal{B})$ are the projection functions $\pi_i(t_1, t_2, \ldots, t_n) = t_i$, for each $i = 1, 2, \ldots, n$.

Note that Spec $F_1(\mathcal{B}) \cong \{0,1\}$ as sets, while as Boolean algebras $\{0,1\}$ is isomorphic with $F_0(\mathcal{B}) = \{[\neg\top], [\top]\}$. Moreover, Spec $\{0,1\} \cong \{[\top]\}$, that is, Spec $\{0,1\}$ is a singleton.

Since classical propositional logic is algebraizable in the sense of [1], all the logical notions introduced in Section 2.2 have a correspondence at algebraic level. In fact the free n-generated Boolean algebra is isomorphic with the Lindenbaum algebra of all classes of logically equivalent formulas over a fixed set of n distinct variables and, hence, to each formula φ we associate its equivalent class modulo logical equivalence $[\varphi] = \{\psi \mid \psi \equiv \varphi\} \in F_n(\mathcal{B})$.

An *endomorphism* of Boolean algebras is a homomorphism mapping a Boolean algebra into itself. A substitution σ on the n-variables $\{v_1, \ldots, v_n\}$ uniquely determines an endomorphism $\bar{\sigma}$ of $F_n(\mathcal{B})$, and *viceversa.* As a matter of fact, the restriction of each endomorphism of $F_n(\mathcal{B})$ to the class of the variables $[v_i]$'s clearly determines one substitution, while each substitution from V into formulas uniquely defines a homomorphism by the very definition of free algebra, $[v_1], \ldots, [v_n]$ being the generators of $F_n(\mathcal{B})$. Valuations $V \to \{0,1\}$ can then be identified with homomorphisms of $F_n(\mathcal{B})$ into $\{0,1\}$. Each min-term α defines the atom $[\alpha]$ of $F_n(\mathcal{B})$, and each atom is the class of a min-term.

Each theory Θ in a language with n variables corresponds uniquely with the filter $\mathfrak{f}_\Theta = \{[\vartheta] \mid \vartheta \in \Theta\}$ of $F_n(\mathcal{B})$, and with the congruence $\bar{\Theta} = \bar{\Theta}_{\mathfrak{f}_\Theta} = \{([\varphi], [\psi]) \mid (\neg\varphi \vee \psi) \wedge (\neg\psi \vee \varphi) \in \Theta\}$. In particular, a theory Θ is prime iff $\mathfrak{f}_\Theta$ is the filter generated by an atom, that is, it is of the form $\uparrow[\alpha] = \{[\psi] \mid [\alpha] \leq [\psi]\}$ for some atom $[\alpha]$. Finally, each min-term $\alpha = \bigwedge_{i=1}^{n} (v_i)^{*(i)}$, and consequently each atom $[\alpha]$ of $F_n(\mathcal{B})$ and each prime theory $\Theta = \langle \alpha \rangle$, uniquely determines a valuation $\rho_\Theta \colon \{v_1, \ldots, v_n\} \to \{0,1\}$ such that ρ_Θ is the unique model of Θ, by associating: $\rho_\Theta(v_i) = *(i)$.

Therefore, prime theories, min-terms, atoms, prime filters and valuations are in 1-1 correspondence: each prime theory Θ is, up to logical equivalence, the deductive closure $\langle \alpha \rangle$ of a unique min-term α; for each min-term α, the class $[\alpha]$ is an atom of $F_n(\mathcal{B})$; each prime filter $\mathfrak{p}$ in $F_n(\mathcal{B})$ is the upward closure $\uparrow[\alpha]$ of a unique atom $[\alpha]$; moreover, $\rho_\Theta(\alpha) = 1$ and $\rho_\Theta(\vartheta) = 1$ for each ϑ in the uniquely determined prime theory $\Theta = \langle \alpha \rangle$.

We shall need the following technical lemma.

Lemma 1. *Let V and V' be two disjoint sets of variables of cardinality n and m respectively. Let Θ be a prime theory in V and φ a formula in V. Then, for every prime theory Θ' in V', the following hold:*

1. *$\langle \Theta \cup \Theta' \rangle$ is prime in $(n+m)$ variables,*
2. *$\langle \Theta \cup \Theta' \rangle \models \varphi$ iff $\Theta \models \varphi$.*

Proof. As we already recalled, each prime theory has exactly one model. Let hence $\rho : \{v_1, \ldots, v_n\} \to \{0,1\}$ and $\rho' : \{v'_1, \ldots, v'_m\} \to \{0,1\}$ be the two models of Θ and Θ' respectively.

1. $\langle \Theta \cup \Theta' \rangle$ is prime in $n+m$ variables. It follows by the observation that its unique model ρ'', in $n+m$ variables, is obtained by the disjoint union of ρ and

ρ'. In details, the map ρ'' is defined on the variables $\{v_1, \ldots, v_n, v'_1, \ldots, v'_m\}$ by setting, for each $v \in V \cup V'$,

$$\rho''(v) = \begin{cases} \rho(v_i) & \text{if } v = v_i \\ \rho'(v'_j) & \text{if } v = v'_j. \end{cases}$$

2. Obviously, if $\Theta \models \varphi$, then $\langle \Theta \cup \Theta' \rangle \models \varphi$ by monotonicity. Conversely, if $\langle \Theta \cup \Theta' \rangle \models \varphi$, the unique model ρ'' of $\langle \Theta \cup \Theta' \rangle$ we described in *1* is also a model of φ. Since φ is written in the variables $v_1, \ldots, v_n$, then, clearly, the restriction $\hat{\rho}$ of ρ'' to $\{v_1, \ldots, v_n\}$ coincides with the unique model of Θ and, moreover, $\hat{\rho}(\varphi) = 1$. Hence $\Theta \models \varphi$. □

Duality. We recall that two categories $\mathbb{C}$ and $\mathbb{D}$ are dually equivalent if there exists a pair of contravariant functors $F\colon \mathbb{C} \to \mathbb{D}$ and $G\colon \mathbb{D} \to \mathbb{C}$ such that both FG and GF are naturally isomorphic with the corresponding identity functors, that is, for each object C in $\mathbb{C}$ and D in $\mathbb{D}$ there are isomorphisms $\eta_C\colon GF(C) \to C$ and $\kappa_D\colon FG(D) \to D$ such that

$$\begin{array}{ccc} GF(C_1) & \xrightarrow{GF(f)} & GF(C_2) \\ \downarrow{\scriptstyle \eta_{C_1}} & & \downarrow{\scriptstyle \eta_{C_2}} \\ C_1 & \xrightarrow[f]{} & C_2 \end{array} \qquad \begin{array}{ccc} FG(D_1) & \xrightarrow{FG(g)} & FG(D_2) \\ \downarrow{\scriptstyle \kappa_{D_1}} & & \downarrow{\scriptstyle \kappa_{D_2}} \\ D_1 & \xrightarrow[g]{} & D_2 \end{array}$$

for each $f\colon C_1 \to C_2$ in $\mathbb{C}$ and $g\colon D_1 \to D_2$ in $\mathbb{D}$.

Let $\mathbb{B}_{fin}$ be the category of finite Boolean algebras and homomorphisms and $\mathbb{Set}_{fin}$ the category of finite sets and functions. The categories $\mathbb{B}_{fin}$ and $\mathbb{Set}_{fin}$ are dually equivalent via the following pair of contravariant functors

$$\text{Spec} : \mathbb{B}_{fin} \to \mathbb{Set}_{fin} \qquad \text{Sub} : \mathbb{Set}_{fin} \to \mathbb{B}_{fin}$$

defined by:

- for each object B of $\mathbb{B}_{fin}$, Spec B is the prime spectrum of B,
- for each $h : B_1 \to B_2 \in \mathbb{B}_{fin}$, Spec h : Spec $B_2 \to$ Spec B_1 is the map $\mathfrak{p} \mapsto h^{-1}(\mathfrak{p})$,
- for each object S of $\mathbb{Set}_{fin}$, Sub $S = (2^S, \cap, \cup, \cdot^c, S)$ is the set of all subsets of S, endowed with intersection, union and complement,
- for each $f : S_1 \to S_2 \in \mathbb{Set}_{fin}$, Sub f : Sub $S_2 \to$ Sub S_1 is the map $X \mapsto f^{-1}(X)$.

Note that for every finite Boolean algebra B, $B \cong$ Sub Spec B. The above duality is the specialization to finitely presented objects of the celebrated *Stone Duality* between the category of all Boolean algebras with homomorphisms, and the category of totally disconnected, compact, Hausdorff spaces with continuous maps.

Remark 1. The following are applications of duality to basic concepts of $\mathbb{Set}_{fin}$ and $\mathbb{B}_{fin}$, which will be used in the sequel.

1. The isomorphisms between $F_n(\mathcal{B})$ and Sub Spec $F_n(\mathcal{B})$ are implemented by the map $\eta_n : F_n(\mathcal{B}) \to \text{Sub Spec } F_n(\mathcal{B})$ and its inverse $\kappa_n : \text{Sub Spec } F_n(\mathcal{B}) \to F_n(\mathcal{B})$:
$$\eta_n : [\varphi] \mapsto \{\uparrow[\alpha] \mid \alpha \text{ min-term}, [\alpha] \leq [\varphi]\}.$$
and
$$\kappa_n : \{\uparrow[\alpha_i] \mid \alpha_i \text{ min-term}, i = 1, \dots, r\} \mapsto [\bigvee_{i=1}^{r} \alpha_i].$$
2. An element of Spec $F_n(\mathcal{B})$ can be clearly identified with a map $\iota\colon \{[\top]\} \to \text{Spec } F_n(\mathcal{B})$, that is, a map $\iota\colon \text{Spec }\{0,1\} \to \text{Spec } F_n(\mathcal{B})$. Dually, Sub $\iota\colon F_n(\mathcal{B}) \to \{0,1\}$, assigns to each class $[\varphi]$ a truth-value in $\{0,1\}$, whence, once we fix the variable set $V = \{v_1, \dots, v_n\}$, the homomorphism
$$\text{Sub } \iota([\varphi]) = \kappa_0(\iota^{-1}(\eta_n([\varphi])))$$
can be identified with a valuation $\rho\colon V \to \{0,1\}$, by setting $\rho(\varphi) = \text{Sub } \iota([\varphi])$.
3. Each function λ mapping Spec $F_n(\mathcal{B})$ on itself dually corresponds to an endomorphism $\bar{\sigma}\colon F_n(\mathcal{B}) \to F_n(\mathcal{B})$ by putting
$$\bar{\sigma}([\varphi]) = \kappa_n(\lambda^{-1}(\eta_n([\varphi]))).$$
4. Each function $\chi\colon \text{Spec } F_n(\mathcal{B}) \to \text{Spec } F_1(\mathcal{B})$ dually corresponds to a homomorphism Sub $\chi\colon F_1(\mathcal{B}) \to F_n(\mathcal{B})$, which is completely determined by the image $[\varphi] \in F_n(\mathcal{B})$ of the generator $[v_1]$ of $F_1(\mathcal{B})$ under $\kappa_n \circ \chi^{-1} \circ \eta_1$.

3 Classical Fortresses

In this section we are going to introduce a logical descriptor for regular languages. Throughout this section, let Σ be a finite alphabet and $V = \{v_1, \dots, v_n\}$ be a finite set of propositional variables.

Definition 1. *A* classical fortress in n variables *over Σ is a triple of the form*
$$\mathcal{F} = (\varphi, \{\sigma_a\}_{a \in \Sigma}, \Theta),$$
where

- *φ is a formula in $Form(V)$,*
- *for each $a \in \Sigma$ the map $\sigma_a\colon V \to Form(V)$ is a substitution,*
- *Θ is a prime theory in the variables V.*

Definition 2. *A classical fortress $\mathcal{F} = (\varphi, \{\sigma_a\}_{a \in \Sigma}, \Theta)$* accepts *a word $w = a_1 \cdots a_k \in \Sigma^*$, denoted by $\mathcal{F} \Vdash w$, if*
$$\Theta \models \varphi[\sigma_{a_1} \circ \cdots \circ \sigma_{a_k}]. \tag{1}$$
The language *of a classical fortress $\mathcal{F}$ is the set of all words accepted by $\mathcal{F}$:*
$$\mathcal{L}(\mathcal{F}) = \{w \in \Sigma^* \mid \mathcal{F} \Vdash w\}.$$

Remark 2. Note that in (1), given a word $w = a_1 \cdots a_k$, the substitutions σ_{a_i} are applied in the converse order with respect to the occurrences of the corresponding letters in w.

Notation. If $w = a_1 \cdots a_k \in \Sigma^*$, when it is convenient, we will denote the substitution $\sigma_{a_1} \circ \cdots \circ \sigma_{a_k}$ simply by σ_w.

Theorem 1. *For every complete DFA A with 2^n states, there exists a classical fortress in n-variables $\mathcal{F}_A$ such that $\mathcal{L}(A) = \mathcal{L}(\mathcal{F}_A)$.*

Proof. Let $A = (S, I, \delta, F)$ be a finite complete deterministic automaton such that $|S| = 2^n$. Note that $S \cong \text{Spec } F_n(\mathcal{B})$ and let us denote by $f : S \to \text{Spec } F_n(\mathcal{B})$ this isomorphism.

We consider $\mathfrak{p}_I = f(I) \in \text{Spec } F_n(\mathcal{B})$, where $I \in S$ is the initial state of A. Remark that $\mathfrak{p}_I \in \text{Spec } F_n(\mathcal{B})$ is uniquely defined by the map

$$\iota_I : \text{Spec } \{0,1\} \to \text{Spec } F_n(\mathcal{B}), \quad \iota_I([\top]) = \mathfrak{p}_I.$$

We define

$$\chi_F : \text{Spec } F_n(\mathcal{B}) \to \text{Spec } F_1(\mathcal{B}), \quad \chi_F(\mathfrak{p}) = \begin{cases} 1, & \text{if } f^{-1}(\mathfrak{p}) \in F \\ 0, & \text{if } f^{-1}(\mathfrak{p}) \notin F \end{cases},$$

for every $\mathfrak{p} \in \text{Spec } F_n(\mathcal{B})$, where F is the set of final states of the automaton A. Furthermore, for every $a \in \Sigma$, we define the endomorphism

$$\lambda_a : \text{Spec } F_n(\mathcal{B}) \to \text{Spec } F_n(\mathcal{B}), \quad \lambda_a(\mathfrak{p}) = f(\delta(f^{-1}(\mathfrak{p}), a)),$$

for every $\mathfrak{p} \in \text{Spec } F_n(\mathcal{B})$. The map λ_a is well-defined since A is a complete deterministic automaton. In conclusion, we defined the following arrows:

$$\text{Spec } \{0,1\} \xrightarrow{\iota_I} \text{Spec } F_n(\mathcal{B}) \xrightarrow{\lambda_a} \text{Spec } F_n(\mathcal{B}) \xrightarrow{\chi_F} \text{Spec } F_1(\mathcal{B})$$

Using Remark 1, we obtain the following maps corresponding to ι_I, λ_a and χ_F, respectively:

$$\{0,1\} \xleftarrow{\quad \rho_\Theta \quad} F_n(\mathcal{B}) \xleftarrow{\quad \bar{\sigma}_a \quad} F_n(\mathcal{B}) \xleftarrow{\quad [\varphi] \quad} F_1(\mathcal{B}),$$

where, by Remark 1 (2) we identify $\rho_\Theta = \kappa_0 \circ \iota_I^{-1} \circ \eta_n$ with a valuation from V to $\{0,1\}$. Further, $\bar{\sigma}_a = \kappa_n \circ \delta_a^{-1} \circ \eta_n$ and finally by $[\varphi]$ we denote the map determined by $[\varphi] = \kappa_n \circ \chi_F^{-1} \circ \eta_1([v_1])$. Let Θ, σ_a and φ be the prime theory, the substitution and the formula respectively corresponding to ρ_Θ, $\bar{\sigma}_a$ and $[\varphi]$. Therefore we consider the classical fortress in n variables

$$\mathcal{F}_A = (\varphi, \{\sigma_a\}_{a \in \Sigma}, \Theta).$$

In the rest of the proof we show that $\mathcal{L}(A) = \mathcal{L}(\mathcal{F}_A)$. Let $w = a_1 \cdots a_k \in \Sigma^*$. Since the automaton A is deterministic and complete, there exists a unique finite sequence of states $s_1, \ldots, s_{k+1}$ such that

$$I = s_1 \xrightarrow{a_1} s_2 \xrightarrow{a_2} s_3 \xrightarrow{a_3} \cdots \xrightarrow{a_k} s_{k+1}.$$

Since $\delta(s_i, a_i) = s_{i+1}$, we obtain that $\lambda_{a_i}(f(s_i)) = f(s_{i+1})$, for every $1 \leq i \leq k$. Also $f(s_1) = \iota_I([\top])$. We have two cases to consider.

- Suppose $w \in \mathcal{L}(A)$. Therefore $s_{k+1} \in F$ and it follows that

  $$\chi_F(\lambda_{a_k}(\cdots(\lambda_{a_1}(\iota_I([\top])))\cdots)) = 1, \text{ that is, } (\chi_F \circ \lambda_{a_k} \circ \cdots \circ \lambda_{a_1})(\mathfrak{p}_I) = 1.$$

  By duality, using Remark 1, we obtain that $\rho_\Theta((\sigma_{a_1} \circ \cdots \circ \sigma_{a_k})(\varphi)) = 1$, which is equivalent with $\Theta \models \varphi[\sigma_{a_1} \circ \cdots \circ \sigma_{a_k}]$. Thus $w \in \mathcal{L}(\mathcal{F}_A)$.
- Suppose $w \notin \mathcal{L}(A)$. Therefore $s_{k+1} \notin F$ and it follows that

  $$\chi_F(\lambda_{a_k}(\cdots(\lambda_{a_1}(\iota_I([\top])))\cdots)) = 0, \text{ that is, } (\chi_F \circ \lambda_{a_k} \circ \cdots \circ \lambda_{a_1})(\mathfrak{p}_I) = 0.$$

  Again, using Remark 1, it follows that $\rho_\Theta((\sigma_{a_1} \circ \cdots \circ \sigma_{a_k})(\varphi)) = 0$, which is equivalent with $\Theta \not\models \varphi[\sigma_{a_1} \circ \cdots \circ \sigma_{a_k}]$, that is, $w \notin \mathcal{L}(\mathcal{F}_A)$.

We have proved that $\mathcal{L}(A) = \mathcal{L}(\mathcal{F}_A)$. □

Remark 3. Note that, without loss of generality, we can fix, once and for all, a min-term α and always assume that the isomorphism $f : S \to \text{Spec } F_n(\mathcal{B})$ is such that $f(I) = \uparrow[\alpha]$, and hence $\Theta = \langle\alpha\rangle$. As the reader can easily verify, this can be safely assumed for all the results in the paper. Then, we could have simplified the definition of classical fortress by omitting Θ. We have preferred the present version, since, as we shall hint in the conclusions, Θ cannot be omitted when generalizing fortresses to other logics.

Theorem 2. *For every classical fortress in n variables $\mathcal{F}$, there exists a complete DFA $A_\mathcal{F}$ with 2^n states such that $\mathcal{L}(\mathcal{F}) = \mathcal{L}(A_\mathcal{F})$.*

Proof. Let $\mathcal{F} = \langle\varphi, \{\sigma_a\}_{a\in\Sigma}, \Theta\rangle$ be a classical fortress in n variables. Referring to Remark 1, we hence have the following uniquely determined arrows:

$$\{0,1\} \xleftarrow{\;\rho_\Theta\;} F_n(\mathcal{B}) \xleftarrow{\;\bar{\sigma}_a\;} F_n(\mathcal{B}) \xleftarrow{\;[\varphi]\;} F_1(\mathcal{B})$$

Reversing the arrows via Remark 1, we obtain

$$\text{Spec } \{0,1\} \xrightarrow{\;\iota_I\;} \text{Spec } F_n(\mathcal{B}) \xrightarrow{\;\lambda_a\;} \text{Spec } F_n(\mathcal{B}) \xrightarrow{\;\chi_F\;} \text{Spec } F_1(\mathcal{B})$$

where $I = \uparrow[\alpha_\Theta]$ for α_Θ being the unique min-term such that $\Theta = \langle\alpha_\Theta\rangle$ and hence $\iota_I([\top]) = I$; $F = \{\uparrow[\alpha] \mid \alpha \text{ min-term}, [\varphi] \in \uparrow[\alpha]\}$ and hence χ_F is the characteristic function of F. Now, we take $S = \text{Spec } F_n(\mathcal{B})$, and $\delta(s, a) = \lambda_a(s)$, for every $s \in S$ and $a \in \Sigma$. Therefore, we can consider the automaton

$$A_\mathcal{F} = (S, I, \delta, F).$$

By the definition of I and δ, $A_\mathcal{F}$ is a complete DFA with 2^n states.

In the sequel, we show that $\mathcal{L}(\mathcal{F}) = \mathcal{L}(A_\mathcal{F})$. Let $w \in \Sigma^*$, $w = a_1 \cdots a_k$. We have two cases to consider.

- Suppose $w \in \mathcal{L}(\mathcal{F})$. Therefore $\Theta \models \varphi[\sigma_{a_1} \circ \cdots \circ \sigma_{a_k}]$ and hence, $\rho_\Theta((\sigma_{a_1} \circ \cdots \circ \sigma_{a_k})(\varphi)) = 1$. Whence, $(\chi_F \circ \lambda_{a_k} \circ \cdots \circ \lambda_{a_1})(\mathfrak{p}_I) = 1$, or equivalently, $\chi_F(\lambda_{a_k}(\cdots(\lambda_{a_1}(\iota_I([\top])))\cdots)) = 1$. Taking into account how I, δ and F were defined, we obtain that there exists a finite sequence of states $s_1, \ldots, s_{k+1}$ such that $s_1 = \iota_I([\top]) = I$, $\delta(s_i, a_i) = s_{k+1}$ and $s_{k+1} \in F$. Thus $w \in \mathcal{L}(A_\mathcal{F})$.
- Suppose $w \notin \mathcal{L}(\mathcal{F})$, that is $\Theta \not\models \varphi[\sigma_{a_1} \circ \cdots \circ \sigma_{a_k}]$. Therefore, $\rho_\Theta((\sigma_{a_1} \circ \cdots \circ \sigma_{a_k})(\varphi)) = 0$. Again by duality, it follows that $(\chi_F \circ \lambda_{a_k} \circ \cdots \circ \lambda_{a_1})(\mathfrak{p}_I) = 0$, or equivalently, $\chi_F(\lambda_{a_k}(\cdots(\lambda_{a_1}(\iota_I([\top])))\cdots)) = 0$. Therefore, there exists a finite sequence of states $s_1, \ldots, s_{k+1}$ such that $s_1 = \iota_I([\top]) = I$, $\delta(s_i, a_i) = s_{k+1}$, but $s_{k+1} \notin F$. Thus $w \notin \mathcal{L}(A_\mathcal{F})$.

We have proved that $\mathcal{L}(\mathcal{F}) = \mathcal{L}(A_\mathcal{F})$. □

The next result shows that classical fortresses are indeed another descriptor for regular languages:

Theorem 3. *A language $\mathcal{L}$ is regular if and only if there is a classical fortress $\mathcal{F}$ such that $\mathcal{L}(\mathcal{F}) = \mathcal{L}$.*

Proof. A language $\mathcal{L}$ is regular if and only if there exists a DFA A such that $\mathcal{L} = \mathcal{L}(A)$. Without loss of generality, we can assume that A is a complete DFA with 2^n states. By virtue of Theorems 1 and 2, we know that complete DFA's and classical fortresses accept the same languages, and our proof is settled. □

The following table gathers together all the ingredients needed to move from finite automata to classical fortresses and backwards.

Deterministic Automaton	**Hidden steps**		**Classical Fortress**
	Dual to algebra	**Algebra**	
Finite set of states $S = 2^n$	$2^n \cong \text{Spec } F_n(\mathcal{B})$	$F_n(\mathcal{B})$	$Form(V)$, $V = \{v_1, \ldots, v_n\}$
Transition relation $\delta : S \times \Sigma \to S$	$\text{Spec } F_n(\mathcal{B}) \xrightarrow{\lambda_a} \text{Spec } F_n(\mathcal{B})$ endomorphism, for each $a \in \Sigma$	$F_n(\mathcal{B}) \xrightarrow{\bar{\sigma}_a} F_n(\mathcal{B})$ endomorphism, for each $a \in \Sigma$	σ_a substitution, for each $a \in \Sigma$
The initial state $I \in S$	$\mathfrak{p}_I \in \text{Spec } F_n(\mathcal{B})$	Prime congruence $\bar{\Theta}$ corresp. to $\mathfrak{p}_I$	Prime theory Θ over V
Set of final states $F \subseteq S$	$\chi_F^{-1}(1)$ $\text{Spec } F_n(\mathcal{B}) \xrightarrow{\chi_F} \text{Spec } F_1(\mathcal{B})$	$[\varphi]$ an element of $F_n(\mathcal{B})$	Formula φ over V
$w = a_1 \cdots a_k$ is accepted if $\delta(I, w) \in F$	$w = a_1 \cdots a_k$ is accepted if $\chi_F((\lambda_{a_k} \circ \cdots \circ \lambda_{a_1})(\mathfrak{p}_I)) = 1$	$w = a_1 \cdots a_k$ is accepted if $(\bar{\sigma}_w([\varphi]), [\top]) \in \bar{\Theta}$	$w = a_1 \cdots a_k$ is accepted if $\Theta \models \varphi[\sigma_w]$

4 Algorithm for Passing from Automata to Fortresses

In this section we present an algorithm that having as input the specification of a deterministic complete automaton $A = (S, I, \delta, F)$ $|S| = 2^n$ builds a fortress $\mathcal{F} = (\varphi, \{\sigma_a\}_{a \in \Sigma}, \Theta)$ in n variables such that $\mathcal{L}(A) = \mathcal{L}(\mathcal{F})$.

Algorithm 1

1. **[The variables V]**
For each $0 \le j \le 2^n - 1$, we represent the state $s_j \in S$ by the binary representation of j, that is, $s_j = \overline{k_1^j \cdots k_n^j}$, where each k_i^j is a bit. We therefore fix the variable set as $V = \{v_1, \ldots, v_n\}$.

2. **[The formula φ]**
For each final state $s \in F$, consider $0 \le j \le 2^n - 1$ such that $s = s_j$ and take the formula
$$\alpha_s = \bigwedge_{i=1}^{n} (v_i)^{*(i)},$$
where $*(i) = k_i^j$, for every $1 \le i \le n$. If $F = \{s_{p_1}, \ldots, s_{p_m}\}$, we take the formula $\varphi = \alpha_{s_{p_1}} \vee \cdots \vee \alpha_{s_{p_m}}$.

3. **[The theory Θ]**
For the initial state $I \in S$, consider $0 \le j \le 2^n - 1$ such that $I = s_j$ and take the formula
$$\alpha_I = \bigwedge_{i=1}^{n} (v_i)^{*(i)},$$
where $*(i) = k_i^j$, for every $1 \le i \le n$. We take the theory $\Theta = \langle \alpha_I \rangle$.

4. **[The substitutions $\{\sigma_a\}_{a \in \Sigma}$]**
Let $a \in \Sigma$. For each $0 \le j \le 2^n - 1$, we have $\delta(s_j, a) = s_{t_j}$ for some $0 \le t_j \le 2^n - 1$. Note that $s_j = \overline{k_1^j \cdots k_n^j}$ and $s_{t_j} = \overline{k_1^{t_j} \cdots k_n^{t_j}}$. For every $1 \le p \le n$, we build the formula β_p^a as follows: Let $L = \{l \mid 0 \le l \le 2^n - 1,\ k_p^{t_l} = 1\}$.
   - For each $l \in L$, we build the formula
   $$\alpha_{p,l} = \bigwedge_{i=1}^{n} (v_i)^{*(i)},$$
   where $*(i) = k_i^l$, for every $1 \le i \le n$.
   - Set $\beta_p^a = \bigvee_{l \in L} \alpha_{p,l}$.

   We define the substitution $\sigma_a(v_p) = \beta_p^a$.

Let us investigate how Algorithm 1 works by the following example:

Example 1. Let us consider the complete deterministic automaton A with 2^2 states depicted as follows:

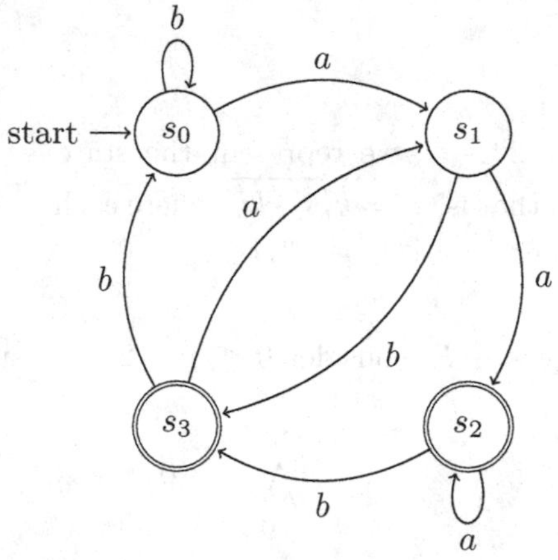

The language accepted by A is $\mathcal{L}(A) = \{(a+b)^*(aa+ab)\}$. We apply Algorithm 1 to find a fortress $\mathcal{F}$ that accepts the language $\mathcal{L}(A)$.

1. **[The variables** V**]** We represent the states as in the following table:

$\mathbf{s_j}$	$\mathbf{k_1^j}$	$\mathbf{k_2^j}$
s_0	0	0
s_1	0	1
s_2	1	0
s_3	1	1

The variables used by the fortress $\mathcal{F}$ are $V = \{v_1, v_2\}$.

2. **[The formula** φ**]** Since the final states of A are s_2 and s_3, we consider the formula $\alpha_{s_2} = v_1 \wedge \neg v_2$ and $\alpha_{s_3} = v_1 \wedge v_2$. We take the formula $\varphi = \alpha_{s_2} \vee \alpha_{s_3}$, that is, $\varphi = (v_1 \wedge \neg v_2) \vee (v_1 \wedge v_2)$.
3. **[The theory** Θ**]** Since the initial state of A is s_0, we consider the formula $\alpha_{s_0} = \neg v_1 \wedge \neg v_2$ and take the theory $\Theta = \langle \neg v_1 \wedge \neg v_2 \rangle$.
4. **[The substitutions** $\{\sigma_a\}_{a \in \Sigma}$**]** We have only two letters in the alphabet:

– *Case of letter* $a \in \Sigma$. The transitions with a in the automaton A are:

$\mathbf{s_j}$	$\delta(\mathbf{s_j}, \mathbf{a}) = \mathbf{s_{t_j}}$
s_0	s_1
s_1	s_2
s_2	s_2
s_3	s_1

$\mathbf{s_j}$		$\mathbf{s_{t_j}}$	
$\mathbf{k_1^j}$	$\mathbf{k_2^j}$	$\mathbf{k_1^{t_j}}$	$\mathbf{k_2^{t_j}}$
0	0	0	1
0	1	1	0
1	0	1	0
1	1	0	1

- For $p = 1$, we obtain $\alpha_{1,1} = \neg v_1 \wedge v_2$ and $\alpha_{1,2} = v_1 \wedge \neg v_2$. We build $\beta_1^a = \alpha_{1,1} \vee \alpha_{1,2}$ and we set $\sigma_a(v_1) = \beta_1^a$, that is,

$$\sigma_a(v_1) = (\neg v_1 \wedge v_2) \vee (v_1 \wedge \neg v_2).$$

- For $p = 2$, we obtain $\alpha_{2,0} = \neg v_1 \wedge \neg v_2$ and $\alpha_{2,3} = v_1 \wedge v_2$. We build $\beta_2^a = \alpha_{2,0} \vee \alpha_{2,3}$ and we set $\sigma_a(v_2) = \beta_2^a$, that is,

$$\sigma_a(v_2) = (\neg v_1 \wedge \neg v_2) \vee (v_1 \wedge v_2).$$

– *Case of letter* $b \in \Sigma$. The transitions with b in the automaton A are:

$\mathbf{s_j}$	$\delta(\mathbf{s_j}, \mathbf{b}) = \mathbf{s_{t_j}}$
s_0	s_0
s_1	s_3
s_2	s_3
s_3	s_0

$\mathbf{s_j}$		$\mathbf{s_{t_j}}$	
$\mathbf{k_1^i}$	$\mathbf{k_2^i}$	$\mathbf{k_1^{t_i}}$	$\mathbf{k_2^{t_i}}$
0	0	0	0
0	1	1	1
1	0	1	1
1	1	0	0

- For $p = 1$, we obtain $\alpha_{1,1} = \neg v_1 \wedge v_2$ and $\alpha_{1,2} = v_1 \wedge \neg v_2$. We build $\beta_1^b = \alpha_{1,1} \vee \alpha_{1,2}$ and we set $\sigma_b(v_1) = \beta_1^b$, that is,

$$\sigma_b(v_1) = (\neg v_1 \wedge v_2) \vee (v_1 \wedge \neg v_2).$$

- For $p = 2$, we obtain $\alpha_{2,1} = \neg v_1 \wedge v_2$ and $\alpha_{2,2} = v_1 \wedge \neg v_2$. We build $\beta_2^b = \alpha_{2,1} \vee \alpha_{2,2}$ and we set $\sigma_b(v_2) = \beta_2^b$, that is,

$$\sigma_b(v_2) = (\neg v_1 \wedge v_2) \vee (v_1 \wedge \neg v_2).$$

Using standard logical equivalences, and the derived connective $\varphi \leftrightarrow \psi := (\neg\varphi \vee \psi) \wedge (\varphi \vee \neg\psi)$, we obtain that

$$\varphi \equiv v_1, \qquad \sigma_a(v_1) = \sigma_b(v_1) = \sigma_b(v_2) \equiv \neg(v_1 \leftrightarrow v_2), \qquad \sigma_a(v_2) \equiv v_1 \leftrightarrow v_2.$$

We have built the fortress $\mathcal{F} = (\varphi, \{\sigma_a\}_{a\in\Sigma}, \Theta)$ in the variables $\{v_1, v_2\}$, where:

$\varphi = v_1$,

	$\sigma_{\mathbf{a}}$	$\sigma_{\mathbf{b}}$
v_1	$\neg(v_1 \leftrightarrow v_2)$	$\neg(v_1 \leftrightarrow v_2)$
v_2	$v_1 \leftrightarrow v_2$	$\neg(v_1 \leftrightarrow v_2)$

, $\Theta = \langle \neg v_1 \wedge \neg v_2 \rangle$.

4.1 Runs in a Fortress

Given a classical fortress $\mathcal{F}$, we can use duality to implement the computation deciding whether a word w is accepted by $\mathcal{F}$. Each substitution σ_a, in fact, corresponds dually to the map λ_a : Spec $F_n(\mathcal{B}) \to$ Spec $F_n(\mathcal{B})$ which moves prime filters of $F_n(\mathcal{B})$ into prime filters of $F_n(\mathcal{B})$. In other words, via the usual 1-1 correspondences between prime filters of $F_n(\mathcal{B})$, atoms of $F_n(\mathcal{B})$, min-terms in n variables and valuations in n variables, each $\bar{\sigma}_a$ can be regarded as the dual of a map moving valuations in valuations, or equivalently, for every $a \in \Sigma$, $\lambda_a : \{0,1\}^n \to \{0,1\}^n$. Given a word $w = a_1 \cdots a_k \in \Sigma^*$, we denote by λ_w the composition map $\lambda_{a_k} \circ \cdots \circ \lambda_{a_1}$ dual to $\bar{\sigma}_w$. Therefore, the lemma below easily follows.

Lemma 2. *Let $\mathcal{F} = (\varphi, \{\sigma_a\}_{a\in\Sigma}, \Theta)$ be a fortress in n-variables and let $w \in \Sigma^*$. Let $\rho_\Theta(V) = \langle \rho_\Theta(v_1), \ldots, \rho_\Theta(v_n)\rangle \in \{0,1\}^n$, and $\chi_\varphi\colon \{0,1\}^n \to \{0,1\}$ be the function $\langle t_1, \ldots, t_n\rangle \mapsto \rho_t(\varphi)$, where $\rho_t(v_i) = t_i$ for each $i = 1, 2, \ldots, n$. Then $\mathcal{F} \Vdash w$ iff $\chi_\varphi \circ \lambda_w \circ \rho_\Theta(V) = 1$.*

Consider the following example:

Example 2. Let $\Sigma = \{a, b\}$ and consider the fortress $\mathcal{F} = (\varphi, \{\sigma_a\}_{a\in\Sigma}, \Theta)$ in 2-variables obtained in Example 1:

$\varphi = v_1,$

	$\sigma_{\mathbf{a}}$	$\sigma_{\mathbf{b}}$
v_1	$\neg(v_1 \leftrightarrow v_2)$	$\neg(v_1 \leftrightarrow v_2)$
v_2	$v_1 \leftrightarrow v_2$	$\neg(v_1 \leftrightarrow v_2)$

, $\Theta = \langle \neg v_1 \wedge \neg v_2\rangle.$

The valuation ρ_Θ, being the unique model of Θ, must map to 1 its generating min-term $\neg v_1 \wedge \neg v_2$. Hence, $\rho_\Theta(v_1) = \rho_\Theta(v_2) = 0$. The maps λ_a and λ_b act on $\{0,1\}^2$ as follows:

$$\lambda_a(x, y) = (\max(\min(x, 1-y), \min(1-x, y)), \min(\max(1-x, y), \max(x, 1-y))),$$
$$\lambda_b(x, y) = (\max(\min(x, 1-y), \min(1-x, y)), \max(\min(x, 1-y), \min(1-x, y))).$$

Then:

– $\mathcal{F} \Vdash abaa$. Indeed,

$$\begin{aligned}\chi_\varphi \circ \lambda_{abaa} \circ \rho_\Theta(v_1, v_2) &= \chi_\varphi \circ \lambda_a \circ \lambda_a \circ \lambda_b \circ \lambda_a(0,0)\\ &= \chi_\varphi \circ \lambda_a \circ \lambda_a \circ \lambda_b(0,1)\\ &= \chi_\varphi \circ \lambda_a \circ \lambda_a(1,1)\\ &= \chi_\varphi \circ \lambda_a(0,1)\\ &= \chi_\varphi(1,0)\\ &= \rho_{(1,0)}(\varphi) = 1.\end{aligned}$$

– $\mathcal{F} \nVdash aba$. Indeed,

$$\begin{aligned}\chi_\varphi \circ \lambda_{aba} \circ \rho_\Theta(v_1, v_2) &= \chi_\varphi \circ \lambda_a \circ \lambda_b \circ \lambda_a(0,0)\\ &= \chi_\varphi \circ \lambda_a \circ \lambda_b(0,1)\\ &= \chi_\varphi \circ \lambda_a(1,1)\\ &= \chi_\varphi(0,1)\\ &= \rho_{(0,1)}(\varphi) = 0.\end{aligned}$$

5 Reduced Fortresses

As we have seen, a fortress corresponds to a DFA with a number of states which is a power of 2. This is not a limitation since, as we have already said, each DFA recognises the same language of a DFA whose number of states is a power of 2. This notwithstanding, it can be interesting to provide a variant of the notion of fortress which naturally associates with any complete DFA, with no constraints on the number of states.

Definition 3. *Let $\mathcal{F} = (\varphi, \{\sigma_a\}_{a\in\Sigma}, \Theta)$ be a classical fortress. Assume there is a theory Γ such that:*

1. *$\Gamma \subseteq \Theta$, that is, $\Theta \models \gamma$ for all $\gamma \in \Gamma$;*
2. *$\Gamma \models \psi$ implies $\Gamma \models \psi[\sigma_a]$ for all $a \in \Sigma$ and all formulas ψ.*

Then the quadruple $\mathcal{F}_\Gamma = (\varphi, \{\sigma_a\}_{a\in\Sigma}, \Theta, \Gamma)$ is called the Γ-reduction of $\mathcal{F}$.

With each Γ-reduction we associate a regular language.

Lemma 3. *Let $\mathcal{F}_\Gamma = (\varphi, \{\sigma_a\}_{a\in\Sigma}, \Theta, \Gamma)$ be the Γ-reduction of a fortress $\mathcal{F}$. Then let $A_\Gamma = (S, I, \delta, F)$ be defined as follows.*

1. *$S = \{\uparrow[\alpha] \mid \Gamma \subseteq \langle\alpha\rangle,\ \alpha$ a min-term$\}$;*
2. *$I = \uparrow[\alpha_\Theta]$ for α_Θ being the unique min-term such that $\Theta = \langle\alpha_\Theta\rangle$;*
3. *$\delta(s, a) = \delta_a(s)$ for each $s \in S$ and each $a \in \Sigma$, where $\delta_a = \mathrm{Spec}\ \bar{\sigma}_a$;*
4. *$F = \{\uparrow[\psi] \mid [\varphi] \in \uparrow[\psi],\ \psi$ a min-term$\} \cap S$.*

Then A_Γ is a complete DFA, and $\mathcal{L}(A_\Gamma) = \mathcal{L}(\mathcal{F})$.

Proof. Notice that Condition 1 in Definition 3 implies that $I = \uparrow[\alpha_\Theta]$ belongs to S. Condition 2, on the other hand, guarantees that each δ_a carries S into S. Indeed, pick $\uparrow[\beta] \in \mathrm{Spec}\ F_n(\mathcal{B}) \setminus S$, for some min-term β. Then $\Gamma \not\subseteq \langle\beta\rangle$. Since $\langle\beta\rangle$ is maximal, the latter entails $\Gamma \models \neg\beta$, and by Condition 2, $\Gamma \models \neg\beta[\sigma_a]$, too. Now, recall that, by duality,

$$\bar{\sigma}_a([\beta]) = [\bigvee\{\alpha \mid \alpha \text{ min-term}, \delta_a(\uparrow[\alpha]) = \uparrow[\beta]\}].$$

Whence, $\Gamma \models \neg\bigvee\{\alpha \mid \alpha \text{ min-term}, \delta_a(\uparrow[\alpha]) = \uparrow[\beta]\}$ and then $\Gamma \models \neg\alpha$ for each min-term α such that $\delta_a(\uparrow[\alpha]) = \uparrow[\beta]$, and, in turns, $\Gamma \not\subseteq \langle\alpha\rangle$ for each such min-term. Then for all min-terms α such that $\delta_a(\uparrow[\alpha]) = \uparrow[\beta]$, $\uparrow[\alpha]$ does not belong to S, thus proving that δ_a maps S into S. We conclude that A_Γ is a well defined complete DFA, with $S \subseteq \mathrm{Spec}\ F_n(\mathcal{B})$, which, in turns, is the set of states of the automaton $A_\mathcal{F}$ built in Theorem 2 from $\mathcal{F}$. Finally, $\mathcal{L}(A_\Gamma) = \mathcal{L}(A_\mathcal{F}) = \mathcal{L}(\mathcal{F})$. □

Notice that the number $|S|$ of states of A_Γ is not constrained to be a power of 2. From Algorithm 1 it is clear that each DFA arises as A_Γ for some theory Γ providing the Γ-reduction of a suitable fortress $\mathcal{F}$.

An interesting consequence is when the Γ-reduction is a fortress as well, in the sense made precise by the proof of the following proposition.

Proposition 1. *Let $\mathcal{F}_\Gamma$ be the Γ-reduction of a fortress $\mathcal{F}$ in n variables. If $F_n(\mathcal{B})/\bar{\Gamma}$ is isomorphic with $F_k(\mathcal{B})$ for some $k \leq n$, then there is a fortress $\mathcal{F}'$ in k variables such that $\mathcal{L}(\mathcal{F}') = \mathcal{L}(\mathcal{F})$.*

Proof. Consider the automaton A_Γ built in the proof of Lemma 3. Notice that A_Γ has exactly 2^k states as $F_n(\mathcal{B})/\bar{\Gamma}$ is isomorphic with $F_k(\mathcal{B})$. Whence, by Theorem 1, there is a fortress $\mathcal{F}'$ in k variables such that $\mathcal{L}(\mathcal{F}') = \mathcal{L}(\mathcal{F})$. □

6 Closure Properties of Regular Languages through Classical Fortresses

In this section we will give alternative proofs for some well-known results on the closure properties of regular languages in the framework of classical fortresses. By these alternative proofs, we show that classical fortresses are also a suitable tool for providing intuitive proofs for some closure properties of regular languages.

Proposition 2. *The complement of a regular language is regular.*

Proof. Let L be a regular language and let $\mathcal{F} = (\varphi, \{\sigma_a\}_{a\in\Sigma}, \Theta)$ be a classical fortress in n variables such that $\mathcal{L}(\mathcal{F}) = L$. We consider the following classical fortress in n variables:

$$\mathcal{F}_c = (\neg\varphi, \{\sigma_a\}_{a\in\Sigma}, \Theta).$$

Let $w \in \Sigma^*$. Since the prime theory Θ is maximal, we have the following:

$$w \in \Sigma^* \setminus L \;\Leftrightarrow\; w \notin \mathcal{L}(\mathcal{F}) \;\Leftrightarrow\; \Theta \not\models \varphi[\sigma_w] \;\Leftrightarrow\; \Theta \models \neg\varphi[\sigma_w] \;\Leftrightarrow\; w \in \mathcal{L}(\mathcal{F}_c).$$

Therefore $\mathcal{F}_c$ accepts the complement of L. □

Proposition 3. *The union of two regular languages is regular.*

Proof. Let L_1 and L_2 be regular languages. Then there is a classical fortress $\mathcal{F}_1 = (\varphi_1, \{\sigma_a\}_{a\in\Sigma}, \Theta_1)$ in n variables $V = \{v_1, \ldots, v_n\}$ such that $\mathcal{L}(\mathcal{F}_1) = L_1$ and a classical fortress $\mathcal{F}_2 = (\varphi_2, \{\tau_a\}_{a\in\Sigma}, \Theta_2)$ in m variables $V' = \{v'_1, \ldots, v'_m\}$ such that $\mathcal{L}(\mathcal{F}_2) = L_2$. Possibly by renaming variables, we can safely assume V and V' are disjoint, so $V \cup V'$ is a set of $n + m$ distinct variables. We consider the following classical fortress:

$$\mathcal{F}_\cup = (\varphi_1 \vee \varphi_2, \{\mu_a\}_{a\in\Sigma}, \langle\Theta_1 \cup \Theta_2\rangle),$$

in the $n + m$ variables $V \cup V'$, where: $\mu_a(v_i) = \sigma_a(v_i)$ for all $i = 1, \ldots, n$ and $\mu_a(v'_j) = \tau_a(v'_j)$ for all $j = 1, \ldots, m$. By our assumptions on V and V', all substitutions μ_a are well defined. Let $w \in \Sigma^*$. By using Lemma 1, we have the following:

$$\begin{aligned} w \in \mathcal{L}(\mathcal{F}_\cup) &\Leftrightarrow \langle\Theta_1 \cup \Theta_2\rangle \models (\varphi_1 \vee \varphi_2)[\mu_w] \\ &\Leftrightarrow \langle\Theta_1 \cup \Theta_2\rangle \models \varphi_1[\mu_w] \vee \varphi_2[\mu_w] \\ &\Leftrightarrow \langle\Theta_1 \cup \Theta_2\rangle \models \varphi_1[\sigma_w] \vee \varphi_2[\tau_w] \\ &\Leftrightarrow \langle\Theta_1 \cup \Theta_2\rangle \models \varphi_1[\sigma_w] \text{ or } \langle\Theta_1 \cup \Theta_2\rangle \models \varphi_2[\tau_w] \\ &\Leftrightarrow \Theta_1 \models \varphi_1[\sigma_w] \text{ or } \Theta_2 \models \varphi_2[\tau_w] \\ &\Leftrightarrow w \in \mathcal{L}(\mathcal{F}_1) \text{ or } w \in \mathcal{L}(\mathcal{F}_2) \\ &\Leftrightarrow w \in \mathcal{L}(\mathcal{F}_1) \cup \mathcal{L}(\mathcal{F}_2) \end{aligned}$$

Therefore, $\mathcal{F}_\cup$ accepts $L_1 \cup L_2$. □

Proposition 4. *The intersection of two regular languages is regular.*

Proof. Let L_1 and L_2 be regular languages. Then there is a classical fortress $\mathcal{F}_1 = (\varphi_1, \{\sigma_a\}_{a\in\Sigma}, \Theta_1)$ in n variables $V = \{v_1, \ldots, v_n\}$ such that $\mathcal{L}(\mathcal{F}_1) = L_1$ and a classical fortress $\mathcal{F}_2 = (\varphi_2, \{\tau_a\}_{a\in\Sigma}, \Theta_2)$ in m variables $V' = \{v'_1, \ldots, v'_m\}$ such that $\mathcal{L}(\mathcal{F}_2) = L_2$. Possibly by renaming variables, we can safely assume V and V' are disjoint, so $V \cup V'$ is a set of $n+m$ distinct variables. We consider the classical fortress

$$\mathcal{F}_\cap = (\varphi_1 \wedge \varphi_2, \{\mu_a\}_{a\in\Sigma}, \langle\Theta_1 \cup \Theta_2\rangle),$$

in the $n+m$ variables $V \cup V'$, where, for all $a \in \Sigma$, the substitution μ_a is defined as in the proof of Proposition 3.

Let $w \in \Sigma^*$. By using Lemma 1, we have the following:

$$\begin{aligned} w \in \mathcal{L}(\mathcal{F}_\cap) &\Leftrightarrow \langle\Theta_1 \cup \Theta_2\rangle \models (\varphi_1 \wedge \varphi_2)[\mu_w] \\ &\Leftrightarrow \langle\Theta_1 \cup \Theta_2\rangle \models \varphi_1[\mu_w] \wedge \varphi_2[\mu_w] \\ &\Leftrightarrow \langle\Theta_1 \cup \Theta_2\rangle \models \varphi_1[\sigma_w] \wedge \varphi_2[\tau_w] \\ &\Leftrightarrow \langle\Theta_1 \cup \Theta_2\rangle \models \varphi_1[\sigma_w] \text{ and } \langle\Theta_1 \cup \Theta_2\rangle \models \varphi_2[\tau_w] \\ &\Leftrightarrow \Theta_1 \models \varphi_1[\sigma_w] \text{ and } \Theta_2 \models \varphi_2[\tau_w] \\ &\Leftrightarrow w \in \mathcal{L}(\mathcal{F}_1) \text{ and } w \in \mathcal{L}(\mathcal{F}_2) \\ &\Leftrightarrow w \in \mathcal{L}(\mathcal{F}_1) \cap \mathcal{L}(\mathcal{F}_2) \end{aligned}$$

Therefore, $\mathcal{F}_\cap$ accepts $L_1 \cap L_2$. □

7 Beyond Classical Logic and Finite Automata

The finite slice of Stone duality is the main ingredient which allows the introduction of classical fortresses from DFA as shown in Section 4. In this section we are going to swap the perspective which allowed us to introduce classical fortresses starting from finite automata, and try to make a step forward through two main generalizations of DFA which will constitute the key arguments of our future work.

Classical fortresses, as objects specified in classical propositional logic on a finite number of variables, allow an easy generalization to any non-classical logical setting. In theoretical terms, in fact, given a propositional logical calculus L, one can easily adapt the definition of classical fortress to the frame of L and introduce a notion of L-fortress and of language accepted by such an object. The converse task, which is not always viable, is to reverse Algorithm 1 and introduce L-automata as the corresponding, computational counterpart of L-fortresses, and deduce from that a characterisation of the class of languages accepted by L-automata. In fact, a logic L allows such a turn-about, only if L enjoys the following, informally stated, properties:

1. L is algebraisable in the sense of [1], its algebraic semantics being $\mathbb{L}$[1].
2. $\mathbb{L}$ is locally finite and, hence, the n-freely generated $\mathbb{L}$-algebras are finite.

[1] Notice that we strongly used the algebraizability of classical logic when passing from the algebraic view to the logical one.

3. There is a Stone-type duality between the finite slice of $\mathbb{L}$ and a target category $\mathbb{C}$ which plays the same rôle as $\mathcal{S}et_{fin}$ does in the classical Stone duality.

In our future work we shall study the following generalizations of DFA.

1. We aim at generalizing the notion of DFA to several logics, starting from Gödel propositional logic [5, §VII], the latter being a non-classical logic which satisfies the above (1)-(3). We explicitly stress that, as we have anticipated in Remark 3, the specific choice of Θ influences the behavior of a *Gödel fortress*, as in general distinct prime congruences give rise to distinct non-isomorphic quotients of the free Gödel algebras.
2. For every cardinal κ, we shall focus on a further generalization of DFA obtained starting from a fortress in 2^{κ} variables and then applying the full Stone duality between Boolean algebras and Stone spaces to derive the corresponding notion of automaton with κ states. We shall try to identify the class of languages recognised by such devices.

References

1. Blok, W., Pigozzi, D.: Algebraizable logics. Memoirs of The American Mathematical Society, vol. 77. American Mathematical Society (1989)
2. Brzozowski, J.A.: Canonical regular expressions and minimal state graphs for definite events. In: Mathematical Theory of Automata. MRI Symposia Series, vol. 12, pp. 529–561. Polytechnic Press, Polytechnic Institute of Brooklyn (1962)
3. Büchi, J.R.: Weak second-order arithmetic and finite automata. Zeitschrift für mathematische Logik und Grundlagen der Mathematik 6, 66–92 (1960)
4. Burris, S., Sankappanavar, H.P.: A course in Universal Algebra. Springer (1981)
5. Cintula, P., Hájek, P., Noguera, C.: Handbook of Mathematical Fuzzy Logic, vol. 2. College Publications (2011)
6. Elgot, C.C.: Decision problems of finite automata design and related arithmetics. Trans. Amer. Math. Soc. 98, 21–51 (1961)
7. Hansen, H.H., Panangaden, P., Rutten, J.J.M.M., Bonchi, F., Bonsangue, M.M., Silva, A.: Algebra-coalgebra duality in Brzozwski's minimization algorithm. ACM Transactions on Computational Logic (to appear)
8. Gehrke, M.: Duality and recognition. In: Murlak, F., Sankowski, P. (eds.) MFCS 2011. LNCS, vol. 6907, pp. 3–18. Springer, Heidelberg (2011)
9. Gehrke, M., Grigorieff, S., Pin, J.-É.: Duality and equational theory of regular languages. In: Aceto, L., Damgård, I., Goldberg, L.A., Halldórsson, M.M., Ingólfsdóttir, A., Walukiewicz, I. (eds.) ICALP 2008, Part II. LNCS, vol. 5126, pp. 246–257. Springer, Heidelberg (2008)
10. Hopcroft, J.E., Motwani, R., Ullman, J.D.: Introduction to Automata Theory, Languages, and Computation, 2nd edn. Addison-Wesley (2000)
11. Johnstone, P.T.: Stone Spaces. Cambridge University Press (1982)
12. Kleene, S.C.: Representation of events in nerve nets and finite automata. In: Shannon, C.E., McCarthy, J. (eds.) Automata Studies, pp. 3–42. Princeton University Press (1956)
13. Weyuker, E.J., Davis, M.E., Sigal, R.: Computability, complexity, and languages: Fundamentals of theoretical computer science. Academic Press, Boston (1994)
14. Trakhtenbrot, B.A.: Finite automata and the logic of oneplace predicates. Siberian Math. J. 3, 103–131 (1962)

On Clock-Aware LTL Properties of Timed Automata

Peter Bezděk, Nikola Beneš*, Vojtěch Havel, Jiří Barnat, and Ivana Černá**

Faculty of Informatics, Masaryk University, Brno, Czech Republic
{xbezdek1,xbenes3,xhavel1,barnat,cerna}@fi.muni.cz

Abstract. We introduce the *Clock-Aware Linear Temporal Logic* (CA-LTL) for expressing linear time properties of timed automata, and show how to apply the standard automata-based approach of Vardi and Wolper to check for the validity of a CA-LTL formula over the continuous-time semantics of a timed automaton. Our model checking procedure employs zone-based abstraction and a new concept of the so called ultraregions. We also show that the Timed Büchi Automaton Emptiness problem is not the problem that the intended automata-based approach to CA-LTL model checking is reduced to. Finally, we give the necessary proofs of correctness, some hints for an efficient implementation, and preliminary experimental evaluation of our technique.

Keywords: Linear Temporal Logic, Timed Automata, Automata-based Model Checking.

1 Introduction

Model checking [1] is a formal verification technique applied to check for logical correctness of discrete distributed systems. While it is often used to prove the unreachability of a bad state (such as an assertion violation in a piece of code), with a proper specification formalism, such as the *Linear Temporal Logic* (LTL), it can also check for many interesting liveness properties of systems, such as repeated guaranteed response, eventual stability, live-lock, etc.

Timed automata have been introduced in [2] and have become a widely accepted framework for modelling and analysis of time-critical systems. The formalism is built on top of the standard finite automata enriched with a set of real-time clocks and allowing the system actions to be guarded with respect to the clock valuations. In the general case, such a timed system exhibits infinite-state semantics (the clock domains are continuous). Nevertheless, when the guards are limited to comparing clock values with integers only, there exists a bisimilar finite state representation of the original infinite-state real-time system referred

* The author has been supported by the MEYS project No. CZ.1.07/2.3.00/30.0009 Employment of Newly Graduated Doctors of Science for Scientific Excellence.
** The author has been supported by the MEYS project No. LH11065 Control Synthesis and Formal Verification of Complex Hybrid Systems.

G. Ciobanu and D. Méry (Eds.): ICTAC 2014, LNCS 8687, pp. 43–60, 2014.

to as the region abstraction. The region abstraction builds on top of the observation that concrete real-time clock valuations that are between two consecutive integers are indistinguishable with respect to the valuation of an action guard. Unfortunately, the size of the region-based abstraction grows exponentially with the number of clocks and the largest integer number used. As a result, the region-based abstraction is difficult to be used in practice for the analysis of more than academic toy examples, even though it has its theoretical value.

A practically efficient abstraction of the infinite-state space came with the so called zones [3]. Unlike the region-based abstraction, a single state in the zone-based abstraction is no more restricted to represent only those clock values that are between two consecutive integers. Therefore, the zone-based abstraction is much coarser and the number of zones *reachable* from the initial state is significantly smaller. This in turns allows for efficient implementation of verification tools for timed automata, see e.g. UPPAAL [4].

In this paper we solve the model checking problem of linear time properties over timed automata. To that end we introduce *Clock-Aware Linear Temporal Logic* (CA-LTL), which is a linear time logic built from the standard boolean operators, the standard LTL operator *Until*, and atomic propositions that are boolean combinations of comparisons of clock valuations against integer constants and guards over variables of an underlying timed automaton.

The ability to use clock-valuation constraints as atomic propositions makes the newly introduced logic rather powerful. Note, for example, that in terms of expressibility, it completely covers the fragment of TCTL as used for specification purposes by UPPAAL model checker. The non-trivial expressive power of CA-LTL is also witnessed with a CA-LTL formula $\mathbf{FG}(x \leq 3)$ expressing that the timed automaton under investigation will eventually come to a stable state where it is guaranteed that from that time on the clock variable x will never exceed the value of 3, i.e. a reset of x is going to happen somewhat regularly.

Regarding model checking of CA-LTL we stress that we are aware of the so called *Timed Büchi Automaton Emptiness* problem [5,6,7]. Timed Büchi Automaton Emptiness could be considered as the solution to the problem of LTL model checking over timed automata provided that the logic used cannot refer to clock valuations. However, for CA-LTL we believe and show later in this paper that the solution of CA-LTL model checking does not reduce to the problem of Timed Büchi Automaton Emptiness.

Contribution. In this paper we define the syntax and the continuous-time semantics of the*Clock-Aware Linear Temporal Logic* (CA-LTL). We then show how to apply the standard automata-based approach to LTL model checking of Vardi and Wolper [8] for a CA-LTL formula and a timed automaton. In particular, we show how to construct a Büchi automaton coming from the CA-LTL specification with a zone-based abstraction of a timed automaton representing the system under verification using the so-called ultraregions. We give the necessary proof of correctness of our construction and list some hints that lead towards an efficient implementation of it. We also report on the practical impact of introducing ultraregions on the size of the zone-base abstracted timed automaton graph.

Outline. The rest of the paper is organised as follows. We first list the preliminaries and define our new CA-LTL in Section 2. Then, we relate CA-LTL with other logics and explain the motivation behind our approach in Section 3. The technical core of the synchronised product of a Büchi automaton and a zone-based abstracted timed automaton is given in Section 4 including the sketch of the proof of correctness. Section 5 gives some details on the implementation of our construction and lists some experimental measurements we did. Finally, Section 6 concludes the paper. Due to space constraints, we did not include the full technically detailed proofs in this paper. These can be found in the technical report [9].

2 Preliminaries and Problem Statement

Let X be a finite set of *clocks*. A *simple guard* is an expression of the form $x \sim c$ where $x \in X$, $c \in \mathbb{N}_0$, and $\sim \in \{<, \leq, \geq, >\}$. A conjunction of simple guards is called a *guard*; the empty conjunction is denoted by the boolean constant **tt**. We use $\mathcal{G}(X)$ to denote the set of all guards over a set of clocks X. A *clock valuation over* X is a function $\eta : X \to \mathbb{R}_{\geq 0}$ assigning non-negative real numbers to each clock. We denote the set of all clock valuations over X by $\mathbb{R}^X_{\geq 0}$ and the valuation that assigns 0 to each clock by $\mathbf{0}$. For a guard g and a valuation η, we say that η *satisfies* g, denoted by $\eta \models g$, if g evaluates to true when all x in g are replaced by $\eta(x)$.

We define two operations on clock valuations. Let η be a clock valuation, d a non-negative real number and $R \subseteq X$ a set of clocks to reset. We use $\eta + d$ to denote the clock valuation that adds the delay d to each clock, i.e. $(\eta + d)(x) = \eta(x) + d$ for all $x \in X$. We further use $\eta[R]$ to denote the clock valuation that resets clocks from the set R, i.e. $\eta[R](x) = 0$ if $x \in R$, $\eta[R](x) = \eta(x)$ otherwise.

Definition 2.1. *A* timed automaton *(TA) is a tuple* $A = (L, l_0, X, \Delta, Inv)$ *where*

- L *is a finite set of locations,*
- $l_0 \in L$ *is an initial location,*
- X *is a finite set of clocks,*
- $\Delta \subseteq L \times \mathcal{G}(X) \times 2^X \times L$ *is a finite transition relation, and*
- $Inv : L \to \mathcal{G}(X)$ *is an invariant function.*

We use $q \xrightarrow{g,R}_{\Delta} q'$ *to denote* $(q, g, R, q') \in \Delta$.

In the following, we assume that the invariants are upper bounds only, i.e. of the form $x < c$ or $x \leq c$. Note that this is without loss of generality, as lower bound invariants may be always moved to guards of incoming transitions.

The semantics of a timed automaton is given as a labelled transition system.

Definition 2.2. *A* labelled transition system *(LTS) over a set of symbols* Σ *is a triple* $(S, s_0, \to)$, *where*

– *S is a set of states,*
– *$s_0 \in S$ is an initial state, and*
– *$\rightarrow \subseteq S \times \Sigma \times S$ is a transition relation.*

We use $s \xrightarrow{a} s'$ to denote $(s, a, s') \in \rightarrow$.

Definition 2.3 (TA Semantics). *Let $A = (L, l_0, X, \Delta, Inv)$ be a TA. The semantics of A, denoted by $[\![A]\!]$, is a LTS $(S, s_0, \rightarrow)$ over the set of symbols $\{act, \infty\} \cup \mathbb{R}_{\geq 0}$, where*

– $S = \{(l, \eta) \in L \times \mathbb{R}^X_{\geq 0} \mid \eta \models Inv(l)\} \cup \{(l, \infty) \mid Inv(l) = \mathbf{tt}\}$,
– $s_0 = (l_0, \mathbf{0})$,
– *the transition relation $\rightarrow$ is specified for all $(q, \eta), (q', \eta') \in S$ such that η is a clock valuation as follows:*
  - $(q, \eta) \xrightarrow{d} (q', \eta')$ *if* $q = q'$, $d \in \mathbb{R}_{\geq 0}$, *and* $\eta' = \eta + d$,
  - $(q, \eta) \xrightarrow{\infty} (q', \eta')$ *if* $q = q'$ *and* $\eta' = \infty$,
  - $(q, \eta) \xrightarrow{act} (q', \eta')$ *if* $\exists g, R : q \xrightarrow{g,R}_{\Delta} q'$, $\eta \models g$, *and* $\eta' = \eta[R]$.

The first two kinds of transitions are called delay transitions, *the latter are called* action transitions.

In the following, we assume that we only deal with deadlock-free timed automata, i.e. that the only states without outgoing transitions in $[\![A]\!]$ are of the form (l, ∞). A deadlock usually signalises a severe error in the model and its (non-)existence may be ascertained in the standard way.

A *proper run* of $[\![A]\!]$ is an alternating sequence of delay and action transitions that begins with a delay transition and is either infinite or ends with a ∞ delay transition. The length of a proper run $|\pi|$ is the number of action transitions it contains. A proper run is called a Zeno run if it is infinite while the sum of all its delays is finite. Zeno runs usually represent non-realistic behaviour and it is thus desirable to ignore them in TA analysis. However, we postpone the question of dealing with Zeno runs until Section 4.

We now define the syntax and semantics of the clock-aware linear temporal logic. The atomic propositions of this logic are going to be of two kinds—those that consider properties of locations and those that consider properties of clocks. The former ones, which we call location propositions are just arbitrary symbols that are assigned to locations via a labelling function. The latter ones are simple guards over the set of clocks.

Definition 2.4 (CA-LTL Syntax). *Let $Ap = Lp \cup G$ where Lp is a set of* location propositions *and G is a set of simple guards. A* clock-aware linear temporal logic *(CA-LTL) formula over Ap is defined as follows:*

$$\varphi ::= l \mid g \mid \neg\varphi \mid \varphi \vee \varphi \mid \varphi\, \mathbf{U}\, \varphi$$

where $l \in Lp$ and $g \in G$.

We also use the standard derived boolean operators such as $\wedge$ and $\Rightarrow$, and the usual derived temporal operators $\mathbf{F}\,\varphi \equiv \mathbf{tt}\,\mathbf{U}\,\varphi$, $\mathbf{G}\,\varphi \equiv \neg\,\mathbf{F}\,\neg\varphi$.

We want our semantics of CA-LTL to reason about continuous linear time. We thus need a notion of a (continuous) suffix of a proper run. For a proper run $\pi = (l_0, \eta_0) \xrightarrow{d_0} (l_0, \eta_0 + d_0) \xrightarrow{act} (l_1, \eta_1) \xrightarrow{d_1} (l_1, \eta_1 + d_1) \xrightarrow{act} \cdots$ we define its suffix $\pi^{k,t}$ as follows:

- if $|\pi| > k$ and $t \le d_k$ then $\pi^{k,t} = (l_k, \eta_k + t) \xrightarrow{d_k - t} (l_k, \eta_k + d_k) \xrightarrow{act} \cdots$,
- if $|\pi| = k$ then $\pi^{k,t} = (l_k, \eta_k + t) \xrightarrow{\infty} (l_k, \infty)$,
- otherwise, $\pi^{k,t}$ is undefined.

Note that the condition $|\pi| = k$ implies that π ends with $\cdots (l_k, \eta_k) \xrightarrow{\infty} (l_k, \infty)$. We further define an ordering on the set of suffixes of π, denoted by $\lessdot_\pi$ as follows: $\pi^{i,t} \lessdot_\pi \pi^{j,s}$ if both $\pi^{i,t}$ and $\pi^{j,s}$ are defined and either $i < j$ or $i = j$ and $t \le s$. (The semantics is that $\pi^{i,t}$ is an "earlier" suffix of π than $\pi^{j,s}$.)

Definition 2.5 (CA-LTL Semantics). *Let $\mathcal{L} : L \to 2^{Lp}$ be a function that assigns a set of location propositions to each location. The semantics of a CA-LTL formula φ on a proper run $\pi = (l_0, \eta_0) \xrightarrow{d} (l_0, \eta_0 + d_0) \xrightarrow{act} (l_1, \eta_1) \xrightarrow{d1} \cdots$ with a labelling $\mathcal{L}$ is given as follows (the semantics of the boolean operators is the usual one and is omitted here):*

$$\begin{aligned}
\pi &\models p &&\iff p \in \mathcal{L}(l_0)\\
\pi &\models g &&\iff \eta_0 \models g\\
\pi &\models \varphi\,\mathbf{U}\,\psi &&\iff \exists k,t : \pi^{k,t} \textit{ is defined}, \pi^{k,t} \models \psi, \textit{ and}\\
& && \qquad \forall j,s \textit{ such that } \pi^{j,s} \lessdot_\pi \pi^{k,t} : \pi^{j,s} \models \varphi \vee \psi
\end{aligned}$$

For a timed automaton A with a location labelling function $\mathcal{L}$, we say that A with $\mathcal{L}$ satisfies φ, denoted by $(A, \mathcal{L}) \models \varphi$ if for all proper runs π of $[\![A]\!]$, $\pi \models \varphi$.

The goal of this paper is to solve the following problem.

CA-LTL Model Checking Problem. Given a timed automaton A, a location labelling function $\mathcal{L}$, and a CA-LTL formula φ, decide whether $(A, \mathcal{L}) \models \varphi$.

3 Related Work and Motivation

There is a plethora of derivatives of linear temporal logics for the specification of properties of real-time systems, timed automata in particular. To name at least some of them, we list TPTL [10], MTL [11], MITL [12], RTTL [13], XCTL [14], CLTL [15], and LTLC [16]. These logics employ various ways of expressing time aspects of underlying systems including one global time clock, time-bounded temporal operators, timing variables with quantifiers, and freeze operators. Some logics are defined with the use of time sampling semantics, which has been shown to be counter-intuitive [17]. The key aspect differentiating our CA-LTL from the logics mentioned above is the ability to properly and intuitively reason about clock values in the classical continuous-time semantics while still preserving practical efficiency of the model checking process.

Similar qualities are found in the branching time logic TCTL [18] a subset of which is actually supported with UPPAAL tool. Our motivation to introduce CA-LTL was to mimic the branching time TCTL in a linear time setting. We stress that CA-LTL is able to reason about values of clocks in timed automata while still being practically simple enough to allow for an efficient model checking procedure. Note that the inclusion of time-bounded operators, such as the *until* operator of TCTL, would lead to the expressive power of at least MTL, model checking of which is considered computationally infeasible. CA-LTL can thus be seen as a practically motivated extension of LTL, which is powerful enough to express the same properties as can be expressed by the specification language of the world-wide leading timed automata verification tool UPPAAL.[1]

Timed automata can be defined with different types of semantics. The standard continuous-time semantics (as used also for the definition of CA-LTL) is in many cases substituted with the so called sampling semantics. However, it has been shown in [17] that cycle detection under the sampling semantics of timed automata with unknown sampling rate is undecidable.

We now use an example of a timed automaton and some CA-LTL formulae to explain the intricacies of our model checking problem.

Example 3.1. Let us consider a timed automaton as given in Fig. 1 with the labelling function $\mathcal{L}$ assigning to each location its own name only, i.e. $\mathcal{L}(l) = \{l\}$ for all l. Let us further consider the CA-LTL formulae

$$\varphi = \mathbf{G}(l_1 \Rightarrow ((x \leq 3 \wedge y \leq 3)\, \mathbf{U}\, (x > 3 \wedge y > 3))) \quad \text{and} \quad \psi = \mathbf{F}\, l_3.$$

Note that while there exists a run satisfying φ and a run satisfying ψ, there is no run satisfying their conjunction, $\varphi \wedge \psi$. The reason is that the runs satisfying φ always perform the reset of y at time 0, while the runs satisfying ψ always perform the reset of y at some other time, to be able to satisfy the guard $y < 6$ together with $x = 6$.

First of all, note that there is no obvious way of combining this TA with a Büchi automaton representing the formula φ (or its negation). The reason is that while staying in l_0, the satisfaction of the guards $x \leq 3$, $y \leq 3$ changes. We could try splitting each location into several ones such that staying in each of these new locations ensures no changes of the guards. However, under the standard TA semantics, such feature is impossible. Indeed, if there were two locations with invariants $x \leq 3$ and $x > 3$, respectively, no transition between them could be enabled at any time. There thus appears to be no direct way of reducing our problem to Timed Büchi Automaton Emptiness.

One way of solving the problem whether a timed automaton satisfies a CA-LTL formula is to evaluate the formula as a standard LTL formula over the automaton's region graph. Suppose that we have the standard region graph construction [19], in which the maximal bounds on each clock also include the

[1] For more details, see Section 6 in [9].

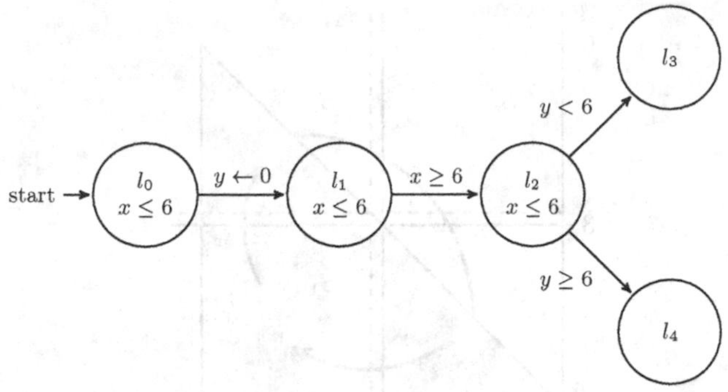

Fig. 1. Timed automaton $A_{3.1}$

bounds appearing in the formula. Then the satisfaction of guards inside a region never changes. This shows that the CA-LTL problem is in the PSPACE complexity class, as both the region graph and the Büchi automaton for the formula may be created on the fly. However, it is well known that the number of regions is impractically large. In the following, we therefore aim to provide a zone-based model checking approach.

Considering the standard zone-based approach [20], the main issue pointed out above remains—the satisfaction of the guards differs for various parts of a zone. Our first idea is to slice (pre-partition) the zones according to the guards of the formula. In our example, this would mean to slice the zones into one of the four "quadrants" $[0,3] \times [0,3]$, $(3,\infty) \times [0,3]$, $(3,\infty) \times (3,\infty)$, $[0,3] \times (3,\infty)$. These are illustrated in Fig. 2 and named S_0 to S_3. Every sliced zone now respects the guards $x \leq 3$, $y \leq 3$. Note, however, that this partitioning also comes with the need of describing new transitions between the newly defined zone slices. These new transitions correspond to the passage of time within the original zone. Now, consider again Example 1 and the zone that is created after the transition from l_0 to l_1 is taken. The zone is defined as the set of all valuations of clocks ν such that $\nu(x) - \nu(y) \geq 0$ and $\nu(x), \nu(y) \in [0,6]$, also illustrated with the greyed area in Fig. 2.

Let us take the S_0 slice of this zone. The next slice is not uniquely determined. One candidate is the S_1 slice as all valuations with $\nu(x) - \nu(y) > 0$ will reach this slice with the passage of time. However, we also have another candidate, the S_2 slice. This is due to the fact that all valuations with $\nu(x) - \nu(y) = 0$ reach the S_2 slice immediately after leaving the S_0 slice. We cannot take both options with a nondeterministic choice. This would introduce incorrect behaviour, as then there would be a run in the zone graph satisfying the conjunction of formulae $\varphi \wedge \psi$. Therefore, we also need to take diagonals into account. The problem here is very similar to the problem that led to the inclusion of diagonals into standard region graphs. Our slicing areas thus somehow resemble regions, only much larger. Also, their count is only dependent on the number of guards appearing in the CA-LTL formula, and may thus be expected to be reasonable. For their similarity with regions, we call these areas *ultraregions*.

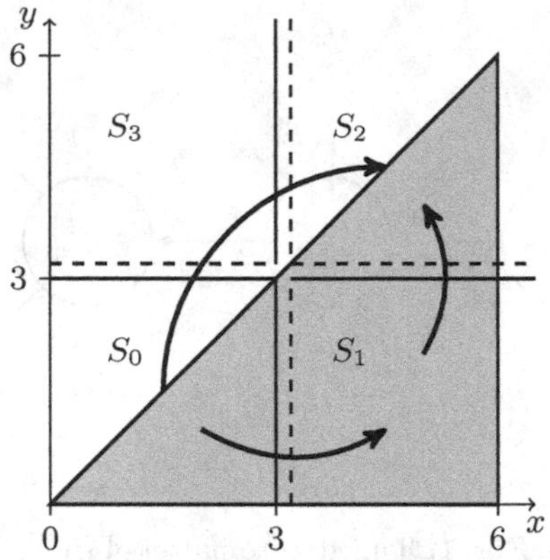

Fig. 2. Illustration of the need to consider diagonals separately

4 Zone-Ultraregion Semantics

In the following, let X be a fixed set of clocks and G a fixed set of simple guards over X. For a clock $x \in X$, we define $\mathcal{I}_x$ to be the coarsest interval partition of $\mathbb{R}_{\geq 0}$ that respects the guards in G, i.e. all values in an interval have to satisfy the same subset of guards of G. Let further B_x denote the set of bounds the clock x is compared against in the guards of G and let $B_{x-y} = \{a - b \mid a \in B_x, b \in B_y\}$. Let then $\mathcal{I}_{x-y} = \{(-\infty, c_0), [c_0, c_0], (c_0, c_1), \dots, (c_k, \infty)\}$ where $B_{x-y} = \{c_0, \dots, c_k\}$ and $c_0 < c_1 < \cdots < c_k$. For a valuation $\eta \in \mathbb{R}^X_{\geq 0}$ we use $\mathcal{I}_x(\eta)$ to denote the interval of $\mathcal{I}_x$ that contains the value $\eta(x)$, similarly for $\mathcal{I}_{x-y}(\eta)$. We say that $\mathcal{I}_x(\eta)$ is *unbounded* if it is of the form $[c, \infty)$ or (c, ∞), otherwise we say that it is *bounded*. We now define an equivalence relation with respect to a set of simple guards G on clock valuations.

Definition 4.1 (Ultraregions). *Let X be a set of clocks, G a set of simple guards over X. We define an equivalence relation $\simeq_G$ on $\mathbb{R}^X_{\geq 0}$ as follows: $\eta \simeq_G \eta'$ if for all x, $\mathcal{I}_x(\eta) = \mathcal{I}_x(\eta')$, and for all y, z such that $\mathcal{I}_y(\eta)$ and $\mathcal{I}_z(\eta)$ are* bounded, *$\mathcal{I}_{y-z}(\eta) = \mathcal{I}_{y-z}(\eta')$. The equivalence classes of $\simeq_G$ are called the* ultraregions *of G.*

Note that every ultraregion is uniquely identified by a choice of intervals from $\mathcal{I}_x$ and $\mathcal{I}_{x-y}$ for all clocks x, y. Also note that a choice in $\mathcal{I}_{x-y}$ always determines a choice in $\mathcal{I}_{y-x}$.

Example 4.2. Let $G = \{x \leq 3, x < 6, y \leq 4\}$. Then $\mathcal{I}_x = \{[0, 3], (3, 6), [6, \infty)\}$, $\mathcal{I}_y = \{[0, 4], (4, \infty)\}$, and $\mathcal{I}_{x-y} = \{(-\infty, -1), [-1, -1], (-1, 2), [2, 2], (2, \infty)\}$.

The ultraregions of G look as follows:

$$U_1 = \{\eta = (x, y) \mid x \in [0,3], y \in [0,4], x - y < -1\}$$
$$U_2 = \{\eta = (x, y) \mid x \in [0,3], y \in [0,4], x - y = -1\}$$
$$U_3 = \{\eta = (x, y) \mid x \in [0,3], y \in [0,4], x - y \in (-1,2)\}$$
$$U_4 = \{\eta = (x, y) \mid x \in [0,3], y \in [0,4], x - y = 2\}$$
$$U_5 = \{\eta = (x, y) \mid x \in [0,3], y \in [0,4], x - y > 2\}$$
$$U_6 = \{\eta = (x, y) \mid x \in [0,3], y > 4\}$$
$$U_7 = \{\eta = (x, y) \mid x \in (3,6), y > 4\}$$
$$U_8 = \{\eta = (x, y) \mid x \in (3,6), y \in [0,4], x - y \in (-1,2)\}$$
$$U_9 = \{\eta = (x, y) \mid x \in (3,6), y \in [0,4], x - y = 2\}$$
$$U_{10} = \{\eta = (x, y) \mid x \in (3,6), y \in [0,4], x - y > 2\}$$
$$U_{11} = \{\eta = (x, y) \mid x \geq 6, y \in [0,4]\}$$
$$U_{12} = \{\eta = (x, y) \mid x \geq 6, y > 4\}$$

These ultraregions are illustrated in Figure 3.

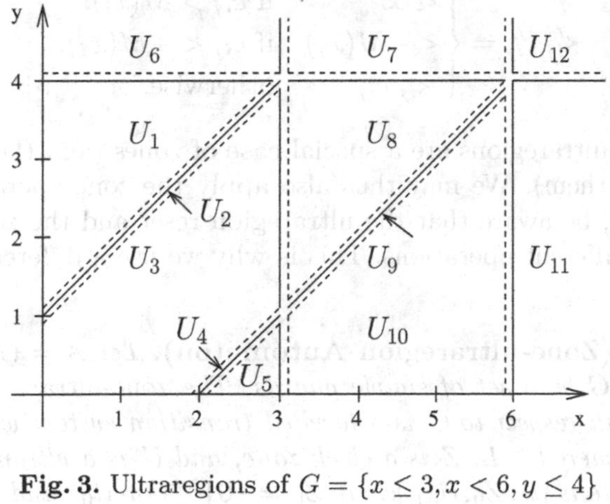

Fig. 3. Ultraregions of $G = \{x \leq 3, x < 6, y \leq 4\}$

Let $U \neq U'$ be ultraregions. We say that U' is a successor of U if for all $\eta \in U$ there exists $d \in \mathbb{R}_{>0}$ such that $\eta + d \in U'$ and $\forall 0 \leq d' \leq d : \eta + d' \in U \cup U'$.

Lemma 4.3. *An ultraregion has at most one successor.*

This allows us to denote the successor of U by *succ*(U). If U has no successor, we additionally define *succ*$(U) = U$. Note that *succ*$(U) = U$ if and only if all clocks are unbounded in U, i.e. for every $\eta \in U$ and every $d \in \mathbb{R}_{\geq 0}$, $\eta + d \in U$.

Let now $R \subseteq X$ be a set of clocks. The *reset* of U with respect to R, denoted by $U\langle R \rangle$, is defined as follows:

$$U\langle R \rangle = \{U' \mid U' \text{ is an ultraregion} \wedge \exists \eta \in U : \eta[R] \in U'\}$$

Example 4.4. Continuing with Example 4.2, we may see that e.g. $succ(U_8) = U_7$, $succ(U_{12}) = U_{12}$, $U_{11}\langle x\rangle = \{U_1, U_2, U_3\}$, $U_9\langle x, y\rangle = \{U_3\}$, and $U_5\langle x\rangle = \{U_3\}$.

We may now define the zone-ultraregion semantics of a timed automaton. We use the standard notion of clock zones here [20]. Every zone is described by a set of diagonal constraints of the form $x_i - x_j \prec_{ij} c_{ij}$ where $c_{ij} \in \mathbb{R}$, $\prec_{ij} \in \{<, \leq\}$ for all clocks x_i, $x_j \in X \cup \{x_0\}$, and x_0 is a special clock that has always the value 0. We use these standard operations on zones: intersection $Z \cap Z'$, reset $Z[R] = \{\eta[R] \mid \eta \in Z\}$, and time passing $Z^{\uparrow} = \{\eta + d \mid \eta \in Z, d \in \mathbb{R}_{\geq 0}\}$. The zones may be efficiently represented using difference bound matrices (DBM) [21,22]. Although there may be different representations of the same zone, it is a standard result that there exists a unique canonical representation in which the bounds $\prec_{ij} c_{ij}$ are as tight as possible.

In order to keep the number of zones finite, we use the standard *k-extrapolation* construction [20,22,23,24]. Let Z be a zone and let $\prec_{ij} c_{ij}$ be the bounds in its canonical representation. Let $M(x)$ be the highest bound in the guards of TA and the guards from G that compare against x. The extrapolated zone $\mathcal{E}(Z)$ is defined by the set of diagonal constraints $x_i - x_j \prec'_{ij} c'_{ij}$ where

$$\prec'_{ij} c'_{ij} = \begin{cases} < \infty & \text{if } c_{ij} > M(x_i), \\ < -M(x_j) & \text{if } c_{ij} < -M(x_j), \\ \prec_{ij} c_{ij} & \text{otherwise.} \end{cases}$$

Note that the ultraregions are a special case of zones (and the extrapolation does not change them). We may thus also apply the zone operations to ultraregions. However, be aware that the ultraregion reset and the zone reset of an ultraregion are different operations. This is why we use a different notation for $U\langle R\rangle$.

Definition 4.5 (Zone-ultraregion Automaton). *Let $A = (L, l_0, X, \Delta, Inv)$ be a TA and let G be a set of simple guards. The* zone-ultraregion automaton *(ZURA) of A with respect to G is a labelled transition system whose states are triples (l, Z, U) where $l \in L$, Z is a clock zone, and U is a ultraregion of G.*

The initial state is (l_0, Z_0, U_0) where $Z_0 = \{\mathbf{0}\}^{\uparrow} \cap Inv(l_0)$ and U_0 is the ultraregion containing the zero valuation $\mathbf{0}$. The transitions are of two kinds:

- *delay transitions: $(l, Z, U) \xrightarrow{\delta} (l, Z, succ(U))$ whenever $Z \cap succ(U) \neq \emptyset$ and $U = succ(U) \implies Z = Z^{\uparrow}$,*
- *action transitions: $(l, Z, U) \xrightarrow{act} (l', \mathcal{E}(Z'), U')$ whenever $l \xrightarrow{g,R}_{\Delta} l'$, $U' \in U\langle R\rangle$, $Z' = ((Z \cap U \cap g)[R] \cap U')^{\uparrow} \cap Inv(l')$, and $Z' \cap U' \neq \emptyset$.*

Example 4.6. Continuing with Example 3.1, Fig. 4 represents the ZURA of the timed automaton $A_{3.1}$ with respect to $G = \{x \leq 3, y \leq 3\}$.

A combination of a ZURA with a location labelling function $\mathcal{L}$ is interpreted as a Kripke structure [1]. The states and transitions of this Kripke structure are

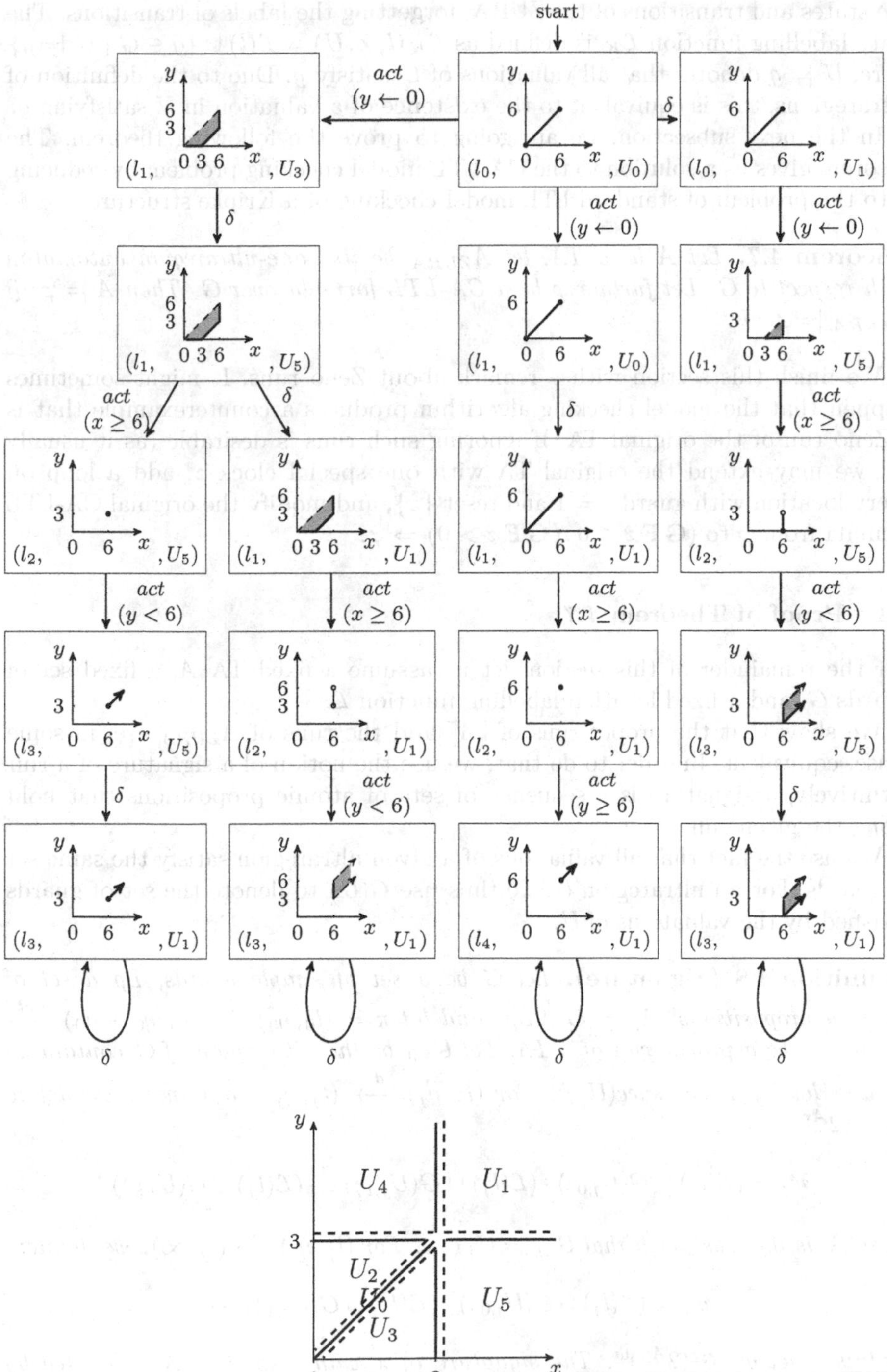

Fig. 4. ZURA state space of timed automaton $A_{3.1}$ with respect to $G = \{x \leq 3, y \leq 3\}$

the states and transitions of the ZURA, forgetting the labels of transitions. The state labelling function $\mathcal{L}_K$ is defined as $\mathcal{L}_K(l, Z, U) = \mathcal{L}(l) \cup \{g \in G \mid U \models g\}$. Here, $U \models g$ denotes that all valuations of U satisfy g. Due to the definition of ultraregions, this is equivalent to the existence of a valuation in U satisfying g.

In the next subsection, we are going to prove the following theorem. The theorem gives us a solution to the CA-LTL model checking problem by reducing it to the problem of standard LTL model checking of a Kripke structure.

Theorem 4.7. *Let A be a TA, let A_{ZURA} be its zone-ultraregion automaton with respect to G. Let further φ be a CA-LTL formula over G. Then $A \models \varphi$ iff $A_{ZURA} \models \varphi$.*

We finish this section with a remark about Zeno runs. It might sometimes happen that the model checking algorithm produces a counterexample that is a Zeno run of the original TA. If ignoring such runs is desirable (as it usually is), we may extend the original TA with one special clock z, add a loop on every location with guard $z = 1$ and reset $\{z\}$, and modify the original CA-LTL formula from φ to $(\mathbf{GF}\, z \leq 0 \wedge \mathbf{GF}\, z > 0) \Rightarrow \varphi$.

4.1 Proof of Theorem 4.7

For the remainder of this section, let us assume a fixed TA A, a fixed set of guards G, and a fixed location labelling function $\mathcal{L}$.

We show that the proper runs of $[\![A]\!]$ and the runs of A_{ZURA} are, in some sense, equivalent. In order to do that, we use the notion of a signature of a run. Intuitively, a signature is a sequence of sets of atomic propositions that hold along the given run.

We use the fact that all valuations of a given ultraregion satisfy the same set of guards. For an ultraregion U, we thus use $G(U)$ to denote the set of guards satisfied by the valuations of U.

Definition 4.8 (Signature). *Let G be a set of simple guards, Lp a set of location propositions, $Ap = G \cup Lp$, and let $\pi = (l_0, \eta_0) \xrightarrow{d_0} (l_0, \eta_0 + d_0) \xrightarrow{act} (l_1, \eta_1) \cdots$ be a proper run of a TA. Let $U_{j,0}$ be the ultraregion of G containing η_j and let $U_{j,i+1} = succ(U_{j,i})$. For $(l_j, \eta_j) \xrightarrow{d_j} (l_j, \eta_j + d_j)$ in π, we define $w_j \in (2^{Ap})^+$:*

$$w_j = (\mathcal{L}(l_j) \cup G(U_{j,0})) \cdot (\mathcal{L}(l_j) \cup G(U_{j,1})) \cdots (\mathcal{L}(l_j) \cup G(U_{j,k}))$$

where k is the least such that $U_{j,k} = U_{j,k+1}$. For $(l_j, \eta_j) \xrightarrow{\infty} (l_j, \infty)$, we define:

$$w_j = (\mathcal{L}(l_j) \cup G(U_{j,0})) \cdot (\mathcal{L}(l_j) \cup G(U_{j,1})) \cdots$$

In this case, $w_j \in (2^{Ap})^\omega$. The signature of π with respect to Ap, denoted by $sig_{Ap}(\pi)$, is defined as the infinite word $w = w_0 w_1 w_2 \cdots w_j$ if π ends with (l_j, ∞), $w = w_0 w_1 w_2 \cdots$ otherwise.

Our first objective is to show that the runs of A_{ZURA} represent exactly the signatures of all proper runs of A. We then show that the CA-LTL satisfaction on a proper run can be reduced to the classic LTL satisfaction on its signature. These results together imply the statement of Theorem 4.7. We only give the main ideas of the proofs here, more detailed proofs and explanation can be found in [9].

Lemma 4.9. *Let π be a proper run of A with signature $sig_{Ap}(\pi)$. Then there exists an infinite run $(l_0, Z_0, U_0) \to (l_1, Z_1, U_1) \to \cdots$ such that $sig_{Ap}(\pi)(i) = \mathcal{L}(l_i) \cup G(U_i)$, where $sig_{Ap}(\pi)(i)$ denotes the ith letter (from 2^{Ap}) of $sig_{Ap}(\pi)$.*

Proof idea. We show that every action transition of $[\![A]\!]$ has its counterpart in A_{ZURA} and that every delay transition of $[\![A]\!]$ has its counterpart as a sequence of δ-transitions in A_{ZURA}. We then build the run of A_{ZURA} inductively.

Lemma 4.10. *Let $(l_0, Z_0, U_0) \xrightarrow{\gamma_0} (l_1, Z_1, U_1) \xrightarrow{\gamma_1} \cdots$ be a run of A_{ZURA} with $\gamma_i \in \{act, \delta\}$. Then there exists a proper run π of A such that $sig_{Ap}(\pi)(i) = \mathcal{L}(l_i) \cup G(U_i)$.*

Proof idea. The proof of this lemma is more involved as we cannot map every transition of A_{ZURA} to a counterpart transition or a sequence of transitions in $[\![A]\!]$ directly in a forward manner. Instead, we use a trick similar to that used in [25]. We consider the region graph of A and show the correspondence between A_{ZURA} and this region graph. We then use the fact that the standard region equivalence is a time-abstracting bisimulation [26] to show the correspondence between runs of the region graph and proper runs of $[\![A]\!]$.

Let us recall that the classic satisfaction relation for LTL on an infinite word from $(2^{Ap})^\omega$ is given by $w \models p \in Ap$ if $p \in w(0)$, $w \models \varphi \,\mathbf{U}\, \psi \iff \exists k : w^k \models \psi$ and for all $j < k : w^j \models \varphi$. Here, $w(0)$ is the first letter of the word w and w^k is the kth suffix of w. We omit the standard boolean operators.

Lemma 4.11. *Let π be a proper run of a TA, let G be a set of simple guards, Lp a set of location propositions, $Ap = G \cup Lp$. Let φ be a CA-LTL formula over Ap. Then $\pi \models \varphi$ iff $sig_{Ap}(\pi) \models \varphi$ when seen as an LTL formula.*

Proof idea. We observe that the suffixes of proper runs correspond to suffixes of their signatures. The lemma is then proved by induction on the structure of φ, with most cases being trivial and the only interesting case (that of the **U** operator) being resolved by the observation about suffixes.

This concludes the proof of Theorem 4.7.

5 Implementation

We first show how to compute the successor of a given fixed ultraregion U. Let $\widehat{X}_U \subseteq X$ be the set of all clocks bounded in U, i.e. $\widehat{X}_U = \{x \in X \mid \exists d \in \mathbb{R}_{\geq 0} : \forall \eta \in U : \eta(x) \leq d\}$.

For a clock $x \in X$ we define $U_x \in \mathcal{I}_x$ to be the interval from $\mathcal{I}_x$ such that there exists a valuation $\eta \in U$ with $\eta(x) \in U_x$. For $x, y \in \widehat{X}_U$ we further define U_{x-y} to be the interval from $\mathcal{I}_{x-y}$ such that there exists a valuation $\eta \in U$ with $\eta(x) - \eta(y) \in U_{x-y}$. Note that the existence and uniqueness of U_x and U_{x-y} follow from the definition of ultraregions, and in the latter case, the fact that we only consider differences of bounded clocks. We further use E_x to denote the right endpoint of U_x.

To establish the successor of U, we need to find out which clocks leave U soonest. We thus define the following relation on $\widehat{X}_U$:

$$x \preccurlyeq_U y \iff \forall \eta \in U : \forall d \in \mathbb{R}_{\geq 0} : \eta(x) + d \in U_x \Rightarrow \eta(y) + d \in U_y$$

It is easy to see that $\preccurlyeq_U$ is reflexive, transitive, and that for all $x, y \in \widehat{X}_U$, either $x \preccurlyeq_U y$ or $y \preccurlyeq_U x$. This means that $\preccurlyeq_U$ is a total preorder. We denote the induced equivalence by $\approx_U$. The following lemma gives us a way of computing $\preccurlyeq_U$. Here, we use the notation $a < I$ with the meaning $\forall b \in I : a < b$, similarly for $>$.

Lemma 5.1. *Let $x, y \in \widehat{X}_U$. Then $x \preccurlyeq_U y$ iff either $(E_x - E_y) < U_{x-y}$ or $(E_x - E_y) \in U_{x-y} \wedge (U_x$ right-closed $\implies U_y$ right-closed$)$.*

We can now show the construction of a successor. Let $\widetilde{X}_U$ be the set of the smallest clocks with respect to $\preccurlyeq_U$, i.e. $\widetilde{X}_U = \{x \in \widehat{X}_U \mid \forall y \in \widehat{X}_U : x \preccurlyeq_U y\}$. For a bounded interval $J \in \mathcal{I}_x$, denote by $J^\uparrow$ the interval in $\mathcal{I}_x$ such that the right endpoint of J is the left endpoint of $J^\uparrow$. For a clock $x \in X$ we then define U'_x as follows:

$$U'_x = \begin{cases} U_x^\uparrow & x \in \widetilde{X}_U \\ U_x & x \notin \widetilde{X}_U \end{cases}$$

We then define $U' = \{\eta \mid \forall x \in X : \eta(x) \in U'_x$ and $\forall x, y \in X : U'_x$ and U'_y bounded $\Rightarrow \eta(x) - \eta(y) \in U_{x-y}\}$. We want to show that $U' = succ(U)$. In order to do that, we first need an auxiliary lemma.

Lemma 5.2. *For every $\eta \in U$ there exists $d \in \mathbb{R}_{>0}$ such that $\forall x \in \widetilde{X}_U : \eta(x) + d \notin U_x$ and $\forall y \in \widehat{X}_U \setminus \widetilde{X}_U : \eta(y) + d \in U_y$.*

Theorem 5.3. *Let U be an ultraregion, let U' be defined as above. Then $U' = succ(U)$.*

We now show the construction of the reset operation on ultraregions. Let U be an ultraregion and let $R \subseteq X$ be a set of clocks. As an ultraregion is a special case of a zone, we may apply the standard zone reset operation on U and then tighten the constraints using the standard Floyd-Warshall algorithm approach. The resulting zone can be written as:

$$\begin{aligned} U[R] = M = \{\eta \mid \forall x \in R : \eta(x) = 0; \forall x \notin R : \eta(x) \in M_x \\ \forall x, y \notin R : \eta(x) - \eta(y) \in M_{x-y}\} \end{aligned}$$

where $M_x \subseteq U_x$ and $M_{x-y} \subseteq U_{x-y}$ for all x, y. The implied constraints are $\eta(x) - \eta(y) = 0$ for $x, y \in R$ and $\eta(x) - \eta(y) \in M_x$ for $x \notin R$ and $y \in R$. We now have to find the set of all ultraregions that intersect M. Clearly, such set is equal to $U\langle R \rangle$. For $x \notin R$, $y \in R$, let $J_{x-y} = \{J \in \mathcal{I}_{x-y} \mid J \cap M_x \neq \emptyset\}$. Let f be a choice function that assigns to x, y an interval in J_{x-y}, i.e. $f(x,y) \in J_{x-y}$. Let further $O_x \in \mathcal{I}_x$ be the interval containing 0, similarly for O_{x-y}. We then define:

$$\begin{aligned} U_f = \{\eta \mid \forall x \in R : \eta(x) \in O_x; \forall x \notin R : \eta(x) \in U_x \\ \forall x, y \notin R : \eta(x) - \eta(y) \in M_{x-y} \\ \forall x, y \in R : \eta(x) - \eta(y) \in O_{x-y} \\ \forall x \notin R, y \in R : \eta(x) - \eta(y) \in f(x,y)\} \end{aligned}$$

Clearly, every ultraregion intersecting M is of the form U_f for some choice function f as define above. However, as we show in the next example, some U_f do not intersect M.

Example 5.4. Let us now extend our ultraregion example 4.2 by adding another clock z with the guard $z \leq 0$. Then $\mathcal{I}_z = \{[0,0], (0,\infty)\}$, $\mathcal{I}_{x-z} = \{(-\infty,3), [3,3], (3,6), [6,6], (6,\infty)\}$, and $\mathcal{I}_{y-z} = \{(-\infty,4), [4,4], (4,\infty)\}$. Let U be the ultraregion

$$U = \{(x,y,z) \mid x \in [0,3], y \in [0,4], z \in (0,\infty), x - y \in (-1,2)\}$$

and let $R = \{y,z\}$. Using the zone reset operation and tightening the bounds gives $M = \{(x,y,z) \mid x \in [0,3], y = z = 0\}$. We then have $J_{x-y} = \{(-1,2), [2,2], (2,\infty)\}$ and $J_{x-z} = \{(-\infty,3), [3,3]\}$, which gives six possible choice functions and thus six candidate ultraregions. It is easy to verify that two of these candidates, namely those corresponding to the choices $(-1,2) \in J_{x-y}$, $[3,3] \in J_{x-z}$ and $[2,2] \in J_{x-y}$, $[3,3] \in J_{x-z}$, do not intersect M. The other four candidates together constitute $U\langle R \rangle$.

The implementation of $U\langle R \rangle$ thus works as follows: We first apply the zone reset operation to U and tighten the bounds. We then create the sets of intervals J_{x-y} and iteratively try all choice functions to get all possible candidates. Each candidate is then intersected with M and checked for emptiness. Those who intersect M are then the result.

We have shown the implementation of the ultraregion reset and the ultraregion successor operations. Note that it is possible to pre-compute the ultraregion automaton with respect to a given G. Such a pre-computed automaton will provide an efficient way of obtaining the result of the successor and reset operations for a given ultraregion. Obviously, all other operations required in the construction of the zone-ultraregion automaton are the standard zone operations. These operations can be computed on the fly, thus allowing to employ efficient on-the-fly model checking algorithms.

5.1 Experiments

We have evaluated our method within the parallel and distributed model checker DIVINE [27]. For the purpose of the evaluation we used several real-time models, namely `2doors.xml`, `bridge.xml`, `fisher.xml`, and `train-gate.xml`, which are distributed with UPPAAL 4.0.13. We have measured the size of ZURA for selected instances of the chosen models with the primary interest in the increase of the state space size with respect to the number of guards used in a specification. Therefore, we work directly with different guard sets instead of CA-LTL formulae. Table 1 shows the growth of zone-ultraregion automaton state space size with the increasing size of the guard set.

We choose guard sets separately for each model as follows, for `2doors.xml` $G_1 = \{x \leq 4\}, G_2 = G_1 \cup \{x \leq 6\}, G_3 = G_2 \cup \{w \leq 0\}$, for `bridge.xml` $G_1 = \{y \leq 10\}, G_2 = G_1 \cup \{y \leq 15\}, G_3 = G_2 \cup \{time \leq 25\}$, for `fisher.xml` $G_1 = \{x \leq 1\}, G_2 = G_1 \cup \{x \leq 2\}, G_3 = G_2 \cup \{x < 2\}$, for `train-gate.xml` $G_1 = \{x \leq 10\}, G_2 = G_1 \cup \{x \leq 5\}, G_3 = G_2 \cup \{x < 15\}$.

Table 1. Experimental evaluation of $ZURA$ state space size

Model	$\lvert G\rvert = 0$		$\lvert G_1\rvert = 1$		$\lvert G_2\rvert = 2$		$\lvert G_3\rvert = 3$	
	states	trans.	states	trans.	states	trans.	states	trans.
2doors	189	343	294	508	364	636	482	817
bridge	723	1851	1446	3702	2169	5553	4617	11334
fischer4	1792	4912	12159	29808	16688	38337	32124	72668
fischer5	15142	45262	157623	426256	219435	556705	420875	1063796
fischer6	140716	453328	2174673	6424394	3070446	8536643	5817098	16279518
train-gate3	610	1153	3689	7486	11286	23023	28066	60422
train-gate4	9977	18233	98366	187327	351388	674504	936973	1915545
train-gate5	200776	359031	2479343	4589462	9662204	18112439	27159806	54271266

6 Conclusion and Future Work

Model checking of CA-LTL over timed automata provides system engineers with another powerful formal verification procedure to check for reliability and correctness of time-critical systems. To our best knowledge we are the first to fully describe and implement the process of zone-based LTL model checking over timed automata for a logic that allows clock constraints as atomic propositions. We again recall that the zone-based solutions to the Timed Büchi Automaton Emptiness problem known so far have not provided the solution to the CA-LTL model checking problem as presented in this paper. Our implementation has been done within the parallel and distributed model checker DIVINE which allows us to employ the aggregate power of multiple computational nodes in order to deal with a single model checking task.

We are currently working on several extensions. One of them is to allow atomic propositions used in CA-LTL formulae to be able to refer to the clock value differences. Another extension deals with different zone extrapolation methods and the last one enhances the logic with actions.

References

1. Clarke, E., Grumberg, O., Peled, D.: Model Checking. MIT Press (1999)
2. Alur, R., Dill, D.L.: A Theory of Timed Automata. Theor. Comput. Sci. 126(2), 183–235 (1994)
3. Daws, C., Tripakis, S.: Model Checking of Real-Time Reachability Properties Using Abstractions. In: Steffen, B. (ed.) TACAS 1998. LNCS, vol. 1384, pp. 313–329. Springer, Heidelberg (1998)
4. Behrmann, G., David, A., Larsen, K.G., Müller, O., Pettersson, P., Yi, W.: UPPAAL - present and future. In: Proc. of 40th IEEE Conference on Decision and Control. IEEE Computer Society Press (2001)
5. Tripakis, S.: Checking timed Büchi Automata Emptiness on Simulation Graphs. TOCL 10(3) (2009)
6. Li, G.: Checking Timed Büchi Automata Emptiness Using LU-Abstractions. In: Ouaknine, J., Vaandrager, F.W. (eds.) FORMATS 2009. LNCS, vol. 5813, pp. 228–242. Springer, Heidelberg (2009)
7. Laarman, A., Olesen, M.C., Dalsgaard, A.E., Larsen, K.G., van de Pol, J.: Multi-core Emptiness Checking of Timed Büchi Automata Using Inclusion Abstraction. In: Sharygina, N., Veith, H. (eds.) CAV 2013. LNCS, vol. 8044, pp. 968–983. Springer, Heidelberg (2013)
8. Vardi, M., Wolper, P.: An automata-theoretic approach to automatic program verification (preliminary report). In: Proceedings, Symposium on Logic in Computer Science (LICS 1986), pp. 332–344. IEEE Computer Society (1986)
9. Bezděk, P., Beneš, N., Havel, V., Barnat, J., Černá, I.: On clock-aware ltl properties of timed automata. Technical report FIMU-RS-2014-04, Faculty of Informatics, Masaryk University, Brno (2014)
10. Alur, R., Henzinger, T.A.: A really temporal logic. Journal of the ACM (JACM) 41(1), 181–203 (1994)
11. Koymans, R.: Specifying real-time properties with metric temporal logic. Real-time Systems 2(4), 255–299 (1990)
12. Alur, R., Feder, T., Henzinger, T.A.: The benefits of relaxing punctuality. J. ACM 43(1), 116–146 (1996)
13. Ostroff, J.S.: Temporal logic for real-time systems, vol. 40. Cambridge Univ. Press (1989)
14. Harel, E., Lichtenstein, O., Pnueli, A.: Explicit clock temporal logic. In: Proceedings of the Fifth Annual IEEE Symposium on e Logic in Computer Science, LICS 1990, pp. 402–413. IEEE (1990)
15. Demri, S., D'Souza, D.: An automata-theoretic approach to constraint LTL. Information and Computation 205(3), 380–415 (2007)
16. Li, G., Tang, Z.: Modelling real-time systems with continuous-time temporal logic. In: George, C.W., Miao, H. (eds.) ICFEM 2002. LNCS, vol. 2495, pp. 231–236. Springer, Heidelberg (2002)
17. Alur, R., Madhusudan, P.: Decision problems for timed automata: A survey. In: Bernardo, M., Corradini, F. (eds.) SFM-RT 2004. LNCS, vol. 3185, pp. 1–24. Springer, Heidelberg (2004)

18. Baier, C., Katoen, J.P.: Principles of Model Checking. Representation and Mind Series. The MIT Press (2008)
19. Alur, R., Courcoubetis, C., Dill, D.: Model-checking for real-time systems. In: Proceedings of the Fifth Annual IEEE Symposium on e Logic in Computer Science, LICS 1990, pp. 414–425. IEEE (1990)
20. Tripakis, S.: Checking timed Büchi automata emptiness on simulation graphs. ACM Transactions on Computational Logic (TOCL) 10(3), 15 (2009)
21. Dill, D.L.: Timing assumptions and verification of finite-state concurrent systems. In: Automatic Verification Methods for Finite State Systems, pp. 197–212. Springer (1990)
22. Bengtsson, J.E., Yi, W.: Timed automata: Semantics, algorithms and tools. In: Desel, J., Reisig, W., Rozenberg, G. (eds.) Lectures on Concurrency and Petri Nets. LNCS, vol. 3098, pp. 87–124. Springer, Heidelberg (2004)
23. Pettersson, P.: Modelling and verification of real-time systems using timed automata: theory and practice, Citeseer (1999)
24. Bouyer, P.: Forward analysis of updatable timed automata. Formal Methods in System Design 24(3), 281–320 (2004)
25. Herbreteau, F., Srivathsan, B., Walukiewicz, I.: Efficient Emptiness Check for Timed Büchi Automata (Extended version). CoRR abs/1104.1540 (2011)
26. Tripakis, S., Yovine, S.: Analysis of timed systems using time-abstracting bisimulations. Formal Methods in System Design 18(1), 25–68 (2001)
27. Barnat, J., Brim, L., Havel, V., Havlíček, J., Kriho, J., Lenčo, M., Ročkai, P., Štill, V., Weiser, J.: DiVinE 3.0 – An Explicit-State Model Checker for Multithreaded C & C++ Programs. In: Sharygina, N., Veith, H. (eds.) CAV 2013. LNCS, vol. 8044, pp. 863–868. Springer, Heidelberg (2013)

Linguistic Mechanisms for Context-Aware Security*

Chiara Bodei, Pierpaolo Degano, Letterio Galletta, and Francesco Salvatori

Dipartimento di Informatica, Università di Pisa
{chiara,degano,galletta}@di.unipi.it, francesco.salvatori@sns.it

Abstract. Adaptive systems improve their efficiency, by modifying their behaviour to respond to changes in their operational environment. Also, security must adapt to these changes and policy enforcement becomes dependent on the dynamic contexts. We extend (the core of) an adaptive functional language with primitives to enforce security policies on the code execution, and we exploit a static analysis to instrument programs. The introduced checks guarantee that no violation of the required security policies occurs.

1 Introduction

Context and Adaptivity. Today's software systems are expected to operate *every time* and *everywhere*. They have to cope with changing environments, and never compromise their intended behaviour or their non-functional requirements, e.g. security or quality of service. Therefore, languages need effective mechanisms to sense the changes in the operational environment, i.e. the *context*, in which the application is plugged in, and to properly *adapt* to changes. At the same time, these mechanisms must maintain the functional and non-functional properties of applications after the adaptation steps.

The context is a key notion for adaptive software. Typically, a context includes different kinds of computationally accessible information coming both from outside (e.g. sensor values, available devices, code libraries offered by the environment), and from inside the application boundaries (e.g. its private resources, user profiles, etc.).

Context Oriented Programming (COP) [9,15,1,17,3] is a recent paradigm that proposes linguistic features to deal with contexts and adaptivity. Its main construct is *behavioural variation*, a chunk of code to be activated depending on the current context hosting the application, to dynamically modify the execution.

Security and Contexts. The combination of security and context-awareness requires to address two aspects. First, security may reduce adaptivity, by adding further constraints on the possible actions of software. Second, new highly dynamic security mechanisms are needed to scale up to adaptive software. In the literature, e.g. in [26,6], this duality is addressed in two ways: *securing context-aware systems* and *context-aware security*.

Securing context-aware systems aims at rephrasing the standard notions and techniques for confidentiality, integrity and availability [24], and at developing techniques for guaranteeing them [26]. The challenge is to understand how to get secure and trusted context information. Context-aware security is dually concerned with the use of context

* Work partially supported by the MIUR-PRIN project Security Horizons.

G. Ciobanu and D. Méry (Eds.): ICTAC 2014, LNCS 8687, pp. 61–79, 2014.

information to dynamically drive security decisions. Consider the usual no flash photography policy in museums. While a standard security policy *never* allows people to use flash, a context-aware security could forbid flash *only* inside particular rooms.

Yet, there is no unifying concept of security, because the two aspects above are often tackled separately. Indeed, mechanisms have been implemented at different levels of the infrastructure, in the middleware [25] or in the interaction protocols [14], that mostly address access control for resources and for smart things (see e.g. [26,16,27], and [2,10]). More foundational issues have been less studied within the programming languages approach we follow; preliminary work can be found, e.g. in [7,23].

Our Proposal. The kernel of our proposal relies on extending ML_{CoDa}, a core ML with COP features introduced in [13]. Its main novelty is to be a two-component language: a declarative part for programming the context and a functional one for computing.

The context in ML_{CoDa} is a knowledge base implemented as a Datalog program (stratified, with negation) [22,19]. To retrieve the state of a resource, programs simply query the context, in spite of the possibly complex deductions required to solve the corresponding goal; context is changed by using the standard *tell*/*retract* constructs.

Programming adaptation is specified through behavioural variations, a first class, higher-order ML_{CoDa} construct. They can be referred to by identifiers, and used as a parameter in functions. This fosters dynamic, compositional adaptation patterns, as well as reusable, modular code. The chunk of a behavioural variation to be run is selected by the *dispatching* mechanism that inspects the actual context and makes the right choices.

Notably, ML_{CoDa}, as it is, offers the features needed for addressing context-aware security issues, in particular for defining and enforcing access control policies. Our version of Datalog is powerful enough to express all relational algebras, is fully decidable, and guarantees polynomial response time [12]. Furthermore, adopting a stratified-negation-model is common and many logical languages for defining access control policies compile in Stratified Datalog, e.g. [5,18,11]. Here, we are only interested in policies imposed by the system, which are unknown at development time. Indeed the policies of the application can be directly encoded by the developer as behavioural variations. The dispatching mechanism then suffices for checking whether a specific policy holds, and for enabling the chunk of behaviour that obeys it. Our language therefore requires no extensions to deal with security policies.

Our aim is to handle, as soon as possible, both failures in adaptation to the current context (*functional failure*) and policy violations (*non-functional failure*). Note that the actual value of some elements in the current context is only known when the application is linked with it at runtime. Actually, we have a sort of *runtime monitor*, natively supported by the dispatching mechanism of ML_{CoDa}, which we switch on and off at need. To specify and implement the runtime monitor, we conservatively extend the two-phase verification of [13]. The first phase is based on a type and effect system that, at compile time, computes a safe over-approximation, call it H, of the application behaviour. Then H is used at loading time to verify that (i) the resources required by the application are available in the actual context, and in its future modifications (as done in [13]); and (ii) to detect within the application where a policy violation may occur, i.e. when the context is modified through *tell* and *retract* actions.

The loading time analysis requires first to build a graph $\mathcal{G}$, that safely approximates which contexts the application will pass through, while running. While building the graph, we also label its edges with the *tell*/*retract* operations in the code, exploiting the approximation H. Before launching the execution, we detect the unsafe operations by checking the policy Φ on each node of $\mathcal{G}$. Our runtime monitor can guard them, where it will be switched off for the remaining actions. Actually, we collect the labels of the risky operations and associate the value on with them, and `off` with all the others.

To make the above effective, the compiler instruments the code by substituting a behavioural variation *bv* for each occurrence of a *tell*/*retract*. At runtime, *bv* checks if Φ holds in the running context, but only when the value of the label is on.

The next section introduces ML_{CoDa} and our proposal, with the help of a running example, along with an intuitive presentation of the various components of our two-phase static analysis, and the way security is dynamically enforced. The formal definitions and the statements of the correctness of our proposal will follow in the remaining sections. The conclusion summarises our results and discusses some future work.

2 Running Example

Consider a multimedia guide to a museum implemented as a smartphone application, starting from the case study in [13]. Assume the museum has a wireless infrastructure exploiting different technologies, like WiFi, Bluetooth, Irda or RFID. When a smartphone is connected, the visitors can access the museum Intranet and its website, from which they download information about the exhibit and further multimedia contents. Each exhibit is equipped with a wireless adapter (Bluetooth, Irda, RFID) and a QR code. They are only used to offer the guide with the URL of the exhibit, retrievable by using one of the above technologies, provided that it is available on the smartphone. If equipped with a Bluetooth adapter, the smartphone connects to that of the exhibit and directly downloads the URL; if the smartphone has a camera and a QR decoder, the guide can retrieve the URL by taking a picture of the code and decoding it.

The smartphone capabilities are stored in the context as Datalog clauses. Consider the following clauses defining when the smartphone can either directly download the URL (the predicate `device(d)` holds when the device `d` $\in$ {`irda`, `bluetooth`, `rfid_reader`} is available), or it can take the URL by decoding a picture (the parameter `x` in the predicate `use_qrcode` is a handle for using the decoder):

```
direct_comm() ← device(irda).
direct_comm() ← device(bluetooth).
direct_comm() ← device(rfid_reader).
use_qrcode(x) ← user_prefer(qr_code),
                qr_decoder(x),
                device(camera).
use_qrcode(x) ← qr_decoder(x),
                device(camera),
                ¬ device(irda),
                ¬ device(rfid_reader),
                ¬ device(bluetooth).
```

Contextual data, like the above predicates `use_qrcode(decoder)` and `direct_comm()`, affect the download. To change the program flow in accordance to the current context, we exploit behavioural variations. Syntactically, they are similar to pattern matching, where Datalog goals replace patterns and parameters can additionally occur. Behavioural variations are similar to functional abstractions, but their application triggers a *dispatching mechanism* that, at runtime, inspects the context and selects the first expression whose goal holds.

In the following function `getExhibitData`, we declare the behavioural variation `url` (with an unused argument "_"), that returns the URL of an exhibit. If the smartphone can directly download the URL, then it does, through the channel returned by the function `getChannel()`; otherwise the smartphone takes a picture of the QR code and decodes it. In the last case, the variables `decoder` and `cam` will be assigned to the handles of the decoder and the one of the camera deduced by the Datalog machinery. These handles are used by the functions `take_picture` and `decode_qr` to interact with the actual smartphone resources.

```
fun getExhibitData () =
  let url = (_){
    ← direct_comm().
        let c = getChannel () in
          receiveData c,
    ← use_qrcode(decoder),camera(cam).
        let p = take_picture cam in
          decode_qr decoder p }
  in getRemoteData #url
```

The behavioural variation (bound to) `url` is applied before the `getRemoteData` call that connects to the corresponding website and downloads the required information (we use here a slightly simplified syntax, for details see Sect. 3).

By applying the function `getExhibitData` to unit and assuming `n` is returned by `getChannel`, we have the following computation ($C, e \to^* C', e'$ says that the expression e in the context C reduces in several steps to e' changing the context in C'):

$$C, \texttt{getExhibitData}() \to^* C, \texttt{getRemoteData\#}u \to^* C, \texttt{getRemoteData}(\texttt{receiveData}\,n)$$

The second configuration above transforms into the third, because C satisfies the goal $\leftarrow$ `direct_comm()`, and so the dispatching mechanism selects the first expression of the behavioural variation u (the one bound to `url` in `getExhibitData`).

To dynamically update the context, we use the constructs *tell* and *retract*, that add and remove Datalog facts. In our example the context stores information about the room in which the user is, through the predicate `current_room`. If the user moves from the *delicate paintings* room to the *sculptures* one, the application updates the context by:

```
retract current_room(delicate_paintings)
tell current_room(sculptures).
```

Assume now that one can take pictures in every room, but that in the rooms with delicate paintings it is forbidden to use the flash so to prevent the exhibits from damages. This policy is specified by the museum (the system) and it must be enforced during the user's tour. Since policies predicate on the context, they are easily expressed as Datalog goals. Let the fact `flash_on` hold when the flash is active and the fact `button_clicked` when the user presses the button of the camera. The above policy intuitively corresponds to the logical condition $current\_room(delicate\_paintings) \Rightarrow (button\_clicked \Rightarrow \neg flash\_on)$ and is expressed in Datalog as the equivalent goal

```
phi ←¬ current_room(delicate_paintings)
phi ←¬ button_clicked
phi ←¬ flash_on
```

Of course, the museum can specify other policies, and we assume that there is a unique global policy Φ (referred in the code as `phi`), obtained by suitably combining them all. The enforcement is obtained by a runtime monitor that checks the validity of Φ right before every context change, i.e. before every *tell/retract*.

An application fails to adapt to a context (*functional failure*), when the dispatching mechanism fails. Consider the evaluation of `getExhibitData` on a smartphone without wireless technology and QR decoder. Since no context will ever satisfy the goals of `url`, it gets stuck. Another kind of failure happens when a *tell/retract* causes a policy violation (*non-functional failure*). If the context includes `current_room(delicate_paintings)`, such a violation occurs when attempting to use the flash.

To avoid functional failures and to optimise policy enforcement, we equip ML_{CoDa} with a two-phase static analysis: a type and effect system, and a control-flow analysis. The analysis checks whether an application will be able to adapt to its execution contexts, and detects which contexts can violate the required policies.

At *compile time*, we associate a type and an effect with an expression *e*. The type is (almost) standard, and the effect is an over-approximation of the actual runtime behaviour of *e*, called the *history expression*. The effect abstractly represents the changes and the queries performed on the context during its evaluation. Consider the expression:

```
eₐ = let x =
      if always_flash
       then let y = tell F¹₁ in tell F²₂
       else let y = tell F³₁ in tell F⁴₃
     in tell F⁵₄
```

For clarity, we show the labels *i* of $\texttt{tell}F_j^i$ in the code, that are inserted by the compiler while parsing (same for `retract`). Let the facts above be $F_1 \equiv$ `camera_on`; $F_2 \equiv$ `flash_on`; $F_3 \equiv$ `mode_museum_activated`; $F_4 \equiv$ `button_clicked`. The type of e_a is `unit`, as well as that of $\texttt{tell}\,F_4$, and its history expression is

$$H_a = (((tell F_1^1 \cdot tell F_2^2)^3 + (tell F_1^4 \cdot tell F_3^5)^6)^7 \cdot tell F_4^8)^9$$

($\cdot$ abstracts sequential composition, $+$ `if-then-else`). Depending upon the value of `always_flash`, that records whether the user wants the flash to be always usable, the expression e_a can either perform the action $\texttt{tell}\,F_1$, followed by $\texttt{tell}\,F_2$, *or* the action $\texttt{tell}\,F_1$, followed by $\texttt{tell}\,F_3$, so recording in the context that the flash is on or off. After that, e_a will perform $\texttt{tell}\,F_4$, no matter what the previous choice was.

The labels of history expressions allow us to link the actions in histories to the corresponding actions of the code, e.g. the first $\texttt{tell}\,F_1^1$ in H_a, corresponds to the first

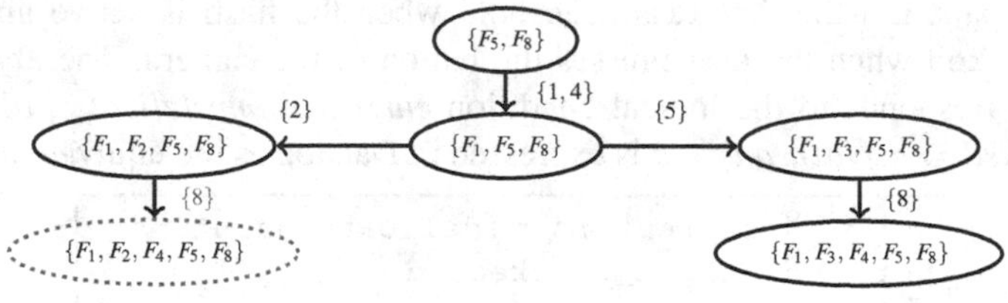

Fig. 1. The evolution graph for the context $\{F_5, F_8\}$ and for the history expression $H_a = (((tell\,F_1^1.tell\,F_2^2)^3 + (tell\,F_1^4.tell\,F_3^5)^6)^7.\underline{tell\,F_4^8})^9$

$\mathtt{tell}\,\mathtt{F}_1$ in e_a, that is also labelled by 1, while the $\mathtt{tell}\,\mathtt{F}_4^8$ in H_a, is linked to the action with label 5 in e_a. All the correspondences are $\{1 \mapsto 1, 2 \mapsto 2, 4 \mapsto 3, 5 \mapsto 4, 8 \mapsto 5\}$ (the abstract labels that do not annotate *tell*/*retract* actions have no counterpart).

Consider now an initial context C that includes the facts F_5 (irrelevant here), and $F_8 \equiv$ `current_room(delicate_paintings)`, but no facts in $\{F_1, F_2, F_3, F_4\}$. Starting from C (and from H_a) our *loading time* analysis builds the graph described in Fig. 1. Nodes represent contexts, possibly reachable at runtime, while edges represent transitions from one context to another. Each edge is annotated with the set of actions in H_a that may cause that transition, e.g. from the context C it is possible to reach a (single) context that also includes the fact F_1, because of the two *tell* operations labelled by 1 and by 4 in H_a. As a matter of fact, an edge can have an annotation including more than one label (e.g. the one labelled $\{1,4\}$). Note also that the same label may occur in the annotation of more than one edge (e.g. the label 8).

By visiting the graph, we observe that the context $\{F_1, F_2, F_4, F_5, F_8\}$ (the dotted node in Fig. 1, red in the pdf) violates our no-flash policy. At runtime the action labelled with 8 (underlined and in red in the pdf), corresponding to $\mathtt{tell}\,\mathtt{F}_4$ must be blocked. For preventing this violation, all we have to do is activate the runtime monitor, right before executing this risky operation.

3 ML$_{\text{CoDa}}$

Below, we survey the syntax and the semantics of ML$_{\text{CoDa}}$; for more details see [13].

Syntax. ML$_{\text{CoDa}}$ consists of two components: Datalog with negation to describe the context, and a core ML extended with COP features. The Datalog part is standard: a program is a set of facts and clauses. We assume that each program is safe, and we adopt *Stratified Datalog*, under the Closed World Assumption to deal with negation [8]. Security policies are simply expressed as Datalog goals, the value of which is true only if the policy holds.

The functional part inherits most of the ML constructs. In addition to the usual ones, our values include Datalog facts F and behavioural variations. Moreover, we introduce the set $\tilde{x} \in DynVar$ of *parameters*, i.e. variables that assume values depending on the properties of the running context, while $x, f \in Var$ are identifiers for standard variables and functions, with the proviso that $Var \cap DynVar = \emptyset$. The syntax of ML$_{\text{CoDa}}$ is below.

$$Va ::= G.e \mid G.e, Va \qquad v ::= c \mid \lambda_f x.e \mid (x)\{Va\} \mid \mathrm{F}$$
$$e ::= v \mid x \mid \tilde{x} \mid e_1\, e_2 \mid let\, x = e_1\, in\, e_2 \mid if\, e_1\, then\, e_2\, else\, e_3 \mid$$
$$dlet\, \tilde{x} = e_1\, when\, G\, in\, e_2 \mid tell(e_1)^l \mid retract(e_1)^l \mid e_1 \cup e_2 \mid \#(e_1, e_2)$$

To facilitate our static analysis (see Sect. 5) we associate each *tell*/*retract* with a label $l \in Lab_C$ (this does not affect the dynamic semantics).

COP-oriented constructs of $\mathrm{ML_{CoDa}}$ include behavioural variations $(x)\{Va\}$, each consisting of a *variation* Va, i.e. a list of expressions $G_1.e_1, \ldots, G_n.e_n$ guarded by Datalog goals G_i (x free in e_i). At runtime, the first goal G_i satisfied by the context selects the expression e_i to be run (*dispatching*). Context-dependent binding is the mechanism to declare variables whose values depend on the context. The *dlet* construct implements the context-dependent binding of a parameter $\tilde{x}$ to a variation Va. The *tell*/*retract* constructs update the context by asserting/retracting facts, provided that the resulting context satisfies the system policy Φ. The append operator $e_1 \cup e_2$ concatenates behavioural variations, so allowing for dynamically composing them. The evaluation of a behavioural variation $\#(e_1, e_2)$ applies e_1 to its argument e_2. To do so, the dispatching mechanism is triggered to query the context and to select from e_1 the expression to run.

Semantics. The Datalog component has the standard top-down semantics [8]. Given a context $C \in Context$ and a goal G, we let $C \vDash G\, with\, \theta$ mean that the goal G, under a ground substitution θ, is satisfied in the context C.

The SOS semantics of $\mathrm{ML_{CoDa}}$ is defined for expressions with no free variables, but possibly with free parameters, thus allowing for openness. To this aim, we have an environment ρ, i.e. a function mapping parameters to variations $DynVar \to Va$. A transition $\rho \vdash C, e \to C', e'$ says that in the environment ρ, the expression e is evaluated in the context C and reduces to e' changing C to C'. The initial configuration is $\rho_0 \vdash C, e_p$, where ρ_0 contains the bindings for all system parameters, and C results from joining the predicates and facts of the system and of the application.

Fig. 2 shows the inductive definitions of the reduction rules for our new constructs; the others ones are standard, and so are the congruence rules that reduce subexpressions, e.g. $\rho \vdash C, \texttt{tell}(e) \to C', \texttt{tell}(e')$ if $\rho \vdash C, e \to C', e'$.

We briefly comment below on the rules displayed. The rules (DLET1) and (DLET2) for the construct *dlet*, and the rule (PAR) for parameters implement our context-dependent binding. For brevity, we assume here that e_1 contains no parameters. The rule (DLET1) extends the environment ρ by appending $G.e_1$ in front of the existent binding for $\tilde{x}$. Then, e_2 is evaluated under the updated environment. Note that the *dlet* does *not* evaluate e_1, but only records it in the environment in a sort of call-by-name style. The rule (DLET2) is standard: the whole *dlet* reduces to the value to which e_2 reduces.

The (PAR) rule looks for the variation Va bound to $\tilde{x}$ in ρ. Then, the dispatching mechanism selects the expression to which $\tilde{x}$ reduces. The dispatching mechanism is implemented by the partial function *dsp*, defined as

$$dsp(C, (G.e, Va)) = \begin{cases} (e, \theta) & \text{if } C \vDash G\, with\, \theta \\ dsp(C, Va) & \text{otherwise} \end{cases}$$

(DLET1)
$$\frac{\rho[(G.e_1,\rho(\tilde{x}))/\tilde{x}] \vdash C,\, e_2 \to C',\, e_2'}{\rho \vdash C,\, dlet\,\tilde{x} = e_1\, when\, G\, in\, e_2 \to C',\, dlet\,\tilde{x} = e_1\, when\, G\, in\, e_2'}$$

(DLET2)
$$\frac{}{\rho \vdash C,\, dlet\,\tilde{x} = e_1\, when\, G\, in\, v \to C,\, v}$$

(PAR)
$$\frac{\rho(\tilde{x}) = Va \qquad dsp(C, Va) = (e, \theta)}{\rho \vdash C,\, \tilde{x} \to C,\, e\theta}$$

(TELL2)
$$\frac{dsp(C \cup \{F\}, phi.()) = ((), \emptyset)}{\rho \vdash C,\, tell(F)^l \to C \cup \{F\},\, ()}$$

(VAAPP3)
$$\frac{dsp(C, Va) = (e, \{\overrightarrow{c}/\overrightarrow{y}\})}{\rho \vdash C,\, \#((x)\{Va\},\, v) \to C,\, e\{v/x,\, \overrightarrow{c}/\overrightarrow{y}\}}$$

(RETRACT2)
$$\frac{dsp(C \setminus \{F\}, phi.()) = ((), \emptyset)}{\rho \vdash C,\, retract(F)^l \to C \setminus \{F\},\, ()}$$

Fig. 2. The reduction rules for the constructs of $\mathrm{ML_{CoDa}}$ concerning adaptation

It inspects a variation from left to right to find the first goal G satisfied by C, under a substitution θ. If this search succeeds, the dispatching returns the corresponding expression e and θ. Then, $\tilde{x}$ reduces to $e\theta$, i.e. to e the variables of which are bound by θ. Instead, if the dispatching fails because no goal holds, the computation gets stuck, because the program cannot adapt to the current context.

Consider the simple conditional expression $\mathtt{if}\,\tilde{x} = F_2\ \mathtt{then}\ 42\ \mathtt{else}\ 51$, in an environment ρ that binds the parameter $\tilde{x}$ to $e' = \mathtt{G_1.F_5, G_2.F_2}$ and in a context C that satisfies the goal G_2, but not G_1:

$$\rho \vdash C,\, \mathtt{if}\,\tilde{x} = F_2\ \mathtt{then}\ 42\ \mathtt{else}\ 51 \to C,\, \mathtt{if}\, F_2 = F_2\ \mathtt{then}\ 42\ \mathtt{else}\ 51 \to C,\, 42$$

where we first retrieve the binding for $\tilde{x}$ (recall it is e'), with $dsp(C, e') = (\mathtt{F_2}, \theta)$, for a suitable substitution θ. Since facts are values, we can bind them to parameters and test their equivalence by a conditional expression.

The application of the behavioural variation $\#(e_1, e_2)$ evaluates the subexpressions until e_1 reduces to $(x)\{Va\}$ and e_2 to a value v. Then, the rule (VAAPP3) invokes the dispatching mechanism to select the relevant expression e from which the computation proceeds after v is substituted for x. Also in this case the computation gets stuck, if the dispatching mechanism fails. Consider the behavioural variation $(x)\{G_1.c_1, G_2.x\}$ and apply it to the constant c in a context C that satisfies the goal G_2, but not G_1. Since $dsp(C, (x)\{G_1.c_1, G_2.x\}) = (x, \theta)$ for some substitution θ, we get

$$\rho \vdash C,\, \#((x)\{G_1.c_1, G_2.x\}, c)\ \to\ C,\, c$$

The rule for $tell(e)^l/retract(e)^l$ evaluates the expression e until it reduces to a fact F, which is a value of $\mathrm{ML_{CoDa}}$. The new context C', obtained from C by adding/removing F, is checked against the security policy Φ. Since Φ is a Datalog goal, we can easily reuse our dispatching machinery, implementing the check as a call to the function dsp where the first argument is C' and the second one is the trivial variation `phi.()`. If this call produces a result, then the evaluation yields the unit value and the new context C'.

The following example shows the reduction of a *retract* construct. Let Φ be the policy of Sect. 2, C be $\{F_3, F_4, F_5\}$, and apply $\mathtt{f} = \lambda\mathtt{x.\,if\,e_1\,then\,F_5\,else\,F_4}$ to unit. If

e_1 evaluates to `false` (without changing the context), the evaluation gets stuck because $dsp(C \smallsetminus \{F_4\}, \mathtt{phi.}())$ fails. Since Φ requires the fact F_4 to always hold, every attempt to remove it from the context violates indeed Φ. If, instead, e_1 reduces to `true`, there is no policy violation and the evaluation reduces to unit.

$$\rho \vdash C, \mathtt{retract(f\ ())}^l \rightarrow^* C, \mathtt{retract(F_4)}^l \not\rightarrow$$
$$\rho \vdash C, \mathtt{retract(f\ ())}^l \rightarrow^* C, \mathtt{retract(F_5)}^l \rightarrow C \smallsetminus \{\mathtt{F_5}\}, ()$$

4 Type and Effect System

We now associate an ML_{CoDa} expression with a type, an abstraction called the *history expression*, and a function called the *labelling environment*. During the verification phase, the virtual machine uses the history expression to ensure that the dispatching mechanism will always succeed at runtime. Then, the labelling environment drives code instrumentation with security checks. First, we briefly present History Expressions and labelling environments, and then the rules of our type and effect system.

History Expressions. A history expression is a term of a simple process algebra that soundly abstracts program behaviour [4]. Here, they approximate the sequence of actions that an application may perform over the context at runtime, i.e. asserting/retracting facts and asking if a goal holds. We assume that a history expression is uniquely labelled on a given set of Lab_H. Labels link static actions in histories to the corresponding dynamic actions inside the code. The syntax of History Expressions follows:

$$H ::= \ni \mid \varepsilon^l \mid h^l \mid (\mu h.H)^l \mid tell\,F^l \mid retract\,F^l \mid (H_1 + H_2)^l \mid (H_1 \cdot H_2)^l \mid \Delta$$
$$\Delta ::= (ask\,G.H \otimes \Delta)^l \mid fail^l$$

The empty history expression abstracts programs which do not interact with the context. For technical reasons, we syntactically distinguish when the empty history expression comes from the syntax (ε^l) and when it is instead obtained by reduction in the semantics ($\ni$). With $\mu h.H$ we represent possibly recursive functions, where h is the recursion variable; the "atomic" history expressions $tell\,F$ and $retract\,F$ are for the analogous constructs of ML_{CoDa}; the non-deterministic sum $H_1 + H_2$ abstracts *if-then-else*; the concatenation $H_1 \cdot H_2$ is for sequences of actions, that arise, e.g. while evaluating applications; Δ mimics our dispatching mechanism, where Δ is an *abstract variation*, defined as a list of history expressions, each element H_i of which is guarded by an $ask\,G_i$.

The history expression of the behavioural variation `url` in `getExhibitData` of Sect. 2, is $H_{url} = ask\,G_1.H_1 \otimes ask\,G_2.H_2 \otimes fail$, where $G_1 = \leftarrow$ `direct_comm()` and $G_2 = \leftarrow$ `use_qrcode (decoder)`, `camera(cam)`, and H_i is the effect of the expression guarded by G_i, for $i = 1, 2$. Intuitively, H_{url} means that at least one goal between G_1 and G_2 must be satisfied by the context to successfully apply the behavioural variation `url`. Given a context C, the behaviour of a history expression H is formalised by the transition system inductively defined in Fig. 3. Transitions $C, H \rightarrow C', H'$ mean that H reduces to H' in the context C and yields the context C'. Most rules are similar to the ones of [4]: below we only comment on those dealing with the context. An action $tell\,F$ reduces to $\ni$ and yields a context C', where the fact F has just been added; similarly for

$$\frac{}{C,(\ni\cdot H)^l \to C,H} \qquad \frac{}{C, tell\,F^l \to C\cup\{F\}, \ni} \qquad \frac{C,H_1 \to C',H_1'}{C,(H_1+H_2)^l \to C',H_1'}$$

$$\frac{}{C,\varepsilon^l \to C, \ni} \qquad \frac{}{C, retract\,F^l \to C\backslash\{F\}, \ni} \qquad \frac{C,H_2 \to C',H_2'}{C,(H_1+H_2)^l \to C',H_2'}$$

$$\frac{C,H_1 \to C',H_1'}{C,(H_1\cdot H_2)^l \to C',(H_1'\cdot H_2)^l} \qquad \frac{C\models G}{C,(ask\,G.H\otimes\Delta)^l \to C,H}$$

$$\frac{}{C,(\mu h.H)^l \to C,H[(\mu h.H)^l/h]} \qquad \frac{C\not\models G}{C,(ask\,G.H\otimes\Delta)^l \to C,\Delta}$$

Fig. 3. Semantics of History Expressions

retract F. Differently from what we do in the semantic rules, here we do not consider the possibility of a policy violation: a history expression approximates how the application would behave in the absence of any kind of check. The rules for Δ scan the abstract variation and look for the first goal G satisfied in the current context; if this search succeeds, the whole history expression reduces to the history expression H guarded by G; otherwise the search continues on the rest of Δ. If no satisfiable goal exists, the stuck configuration *fail* is reached, meaning that the dispatching mechanism fails.

We assume we are given the function $h : Lab_H \to \mathbb{H}$ that recovers a construct in a given history expression $h \in \mathbb{H}$ from a label l. Below, we specify the link between a *tell/retract* in a history expression and its corresponding operation in the code, labelled on Lab_C (see Sect 3). Consider, e.g. the history expression H_a of Sect. 2, and the label correspondence given there: $\{1 \mapsto 1, 2 \mapsto 2, 4 \mapsto 3, 5 \mapsto 4, 8 \mapsto 5\}$.

Definition 1 (Labelling Environment). *A* labelling environment *is a (partial) function* $\Lambda : Lab_H \to Lab_C$*, defined only if* $h(l) \in \{tell(F), retract(F)\}$.

Typing Rules. We assume that each Datalog predicate has a fixed arity and a type (see [20]). From here onwards, we also assume that there exists a Datalog typing function γ that, given a goal G, returns a list of pairs (x, type-of-x), for all variables $x \in G$.

The rules of our type and effect systems have:

- the usual environment $\Gamma ::= \emptyset \mid \Gamma, x : \tau$, binding the variables of an expression; $\emptyset$ denotes the empty environment, and $\Gamma, x : \tau$ denotes an environment having a binding for the variable x (x does not occur in Γ).
- a further environment $K ::= \emptyset \mid K, (\tilde{x}, \tau, \Delta)$, that maps a parameter $\tilde{x}$ to a pair consisting of a type and an abstract variation Δ, used to solve the binding for $\tilde{x}$ at runtime; $K, (\tilde{x}, \tau, \Delta)$ denotes an environment with a binding for the parameter $\tilde{x}$ (not in K).

Our typing judgements $\Gamma; K \vdash e : \tau \triangleright H; \Lambda$, express that in the environments Γ and K the expression e has type τ, effect H and yields a labelling environment Λ. We have basic types $\tau_c \in \{int, bool, unit, \ldots\}$, functional types, behavioural variations types, and facts:

$$\tau ::= \tau_c \mid \tau_1 \xrightarrow{K|H} \tau_2 \mid \tau_1 \xRightarrow{K|\Delta} \tau_2 \mid fact_\phi \qquad\qquad \phi \in \wp(Fact)$$

(TFACT)

$$\Gamma; K \vdash F : fact_{\{F\}} \triangleright \varepsilon; \bot$$

(TTELL/TRETRACT)

$$\frac{\Gamma; K \vdash e : fact_{\phi} \triangleright H; \Lambda \qquad op \in \{tell, retract\}}{\Gamma; K \vdash op(e)^{l} : unit \triangleright \left(H \cdot \left(\sum_{F_i \in \phi} opF_i^{l_i}\right)\right)^{l} ; \Lambda \biguplus_{F_i \in \phi} [l_i \mapsto l]}$$

(TLET)

$$\frac{\Gamma; K \vdash e_1 : \tau_1 \triangleright H_1; \Lambda_1 \qquad \Gamma, x : \tau_1; K \vdash e_2 : \tau_2 \triangleright H_2; \Lambda_2}{\Gamma; K \vdash let\ x = e_1\ in\ e_2 : \tau_2 \triangleright H_1 \cdot H_2; \Lambda_1 \uplus \Lambda_2}$$

(TVARIATION)

$$\frac{\begin{array}{cc} \forall i \in \{1, \ldots, n\} & \gamma(G_i) = \vec{y_i} : \vec{\tau_i} \\ \Gamma, x : \tau_1, \vec{y_i} : \vec{\tau_i}; K' \vdash e_i : \tau_2 \triangleright H_i; , \Lambda_i & \Delta = ask G_1.H_1 \otimes \cdots \otimes ask G_n.H_n \otimes fail \end{array}}{\Gamma; K \vdash (x)\{G_1.e_1, \ldots, G_n.e_n\} : \tau_1 \xrightarrow{K'|\Delta} \tau_2 \triangleright \varepsilon; \quad \biguplus_{i \in \{1, \ldots, n\}} \Lambda_i}$$

(TDLET)

$$\frac{\begin{array}{c} \Gamma, \vec{y} : \vec{\tau}; K \vdash e_1 : \tau_1 \triangleright H_1; \Lambda_1 \\ \Gamma; K, (\tilde{x}, \tau_1, \Delta') \vdash e_2 : \tau \triangleright H; \Lambda_2 \end{array}}{\Gamma; K \vdash dlet\, \tilde{x} = e_1\, when G in e_2 : \tau \triangleright H; \Lambda_1 \uplus \Lambda_2} \quad \begin{array}{l} \text{where } \gamma(G) = \vec{y} : \vec{\tau} \\ \text{if } K(\tilde{x}) = (\tau_1, \Delta) \text{ then } \Delta' = G.H_1 \otimes \Delta \\ \text{else (if } \tilde{x} \notin K \text{ then } \Delta' = G.H_1 \otimes fail) \end{array}$$

Fig. 4. Typing rules for the new constructs implementing adaptation

Some types are annotated for analysis reasons. In $fact_{\phi}$, the set ϕ contains the facts that an expression can be reduced to at runtime (see the semantics rules (TELL2) and (RETRACT2)). Here, K stores the types and the abstract variations of the parameters occurring inside the body of f. The history expression H is the latent effect of f, i.e. the sequence of actions which may be performed over the context during the function evaluation. Similarly, in $\tau_1 \xrightarrow{K|\Delta} \tau_2$ associated with the behavioural variation $bv = (x)\{Va\}$, K is a precondition for applying bv, while Δ is an abstract variation, that represents the information used at runtime by the dispatching mechanism to apply bv.

We now introduce the *partial orderings* $\sqsubseteq_H, \sqsubseteq_\Delta, \sqsubseteq_K, \sqsubseteq_\Lambda$ on H, Δ, K and Λ, resp. (often omitting the indexes when unambiguous):

- $H_1 \sqsubseteq_H H_2$ iff $\exists H_3$ such that $H_2 = H_1 + H_3$;
- $\Delta_1 \sqsubseteq_\Delta \Delta_2$ iff $\exists \Delta_3$ such that $\Delta_2 = \Delta_1 \otimes \Delta_3$ (note that Δ_2 has a single trailing *fail*);
- $K_1 \sqsubseteq_K K_2$ iff $((\tilde{x}, \tau_1, \Delta_1) \in K_1$ implies $(\tilde{x}, \tau_2, \Delta_2) \in K_2$ and $\tau_1 \leq \tau_2 \wedge \Delta_1 \sqsubseteq_\Delta \Delta_2)$;
- $\Lambda_1 \sqsubseteq_\Lambda \Lambda_2$ iff $\exists \Lambda_3$ such that $dom(\Lambda_3) \cap dom(\Lambda_1) = \emptyset$ and $\Lambda_2 = \Lambda_1 \uplus \Lambda_3$.

Most of the rules of our type and effect system are inherited from ML, and those for the new constructs are in Fig. 4. A few comments are in order.

The rule (TFACT) gives a fact F type *fact* annotated with $\{F\}$ and the empty effect. The rule (TTELL)/(TRETRACT) asserts that the expression $tell(e)/retract(e)$ has type *unit*, provided that the type of e is $fact_{\phi}$. The overall effect is obtained by combining the effect of e with the nondeterministic summation of $tell F/retract F$, where F is any of the facts in the type of e. In rule (TVARIATION) we determine the type for each subexpression e_i under K', and the environment Γ, extended by the type of x and of the variables $\vec{y_i}$ occurring in the goal G_i (recall that the Datalog typing function γ returns a list of pairs (z, type-of-z) for all variables z of G_i). Note that all subexpressions e_i have the same type τ_2. We also require that the abstract variation Δ results from concatenating $ask\,G_i$ with the effect computed for e_i. The type of the behavioural variation is annotated by K' and Δ. Consider the behavioural variation $bv_1 = (x)\{G_1.e_1, G_2.e_2\}$. Assume that the two cases of this behavioural variation have type τ and effects H_1 and H_2, respectively, under the environment $\Gamma, x : int$ (goals have no variables) and the guessed environment K'. Hence, the type of bv_1 will be $int \xRightarrow{K'|\Delta} \tau$ with $\Delta = ask\,G_1.H_1 \otimes ask\,G_2.H_2 \otimes fail$ and the effect will be empty. The rule (TDLET) requires that e_1 has type τ_1 in the environment Γ extended with the types for the variables $\overrightarrow{y}$ of the goal G. Also, e_2 has to type-check in an environment K, extended with the information for parameter $\tilde{x}$. The type and the effect for the overall *dlet* expression are the same as e_2. The labelling environment generated by the rules (TFACT) is $\bot$, because there is no *tell* or *retract*. Instead both (TTELL) and (TRETRACT) update the current environment Λ by associating all the labels of the facts which e can evaluate to, with the label l of the $tell(e)$ ($retract(e)$, resp.) being typed. The rule (TLET) produces an environment Λ that contains all the correspondences of Λ_1 and Λ_2 coming from e_1 and e_2; note that unicity of the labelling is guaranteed by the condition $dom(\Lambda_1) \cap dom(\Lambda_2) = \emptyset$.

The correspondence between the labels in the expression e_a and those of its history expression H_a of Sect. 2 are $\{1 \mapsto 1, 2 \mapsto 2, 4 \mapsto 3, 5 \mapsto 4, 8 \mapsto 5\}$, and the other labels are mapped to $\bot$. Note that a labelling environment need not be injective.

Soundness. Our type and effect system is sound with respect to the operational semantics of $\mathrm{ML_{CoDa}}$. First, we introduce the typing dynamic environment and an ordering on history expressions. Intuitively, the history expression H_1 could be obtained from H_2 by evaluation.

Definition 2 (Typing Dynamic Environment). *Given the type environments Γ and K, we say that the dynamic environment ρ has type K under Γ (in symbols $\Gamma \vdash \rho : K$) iff $dom(\rho) \subseteq dom(K)$ and $\forall \tilde{x} \in dom(\rho).\ \rho(x) = G_1.e_1, \ldots, G_n.e_n\ K(\tilde{x}) = (\tau, \Delta)$ and $\forall i \in \{1, \ldots, n\}.\ \gamma(G_i) = \overrightarrow{y_i} : \overrightarrow{\tau_i}\ \Gamma, \overrightarrow{y_i} : \overrightarrow{\tau_i}; K_{\tilde{x}} \vdash e_i : \tau' \triangleright H_i$ and $\tau' \leq \tau$ and $\bigotimes_{i \in \{1,\ldots,n\}} G_i.H_i \sqsubseteq \Delta$.*

Definition 3. *Given H_1, H_2 then $H_1 \preccurlyeq H_2$ iff one of the following cases holds*

(a) $H_1 \sqsubseteq H_2$; *(b)* $H_2 = H_3 \cdot H_1$ *for some* H_3;
(c) $H_2 = \bigotimes_{i \in \{1,\ldots,n\}} ask\,G_i.H_i \otimes fail \wedge H_1 = H_i,\ i \in [1..n]$.

Theorem 1 (Preservation). *Let e_s be a closed expression; and let ρ be a dynamic environment such that $dom(\rho)$ includes the set of parameters of e_s and such that $\Gamma \vdash \rho : K$. If $\Gamma; K \vdash e_s : \tau \triangleright H_s; \Lambda_s$ and $\rho \vdash C, e_s \rightarrow C', e_s'$ then $\Gamma; K \vdash e_s' : \tau \triangleright H_s'; \Lambda_s'$ and $\exists \overline{H}$, s.t. $\overline{H} \cdot H_s' \preccurlyeq H_s$ and $C, \overline{H} \cdot H_s' \rightarrow^+ C', H_s'$ and $\Lambda_s' \sqsubseteq \Lambda_s$.*

The Progress Theorem assumes that the effect H is *viable*, i.e. it does not reach *fail*, meaning that the dispatching mechanism succeeds at runtime. The control flow analysis of Sect. 5 guarantees viability (below $\rho \vdash C, e \nrightarrow$ means no transition from C, e). The next corollary ensures that the effect computed for e soundly approximates the actions that may be performed over the context during the evaluation of e.

Theorem 2 (Progress). *Let e_s be a closed expression such that $\Gamma;K \vdash e_s : \tau \triangleright H_s;\Lambda_s$; and let ρ be a dynamic environment such that $dom(\rho)$ includes the set of parameters of e_s, and such that $\Gamma \vdash \rho : K$. If $\rho \vdash C, e_s \not\rightarrow$ and H is viable for C (i.e. $C, H_s \not\rightarrow^+ C', fail$) and there is no policy violation then e_s is a value.*

Corollary 1 (Over-Approximation). *Let e be a closed expression.*
If $\Gamma;K \vdash e : \tau \triangleright H;\Lambda_s \wedge \rho \vdash C, e \rightarrow^* C', e'$, for some ρ such that $\Gamma \vdash \rho : K$, then there exists a computation $C, H \rightarrow^* C', H'$, for some H'.

Note that the type of e' is the same of e, because of Theorem 1, and the obtained label environment is included in Λ_s.

5 Loading-Time Analysis

Our execution model for $\mathrm{ML_{CoDa}}$ extends the one in [13]: the compiler produces a quadruple $(C_p, e_p, H_p, \Lambda_p)$ given by the application context, the object code, the history expression over-approximating the behaviour of e_p; and the labelling environment associating labels of H_p with those in the code. Given the quadruple, at loading time, the virtual machine performs the following two phases:

- *linking*: to resolve system variables and constructs the initial context C (combining C_p and the system context); and
- *verification*: to build from H_p a graph $\mathcal{G}$ that describes the possible evolutions of C.

Technically, we compute $\mathcal{G}$ through a static analysis, specified in terms of Flow Logic [21]. To support the formal development, we assume below that all the bound variables occurring in a history expression are distinct. So we can define a function $\mathbb{K}$ mapping a variable h^l to the history expression $(\mu h.H_1^{l_1})^{l_2}$ that introduces it.

The static approximation is represented by a pair $(\Sigma_\circ, \Sigma_\bullet)$, called *estimate* for H, with $\Sigma_\circ, \Sigma_\bullet : Lab_H \to \wp(Context \cup \{\bullet\})$, where $\bullet$ is the distinguished "failure" context representing a dispatching failure. For each label l,

- the *pre-set* $\Sigma_\circ(l)$ contains the contexts possibly arising *before* evaluating H^l;
- the *post-set* $\Sigma_\bullet(l)$ contains the contexts possibly resulting *after* evaluating H^l.

The analysis is specified by a set of clauses upon judgements $(\Sigma_\circ, \Sigma_\bullet) \vDash H^l$, where $\vDash \subseteq \mathcal{AE} \times \mathbb{H}$ and $\mathcal{AE} = (Lab_H \to \wp(Context \cup \{\bullet\}))^2$ is the domain of the results of the analysis and $\mathbb{H}$ the set of history expressions. The judgement $(\Sigma_\circ, \Sigma_\bullet) \vDash H^l$ says that $\Sigma_\circ$ and $\Sigma_\bullet$ form an acceptable analysis estimate for the history expression H^l.

We will use the notion of acceptability to check whether the history expression H_p, hence the expression e it is an abstraction of, will never fail in a given initial context C.

In Fig. 5, we give the set of inference rules that validate the correctness of a given estimate $\mathcal{E} = (\Sigma_\circ, \Sigma_\bullet)$. Intuitively, the checks in the clauses mimic the semantic evolution of the history expression in the given context, by modelling the semantic preconditions and the consequences of the possible reductions.

In the rule (ATELL), the analysis checks whether the context C is in the pre-set, and $C \cup \{F\}$ is in the post-set; similarly for(ARETRACT), where $C \backslash \{F\}$ should be in the

$$\text{(ANIL)}\ \frac{}{(\Sigma_\circ,\Sigma_\bullet)\models \ni} \qquad \text{(ATELL)}\ \frac{\forall C\in\Sigma_\circ(l)\ \ C\cup\{F\}\in\Sigma_\bullet(l)}{(\Sigma_\circ,\Sigma_\bullet)\models \mathit{tell}\,F^l} \qquad \text{(ARETRACT)}\ \frac{\forall C\in\Sigma_\circ(l)\ \ C\setminus\{F\}\in\Sigma_\bullet(l)}{(\Sigma_\circ,\Sigma_\bullet)\models \mathit{retract}\,F^l}$$

$$\text{(ASEQ1)}\ \frac{\begin{array}{ll}(\Sigma_\circ,\Sigma_\bullet)\models H_1^{l_1} & \Sigma_\circ(l)\subseteq\Sigma_\circ(l_1)\\ (\Sigma_\circ,\Sigma_\bullet)\models H_2^{l_2} & \Sigma_\bullet(l_1)\subseteq\Sigma_\circ(l_2)\\ & \Sigma_\bullet(l_2)\subseteq\Sigma_\bullet(l)\end{array}}{(\Sigma_\circ,\Sigma_\bullet)\models (H_1^{l_1}\cdot H_2^{l_2})^l} \qquad \text{(ASEQ2)}\ \frac{(\Sigma_\circ,\Sigma_\bullet)\models H_2^{l_2} \quad \begin{array}{l}\Sigma_\circ(l)\subseteq\Sigma_\circ(l_2)\\ \Sigma_\bullet(l_2)\subseteq\Sigma_\bullet(l)\end{array}}{(\Sigma_\circ,\Sigma_\bullet)\models (\ni\cdot H_2^{l_2})^l}$$

$$\text{(AEPS)}\ \frac{\Sigma_\circ(l)\subseteq\Sigma_\bullet(l)}{(\Sigma_\circ,\Sigma_\bullet)\models \varepsilon^l} \qquad \text{(ASUM)}\ \frac{\begin{array}{lll}(\Sigma_\circ,\Sigma_\bullet)\models H_1^{l_1} & \Sigma_\circ(l)\subseteq\Sigma_\circ(l_1) & \Sigma_\bullet(l_1)\subseteq\Sigma_\bullet(l)\\ (\Sigma_\circ,\Sigma_\bullet)\models H_2^{l_2} & \Sigma_\circ(l)\subseteq\Sigma_\circ(l_2) & \Sigma_\bullet(l_2)\subseteq\Sigma_\bullet(l)\end{array}}{(\Sigma_\circ,\Sigma_\bullet)\models (H_1^{l_1}+H_2^{l_2})^l}$$

$$\text{(AASK1)}\ \frac{\begin{array}{l}\forall C\in\Sigma_\circ(l)\ \ (C\models G \Longrightarrow (\Sigma_\circ,\Sigma_\bullet)\models H^{l_1}\ \ \Sigma_\circ(l)\subseteq\Sigma_\circ(l_1)\ \ \Sigma_\bullet(l_1)\subseteq\Sigma_\bullet(l))\\ \qquad\qquad\ \ (C\not\models G \Longrightarrow (\Sigma_\circ,\Sigma_\bullet)\models \Delta^{l_2}\ \ \Sigma_\circ(l)\subseteq\Sigma_\circ(l_2)\ \ \Sigma_\bullet(l_2)\subseteq\Sigma_\bullet(l))\end{array}}{(\Sigma_\circ,\Sigma_\bullet)\models (\mathit{ask}G.H^{l_1}\otimes\Delta^{l_2})^l} \qquad \text{(AASK2)}\ \frac{\bullet\in\Sigma_\bullet(l)}{(\Sigma_\circ,\Sigma_\bullet)\models \mathit{fail}^l}$$

$$\text{(AREC)}\ \frac{(\Sigma_\circ,\Sigma_\bullet)\models H^{l_1} \quad \begin{array}{l}\Sigma_\circ(l)\subseteq\Sigma_\circ(l_1)\\ \Sigma_\bullet(l_1)\subseteq\Sigma_\bullet(l)\end{array}}{(\Sigma_\circ,\Sigma_\bullet)\models (\mu h.H^{l_1})^l} \qquad \text{(AVAR)}\ \frac{\mathbb{K}(h)=(\mu h.H^{l_1})^{l'} \quad \begin{array}{l}(\Sigma_\circ(l)\subseteq\Sigma_\circ(l')\\ \Sigma_\bullet(l')\subseteq\Sigma_\bullet(l)\end{array}}{(\Sigma_\circ,\Sigma_\bullet)\models h^l}$$

Fig. 5. Specification of the analysis for History Expressions

post-set. The rule (ANIL) says that every pair of functions is an acceptable estimate for the "semantic" empty history expression $\ni$. The estimate $\mathcal{E}$ is acceptable for the "syntactic" ε^l if the pre-set is included in the post-set (rule (AEPS)). The rules (ASEQ1) and (ASEQ2) handle the sequential composition of history expressions. The first rule states that $(\Sigma_\circ,\Sigma_\bullet)$ is acceptable for $H=(H_1^{l_1}\cdot H_2^{l_2})^l$ if it is valid for both H_1 and H_2. Moreover, the pre-set of H_1 must include the pre-set of H and the pre-set of H_2 includes the post-set of H_1; finally, the post-set of H includes that of H_2. The second rule states that $\mathcal{E}$ is acceptable for $H=(\ni\cdot H_1^{l_2})^l$ if it is acceptable for H_1 and the pre-set of H_1 includes that of H, while the post-set of H includes that of H_1. The rules (AASK1) and (AASK2) handle the abstract dispatching mechanism. The first states that $\mathcal{E}$ is acceptable for $H=(\mathit{ask}G.H_1^{l_1}\otimes\Delta^{l_2})^l$, provided that, for all C in the pre-set of H, if the goal G succeeds in C then the pre-set of H_1 includes that of H and the post-set of H includes that of H_1. Otherwise, the pre-set of Δ^{l_2} must include the pre-set of H and the post-set of Δ^{l_2} is included in that of H. The second requires $\bullet$ to be in the post-set of fail^l. By the rule (ASUM), $\mathcal{E}$ is acceptable for $H=(H_1^{l_1}+H_2^{l_2})^l$ if it is valid for each H_1 and H_2; the pre-set of H is included in the pre-sets of H_1 and H_2; and the post-set of H includes those of H_1 and H_2. By the rule (AREC), $\mathcal{E}$ is acceptable for $H=(\mu h.H_1^{l_1})^l$ if it is valid for $H_1^{l_1}$, the pre-set of H_1 includes that of H; and the post-set of H includes that of H_1.

The rule (AVAR) says that a pair $(\Sigma_\circ, \Sigma_\bullet)$ is an acceptable estimate for a variable h^l if the pre-set of the history expression introducing h, namely $\mathbb{K}(h)$, is included in that of h^l, and the post-set of h^l includes that of $\mathbb{K}(h)$.

Semantic Properties. We now formalise the notion of valid estimate for a history expression; we prove that there always exists a minimal valid analysis estimate; and that a valid estimate is correct w.r.t. the operational semantics of history expressions.

Definition 4 (Valid Analysis Estimate). *Given $H_p^{l_p}$ and an initial context C, we say that a pair $(\Sigma_\circ, \Sigma_\bullet)$ is a* valid analysis estimate *for H_p and C iff $C \in \Sigma_\circ(l_p)$ and $(\Sigma_\circ, \Sigma_\bullet) \models H_p^{l_p}$.*

Theorem 3 (Existence of Estimates). *Given H^l and an initial context C, the set $\{(\Sigma_\circ, \Sigma_\bullet) \mid (\Sigma_\circ, \Sigma_\bullet) \models H^l\}$ of the acceptable estimates of the analysis for H^l and C is a Moore family; hence, there exists a minimal valid estimate.*

Theorem 4 (Subject Reduction). *Let H^l be a closed history expression s.t. $(\Sigma_\circ, \Sigma_\bullet) \models H^l$. If $\forall C \in \Sigma_\circ(l)$, $C, H^l \to C', H'^{l'}$ then $(\Sigma_\circ, \Sigma_\bullet) \models H'^{l'}$, $\Sigma_\circ(l) \subseteq \Sigma_\circ(l')$, and $\Sigma_\bullet(l') \subseteq \Sigma_\bullet(l)$.*

Viability of history expressions. We now define when a history expression H_p is viable for an initial context C, i.e. when it passes the verification phase. Below, let *lfail*(H) be the set of labels of the *fail* sub-terms in H:

Definition 5 (Viability). *Let H_p be a history expression and C be an initial context. We say that H_p is* viable *for C if there exists the minimal valid analysis estimate $(\Sigma_\circ, \Sigma_\bullet)$ such that $\forall l \in dom(\Sigma_\bullet) \backslash lfail(H_P)\ \bullet \notin \Sigma_\bullet(l)$.*

To illustrate how viability is checked, consider the following history expressions:

$$H_p = ((tell\, F_1^1 \cdot retract\, F_2^2)^3 + (ask\, F_5.retract\, F_8^5 \otimes (ask\, F_3.retract\, F_4^6 \otimes fail^7)^8)^4)^9$$
$$H_p' = ((tell\, F_1^1 \cdot retract\, F_2^2)^3 + (ask\, F_3.retract\, F_4^5 \otimes fail^6)^4)^7$$

and the initial context $C = \{F_2, F_5, F_8\}$, only consisting of facts.

The left part of Fig. 6 shows the values of $\Sigma_\circ^1(l)$ and $\Sigma_\bullet^1(l)$ for H_p. Notice that the pre-set of $tell\, F_1^1$ includes $\{F_2, F_5, F_8\}$, and the post-set also includes $\{F_1\}$. Also, the pre-set of $retract\, F_8^5$ includes $\{F_2, F_5, F_8\}$, while the post-set includes $\{F_2, F_5\}$. The column describing $\Sigma_\bullet$ contains $\bullet$ only for $l = 7$, the label of *fail*, so H_p is viable for C. However, the history expression H_p' fails to pass the verification phase, when put in the initial context C. Since the goal F_3 does not hold in C, H_p' is not viable. This is reflected by the occurrences of $\bullet$ in $\Sigma_\bullet^2(4)$ and $\Sigma_\bullet^2(7)$, as shown in the right part of Fig. 6.

Now, we exploit the result of the above analysis to build up the evolution graph $\mathcal{G}$. It describes how the initial context C will evolve at runtime, paving the way to security enforcement.

	$\Sigma_\circ^1$	$\Sigma_\bullet^1$
1	$\{\{F_2,F_5,F_8\}\}$	$\{\{F_1,F_2,F_5,F_8\}\}$
2	$\{\{F_1,F_2,F_5,F_8\}\}$	$\{\{F_1,F_5,F_8\}\}$
3	$\{\{F_2,F_5,F_8\}\}$	$\{\{F_1,F_5,F_8\}\}$
4	$\{\{F_2,F_5,F_8\}\}$	$\{\{F_2,F_5\}\}$
5	$\{\{F_2,F_5,F_8\}\}$	$\{\{F_2,F_5\}\}$
6	$\emptyset$	$\emptyset$
7	$\emptyset$	$\{\bullet\}$
8	$\emptyset$	$\emptyset$
9	$\{\{F_2,F_5,F_8\}\}$	$\{\{F_1,F_5,F_8\},\{F_2,F_5\}\}$

	$\Sigma_\circ^2$	$\Sigma_\bullet^2$
1	$\{\{F_2,F_5,F_8\}\}$	$\{\{F_1,F_2,F_5,F_8\}\}$
2	$\{\{F_1,F_2,F_5,F_8\}\}$	$\{\{F_1,F_5,F_8\}\}$
3	$\{\{F_2,F_5,F_8\}\}$	$\{\{F_1,F_5,F_8\}\}$
4	$\{\{F_2,F_5,F_8\}\}$	$\{\bullet\}$
5	$\emptyset$	$\emptyset$
6	$\{\{F_2,F_5,F_8\}\}$	$\{\bullet\}$
7	$\{\{F_2,F_5,F_8\}\}$	$\{\{F_1,F_5,F_8\},\bullet\}$

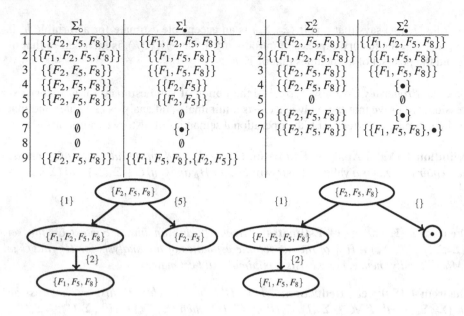

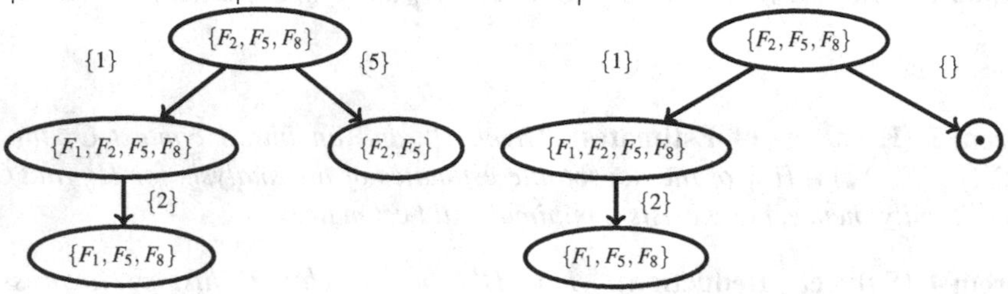

Fig. 6. The analysis results (top) and the evolution graphs $\mathcal{G}_p$ (bottom left) and $\mathcal{G}'_p$ (bottom right) for the initial context $C = \{F_2, F_5, F_8\}$, and for the history expressions $H_p = ((tell\,F_1^1 \cdot retract\,F_2^2)^3 + (ask\,F_5.retract\,F_8^5 \otimes ask\,F_3.retract\,F_4^6 \otimes fail^7)^4)^8$ and $H'_p = ((tell\,F_1^1 \cdot retract\,F_2^2)^3 + (ask\,F_3.retract\,F_4^5 \otimes fail^6)^4)^7$, respectively

Definition 6 (Evolution Graph). *Let H_p be a history expression, C be a context, and $(\Sigma_\circ, \Sigma_\bullet)$ be a valid analysis estimate. The evolution graph of C is $\mathcal{G} = (N, E, L)$, where*

$$\begin{aligned} N &= \bigcup_{l \in Lab_H^*} (\Sigma_\circ(l) \cup \Sigma_\bullet(l)) \\ E &= \{(C_1, C_2) \mid \exists F \in Fact^*,\ l \in Lab_H^*\ s.t.\ C_1 \in \Sigma_\circ(l) \wedge C_2 \in \Sigma_\bullet(l) \wedge \\ &\qquad (h(l) \in \{tell(F), retract(F)\} \vee (C_2 = \bullet))\} \\ L &: E \to \mathcal{P}(Labels) \\ \forall t &= (C_1, C_2) \in E,\ l \in L(t)\ iff\ C_1 \in \Sigma_\circ(l) \wedge C_2 \in \Sigma_\bullet(l) \wedge h(l) \neq fail \end{aligned}$$

Intuitively, the nodes of $\mathcal{G}$ are sets of contexts, and an edge between two nodes C_1 and C_2 records that C_2 is obtained from C_1, through a $tell/retract$. Using the labels of arcs we can locate the abstract $tell/retract$ that may lead to a context violating a given policy Φ. By putting guards on the corresponding risky actions in the code (via Λ), we can enforce Φ. In the following, let $Fact^*$ and Lab_H^* be the set of facts and the set of labels occurring in H_p, i.e. the history expression under verification.

Consider again the history expressions H_p and H'_p and their evolution graphs $\mathcal{G}_p$ and $\mathcal{G}'_p$ (Fig. 6, bottom). In $\mathcal{G}_p$, from the initial context C there is an arc labelled $\{1\}$ to $C \cup \{F_1\}$, because of $tell\,F_1^1$, and there is an arc labelled $\{5\}$ to the $C \setminus F_8$, because of $retract\,F_8^5$. It is easy to see that H_p is viable for C, because the node $\bullet$ is not reachable from the initial context C in $\mathcal{G}_p$, However H'_p is not, because $\bullet$ is reachable in $\mathcal{G}'_p$.

6 Code Instrumentation

Once we detected the potentially risky operations through the evolution graph $\mathcal{G}$, we can instrument the code of an application e and only switch on our *runtime monitor* to guard them. First, since a node n of $\mathcal{G}$ represents a context reachable while executing e, we *statically* verify whether n satisfies Φ. If this is not the case, we consider all the edges with target n and the set R of their labels. The labelling environment Λ, computed while type checking e, determines those actions in the code that require monitoring during the execution, indexed by the set $Risky = \Lambda(R)$.

In fact we will guard all the *tell*/*retract* actions in the code, but our runtime monitor will only be invoked on the risky ones. To do that, the compiler labels the source code as said in Sect. 2 and generates specific calls to *monitoring* procedures. We offer a lightweight form of code instrumentation that does not operate on the object code, differently from standard instrumentation. In more detail, we define a procedure, called `check_violation(l)`, for verifying if the policy Φ is satisfied. It takes a label `l` as parameter and returns the type *unit*. At loading time, we assign a global mask `risky[l]` for each label `l` in the source code, by using the information in the set *Risky*.
The procedure code in a pseudo ML_{CoDa} and the definition of `risky[l]` are as follows:

```
fun check_violation l =
    if risky[l] then ask phi.() else ()
```

$$\text{where } \texttt{risky[l]} = \begin{cases} \texttt{true} & \text{if } \texttt{l} \in Risky \\ \texttt{false} & \text{otherwise} \end{cases}$$

If `risky[l]` is `false`, then the procedure returns to the caller and the execution goes on normally. Otherwise, it calls for a check on Φ, by triggering with the call `ask phi.()` the dispatching mechanism: if the call fails then a policy violation is about to occur. In this case the computation is aborted or a recovery mechanism is possibly invoked.

Our compilation schema needs to replace every $tell(e)^l$ (similarly for $retract(e)^l$) in the source code with the following, where `z` is fresh.:

```
let z = tell(e) in check_violation(l)
```

An easy optimisation is possible when *Risky* is empty, i.e. when the analysis ensures that all the *tell*/*retract* actions are safe and so no execution paths lead to a policy violation. To do this, we introduce the flag `always_ok`, whose value will be computed at linking time: if it is true, no check is needed. The previous compilation schema is simply refined by testing `always_ok` before calling `check_violation`.

7 Conclusions

We have addressed security issues in an adaptive framework, by extending ML_{CoDa}, a functional language introduced in [13] for adaptive programming. Our main contributions can be summarised as follows:

- We have expressed and enforced *context-dependent* security policies in Datalog, originally used by ML_{CoDa} to deal with contexts.
- We have extended the ML_{CoDa} type and effect system for computing a type and a labelled abstract representation of the overall behaviour of an application. Actually, an effect over-approximates the sequences of the possible dynamic actions over the context, and labels link the security-critical operations of the abstraction with those in the code of the application.
- We have enhanced the static analysis of [13] to identify the operations that may affect contexts and violate the required *policy*, besides verifying that the application can adapt to all the possible contexts arising at runtime.
- Based on the results of the static analysis, we have defined a way to instrument the code of an application *e*, so as to introduce an *adaptive runtime monitor* that stops *e* when about to violate the policy to be enforced, and is switched on and off at need.

We plan to investigate richer forms of policies, in particular those having an additional dynamic scope, and those which are history dependent [4], and study their impact on adaptivity. A long term goal is extending these policies with quantitative information, e.g. statistical information about the usage of contexts, reliability of resources therein, etc. Finally, we are thinking of providing a kind of recovery mechanism for behavioural variations, to allow the user to undo some actions considered risky or sensible, and force the dispatching mechanism to make different, alternative choices.

References

1. Achermann, F., Lumpe, M., Schneider, J., Nierstrasz, O.: PICCOLA-a small composition language. In: Formal Methods for Distributed Processing. Cambridge University Press (2001)
2. Al-Neyadi, F., Abawajy, J.H.: Context-based E-health system access control mechanism. In: Park, J.H., Zhan, J., Lee, C., Wang, G., Kim, T.-h., Yeo, S.-S. (eds.) ISA 2009. CCIS, vol. 36, pp. 68–77. Springer, Heidelberg (2009)
3. Appeltauer, M., Hirschfeld, R., Haupt, M., Masuhara, H.: ContextJ: Context-oriented programming with Java. Computer Software 28(1) (2011)
4. Bartoletti, M., Degano, P., Ferrari, G.L., Zunino, R.: Local policies for resource usage analysis. ACM Trans. Program. Lang. Syst. 31(6) (2009)
5. Bonatti, P., De Capitani Di Vimercati, S., Samarati, P.: An algebra for composing access control policies. ACM Transactions on Information and System Security 5(1), 1–35 (2002)
6. Campbell, R., Al-Muhtadi, J., Naldurg, P., Sampemane, G., Mickunas, M.D.: Towards security and privacy for pervasive computing. In: Okada, M., Babu, C. S., Scedrov, A., Tokuda, H. (eds.) ISSS 2002. LNCS, vol. 2609, pp. 1–15. Springer, Heidelberg (2003)
7. Cardelli, L., Gordon, A.D.: Mobile ambients. Theor. Comput. Sci. 240(1), 177–213 (2000)
8. Ceri, S., Gottlob, G., Tanca, L.: What you always wanted to know about datalog (and never dared to ask). IEEE Trans. on Knowl. and Data Eng. 1(1), 146–166 (1989)
9. Costanza, P.: Language constructs for context-oriented programming. In: Proc. of the Dynamic Languages Symposium, pp. 1–10. ACM Press (2005)
10. Deng, M., Cock, D.D., Preneel, B.: Towards a cross-context identity management framework in e-health. Online Information Review 33(3), 422–442 (2009)
11. DeTreville, J.: Binder, a Logic-Based Security Language. In: Proc. of the 2002 IEEE Symposium on Security and Privacy, SP 2002, pp. 105–113. IEEE Computer Society (2002)

12. Eiter, T., Gottlob, G., Mannila, H.: Disjunctive datalog. ACM Transactions on Database Systems 5(1), 1–35 (1997)
13. Galletta, L.: Adaptivity: linguistic mechanisms and static analysis techniques. Ph.D. thesis, University of Pisa (2014), `http://www.di.unipi.it/~galletta/phdThesis.pdf`
14. Heer, T., Garcia-Morchon, O., Hummen, R., Keoh, S., Kumar, S., Wehrle, K.: Security challenges in the IP-based internet of things. Wireless Personal Communications, 1–16 (2011)
15. Hirschfeld, R., Costanza, P., Nierstrasz, O.: Context-oriented programming. Journal of Object Technology 7(3), 125–151 (2008)
16. Hulsebosch, R., Salden, A., Bargh, M., Ebben, P., Reitsma, J.: Context sensitive access control. In: Proc. of the ACM Symposium on Access Control Models and Technologies, pp. 111–119 (2005)
17. Kamina, T., Aotani, T., Masuhara, H.: Eventcj: a context-oriented programming language with declarative event-based context transition. In: Proc. of the 10th International Conference on Aspect-Oriented Software Development (AOSD 2011), pp. 253–264. ACM (2011)
18. Li, N., Mitchell, J.C.: DATALOG with Constraints: A Foundation for Trust Management Languages. In: Dahl, V. (ed.) PADL 2003. LNCS, vol. 2562, pp. 58–73. Springer, Heidelberg (2002)
19. Loke, S.W.: Representing and reasoning with situations for context-aware pervasive computing: a logic programming perspective. Knowl. Eng. Rev. 19(3), 213–233 (2004)
20. Mycroft, A., O'Keefe, R.A.: A polymorphic type system for prolog. Artificial Intelligence 23(3), 295–307 (1984)
21. Riis Nielson, H., Nielson, F.: Flow logic: a multi-paradigmatic approach to static analysis. In: Mogensen, T.Æ., Schmidt, D.A., Sudborough, I.H. (eds.) The Essence of Computation. LNCS, vol. 2566, pp. 223–244. Springer, Heidelberg (2002)
22. Orsi, G., Tanca, L.: Context modelling and context-aware querying. In: de Moor, O., Gottlob, G., Furche, T., Sellers, A. (eds.) Datalog 2010. LNCS, vol. 6702, pp. 225–244. Springer, Heidelberg (2011)
23. Pasquale, L., Ghezzi, C., Menghi, C., Tsigkanos, C., Nuseibeh, B.: Topology Aware Adaptive Security (to appear in SEAMS 2014)
24. Pfleeger, C., Pfleeger, S.: Security in computing. Prentice Hall (2003)
25. Román, M., Hess, C., Cerqueira, R., Ranganathan, A., Campbell, R., Nahrstedt, K.: Gaia: a middleware platform for active spaces. ACM SIGMOBILE Mobile Computing and Communications Review 6(4), 65–67 (2002)
26. Wrona, K., Gomez, L.: Context-aware security and secure context-awareness in ubiquitous computing environments. In: XXI Autumn Meeting of Polish Information Processing Society (2005)
27. Zhang, G., Parashar, M.: Dynamic context-aware access control for grid applications. In: Proc. of Fourth International Workshop on Grid Computing, pp. 101–108. IEEE (2003)

Partial Models and Weak Equivalence

Adilson Luiz Bonifacio[1,*] and Arnaldo Vieira Moura[2]

[1] Computing Department, University of Londrina, Londrina, Brazil
bonifacio@uel.br

[2] Computing Institute, University of Campinas, Campinas, Brazil
arnaldo@ic.unicamp.br

Abstract. One of the important tasks in model-based testing is checking completeness of test suites. In this paper we first extend some known sufficient conditions for test suite completeness by also allowing partial implementations. We also study a new notion of equivalence, and show that the same conditions are still sufficient when treating complete implementations. But when we also allow for partial implementations under this new notion of equivalence such conditions are not sufficient anymore for the completeness of test suites.

Keywords: test suite completeness, partial models, weak equivalence, confirmed sets.

1 Introduction

Automatic test suite generation for Finite State Machine (FSM) models has been widely investigated in the literature [1, 3, 4, 7]. Several approaches proposed conditions and techniques to generate test suites based on FSMs with complete fault coverage [2, 5, 8–10]. Some of these works have shown sufficient conditions that guarantee the completeness of the test suites [4, 12].

Simao and Petrenko [11] proposed sufficient conditions for checking test suite completeness based on a notion of "confirmed sets". Informally, a set of input sequences T is confirmed when any two of its sequences lead to a same state in a specification, and these same sequences also lead to a common state in any implementation that is T-equivalent to the specification. However, in that approach, implementation candidates are assumed to be complete models. Further, specifications and implementations are also required to be reduced and initially connected FSMs with the same number of states.

In this work we first remove the restriction of implementation machines being complete models. We also show that the existence of confirmed sets is enough to guarantee test suite completeness even when the implementations have any number of states. These relaxations seem natural, given that implementations are usually treated as black boxes. We proceed to explore a new notion of equivalence, called *weak-equivalence*, where we treat test cases that may not run to completion in one or both of the models. This further improves the treatment of

* Supported by FAPESP, process 2012/23500-6.

G. Ciobanu and D. Méry (Eds.): ICTAC 2014, LNCS 8687, pp. 80–96, 2014.

implementations as real black boxes. It is shown that confirmed sets, under the new notion of equivalence, are also sufficient to guarantee test suite completeness when implementations are complete models. By contrast, the existence of confirmed sets is no longer sufficient to ascertain test suite completeness when implementations may be partial machines under the new notion of equivalence.

This paper is organized as follows. Section 2 gives some important definitions. In Section 3 we show that confirmed sets are sufficient for test suite completeness under the classical notion of equivalence, and even when implementations are partial machines. We present an example with partial implementations in Section 4. In Section 5 we extend this result under a new notion of equivalence, where implementations are still restricted to being complete FSMs. Test suite completeness under the new notion of equivalence, but now allowing for partial implementations, is studied in Section 6. Section 7 ends with some concluding remarks.

2 Definitions

Let $\mathcal{I}$ be an alphabet. We denote by $\mathcal{I}^\star$ the set of all finite sequences of symbols from $\mathcal{I}$. When we write $\sigma = x_1x_2\cdots x_n \in \mathcal{I}^\star$ $(n \geq 0)$ we mean $x_i \in \mathcal{I}$ $(1 \leq i \leq n)$, unless noted otherwise. The length of any finite sequence of symbols α over $\mathcal{I}$ is indicated by $|\alpha|$. The empty sequence will be indicated by ε, with $|\varepsilon| = 0$. Given any two sets of sequences $A, B \subseteq \mathcal{I}^\star$, their symmetric difference will be indicated by $A \ominus B$.

Remark 1. *$A \ominus B = \emptyset$ iff*[1] *$A = B$.*

Next, we define Finite State Machines (FSMs) [6, 11].

Definition 1. *A FSM is a tuple $(S, s_0, \mathcal{I}, \mathcal{O}, D, \delta, \lambda)$ where*

1. *S is a finite set of* states
2. *$s_0 \in S$ is the initial state*
3. *$\mathcal{I}$ is a finite set of* input actions *or* input events
4. *$\mathcal{O}$ is a finite set of* output actions *or* output events
5. *$D \subseteq S \times \mathcal{I}$ is a* specification domain
6. *$\delta : D \to S$ is the* transition function
7. *$\lambda : D \to \mathcal{O}$ is the* output function.

All FSMs treated here are deterministic, since δ and λ are functions.

In the sequel, M and N will always denote the FSMs $(S, s_0, \mathcal{I}, \mathcal{O}, D, \delta, \lambda)$ and $(Q, q_0, \mathcal{I}, \mathcal{O}', D', \mu, \tau)$, respectively. Let $\sigma = x_1x_2\cdots x_n \in \mathcal{I}^\star$, $\omega = a_1a_2\cdots a_n \in \mathcal{O}^\star$ $(n \geq 0)$. If there are states $r_i \in S$ $(0 \leq i \leq n)$ such that $(r_{i-1}, x_i) \in D$, with $\delta(r_{i-1}, x_i) = r_i$ and $\lambda(r_{i-1}, x_i) = a_i$ $(1 \leq i \leq n)$, we may write $r_0 \overset{\sigma/\omega}{\to} r_n$. When σ, or ω, or both, is not important, we may write $r_0 \overset{\sigma/}{\to} r_n$, or $r_0 \overset{/\omega}{\to} r_n$, or $r_0 \to r_n$, respectively. We can also drop the target state, when it is not important, *e.g.*

[1] Here, 'iff' is short for 'if and only if'.

$r_0 \overset{\sigma/\omega}{\to}$ or $r_0 \to$. It will be useful to extend the functions δ and λ to pairs $(s,\sigma) \in S \times \mathcal{I}^\star$. Let $\widehat{D} = \{(s,\sigma) \mid s \overset{\sigma/}{\to}\}$. Define the extensions $\widehat{\delta} : \widehat{D} \to S$ and $\widehat{\lambda} : \widehat{D} \to \mathcal{O}^\star$ by letting $\widehat{\delta}(s,\sigma) = r$ and $\widehat{\lambda}(s,\sigma) = \omega$ whenever $s \overset{\sigma/\omega}{\to} r$. When there is no reason for confusion, we may write D, δ and λ instead of $\widehat{D}$, $\widehat{\delta}$ and $\widehat{\lambda}$, respectively. Also, let $U(s) = \{\sigma \mid (s,\sigma) \in \widehat{D}\}$ for all $s \in S$.

Remark 2. *We always have $\varepsilon \in U(s)$, for all $s \in S$.*

Next we define reachability and completeness for FSMs.

Definition 2. *Let M be a FSM, and $s, r \in S$. Then r is* reachable from s *iff $s \overset{\sigma/}{\to} r$ for some $\sigma \in \mathcal{I}^\star$. We say that r is* reachable *iff it is reachable from s_0. Also, M is* complete *iff for all reachable $s \in S$ and all $x \in \mathcal{I}$ we have $s \overset{x/}{\to}$.*

From Definition 1, FSMs need not be complete, that is, we may have $D \neq S \times \mathcal{I}$. On the other hand, if M is complete then $U(s) = \mathcal{I}^\star$ for all $s \in S$.

Observe that when M and N are not complete then there are cases when we can differentiate two states $s \in S$ and $q \in Q$ if there is $\sigma \in U(s) \ominus U(q)$. In this case, by running σ starting at s and q, one would observe a longer output sequence from either M or N. The notion of weak equivalence captures this effect: two FSMs are weakly equivalent when any input sequence that runs in one also runs in the other. The notion of equivalence is the classical one: any input sequence that runs on both machines must yield the same behavior.

Definition 3. *Let M and N be FSMs and $s \in S$, $q \in Q$. Let $C \subseteq \mathcal{I}^\star$. We say that s and q are*

1. *C-*weakly-distinguishable *iff $\big(U(s) \ominus U(q)\big) \cap C \neq \emptyset$, denoted $s \not\sim_C q$. Otherwise they are C-*weakly-equivalent*, denoted $s \sim_C q$.*
2. *C-*distinguishable *iff $\lambda(s,\sigma) \neq \tau(q,\sigma)$ for some $\sigma \in U(s) \cap U(q) \cap C$, denoted $s \not\approx_C q$. Otherwise, they are C-*equivalent*, denoted $s \approx_C q$.*

Two FSMs M and N are *C-weakly-distinguishable* or *C-distinguishable* iff $s_0 \not\sim q_0$ or $s_0 \not\approx q_0$, respectively. Otherwise M and N are *C-weakly-equivalent* or *C-equivalent*, respectively. We will use the same notation for FSM equivalence as we did for states, *e.g.*, $M \approx_C N$ when $s_0 \approx_C q_0$. When C is not important, or it is clear from the context, we might drop the index. When there is no mention to C, we mean $C = \mathcal{I}^\star$.

Remark 3. *Weak-distinguishability and weak-equivalence are purely structural notions. They do not mention the output sequence of the FSMs in any way.*

Next we say when FSMs are reduced.

Definition 4. *We say that a FSM M is* reduced *iff every state is reachable and every two distinct states are distinguishable.*

Now we define test cases and test suites.

Definition 5. *Let M be a FSM. A* test suite *for M is any finite subset of $\mathcal{I}^\star$. Any element of a test suite is a* test case.

Consider a specification M, a test suite T for M, and the notion of T-equivalence. In many situations, one considers only test suites whose test cases run in M, that is, with $T \subseteq U(s_0)$. It is also common to assume that implementations N are complete machines, that is, $U(q_0) = \mathcal{I}^\star$. Under these conditions we get $T \cap U(s_0) \cap U(q_0) = T$, and T-equivalence reduces to $\lambda(s_0, \alpha) = \tau(q_0, \alpha)$, for all $\alpha \in T$, that is M and N display the same behavior under all tests in T. Another common constraint is to restrict implementation N to have at most as many states as a specification M. However, when implementation N is treated as a black box, usually one cannot guarantee completeness of N, nor does one control the number of states in N. Many of the results that follow will establish a contrast between this simpler state of affairs and other, more realistic scenarios, where test suites can be more general sets and implementations may be partial models with any number of states.

Next, we recall the concept of n-completeness for test suites, under the two notions of equivalence.

Definition 6. *Let M be a FSM, let T be a test suite for M and take $n \geq 1$. Then T is*

1. weakly-n-complete *for M iff for any FSM N with $|Q| \leq n$, if $M \sim_T N$ then $M \sim N$.*
2. n-complete *for M iff for any FSM N with $|Q| \leq n$, if $M \approx_T N$ then $M \approx N$.*

3 Equivalence and Partial Implementations

Simão and Petrenko [11] proposed sufficient conditions for checking test suite n-completeness under a number of restrictions over the specifications and the implementation models, such as reducibility of both machines and completeness of the implementations. In this section we extend that result, by showing that those conditions are also sufficient for checking test suite completeness even when implementations are not complete. We also do not require that implementations have no more states than the specifications, nor do we require that both models be reduced FSMs. We remark that in this section we will be treating only the classical definition of equivalence. See definition 3(2).

For the ease of reference, we repeat part of [11]. If M is a FSM and T is a test suite for M, let $\Im_T(M)$ be the set of all FSMs with the same number of states as M, and that are reduced, complete and T-equivalent to it, with this notation as in [11]. Since we will not enforce all of these restrictions, we let $\mathcal{E}_T(M)$ be the larger set of all FSMs that are just T-equivalent to M.

Definition 7 ([11]). *Let M be a FSM, let T be a test suite for M and $K \subseteq T$. The set K is* confirmed *iff for all $s \in S$ there is some $\alpha \in K$ such that $\delta(s_0, \alpha) = s$ and, further, for each $N \in \mathcal{E}_T(M)$, it holds that*

1. *$K \subseteq U(s_0) \cap U(q_0)$;*
2. *For all α, $\beta \in K$, $\mu(q_0, \alpha) = \mu(q_0, \beta)$ if and only if $\delta(s_0, \alpha) = \delta(s_0, \beta)$.*

Remark 4. *We note that condition (1) is not explicitly stated in Definition 4 of [11]. Rather, in [11] it is implicitly assumed the slightly stronger condition that* $T \subseteq U(s_0) \cap U(q_0)$ *or, equivalently* $T \subseteq U(s_0)$*, given that implementations are always taken as complete models in [11]. Also, we required* $N \in \mathcal{E}_T(M)$ *in our Definition 7, and not* $N \in \Im_T(M)$ *as in Definition 4 of [11], since we are not restricted to only complete and reduced implementations with the same number of states as* M*.*

Theorem 1 ([11]). *Let* M *be a reduced FSM with* n *states. When restricted to reduced and complete implementations,* T *is* n*-complete for* M *if there exists a confirmed set* $K \subseteq T$ *such that* $\varepsilon \in K$ *and for each* $(s, x) \in D$ *there exist* α*,* $\alpha x \in K$ *such that* $\delta(s_0, \alpha) = s$*.*

We start by showing that confirmed sets induce certain injective functions between the states of a specification and an implementation, given that they are T-equivalent.

Lemma 1. *Let* M*,* N *be FSMs and let* T *be a test suite for* M*. Assume that* $M \approx_T N$*, and that* $K \subseteq T$ *is a confirmed set. Let* $f = \{(\delta(s_0, \alpha), \mu(q_0, \alpha)) \mid \alpha \in K\}$*. Then* f *is an injective function.*

Proof. Assume that (s, q_1), $(s, q_2) \in f$. Then, we must have α_1, $\alpha_2 \in K$ with $\delta(s_0, \alpha_1) = s = \delta(s_0, \alpha_2)$ and $q_1 = \mu(q_0, \alpha_1)$, $q_2 = \mu(q_0, \alpha_2)$. But since K is confirmed, $M \approx_T N$ and α_1, $\alpha_2 \in K$ we get $\mu(q_0, \alpha_1) = \mu(q_0, \alpha_2)$. This gives $q_1 = q_2$, showing that f is a function. Let now (s_1, q), $(s_2, q) \in f$. This gives $\alpha_1, \alpha_2 \in K$ with $\delta(s_0, \alpha_1) = s_1$, $\delta(s_0, \alpha_2) = s_2$ and $\mu(q_0, \alpha_1) = q = \mu(q_0, \alpha_2)$. Again, since K is confirmed, $M \approx_T N$ and α_1, $\alpha_2 \in K$ we get $\delta(s_0, \alpha_1) = \delta(s_0, \alpha_2)$. Hence, $s_1 = s_2$ and f is injective. □

Remark 5. *The bi-directional requirement in Definition 7(2) was crucial to show that* f *is an injective function.*

If we take only complete implementations with no more states than the specification, then f is, in fact, a bijection.

Lemma 2. *Let* M*,* N *be FSMs and let* T *be a test suite for* M*. Assume that* $M \approx_T N$ *and that* $K \subseteq T$ *is a confirmed set. Let* f *be as in Lemma 1. Then,* f *is a bijection if* N *is complete and* $|Q| \leq |S|$*.*

Proof. Let $s \in S$. Since K is confirmed, there is some $\alpha \in K$ such that $\delta(s_0, \alpha) = s$. Since N is complete, we know that $\alpha \in U(q_0)$. The definition of f gives $(s, \mu(q_0, \alpha)) \in f$, and so f is a total relation. Since, by Lemma 1, $f \subseteq S \times Q$ is also an injective function, we get $|S| \leq |Q|$. Thus, $|S| = |Q|$ and so f is also onto. We conclude that f is a bijection. □

The next result says that if f is a function and satisfies the hypothesis of Theorem 1, then its domain includes all of $U(s_0) \cap U(q_0)$.

Lemma 3. *Let* M*,* N *be FSMs and let* T *be a test suite for* M *with* $M \approx_T N$*. Let* K *be a confirmed set satisfying the hypothesis of Theorem 1. Assume that* f*, as defined in Lemma 1, is a function. Then* $(\delta(s_0, \rho), \mu(q_0, \rho)) \in f$ *when* $\rho \in U(s_0) \cap U(q_0)$*.*

Proof. We go by induction on $|\rho| \geq 0$. When $|\rho| = 0$ we get $\rho = \varepsilon$. Hence, by the hypothesis in Theorem 1, $\rho \in K$. Clearly, we also have $\rho \in U(s_0) \cap U(q_0)$ and the result follows. Inductively, assume the result for all ρ with $|\rho| \leq n$. Take ρx, with $x \in \mathcal{I}$, $|\rho| = n$ and such that $\rho x \in U(s_0) \cap U(q_0)$. Then $\rho \in U(s_0) \cap U(q_0)$ and the induction hypothesis gives $(s, q) \in f$, where $\delta(s_0, \rho) = s$ and $\mu(q_0, \rho) = q$. Since $\rho x \in U(s_0)$, we have $(s, x) \in D$. By the hypothesis of Theorem 1 we know that K is confirmed, which gives α, $\alpha x \in K$ with $\delta(s_0, \alpha) = s$. Because $\alpha x \in K$, the definition of f gives $(\delta(s_0, \alpha x), \mu(q_0, \alpha x)) \in f$. We conclude the argument by showing that $\delta(s_0, \alpha x) = \delta(s_0, \rho x)$ and $\mu(q_0, \alpha x) = \mu(q_0, \rho x)$, thus extending the induction. The former follows immediately since $\delta(s_0, \rho) = s = \delta(s_0, \alpha)$. For the latter, since $\alpha \in K$, from the definition of f we also get $(\delta(s_0, \alpha), \mu(q_0, \alpha)) \in f$. But we already know that $(s, q) = (\delta(s_0, \alpha), q)$ is also in f and so, because f is a function, we conclude that $\mu(q_0, \alpha) = q = \mu(q_0, \rho)$. Hence, $\mu(q_0, \alpha x) = \mu(q_0, \rho x)$, as desired. □

The next result essentially says that states mapped under f will display the same behavior.

Lemma 4. *Let M, N be FSMs, let T be a test suite for M and let $K \subseteq T$ be a confirmed set satisfying the sufficiency conditions of Theorem 1. Let $s \in S$, $x \in \mathcal{I}$ with $(s, x) \in D$, and assume that f, as in Lemma 1, is a function. Then $\lambda(s, x) = \tau(f(s), x)$ whenever $M \approx_T N$.*

Proof. The sufficiency conditions in Theorem 1 give α, $\alpha x \in K$ such that $\delta(s_0, \alpha) = s$. From Definition 7 and $K \subseteq T$, we get $\alpha x \in U(s_0) \cap U(q_0) \cap T$ and so, because $M \approx_T N$, we get $\lambda(s_0, \alpha x) = \tau(q_0, \alpha x)$ using Definition 3. But

$$\lambda(s_0, \alpha x) = \lambda(s_0, \alpha)\lambda(\delta(s_0, \alpha), x) = \lambda(s_0, \alpha)\lambda(s, x) \quad \text{and}$$
$$\tau(q_0, \alpha x) = \tau(q_0, \alpha)\tau(\mu(q_0, \alpha), x).$$

Hence, $\lambda(s, x) = \tau(\mu(q_0, \alpha), x)$. Because $\alpha \in K$, the definition of f now yields $(\delta(s_0, \alpha), \mu(q_0, \alpha)) = (s, \mu(q_0, \alpha))$ is in f. Since f is a function, $f(s) = \mu(q_0, \alpha)$. Thus, $\lambda(s, x) = \tau(f(s), x)$. □

We are now in a position to extend Theorem 1. We show that the conditions there required are sufficient to guarantee test suite completeness even when implementations are not complete and have any number of states.

Theorem 2. *Let M be a FSM and let T be a test suite for M. Let $K \subseteq T$ be a confirmed set such that $\varepsilon \in K$, and for each $(s, x) \in D$ there exist α, $\alpha x \in K$ such that $\delta(s_0, \alpha) = s$. Then T is n-complete for M, for all $n \geq 1$.*

Proof. Assume that T is not n-complete for M, for some $n \geq 1$. Then, by Definition 6, there is a FSM N, with $|Q| \leq n$ and such that $M \not\approx N$ and $M \approx_T N$. Using Definition 3 we get $\rho \in \mathcal{I}^\star$, $x \in \mathcal{I}$ with $\rho x \in U(s_0) \cap U(q_0)$, and such that $\lambda(s_0, \rho) = \tau(q_0, \rho)$ and $\lambda(s_0, \rho x) \neq \tau(q_0, \rho x)$. Let $s = \delta(s_0, \rho)$ and $q = \mu(q_0, \rho)$, so that $\lambda(s, x) \neq \tau(q, x)$. Let f be the relation defined in Lemma 1. Since $M \approx_T N$ and K is confirmed, Lemma 1 says that f is an injective function. Lemma 3 says that $f(s) = f(\delta(s_0, \rho)) = \mu(q_0, \rho) = q$. Clearly. $(\delta(s_0, \rho), x) = (s, x) \in D$, and so Lemma 4 gives $\lambda(s, x) = \tau(f(s), x)$, that is, $\lambda(s, x) = \tau(q, x)$, which is a contradiction. □

Remark 6. *For this proof it is important that f is simply a* function, *and not necessarily a* bijection, *as required in [11]. Also, the fact that* $|Q| \leq n$ *was not important for the proof. This assumption is need in Lemma 2, but not in Lemma 1.*

4 An Example with Partial Implementations

In this section we show an application of Theorem 2, using a simple example where implementations are partial FSMs.

Let M be the FSM specification depicted in Figure 1, and let $T = \{\epsilon, 1, 10, 11, 111, 100, 101, 10110, 10010\}$ be a test suite for M. Note that M is a partial specification with three states, and assume that we are treating implementation machines with up to three states.

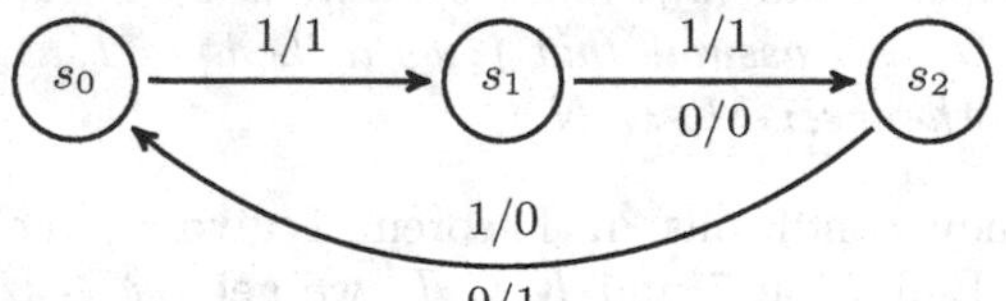

Fig. 1. FSM specification M

We have systematically constructed all FSMs T-equivalent to M with up to three states. We obtained seven possible implementation FSMs, named N_0 to N_6, that is $\mathcal{E}_T(M) = \{N_i | 0 \leq i \leq 6\}$. Machine N_0 is isomorphic to M, and so it is also a partial model. Machines N_1 and N_4 are the completely specified FSMs depicted in Figure 2. They are distinguishable only by the output of the loop transition at q_0. For N_1 we let $a = 0$, and for N_4 we let $a = 1$. Similarly, N_2 and N_5 are the complete FSMs depicted in Figure 3, where $a = 0$ in N_2, and $a = 1$ in N_5. Machines N_3 and N_6 are depicted in Figure 4, where $a = 0$ and $a = 1$, respectively.

Now let $K = \{\epsilon, 1, 10, 11, 100, 101\}$. It is a simple matter to check that $K \subseteq T$, and also that for any $(s, x) \in D$ there are $\alpha, \alpha x \in K$ such that $\delta(s_0, \alpha) = s$. In order to verify that K is a confirmed set for M we need the following conditions, for each $N \in \mathcal{E}_T(M)$:

1. $K \subseteq U(s_0) \cap U(q_0)$; and
2. for all $\alpha, \beta \in K$, $\mu(q_0, \alpha) = \mu(q_0, \beta)$ if and only if $\delta(s_0, \alpha) = \delta(s_0, \beta)$.

Note that we have $T \subseteq U(s_0)$, and also $U(s_0) \subseteq U(q_0)$, since all input sequences that run in the specification M also run in any implementation N_i

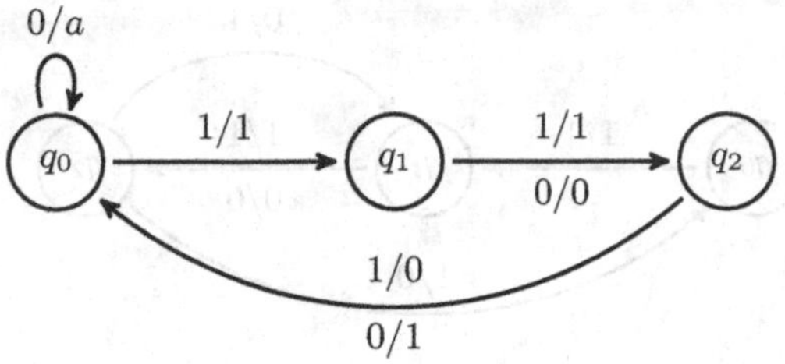

Fig. 2. FSM implementations N_1 and N_4

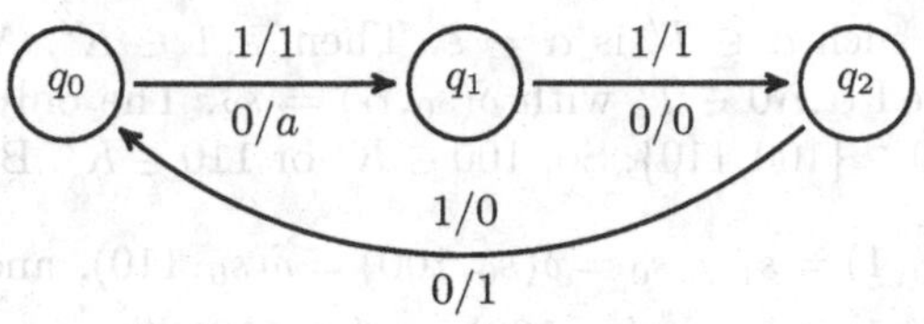

Fig. 3. FSM implementations N_2 and N_5

$(0 \leq i \leq 6)$, because they are extensions of M. The second condition is also easily checked, and so K is a confirmed set for M. We can now apply Theorem 2 and conclude that T is a 3-complete test suite for M.

We also know that if a test suite T is not 3-complete for M then the sufficiency conditions in Theorem 2 do not hold, that is, there will not be a confirmed set K for M. To illustrate this, let $T' = \{\epsilon, 1, 10, 11, 111, 100, 101, 110, 10110\}$ be a new test suite for M. We know that T' is not 3-complete for M since there exists a machine N^*, depicted in Figure 5, that is T'-equivalent to M but N^* is not equivalent to M. In order to see this, we can easily check that both machines give the same output behavior when we apply all test cases in T' to M and N^*. But, if we apply 10010 to both machines, we get 10110 in M and 10111 in N^*.

Now, suppose that there is $K' \subseteq T'$ that satisfies the conditions at Theorem 2. Then, for any $(s, x) \in D$ there must be some $\alpha, \alpha x \in K'$ such that $\delta(s_0, \alpha) = s$. Since we have $(s_0, 1) \in D$ so we will need $\alpha \in T'$ such that $\delta(s_0, \alpha) = s_0$ and

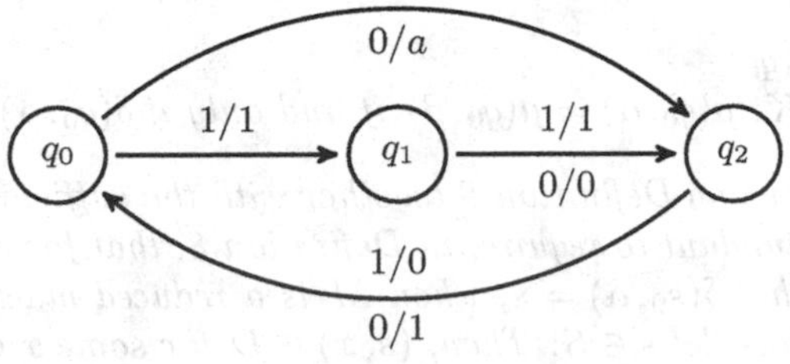

Fig. 4. FSM implementations N_3 and N_6

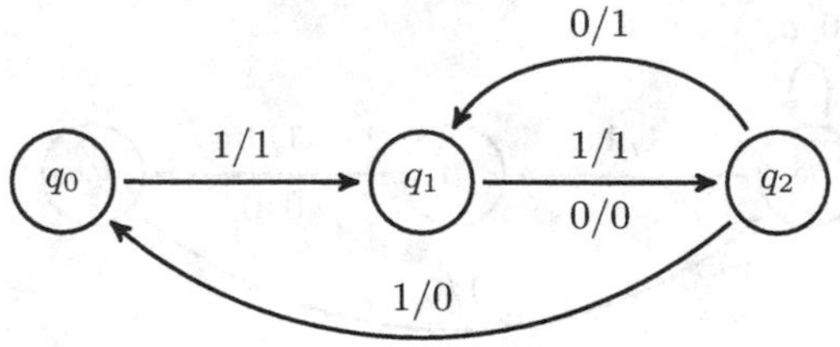

Fig. 5. FSM implementation N^*

$\alpha 1 \in T'$. The only such $\alpha \in T'$ is $\alpha = \varepsilon$. Then, $\varepsilon, 1 \in K'$. Also, $(s_2, 0) \in D$. Then, we would need $\alpha, \alpha 0 \in T'$ with $\delta(s_0, \alpha) = s_2$. The only possibilities are $\alpha \in \{10, 11\}$ and $\alpha 0 \in \{100, 110\}$. So, $100 \in K'$ or $110 \in K'$. But

$$\delta(s_0, 1) = s_1 \neq s_0 = \delta(s_0, 100) = \delta(s_0, 110), \text{ and}$$
$$\mu(q_0, 1) = q_1 = \mu(q_0, 100) = \mu(s_0, 110).$$

So, there is no $K' \subseteq T'$ that satisfies the conditions of Theorem 2.

5 Weak-Equivalence and Complete Implementations

In this section we treat the completeness of test suites using the notion of weak equivalence, when test cases may not run to completion in the specification or in the implementation models. Here we consider only complete implementations models.

Let M be a FSM with n states. We denote by $\mathcal{W}(M)$ and by $\mathcal{WC}(M)$ the sets of all FSMs and all complete FSMs, respectively, with at most n states. Let T be a test suite for M. We denote by $\mathcal{W}_T(M)$, and by $\mathcal{WC}_T(M)$, the sets of all FSMs in $\mathcal{W}(M)$, and in $\mathcal{WC}(M)$, respectively, that are T-weakly-equivalent to M. See Definition 6 (1).

Next we introduce the definition of confirmed sets over the new notion of weak equivalence.

Definition 8. *Let M be a FSM, let T be a test suite for M and $K \subseteq T$. The set K is* weak-confirmed *for M and T iff for any $s \in S$ there is some $\alpha \in K$ such that $\delta(s_0, \alpha) = s$ and, further, for each $N \in \mathcal{W}_T(M)$, it holds that*

1. *$K \subseteq U(s_0) \cap U(q_0)$;*
2. *For all α, $\beta \in K$, $\mu(q_0, \alpha) = \mu(q_0, \beta)$ if and only if $\delta(q_0, \alpha) = \delta(q_0, \beta)$.*

Remark 7. *When using Definition 8 together with the sufficiency conditions in Theorem 1, it is redundant to require, in Definition 8, that for any $s \in S$ there is some $\alpha \in K$ such that $\delta(s_0, \alpha) = s$, when M is a reduced machine with at least two states. To see this, let $s \in S$. Then, $(s, x) \in D$ for some $x \in \mathcal{I}$ because M is reduced. The sufficiency conditions then give some $\alpha \in K$ such that $\delta(s_0, \alpha) = s$, as desired.*

Suppose now that we are treating only complete implementations. In particular, in Definition 8 we replace $\mathcal{W}_T(M)$ by $\mathcal{WC}_T(M)$. The next lemma then states some important properties that will be useful later.

Lemma 5. *Let M and N be FSMs, with N complete, and let T be a test suite for M. Then*

1. $M \sim_T N$ *iff* $T \subseteq U(s_0)$.
2. $M \not\sim N$ *iff* $U(s_0) \neq \mathcal{I}^\star$.

Proof. Follow directly from the definitions. □

The following result states similar sufficiency conditions as does Theorem 1, but now under weak-equivalence.

Theorem 3. *Assume that we only allow complete implementations. Let M be a FSM with $n \geq 2$ states, and let T be a test suite for M. If there is a weakly-confirmed set $K \subseteq T$ that also satisfies the sufficiency conditions of Theorem 1, then T is weak-n-complete for M.*

Proof. First assume that $T \not\subseteq U(s_0)$. For the sake of contradiction, assume the result is false. Then we have a FSM M and a test suite T for M such that there is a weakly-confirmed $K \subseteq T$ that satisfies the sufficiency conditions of Theorem 1. Further, we know that $M \sim_T N$ and $M \not\sim N$. But Lemma 5 immediately gives $M \not\sim_T N$, a contradiction.

Now let $T \subseteq U(s_0)$. Lemma 5 immediately gives that $M \sim_T N$ for any complete FSM N. We construct a complete FSM N with $Q = S$ as follows. Since M has $n \geq 2$ states, let $s_0 \overset{x/}{\underset{M}{\rightarrow}} s_1$, for some $s_1 \in S$. Fix some a, $b \in \mathcal{O}$ with $a \neq b$. Since we also want N to be reduced, construct the cycle $s_i \overset{y/a}{\underset{N}{\rightarrow}} s_{i+1}$, $0 \leq i < n$, and close the cycle with $s_n \overset{y/b}{\underset{N}{\rightarrow}} s_0$, for all $y \in \mathcal{I}$ and $y \neq x$. Terminate the construction with $s_i \overset{x/a}{\underset{N}{\rightarrow}} s_0$, $0 \leq i \leq n$. Clearly, N is also complete. It is easy to see that N is reduced, since the input sequence y^n starting at state s_i gives $a^{n-1-i}ba^i$, and $a^{n-1-i}ba^i \neq a^{n-1-j}ba^j$ when $i \neq j$. Using Lemma 5 we get $N \sim_T M$ and so $N \in \mathcal{WC}_T(M)$. Let $K \subseteq T$ be a weakly-confirmed set that also satisfies the sufficiency conditions of Theorem 1. Then, since $(s_0, x) \in D$, there are α, $\alpha x \in K$ with $\delta(s_0, \alpha) = s_0$. So, $\delta(s_0, \alpha x) = s_1$. We also have $\varepsilon \in K$. Then, ε, $\alpha x \in K$, with $\delta(s_0, \varepsilon) = s_0 \neq s_1 = \delta(s_0, \alpha x)$ in M. But the construction of N gives $\mu(s_0, \varepsilon) = s_0 = \mu(s_0, \alpha x)$. This contradicts K being weakly-confirmed. □

6 Weak-Equivalence and Partial Implementations

In this section we show that Theorem 1 does not hold under weak equivalence when we also allow for partial implementations. We want a (partial) specification FSM M and a test suite T for M such that T contains a confirmed subset $K \subseteq T$ which also satisfies the sufficiency conditions of Theorem 1. We then construct a (partial) implementation FSM N such that $M \sim_T N$, but $M \not\sim N$, thus establishing that Theorem 1 does not hold under these conditions, as desired.

Remark 8. *Since, in this section, we are treating weak-equivalence only, we will drop output symbols in transitions. Output symbols could easily be injected so that all FSM considered here are reduced.*

We start with a specification M with $n+1$ states, where $n \geq 2$, that is we let $S = \{s_0, s_1, \ldots, s_n\}$. The simpler cases when $n = 1$ or $n = 0$ will be dealt with later. We define machine M as in Figure 6. More formally, we let $\mathcal{I} = \{0, 1\}$,

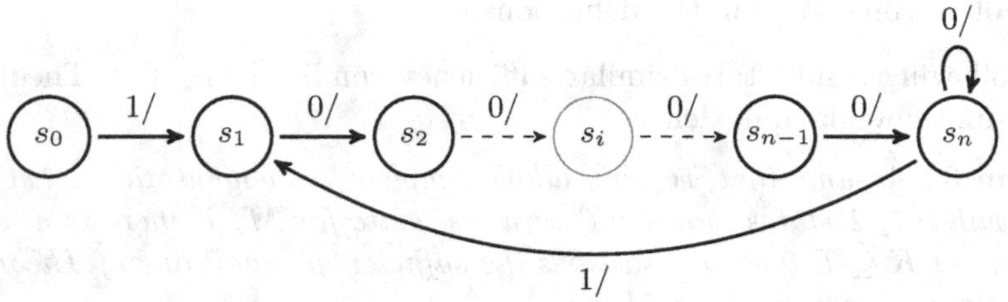

Fig. 6. Speficication M

$S = \{s_0, s_1, \cdots, s_n\}$ and

$$\begin{aligned} & s_0 \overset{1/}{\to} s_1 \overset{0/}{\to} s_2 \overset{0/}{\to} s_3 \overset{0/}{\to} \cdots \overset{0/}{\to} s_{n-1} \overset{0/}{\to} s_n \overset{1/}{\to} s_1 \\ & s_n \overset{0/}{\to} s_n. \end{aligned} \tag{1}$$

Next we define a test suite T and a subset $K \subseteq T$. For the ease of reference, we let $K = K_1 \cup K_2$ with

$$\begin{aligned} K_1 &= \{\varepsilon\} \cup \{10^p \mid 0 \leq p \leq n\}, \quad \text{and} \\ K_2 &= \{10^{n-1}1, 10^{n+1}1, 10^{n-1}10, 10^n10\}. \end{aligned} \tag{2}$$

The test suite T is now given by $T = K \cup T_1$, where

$$T_1 = \{0\} \cup \{10^p10, 10^{n-1}10^p10 \mid 0 \leq p \leq n-2\}. \tag{3}$$

Note that since $n \geq 2$, K and T are well defined. Further, we clearly have $T_1 \cap K = \emptyset$.

Using K_1 it is easy to see that for all states $s \in S$, there is some $\alpha \in K$ such that $\delta(s_0, \alpha) = s$, so that K satisfies the reachability condition in Definition 8. Using K_1 again and $10^{n-1}1 \in K_2$ it is also a simple matter to verify that K satisfies the sufficiency conditions of Theorem 1.

To ease the notation, given a FSM M, some $s \in S$ and some $\alpha \in \mathcal{I}^\star$, we will say that s *runs* α if $\alpha \in U(s)$, and we will say that M *runs* α if s_0 runs α.

Remark 9. *It is easy to verify that $K \subseteq U(s_0)$, that is M runs all $\alpha \in K$. It is also easy to verify that $T_1 \cap U(s_0) = \emptyset$, that is, M does not run any $\alpha \in T_1$.*

Next, in order to verify that K is indeed confirmed, according to Definition 8, we need to investigate which FSMs are in $\mathcal{W}_T(M)$. Note that, according to Definition 3 and Remark 9, for N to be in $\mathcal{W}_T(M)$ it is necessary and sufficient that N runs all $\alpha \in K$, and that N does not run any $\alpha \in T_1$. We will show that any FSM in $\mathcal{W}_T(M)$ has $n+1$ states and must also have, at least, all transitions that are in M. More specifically, if N is a FSM in $\mathcal{W}_T(M)$, then the states in N can be listed as $Q = \{q_0, q_1, \ldots, q_n\}$, with the property that if $s_i \xrightarrow{x/} s_j$ in M, then we also have $q_i \xrightarrow{x/} q_j$ in N, for all $x \in \mathcal{I}^\star$. We establish this result by proving a series of claims.

Let N be a FSM such that $N \sim_T M$.

Claim 1: We have $q_0 \xrightarrow{1/} q_1 \xrightarrow{0/} q_2$, with q_0, q_1, q_2 distinct.

Proof. First note that since M does not run 0, then there is no outgoing transition on 0 from q_0. That is, N does not run 0. Since M runs $100^{n-2}1$, then N will also run 1, but we cannot have $q_0 \xrightarrow{1/} q_0$ because, since N will also run 10, then it would also run 0, which is a contradiction. We conclude that we must have $q_0 \xrightarrow{1/} q_1$, for some $q_1 \in Q$, $q_0 \neq q_1$.
Now, since M runs $1000^{n-1}1$, so does N. This means that q_1 runs $000^{n-1}1$. Clearly, we cannot have $q_1 \xrightarrow{0/} q_0$ because this will make q_0 run 0, which cannot happen.
Next, we show that we cannot have a self-loop on 0 at q_1. Since M runs $100^{n-2}1$ so would N. Then, q_1 would run 1. There are three cases.
Assume that $q_1 \xrightarrow{1/} q_2$, for some $q_2 \in Q$. Since M runs $10^{n-1}10$, so does N. But then N would also run $\alpha = 110 = 10^p10$, with $p = 0$. Since $\alpha \in T_1$, we get a contradiction. Assume that $q_1 \xrightarrow{1/} q_1$. Then, as before, N would also run 110, which cannot happen. Finally, assume that $q_1 \xrightarrow{1/} q_0$. Then, since $\alpha = 10^{n-1}10 \in K_2$, N would run α, and so N would also run $0 \in T_1$, a contradiction.
We conclude that $q_1 \xrightarrow{0/} q_2$, for some $q_2 \in Q$, distinct from q_0 and q_1. △

From Claim 1, we have $q_0 \xrightarrow{1/} q_1 \xrightarrow{0/} q_2$ in N, with $|\{q_0, q_1, q_2\}| = 3$. Next, we show that this argument can be extended inductively.

Claim 2: Assume that we have q_j $(0 \leq j \leq k)$ as distinct states in Q, where $2 \leq k < n$, and such that $q_0 \xrightarrow{1/} q_1$ and

$$q_1 \xrightarrow{0/} q_2 \xrightarrow{0/} q_3 \xrightarrow{0/} \cdots \xrightarrow{0/} q_k.$$

Then there is a new distinct state $q_{k+1} \in Q$ with $q_k \xrightarrow{0/} q_{k+1}$.

Proof. Recall that $10^{n+1}1 \in K_2$, and so N runs $10^{n+1}1$. Since $n > k$, this means that q_k runs $000^{n-k}1$. We cannot have $q_k \xrightarrow{0/} q_0$ because then N would run $0 \in T_1$, which would be a contradiction. There are two other cases: either $q_k \xrightarrow{0/} q_j$, where $1 \leq j \leq k$, or $q_k \xrightarrow{0/} q_{k+1}$, where $q_{k+1} \in Q$ is a new distinct state. We show that the former cannot happen, and so the latter must be true, establishing the claim.

For the sake of contradiction, assume that $q_k \xrightarrow{0/} q_j$, where $1 \leq j \leq k$. Since $10^{n-1}1 \in K_2$, then N must run $10^{n-1}1$. We have $n-1-(j-1) = n-j \geq 1$ and so q_j must run $0^{n-j}1$. Also, $n - j \geq (k+1) - j = k - j + 1 = \ell$, and so we have the cycle $\mu(q_j, 0^\ell) = q_j$. We can write $n - j = c\ell + r$, for some $c \geq 1$ and some $0 \leq r < \ell$. So, because of the cycle, we have

$$\begin{aligned}\mu(q_0, 10^{n-1}) &= \mu(q_j, 0^{n-j}) = \mu(q_j, 0^{c\ell+r}) \\ &= \mu(q_j, 0^r) = q_{j+r}.\end{aligned}$$

Note that $1 \leq r+j < \ell+j = k+1$, so that $1 \leq j+r \leq k$. Since $\mu(q_0, 10^{n-1}) = q_{j+r}$ and $10^{n-1}10 \in K_2$, we know that $q_{j+r} \xrightarrow{1/} q$, for some $q \in Q$. If $q = q_0$ then N would run $0 \in T_1$, which cannot happen. Then, $q_{j+r} \xrightarrow{1/} q_i$, for some $1 \leq i \leq k+1$, where $q_{k+1} \in Q$ is a new distinct state. If this is the case, then we would get

$$\mu(q_0, 10^{n-1}10) = \mu(q_{j+r}, 10) = \mu(q_i, 0).$$

So, q_i runs 0. Since $j + r \leq k$, we also have $\mu(q_0, 10^{j+r-1}10) = \mu(q_{j+r}, 10) = \mu(q_i, 0)$, and so N would run $10^{j+r-1}10 = 10^p10$, with $p = j + r - 1$. Since $1 \leq j + r \leq k$ we get $0 \leq p \leq k - 1 \leq n - 2$, because $k < n$. This shows that N runs $10^p10 \in T_1$, which is a contradiction. △

Now, using Claim 1 as the basis and Claim 2 as the induction step, we have

$$q_0 \xrightarrow{1/} q_1 \xrightarrow{0/} q_2 \xrightarrow{0/} q_3 \xrightarrow{0/} \cdots \xrightarrow{0/} q_{n-1} \xrightarrow{0/} q_n.$$

But $10^{n+1}1 \in K_2$, and we conclude that q_n runs 001.

Claim 3: We must have $q_n \xrightarrow{0/} q_n$.

Proof. If $q_n \xrightarrow{0/} q_0$ we would immediately have N running $0 \in T_1$, which cannot happen.

Now assume that $q_n \xrightarrow{0/} q_i$ with $1 \leq i < n$. Recall that because $10^n10 \in K_2$, then N runs 10^n10. Hence,

$$\mu(q_0, 10^{n-1}010) = \mu(q_n, 010) = \mu(q_i, 10),$$

and so q_i runs 1. We cannot have $q_i \xrightarrow{1/} q_0$ because q_0 would run $0 \in T_1$, which is not allowed. If $q_i \xrightarrow{1/} q_j$, with $2 \leq j \leq n$, then we would have

$$\mu(q_0, 10^n10) = \mu(q_i, 10) = \mu(q_j, 0),$$

and so q_j would run 0. But we also have $\mu(q_0, 10^{i-1}10) = \mu(q_i, 10) = \mu(q_j, 0)$, and so N would also run $10^{i-1}10$. Because $0 \leq i-1 \leq n-2$, we get $10^{i-1}10 \in T_1$, a contradiction.

The only other possibility is to have $q_n \xrightarrow{0/} q_n$, thus establishing the claim. △

Using Claim 3, we now have

$$q_0 \overset{1/}{\to} q_1 \overset{0/}{\to} q_2 \overset{0/}{\to} q_3 \overset{0/}{\to} \cdots \overset{0/}{\to} q_{n-1} \overset{0/}{\to} q_n, \text{ and } q_n \overset{0/}{\to} q_n.$$

CLAIM 4: We must have $q_n \overset{1/}{\to} q_1$.

PROOF. Observe that $10^{n-1}10 \in K_2$, and so N must run $10^{n-1}10$. Since $\mu(q_0, 10^{n-1}10) = \mu(q_n, 10)$, we conclude that q_n runs 10. If $q_n \overset{1/}{\to} q_0$ we immediately get that q_0 runs $0 \in T_1$, which cannot happen.
Assume now that $q_n \overset{1/}{\to} q_j$, with $2 \leq j \leq n$. Then, q_j runs 0. Also,

$$\begin{aligned}\mu(q_0, 10^{n-1}10^{n-j}10) &= \mu(q_n, 10^{n-j}10) = \mu(q_j, 0^{n-j}10)\\ &= \mu(q_n, 10) = \mu(q_j, 0).\end{aligned}$$

Thus, N runs $\alpha = 10^{n-1}10^{n-j}10$. Since $2 \leq j \leq n$, we have $0 \leq n - j \leq n - 2$ and so $\alpha \in T_1$, contradicting N running α.
Since N can have at most $n+1$ states, because $N \in \mathcal{W}_T(M)$, the only possibility left is $q_n \overset{1/}{\to} q_1$, thus establishing the claim. △

We now have reached the desired result.
CLAIM 5: Let N be a FSM. If $N \sim_T M$ then N has $n+1$ states $\{q_0, q_1, \ldots, q_n\}$ satisfying

$$\begin{aligned}&q_0 \overset{1/}{\to} q_1 \overset{0/}{\to} q_2 \overset{0/}{\to} q_3 \overset{0/}{\to} \cdots \overset{0/}{\to} q_{n-1} \overset{0/}{\to} q_n \overset{1/}{\to} q_1\\ &q_n \overset{0/}{\to} q_n.\end{aligned} \tag{4}$$

PROOF. Use Claims 1–4. △

We can now argue that K is confirmed for M and T.
CLAIM 6: Let M, T and K be as given in (1)–(3). Then, K is confirmed for M and T.

PROOF. Let N be a FSM. From Remark 9, we get $K \subseteq U(s_0)$. Let $N \in \mathcal{W}_T(M)$. Then, by Claim 5, N satisfies (4), and we clearly get $U(s_0) \subseteq U(q_0)$. Hence, $K \subseteq U(s_0) \cap U(q_0)$, satisfying the first condition at Definition 8. It is also easily seen that we have $\delta(s_0, \alpha) = s_i$ if and only if $\mu(q_0, \alpha) = q_i$, for all $\alpha \in K$ and all $0 \leq i \leq n$. From this, it follows easily that $\delta(s_0, \alpha) = \delta(s_0, \beta)$ if and only if $\mu(q_0, \alpha) = \mu(q_0, \beta)$, for all $\alpha, \beta \in K$. Thus, the second condition at Definition 8 is satisfied and so K is weakly-confirmed for M and T. △

Finally, we construct the counter-example to Theorem 1. Let N be a FSM with $n+1$ states and that satisfies (4). To conclude the construction of N, add the transition $q_1 \overset{1/}{\to} q_0$ as shown in Figure 7.
Now, we check that N runs any string in K.
CLAIM 7: N runs all $\alpha \in K$.

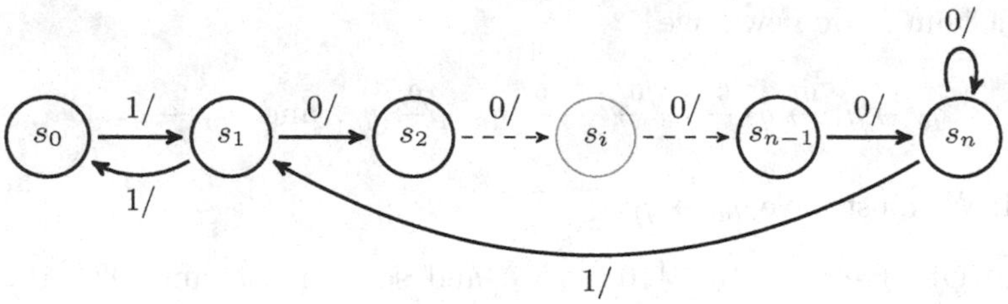

Fig. 7. Implementation N

PROOF. Follows from (1) and (4). △

Next, we check that N does not run any string in T_1.

CLAIM 8: N does not run α, for any $\alpha \in T_1$.

PROOF. Clearly, N does not run 0.
Take $10^p10 \in T_1$, with $0 \le p \le n-2$. Then,

$$\mu(q_0, 10^p10) = \mu(q_1, 0^p10) = \mu(q_{p+1}, 10).$$

There are two cases. If $1 \le p \le n-2$, then $2 \le p+1 \le n-1$ and from (4) we get that q_{p+1}. does not run 10, and so N does not run 10^p10 either. If $p = 0$ we get

$$\mu(q_0, 10^p10) = \mu(q_1, 10) = \mu(q_0, 0).$$

Again, from (4), q_0 does not run 0 and so N does not run 10^p10.
Finally, take $10^{n-1}10^p10 \in T_1$, with $0 \le p \le n-2$. Now we have

$$\mu(q_0, 10^{n-1}10^p10) = \mu(q_1, 0^{n-1}10^p10) = \mu(q_n, 10^p10) = \mu(q_1, 0^p10),$$

and we can repeat the previous argument.
This exhausts all of T_1 and the claim holds. △

We have now our negative result.

Theorem 4. *Assume that specifications are partial FSMs with $n \ge 3$ states, and implementations are partial FSMs with at most n states. Then Theorem 1 does not hold under weak-equivalence.*

Proof. From Claims 7 and 8 we know that $N \sim_T M$. But we can easily check that N runs 11 and M does not run 11. Hence $N \not\sim M$. Since, by Claim 6, K is confirmed for M and T, we conclude that Theorem 1 does not hold. □

The particular cases when specifications have 1 or 2 states are easy to treat separately. First, let M be a two-state FSM with $S = \{s_0, s_1\}$, $\mathcal{I} = \{0, 1\}$ and

$$s_0 \xrightarrow{1/} s_1 \xrightarrow{0/} s_1. \tag{5}$$

Let $T = K \cup T'$ where

$$K = \{\varepsilon, 1, 10, 100\}, \quad T' = \{0\}. \tag{6}$$

Clearly, any state in M is reachable using a string in K. Moreover, it is easy to check that K satisfies the sufficiency condition at Theorem 1. Let N be a FSM with at most 2 states and such that $N \sim_T M$. Since M runs 10, so does N. If $q_0 \xrightarrow{1/} q_0$ then q_0 would also run 0. But this is a contradiction since $0 \in T'$ and M does not run 0. Hence, we must have $q_0 \xrightarrow{1/} q_1$, with $q_0 \neq q_1$. Since M also runs 100, so does N. If we had $q_1 \xrightarrow{0/} q_0$, then q_0 would run 0, which can not happen. We conclude that in N we have

$$q_0 \xrightarrow{1/} q_1 \xrightarrow{0/} q_1. \tag{7}$$

But now it is immediate that $\delta(s_0, \alpha) = \delta(s_0, \beta)$ if and only if $\mu(q_0, \alpha) = \mu(q_0, \beta)$, for all α, $\beta \in K$. Since N was any machine in $\mathcal{W}_T(M)$, we thus conclude that K is confirmed for M and K. For the desired counter-example, let N' be as in (7) with the added transition $q_1 \xrightarrow{1/} q_1$. Clearly, N' runs all $\alpha \in K$ and does not run $0 \in T'$. Hence, $N' \sim_T M$. But N' runs 11 and M does not, so that $N' \not\sim M$. Because K was shown to be weakly-confirmed for M and T, we conclude that Theorem 1 also does not hold for two-state FSMs under weak-equivalence and when we allow for partial specifications and partial implementations.

For completeness, we also look at one-state machines. Take M with $S = \{s_0\}$ and no transitions. Let $T = K = \{\varepsilon\}$. It is easily checked that K satisfies the sufficiency conditions at Theorem 1. Since M is a one-state FSM and $T = \{\varepsilon\}$ we immediately conclude that any one-state FSM is T-weakly-equivalent to M. From this, and since $K = \{\varepsilon\}$, it follows easily that K is weakly-confirmed for M and T. Now take the one-state FSM N where we just let $q_0 \xrightarrow{0/} q_0$. Then, $N \sim_T M$ but $N \not\sim M$ because N runs 0 and M does not. Again, Theorem 1 does not hold for one-state FSMs.

We can now state a more general result, that is Theorem 1 does not hold under weak-equivalence and for partial implementation models.

Corollary 1. *Assume that specifications are partial FSMs with n states, and implementations are partial FSMs with at most n states. Then Theorem 1 does not hold under weak-equivalence.*

Proof. From Theorem 4 and the preceding discussion. □

7 Conclusions

In this work we showed that the notion of confirmed sets can be applied to less restrictive FSM models to check test suite completeness, when we do not require implementations to be completely specified, an important relaxation when treating implementations as true black boxes. Further, we showed that neither specifications nor implementations need to be reduced FSMs and also implementation models can have any number of states.

We also investigated the problem of checking test suite completeness under the notion of weak-equivalence, that is when we have test suites whose test cases that may not run to completion in specifications or in implementations. This further strengths the black box characteristic of implementations. We showed that confirmed sets, in the presence of weak-equivalence, are still sufficient for checking test suite completeness when we allow for complete implementations only. In contrast, we also showed that confirmed sets are not sufficient anymore for checking test suite completeness when we also consider partial implementations.

All statements were proved correct by rigorous arguments.

References

1. Bonifacio, A.L., Moura, A.V., da, S., Simao, A.: A generalized model-based test generation method. In: Cerone, A., Gruner, S. (eds.) Sixth IEEE International Conference on Software Engineering and Formal Methods, SEFM, November 10-14, pp. 139–148. IEEE Computer Society, Cape Town (2008)
2. Bonifacio, A.L., Moura, A.V., da Silva Simão, A.: Model partitions and compact test case suites. Int. J. Found. Comput. Sci. 23(1), 147–172 (2012)
3. Chow, T.S.: Testing software design modeled by finite-state machines. IEEE Transactions on Software Engineering 4(3), 178–187 (1978)
4. Dorofeeva, R., El-Fakih, K., Yevtushenko, N.: An improved conformance testing method. In: Wang, F. (ed.) FORTE 2005. LNCS, vol. 3731, pp. 204–218. Springer, Heidelberg (2005)
5. Fujiwara, S., Bochmann, G.V., Khendek, F., Amalou, M., Ghedamsi, A.: Test selection based on finite state models. IEEE Transactions on Software Engineering 17(6), 591–603 (1991)
6. Gill, A.: Introduction to the theory of finite-state machines. McGraw-Hill, New York (1962)
7. Hennie, F.C.: Fault detecting experiments for sequential circuits. In: Proceedings of the Fifth Annual Symposium on Switching Circuit Theory and Logical Design, Princeton, New Jersey, USA, November 11-13, pp. 95–110. IEEE (1964)
8. Hierons, R.M.: Separating sequence overlap for automated test sequence generation. Automated Software Engg. 13(2), 283–301 (2006)
9. Hierons, R.M., Ural, H.: Reduced length checking sequences. IEEE Trans. Comput. 51(9), 1111–1117 (2002), `http://dx.doi.org/10.1109/TC.2002.1032630`
10. Hierons, R.M., Ural, H.: Optimizing the length of checking sequences. IEEE Trans. Comput. 55(5), 618–629 (2006), `http://dx.doi.org/10.1109/TC.2006.80`
11. da Simao, A.S., Petrenko, P.: Checking completeness of tests for finite state machines. IEEE Trans. Computers 59(8), 1023–1032 (2010)
12. Ural, H., Wu, X., Zhang, F.: On minimizing the lengths of checking sequences. IEEE Trans. Comput. 46(1), 93–99 (1997), `http://dx.doi.org/10.1109/12.559807`

Probabilistic Recursion Theory and Implicit Computational Complexity*

Ugo Dal Lago[1,2] and Sara Zuppiroli[1,2]

[1] Università di Bologna, Italian
[2] INRIA, France
{dallago,zuppirol}@cs.unibo.it

Abstract. We show that probabilistic computable functions, i.e., those functions outputting distributions and computed by probabilistic Turing machines, can be characterized by a natural generalization of Church and Kleene's partial recursive functions. The obtained algebra, following Leivant, can be restricted so as to capture the notion of polytime sampleable distributions, a key concept in average-case complexity and cryptography.

1 Introduction

Models of computation as introduced one after the other in the first half of the last century were all designed around the assumption that *determinacy* is one of the key properties to be modeled: given an algorithm and an input to it, the sequence of computation steps leading to the final result is *uniquely* determined by the way an *algorithm* describes the state evolution. The great majority of the introduced models are *equivalent*, in that the classes of functions (on, say, natural numbers) they are able to compute are the same.

The second half of the 20th century has seen the assumption above relaxed in many different ways. Nondeterminism, as an example, has been investigated as a way to abstract the behavior of certain classes of algorithms, this way facilitating their study without necessarily changing their expressive power: think about how NFAs [15] make the task of proving closure properties of regular languages easier.

A relatively recent step in this direction consists in allowing algorithms' internal state to evolve probabilistically: the next state is not *functionally* determined by the current one, but is obtained from it by performing a process having possibly many outcomes, each with a certain probability. Again, probabilistically evolving computation can be a way to abstract over determinism, but also a way to model situations in which algorithms have access to a source of true randomness.

Probabilistic models are nowadays more and more pervasive. Not only are they a formidable tool when dealing with uncertainty and incomplete information, but they sometimes are a *necessity* rather than an option, like in computational cryptography (where, e.g., secure public key encryption schemes need to be probabilistic [9]). A nice way to deal computationally with probabilistic

* This work is partially supported by the ANR project 12IS02001 PACE.

G. Ciobanu and D. Méry (Eds.): ICTAC 2014, LNCS 8687, pp. 97–114, 2014.

models is to allow probabilistic choice as a primitive when designing algorithms, this way switching from usual, deterministic computation to a new paradigm, called probabilistic computation.

But what does the presence of probabilistic choice give us in terms of expressivity? Are we strictly more expressive than usual, deterministic, computation? And how about efficiency: is it that probabilistic choice permits to solve computational problems more efficiently? These questions have been among the most central in the theory of computation, and in particular in computational complexity, in the last forty years (see below for more details about related work). Roughly, while probability has been proved not to offer any advantage in the absence of resource constraints, it is not known whether probabilistic classes such as **BPP** or **ZPP** are different from **P**.

This work goes in a somehow different direction: we want to study probabilistic computation without necessarily *reducing* or *comparing* it to deterministic computation. The central assumption here is the following: a probabilistic algorithm computes what we call a *probabilistic function*, i.e. a function from a discrete set (e.g. natural numbers or binary strings) to *distributions* over the same set. What we want to do is to study the set of those probabilistic functions which can be computed by algorithms, possibly with resource constraints.

We give some initial results here. First of all, we provide a characterization of computable probabilistic functions by the natural generalization of Kleene's partial recursive functions, where among the initial functions there is now a function corresponding to tossing a fair coin. In the non-trivial proof of completeness for the obtained algebra, Kleene's minimization operator is used in an unusual way, making the usual proof strategy for Kleene's Normal Form Theorem (see, e.g., [18]) useless. We later hint at how to recover the latter by replacing minimization with a more powerful operator. We also mention how probabilistic recursion theory offers characterizations of concepts like the one of a computable distribution and of a computable real number.

The second part of this paper is devoted to applying the aforementioned recursion-theoretical framework to polynomial-time computation. We do that by following Bellantoni and Cook's and Leivant's works [1,12], in which polynomial-time deterministic computation is characterized by a restricted form of recursion, called *predicative* or *ramified* recursion. Endowing Leivant's ramified recurrence with a random base function, in particular, is shown to provide a characterization of polynomial-time computable distributions, a key notion in average-case complexity [2].

Related Work. This work is rooted in the classic theory of computation, and in particular in the definition of partial computable functions as introduced by Church and later studied by Kleene [11]. Starting from the early fifties, various forms of automata in which probabilistic choice is available have been considered (e.g. [14]). The inception of probabilistic choice into an universal model of computation, namely Turing machines, is due to Santos [16,17], but is (essentially) already there in an earlier work by De Leeuw and others [5]. Some years later, Gill [6] considered probabilistic Turing machines with bounded complexity: his work has been the starting point of a florid research about the interplay between computational complexity and randomness. Among the many side effects of this research one can of course mention modern cryptography [10], in which algorithms (e.g. encryption schemes, authentication schemes, and adversaries for them) are almost invariably assumed to work in probabilistic polynomial time.

Implicit computational complexity (ICC), which studies machine-free characterizations of complexity classed based on mathematical logic and programming language theory, is a much younger research area. Its birth is traditionally made to correspond with the beginning of the nineties, when Bellantoni and Cook [1] and Leivant [12] independently proposed function algebras precisely characterizing (deterministic) polynomial time computable functions. In the last twenty years, the area has produced many interesting results, and complexity classes spanning from the logarithmic space computable functions to the elementary functions have been characterized by, e.g., function algebras, type systems [13], or fragments of linear logic [7]. Recently, some investigations on the interplay between implicit complexity and probabilistic computation have started to appear [3]. There is however an intrinsic difficulty in giving *implicit* characterizations of probabilistic classes like **BPP** or **ZPP**: the latter are semantic classes defined by imposing a polynomial bound on time, but also appropriate bounds on the probability of error. This makes the task of enumerating machines computing problems in the classes much harder and, ultimately, prevents from deriving implicit characterization of the classes above. Again, our emphasis is different: we do not see probabilistic algorithms as artifacts computing functions of the same kind as the one deterministic algorithms compute, but we see probabilistic algorithms as devices outputing distributions.

2 Probabilistic Recursion Theory

In this section we provide a characterization of the functions computed by a Probabilistic Turing Machine (PTM) in terms of a function algebra *à la* Kleene. We first define *probabilistic recursive functions*, which are the elements of our algebra. Next we define formally the class of probabilistic functions computed by a PTM. Finally, we show the equivalence of the two introduced classes. In the following, $\mathbb{R}_{[0,1]}$ is the unit interval.

Since PTMs compute probability distributions, the functions that we consider in our algebra have domain $\mathbb{N}^k$ and codomain $\mathbb{N} \to \mathbb{R}_{[0,1]}$ (rather than $\mathbb{N}$ as in the classic case). The idea is that if $f(x)$ is a function which returns $r \in \mathbb{R}_{[0,1]}$ on input $y \in \mathbb{N}$, then r is the probability of getting y as the output when feeding f with the input x. We note that we could extend our codomain from $\mathbb{N} \to \mathbb{R}_{[0,1]}$ to $\mathbb{N}^m \to \mathbb{R}_{[0,1]}$, however we use $\mathbb{N} \to \mathbb{R}_{[0,1]}$ in order to simplify the presentation.

Definition 1 (Pseudodistributions and Probabilistic Functions). *A* pseudodistribution *on* $\mathbb{N}$ *is a function* $\mathcal{D} : \mathbb{N} \to \mathbb{R}_{[0,1]}$ *such that* $\sum_{n\in\mathbb{N}} \mathcal{D}(n) \leq 1$. $\sum_{n\in\mathbb{N}} \mathcal{D}(n)$ *is often denoted as* $\sum \mathcal{D}$. *Let* $\mathbb{P}_{\mathbb{N}}$ *be the set of all pseudodistributions on* $\mathbb{N}$. *A* probabilistic function *(PF) is a function from* $\mathbb{N}^k$ *to* $\mathbb{P}_{\mathbb{N}}$, *where*

$\mathbb{N}^k$ stands for the set of k-tuples in $\mathbb{N}$. We use the expression $\{n_1^{p1}, \ldots, n_k^{pk}\}$ to denote the pseudodistribution $\mathcal{D}$ defined as $\mathcal{D}(n) = \sum_{n_i=n} p_i$. Observe that $\sum \mathcal{D} = \sum_{i=1}^{k} p_i$. When this does not cause ambiguity, the terms distribution and pseudodistribution will be used interchangeably.

Please notice that probabilistic functions are always *total* functions, but their codomain is a set of distributions which do not necessarily sum to 1, but rather to a real number *smaller* or equal to 1, this way modeling the probability of divergence. For example, the nowhere-defined partial function $\Omega : \mathbb{N} \rightharpoonup \mathbb{N}$ of classic recursion theory becomes a probabilistic function which returns the empty distributions $\emptyset$ on any input. The first step towards defining our function algebra consists in giving a set of functions to start from:

Definition 2 (Basic Probabilistic Functions). *The* basic probabilistic functions *(BPFs) are as follows:*

- *The* zero function $z : \mathbb{N} \to \mathbb{P}_{\mathbb{N}}$ *defined as:* $z(n)(0) = 1$ *for every* $n \in \mathbb{N}$*;*
- *The* successor function $s : \mathbb{N} \to \mathbb{P}_{\mathbb{N}}$ *defined as:* $s(n)(n+1) = 1$ *for every* $n \in \mathbb{N}$*;*
- *The* projection function $\Pi_m^n : \mathbb{N}^n \to \mathbb{P}_{\mathbb{N}}$ *defined as:* $\Pi_m^n(k_1, \cdots, k_n)(k_m) = 1$ *for every positive* $n, m \in \mathbb{N}$ *such that* $1 \leq m \leq n$*;*
- *The* fair coin function $r : \mathbb{N} \to \mathbb{P}_{\mathbb{N}}$ *that is defined as:*

$$r(x)(y) = \begin{cases} 1/2 \textit{ if } y = x \\ 1/2 \textit{ if } y = x+1 \end{cases}$$

The first three BPFs are the same as the basic functions from classic recursion theory, while r is the only truly probabilistic BPF.

The next step consists in defining how PFs *compose*. Function composition of course cannot be used here, because when composing two PFs g and f the codomain of g does not match with the domain of f. Indeed g returns a distribution $\mathbb{N} \to \mathbb{R}_{[0,1]}$ while f expects a natural number as input. What we have to do here is the following. Given an input $x \in \mathbb{N}$ and an output $y \in \mathbb{N}$ for the composition $f \bullet g$, we apply the distribution $g(x)$ to any value $z \in \mathbb{N}$. This gives a probability $g(x)(z)$ which is then multiplied by the probability that the distribution $f(z)$ associates to the value $y \in \mathbb{N}$. If we then consider the sum of the obtained product $g(x)(z) \cdot f(z)(y)$ on all possible $z \in \mathbb{N}$ we obtain the probability of $f \bullet g$ returning y when fed with x. The sum is due to the fact that two different values, say $z_1, z_2 \in \mathbb{N}$, which provide two different distributions $f(z_1)$ and $f(z_2)$ must both contribute to the same probability value $f(z_1)(y) + f(z_2)(y)$ for a specific y. In other words, we are doing nothing more than lifting f to a function from distributions to distributions, then composing it with g. Formally:

Definition 3 (Composition). *We define the* composition $f \bullet g : \mathbb{N} \to \mathbb{P}_{\mathbb{N}}$ *of two functions* $f : \mathbb{N} \to \mathbb{P}_{\mathbb{N}}$ *and* $g : \mathbb{N} \to \mathbb{P}_{\mathbb{N}}$ *as:*

$$((f \bullet g)(x))(y) = \sum_{z \in \mathbb{N}} g(x)(z) \cdot f(z)(y).$$

The previous definition can be generalized to functions taking more than one parameter in the expected way:

Definition 4 (Generalized Composition). *We define the* generalized composition *of functions* $f : \mathbb{N}^n \to \mathbb{P}_\mathbb{N}$, $g_1 : \mathbb{N}^k \to \mathbb{P}_\mathbb{N}, \ldots, g_n : \mathbb{N}^k \to \mathbb{P}_\mathbb{N}$ *as the function* $f \odot (g_1, \ldots, g_n) : \mathbb{N}^k \to \mathbb{P}_\mathbb{N}$ *defined as follows:*

$$((f \odot (g_1, \ldots, g_n))(\mathbf{x}))(y) = \sum_{z_1,\ldots,z_n \in \mathbb{N}} \left(f(z_1, \ldots, z_n)(y) \cdot \prod_{1 \le i \le n} g_i(\mathbf{x})(z_i) \right).$$

With a slight abuse of notation, we can treat probabilistic functions as ordinary functions when forming expressions. Suppose, as an example, that $x \in \mathbb{N}$ and that $f : \mathbb{N}^3 \to \mathbb{P}_\mathbb{N}$, $g : \mathbb{N} \to \mathbb{P}_\mathbb{N}$, $h : \mathbb{N} \to \mathbb{P}_\mathbb{N}$. Then the expression $f(g(x), x, h(x))$ stands for the distribution in $\mathbb{P}_\mathbb{N}$ defined as follows: $(f \odot (g, id, h))(x)$, where $id = \Pi_1^1$ is the identity PF.

The way we have defined probabilistic functions and their composition is reminiscent of, and indeed inspired by, the way one defines the Kleisli category for the Giry monad, starting from the category of partial functions on sets. This categorical way of seeing the problem can help a lot in finding the right definition, but by itself is not adequate to proving the existence of a correspondence with machines like the one we want to give here.

Primitive recursion is defined as in Kleene's algebra, provided one uses composition as previously defined:

Definition 5 (Primitive Recursion). *Given functions* $g : \mathbb{N}^{k+2} \to \mathbb{P}_\mathbb{N}$, *and* $f : \mathbb{N}^k \to \mathbb{P}_\mathbb{N}$, *the function* $h : \mathbb{N}^{k+1} \to \mathbb{P}_\mathbb{N}$ *defined as*

$$h(\mathbf{x}, 0) = f(\mathbf{x}); \qquad h(\mathbf{x}, y+1) = g(\mathbf{x}, y, h(\mathbf{x}, y));$$

is said to be defined by primitive recursion *from* f *and* g, *and is denoted as* $rec(f, g)$.

We now turn our attention to the minimization operator which, as in the deterministic case, is needed in order to obtain the full expressive power of (P)TMs. The definition of this operator is in our case delicate and requires some explanation. Recall that, in the classic case, the minimization operator allows from a partial function $f : \mathbb{N}^{k+1} \rightharpoonup \mathbb{N}$, to define another partial function, call it μf, which computes from $\mathbf{x} \in \mathbb{N}^k$ the least value of y such that $f(\mathbf{x}, y)$ is equal to 0, if such a value exists (and is undefined otherwise). In our case, again, we are concerned with distributions, hence we cannot simply consider the least value on which f returns 0, since functions return 0 *with a certain probability*. The idea is then to define the minimization μf as a function which, given an input $\mathbf{x} \in \mathbb{N}^k$, returns a distribution associating to each natural y the probability that the result of $f(\mathbf{x}, y)$ is 0 *and* the result of $f(\mathbf{x}, z)$ is positive for every $z < y$. Formally:

Definition 6 (Minimization). *Given a PF* $f : \mathbb{N}^{k+1} \to \mathbb{P}_\mathbb{N}$, *we define another PF* $\mu f : \mathbb{N}^k \to \mathbb{P}_\mathbb{N}$ *as follows:*

$$\mu f(\mathbf{x})(y) = f(\mathbf{x}, y)(0) \cdot (\prod_{z<y} (\sum_{k>0} f(\mathbf{x}, z)(k))).$$

We are finally able to define the class of functions we are interested in as follows.

Definition 7 (Probabilistic Recursive Functions). *The class* $\mathscr{PR}$ *of* probabilistic recursive functions *is the smallest class of probabilistic functions that contains the BPFs (Definition 2) and is closed under the operation of General Composition (Definition 4), Primitive Recursion (Definition 5) and Minimization (Definition 6).*

It is easy to show that $\mathscr{PR}$ includes all partial recursive functions, seen as probabilistic functions: first, for every partial function $f : \mathbb{N}^k \rightharpoonup \mathbb{N}$, define $p_f : \mathbb{N}^k \to \mathbb{P}_{\mathbb{N}}$ by stipulating that $p_f(\mathbf{x})(y) = 1$ whenever $y = f(\mathbf{x})$, and $p_f(\mathbf{x})(y) = 0$ otherwise; then, by an easy induction, $p_f \in \mathscr{PR}$ whenever f is partial recursive.

Example 1. The following are examples of probabilistic recursive functions:

- The *identity function* $id : \mathbb{N} \to \mathbb{P}_{\mathbb{N}}$, defined as $id(x)(x) = 1$. For all $x, y \in \mathbb{N}$ we have that
$$id(x)(y) = \begin{cases} 1 \text{ if } y = x \\ 0 \text{ otherwise} \end{cases}$$
as a consequence $id = \Pi_1^1$, and, since the latter is a BPF (Definition 2) id is in $\mathscr{PR}$.
- The probabilistic funtion $rand : \mathbb{N} \to \mathbb{P}_{\mathbb{N}}$ such that for every $x \in \mathbb{N}$, $rand(x)(0) = \frac{1}{2}$ and $rand(x)(1) = \frac{1}{2}$ can be easily shown to be recursive, since $rand = r \odot z$ (and we know that both r and z are BPF). Actually, $rand$ could itself be taken as the only genuinely probabilistic BPF, i.e., r can be constructed from $rand$ and the other BPF by composition and primitive recursion. We proceed by defining $g : \mathbb{N}^3 \to \mathbb{P}_{\mathbb{N}}$ as follow:
$$g(x_1, x_2, z)(y) = \begin{cases} 1 \text{ if } y = z + 1 \\ 0 \text{ otherwise} \end{cases}$$
g is in $\mathscr{PR}$ because $g = s \odot (\Pi_3^3)$. Now we observe that the function add defined by $add(x, 0) = id(x)$ and $add(x_1, x_2 + 1) = g(x_1, x_2, add(x_1, x_2))$ is a probabilistic recursive function, since it can be obtained from basic functions using composition and primitive recursion. We can conclude by just observing that $r = add \odot (\Pi_1^1, rand)$.
- All functions we have proved recursive so far have the property that the returned distribution is *finite* for any input. Indeed, this is true for every probabilistic *primitive* recursive function, since minimization is the only way to break this form of finiteness. Consider the function $f : \mathbb{N} \to \mathbb{P}_{\mathbb{N}}$ defined as $f(x)(y) = \frac{1}{2^{y-x+1}}$ if $y \geq x$, and $f(x)(y) = 0$ otherwise. We define another function $h : \mathbb{N} \to \mathbb{P}_{\mathbb{N}}$ by stipulating that $h(x)(y) = \frac{1}{2^{y+1}}$ for every $x, y \in \mathbb{N}$. h is a probabilistic recursive function; indeed consider the function $k : \mathbb{N}^2 \to \mathbb{P}_{\mathbb{N}}$ defined as $rand \odot \Pi_1^2$ and build $\mu\, k$. By definition,
$$(\mu\, k)(x)(y) = k(x, y)(0) \cdot (\prod_{z<y} (\sum_{q>0} k(x, z)(q))). \tag{1}$$

Then observe that $(\mu\ k)(x)(y) = \frac{1}{2^{y+1}}$: by (1), $(\mu\ k)(x)(y)$ unfolds into a product of exactly $y+1$ copies of $\frac{1}{2}$, each "coming from the flip of a distinct coin". Hence, $h = \mu\ k$. Then we observe that

$$(add \odot (\mu\ k, id))(x)(y) = \sum_{z_1, z_2} add(z_1, z_2)(y) \cdot ((\mu\ k)(x)(z_1) \cdot id(x)(z_2)).$$

But notice that $id(x)(z_2) = 1$ only when $z_2 = x$ (and in the other cases $id(x)(z_2) = 0$), $(\mu\ k)(x)(z_1) = \frac{1}{2^{z_1+1}}$, and $add(z_1, z_2)(y) = 1$ only when $z_1 + z_2 = y$ (and in the other cases, $add(z_1, z_2)(y) = 0$). This implies that the term in the sum is different from 0 only when $z_2 = x$ and $z_1 + z_2 = y$, namely when $z_1 = y - z_2 = y - x$, and in that case its value is $\frac{1}{2^{y-x+1}}$. Thus, we can claim that $f = (add \odot (\mu\ k, id))$, and that f is in $\mathscr{PR}$.

2.1 Probabilistic Turing Machines and Computable Functions

In this section we introduce computable functions as those probabilistic functions which can be computed by Probabilistic Turing Machines. As previously mentioned, probabilistic computation devices have received a wide interest in computer science already in the fifties [5] and early sixties [14]. A natural question which arose was then to see what happened if random elements were allowed in a Turing machine. This question led to several formalizations of probabilistic Turing machines (PTMs in the following) [5,16] — which, essentially, are Turing machines which have the ability to flip coins in order to make random decisions — and to several results concerning the computational complexity of problems when solved by PTMs [6].

Following [6], a Probabilistic Turing Machine (PTM) M can be seen as a Turing Machine with two transition functions δ_0, δ_1. At each computation step, either δ_0 or δ_1 can be applied, each with probability 1/2. Then, in a way analogous to the deterministic case, we can define a notion of a (initial, final) configuration for a PTM M. In the following, Σ_b denotes the set of possible symbols on the tape, including a blank symbol $\Box$; Q denotes the set of states; $Q_f \subseteq Q$ denotes the set of final states and $q_s \in Q$ denotes the initial state.

Definition 8 (Probabilistic Turing Machine). *A Probabilistic Turing Machine (PTM) is a Turing machine endowed with two transition functions δ_0, δ_1. At each computation step the transition function δ_0 can be applied with probability 1/2 and the transition δ_1 can be applied with probability 1/2.*

Definition 9 (Configuration of a PTM). *Let M be a PTM. We define a PTM* configuration *as a 4-tuple $\langle s, a, t, q\rangle \in \Sigma_b^* \times \Sigma_b \times \Sigma_b^* \times Q$ such that:*

- *The first component, $s \in \Sigma_b^*$, is the portion of the tape lying on the left of the head.*
- *The second component, $a \in \Sigma_b$, is the symbol the head is reading.*
- *The third component, $t \in \Sigma_b^*$, is the portion of the tape lying on the right of the head.*
- *The fourth component, $q \in Q$ is the current state.*

Moreover we define the set of all configurations as $\mathcal{C}_M = \Sigma_b^* \times \Sigma_b \times \Sigma_b^* \times Q$.

Definition 10 (Initial and Final Configurations of a PTM). *Let M be a PTM. We define the* initial configuration *of M for the string s as the configuration in the form $\langle \varepsilon, a, v, q_s \rangle \in \Sigma_b^* \times \Sigma_b \times \Sigma_b^* \times Q$ such that $s = a \cdot v$ and the fourth component, $q_s \in Q$, is the initial state. We denote it with $\mathcal{IN}_M^s$. Similarly, we define a* final configuration *of M for s as a configuration $\langle s, \square, \varepsilon, q_f \rangle \in \Sigma_b^* \times \Sigma_b \times \Sigma_b^* \times Q_f$. The set of all such final configurations for a PTM M is denoted by $\mathcal{FC}_M^s$.*

For a function $T : \mathbb{N} \to \mathbb{N}$, we say that a PTM M *runs in time bounded by* T if for any input x, M halts on input x within $T(|x|)$ steps *independently* of the random choices it makes. Thus, M *works in polynomial time* if it runs in time bounded by P, where P is any polynomial.

Intuitively, the function computed by a PTM M associates to each input s, a pseudodistribution which indicates the probability of reaching a final configuration of M from $\mathcal{IN}_M^s$. It is worth noticing that, differently from the deterministic case, since in a PTM the same configuration can be obtained by different computations, the probability of reaching a given final configuration is the *sum* of the probabilities of reaching the configuration along all computation paths, of which there can be (even infinitely) many. It is thus convenient to define the function computed by a PTM through a fixpoint construction, as follows. First, we can define a partial order on the string distributions as follows.

Definition 11. *A* string pseudodistribution *on Σ^* is a function $\mathcal{D} : \Sigma^* \to \mathbb{R}_{[0,1]}$ such that $\sum_{s \in \Sigma^*} \mathcal{D}(s) \leq 1$. $\mathbb{P}_{\Sigma^*}$ denotes the set of all string pseudodistributions on Σ^*. The relation $\sqsubseteq_{\mathbb{P}_{\Sigma^*}} \subseteq \mathbb{P}_{\Sigma^*} \times \mathbb{P}_{\Sigma^*}$ is defined as the pointwise extension of the usual partial order on $\mathbb{R}$.*

It is easy to show that the relation $\sqsubseteq_{\mathbb{P}_{\Sigma^*}}$ from Definition 11 is a partial order. Next, we can define the domain $\mathcal{CEV}$ of those functions computed by a PTM M from a given configuration, i.e., the set of those functions f such that $f : \mathcal{C}_M \to \mathbb{P}_{\Sigma^*}$. Inheriting the structure from $\mathbb{P}_{\Sigma^*}$, we can obtain a poset $(\mathcal{CEV}, \sqsubseteq_{\mathcal{CEV}})$, again by defining $\sqsubseteq_{\mathcal{CEV}}$ pointwise. Moreover, it is also easy to show that the two introduced posets are ω**CPO**s.

We can now define a functional F_M on $\mathcal{CEV}$ which will be used to define the function computed by M via a fixpoint construction. Intuitively, the application of the functional F_M describes *one* computation step. Formally:

Definition 12. *Given a PTM M, we define a functional $F_M : \mathcal{CEV} \to \mathcal{CEV}$ as:*

$$F_M(f)(C) = \begin{cases} \{s^1\} & \text{if } C \in \mathcal{FC}_M^s; \\ \frac{1}{2} f(\delta_0(C)) + \frac{1}{2} f(\delta_1(C)) & \text{otherwise.} \end{cases}$$

One can show that the functional F_M from Definition 12 is continuous on $\mathcal{CEV}$. A classic fixpoint theorem ensures that F_M has a least fixpoint. Such a least fixpoint is, once composed with a function returning $\mathcal{IN}_M^s$ from s, the *function computed by the machine* M, which is denoted as $\mathcal{IO}_M : \Sigma^* \to \mathbb{P}_{\Sigma^*}$. The set of those functions which can be computed by any PTM is denoted as $\mathscr{PC}$, while $\mathscr{PPC}$ is the set of probabilistic functions which can be computed by a PTM working in *polynomial* time. The notion of a computable probabilistic function subsumes other key notions in probabilistic and real-number computation. As an example, *computable distributions* can be characterized as those distributions on Σ^* which can be obtained as the result of a function in $\mathscr{PC}$ on a *fixed* input. Analogously, *computable real numbers* from the unit interval $[0,1]$ can be seen as those elements of $\mathbb{R}$ in the form $f(0)(0)$ for a computable function $f \in \mathscr{PC}$.

2.2 Probabilistic Recursive Functions Equals Functions Computed by Probabilistic Turing Machines

In this section we prove that probabilistic *recursive* functions are the same as probabilistic *computable* functions, modulo an appropriate bijection between strings and natural numbers which we denote (as its inverse) with $\overline{(\cdot)}$.

In order to prove the equivalence result we first need to show that a probabilistic recursive function can be computed by a PTM. This result is not difficult and, analogously to the deterministic case, is proved by exhibiting PTMs which simulate the basic probabilistic recursive functions and by showing that $\mathscr{PC}$ is closed by composition, primitive recursion, and minimization. We omit the details, which can be found in [4].

The most difficult part of the equivalence proof consists in proving that each probabilistic computable function is actually *recursive*. Analogously to the classic case, a good strategy consists in representing configurations as natural numbers, then encoding the transition functions of the machine at hand, call it M, as a (recursive) function on $\mathbb{N}$. In the classic case the proof proceeds by making essential use of the minimization operator by which one determines the *number* of transition steps of M necessary to reach a final configuration, if such number exists. This number can then be fed into another function which simulates M (on an input) a given number of steps, and which is primitive recursive. In our case, this strategy does not work: the number of computation steps can be infinite, even when the convergence probability is 1. Given our definition of minimization which involves distributions, this is delicate, since we have to define a suitable function on the PTM computation tree to be minimized.

In order to adapt the classic proof, we need to formalize the notion of a *computation tree* which represents all computation paths corresponding to a given input string x. We define such a tree as follows. Each node is labelled by a configuration of the machine and each edge represents a computation step. The root is labelled with $\mathcal{IN}_M^x$ and each node labelled with C has either no child (if C is final) or 2 children (otherwise), labelled with $\delta_0(C)$ and $\delta_1(C)$. Please notice that the same configuration may be duplicated across a single level of the tree as well as appear at different levels of the tree; nevertheless we represent each such appearance by a separate node.

We can naturally associate a probability with each node, corresponding to the probability that the node is reached in the computation: it is $\frac{1}{2^n}$, where n is the height of the node. The probability of a particular *final* configuration is the sum of the probabilities of all leaves labelled with that configuration. We also enumerate nodes in the tree, top-down and from left to right, by using binary strings in the following way: the root has associated the number ε. Then if b is the binary string representing the node N, the left child of N has associated the string $b \cdot 0$ while the right child has the number $b \cdot 1$. Note that from this definition it follows that each binary number associated to a node N indicates a path in the tree from the root to N. The computation tree for x will be denoted as $CT_M(x)$

We give now a more explicit description of the constructions described above. First we need to encode the rational numbers $\mathbb{Q}$ into $\mathbb{N}$. Let $pair : \mathbb{N} \times \mathbb{N} \to \mathbb{N}$ be any recursive bijection between pairs of natural numbers and natural numbers such that *pair* and its inverse are both computable. Let then *enc* be just p_{pair}, i.e. the function $enc : \mathbb{N} \times \mathbb{N} \to \mathbb{P}_{\mathbb{N}}$ defined as follows

$$enc(a,b)(q) = \begin{cases} 1 & \text{if } q = pair(a,b) \\ 0 & \text{otherwise} \end{cases}$$

The function *enc* allows to represent positive rational numbers as pairs of natural numbers in the obvious way and is recursive.

It is now time to define a few notions on computation trees

Definition 13 (Computation Trees and String Probabilities). *The function* $PT_M : \mathbb{N} \times \mathbb{N} \to \mathbb{Q}$ *is defined by stipulating that* $PT_M(x,y)$ *is the probability of observing the string* $\overline{y}$ *in the tree* $CT_M(x)$*, namely* $\frac{1}{2^{|\overline{y}|}}$.

Of course, PT_M is partial recursive, thus p_{PT_M} is probabilistic recursive. Since the same configuration C can label more than one node in a computation tree $CT_M(x)$, PT_M does not indicate the probability of reaching C, even when C is the label of the node corresponding to the second argument. Such a probability can be obtained by summing the probability of all nodes labelled with the configuration at hand:

Definition 14 (Configuration Probability). *Suppose given a PTM* M*. If* $x \in \mathbb{N}$ *and* $z \in \mathcal{C}_M$*, the subset* $CC_M(x,z)$ *of* $\mathbb{N}$ *contains precisely the indices of nodes of* $CT_M(x)$ *which are labelled by* z*. The function* $PC_M : \mathbb{N} \times \mathbb{N} \to \mathbb{Q}$ *is defined as follows:*

$$PC_M(x,z) = \Sigma_{y \in CC_M(x,z)} PT_M(x,y)$$

Contrary to PT_M, there is nothing guaranteeing that PC_M is indeed computable. In the following, however, what we do is precisely to show that this is the case.

In Figure 1 we show an example of computation tree $CT_M(x)$ for an hypothetical PTM M and an input x. The leaves, depicted as red nodes, represent the final configurations of the computation. So, for example, $PC_M(x,C) = 1$, while $PC_M(x,E) = \frac{3}{4}$. Indeed, notice that there are three nodes in the tree which are labelled with E, namely those corresponding to the binary strings 00,

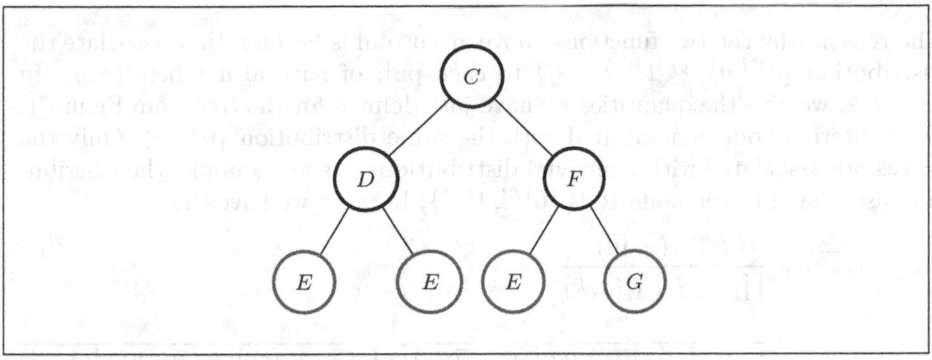

Fig. 1. An Example of a Computation Tree

01, and 10. As we already mentioned, our proof separates the classic part of the computation performed by the underlying PTM, which essentially computes the configurations reached by the machine in different paths, from the probabilistic part, which instead computes the probability values associated to each computation by using minimization. These two tasks are realized by two suitable probabilistic recursive functions, which are then composed to obtain the function computed by the underlying PTM. We start with the probabilistic part, which is more complicated.

We need to define a function, which returns the *conditional* probability of terminating at the node corresponding to the string $\overline{y}$ in the tree $CT_M(x)$, given that all the nodes $\overline{z}$ where $z < y$ are labelled with non-final configurations. This is captured by the following definition:

Definition 15. *Given a PTM M, we define $PT^0_M : \mathbb{N} \times \mathbb{N} \to \mathbb{Q}$ and $PT^1_M : \mathbb{N} \times \mathbb{N} \to \mathbb{Q}$ as follows:*

$$PT^1_M(x,y) = \begin{cases} 1 & \textit{if } y \textit{ is not a leaf of } CT_M(x); \\ 1 - PT^0_M(x,y) & \textit{otherwise;} \end{cases}$$

$$PT^0_M(x,y) = \begin{cases} 0 & \textit{if } y \textit{ is not a leaf of } CT_M(x); \\ \frac{PT_M(x,y)}{\prod_{k<y} PT^1_M(x,k)} & \textit{otherwise;} \end{cases}$$

Note that, according to previous definition, $PT^1_M(x,y)$ is the probability of *not* terminating the computation in the node y, while $PT^0_M(x,y)$ represents the probability of terminating the computation in the node y, both *knowing* that the computation has not terminated in any node k preceding y.

Proposition 1. *The functions $PT^0_M : \mathbb{N} \times \mathbb{N} \to \mathbb{Q}$ and $PT^1_M : \mathbb{N} \times \mathbb{N} \to \mathbb{Q}$ are partial recursive.*

Proof. Please observe that PT_M is partial recursive and that the definitions above are mutually recursive, but the underlying order is well-founded. Both functions are thus intuitively computable, thus partial recursive by the Church-Turing thesis. □

The reason why the two functions above are useful is because they associate the distribution $\{0^{PT^1_M(x,y)}, 1^{PT^0_M(x,y)}\}$ to each pair of natural numbers (x, y). In Figure 2, we give the quantities we have just defined for the tree from Figure 1. Each internal node is associated with the same distribution $\{0^0, 1^1\}$. Only the leaves are associated with nontrivial distributions. As an example, the distribution associated to the node 10 is $\{0^{1/2}, 1^{1/2}\}$, because we have that

$$PT^0_M(x,\overline{10}) = \frac{PT_M(x,\overline{10})}{\prod_{k<\overline{10}} PT^1_M(x,k)}$$
$$= \frac{1}{4 \cdot PT^1_M(x,\overline{01}) \cdot PT^1_M(x,\overline{00}) \cdot PT^1_M(x,\overline{1}) \cdot PT^1_M(x,\overline{0}) \cdot PT^1_M(x,\overline{\varepsilon})}$$
$$= \frac{1}{4 \cdot PT^1_M(x,\overline{01}) \cdot PT^1_M(x,\overline{00})}.$$

As it can be easily verified, $PT^1_M(x,\overline{00}) = \frac{3}{4}$, while $PT^1_M(x,\overline{01}) = \frac{2}{3}$. Thus, $PT^0_M(x,\overline{10}) = \frac{1}{2}$.

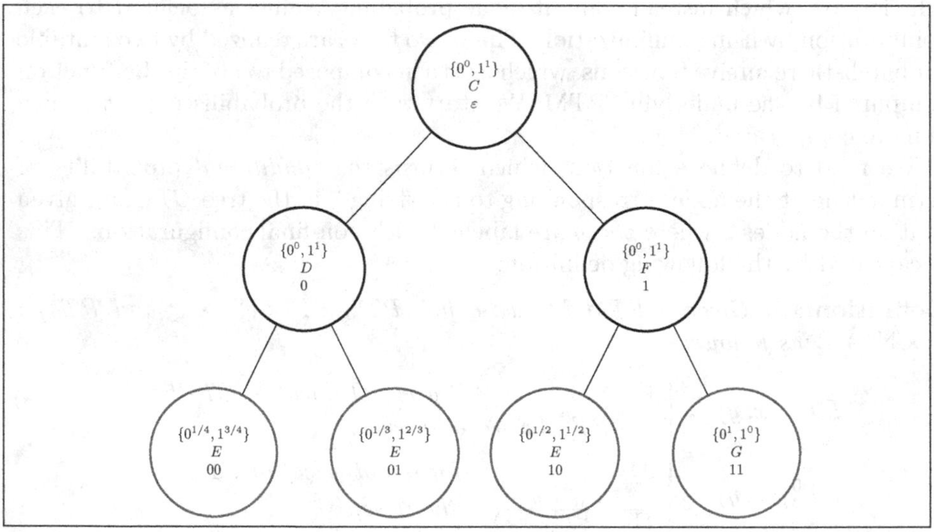

Fig. 2. The Conditional Probabilities for the Computation Tree from Figure 1

We now need to go further, and prove that the probabilistic function returning, on input (x, y), the distribution $\{0^{PT^1_M(x,y)}, 1^{PT^0_M(x,y)}\}$ is recursive. This is captured by the following definition:

Definition 16. *Given a PTM M, the function $PTC_M : \mathbb{N} \times \mathbb{N} \to \mathbb{P}_{\mathbb{N}}$ is defined as follows*

$$PTC_M(x,y)(z) = \begin{cases} PT^0_M(x,y) & \textit{if } z = 0; \\ PT^1_M(x,y) & \textit{if } z = 1; \\ 0 & \textit{otherwise-} \end{cases}$$

The function PTC_M is really the core of our encoding. On the one hand, we will show that it is indeed recursive. On the other, minimizing it is going to provide us exactly with the function we need to reach our final goal, namely proving that the probabilistic function computed by M is itself recursive. But how should we proceed if we want to prove PTC_M to be recursive? The idea is to compose $p_{PT^1_M}$ with a function that turns its input into the probability of returning 1. This is precisely what the following function does:

Definition 17. *The function* $I2P : \mathbb{Q} \to \mathbb{P}_{\mathbb{N}}$ *is defined as follows*

$$I2P(x)(y) = \begin{cases} x & \textit{if } (0 \le x \le 1) \wedge (y = 1) \\ 1 - x & \textit{if } (0 \le x \le 1) \wedge (y = 0) \\ 0 & \textit{otherwise} \end{cases}$$

Please observe how the input to *I2P* is the set of rational numbers, as usual encoded by pairs of natural numbers. Previous definitions allow us to treat (rational numbers representing) probabilities in our algebra of functions. Indeed:

Proposition 2. *The probabilistic function I2P is recursive.*

Proof. We first observe that $h : \mathbb{N} \to \mathbb{P}_{\mathbb{N}}$ defined as

$$h(x)(y) = 1/2^{y+1}$$

is a probabilistic recursive function, because $h = \mu\ (rand \odot \Pi^2_1)$. Next we observe that every $q \in \mathbb{Q} \cap [0,1]$ can be represented in binary notation as:

$$q = \sum_{i \in \mathbb{N}} \frac{c^q_i}{1/2^{i+1}}$$

where $c^q_i \in \{0,1\}$ (i.e., c^q_i is the i-th element of the binary representation of q). Moreover, a function computing such a c^q_i from q and i is partial recursive. Hence we can define $b : \mathbb{N} \times \mathbb{N} \to \mathbb{P}_{\mathbb{N}}$ as follows

$$b(q,i)(y) = \begin{cases} 1 \text{ if } y = c^q_i \\ 0 \text{ otherwise} \end{cases}$$

and conclude that b is indeed a probabilistic recursive function (because $\mathscr{PR}$ includes all the partial recursive functions, seen as probabilistic functions). Observe that:

$$b(q,i)(y) = \begin{cases} c^q_i & \text{if } y = 1 \\ 1 - c^q_i & \text{if } y = 0 \end{cases}$$

From the definition of composition, it follows that

$$\begin{aligned} (b \odot (id, h))(q)(y) &= \sum_{x_1, x_2} b(x_1, x_2)(y) \cdot id(q)(x_1) \cdot h(q)(x_2) \\ &= \sum_{x_2} b(q, x_2)(y) \cdot h(q)(x_2) = \sum_{x_2} b(q, x_2)(y) \cdot \frac{1}{2^{x_2+1}} \\ &= \begin{cases} \sum_{x_2} \frac{c^q_{x_2}}{2^{x_2+1}} \text{ if } y = 1 \\ \sum_{x_2} \frac{1 - c^q_{x_2}}{2^{x_2+1}} \text{ if } y = 0 \end{cases} = \begin{cases} q & \text{if } y = 1 \\ 1 - q \text{ if } y = 0 \end{cases} \end{aligned}$$

This shows that

$$I2P = b \odot (id, h),$$

and hence that $I2P$ is probabilistic recursive. □

The following is an easy corollary of what we have obtained so far:

Proposition 3. *The probabilistic function PTC_M is recursive.*

Proof. Just observe that $PTC_M = I2P \odot p_{PT^1_M}$. □

The probabilistic recursive function obtained as the minimization of PTC_M allows to compute a probabilistic function that, given x, returns y with probability $PT_M(x,y)$ *if* y is a leaf (and otherwise the probability is just 0).

Definition 18. *The function $\mathcal{CF}_M : \mathbb{N} \rightarrow \mathbb{P}_\mathbb{N}$ is defined as follows*

$$\mathcal{CF}_M(x)(y) = \begin{cases} PT_M(x,y) & \textit{if } y \textit{ corresponds to a leaf} \\ 0 & \textit{otherwise.} \end{cases}$$

Proposition 4. *The probabilistic function $\mathcal{CF}_M$ is recursive.*

Proof. The probabilistic function $\mathcal{CF}_M$ is just the function obtained by minimizing PTC_M, which we already know to be recursive. Indeed, if z corresponds to a leaf, then:

$$\begin{aligned}
(\mu PTC_M)(x)(z) &= PTC_M(x,z)(0) \cdot \prod_{y<z} \sum_{k>0} PTC_M(x,y)(k) \\
&= PTC_M(x,z)(0) \cdot \prod_{y<z} PTC_M(x,y)(1) \\
&= PT^0_M(x,z) \cdot \prod_{y<z} PT^1_M(x,y) \\
&= \frac{PT_M(x,z)}{\prod_{y<z} PT^1_M(x,y)} \cdot \prod_{y<z} PT^1_M(x,y) = PT_M(x,z).
\end{aligned}$$

If, however, z does not correspond to a leaf, then:

$$\begin{aligned}
(\mu PTC_M)(x)(z) &= PTC_M(x,z)(0) \cdot \prod_{y<z} \sum_{k>0} PTC_M(x,y)(k) \\
&= PT^0_M(x,z)(0) \cdot \prod_{y<z} \sum_{k>0} PTC_M(x,y)(k) = 0.
\end{aligned}$$

This concludes the proof. □

We are almost ready to wrap up our result, but before proceeding further, we need to define the function $SP_M : \mathbb{N} \times \mathbb{N} \rightarrow \mathbb{N}$ that, given in input a pair (x,y) returns the (encoding) of the string found in the configuration labeling the node y in $CT_M(x)$. We can now prove the desired result:

Theorem 1. $\mathscr{PC} \subseteq \mathscr{PR}$.

Proof. It suffices to note that, given any PTM M, the function computed by M is nothing more than

$$psp_M \odot (id, \mathcal{CF}_M).$$

Indeed, one can easily realize that a way to simulate M consists in generating, from x, all strings corresponding to the leaves of $CT_M(x)$, each with an appropriate probability. This is indeed what $\mathcal{CF}_M$ does. What remains to be done is simulating psp_M along paths leading to final configurations. □

We are finally ready to prove the main result of this Section:

Corollary 1. $\mathscr{PR} = \mathscr{PC}$

Proof. Immediate from Theorem 1, observing that $\mathscr{PR} \subseteq \mathscr{PC}$ (this implication is easy to prove). □

The way we prove Corollary 1 implies that we cannot deduce Kleene's Normal Form Theorem from it: minimization has been used many times, some of them "deep inside" the construction. A way to recover Kleene's Theorem consists in replacing minimization with a more powerful operator, essentially corresponding to computing the fixpoint of a given function (see [4] for more details).

3 Characterizing Probabilistic Complexity by Tiering

In this section we provide a characterization of the probabilistic functions which can be computed in polynomial time by an algebra of functions acting on word algebras. More precisely, we define a type system inspired by Leivant's notion of tiering [12], which permits to rule out functions having a too-high complexity, thus allowing to isolate the class of *predicative probabilistic functions*. Finally, we give a hint at how the equivalence between polytime probabilistic functions and predicative probabilistic functions can be proved (more details are in [4]).

The constructions from Section 2 can be easily generalized to a function algebra on strings in a given alphabet Σ, which themselves can be seen as a *word algebra* $\mathbb{W}$. Base functions include a function computing the empty string, called ε, and concatenation with any character $a \in \Sigma$, called c_a. Projections remain of course available, while the only truly random functions concatenate a symbol $a \in \Sigma$ to the input, with probability $\frac{1}{2}$, or leave it unchanged, with probability $\frac{1}{2}$. Such a function is denoted as r_a. Composition and primitive recursion are available, although the latter takes the form of recursion *on notation.* We do not need minimization: the distribution a polytime computable probabilistic function returns (on any input) is always finite, and primitive recursion is anyway powerful enough for our purposes.

The following construction is redundant in presence of primitive recursion, but becomes essential when predicatively restricting it:

Definition 19 (Case Distinction). *If* $g_\varepsilon : \mathbb{W}^k \to \mathbb{P}_\mathbb{W}$ *and for every* $a \in \Sigma$, $g_a : \mathbb{W}^{k+1} \to \mathbb{P}_\mathbb{W}$, *the function* $h : \mathbb{W}^{k+1} \to \mathbb{P}_\mathbb{W}$ *such that* $h(\varepsilon, \mathbf{y}) = g_\varepsilon(\mathbf{y})$ *and* $h(a \cdot w, \mathbf{y}) = g_a(w, \mathbf{y})$ *is said to be defined by* case distinction *from* g_ε *and* $\{g_a\}_{a\in\Sigma}$ *and is denoted as* $case(g_\varepsilon, \{g_a\}_{a\in\Sigma})$.

The idea behind tiering consists in working with denumerable many copies of the underlying algebra $\mathbb{W}$, each indexed by a natural number $n \in \mathbb{N}$ and denoted by $\mathbb{W}_n$. Judgments take the form $f \rhd \mathbb{W}_{n_1} \times \ldots \times \mathbb{W}_{n_k} \to \mathbb{W}_m$, where $f : \mathbb{W}^k \to \mathbb{W}$. In the following, with slight abuse of notation, $\mathbf{W}$ stands for any expression in the form $\mathbb{W}_{i_1} \times \cdots \times \mathbb{W}_{i_j}$.

Typing rules are in Figure 3. The idea here is that, when generating functions

$$\frac{}{\varepsilon \rhd \mathbb{W}_k \to \mathbb{W}_k} \qquad \frac{}{r_a \rhd \mathbb{W}_k \to \mathbb{W}_k} \qquad \frac{}{c_a \rhd \mathbb{W}_k \to \mathbb{W}_k} \qquad \frac{}{\Pi^k_m \rhd \mathbb{W}_{n_1} \times \cdots \times \mathbb{W}_{n_k} \to \mathbb{W}_{n_m}}$$

$$\frac{\{g_i \rhd \mathbb{W}_{s_1} \times \cdots \times \mathbb{W}_{s_r} \to \mathbb{W}_{m_i}\}_{1\leq i\leq p} \quad f \rhd \mathbb{W}_{m_1} \times \cdots \times \mathbb{W}_{m_p} \to \mathbb{W}_l}{f \odot (g_1, \ldots, g_l) \rhd \mathbb{W}_{s_1} \times \cdots \times \mathbb{W}_{s_r} \to \mathbb{W}_l}$$

$$\frac{g_\varepsilon \rhd \mathbf{W} \to \mathbb{W}_l \qquad \{g_a \rhd \mathbb{W}_k \times \mathbf{W} \to \mathbb{W}_l\}_{a\in\Sigma}}{case(g_\varepsilon, \{g_a\}_{a\in\Sigma}) \rhd \mathbb{W}_k \times \mathbf{W} \to \mathbb{W}_l} \qquad \frac{g_\varepsilon \rhd \mathbf{W} \to \mathbb{W}_k \quad m > k \qquad \{g_a \rhd \mathbb{W}_k \times \mathbb{W}_m \times \mathbf{W} \to \mathbb{W}_k\}_{a\in\Sigma}}{rec(g_\epsilon, \{g_a\}_{a\in\Sigma}) \rhd \mathbb{W}_m \times \mathbf{W} \to \mathbb{W}_k}$$

Fig. 3. Tiering as a Typing System

by primitive recursion, one goes from a level (tier) m for the domain to a *strictly* lower level k for the result. This predicative constraint ensures that recursion does not cause any exponential blowup, simply because the way one can *nest* primitive recursive definitions one inside the other is severely restricted. Please notice that case distinction, although being typed in a similar way, does *not* require the same constraints.

Those probabilistic functions $f : \mathbb{W}^k \to \mathbb{P}_\mathbb{W}$ such that f can be given a type through the rules in Figure 3 are said to be *predicatively recursive.* The class of all predicatively recursive functions is $\mathscr{PT}$. Actually, the class coincides with the one of probabilistic functions which can be computed by PTMs in polynomial time:

Theorem 2. $\mathscr{PT} = \mathscr{PPC}$.

We don't have enough space to give the details of the proof of Theorem 2. It however proceeds essentially by showing the following four lemmas, from which the thesis can be easily inferred:

- On the one hand, one can prove, by a careful encoding, that a form of *simultaneous primitive recursion* is available in predicative recursion.
- On the other hand, PTMs can be shown equivalent, in terms of expressivity, to probabilistic *register* machines; going through register machines has the advantage of facilitating the last two steps.
- Thirdly, any function definable by predicative recurrence can be proved computable by a polytime probabilistic register machine.
- Lastly, one can give an embedding of any polytime probabilistic register machine into a predicatively recursive function, making use of simultaneous recurrence.

Characterizing complexity classes of *probabilistic* functions allows to deal implicitly with concepts like that of a *polynomial time samplable* distribution [2,8], which is a family $\{\mathcal{D}_n\}_{n\in\mathbb{N}}$ of distributions on strings such that a polytime randomized algorithm produces $\mathcal{D}_n$ when fed with the string 1^n. By Theorem 2, each of them is computed by a function in $\mathscr{PT}$ and, conversely, any predicatively recursive probabilistic function computes one such family.

4 Conclusions

In this paper we make a first step in the direction of characterizing probabilistic computation in itself, from a recursion-theoretical perspective, without reducing it to deterministic computation. The significance of this study is genuinely foundational: working with probabilistic functions allows us to better understand the nature of probabilistic computation on the one hand, but also to study the implicit complexity of a generalization of Leivant's predicative recurrence, all in a unified framework.

More specifically, we give a characterization of computable probabilistic functions by a natural generalization of Kleene's partial recursive functions which includes, among initial functions, one that returns the uniform distribution on $\{0, 1\}$. We then prove the equi-expressivity of the obtained algebra and the class of functions computed by PTMs. In the the second part of the paper, we investigate the relations existing between our recursion-theoretical framework and sub-recursive classes, in the spirit of ICC. More precisely, endowing predicative recurrence with a random base function is proved to lead to a characterization of polynomial-time computable probabilistic functions.

An interesting direction for future work could be the extension of our recursion-theoretic framework to *quantum* computation. In this case one should consider transformations on Hilbert spaces as the basic elements of the computation domain. The main difficulty towards obtaining a completeness result for the resulting algebra and proving the equivalence with quantum Turing machines seems to be the definition of suitable recursion and minimization operators generalizing the ones described in this paper, given that qubits (the quantum analogues of classical bits) cannot be copied nor erased.

References

1. Bellantoni, S., Cook, S.: A new recursion-theoretic characterization of the polytime functions. Computational Complexity 2(2), 97–110 (1992)
2. Bogdanov, A., Trevisan, L.: Average-case complexity. Foundations and Trends in Theoretical Computer Science 2(1) (2006)
3. Dal Lago, U., Parisen Toldin, P.: A higher-order characterization of probabilistic polynomial time. In: Peña, R., van Eekelen, M., Shkaravska, O. (eds.) FOPARA 2011. LNCS, vol. 7177, pp. 1–18. Springer, Heidelberg (2012)
4. Dal Lago, U., Zuppiroli, S.: Probabilistic recursion theory and implicit computational complexity (long version) (2014), `http://arxiv.org/abs/1406.3378`
5. De Leeuw, K., Moore, E.F., Shannon, C.E., Shapiro, N.: Computability by probabilistic machines. Automata Studies 34, 183–198 (1956)
6. Gill, J.: Computational complexity of probabilistic Turing machines. SIAM Journal on Computing 6(4), 675–695 (1977)
7. Girard, J.-Y.: Light linear logic. Inf. Comput. 143(2), 175–204 (1998)
8. Goldreich, O.: Foundations of Cryptography: Basic Tools. Cambridge University Press (2000)
9. Goldwasser, S., Micali, S.: Probabilistic encryption. Journal of Computer and System Sciences 28(2), 270–299 (1984)
10. Katz, J., Lindell, Y.: Introduction to Modern Cryptography. Chapman & Hall/Crc Cryptography and Network Security Series. Chapman & Hall/CRC (2007)
11. Kleene, S.C.: General recursive functions of natural numbers. Mathematische Annalen 112(1), 727–742 (1936)
12. Leivant, D.: Ramified recurrence and computational complexity I: Word recurrence and poly-time. In: Feasible Mathematics II, pp. 320–343. Springer (1995)
13. Leivant, D., Marion, J.-Y.: Lambda calculus characterizations of poly-time. Fundam. Inform. 19(1/2), 167–184 (1993)
14. Rabin, M.O.: Probabilistic automata. Information and Control 6(3), 230–245 (1963)
15. Rabin, M.O., Scott, D.: Finite automata and their decision problems. IBM J. Res. Dev. 3(2), 114–125 (1959)
16. Santos, E.S.: Probabilistic Turing machines and computability. Proceedings of the American Mathematical Society 22(3), 704–710 (1969)
17. Santos, E.S.: Computability by probabilistic turing machines. Transactions of the American Mathematical Society 159, 165–184 (1971)
18. Soare, R.I.: Recursively enumerable sets and degrees: a study of computable functions and computably generated sets. Perspectives in mathematical logic. Springer (1987)

Heterogeneous Timed Machines

Benoît Delahaye[1], José Luiz Fiadeiro[2], Axel Legay[2,3], and Antónia Lopes[4]

[1] Université de Nantes, LINA, France
benoit.delahaye@univ-nantes.fr
[2] Dep. of Computer Science, Royal Holloway University of London, UK
jose.fiadeiro@rhul.ac.uk
[3] INRIA/IRISA, Rennes, France
axel.legay@irisa.fr
[4] Dep. of Informatics, Faculty of Sciences, University of Lisbon, Portugal
mal@di.fc.ul.pt

Abstract. We present an algebra of discrete timed input/output automata that execute in the context of different clock granularities — timed machines — as models of systems that can be dynamically interconnected at run time in a heterogeneous context. We show how timed machines can be refined to a lower granularity of time and how timed machines with different clock granularities can be composed. We propose techniques for checking whether timed machines are consistent or feasible. Finally, we investigate how consistency and feasibility of composition can be proved at run-time without computing products of automata.

1 Introduction

Many software applications operating in cyberspace need to connect, dynamically, to other software systems. For example, in the domain of intelligent transportation, systems for congestion avoidance or coordination of self-driven convoys of cars need to be able to accommodate interconnections that are established at run time between components that cannot be pre-determined at design time.

Applications such as these often have real-time requirements, i.e., their correctness depends not only on what outputs are returned to given inputs, but also on the time at which inputs are received and corresponding outputs are produced and communicated. When components of such software applications, usually written in a high-level programming language and relying on particular time abstractions, are executed in a given execution platform, their real-time behaviour is additionally restricted by the clock period of that platform. Components interconnected at run time will be likely to operate over different clock periods, resulting in a timed heterogeneous system.

Existing formalisms for modeling time-constrained systems focus mainly on mono-periodic systems, i.e., they assume that all system components will operate over a shared clock period. These models can still be used for timed heterogeneous systems whose structure is fixed and known *a-priori* by modeling the system components in terms of a global clock that is the least common multiple

G. Ciobanu and D. Méry (Eds.): ICTAC 2014, LNCS 8687, pp. 115–132, 2014.

of all local clocks. In the case of systems whose structure is dynamic and defined at run time, this is no longer possible [3].

In this paper, we propose a formal model for timed heterogeneous systems that does not require *a-priori* knowledge of their composition structure. Our model is based on input/output automata and supports *run-time compositionality* in the following sense: it is possible to ensure that components can work together as interconnected over heterogeneous local clocks by relying only on properties of models of those components, with no need for calculating their composition. More specifically, we provide the means to determine if the interconnection of two automata is consistent (there is at least a joint execution) or feasible (there is at least a joint execution no matter what inputs the components receive from their environment) not by calculating and checking properties of their product at run time but by relying on properties of the individual automata that can be established at design or composition time. Those properties ensure that the automata are able to co-operate at run time without modifying their time domains.

Our starting point is the homogeneous timed component algebra that we proposed in [8] for services. The extension from a homogenous to a heterogeneous setting is not trivial (which justifies this paper) because, where the algebraic properties of composition in a homogenous-time domain generalise those of the un-timed domain, interconnection in a heterogeneous setting is not even always admissible. For that reason, the algebra that we propose in Sec. 3 separates the space of discrete timed input/output automata (TIOA) [14,7] from that of their executions over a given clock: the components of our algebra are pairs of a TIOA and a clock granularity, what we call timed machines. Two operations are defined over timed machines: *heterogeneous composition*, which extends the traditional product of TIOA to the situation in which the granularities of the two machines are not the same, and *refinement*, which extends a machine with new states and transitions in order to accommodate a finer clock granularity as required to interoperate with other machines. Still, refinement does not reduce heterogeneous composition to the homogeneous setting, which leads us to define a notion of 'best approximation' through which we can characterise classes of timed machines that can be used to reason about or simulate interconnections of timed machines with commensurable clock granularities.

In Sec. 4, we study two important properties when modelling systems: *consistency*, in the sense that a machine can be ensured to generate a non-empty language, and *feasibility*, in the sense that a machine can be ensured to generate a non-empty language no matter what inputs it receives. Finally, we prove two compositionality results, one for consistency and the other for feasibility. Those results rely on a number of properties that can be checked, at design time, over given timed machines to ensure that their interconnection will be consistent or feasible without actually having to calculate the product of the corresponding automata at run time. Those properties ensure that components that implement the timed machines can work together across different clock granularities.

2 Preliminaries

2.1 Timed Traces

Although transition systems are typically used as operational semantics of automata (including timed transitions systems for timed automata as in [13]), we use instead a trace semantics because the topological properties of trace domains allow us to provide a finer characterisation of properties such as consistency and feasibility (cf. Sec. 4). For example, existing transition-system semantics such as [7] offer a weaker notion of consistency for timed automata because it fails to enforce time progression and, therefore, an automaton that does not accept any non-Zeno timed sequence can still be consistent. The proposed operational semantics is also much closer to the one that we used in the homogeneous-timed [8] and un-timed [9] domains, thus making it easier to understand the challenges raised by a heterogeneous domain.

We start by recalling a few concepts related to traces. Given a set A, a *trace* λ over A is an element of A^ω, i.e., an infinite sequence of elements of A. We denote by $\lambda(i)$ the $(i+1)$-th element of λ. A *segment* π is an element of A^*, i.e., a finite sequence of elements of A.

In our timed model, a trace consists of an infinite sequence of pairs of an instant of time and of the set of actions that are observed at that instant of time. Every such set of actions can be empty so that, on the one hand, we can model components that stop executing after a certain point in time while still part of a system and, on the other hand, we can model observations that are triggered by actions performed by components outside the system.

Definition 1 (Timed Traces). *Let A be a set (of actions).*

- *A* time sequence τ *is a trace over $\mathbb{R}_{\geq 0}$ such that:*
  - $\tau(0) = 0$;
  - *for every $i \in \mathbb{N}$, $\tau(i) < \tau(i+1)$;*
  - *the set $\{\tau(i) : i \in \mathbb{N}\}$ is unbounded, i.e., time progresses (also called the 'non-Zeno' condition).*
- *An* action sequence σ *is a trace over 2^A, i.e., an infinite sequence of sets of actions, such that $\sigma(0) = \emptyset$.*
- *A* timed trace *over A is a pair $\lambda = \langle\sigma, \tau\rangle$ of an action and a time sequence. We denote by $\Lambda(A)$ the set of timed traces over A and we call any $\Lambda \subseteq \Lambda(A)$ a* timed property.
- *Given $\delta \in \mathbb{R}_{>0}$, the* δ-time sequence τ_δ *consists of all multiples of δ — for every $i \in \mathbb{N}$, $\tau_\delta(i) = i \cdot \delta$. A* δ-timed trace *over A is a timed trace $\langle\sigma, \tau_\delta\rangle$.*

That is, in δ-timed traces, actions occur according to a fixed period (δ). These traces are useful to capture the behaviour of discrete systems that execute according to a fixed clock granularity.

In order to address heterogeneity, we need a notion of time refinement:

Definition 2 (Time Refinement). *Let $\rho : \mathbb{N} \to \mathbb{N}$ be a monotonically increasing function that satisfies $\rho(0) = 0$.*

- *Let τ, τ' be two time sequences. We say that τ'* refines *τ through ρ ($\tau' \preceq_\rho \tau$) iff, for every $i \in \mathbb{N}$, $\tau(i) = \tau'(\rho(i))$. We say that τ'* refines *τ ($\tau' \preceq \tau$) iff $\tau' \preceq_\rho \tau$ for some ρ.*
- *Let $\lambda = \langle\sigma, \tau\rangle$, $\lambda' = \langle\sigma', \tau'\rangle$ be two timed traces. We say that λ' refines λ through ρ ($\lambda' \preceq_\rho \lambda$) iff*
  - $\tau' \preceq_\rho \tau$,
  - *$\sigma(i) = \sigma'(\rho(i))$ for every $i \in \mathbb{N}$, and*
  - *$\sigma'(j) = \emptyset$ for every $\rho(i) < j < \rho(i+1)$.*

  We also say that λ' refines *λ ($\lambda' \preceq \lambda$) iff $\lambda' \preceq_\rho \lambda$ for some ρ.*
- *The* r-closure *of a timed property Λ is $\Lambda^r = \{\lambda' : \exists\lambda{\in}\Lambda(\lambda' \preceq \lambda)\}$.*
- *We say that Λ is* closed under time refinement, *or r-closed, iff $\Lambda^r \subseteq \Lambda$.*

We extend the notion of refinement to timed properties:

- *A timed property Λ' refines a timed property Λ ($\Lambda' \preceq \Lambda$) iff, for every $\lambda'{\in}\Lambda'$, there exists $\lambda{\in}\Lambda$ such that $\lambda' \preceq \lambda$.*
- *A timed property Λ'* approximates *a timed property Λ ($\Lambda' \precapprox \Lambda$) iff $\Lambda' \preceq \Lambda$ and, for every $\lambda{\in}\Lambda$, there exists $\lambda'{\in}\Lambda'$ such that $\lambda' \preceq \lambda$.*

That is, a time sequence refines another if the former interleaves time observations between any two time observations of the latter. For instance,

$$\langle\emptyset \cdot \{a, b\} \cdot \{b, c\} \ldots, 0 \cdot 2 \cdot 4 \cdots\rangle$$

is refined by

$$\langle\emptyset \cdot \emptyset \cdot \{a, b\} \cdot \emptyset \cdot \{b, c\} \ldots, 0 \cdot 1 \cdot 2 \cdot 3 \cdot 4 \cdots\rangle$$

Refinement extends to traces by requiring that no actions be observed in the finer trace between two consecutive times of the coarser. A timed property Λ' refines Λ if all traces of Λ' refine some trace of Λ. If all the traces of Λ have a refinement in Λ', then Λ' approximates Λ.

2.2 Timed Input/Output Automata

In order to model machines, we use timed I/O automata as in [7] except that transitions perform sets of actions instead of single actions. Working with sets of actions simplifies the treatment of interconnections by introducing synchronisation sets and gives us for free the empty set as an abstraction of actions performed by the environment that an automaton can observe without being directly involved.

A timed automaton is defined in terms of a finite set $\mathbb{C}$ of clocks. A *condition* over $\mathbb{C}$ is a finite conjunction of expressions of the form $c \bowtie n$ where $c \in \mathbb{C}$, $\bowtie{\in}\{\leq, \geq\}$ and $n{\in}\mathbb{N}$. We denote by $\mathcal{B}(\mathbb{C})$ the set of conditions over $\mathbb{C}$.

Definition 3 (TIOA). *A timed I/O automaton $\mathcal{A}$ (TIOA) is a tuple*

$$\mathcal{A} = \langle Loc, q_0, \mathbb{C}, E, Act, Inv\rangle$$

where:

- *Loc is a finite set of locations;*
- *$q_0 \in Loc$ is the initial location;*
- *$\mathbb{C}$ is a finite set of clocks;*
- *$E \subseteq Loc \times 2^{Act} \times \mathcal{B}(\mathbb{C}) \times 2^{\mathbb{C}} \times Loc$ is a finite set of edges;*
- *$Act = Act^I \cup Act^O \cup Act^\tau$ is a finite set of actions partitioned into inputs, outputs and internal actions, respectively;*
- *Inv: $Loc \to \mathcal{B}(\mathbb{C})$ is a mapping that associates an invariant with every location.*

In addition, we impose that every TIOA is open *in the sense of not interfering with the ability of the environment to make progress: for all $l \in Loc$, there is an edge $(l, \emptyset, \phi, \emptyset, l') \in E$ for some location l' such that $Inv(l')$ is implied by $Inv(l)$ and for some tautology ϕ.*

Given an edge (l, S, C, R, l'), l is the source location, l' is the target location, S is the set of actions executed during the transition, C is a guard (a condition that determines if the transition can be performed), and R is the set of clocks that are reset by the transition. The requirement that every location is the source of a transition labelled by $\emptyset$ that is always enabled means that the behavior of $\mathcal{A}$ is always open to the execution of actions in which it is not involved.

A *clock valuation* over a set $\mathbb{C}$ of clocks is a mapping v: $\mathbb{C} \to \mathbb{R}_{\geq 0}$. Given $d \in \mathbb{R}_{\geq 0}$ and a valuation v, we denote by $v+d$ the valuation defined by, for any clock $c \in \mathbb{C}$, $(v+d)(c) = v(c)+d$. Given $R \subseteq \mathbb{C}$ and a clock valuation v, we denote by $v^{\mathbf{R}}$ the valuation where clocks from R are reset, i.e., such that $v^{\mathbf{R}}(c)=0$ if $c \in R$ and $v^{\mathbf{R}}(c)=v(c)$ otherwise. Given a condition C in $\mathcal{B}(\mathbb{C})$, we use $v \models C$ to express that C holds for the clock valuation v.

Definition 4 (Execution). *Let $\mathcal{A} = \langle Loc, q_0, \mathbb{C}, E, Act, Inv \rangle$ be a TIOA. An* execution *of $\mathcal{A}$ starting in l_0 and valuation v_0 is a sequence*

$$(l_0, v_0, d_0) \xrightarrow{S_0, R_0} (l_1, v_1, d_1) \xrightarrow{S_1, R_1} \ldots$$

where, for all i: (1) $l_i \in Loc$, v_i is a clock valuation over $\mathbb{C}$ and $d_i \in \mathbb{R}_{>0}$; (2) $S_i \subseteq Act$ and $R_i \subseteq \mathbb{C}$; (3) for all $0 \leq t \leq d_i$, $v_i + t \models Inv(l_i)$; (4) $v_{i+1}=(v_i + d_i)^{\mathbf{R_i}}$; and (5) there is $(l_i, S_i, C, R_i, l_{i+1}) \in E$ such that $v_i + d_i \models C$. A partial execution *is of the form*

$$(l_0, v_0, d_0) \xrightarrow{S_0, R_0} \ldots \xrightarrow{S_{n-1}, R_{n-1}} (l_n, v_n, d_n)$$

where (1) and (3) hold for all $i \in [0..n]$, and (2), (4) and (5) for all $i \in [0..n-1]$.

That is, each triple (l_i, v_i, d_i) consists of a location, the value of the clocks when that location is reached at that point of the execution, and the duration for which the automaton remains at that location before the next transition (which can leave the automaton in the same location). During this time, the invariant $Inv(l_i)$ must hold. A transition out of (l_i, v_i, d_i) happens at the end of d_i units of time and needs to be made by an edge whose guard C_i holds at that time and leads to a location whose invariant is satisfied. As a result of the transition, the clocks are updated to $(v_i + d_i)^{\mathbf{R_i}}$.

A pair (l, v) where l is a location and v is a clock valuation is said to be *reachable at time* $T \in \mathbb{R}_{\geq 0}$ if either (a) $(l, v) = (q_0, 0)$, $T = 0$ and, there exists $d_0 > 0$ such that $t \models Inv(q_0)$ for all $0 \leq t \leq d_0$; or (b) there exists a partial execution that starts at $(q_0, 0)$ and ends at $(l_n, v_n) = (l, v)$, and $T = \sum_{i=0\cdots n-1} d_i$.

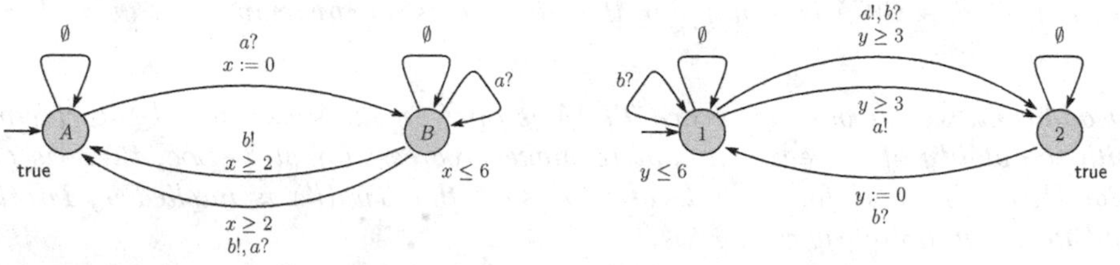

Fig. 1. Two TIOAs: $\mathcal{A}^x$ (left) and $\mathcal{A}^y$ (right)

Example 5. *Consider the TIOAs in Fig. 1:* $\mathcal{A}^x = \langle \{A, B\}, A, \{x\}, E^x, Act_x, Inv_x \rangle$ *with* $Act_x^I = \{a\}$ *and* $Act_x^O = \{b\}$, *and* $\mathcal{A}^y = \langle \{1, 2\}, 1, \{y\}, E^y, Act_y, Inv_y \rangle$ *with* $Act_y^I = \{b\}$ *and* $Act_y^O = \{a\}$ *(for clarity, inputs are decorated with ? and outputs with !).*

- $\mathcal{A}^y$ *starts by sending an* a *within six time units but not before three units have passed; it then waits for receiving a* b *to start again and send another* a. *More* b*'s can be received meanwhile (even while sending an* a*), but they are all ignored.*
- $\mathcal{A}^x$ *waits for receiving an* a, *after which it sends a* b *within six time units but not before two times units have passed (all* a*'s received in the meanwhile being ignored); then,* $\mathcal{A}^x$ *waits for receiving another* a.

An example of a partial execution of $\mathcal{A}^x$ *is*

$$(A, 0, 2) \xrightarrow{\{a\},\{x\}} (B, 0, 3) \xrightarrow{\{b\},\emptyset} (A, 3, 5) \xrightarrow{\{a\},\{x\}} (B, 0, 2)$$

which shows that $(B, 0)$ *is reachable at times* 2 *and* 10.

We now recall the classical definition of composition of *compatible* TIOAs, which captures partial synchronisation.

Definition 6 (Compatibility). *Two TIOAs* $\mathcal{A}_i = \langle Loc^i, q_0^i, \mathbb{C}^i, E^i, Act_i, Inv^i \rangle$ *are compatible iff* $\mathbb{C}^1 \cap \mathbb{C}^2 = Act_1^I \cap Act_2^I = Act_1^O \cap Act_2^O = Act_1^\tau \cap Act_2 = Act_2^\tau \cap Act_1 = \emptyset$.

Definition 7 (Composition). *The* composition *of two compatible TIOAs* $\mathcal{A}_i = \langle Loc^i, q_0^i, \mathbb{C}^i, E^i, Act_i, Inv^i \rangle$ *is*

$$\mathcal{A}_1 \parallel \mathcal{A}_2 = \langle Loc^1 \times Loc^2, (q_0^1, q_0^2), \mathbb{C}^1 \cup \mathbb{C}^2, E, Act, Inv \rangle$$

where:

– $Act^I = (Act_1^I \backslash Act_2^O) \cup (Act_2^I \backslash Act_1^O)$,
– $Act^O = (Act_1^O \backslash Act_2^I) \cup (Act_2^O \backslash Act_1^I)$,
– $Act^\tau = Act_1^\tau \cup Act_2^\tau \cup (Act_1^I \cap Act_2^O) \cup (Act_1^O \cap Act_2^I)$, *and*
– *for all* $(q_1, q_2) \in Loc^1 \times Loc^2$:
  • $Inv((q_1, q_2)) = Inv^1(q_1) \wedge Inv^2(q_2)$;
  • $((q_1, q_2), S, C, R, (q_1', q_2')) \in E$ *iff* $(q_1, S_1, C_1, q_1') \in E^1$, $(q_2, S_2, C_2, q_2') \in E^2$, $C = C_1 \wedge C_2$, $S_i = S \cap Act_i$ $(i = 1, 2)$ *and* $R = R_1 \cup R_2$.

Notice that, because the guards of transitions are conjoined, for the TIOA that results from the composition to be open (cf. Def. 3) we need to require the existence of a transition labeled with $\emptyset$ and a tautological guard (instead of simply *true*). Notice also that, by construction, whenever $S \cap Act_1 \neq \emptyset$ and $S \cap Act_2 \neq \emptyset$, all actions on which $\mathcal{A}_1$ and $\mathcal{A}_2$ synchronise (those in $S \cap Act_1 \cap Act_2$) are necessarily inputs on one side and outputs on the other; the composition makes those actions internal. Finally, transitions such that $S \cap Act_i = \emptyset$, which are usually considered as non-synchronising, are in our case handled as synchronising transitions where $\mathcal{A}_i$ performs the empty set of actions (which corresponds to an open semantics).

3 Timed Machines: Definition and Operations

In order to model systems where applications execute over specific platforms, which implies that they are subject to the clock granularity of the platform, we extend TIOAs to what we call timed machines.

3.1 Timed Machines

A timed machine is a TIOA that executes in the context of a clock granularity δ, i.e., its actions are always executed at instant times that are multiples of δ.

Definition 8 (DTIOM). *A* discrete timed I/O machine *(DTIOM) is a pair*

$$\mathcal{M} = \langle \delta_{\mathcal{M}}, \mathcal{A}_{\mathcal{M}} \rangle$$

where $\delta_{\mathcal{M}} \in \mathbb{R}_{>0}$ *and* $\mathcal{A}_{\mathcal{M}}$ *is a TIOA.*

The executions *and* partial executions *of* $\mathcal{M}$ *are those of* $\mathcal{A}_{\mathcal{M}}$ *restricted to transitions at every* $\delta_{\mathcal{M}}$, *i.e.*,

$$(l_0, v_0, d_0) \xrightarrow{S_0, R_0} (l_1, v_1, d_1) \xrightarrow{S_1, R_1} \ldots$$

such that all the durations d_i *are* $\delta_{\mathcal{M}}$. *Therefore, we represent executions of DTIOMs as sequences*

$$(l_0, v_0) \xrightarrow{S_0, R_0} (l_1, v_1) \xrightarrow{S_1, R_1} \ldots$$

and call each pair (l_i, v_i) *an execution state.*

The behaviour $[\![\mathcal{M}]\!]$ *of* $\mathcal{M}$ *is the set of executions such that* $l_0{=}q_0$ *and* $v_0(c){=}0$ *for all* $c{\in}\mathbb{C}$*, i.e., those that start in the initial location with all clocks set to* 0.

Every execution of a DTIOM $\mathcal{M}$ *defines the* $\delta_{\mathcal{M}}$*-timed trace* $\lambda{=}\langle\sigma, \tau_{\delta_{\mathcal{M}}}\rangle$ *over* Act *where* $\sigma(0){=}\emptyset$ *and, for* $i \geq 0$, $\sigma(i+1) = S_i$*. We denote by* $\Lambda_{\mathcal{M}}$ *the r-closure of the set of timed traces defined by* $[\![\mathcal{M}]\!]$*, which we call its* language.

The fact that the language of a DTIOM is r-closed means that it contains all possible interleavings of empty observations, thus capturing the behaviour of the DTIOM in any possible environment. This notion of closure can be related to mechanisms that, such as stuttering [1], ensure that components do not constrain their environment.

Example 9. *Consider* $\mathcal{M}^x = \langle\delta_x, \mathcal{A}^x\rangle$*, and* $\mathcal{M}^y = \langle\delta_y, \mathcal{A}^y\rangle$*, with* $\delta_x = 2$, $\delta_y = 1$ *and* $\mathcal{A}^x$ *and* $\mathcal{A}^y$ *as in Ex. 5. Notice that the partial execution of* $\mathcal{A}^x$ *given in Ex. 5 is not a partial execution of* $\mathcal{M}^x$ *as it does not respect the granularity* $\delta_x = 2$*. An example of a partial execution of* $\mathcal{M}^x$ *is*

$$(A, 0) \xrightarrow{\{a\},\{x\}} (B, 0) \xrightarrow{\emptyset,\emptyset} (B, 2) \xrightarrow{\{b\},\emptyset} (A, 4)$$

Note that this means that a *was executed at time* 2*, nothing was executed at time* 4 *and* b *was executed at time* 6.

3.2 Composition and Refinement of Timed Machines

Composition of DTIOMs with the same clock granularity is as for TIOA:

Definition 10 (Composition). *Given two TIOAs* $\mathcal{A}_1$ *and* $\mathcal{A}_2$ *that are compatible, we define the composition* $\langle\delta, \mathcal{A}_1\rangle \| \langle\delta, \mathcal{A}_2\rangle = \langle\delta, \mathcal{A}_1 \| \mathcal{A}_2\rangle$.

It is not difficult to prove that the language $\Lambda_{\langle\delta,\mathcal{A}_1\|\mathcal{A}_2\rangle}$ of the composition is the intersection

$$\Lambda^{\iota_1}_{\langle\delta,\mathcal{A}_1\rangle} \cap \Lambda^{\iota_2}_{\langle\delta,\mathcal{A}_2\rangle}$$

where

- ι_i is the inclusion of Act_i in $Act_1 \cup Act_2$ and
- $\Lambda^{\iota_i}_{\langle\delta,\mathcal{A}_i\rangle} = \{\lambda : \iota_i^{-1}(\lambda) \in \Lambda_{\langle\delta,\mathcal{A}_i\rangle}\}$ are the projections of the languages of the machines to the alphabet of the composition defined by, for every k, $\iota_i^{-1}(\lambda)(k) = \iota_i^{-1}(\lambda(k))$.

That is, the language of the composition consists of the timed traces that project to timed traces of languages of the component DTIOMs. This is what is usually taken to be the joint behaviour of a system of components in a trace-based semantic domain, meaning that $\langle\delta, \mathcal{A}_1 \| \mathcal{A}_2\rangle$ provides a model of the joint behaviour of two systems of which $\langle\delta, \mathcal{A}_1\rangle$ and $\langle\delta, \mathcal{A}_2\rangle$ are models.

If $\langle\delta_1, \mathcal{A}_1\rangle$ and $\langle\delta_2, \mathcal{A}_2\rangle$ have different clock granularities, we can still calculate the intersection $\Lambda^{\iota_1}_{\langle\delta,\mathcal{A}_1\rangle} \cap \Lambda^{\iota_2}_{\langle\delta,\mathcal{A}_2\rangle}$, which is the joint behaviour of the two machines synchronising on shared inputs and outputs at times that are multiple of both

δ_1 and δ_2. If no such multiples exist, the two machines cannot synchronise and, therefore, either they do not have liveness requirements, in which case they can agree on timed traces that only execute the empty set of actions, or they cannot agree on any timed trace — their interconnection is inconsistent.

If δ_1 and δ_2 admit a common multiple, i.e., $\delta_1 \cdot n = \delta_2 \cdot m$ for given $n, m \in \mathbb{N}_{>0}$, then they are *commensurable*, i.e., they admit a common divisor ($\delta_1/m = \delta_2/n$) – again, a real number. This is the situation that we characterise now. More precisely, our aim is to construct a machine $\mathcal{M}$ that, although it may not generate the full set of joint behaviours, i.e., be such that $\Lambda_{\mathcal{M}} = \Lambda^{\iota_1}_{\mathcal{M}_1} \cap \Lambda^{\iota_2}_{\mathcal{M}_2}$, it will be the 'best' approximation of that set in the sense that $\Lambda_{\mathcal{M}} \precapprox \Lambda^{\iota_1}_{\mathcal{M}_1} \cap \Lambda^{\iota_2}_{\mathcal{M}_2}$ and, for any other machine $\mathcal{M}'$ such that $\Lambda_{\mathcal{M}'} \precapprox \Lambda^{\iota_1}_{\mathcal{M}_1} \cap \Lambda^{\iota_2}_{\mathcal{M}_2}$, $\Lambda_{\mathcal{M}'} \precapprox \Lambda_{\mathcal{M}}$. Having a best approximation is important so that properties of the global behaviour of the system (such as consistency and feasibility, discussed in Sec. 4) can be inferred from that of the composed machine or that the behaviour of the system can be simulated through a machine.

The idea is to refine the timed machines to a common clock granularity and then compose the refinements: intuitively, given a timed machine $\mathcal{M} = \langle \delta, \mathcal{A} \rangle$, we define its *k-refinement* $\mathcal{M}_k = \langle \delta/k, \mathcal{A}_k \rangle$ by dividing both the clock granularity and the TIOA $\mathcal{A}$ by k so as to produce a TIOA $\mathcal{A}_k$ that divides every state in k copies such that the original transitions are performed in the last 'tick', all previous 'ticks' performing no actions and, therefore, being open for synchronisation with a machine that ticks with a granularity δ/k.

Definition 11 (Refinement). *Given a TIOA $\mathcal{A} = \langle Loc, q_0, \mathbb{C}, E, Act, Inv \rangle$ and $k \in \mathbb{N}_{>0}$, its k-*refinement *is the TIOA $\mathcal{A}_k = \langle Loc_k, q_{k0}, \mathbb{C}, E_k, Act, Inv_k \rangle$ where:*

- $Loc_k = Loc \times [0..k-1]$;
- $q_{k0} = (q_0, 0)$;
- $Inv_k(l,i) = Inv(l)$;
- *for every (l, S, C, R, l') of E, E_k has the edge $((l, k-1), S, C, R, (l', 0))$ and all edges of the form $((l,i), \emptyset, true, \emptyset, (l, i+1))$, $i \in [0..k-2]$.*

It is easy to see that $\Lambda_{\mathcal{M}_k} \preceq \Lambda_{\mathcal{M}}$, i.e., the language of $\mathcal{M}_k$ refines that of $\mathcal{M}$; in fact, because the languages are r-closed, $\Lambda_{\mathcal{M}_k} \subseteq \Lambda_{\mathcal{M}}$. Because every execution of $\mathcal{M}$ defines a (unique) execution of $\mathcal{M}_k$, the language of $\mathcal{M}_k$ actually approximates that of $\mathcal{M}$, i.e., $\Lambda_{\mathcal{M}_k} \precapprox \Lambda_{\mathcal{M}}$, meaning that all possible behaviours of $\mathcal{M}$ can be accounted for in $\mathcal{M}_k$ through a refinement: we say that $\mathcal{M}_k$ *approximates* $\mathcal{M}$ and write $\mathcal{M}_k \precapprox \mathcal{M}$. More generally, for arbitrary DTIOMs $\mathcal{M}$ and $\mathcal{M}'$ that have a common alphabet (i.e., $Act_{\mathcal{M}'} = Act_{\mathcal{M}}$), we define $\mathcal{M}' \precapprox \mathcal{M}$ to mean that $\delta_{\mathcal{M}}$ is a multiple of $\delta_{\mathcal{M}'}$ and $\Lambda_{\mathcal{M}'} \precapprox \Lambda_{\mathcal{M}}$.

Example 12. *Consider the TIOA $\mathcal{A}^x$ in Fig. 1 and the corresponding DTIOM $\mathcal{M}^x$ defined in Ex. 9, which has granularity 2. Its refinement to a DTIOM with granularity 1 is $\mathcal{M}^x_2 = \langle 1, \mathcal{A}^x_2 \rangle$, with $\mathcal{A}^x_2$ given in Fig. 2. The refinement of the partial execution of $\mathcal{M}^x$ given in Ex. 9 is:*

$$((A,0),0) \xrightarrow{\emptyset,\emptyset} ((A,1),1) \xrightarrow{\{a\},\{x\}} ((B,0),0) \xrightarrow{\emptyset,\emptyset} ((B,1),1) \xrightarrow{\emptyset,\emptyset} ((B,0),2) \xrightarrow{\emptyset,\emptyset} ((B,1),3) \xrightarrow{\{b\},\emptyset} ((A,0),4)$$

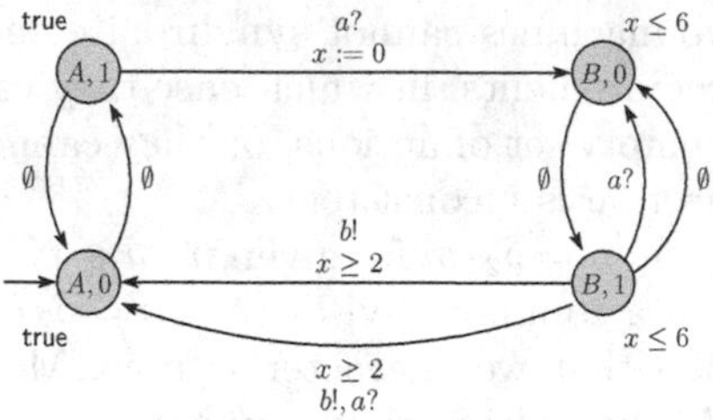

Fig. 2. The refinement $\mathcal{A}_2^x$ of $\mathcal{A}^x$

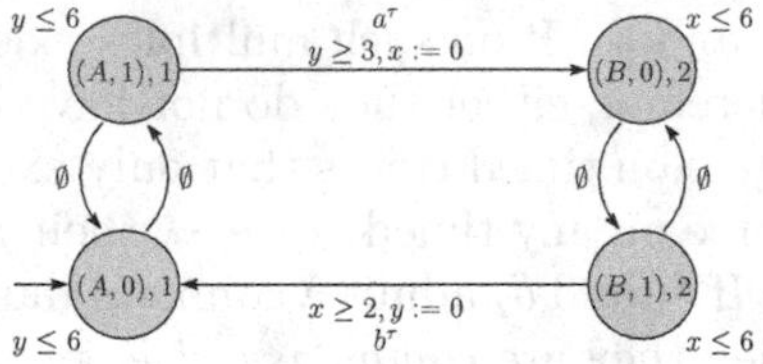

Fig. 3. The TIOA $\mathcal{A}^{x,y}$ of $\mathcal{M}^x \| \mathcal{M}^y$

We can now extend the composition of two timed machines to the case where their clock granularities are commensurable (have a common divisor):

Definition 13 (Heterogeneous Compatibility). *Two DTIOMs* $\mathcal{M}_i = \langle \delta_i, \mathcal{A}_i \rangle$, $i = 1, 2$, *are said to be* δ-compatible *(where* $\delta \in \mathbb{R}_{>0}$*) if (a)* $\mathcal{A}_1$ *and* $\mathcal{A}_2$ *are compatible, and (b)* δ *is a common divisor of* δ_1 *and* δ_2*. They are said to be* compatible *if they are* δ-compatible *for some* δ.

Definition 14 (Heterogeneous Composition). *The* δ-composition *of two* δ-*compatible DTIOMs is*

$$\mathcal{M}_1 \|_\delta \mathcal{M}_2 = \mathcal{M}_{1(\delta_1/\delta)} \| \mathcal{M}_{2(\delta_2/\delta)} = \langle \delta, \mathcal{A}_{1(\delta_1/\delta)} \| \mathcal{A}_{2(\delta_2/\delta)} \rangle$$

If δ *is the greatest common divisor of* δ_1 *and* δ_2*, we use the notation* $\mathcal{M}_1 \| \mathcal{M}_2$ *and simply refer to* the composition *of* $\mathcal{M}_1$ *and* $\mathcal{M}_2$.

Notice that if $\mathcal{A}_1$ and $\mathcal{A}_2$ are compatible, so are $\mathcal{A}_{1(\delta_1/\delta)}$ and $\mathcal{A}_{2(\delta_2/\delta)}$.

Example 15. *Consider DTIOMs* $\mathcal{M}^x$ *and* $\mathcal{M}^y$ *from Ex. 9. Because* $\mathcal{A}^x$ *and* $\mathcal{A}^y$ *are compatible and* δ_x *and* δ_y *have a common divisor (*$\delta = 1$*), we can compute their composition. The first step consists in refining* $\mathcal{A}^x$ *into* $\mathcal{A}_2^x$ *(Fig. 2). The composition* $\mathcal{M}^x \| \mathcal{M}^y$ *is* $\langle 1, \mathcal{A}^{x,y} \rangle$ *where* $\mathcal{A}^{x,y} = \mathcal{A}_2^x \| \mathcal{A}^y$ *is given in Fig. 3. Notice that actions a and b are synchronised and, hence, made internal in the composition, which we denote by* a^τ *and* b^τ*, respectively.*

The language of a heterogeneous composition is not necessarily the intersection of the languages of the components. However, if $\mathcal{M}_1$ and $\mathcal{M}_2$ can be composed, the machine $\mathcal{M}_1 \| \mathcal{M}_2$ approximates $\Lambda^{\iota_1}_{\mathcal{M}_1} \cap \Lambda^{\iota_2}_{\mathcal{M}_2}$, and is a best approximation:

Theorem 16. *Let* $\mathcal{M}_i$, $i = 1, 2$, *be two compatible DTIOMs. The composition* $\mathcal{M}_1 \| \mathcal{M}_2$ *is the machine that best approximates* $\Lambda = \Lambda^{\iota_1}_{\mathcal{M}_1} \cap \Lambda^{\iota_2}_{\mathcal{M}_2}$*, i.e.,*

- $\Lambda_{\mathcal{M}_1 \| \mathcal{M}_2} \precapprox \Lambda$ *and,*
- *for any other machine* $\mathcal{M}$ *such that* $\Lambda_{\mathcal{M}} \precapprox \Lambda$, $\mathcal{M} \precapprox \mathcal{M}_1 \| \mathcal{M}_2$.

3.3 Büchi Representation of Timed Machines

In order to check different structural properties of DTIOMs, namely properties formulated in terms of reachable states, it is useful to be able to construct Büchi-automata "equivalent".

Let $\mathcal{A} = \langle Loc, l_0, \mathbb{C}, E, Act, Inv\rangle$ be a TIOA. Given a clock c, let $\mathsf{Max}^{\mathcal{A}}(c)$ denote the maximal constant with which c is compared in the guards and invariants of $\mathcal{A}$. Let $\mathcal{M} = \langle \delta, \mathcal{A}\rangle$ and $\mathcal{B}_{\mathcal{M}} = \langle Q, q_0, 2^{Act}, \rightarrow, Q\rangle$ be the Büchi automaton such that:

- $Q = Loc \times \prod_{c\in\mathbb{C}}[0 \mathinner{..} \lfloor \frac{\mathsf{Max}^{\mathcal{M}}(c)}{\delta}\rfloor + 1]$ (i.e., states consist of a location l and a natural number $n_c \leq \lfloor \frac{\mathsf{Max}^{\mathcal{M}}(c)}{\delta}\rfloor + 1$, for every $c \in \mathbb{C}$);
- $q_0 = (l_0, \mathbf{0})$;
- $(l, \boldsymbol{\nu}) \xrightarrow{S} (l', \boldsymbol{\nu}')$ iff there exists a transition $(l, S, C, R, l') \in E$ such that:
  (i) for all $0 \leq t \leq \delta$, $\boldsymbol{\nu}\cdot\delta + t \models Inv(l)$,
  (ii) $\boldsymbol{\nu}\cdot\delta + \delta \models C$,
  (iii) for all $c \in \mathbb{C}$, $\boldsymbol{\nu}'(c) = \left\{\begin{array}{ll} 0 & \text{if } c \in R \\ \boldsymbol{\nu}(c) & \text{if } c \notin R \text{ and } \boldsymbol{\nu}(c) = \lfloor \frac{\mathsf{Max}^{\mathcal{A}}(c)}{\delta}\rfloor + 1 \\ \boldsymbol{\nu}(c) + 1 & \text{otherwise} \end{array}\right\}$
  (iv) $\boldsymbol{\nu}'\cdot\delta \models Inv(l')$.

Notice that Q involves only natural numbers. The size of $\mathcal{B}_{\mathcal{M}}$ is in $O(|Loc| \cdot (\lfloor \frac{\mathsf{Max}}{\delta}\rfloor + 2)^{|\mathbb{C}|})$, where $|Loc|$ and $|\mathbb{C}|$ are the size of Loc and the number of clocks, respectively, and $\mathsf{Max} = \max\{\mathsf{Max}^{\mathcal{A}}(c) \mid c \in \mathbb{C}\}$ is the maximal constant considered in all constraints and invariants of $\mathcal{M}$.

The Büchi automaton $\mathcal{B}_{\mathcal{M}}$ is equivalent to $\mathcal{M}$ in the following sense:

Theorem 17. *For all action sequences σ over Act, $\langle\sigma, \tau_\delta\rangle \in [\![\mathcal{M}]\!]$ iff the infinite sequence $\sigma(1)\sigma(2)\ldots$ is in the language of $\mathcal{B}_{\mathcal{M}}$.*

4 Consistency and Feasibility of Timed Machines

In this section, we investigate two important properties of DTIOMs as models of systems: consistency (in the sense that they generate a non-empty language) and feasibility (in the sense that they generate a non-empty language no matter what inputs they receive). We are especially interested in conditions under which consistency/feasibility are preserved by composition. This is because, in order to support run-time interconnections, one should be able to guarantee that a composition of DTIOMs is consistent/feasible *without* having to compose them.

4.1 Consistency

Definition 18 (Consistency). *A DTIOM $\mathcal{M}$ is said to be* consistent *if $\Lambda_{\mathcal{M}} \neq \emptyset$.*

Notice that consistency is preserved by refinement:

Proposition 19. *Let $k \in \mathbb{N}_{>0}$. A DTIOM $\mathcal{M}$ is consistent iff its k-refinement $\mathcal{M}_k$ is consistent. More generally, for arbitrary DTIOM $\mathcal{M}$ and $\mathcal{M}'$,*

- *if $\mathcal{M}' \preceq \mathcal{M}$ and $\mathcal{M}'$ is consistent, then so is $\mathcal{M}$, and*
- *if $\mathcal{M}' \precapprox \mathcal{M}$, then $\mathcal{M}'$ is consistent iff $\mathcal{M}$ is consistent.*

A sufficient condition for consistency is that the DTIOM is initializable and makes independent progress. A DTIOM is initializable if it can stay in the initial state until the first tick of the clock:

Definition 20 (Initializable). *A DTIOM $\mathcal{M}$ is said to* be initializable *if, for all $0 \leq t \leq \delta_{\mathcal{M}}$, $(q_0, t) \models Inv(q_0)$.*

For a machine to make independent progress (which we adapt from [7]), it needs to make a transition from any reachable state without forcing the environment to provide any input:

Definition 21 (Independent Progress). *A DTIOM $\mathcal{M}$ is said to* make independent progress *if, for every reachable state (l, v), there is an edge (l, A, C, R, l') such that: (a) $A \subseteq Act^O_{\mathcal{M}} \cup Act^{\tau}_{\mathcal{M}}$, (b) $v + \delta_{\mathcal{M}} \models C$, and (c) for all $0 \leq t \leq \delta_{\mathcal{M}}$, $(v + \delta_{\mathcal{M}})^{\mathbf{R}} + t \models Inv(l')$.*

As an example, both $\mathcal{M}^x$ and $\mathcal{M}^y$ as in Ex. 9 are initializable and make independent progress.

Proposition 22. *Any initializable DTIOM that makes independent progress is consistent.*

Notice that checking that a timed machine makes independent progress requires only the analysis of properties of its reachable states. In practice, this can be done using a syntactic check on the Büchi automaton as constructed in Sec. 3.3: a given DTIOM $\mathcal{M}$ makes independent progress iff all reachable states $(l, \boldsymbol{\nu})$ of the equivalent Büchi automaton $\mathcal{B}_{\mathcal{M}}$ have at least one outgoing transition $(l, \boldsymbol{\nu}) \xrightarrow{A} (l', \boldsymbol{\nu}')$ with $A \subseteq Act^O_{\mathcal{M}} \cup Act^{\tau}_{\mathcal{M}}$. $\mathcal{B}_{\mathcal{M}}$ has only finitely many states, denoted by $|\mathcal{B}_{\mathcal{M}}|$, and finitely many transitions, denoted by $|E_{\mathcal{M}}|$, and, hence, this can be checked in time $O(|\mathcal{B}_{\mathcal{M}}| \cdot |E_{\mathcal{M}}|)$.

4.2 Compositional Consistency Checking

In order to investigate conditions that can guarantee compositionality of consistency checking, we start by remarking that the fact that two DTIOMs $\mathcal{M}_1$ and $\mathcal{M}_2$ are such that δ_1 and δ_2 are commensurate simply means that we can find a clock granularity in which we can accommodate the transitions that the two DTIOMs perform: by itself, this does not ensure that the two DTIOMs can jointly execute their input/output synchronisation pairs. For example, if $\delta_1 = 2$ and $\delta_2 = 3$ and $\mathcal{M}_2$ only performs non-empty actions at odd multiples of 3, the two machines will not be able to agree on their input/output synchronisation pairs. For the DTIOMs to actually interact with each other it is necessary that their input/output synchronisation pairs can be performed on a common multiple of δ_1 and δ_2.

Definition 23 (Cooperative). *A DTIOM $\mathcal{M}$ is said to be* cooperative *in relation to $Q \subseteq Act_{\mathcal{M}}$ and a multiple δ of $\delta_{\mathcal{M}}$ if the following holds for every (l, v) reachable at a time T such that $(T + \delta_{\mathcal{M}})$ is not a multiple of δ:*

> *for every edge $(l, A, C, R, l') \in E_{\mathcal{M}}$ such that $v + \delta_{\mathcal{M}} \models C$ and $(v + \delta_{\mathcal{M}})^{\mathbf{R}} + t \models Inv_{\mathcal{M}}(l')$ for all $0 \le t \le \delta_{\mathcal{M}}$ — i.e., the machine makes a transition at a time that is not a multiple of δ — there exists an edge $(l, A \backslash Q, C', R', l'')$ such that $v + \delta_{\mathcal{M}} \models C'$; for all $0 \le t \le \delta_{\mathcal{M}}$, $(v + \delta_{\mathcal{M}})^{\mathbf{R}'} + t \models Inv_{\mathcal{M}}(l'')$ — i.e., the machine can make an alternative transition that does not perform any actions in Q.*

Essentially, being cooperative in relation to Q and δ means that the machine will not force transitions that perform actions in Q at times that are not multiples of δ. In practice, this can be verified using a syntactic check on the states of the equivalent Büchi automaton that can be reached with a number of transitions n such that $n + 1$ is not a multiple of $\delta/\delta_{\mathcal{M}}$. This can be done in time in $O(\frac{\delta}{\delta_{\mathcal{M}}} \cdot |\mathcal{B}_{\mathcal{M}}| \cdot |E_{\mathcal{M}}|^2)$, with $|\mathcal{B}_{\mathcal{M}}|$ the size of the Büchi automaton $\mathcal{B}_{\mathcal{M}}$ defined in Sec. 3.3.

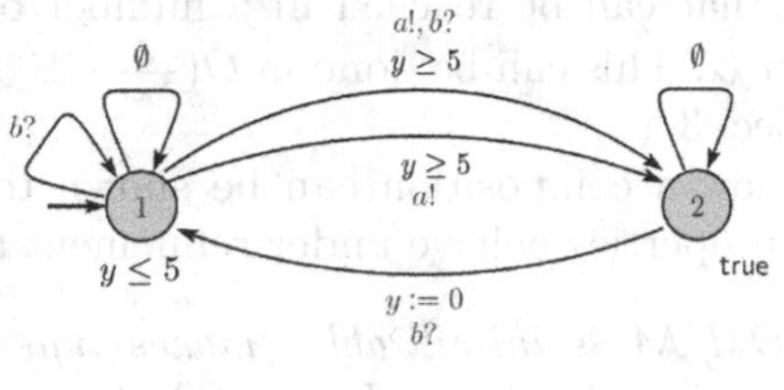

Fig. 4. The TIOA $\mathcal{A}'$

Example 24. *$\mathcal{M}^y$ from Ex. 9 is cooperative in relation to $\{a, b\}$ and $\delta = 2$. In constrast, the machine $\mathcal{M}'$ with $\delta' = 1$ and the TIOA $\mathcal{A}'$ presented in Fig. 4 is not cooperative in relation to $\{a, b\}$ and $\delta = 2$. Indeed, the fact that the state corresponding to the location 1 is reached at time 4 enables the transition $(1, a, y \ge 5, \emptyset, 2)$, which cannot be replaced by $(1, \emptyset, \mathsf{true}, \emptyset, 1)$ because the last condition — for all $0 \le t \le 1 = \delta_y$, $5 + t \le 5$ — is violated. Because the machine $\mathcal{M}'$ forces the output of* a *at time* 5*, it is easy to conclude that its composition with the machine $\mathcal{M}^x$ from Ex. 9 (which has a clock granularity $\delta^x = 2$) results in a inconsistent DTIOM.*

In relation to the composition of $\mathcal{M}_1$ and $\mathcal{M}_2$, the idea is to require that a common multiple of δ_1 and δ_2 exists such that both DTIOMs are cooperative in relation to $Act_{\mathcal{M}_1} \cap Act_{\mathcal{M}_2}$. However, this is not enough to guarantee that the two DTIOMs can actually work together: we need to ensure that if, say, $\mathcal{M}_1$ wants to output an action, $\mathcal{M}_2$ can accept it.

Definition 25 (DP-enabled). *A DTIOM $\mathcal{M}$ is said to be* DP-enabled *in relation to $J \subseteq Act^I_{\mathcal{M}}$ and δ multiple of $\delta_{\mathcal{M}}$ if the following property holds for every $B \subseteq J$ and state (l, v) reachable at a time T such that $(T + \delta_{\mathcal{M}})$ is a multiple of δ:*

for every edge $(l, A, C, R, l') \in E_{\mathcal{M}}$ *such that* $v + \delta_{\mathcal{M}} \models C$ *and, for all* $0 \leq t \leq \delta_{\mathcal{M}}$, $(v + \delta_{\mathcal{M}})^{\mathbf{R}} + t \models Inv_{\mathcal{M}}(l')$ *— i.e., the machine can make a transition — there exists an edge* $(l, B \cup (A \backslash J), C', R', l'')$ *such that* $v + \delta_{\mathcal{M}} \models C'$ *and, for all* $0 \leq t \leq \delta_{\mathcal{M}}$, $(v + \delta_{\mathcal{M}})^{\mathbf{R}'} + t \models Inv_{\mathcal{M}}(l'')$ *— i.e., the machine can make an alternative transition that accepts instead B as inputs and still performs the same outputs (and inputs outside J).*

That is, a DTIOM is DP-enabled in relation to a set of inputs J and a multiple δ of its clock granularity if, whenever it leaves a reachable state at a multiple of δ, it can do so by accepting any subset of J, and if its outputs are independent of the inputs in J that it receives. Both $\mathcal{M}^x$ and $\mathcal{M}^y$ from Ex. 9 are DP-enabled in relation to the set of input actions (resp., $\{a\}$ and $\{b\}$) and $\delta_x = 2$.

Notice that being DP-enabled is different from being input-enabled [14] in that we work with sets of actions (synchronisation sets), not just individual actions in the edges. Because, in our case, inputs and outputs can occur simultaneously, we need to ensure that there is no dependency between those that are included in the same synchronisation set.

DP-enabledness can be verified using a syntactic check on states of the equivalent Büchi automaton that can be reached in a number of steps n such that $n+1$ is a multiple of $\delta/\delta_{\mathcal{M}}$. This can be done in $O(\frac{\delta}{\delta_{\mathcal{M}}} \cdot |\mathcal{B}_{\mathcal{M}}| \cdot |E_{\mathcal{M}}|^2 \cdot 2^{|Act^I_{\mathcal{M}}|})$, with $|\mathcal{B}_{\mathcal{M}}|$ as given in Sec. 3.3.

We now investigate how a composition can be shown to be consistent. We start by analysing how properties behave under refinement and composition.

Lemma 26. *If a DTIOM* $\mathcal{M}$ *is initializable (makes independent progress, is DP-enabled / cooperative in relation to* J *and* δ'*), then so does* $\mathcal{M}_k$ *for all* $k \in \mathbb{N}_{>0}$.

That is, refinement preserves initializability, independent progress, DP-enabledness and cooperativeness.

Lemma 27. *Let* $\mathcal{M}_i = \langle \delta_i, \mathcal{A}_i \rangle$, $i = 1, 2$, *be two* δ*-compatible DTIOMs and* δ'_1 *a multiple of* δ_1.

(a) If $\mathcal{M}_1$ *and* $\mathcal{M}_2$ *are initializable so is* $\mathcal{M}_1 \parallel_\delta \mathcal{M}_2$.
(b) If $\mathcal{M}_1$ *is DP-enabled in relation to* $J \subseteq Act^I_1$ *and* δ'_1, *then* $\mathcal{M}_1 \parallel_\delta \mathcal{M}_2$ *is DP-enabled in relation to* $J \setminus Act^O_2$ *and* δ'_1.
(c) If $\mathcal{M}_1$ *is cooperative in relation to* $Q \subseteq Act^O_1 \setminus Act^I_2$ *and* δ'_1, *then* $\mathcal{M}_1 \parallel_\delta \mathcal{M}_2$ *is cooperative in relation to* Q *and* δ'_1.

Notice that in the preservation of DP-enabledness, we need to remove from J any actions that were used for synchronisation with $\mathcal{M}_2$, which are necessarily in Act^O_2. This is because they become internal to the composition and, therefore, are no longer available for synchronisation. The preservation of cooperativeness is relative to set of actions that are not used for synchronisation.

Theorem 28 (Compositionality). *Let* $\mathcal{M}_i$, $i = 1, 2$, *be two* δ*-compatible and initializable DTIOMs that can make independent progress. Let* δ' *be a common* *multiple of δ_1 and δ_2. If $\mathcal{M}_1$ is DP-enabled in relation to $Act_1^I \cap Act_2^O$ and δ', $\mathcal{M}_2$ is DP-enabled in relation to $Act_2^I \cap Act_1^O$ and δ', and both $\mathcal{M}_1$ and $\mathcal{M}_2$ are δ'-cooperative in relation to $Act_1 \cap Act_2$, then $\mathcal{M}_1 \parallel_\delta \mathcal{M}_2$ is initializable and can make independent progress (and, hence, by Prop. 22, is consistent).*

This result allows us to conclude that the machines $\mathcal{M}^x$ and $\mathcal{M}^y$ presented in Ex. 9 can work together (i.e., $\mathcal{M}^x \parallel \mathcal{M}^y$ is consistent). This is because, as noted before, $\mathcal{M}^x$ and $\mathcal{M}^y$ are DP-enabled in relation to $\delta' = 2$ and $\{a\}$ and $\{b\}$, respectively, and are cooperative in relation to $\{a, b\}$ and $\delta' = 2$.

Notice that, from Lemma 27, if $\mathcal{M}_i$ is DP-enabled in relation to $J \subseteq Act_i^I$ and δ_i' is a multiple of δ_i, the composition $\mathcal{M}_1 \parallel_\delta \mathcal{M}_2$ is DP-enabled in relation to $J \setminus Act_{\bar{i}}^O$ and δ_i', with $\bar{1} = 2$ and $\bar{2} = 1$. Moreover, if $\mathcal{M}_i$ is cooperative in relation to $Q \subseteq Act_i^O \setminus Act_{\bar{i}}^I$ and δ_i' multiple of δ_i, $\mathcal{M}_1 \parallel_\delta \mathcal{M}_2$ is cooperative in relation to Q and δ_i' (also a multiple of δ). This implies that, in order to ensure that the composition of $\mathcal{M}_1 \parallel_\delta \mathcal{M}_2$ with a third machine $\mathcal{M}_3$ (which can itself be the result of a composition) is consistent, we can verify the required properties (being initializable, DP-enabled and cooperative) over the component machines: we do not need to make checks over the machines resulting from the compositions (compositionality).

This result is also important to certify that the behaviour of a system of interacting components — $\Lambda = \Lambda_{\mathcal{M}_1}^{\iota_1} \cap \Lambda_{\mathcal{M}_2}^{\iota_2}$ in the case of two components that implement $\mathcal{M}_1$ and $\mathcal{M}_2$ — is not empty and, hence, the components can indeed operate together. This is because, by Theo. 16, $\Lambda_{\mathcal{M}_1 \parallel \mathcal{M}_2} \precapprox \Lambda$ and, hence, if $\mathcal{M}_1 \parallel \mathcal{M}_2$ is consistent, $\Lambda_{\mathcal{M}_1}^{\iota_1} \cap \Lambda_{\mathcal{M}_2}^{\iota_2}$ is not empty.

4.3 Feasibility

The property of being DP-enabled is related to a stronger notion of consistency called 'feasibility': whereas consistency guarantees the existence an execution, feasibility requires that, no matter what inputs the machine receives from its environment, it can produce an execution.

Definition 29 (Feasible). *A DTIOM $\mathcal{M}$ is said to be* feasible in relation to *$J \subseteq Act_{\mathcal{M}}^I$ and a multiple δ of $\delta_{\mathcal{M}}$ if, for every δ-timed trace λ over J and state (l, v) reachable at a time T such that $(T + \delta_{\mathcal{M}})$ is a multiple of δ, there is an execution starting at (l, v) that generates a $\delta_{\mathcal{M}}$-timed trace λ' such that $\lambda'|_J \preceq \lambda$, where $\lambda'|_J$ is the timed trace obtained from λ' by forgetting the elements in $Act_{\mathcal{M}} \setminus J$ from the underlying action sequence. A DTIOM $\mathcal{M}$ is said to be* feasible *if it is feasible in relation to $Act_{\mathcal{M}}^I$ and $\delta_{\mathcal{M}}$.*

This notion of feasibility is similar to the one use, for example, in [14], which we have relativised to given sets of input actions in order to account for structured interactions with the environment.

Proposition 30. *A DTIOM $\mathcal{M}$ that makes independent progress and is DP-enabled in relation to $J \subseteq Act_{\mathcal{M}}^I$ and a multiple δ of $\delta_{\mathcal{M}}$ is feasible in relation to J and δ.*

In relation to the compositionality of feasibility, we can prove:

Theorem 31. *Let $\mathcal{M}_i$, $i = 1, 2$, be two δ-compatible DTIOMs that can make independent progress. Let δ' be a common multiple of δ_1 and δ_2 and δ'_1 a multiple of δ_1 and $J \subseteq Act_1^I$. If (a) $\mathcal{M}_1$ is DP-enabled in relation to J and δ'_1, (b) $\mathcal{M}_1$ is DP-enabled in relation to $Act_1^I \cap Act_2^O$ and δ', (c) $\mathcal{M}_2$ is DP-enabled in relation to $Act_2^I \cap Act_1^O$ and δ', and (d) both $\mathcal{M}_1$ and $\mathcal{M}_2$ are δ'-cooperative in relation to $Act_1 \cap Act_2$, then $\mathcal{M}_1 \parallel_\delta \mathcal{M}_2$ is feasible in relation to $J \setminus Act_2^O$ and δ'_1.*

5 Related Work

Several researchers have recently addressed discrete timed systems with heterogeneous clock granularities. However, the main focus has been either on specification or on modelling and simulation, not so much on the challenges that heterogeneity raises on run-time interconnection of systems. For example, Forget et al. propose in [10] a synchronous data-flow language that supports the modelling of multi-periodic systems. In this setting, each system has its own discrete periodic clock granularity; composition is supported by a formal clock calculus that allows, in particular, for the refinement of clock granularities in a way that is similar to what we propose in Sec. 3. Aside from the fact that we adopt an automata-based representation, the main difference with our work is that they leave open the question of component-based verification of properties such as consistency.

Similarly, in [6], the authors introduce a formal communication model of behaviour for the composition of heterogeneous real-time cyber-physical systems based on logical clock constraints. Although this model supports the combination of heterogeneous timed systems, the authors do not consider the particular case of discrete periodic systems. In [17], the authors present a methodology (ForSyDe) for high-level modelling and refinement of heterogeneous embedded systems; whilst the semantics they propose, and the notion of clock-refinement they introduce, are similar in essence to ours, their main focus is again on modelling and simulation, whereas ours is on the structures that support compositional reasoning over properties of interconnected systems.

To cope with heterogenous time scales, several approaches to the specification of real-time systems, notably the Timebands Framework [5], have also adopted an explicit representation of time granularity. That framework, unlike others, does not require that all descriptions be transformed into the finest granularity.

Some attempts have also been made at addressing compositionality, for example in [15] that exploits the concept of tag machines [2]. However, the notion of composition of systems introduced by the authors (using tag morphisms) is more relaxed than ours in that it allows for the delay between events to be modified in given tag machines. A consequence of this generality is that the language resulting from a composition is not an approximation of the intersection of the original languages, which, as argued in our paper, is essential for addressing global properties of interconnected systems as implemented.

From a practical point of view, some tools have been developed for modelling and simulating heterogeneous systems. For example, Ptolemy Classic [4] introduced the concept of heterogeneous combinations of semantics such as asynchronous models with timed discrete-events models. The concept was picked up in other tools such as System C [12], Metropolis [11] and Ptolemy II [16]. The common characteristics of these tools is that (1) they are based on a model that is more general than the one we propose in this paper, and (2) they do not consider composition of discrete timed systems with different periodic clocks. As a consequence, they are not able to provide results as strong as ours when it comes to reasoning about specific global properties of interconnected systems.

6 Concluding Remarks

This paper proposes a new theoretical framework for the compositional design of timed heterogeneous systems based on an extension of timed input/output automata [14,7] where automata are assigned a clock granularity (what we call timed machines). Composition is thus extended to cater for automata that operate over different clock granularities.

One key aspect of our work is that we support the design of heterogeneous timed systems whose clock granularities can be made compatible without modifying the time domains of the individual components. This is important so that components can be interconnected at run time, not design time, which is essential for addressing the new generation of systems that are operating in cyberspace, where they need to be interconnected, on the fly, to other systems. Our approach is truly compositional in that we can obtain properties of a whole system of interconnected components without having to compute their composition.

The main properties that we address are consistency (there exists at least a joint trace on which all components can agree) and feasibility (there exists at least a joint trace on which all components can agree no matter what input they receive from their environment). The technical results that support compositional verification of consistency and feasibility are based on new notions of time refinement and of cooperation conditions through which timed components can be ensured to be open to interactions with other components across different time granularities.

There are two main directions for future work. The first is to implement and evaluate our approach on concrete case studies. A possibility would be to implement the framework as an extension of Ptolemy [4], which would give us access to industrial-size case studies. The second aims at extending our work to networks of heterogeneous timed systems that communicate asynchronously by building on [8] and [9].

Acknowledgments. This work was partially supported by the Royal Society International Exchange grant IE130154.

References

1. Abadi, M., Lamport, L.: The existence of refinement mappings. Theor. Comput. Sci. 82(2), 253–284 (1991)
2. Benveniste, A., Caillaud, B., Carloni, L.P., Sangiovanni-Vincentelli, A.L.: Tag machines. In: EMSOFT, pp. 255–263. ACM (2005)
3. Broy, M., Stølen, K.: Specification and Development of Interactive Systems: Focus on Streams, Interfaces, and Refinement. Springer-Verlag New York, Inc., Secaucus (2001)
4. Buck, J.T., Ha, S., Lee, E.A., Messerschmitt, D.G.: Ptolemy: A framework for simulating and prototyping heterogenous systems. Int. Journal in Computer Simulation 4(2), 155–182 (1994)
5. Burns, A., Hayes, I.J.: A timeband framework for modelling real-time systems. Real-Time Syst. 45(1-2), 106–142 (2010)
6. Chen, Y., Chen, Y., Madelaine, E.: Timed-pNets: A formal communication behavior model for real-time CPS system. In: Trustworthy Cyber-Physical Systems, Newcastle University. School of Computing Science Technical Report Series (2012)
7. David, A., Larsen, K.G., Legay, A., Nyman, U., Wasowski, A.: Timed I/O automata: a complete specification theory for real-time systems. In: HSCC, pp. 91–100. ACM (2010)
8. Delahaye, B., Fiadeiro, J.L., Legay, A., Lopes, A.: A timed component algebra for services. In: Beyer, D., Boreale, M. (eds.) FORTE 2013 and FMOODS 2013. LNCS, vol. 7892, pp. 242–257. Springer, Heidelberg (2013)
9. Fiadeiro, J.L., Lopes, A.: An interface theory for service-oriented design. Theor. Comput. Sci. 503, 1–30 (2013)
10. Forget, J., Boniol, F., Lesens, D., Pagetti, C.: A multi-periodic synchronous dataflow language. In: HASE, pp. 251–260. IEEE Computer Society (2008)
11. Gößler, G., Sangiovanni-Vincentelli, A.L.: Compositional modeling in metropolis. In: Sangiovanni-Vincentelli, A.L., Sifakis, J. (eds.) EMSOFT 2002. LNCS, vol. 2491, pp. 93–107. Springer, Heidelberg (2002)
12. Grötker, T.: System Design with SystemC. Springer (2002)
13. Henzinger, T.A., Manna, Z., Pnueli, A.: Timed transition systems. In: Huizing, C., de Bakker, J.W., Rozenberg, G., de Roever, W.-P. (eds.) REX 1991. LNCS, vol. 600, pp. 226–251. Springer, Heidelberg (1992)
14. Kaynar, D.K., Lynch, N., Segala, R., Vaandrager, F.: The Theory of Timed I/O Automata. Morgan & Claypool Publishers (2006)
15. Le, T.T.H., Passerone, R., Fahrenberg, U., Legay, A.: A tag contract framework for heterogeneous systems. In: Canal, C., Villari, M. (eds.) ESOCC 2013. CCIS, vol. 393, pp. 204–217. Springer, Heidelberg (2013)
16. Lee, E.A., Zheng, H.: Leveraging synchronous language principles for heterogeneous modeling and design of embedded systems. In: EMSOFT, pp. 114–123. ACM (2007)
17. Sander, I., Jantsch, A.: System modeling and transformational design refinement in ForSyDe [formal system design]. IEEE Trans. on CAD of Integrated Circuits and Systems 23(1), 17–32 (2004)

Refinement of Structured Interactive Systems*

Denisa Diaconescu[1], Luigia Petre[2], Kaisa Sere†, and Gheorghe Stefanescu[1]

[1] Department of Computer Science, University of Bucharest, Romania
{ddiaconescu,gheorghe.stefanescu}@fmi.unibuc.ro

[2] Department of Information Technologies, Åbo Akademi University, Finland
lpetre@abo.fi

Abstract. The refinement concept provides a formal tool for addressing the complexity of software-intensive systems, by verified stepwise development from an abstract specification towards an implementation. In this paper we propose a novel notion of refinement for a structured formalism dedicated to interactive systems, that combines a data-flow with a control-oriented approach. Our notion is based on *scenarios*, extending to two dimensions the trace-based definition for the refinement of classical sequential systems. We illustrate our refinement notion with a simple example and outline several extensions to include more sophisticated distributed techniques.

Keywords: scenario-based refinement, interactive systems, integration of data flow and control flow, coordination programming languages, trace semantics, stuttering equivalence.

1 Introduction

Using the *refinement* approach, a system can be described at different levels of *abstraction* [3], and the consistency in and between levels can be proved mathematically. Abstraction is a fundamental tool in addressing the ever increasing complexity of contemporary software-intensive systems and it becomes very attractive in the current context of various platforms (e.g., [2]) that allow the formal modelling and (semi-) automatic proving of refinement-based system development.

Interactive computation [32] is an important component of the software-intensive infrastructures of our society. Often, the term is related to HCI (Human-Computer Interaction), the particular case when one of the interacting entities is human. While able to deal with such cases as well, our approach here is process-to-process interaction oriented.

Classical models for process interaction include, among many other models, process algebra models [9], Petri nets [33], dataflow networks [11], etc. In these models, process synchronization is a key feature. For instance, in process algebra models, synchronization is achieved by handshake communications, while in

* Part of this research has been sponsored by the EU funded FP7 project 214158 (DEPLOY).

G. Ciobanu and D. Méry (Eds.): ICTAC 2014, LNCS 8687, pp. 133–150, 2014.

Petri nets and dataflow models, explicit transitions and dataflow nodes are respectively used. These models treats interaction as a primary feature, considering sequential computation to be either derived from communication or implicitly included in the dataflow node behaviour. A more recent proposal falling into this class is the ORC programming language [21], based on name-free processes and structured interaction.

In this paper we propose an extension of the notion of trace refinement [6] in a novel interaction-oriented formalism. The formalism is called *register-voice interactive systems* [27–29, 25, 17, 18] (*rv-IS*) and is a recent approach for developing software systems using both structural state-based as well as interaction-based composition operators. One of the most interesting feature of this formalism is the structuring of the component interactions.

Our work on refinement started from an attempt to get a bidirectional translation between the Event-B and rv-IS formalisms. Event-B [3] is a state-based formal method dedicated to the stepwise development of correct systems, extending the Action Systems [4, 31] and the B-Method [1] formalisms. A central advantage of Event-B is the associated Rodin tool platform [30, 2] employed in discharging the proof obligations that ensure this correct development. Event-B is currently successfully integrated in several industrial developments, for instance at Space Systems Finland [19] and at SAP [13].

Our approach for the integration of the Event-B and rv-IS formalisms is based on the following working plan: (1) define a notion of refinement for rv-IS models combining the refinement of state-based systems with Broy-like refinement of dataflow-based systems; (2) define/use refinement preserving translation between rv-IS and Event-B models; (3) use these translations, on one hand, to get tool support to develop and analyse rv-IS models and, on the other hand, to improve the discharging of proof obligations in Event-B using rv-IS structured decomposition techniques.

A key element in this approach is to find an appropriate definition of refinement for interactive rv-IS systems, which is the main topic of this paper.

Refinement can be defined in a multitude of ways, e.g., [1, 3, 5, 6, 10, 22]. The refinement definition in this paper is based on the idea of a trace-based refinement. Traces may be obtained by flattening (two-dimensional) scenarios used for describing the semantics of rv-IS programs, but they do not faithfully characterize the execution of these interactive programs. To address this problem, we first define *scenario*-equivalence to generalize the "up-to-stuttering trace equivalence" in two dimensions. Then, we propose a definition for scenario-based refinement of rv-IS systems.

The paper is organized as follows. In Section 2 we overview the rv-IS formalism to the extent needed in this paper. In Section 3 we tackle the scenario equivalence problem, in particular defining a scenario stuttering equivalence. In Section 4 we introduce the scenario-based notion of refinement for register-voice interactive systems and discuss the applicability of our approach. Section 5 concludes the paper.

	x=6				.	
.	IL	tx=6	IA	tx=6	IR	.
	x=3		y=6		z=6	
.	P	tx=3	D	tx=3	M	.
	x=2		y=6		z=3	
.	P	tx=2	D	tx=2	M	.
	x=1		y=6		z=1	
.	P	tx=1	D	tx=1	M	.
	x=0		y=6		z=0	
.	TL	tx=-1	TA	tx=-1	TR	.
	.		.		z=0	

Fig. 1. A scenario $S1$ for the **Perfect1** program

2 Register-Voice Interactive Systems

The rv-IS formalism is built on top of register machines, closing them with respect to a space-time duality transformation [28, 25, 17, 8]. In the following, we shortly overview the approach, following [14, 15].

Scenarios. A *scenario* is a two-dimensional rectangular area[1] filled in with identifiers and enriched with data around each identifier. In our interpretation the columns correspond to processes, the top-to-bottom order describing their progress in time. The left-to-right order corresponds to process interaction in a nonblocking message passing discipline. This means that a process sends a message to the right, then it resumes its execution. *(Memory) states* are placed at the north and at the south borders of the identifiers and *(interaction) classes* are placed at the west and at the est borders of the identifiers. In Fig. 1 we illustrate an rv-IS scenario for deciding whether the number 6 is a perfect number (i.e., it is equal to the sum of its proper divisors). More on these scenarios and an rv-IS program generating them are included in the last part of this section.

Spatio-Temporal Specifications. A spatio-temporal specification combines constraints on both spatial and temporal data. Spatial data are placed on north and south borders of the scenario cells, while temporal data on the west and east borders. The later data are pieces of streams recording the communication messages between processes.

For spatial data we use the common data structures and their natural representations in memory. For representing temporal data we use streams: a *stream* is a sequence of data ordered in time and is denoted as $a_0 \frown a_1 \frown \ldots$, where $a_0, a_1, \ldots$ are the data laying on the stream at time $0, 1, \ldots$, respectively. Most of the usual data structures have natural temporal representations. Voices are the temporal dual corresponding to registers [27, 28]. In the following, we will use spatial and temporal integer types, only.

[1] Scenarios of arbitrary shapes may be used, as well; see [7, 8].

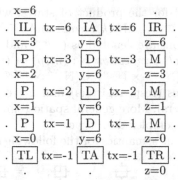

The notation $\otimes$ is used for the product of states, while $\frown$ for the product of classes; $\mathbb{N}^{\otimes k}$ denotes $\mathbb{N} \otimes \ldots \otimes \mathbb{N}$ (k terms) and $\mathbb{N}^{\frown k}$ denotes $\mathbb{N}^{\frown} \ldots {}^{\frown}\mathbb{N}$ (k terms); the "star" operations are denoted as $(\_^{\otimes})^*$ and $(\_^{\frown})^*$.

A simple *spatio-temporal specification* $S : (m, p) \to (n, q)$, using only integer types, is a relation $S \subseteq (\mathbb{N}^{\frown m} \times \mathbb{N}^{\otimes p}) \times (\mathbb{N}^{\frown n} \times \mathbb{N}^{\otimes q})$, where m (resp. p) is the number of input voices (resp. registers) and n (resp. q) is the number of output voices (resp. registers). More general spatio-temporal specifications may be introduced using more complex data types.

As an example, we give the semantics for the following simple wiring constants

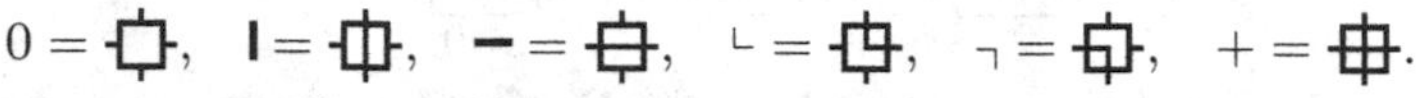

The natural relational interpretation of these constants is:

$0 = \emptyset$ (*empty cell*); $+ = \{(x, y, x, y) : x \in T\}$ (*double identity*, or *cross*)
I $= \{(., x, ., x) : x \in T\}$ (*vertical identity*);
— $= \{(x, ., x, .) : x \in T\}$ (*horizontal identity*);
└ $= \{(., x, x, .) : x \in T\}$ (*speaker*, or *space-to-time converter*);
┐ $= \{(x, ., ., x) : x \in T\}$ (*recorder*, or *time-to-space converter*).

where T is the used type of data and '.' denotes the empty data on a nil interface.

Syntax of structured rv-programs. The syntax of structured rv-programs is:

```
P ::= X | P % P | P # P | P $ P | if(C) then {P} else {P}
        | while_t(C) {P} | while_s(C) {P} | while_st(C) {P}
```

It uses modules X, if statements, vertical (or temporal) %, horizontal (or spatial) #, diagonal (or spatio-temporal) $ compositions and their iterated versions.

The starting blocks for the construction of structured rv-programs are called *modules*. The syntax of a module is given as follows:

```
module module_name
{listen temporal_variables}{read spatial_variables}{
  code
}{speak temporal_variables}{write spatial_variables}
```

The operations on structured rv-programs are briefly described below. More details and examples may be found in [27, 28, 17, 18].

The *type* of a *structured rv-program* $P : (w(P), n(P)) \to (e(P), s(P))$ collects the types at the west, north, east, and south borders of its scenarios.

1. **Composition:** Due to their two dimensional structure, programs may be composed horizontally and vertically, as long as their types agree. They can also be composed diagonally by mixing the horizontal and vertical composition.
   (a) For two programs $P_i : (w_i, n_i) \to (e_i, s_i)$, $i = 1, 2$, the *horizontal composition* $P_1 \# P_2$ is well defined only if $e_1 = w_2$; the type of the composite is $(w_1, n_1 \otimes n_2) \to (e_2, s_1 \otimes s_2)$.
   (b) Similarly, the *vertical composition* $P_1 \% P_2$ is defined only if $s_1 = n_2$; the type of the composite is $(w_1 {}^\frown w_2, n_1) \to (e_1 {}^\frown e_2, s_2)$.

   (c) The *diagonal composition* $P_1\$P_2$ is a derived operation - it connects the east border of P_1 to the west border of P_2 and the south border of P_1 to the north border of P_2; it is defined only if $e_1 = w_2$ and $s_1 = n_2$; the type of the composite is $(w_1, n_1) \to (e_2, s_2)$.
2. **If:** For the "if" operation, given two programs with the same type $P, Q : (w, n) \to (e, s)$, a new program **if(C) then** {P} **else** {Q} : $(w, n) \to (e, s)$ is constructed, for a condition C involving both, the temporal variables in w and the spatial variables in n.
3. **While:** There are three while statements, each being the iteration of the corresponding composition operation.
   (a) For a program $P : (w, n) \to (e, s)$, the statement `while_t(C){P}` is defined if $n = s$ and C is a condition on the variables in $w \cup n$. The type of the result is $((w^\frown)^*, n) \to ((e^\frown)^*, n)$.
   (b) The case of *spatial while* `while_s(C){P}` is similar.
   (c) If $P : (w, n) \to (e, s)$, the statement `while_st(C){P}` is defined if $w = e$ and $n = s$ and C is a condition on $w \cup n$. The type of the result is $(w, n) \to (e, s)$.

Derived statements. Many usual programming idioms may be naturally extended in this setting; e.g., temporal/spatial `for` statements, see [15].

Operational semantics of structured rv-programs. The operational semantics is given in terms of scenarios. Scenarios are built up with the following procedure:

1. Each cell of the scenario has as label a module name.
2. An area around a cell may have additional information. For example, if a cell has the information x = 2, then in that area x is updated to be 2.
3. The scenario is built from the current rv-program by reducing it to simple compositions of spatio-temporal specifications w.r.t. the syntax of the program, until we reach basic blocks, e.g. modules (see, e.g. [7] for more details on this relational semantics).

Example. We illustrate the operational semantics with the following structured rv-program **Perfect1** verifying if a number n is perfect - its modules are listed in Table 1:

```
(IL # IA # IR) % while_t(x > 0){P # D # M} % (TL # TA # TR)
```

In our rv-program we can imagine that we have three processes: one generates all the numbers in the set $\{n/2, \ldots, 1\}$ (module P), one checks if a number is a divisor of n (module D) and the last one updates a variable z (module M). Modules IL, IA and IR are used for initializations and TL, TA and TR for termination. At the end of the program, if the variable z is 0, then the number n is perfect.

The scenario for $n = 6$ is presented in Fig. 1. In the first line of the scenario we initialize the processes with the needed informations: module IL is reading the value $n = 6$ and provides the first process with $x = 3$ $(= n/2)$ and declare a temporal variant of n, namely $tn = 6$, that will be used by modules IA and IR for the other initializations; modules IA and IR use the temporal variable tn for initializing the other two processes with the initial value of n, namely $y = 6$, $z = 6$, respectively. In the next step, module P produces a temporal data $tx = 3$

Table 1. The modules of the **Perfect1** rv-program

`module IL`
`{listen nil}{read n}{`
`  tn:tInt; x:Int;`
`  tn = n; x = n/2;`
`}{speak tn}{write x}`
	`module IA`
	`{listen tn}{read nil}{`
	`  y:Int;`
	`  y = tn;`
	`}{speak tn}{write y}`
		`module IR`
		`{listen tn}{read nil}{`
		`  z:Int;`
		`  z = tn;`
		`}{speak nil}{write z}`

`module P`
`{listen nil}{read x}{`
`  tx:tInt;`
`  tx = x; x = x-1;`
`}{speak tx}{write x}`
	`module D`
	`{listen tx}{read y}{`
	`  if(y % tx !=0){`
	`    tx = 0;};`
	`}{speak tx}{write y}`
		`module M`
		`{listen tx}{read z}{`
		`  z = z - tx;`
		
		`}{speak nil}{write z}`

`module TL`
`{listen nil}{read x}{`
`  tx:tInt; tx = -1;`
`}{speak tx}{write nil}`
	`module TA`
	`{listen tx}{read y}{`
	
	`}{speak tx}{write nil}`
		`module TR`
		`{listen tx}{read z}{`
		
		`}{speak nil}{write z}`

and decrease x. Module `D` verifies if tx is a divisor of y and, if no, it resets the value of tx to 0. Finally, module `M` decreases the value of z by tx. We continue this steps until the variable x becomes 0. A final line contains terminating modules that rearrange some interfaces, keeping only the relevant result z.

3 Scenario Equivalence and Refinement

The notion of running scenario for an rv-program is a natural two-dimensional extension of the notion of running path (or trace) of a sequential program. The stuttering relation on traces, consisting in state repetition, is easy to understand. However, defining scenario equality up to a kind of "stuttering scenario equivalence", where memory state and interaction class repetitions are allowed, is more challenging. It is not only the case that memory states or interaction classes may stutter, but also more complex phenomena such as process migration have to be taken into account.

In this section we introduce a definition for scenario equivalence. This scenario equivalence obeys the following:

> *Correctness criterion for scenario equivalence:* If two scenarios $S1$ and $S2$ are equivalent, then the associated sets of traces $\flat(S1)$ and $\flat(S2)$, obtained using the flattening operator[2] $\flat$ [27], are stuttering equivalent.

This relation is later used to define scenario refinement between two scenarios $S1$ and $S2$; roughly speaking, this means the set of traces associated to $S1$ includes, up to the stuttering equivalence, the set of traces associated to $S2$.

[2] For a scenario S, $\flat(S)$ consists of all traces formed by the cells of S such that the occurrence of each cell in a trace is preceded in that trace by its left neighbour in the row and its top neighbour in the column.

Structural Dependencies. As a first step, we consider the dependence of the cells in a scenario induces by a set of "wiring constants".

The components of an atomic scenario cell $\mathtt{v} = \mathtt{A}\ \overset{\mathtt{s}}{\underset{\mathtt{t}}{\boxed{\mathtt{U}}}}\ \mathtt{B}$ are denoted as follows (using the north, west, east, south, and center positions):

$$\mathtt{v.n=s,\ v.w=A,\ v.e=B,\ v.s=t,\ v.c=\ U.}$$

Let E be a set of *wiring constants* (examples will be given below). For a scenario w, we build up an associated *E-dependencies graph* $D_E(w)$ representing its cells and their connections as follows:

1. The graph $D_E(w)$ has as nodes the inputs of w, the outputs of w, and the non-constants cells - the latter are the cells with labels not in the given set E of wiring constants. Each node C corresponding to a cell has four connecting ports $C.\alpha$ with $\alpha \in \{w, n, e, s\}$.
2. In $D_E(w)$ there is an edge between $C1.i$ and $C2.j$ if the i-th border of $C1$ is connected with the j-th border of $C2$ by a wire built up from a chain of constants in E.

An example is shown below:

$w =$
```
XY———
└+┐00
—+Z┐0  ;
0└++┐
00|└W
```

$D_E(w) : (2,2) \to (2,2)$ is the graph with 2 inputs on each north and west borders, 2 outputs on the east and south borders, and the edges: $\{(in1.n, X.n), (in2.n, Y.n), (in1.w, X.w), (in2.w, Z.w), (X.e, Y.w), (X.s, Z.n), (Y.s, W.n), (Z.e, W.w), (Y.e, out1.e), (W.e, out2.e), (Z.s, out1.s), (W.s, out2.s)\}$.

$D_E(w)$ is

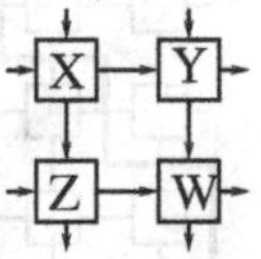

The constants used in this example are: empty cell, vertical, horizontal, cross, speaker and recorder identities - they are formally defined in Sec. 2. For instance, 4 crosses, a speaker and a recorder connect $Y.s$ with $W.n$ in the example above. We note that we cannot have the edge $(Y.s, Z.n)$ in the graph, because the cross identity only acts as a (double) skip statement. Similarly, the vertical and horizontal identities are skip statements, while speaker and recorder identities change the data representation from space to time and conversely.

Definition 1. Let E be a set of constant cells used for wiring. We say two scenarios v, w are $=^E_{struc}$*-equivalent* (or *E-structurally equivalent*) if the associated E-dependencies graphs $D_E(v)$ and $D_E(v)$ are isomorphic. □

The definition may be instantiated upon the set of specified connectors E, in particular for the sets *Connect1* and *Connect2* below.

The basic set *Connect1* consists of the (bijective) connectors in Table 2(a). In this table, the first column shows the symbols, the next the names, and the last the relational semantics described as (w, n, e, s) tuples, where $w/n/e/s$ are the values on the west/north/east/south borders. The extended set *Connect2* is *Connect1* completed with the "branching connectors" in Table 2(b).

Table 2. The set of constants Connect1 and Connect2

(a) The set of (bijective) wiring constants "Connect1"

—	*horizontal line* (the message passes);	$\{(x,.,x,.) : x \in T\}$
|	*vertical line* (the process stays active, doing nothing);	$\{(.,x,.,x) : x \in T\}$
+	*cross* (the message passes and the process does nothing);	$\{(x,y,x,y) : x \in T\}$
0	*empty cell* (the process is terminated and no message passes);	$\{(.,.,.,.) : x \in T\}$
└	*speaker* (the process passes its state as a message and then it terminates);	$\{(.,x,x,.) : x \in T\}$
┐	*recorder* (a terminated process grabs a message and it becomes active starting from the state specified by the message).	$\{(x,.,.,x) : x \in T\}$

(b) The set of additional branching wiring constants for "Connect2"

⊢	*active speaker* (the process passes its state as a message and it remains active);	$\{(.,x,x,x) : x \in T\}$
⊤	*transparent recorder* (a terminated process sees a message, then it becomes active starting from the state specified by the message, and it also lets the message to pass);	$\{(x,.,x,x) : x \in T\}$
⊥	*terminate a process*;	$\{(.,x,.,.) : x \in T\}$
⊣	*block a message*.	$\{(x,.,.,.) : x \in T\}$

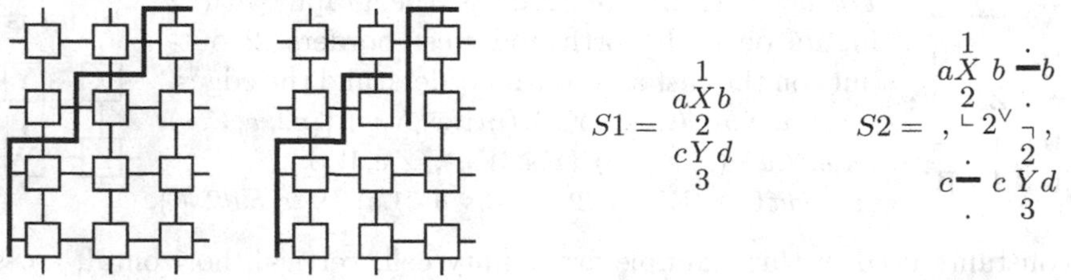

(a) Configuration (b) Empty interfaces (c) Detailed scenario (d) Process migration

Fig. 2. Scenarios and configurations

Getting traces from scenarios. In order to compare traces and scenarios, we recall that a scenario execution starts with the input data placed on the top and on the left borders and follows the execution of the scenario actions going top-to-down on columns and left-to-right on rows. A *sequential execution* $c_0 \xrightarrow{X_1} c_1 \xrightarrow{X_2} c_2 \xrightarrow{X_3} \ldots$ is one where no more than one cell is executed at any time. A *configuration* c of the system is a sequence of states and classes, *displayed from bottom-left to top-right*, that results after a number of steps in the execution of the scenario have already been applied. An example is shown in Fig. 2(a). A *running step* consists of the application of a scenario cell – it changes the current configuration to a new one. Projecting this sequential execution on the actions X_i one gets an *action trace* $X_1X_2X_3\ldots$; by projection on configurations c_i one gets a *class-and-state trace* $c_0c_1c_2\ldots$. The set of action traces associated to a scenario S is denoted by $\flat a(S)$, while the set of class-and-state traces by $\flat cs(S)$.

Example: Process migration. An example of structurally equivalent scenarios is shown in Fig. 2: the scenarios in (c) and (d) are *Connect1*-structurally equivalent.

For technical reasons, we distinguish here between the empty state interface '.' and the empty class interface ','. The recorder and speaker constants pass the information from time to space and conversely and it is convenient to denote by '$\_^{\vee}$' this transformation. E.g., $2^{\vee}$ denotes a class holding in a temporal form the spatial information of state 2. This transformation is an involution, i.e., $x^{\vee\vee} = x$, for all x.

There is only one sequential execution associated to $S1$

(1) $ca1 \xrightarrow{X} c2b \xrightarrow{Y} 3db.$

For $S2$, denoting by $\mathbf{-}_1$ and $\mathbf{-}_2$ the top-right and bottom-left occurrences of $\mathbf{-}$ in $S2$, there are 5 sequential executions

(2) $c, a1. \xrightarrow{X} c, 2b. \xrightarrow{\mathbf{-}_1} c, 2.b \xrightarrow{\llcorner} c.2^{\vee}.b \xrightarrow{\urcorner} c.2, b \xrightarrow{\mathbf{-}_2} .c2, b \xrightarrow{Y} .3d, b$

(3) $c, a1. \xrightarrow{X} c, 2b. \xrightarrow{\mathbf{-}_1} c, 2.b \xrightarrow{\llcorner} c.2^{\vee}.b \xrightarrow{\mathbf{-}_2} .c2^{\vee}.b \xrightarrow{\urcorner} .c2, b \xrightarrow{Y} .3d, b$

(4) $c, a1. \xrightarrow{X} c, 2b. \xrightarrow{\llcorner} c.2^{\vee}b. \xrightarrow{\mathbf{-}_1} c.2^{\vee}.b \xrightarrow{\urcorner} c.2, b \xrightarrow{\mathbf{-}_2} .c2, b \xrightarrow{Y} .3d, b$

(5) $c, a1. \xrightarrow{X} c, 2b. \xrightarrow{\llcorner} c.2^{\vee}b. \xrightarrow{\mathbf{-}_1} c.2^{\vee}.b \xrightarrow{\mathbf{-}_2} .c2^{\vee}.b \xrightarrow{\urcorner} .c2, b \xrightarrow{Y} .3d, b$

(6) $c, a1. \xrightarrow{X} c, 2b. \xrightarrow{\llcorner} c.2^{\vee}b. \xrightarrow{\mathbf{-}_2} .c2^{\vee}b. \xrightarrow{\mathbf{-}_1} .c2^{\vee}.b \xrightarrow{\urcorner} .c2, b \xrightarrow{Y} .3d, b.$

When comparing traces (2)-(6) and (1), the empty state and class interfaces (i.e., '.' and ',') are removed from configurations. Notice that the stuttering in the configurations may change a spatial representation of a value into a temporal representation or conversely. For instance, after removing the empty interfaces, (2) gives a class-and-state trace

$$(ca1)(c2b)(c2b)(c2^{\vee}b)(c2b)(c2b)(3db),$$

which is a stuttering of $(ca1)(c2b)(3db)$, this being the trace associated to (1).

The class-and-state trace associated to (1) and those resulting from the sequential executions described in (2)-(6) are the same up to the following transformations

(i) removing of empty interfaces '.' and ',';
(ii) swapping via $\_^{\vee}$ between spatial and temporal information;
(iii) repeating/removing information in a configuration; and
(iv) stuttering in a trace.

Example: Read-only variables. If the state variables of a module M are read-only used in the module code, then $M = \overset{\vdash\;\urcorner}{\underset{\perp\;|}{M\,+}}$. This type of property needs constants in the extended set *Connect2*. A similar analysis as in the previous example shows this transformation is correct.

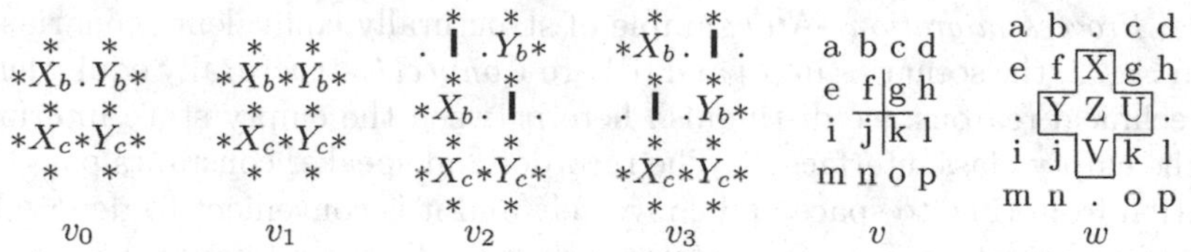

Fig. 3. Examples for scenario equivalence

Stuttering Equivalence. Next, we consider scenarios stuttering. This allows to insert or remove sub-scenarios of a scenario, provided the state and class values on the connecting interfaces of the sub-scenario with the remaining scenario are preserved.

Definition 2. Let v be a scenario and c a "hole" in v, i.e., a path via neighbouring states and classes with an empty contents. Let w be a scenario with the same interface as c. By replacing in v the hole c by w we get a scenario $v\{w/c\}$ considered to be in a *1-step stuttering relation* with v, denoted $v =_{1ss} v\{w/c\}$. The *scenario stuttering equivalence* $=^E_{ss}$ is the equivalence relation generated by the 1-step stuttering relation $=_{1ss}$ and $=^E_{struc}$. □

In fact, the stuttering equivalence $=^E_{ss}$ can be obtained as the symmetric, reflexive, and transitive closure of $=^E_{1ss} \cup =^E_{struc}$. If E is not specified, then it should be clear form the context if *Connect1* or *Conncet2* is used.

Example. We consider the scenarios v and w described in the pictures in Fig. 3 (for the sake of simplicity, the data around the cells are omitted). The "hole" in v consists of a circular line passing the lines inserted in the picture; one possible description, starting from the center, is: up-down-right-left-down-up-left-right. Suppose the cross represented by `X,Y,Z,U,V` in w is such that:

1. `X` and `V` have the data on the eastern borders equal to that of their western borders and, moreover, the northern interface of `X` and the southern interface of `V` are `nil`; and
2. `Y` and `U` have the data on the southern borders equal to that of their northern borders and the western interface of `Y` and the eastern interface of `U` are `nil`.

Then, w and v are in the 1-step stuttering relation, hence they are stuttering equivalent.

Scenario Refinement. We now consider the scenario comparison with respect to a kind of "structural extension" relation, i.e., preparing for the case of more internal causality structure in a refined model. For instance, a scenario in the refined model may have a dependency between two cells via a variable which is hidden in the abstract model. And more dependencies mean less traces.

Definition 3. We say $S2$ *E-structurally extends* $S1$, denoted $S1 >^E_{ref} S2$, if the dependencies graph $D_E(S2)$ includes $D_E(S1)$ and, moreover, $D_E(S2)$ has the same nodes and the same inputs and outputs as $D_E(S1)$.

We say S'' is an *E-refinement* of S', denoted $S' >^E_{struc} S''$ if there is a an alternating chain of $=^E_{struc}$ and $>^E_{ref}$ relations $S' =^E_{struc} S1 >^E_{ref} S2 =^E_{struc} S3 >^E_{ref} S4 \ldots Sn =^E_{struc} S''$ connecting them. □

In other words, while having the same non-constants cells (i.e., cells not in E) and the same inputs and outputs, the "refined" scenario S'' may have more internal connections than S'. Notice that $f =^E_{struc} g$ implies $f >^E_{ref} g$ and $g >^E_{ref} f$.

As an example, v_0 and v_1 in Fig. 3 satisfy $v_0 >^E_{ref} v_1$. Indeed, the dependency graph of v_1 includes the dependency graph of v_0.

Proposition 1. *Scenarios with more dependences have less class-and-state traces. Formally, if $S1 >^E_{struc} S2$, then, up to the described trace stuttering equivalence, $\flat cs(S1) \supseteq \flat cs(S2)$.*

Definition 4. As above, we can combine this relation $>^E_{ref}$ with the sub-scenario stuttering equivalence $=^E_{ss}$ getting a larger scenario refinement relation $>^E_{ss}$, called *E(ss)-refinement.* □

Section Conclusion. To summarize this section, two scenarios are in the *E(ss)-refinement* relation if and only if one only adds connecting interfaces where there were none and/or one adds local processing where there was no processing, communication or interfacing, while keeping the values on the existing interfaces unchanged. This relation obeys the correctness criterion that the refined scenario has less associated class-and-state traces. We employ this characterization of the scenario equivalence in the definition of refinement that we propose in the following section.

4 Refinement of Register-Voice Interactive Systems

The rv-IS model is a combination of state-based and interactive dataflow computations. The rv-IS refinement notion, to be defined below, has the following properties:

(1) By restriction to systems with no interaction classes it reduces itself to the usual refinement of classical state-based systems (a presentation of this type of refinement may be found in [3]);
(2) Similarly, by restriction to systems with no states it produces a refinement for interactive dataflow networks as the one used in [10].

For usual sequential systems there are several approaches to define refinement relations. A simple approach is to use traces [3, 6]: in terms of traces, except for some additional technical conditions, a concrete system C is a refinement of an abstract system A if the traces of C represent a subset of the traces of A, modulo a relation connecting the states of C to those of A.

In this section we present a notion of refinement of rv-IS systems in terms of associated scenarios. Scenarios represent an extension of traces to two dimensions, hence this approach directly lifts the former refinement definition to the level of rv-IS systems.

4.1 Refinement in State-Based Computing Systems

Refinement of classical sequential specifications/programs has a long history; see [1, 22, 10, 6, 5, 3], to mention just a few pointers to this active topic of research. We outline here the key features of the approach following Abrial's book [3].

Refinement may be defined using traces. We say a concrete system C is a *refinement* of an abstract system A if the following conditions hold:

(TI) - *trace inclusion*,
(SI) - *stuttering invariance*, and
(RDF) - *relative deadlock freedom*.

The basic constraint on refinement is (TI), stating that the traces of the refined system C are a subset of the traces of the abstract system A. This property is the cornerstone of the refinement method producing a strategy to develop correct-by-construction implementations. The approach starts with a general, abstract, and often non-deterministic, specification. Gradually, refined models of the system with less degree of non-determinism are produced till the very end when, hopefully, a deterministic and easy to implement model is obtained.

This straightforward relation has to be extended to cope with data refinement. To this end, trace equality is to be considered up to a *stuttering relation* (i.e., state repetition in the traces); this is condition (SI). A simple example is when a distinction makes sense between "internal" and "external" variables. The traces in the concrete system C may have details presented in terms of internal variables, while in the abstract system A one can only see the external ones. During the projection from C to A a sequence of states in a C trace may have no visible changes in terms of external variables, hence stuttering states in the associated A trace may occur.

Technically, one more condition is needed: (RDF). In terms of traces, it says that whenever a concrete trace is related to an abstract trace and the latter may be extended in the abstract system, then the former may be correspondingly extended in the concrete system. Its role is to avoid having deadlock in a state of C corresponding to a state of A with no deadlock at all.

4.2 Refinement in State- and Interaction-Based Systems

To get a notion of refinement in (state- and interaction-based) rv-IS sytems we combine the above notion of refinement with a technique used in [10] to define refinement of dataflow interactive systems. Compared with the independent state-based or interaction-based refinement, our combined approach here has a few advantages:

- it gives a better (structured, compositional) way to handle shared events or shared variables in classical systems using a dataflow-like interaction model;
- it increases the expressivity power of dataflow-like interaction systems by including complex, structural, state-based control mechanisms.

Refinement. The definition below of rv-IS systems refinement is a natural extension of the trace-based definition of state-based systems refinement. Stuttering equivalence on traces is replaced by the stuttering scenario equivalence introduced in the previous section.

Definition 5. Let E be a set of wiring constants. Given two rv-IS systems AIS and CIS, we say CIS is an *E-refinement* of AIS if the following conditions hold:

(TI+SI) The scenarios of CIS are a *subset* of the scenarios of AIS, under the following assumptions:
  1. The scenario equality is up to the *E-stuttering equivalence* relation defined in the previous section.
  2. The scenarios are *projected on classes and states*, hence the cells' labels do not matter.
  3. For comparison one uses a *gluing correspondence* between the state and the class variables of the CIS and the AIS systems.

(RDF) A scenario in CIS, corresponding to as abstract scenario in AIS, *can be correspondingly extended* in CIS, whenever the abstract one can be extended in AIS. □

We note the condition 2. above, that essentially shows that our refinement notion is focused on the trace, in terms of states and classes, of a system. We do not concern ourselves with the code of the modules, but instead with the data and messages flowing and transmitted throughout our model (together with the overall structure of our models). The meaning of the condition 3. above is the following. Suppose $G(var_A, var_C)$ is a relation connecting the variables in AIS and CIS. Then, from a scenario in CIS we get a scenario in AIS by replacing the values α of the variables var_C with values β of var_A such that $G(\beta, \alpha)$ holds true.[3]

4.3 A Refinement of the Perfect1 Program

We can refine the **Perfect1** program, presented in Section 2, when interpreting the process corresponding to the second column in Fig. 1 as a "divisibility checking service" agent: given a number and an input, it returns that number, if it is a divisor of the given input, otherwise zero. Then, we consider the whole system as consisting of a leader (the first process), a worker agent (the second process) and a result collecting process (the third process).

Based on this idea, in the following we refine the working agent by replacing it with a pool of `k` working agents to be used for repeated requests of the divisibility checking service. The program below implements a *round-robin distribution protocol* managed by the *turn* variable `t` (and its temporal version `tt`), i.e., the tasks are sent to the agents in the order `1,2,..,k,1,2...`.

[3] Often, the states and the classes of the AIS system are also present in the CIS system. In such a case, the correspondence mentioned in the definition is a simple projection. This means, by ignoring the new state and class variables used in the CIS scenarios one gets scenarios written with the AIS variables.

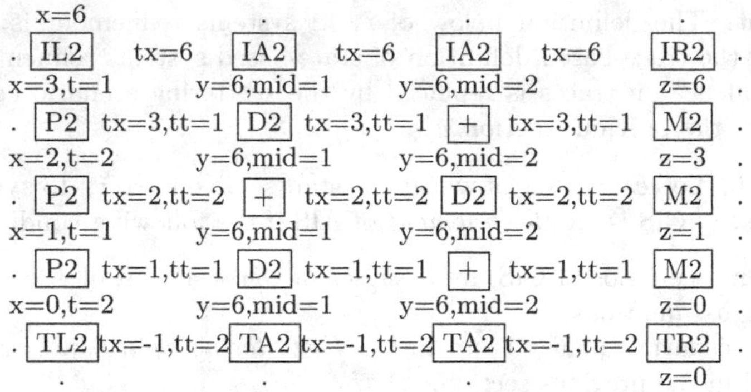

Fig. 4. A scenario $S2$ for **Perfect2** program ($k = 2$)

Table 3. The modules of the **Perfect2** program

`module IL2`
`{listen nil}{read n}{`
`tn:tInt; x,t:Int;`
`tn = n; x = n/2; t = 1;`
`}{speak tn}{write x,t}`
	`module IA2`
	`{listen tn,ti}{read nil}{`
	`y,mid:Int;`
	`y = tn; mid = ti;`
	`}{speak tn,ti}{write y,mid}`
		`module IR2`
		`{listen tn}{read nil}{`
		`z:Int;`
		`z = tn;`
		`}{speak nil}{write z}`

`module P2`
`{listen nil}{read x,t}{`
`tx,tt:tInt;`
`tx = x; tt = t; x = x-1;`
`t = t+1; if(t>k){t=1};`
`}{speak tx,tt}{write x,t}`
	`module D2`
	`{listen tx,tt,ti}{read y,mid}{`
	`if(y % tx !=0){`
	`tx = 0;`
	`};`
	`}{speak tx,tt,ti}{write y,mid}`
		`module M2`
		`{listen tx,tt}{read z}{`
		`z = z - tx;`
		`}{speak nil}{write z}`

`module TL2`
`{listen nil}{read x,t}{`
`tx,tt:tInt; tx = -1; tt = t;`
`}{speak tx,tt}{write nil}`
	`module TA2`
	`{listen tx,tt,ti}{read y,mid}{`
	`}{speak tx,tt,ti}{write nil}`
		`module TR2`
		`{listen tx,tt}{read z}{`
		`}{speak nil}{write z}`

The program **Perfect2** is

```
  (IL2 # for_s(tInt ti=1;ti=<k;ti++){IA2} # IR2)
% while_t(x > 0){(P2 # for_s(tInt ti=1;ti=<k;ti++){if(tt=mid){D2}} # M2)}
% (TL2 # for_s(tInt ti=1;ti=<k;ti++){TA2} # TR2)
```

The modules of **Perfect2** are shown in Table 3 and a scenario in Fig. 4.

Verifying refinement conditions. We consider the (typical) scenarios $S1$ in Fig. 1 and $S2$ in Fig. 4 of `Perfect1` and `Perfect2` programs, denoted $P1$ and $P2$. For $P2$, the particular case of two agents (i.e., `k = 2`) is presented in Fig. 4.

Proof idea: The scenarios $S1$ and $S2$ are $=_{ss}^{Connect2}$-equivalent, when considering the following transformations:

1. the module D uses its state variables in a read-only way, hence it can be replaced by $D+$;
2. we consider only the variables in $P1$, hence the new variables `mid,t,tt`, used in the refined model are ignored.

Table 4. Related `Perfect1` and `Perfect2` scenarios

$$
S1 = \begin{matrix} IL & IA & IR \\ P & D & M \\ P & D & M \\ P & D & M \\ TL & TA & TR \end{matrix} = \begin{matrix} IL & \top{-}{-}{-} & IR \\ & \vdash \quad . & \\ P & D+ {-}{-} & M \\ & \bot\vdash \quad & \\ P & {-}D+ {-} & M \\ & . \; . \; \vdash & \\ P & {-}{-}D+ & M \\ & . \; \bot \; \bot & \\ TL & {-}{-}{-} & TA\,TR \end{matrix}
\quad \text{and} \quad
S2' = \begin{matrix} IL & IA & IA & IR \\ P & D & + & M \\ P & + & D & M \\ P & D & + & M \\ TL & TA & TA & TR \end{matrix} = \begin{matrix} IL & \top{-} & \top & IR \\ & \vdash & & \\ P & D+ & + & M \\ & \bot & & \\ P & {-} & + & D\,M \\ & . & & \\ P & {-} & D & +\,M \\ & & & \\ TL & {-} & TA\,TA & TR \end{matrix}
$$

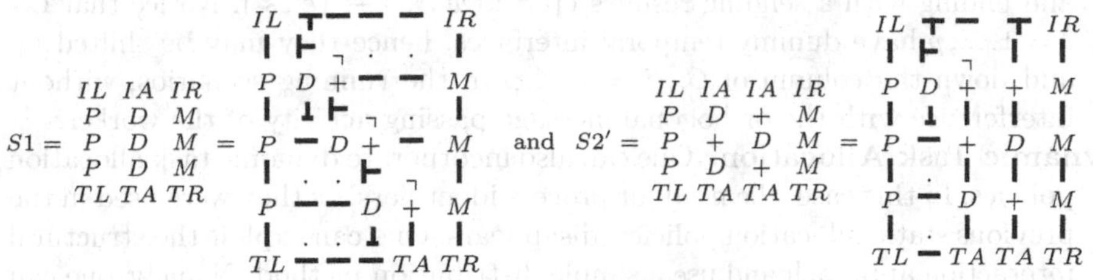

Indeed, after these transformations, the resulting scenarios have the same *Connect2*-dependency graph, so they are *Connect2*-stuttering equivalent.

The details are included below. Consider the scenarios in Table 4. In the transformation of $S1$, the module D is replaced by $D1 = D+$ and IA by $\top$. After these transformations, one can see that all instances of the D module have equal values for the state variables at their northern borders. Moreover, these values are a spatial version of the values on the eastern interface of IL.

The projection of $S2$ (in Fig. 4) on the variables in $P1$ is $S2'$ in Table 4. Indeed, via this projection, for an X in $\{IL, IA, IR, P, D, M, TL, TA, TR\}$ a cell $X2$ in $P2$ is identical with a corresponding cell X in $P1$, hence $S2'$ has the above format. Then, the occurrences of IA's are replaced by $\top$ and the first occurrence of D in the second process by $D1$.

After these transformations of $S1$ and $S2'$, all D's in the corresponding lines in $S1$ and $S2'$ have equal values at their interfaces, hence the *Connect2*-dependency graphs of $S1$ and $S2'$ are isomorphic. This concludes the check of the stuttering equivalence of $S1$ and $S2'$.

The proof above is generic and can be extended to all scenarios of $P1$ and $P2$. The last condition (RDF) in Definition 5 can be easily proved, hence $P2$ is a refinement of $P1$, indeed.

4.4 Refinement Strategies

The "divisibility task" used in `Perfect1`/`Perfect2` programs was chosen to illustrate the method. However, the method may be applied to similar problems for which a parallel execution may be mandatory for an efficient execution.

Other Static Distribution Policies. The round-robin method used for task distribution may be replaced by other static master/slave task distribution policies. The refinement proof is similar, with a slightly more complicated program and corresponding scenarios, resulting from the implementation of the chosen distribution method.

Task Refinement. As a further refinement we may consider the case when, within each worker, the task processing is a sequence $R; C_1; \ldots; C_p; S$ of refined activities: a receiving task step $R : (w, n) \to (e, s)$, followed by a

sequence of internal computations $C_1; \ldots; C_p$, with $C_i : (nil, n_i) \to (nil, s_i)$, and ending with a sending result step $S : (w', n') \to (e', s')$. Notice that C_i, $i = 1, ..., p$ have dummy temporal interfaces, hence they may be shifted up and down the column of C_i, $i = 1, ..., p$ in the running scenarios, without interference with the orthogonal message passing activity of the worker.

Dynamic Task Allocation. One can also incorporate dynamic task allocation policies. In this case, the need for process identifiers, as they were used in the previous static allocation policies, disappears. One can exploit the structural interaction approach and use a simple distribution method. Namely, one can repeat sending the task list via the interaction interfaces to all the workers ordered in a row; free workers will grab tasks from the list and busy workers will put tasks into the list.

Termination Detection Protocols. The dynamic task allocation described above requires more sophisticated termination detection protocols to be implemented and added to the design.

Load Balancing. When the jobs have quite a different time complexity, migrating jobs from a process to another may be worthwhile for getting a better overall execution. The migration process may be modelled as the one presented in Section 3, perhaps using the process names as in **Perfect2** program.

5 Conclusions

Refinement is a fundamental concept in computer science. It has been studied in numerous contexts, essentially starting from Dijkstra's work with weakest precondition calculus [16]. A thorough foundation of refinement calculus is presented in [5] by Back and von Wright, where its mathematical foundations are laid in terms of lattice theory as well as formalized in higher-order logic. Morgan [22] has promoted refinement as a practical method to programming problems, based on pre- and post-conditions as a contract between customers and programmers.

However, refinement is not the only method for establishing a relation between different versions (abstraction-wise) of a specification. For instance, in *pi*-calculus [20], the notion of weak or strong *bisimilarity* – a behavioral equivalence – is used to check that a specification and a concrete system correspond to each other. In Petri nets [24] one can analyze the current model, so that errors found are solved, resulting in a new model; this process is repeated iteratively, until the final model is correct with respect to the initial specification.

As can be seen, a stepwise iterative process is fundamental for software development, independently of the approach taken. As a concept, refinement proves its usability in current popular development methods for software, such as in Agile development [23]: small, iterative steps are taken, so that the next step always starts from the assumption that the previous one produced a system that agrees to some specification.

The main contribution of this paper consists in the introduction of a scenario-based definition of refinement in the register-voice interactive systems formalism.

There are several advantages brought by our approach. First, we include the classical state-base systems as a particular case of our model. In this way, we inherit a very rich set of results, techniques and applications. As a second strength, we note the strong mathematical foundation of the model, including natural operational and relational semantics, program transformations, Floyd-Hoare verification techniques, direct connections with mature mathematical fields such as logic, regular expressions[4], cellular automata, etc. Quite importantly, the increased modularity provided by the structural interaction paradigm has direct implications in the design of multicore, parallel, component, and service-oriented systems at a larger scale.

For our future research plans, directly related to the research presented in this paper, we plan to: (1) propose a concise definition of rv-IS refinement based on the rv-IS models themselves, not on the associated scenarios; (2) develop an algebra for representing rv-IS systems and check its compatibility with the proposed refinement; and (3) provide an automatic translation from rv-IS to Event-B and use Rodin tool platform for automatic reasoning on the correctness of the refinement steps.

References

1. Abrial, J.-R.: The B-Book: Assigning Programs to Meanings. Cambridge University Press (1996)
2. Abrial, J.-R., Butler, M., Hallerstede, S., Hoang, T.S., Mehta, F., Voisin, L.: Rodin: An Open Toolset for Modelling and Reasoning in Event-B. International Journal on Software Tools for Technology Transfer 6, 447–466 (2010)
3. Abrial, J.-R.: Modeling in Event-B: System and Software Design. Cambridge University Press (2010)
4. Back, R.J., Kurki-Suonio, R.: Decentralization of process nets with centralized control. In: PODC 1983, pp. 131-142. ACM (1983)
5. Back, R.J., Wright, J.V.: Refinement Calculus: A Systematic Introduction. Springer, Heidelberg (1998)
6. Back, R.J., Wright, J.V.: Trace Refinement of Action Systems. In: Jonsson, B., Parrow, J. (eds.) CONCUR 1994. LNCS, vol. 836, pp. 367–384. Springer, Heidelberg (1994)
7. Banu-Demergian, I.T., Paduraru, C.I., Stefanescu, G.: A new representation of two-dimensional patterns and applications to interactive programming. In: Arbab, F., Sirjani, M. (eds.) FSEN 2013. LNCS, vol. 8161, pp. 183–198. Springer, Heidelberg (2013)
8. Banu-Demergian, I.T., Stefanescu, G.: Towards a formal representation of interactive systems. Fundamenta Informaticae 131, 313–336 (2014)
9. Bergstra, J.A., Ponse, A., Smolka, S.A. (eds.): Handbook of Process Algebra. Elsevier (2001)

[4] Recently, a new approach for getting two-dimensional regular expressions have been introduced in [7, 8]. They use scenarios of arbitrary shape and powerful control mechanisms on scenario compositions. It seems these arbitrary shape scenarios are mandatory for an appropriate modelling of spatial and temporal pointers in Agapia/rv-IS systems.

10. Broy, M.: Compositional refinement of interactive systems. Journal of the ACM 44, 850–891 (1997)
11. Broy, M., Olderog, E.R.: Trace-oriented models of concurrency. In: [9], pp. 101–196
12. Broy, M., Stefanescu, G.: The algebra of stream processing functions. Theoretical Compututer Science 258, 99–129 (2001)
13. Bryans, J.W., Wei, W.: Formal Analysis of BPMN Models Using Event-B. In: Kowalewski, S., Roveri, M. (eds.) FMICS 2010. LNCS, vol. 6371, pp. 33–49. Springer, Heidelberg (2010)
14. Diaconescu, D., Leustean, I., Petre, L., Sere, K., Stefanescu, G.: Refinement Preserving Translation from Event-B to Register-Voice Interactive Systems. TUCS Technical Reports No. 1028 (December 2011), `http://tucs.fi/publications/view/?pub$_$id=tDiLePeSeSt11a`
15. Diaconescu, D., Leustean, I., Petre, L., Sere, K., Stefanescu, G.: Refinement-Preserving Translation from Event-B to Register-Voice Interactive Systems. In: Derrick, J., Gnesi, S., Latella, D., Treharne, H. (eds.) IFM 2012. LNCS, vol. 7321, pp. 221–236. Springer, Heidelberg (2012)
16. Dijkstra, E.W.: A Discipline of Programming. Prentice-Hall International (1976)
17. Dragoi, C., Stefanescu, G.: AGAPIA v0.1: A programming language for interactive systems and its typing systems. In: FINCO 2007. ENTCS, vol. 203, pp. 69–94 (2008)
18. Dragoi, C., Stefanescu, G.: On compiling structured interactive programs with registers and voices. In: Geffert, V., Karhumäki, J., Bertoni, A., Preneel, B., Návrat, P., Bieliková, M. (eds.) SOFSEM 2008. LNCS, vol. 4910, pp. 259–270. Springer, Heidelberg (2008)
19. Salehi Fathabadi, A., Rezazadeh, A., Butler, M.: Applying Atomicity and Model Decomposition to a Space Craft System in Event-B. In: Bobaru, M., Havelund, K., Holzmann, G.J., Joshi, R. (eds.) NFM 2011. LNCS, vol. 6617, pp. 328–342. Springer, Heidelberg (2011)
20. Milner, R., Parrow, J., Walker, D.: A Calculus of Mobile Processes I and II. Information and Computation 100, 1–77 (1992)
21. Misra, J., Cook, W.R.: Computation Orchestration. Software and System Modeling 6, 83–110 (2007)
22. Morgan, C.: Programming from Specifications. Prentice-Hall (1998)
23. Olszewski, M., Back, R.-J.: Agile Development with Stepwise Feature Introduction. In: ENASE 2012, pp. 161–166 (2012)
24. Petri, C.A., Reisig, W.: Petri net. Scholarpedia 3(4), 6477
25. Popa, A., Sofronia, A., Stefanescu, G.: High-level structured interactive programs with registers and voices. Journal of Universal Computer Science 13, 1722–1754 (2007)
26. Stefanescu, G.: Network algebra. Springer (2000)
27. Stefanescu, G.: Interactive systems with registers and voices. Draft, School of Computing, National University of Singapore (July 2004)
28. Stefanescu, G.: Interactive systems with registers and voices. Fundamenta Informaticae 73, 285–306 (2006)
29. Stefanescu, G.: Towards a Floyd logic for interactive rv-systems. In: ICCP 2006, Technical University of Cluj-Napoca, pp. 169–178 (2006)
30. RODIN tool platform, `http://www.event-b.org/platform.html`
31. Walden, M., Sere, K.: Reasoning About Action Systems Using the B-Method. Formal Methods in Systems Design 13, 5–35 (1998)
32. Wegner, P.: Interactive foundations of computing. Theoretical Computer Science 192, 315–351 (1998)
33. URL, `http://www.petrinets.info/`

Reasoning Algebraically About Refinement on TSO Architectures

Brijesh Dongol[1], John Derrick[1], and Graeme Smith[2]

[1] Department of Computing, University of Sheffield, UK
[2] School of Information Technology and Electrical Engineering, The University of Queensland, Australia

Abstract. The Total Store Order memory model is widely implemented by modern multicore architectures such as x86, where local buffers are used for optimisation, allowing limited forms of instruction reordering. The presence of buffers and hardware-controlled buffer flushes increases the level of non-determinism from the level specified by a program, complicating the already difficult task of concurrent programming. This paper presents a new notion of refinement for weak memory models, based on the observation that pending writes to a process' local variables may be treated as if the effect of the update has already occurred in shared memory. We develop an interval-based model with algebraic rules for various programming constructs. In this framework, several decomposition rules for our new notion of refinement are developed. We apply our approach to verify the spinlock algorithm from the literature.

1 Introduction

Logics for reasoning about concurrency in shared memory systems are based on the assumption that hardware is *sequentially consistent* [18], guaranteeing that instructions within each process are never executed out of order in memory. However, modern processors have abandoned sequential consistency in favour of weaker memory guarantees, using local buffers to offer greater scope for optimisation. There are several different weak memory models [1,2,23]; in this paper, we focus on the most restricted of these: the *Total Store Order* (TSO) memory model, which is implemented by architectures such as x86 (see Fig. 1). Under TSO, instead of committing writes immediately to main memory, the process executing the write stores it as a pending write in its local buffer. Pending writes are not visible to other processors until they are *flushed*, which commits the write to shared memory. A flush is either programmer controlled (via instructions such as `fence` or `lock`) or hardware controlled. Programmer-controlled flushes are ultimately expensive (and inefficient), hence, one would like to keep these to a minimum. On the other hand, reasoning about hardware-controlled flushes is difficult due to the increase in non-determinism of a program's behaviour.

Several approaches to program verification under TSO have been developed; we provide a brief survey. Researchers have considered direct methods, such as executable memory models [22], theorems for reduction [9], and identification of race conditions [20]. Others have linked programs under TSO executions to an abstract specification using linearizability [7,16], however, these use abstract specifications different from the

G. Ciobanu and D. Méry (Eds.): ICTAC 2014, LNCS 8687, pp. 151–168, 2014.

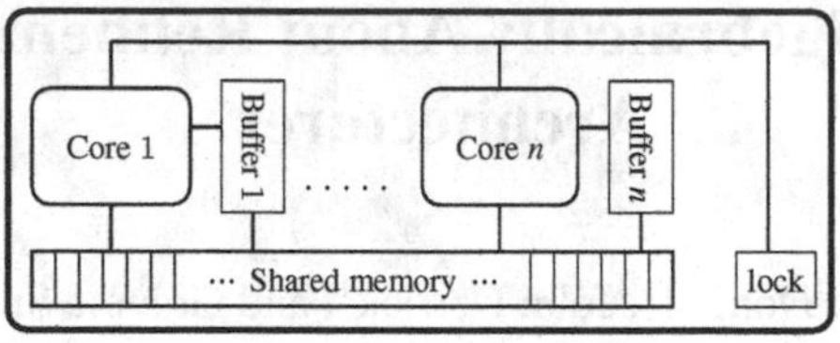

Fig. 1. TSO hardware overview

natural abstractions one would expect; [7] requires buffers to be present in the abstract specification, while [16] uses a non-deterministic abstract specification.

An issue with many existing approaches is that program semantics is given at a low level of abstraction of individual read and writes, which means programs must be understood and analysed using a verbose representation. Our work is based on the desire to lift reasoning to higher levels of abstraction [15], which in turn improves scalability. To this end, we develop an interval-based semantics by adapting Interval Temporal Logic [19]. Such an approach has two distinct advantages: (a) it allows one to define *truly concurrent* executions [10,11], providing a more accurate model of TSO-based hardware; and (b) it is amenable to algebraic reasoning [3,13], which enables one to develop *algebraic laws* for syntactically manipulating formulas representing program behaviour. In this paper, we develop algebraic rules to verify *refinement* between a concrete program and its abstract representation. The development of algebraic laws is non-trivial. However, once available, they provide high-level reusable theories for verification. We do not claim to have a complete set of laws (this is a topic of future work), instead, we provide a set of rules that are required for proving the spinlock example we verify.

Within this interval-based logic, we develop a framework for reasoning on TSO, simplifying our existing semantics [15] and introducing enhancements specifically designed for reasoning about buffer-based programs. This includes a simplified permission framework (Section 3.1), a novel methodology for evaluating expressions in the presence of local buffers (Section 4.3) and a novel notion of *local buffer refinement* (Section 5.2). Local buffer refinement is based on the observation that: To show a command C refines another command A with respect to a process p, the pending writes to local variables of p may be treated as if they have already taken effect in A. Thus, local updates at the concrete level may be treated as if they occur in their program order (without waiting for their flush to occur). This benefits verification because the non-determinism from flushes of local variables is resolved earlier. We develop a number of algebraic transformation laws for both refinement and local buffer refinement.

2 Background

2.1 Total Store Order Example

Total Store Order (TSO) memory allows a process to store a write in its local buffer and continue processing without waiting for this write to be commited to memory (i.e., while the write is pending). The values in the buffer are flushed in a FIFO order. To see

```
word x = 1;
  void acquire() {
a1  while(1) {
a2    lock;
a3    if (x = 1) {
a4      x := 0;
a5      unlock;
a6      return }
a7    unlock;
a8    while(x = 0);
  } }

  void release() {
r1  x := 1; }

  bool tryacquire() {
t1  lock;
t2  if (x = 1) {
t3    x := 0;
t4    unlock;
t5    return true }
t6  unlock;
t7  return false; }
```

Fig. 2. Spinlock algorithm

```
Noncritical section ;
acquire();
Critical section ;
release();
```

Fig. 3. Spinlock client (a)

```
Noncritical section;
if tryacquire() {
  Critical section ;
  release() ;
}
```

Fig. 4. Spinlock client (b)

the effect of this, consider the following classic example with processes p and q that modify shared variables x and y, which are initialised to 0. In this paper, we assume maximum parallelism and that each thread resides in exactly one core, therefore, the words *process* and *core* are used synonymously.

```
word x=0, y=0;
p { p1:  x := 1 ;          q { q1:  y := 1 ;
    p2:  r1 := y }             q2:  r2 := x }
```

Under sequentially consistent memory, at the end of execution, at least one of `r1` or `r2` would have a value 1. However, in TSO memory, it is possible to end execution so that both `r1` and `r2` read the original values of x and y, i.e., both `r1` and `r2` are 0 at termination. One such execution is ⟨`p1`, `p2`, `q1`, `q2`, `flush(p)`, `flush(p)`, `flush(q)`, `flush(q)`⟩, where `flush(p)` denotes a (hardware-controlled) flush event for process *p*. The write to x at `p1` is not seen by process q until p's buffer is flushed, and symmetrically for the write to y at `q1`. Hence, it is possible for q to read a value 0 for x at `q2` even though `q2` is executed after `p1`.

In addition to the above behaviour, each TSO process reads pending writes from its own buffer if possible, and hence, may obtain values that are not yet globally visible to other processes, e.g., if p2 is replaced with `r1 := x`, process p would read `x = 1` even if the write to x is pending. If there are multiple pending writes to the same location, then the write value corresponding to the last pending write is returned.

2.2 Case Study: Spinlock

Spinlock [4] is a locking mechanism designed to avoid operating system overhead associated with process scheduling and context switching. A typical implementation of spinlock is shown in Fig. 2, where a global variable x represents the lock and is set to `0` when the lock is held by a process, and 1 otherwise. The lock x is itself acquired using a secondary hardware lock (see Fig. 1), and this hardware lock is acquired/released using `lock`/`unlock` instructions. A process trying to acquire the lock x *spins*, i.e., waits in a loop and repeatedly checks the lock for availability.

Operation `acquire` only terminates if it successfully obtains the lock x. It will first lock the hardware so that no other process can access `x`. If, another process has already acquired x (i.e., `x = 0`) then it will release the hardware lock at `a7` and spin at `a8`, i.e., loop in the while-loop until x becomes free, before starting over from `a2`. Otherwise, it acquires the lock at `a4` by setting `x` to `0`, releases the hardware lock at `a5` and returns at `a6`. The operation `release` releases the lock by setting `x` to 1. The operation `tryacquire` is similar to `acquire`, but unlike `acquire` it only makes one attempt to acquire the lock. If this is successful it returns `true`, otherwise it returns `false`. Under TSO, a process p executing an assignment (e.g., `x := 0`) places a pending write in p's local buffer, which is not visible to other processes until the buffer is flushed.

We refer to processes that use spinlock to provide mutual exclusion to a critical section of code as its *clients*. Here, as in [22], we assume that clients of the spinlock behave either as the program in Fig. 3 or Fig. 4. Thus, one can assume that a client only calls a `release` operation when it holds the lock.[1] Note however, that the behaviours in Fig. 3 and Fig. 4 are not exhaustive. To admit other behaviours, one may formalise the additional client code, then apply our proof methods in this paper to verify this additional behaviour.

Clients can ensure mutual exclusion in the critical sections if in place of `acquire`, `release` and `tryacquire`, they use *abstract operations* `AAcq`, `ARel` and `ATry` below, respectively, which do not use buffers. We will refer to such clients as *abstract clients*.

```
word x = 1;

void AAcq() {            void ARel() {          bool ATry() {
  await (x = 1) {          x <== 1                if CAS(x, 1, 0) {
    x <== 0              }                          return true
} }                                               else return false } }
```

Here, statement `await` denotes a blocking atomic test-and-set statement, e.g., `AAcq()` can only execute if `x = 1` holds, and its execution atomically sets the value of `x` to `0`. Unlike the concrete program in Fig. 2, all reads and writes occur directly with main memory; we use assignments of the form `x <== 0` (which directly updates the value of x in memory) to distinguish this difference. If `x = 0`, then `AAcq()` blocks and cannot execute further until its guard `x = 1` is set to true by another process. Operation `ATry()` attempts to update `x` to `0` using a non-blocking atomic compare-and-swap operation `CAS`, and returns 1 if the operation is successful and `0`, otherwise.[2]

Our notion of correctness of the spinlock will be to show that every possible execution of a spinlock client is a possible execution of an abstract client. To this end, we prove *refinement* between the behaviour of the two executions (see Section 5.3). Proving refinement under TSO is difficult; one must not only verify concurrency effects, but additionally consider the effect of accessing the buffer during a program's execution. Furthermore, the level of atomicity at which these effects are visible is fine-grained, occuring at the level of individual reads and writes. This paper develops a high-level approach

[1] Such restrictions on client behaviour must be made due to the simplicity of the spinlock algorithm in Fig. 2. Arbitrary client behaviour e.g., two consecutive calls to `release` without acquiring the lock will result in incorrect behaviour under TSO.

[2] `CAS(a, b, c)` is equivalent to `atomic { if a = b then a := c ; return true else return false}`.

for proving refinement that avoids the need to consider low-level (fine-grained) effects whenever possible by developing an interval-based semantics for programs under TSO. This allows one to view the concurrent execution of two processes as the conjunction of their behaviours over an interval (as opposed to an interleaving of their traces), reducing the impact of non-determinism due to concurrency.

3 Interval-Based Reasoning

3.1 Permission Monitoring

Using an interleaved execution semantics, one can guarantee that a variable will not be simultaneously written, or read and written as part of the same transition. This is not true for shared-memory true concurrency, where one must model variable access by the different processes (e.g., two processes simultaneousy modifying variable **x** in Fig. 2).

Our solution is to explicitly define read/write permissions. To this end, we assume that programs are executed by processes from a set *Proc*; each process represents a concurrent thread which modifies a set of variables from a set *Var*. The TSO architecture uses sophisticated coherence protocols to provide an illusion of shared memory. One may assume the following about read and write instructions:

- Two simultaneous writes (by different processes) to the same variable do not occur.
- A simultaneous read and write of the same variable does not occur.
- A process never has access permission to the local variable of another process.

As we shall see in Sections 4.2 and 4.4, permissions also provide a convenient mechanism for formalising the effect of a `lock-unlock` block.

In previous work [11], we have modelled permissions using a fractional encoding (inspired by [5]). Here, we simplify these general notions and define the *permission space* as $Perm \mathrel{\widehat{=}} Proc \to Var \to \mathbb{P}\{\mathsf{wr}, \mathsf{rd}\}$, where wr and rd denote write and read permission, respectively. Using '.' for function application, given $\pi \in Perm$, we interpret $\mathsf{wr} \in \pi.p.v$ (resp., $\mathsf{rd} \in \pi.p.v$) as $p \in Proc$ has permission to write to (resp., read the value of) $v \in Var$.

A system at any time is described by a state of type $State \mathrel{\widehat{=}} Var \to Val$, where *Val* is the set of values. The system over time is formalised by a *stream*, which is a total function of type $Stream \mathrel{\widehat{=}} \mathbb{Z} \to (State \times Perm)$. Therefore, for each time in $\mathbb{Z}$, a stream formalises the state of the system and the permissions for each process and variable. Properties of a system are given by *predicates*; a predicate of type T is a member of $\mathcal{P}T \mathrel{\widehat{=}} T \to \mathbb{B}$, e.g., $\mathcal{P}State$, $\mathcal{P}Stream$, and $\mathcal{P}Perm$ are state, stream, and permission predicates, respectively. We assume pointwise lifting of boolean operators over predicates in the normal manner.

If $\pi \in Perm$, then $\mathcal{W}.p.v.\pi \mathrel{\widehat{=}} (\mathsf{wr} \in \pi.p.v)$, $\mathcal{R}.p.v.\pi \mathrel{\widehat{=}} (\mathsf{rd} \in \pi.p.v)$ and $\mathcal{N}.p.v.\pi \mathrel{\widehat{=}} (\pi.p.v = \varnothing)$ denote permission predicates that hold iff process p has write, read or no access to v in the permission space π, respectively.

Example 1. Suppose $Var = \{u, v\}$, $Proc = \{p, q\}$ and

$$\pi \mathrel{\widehat{=}} \{p \mapsto \{u \mapsto \{\mathsf{rd}, \mathsf{wr}\}, v \mapsto \{\mathsf{rd}\}\}, q \mapsto \{u \mapsto \varnothing, v \mapsto \{\mathsf{rd}\}\}\}$$

Then, $\mathcal{W}.p.u.\pi$ (p has write permission to u), $(\mathcal{R}.p \wedge \mathcal{R}.q).v.\pi$ (both p and q have read permission to v) and $\mathcal{N}.q.u.\pi$ (q has no permission to u) in space π. Note that due to pointwise lifting $(\mathcal{R}.p \wedge \mathcal{R}.q).v.\pi = \mathcal{R}.p.v.\pi \wedge \mathcal{R}.q.v.\pi$. □

The assumptions on reads and writes above are then taken into account by assuming that each *valid* permission space π satisfies the following, where p and q such that $p \neq q$ are processes, v is a variable, and u_p is a local variable of process p.

$$(\mathcal{W}.p.v.\pi \Rightarrow \mathcal{N}.q.v.\pi) \wedge (\mathcal{R}.p.v.\pi \Rightarrow \neg\mathcal{W}.q.v.\pi) \wedge \pi.p.u_p = \{\mathsf{wr}, \mathsf{rd}\} \qquad (1)$$

Note that the third conjunct combined with the first ensures that q does not have read nor write permission to any local variable of p. This is lifted to the level of streams by defining a *valid* stream to be one in which each $s.t$ is valid for $t \in \mathbb{Z}$. For the rest of the paper, we assume each stream is valid.

To simplify the notation, for a state predicate b and permission predicate z, we assume $b.(\sigma, \pi) = b.\sigma$ and $z.(\sigma, \pi) = z.\pi$, where σ and π are a state and a permission state, respectively. We assume '$\upharpoonright$' denotes a projection operator, e.g., $(x, y) \upharpoonright 1 = x$.

3.2 Interval Predicates

In this section, we provide the basics of interval predicates, which forms the logical basis of our program semantics. Our logic is an adaptation of Interval Temporal Logic [19]. An *interval* is a contiguous set of integers (denoted $\mathbb{Z}$), and hence the set of all intervals is $Intv \mathrel{\widehat{=}} \{\Delta \subseteq \mathbb{Z} \mid \forall t_1, t_2{:}\, \Delta \bullet \forall t{:}\, \mathbb{Z} \bullet t_1 \leq t \leq t_2 \Rightarrow t \in \Delta\}$.

We let $\mathsf{lub}.\Delta$ and $\mathsf{glb}.\Delta$ denote the *least upper* and *greatest lower* bounds of an interval Δ, respectively. Furthermore, we define $\mathsf{inf}.\Delta \mathrel{\widehat{=}} (\mathsf{lub}.\Delta = \infty)$, $\mathsf{fin}.\Delta \mathrel{\widehat{=}} \neg\mathsf{inf}.\Delta$, and $\varepsilon.\Delta \mathrel{\widehat{=}} (\Delta = \varnothing)$. We define an ordering $\Delta_1 < \Delta_2 \mathrel{\widehat{=}} \forall t_1{:}\, \Delta_1, t_2{:}\, \Delta_2 \bullet t_1 < t_2$. To facilitate reasoning about specific parts of a stream, we use *interval predicates*, which have type $IntvPred \mathrel{\widehat{=}} Intv \rightarrow \mathcal{P}Stream$ [11,13].

Example 2. Given Var, $Proc$ and π as defined in Example 1, we define

$$\sigma_1 \mathrel{\widehat{=}} \{u \mapsto 500, v \mapsto 42\} \qquad s \mathrel{\widehat{=}} \lambda t \bullet \text{ if } t \geq 10 \text{ then } (\sigma_1, \pi) \text{ else } (\sigma_2, \pi)$$
$$\sigma_2 \mathrel{\widehat{=}} \{u \mapsto 0, v \mapsto 1\} \qquad g \mathrel{\widehat{=}} \lambda \Delta \bullet \lambda s \bullet \forall t{:}\, \Delta \bullet ((s.t) \upharpoonright 1).u \geq 300$$
$$b \mathrel{\widehat{=}} \lambda \sigma \bullet \sigma.u < \sigma.v$$

Then, σ_1, σ_2 are states, b a state predicate, s is a stream and g is an interval predicate. Each of $(\neg b).(\sigma_1, \pi)$, $b.(\sigma_2, \pi)$, $\neg g.[-3, 3].s$ and $g.[10, 100).s$ hold.[3] □

We define *universal implication* $g_1 \Rrightarrow g_2 \mathrel{\widehat{=}} \forall \Delta{:}\, Intv, s{:}\, Stream \bullet g_1.\Delta.s \Rightarrow g_2.\Delta.s$ for interval predicates g_1 and g_2, and write $g_1 \equiv g_2$ iff both $g_1 \Rrightarrow g_2$ and $g_2 \Rrightarrow g_1$ hold.

4 Concurrent Programming with Intervals

4.1 Operators to Model Programming Constructs

In this section, we introduce interval predicate operators used to formalise common programming constructs: sequential composition, branching, loops, and parallel composition. To model sequential composition, we define the *chop* operator [19,13]. Unlike Interval Temporal Logic, which requires adjoining intervals to overlap at a single point, adjoining intervals in our logic are disjoint.

[3] Here, $[-3, 3]$ is the closed interval from -3 to 3 and $[10, 100)$ is the right-open interval from 10 to 100.

$$(g_1 \,;\, g_2).\Delta.s \mathrel{\widehat{=}} \left(\exists \Delta_1, \Delta_2 \bullet (\Delta_1 \cup \Delta_2 = \Delta) \wedge \Delta_1 < \Delta_2 \wedge g_1.\Delta_1.s \wedge g_2.\Delta_2.s\right) \vee (\mathsf{inf} \wedge g_1).\Delta.s$$

Thus, $(g_1 \,;\, g_2).\Delta.s$ holds iff either interval Δ may be split into two adjoining parts Δ_1 and Δ_2 so that g_1 holds for Δ_1 and g_2 holds for Δ_2 in s, or the least upper bound of Δ is ∞ and g_1 holds for Δ in s. Inclusion of the second disjunct $(\mathsf{inf} \wedge g_1).\Delta.s$ enables g_1 to model an infinite (divergent or non-terminating) program. We assume that ';' binds tighter than all other binary operators, e.g., $g_1 \,;\, g_2 \vee h = (g_1 \,;\, g_2) \vee h$.

Example 3. For b, g and s as defined in Example 2, if $h \mathrel{\widehat{=}} \lambda \Delta \bullet \lambda s \bullet \exists t{:}\, \Delta \bullet b.(s.t)$, then $(h \,;\, g).[0, 100).s$ holds because both $h.[0, 10).s$ and $g.[10, 100).s$ hold. Note that there may be more than one possible way to split up an interval when applying the definition of chop. □

Non-deterministic choice is modelled by (lifted) disjunction, and hence, for example, the behaviour of `if b then S1 else S2` is given by *test*.b ; *beh*.S1 $\vee$ *test*.($\neg$b) ; *beh*.S2, where *test*.b and *beh*.S1 are interval predicate formalisations of evaluating b and executing S1, respectively. The precise value of *test*.b depends on the atomicity assumptions of the program under consideration, and hence, the interpretation of *test*.b is non-trivial (see [17,13,11]). The value of *test*.b is modelled by command $[b]$ (see Section 4.2), whereas at the concrete level its value is formalised by a different command $[\![b]\!]$, which takes the effect of the buffer into account (see Section 4.3).

Iteration g^* and g^ω are the least and greatest fixed points of $\lambda z \bullet gz \vee \varepsilon$, respectively [14], where g^* allows empty and finite iterations and g^ω allows empty, finite and infinite iterations of g. We also define strictly finite and possibly infinite *positive* iterations.

$$g^* \mathrel{\widehat{=}} \mu z \bullet ((g \,;\, z) \vee \varepsilon) \qquad g^+ \mathrel{\widehat{=}} g \,;\, g^*$$
$$g^\omega \mathrel{\widehat{=}} \nu z \bullet ((g \,;\, z) \vee \varepsilon) \qquad g^{\omega+} \mathrel{\widehat{=}} g \,;\, g^\omega$$

A thorough algebraic treatment of loops using iteration is given in [3]. For example, program code `while b do S` is modelled by $(\mathit{test}.b \,;\, \mathit{beh}.S)^\omega \,;\, \mathit{test}.(\neg b)$. In this paper, we use the following rule.

Law 1 (Leapfrog [3]). *For interval predicates g and h, both* $g \,;\, (h \,;\, g)^\omega \equiv (g \,;\, h)^\omega \,;\, g$ *and* $g \,;\, (h \,;\, g)^* \equiv (g \,;\, h)^* \,;\, g$ *hold.*

We are interested in modelling true concurrency and therefore simply treat the parallel composition of two or more processes using (lifted) logical conjunction. For example, the behaviour of $g_1 \,;\, g_2$ in parallel with $h_1 \,;\, h_2$ over an interval Δ in stream s is given by $(g_1 \,;\, g_2 \wedge h_1 \,;\, h_2).\Delta.s$. Using pointwise lifting, this is equivalent to $(g_1 \,;\, g_2).\Delta.s \wedge (h_1 \,;\, h_2).\Delta.s$, which holds iff (a) Δ can be split into adjoining intervals Δ_1 and Δ_2 such that $g_1.\Delta_1.s \wedge g_2.\Delta_2.s$ holds; and (b) Δ can also be split into adjoining intervals Δ_3 and Δ_4 such that $h_1.\Delta_3.s \wedge h_2.\Delta_4.s$. Note that there is no immediate correlation between the lengths of Δ_1 and Δ_3, i.e. g_1 could terminate earlier than h_1, and vice versa.

Modelling Tests. Interval predicates provide a flexible approach to non-deterministic state predicate evaluation [17], where expression evaluation is assumed to take time (as opposed to being instantaneous). In this paper, guards and assignments are restricted to contain at most one shared variable.[4] Given that c is either a state or permission predicate, and Δ and s are an interval and stream, respectively, we define:

$$(\boxdot\, c).\Delta.s \mathrel{\widehat{=}} \neg\varepsilon.\Delta \wedge \forall t{:}\, \Delta \bullet c.(s.t)$$

Thus, $(\boxdot\, c).\Delta.s$ holds iff Δ is non-empty and c holds for each state of s within Δ. For example, $\neg\varepsilon \wedge g \equiv \boxdot(u \geq 300)$, where g is the interval predicate defined in Example 2.

Reasoning About Pre/Post Assertions. One may define several additional interval predicate operators [13]. For the purposes of this paper, we find it useful to reason about properties that hold in the immediately preceding interval. We therefore define

$$(\ominus g).\Delta.s \mathrel{\widehat{=}} \neg\varepsilon.\Delta \wedge \mathsf{glb}.\Delta \neq -\infty \wedge g.(\mathrm{prev}.\Delta).s$$

where $\mathrm{prev}.\Delta \mathrel{\widehat{=}} \{t{:}\, \mathbb{Z} \mid \forall u{:}\, \Delta \bullet t < u\}$ is the interval of all times before Δ. If c is a state or permission predicate, we use notation $\overrightarrow{c} \mathrel{\widehat{=}} \mathit{true}\,;\, \boxdot\, c$, where $\overrightarrow{c}.\Delta$ states that c holds at the end of Δ whenever $\mathsf{inf}.\Delta \neq \infty$. Additionally, we define the following notation to reason about assertions that immediately precede, or are a result of a computation.

$$\{c\}g \mathrel{\widehat{=}} \ominus\overrightarrow{c} \wedge g \qquad\qquad g\{c\} \mathrel{\widehat{=}} g \wedge \overrightarrow{c}$$

Such a definition of a pre-assertion is necessary because we assume adjoining intervals do not overlap (unlike [19]). We have the following useful properties, which can be proved in a straightforward manner.

$$\{c\}(g_1 \vee g_2) \equiv (\{c\}g_1) \vee (\{c\}g_2) \tag{2}$$

$$g_1\{c\}\,;\, g_2 \equiv g_1\,;\, \{c\}g_2 \qquad \text{provided } g_1 \vee g_2 \Rrightarrow \neg\varepsilon \tag{3}$$

4.2 Abstract Commands

Using the interval-based semantic basis from the previous sections, we formalise *commands*, which describe the behaviours of the system processes. Formally, a command is of type $\mathit{Cmd} \mathrel{\widehat{=}} \mathbb{P}_1\, \mathit{Proc} \to \mathit{IntvPred}$, mapping non-empty sets of processes to an interval predicate representing their behaviour. We use $C.p$ as shorthand for $C.\{p\}$, where C is a command and p is a process.

The semantics of sequential composition, iteration, non-deterministic choice and parallel composition of commands are defined pointwise lifting of the interval predicate operators, and hence, are given in the same syntax, e.g., $(C_1\,;\,C_2).p = C_1.p\,;\,C_2.p$. What remains is to define the commands to model, say, guard evaluation and assignment.

We first present some basic commands that may be used to models the abstract (sequentially consistent) specification. In particular, we define *idling* (denoted id), *abstract guard evaluation* (denoted $[b]$), *memory update* (denoted $\overline{v} \Lleftarrow \overline{e}$) and *locked access* (denoted $\boxed{\overline{v} \bullet C}$), where $\overline{v}$ and $\overline{e}$ denote vectors of variables and expressions, respectively. We define $\mathit{Deny}.\overline{v}.p \mathrel{\widehat{=}} \boxdot(\forall q{:}\, \mathit{Proc}\backslash\{p\}, v{:}\, \overline{v} \bullet (\neg\mathcal{W} \wedge \neg\mathcal{R}).q.v)$, which states that the variables in $\overline{v}$ are not accessed by processes other than p. We assume that $\mathit{vars}.b$ denotes the free variables in b.

$$\begin{aligned}
\mathsf{nid}.p &\mathrel{\widehat{=}} \forall v{:}\, \mathit{Var} \bullet \boxdot\neg\mathcal{W}.p.v \\
\mathsf{id}.p &\mathrel{\widehat{=}} \neg\varepsilon \Rightarrow \mathsf{nid}.p \\
[b].p &\mathrel{\widehat{=}} \boxdot b \wedge \mathsf{nid}.p \wedge \\
&\qquad \forall v{:}\, \mathit{vars}.b \bullet \boxdot\,\mathcal{R}.p.v
\end{aligned}
\qquad
\begin{aligned}
(\overline{v} \Lleftarrow \overline{e}).p &\mathrel{\widehat{=}} \exists \overline{k} \bullet\ \{\overline{e} = \overline{k}\} \\
&\qquad \boxdot(\overline{v} = \overline{k}) \wedge \boxdot(\forall v{:}\, \overline{v} \bullet \mathcal{W}.p.v) \\
\boxed{\overline{v} \bullet C}.p &\mathrel{\widehat{=}} (\overline{v} \neq \varnothing \Rightarrow \mathit{Deny}.\overline{v}.p) \wedge C.p \wedge \\
&\qquad (\forall v{:}\, \overline{v} \bullet \boxdot(\mathcal{W} \wedge \mathcal{R}).p.v)
\end{aligned}$$

[4] This can be extended to handle more complex expressions [17].

Thus $\mathsf{nid}.p$ states that p is *write idle* i.e., p does not have write access to any variable during the given (non-empty) interval; $\mathsf{id}.p$ states that either p is write idle or the interval under consideration is empty; $[b].p$ holds iff b holds throughout the given interval and p is idle; $\overline{v} \Leftarrow \overline{e}$ denotes an instantaneous update, which holds iff $\overline{e}$ evaluates to a vector of values $\overline{k}$ as a pre-assertion, and $\overline{v}$ is updated to $\overline{k}$, where p has write permission to each $v \in \overline{v}$; and $\boxed{\overline{v} \bullet C}.p$ holds iff $C.p$ holds and additionally no process other than p has permission to access $v \in \overline{v}$.

Example 4. The abstract specification is formalised as follows, where *AAcq* and *ARel* specify operations `AAcq` and `ARel`, respectively, while *ATryOK* and *ATryFl* specify execution of the `ATry` operation that succeed and fail to acquire the lock, respectively. We abbreviate `x = 1` and `x = 0` to x and $\neg x$, respectively. The return value of an execution of `tryacquire` in process p is modelled by a local variable r_p.

$$AAcq.p \mathrel{\widehat=} \boxed{x \bullet [x]\ ;\ (x \Leftarrow 0)} \qquad ATryOK.p \mathrel{\widehat=} AAcq.p\ ;\ \mathsf{id}\ ;\ (r_p \Leftarrow true)$$

$$ARel.p \mathrel{\widehat=} x \Leftarrow 1 \qquad ATryFl.p \mathrel{\widehat=} \boxed{x \bullet [\neg x]}\ ;\ \mathsf{id}\ ;\ (r_p \Leftarrow false)$$

$$AExec.p \mathrel{\widehat=} ((AAcq\ ;\ \mathsf{id}\ ;\ ARel) \vee (ATryOk\ ;\ \mathsf{id}\ ;\ ARel) \vee ATryFl).p$$

$$Spec.P \mathrel{\widehat=} \{x\} \bigwedge_{p:P}((\mathsf{id}\ ;\ AExec)^{\omega}\ ;\ \mathsf{id}).p$$

The concurrent execution of abstract clients is modelled by *Spec*, which begins in a state in which the lock `x` is available (i.e., x holds) and consists of a number of (truly) parallel processes. We assume that each client of the spinlock behaves as either Fig. 3 or Fig. 4, and furthermore, that the critical and non-critical sections do not modify variables x and r_p, and hence, both the critical and non-critical sections are modelled by id. Therefore, id ; *AExec* models a single call to the abstract spinlock. Each process may make multiple (zero or more) calls, followed by no calls, and hence, all possible behaviours of an abstract client is given by $(\mathsf{id}\ ;\ AExec)^{\omega}\ ;\ \mathsf{id}$.

We now explain how each operation is modelled. If $AAcq.p.\Delta.s$ holds for interval Δ and stream s, then only process p has access to x (i.e., no process $q \neq p$ may read or write to x) and either (i) Δ can be partitioned into Δ_1 and Δ_2 with $\Delta_1 < \Delta_2$ such that x holds in s throughout Δ_1 and x is updated to 0 in s within Δ_2, or (ii) Δ is infinite and x (i.e., `x = 1`) holds in s throughout Δ. Because `await b` *blocks* until *test*.b becomes true, there are no behaviours for *AAcq.p* when *test*.($\neg$b) holds. Operation *ARel.p* immediately sets x to 1, and by the definition of $\Leftarrow$ together with assumption (1), we have that no other process reads or writes to x while this update occurs. Operation *ATryOK.p* behaves as *AAcq.p*, performs some idling, then updates r_p to *true*. The idling between *AAcq.p* and update to r_p provides scope for potential stuttering at the concrete level. Operation *ATryFl.p* starts by behaving as $\boxed{x \bullet [\neg x]}$, which implies that x is not accessed by any process $q \neq p$ and that $\neg x$ holds throughout the given interval. Then, *ATryFl.p* performs some idling and updates the return value r_p to *false*. □

4.3 Reading Variables for Expression Evaluation with Buffer Effects

Section 4.2 provided an interval-based semantics for commands without buffers, which were in turn used to model the abstract specification. The concrete program executes under TSO memory and contains local buffers, whose effects on the program's behaviour must be formalised. In this section, we present a method for evaluating expressions, i.e., when processes read variables, in the presence of local buffers. In particular, we formalise the fact that a TSO process first checks its buffer for pending writes; if a pending write exists, the last pending value is returned, and otherwise the value from memory is returned. Using interval-based methods enables one to formalise the effects of a buffer on the value of an expression at a high level of abstraction [15].

We assume that $\mathrm{B}_p \in \mathit{Var}$ denotes the buffer for process p, whose value is of type $\mathrm{seq}.(\mathit{Var} \times \mathit{Val})$, representing a pending write. Each buffer may contain multiple pending writes to the same location, and hence, we define a function *cover* that returns a set of mappings to the last pending write in a given buffer. Because $\mathrm{seq}.X$ is a partial function of type $\mathbb{N} \nrightarrow X$, we may use $\mathrm{dom}.z$ to refer to the indices of $z \in \mathrm{seq}.X$.

$$\mathit{cover}.B \mathrel{\widehat{=}} \{B.i \mid i{:}\, \mathrm{dom}.B \wedge \forall j{:}\, \mathrm{dom}.B \bullet j > i \Rightarrow (B.i \upharpoonright 1) \neq (B.j \upharpoonright 1)\}$$

When a process evaluates an expression, the values of pending writes in a process' buffer *mask* those in memory, which is modelled formally using functional override '$\oplus$' (see [24] for a formal definition).

Example 5. Suppose B and BB are buffers (BB is not shown below), p and q are processes, u, v, and w are variables, and σ is a state such that

$$B \mathrel{\widehat{=}} \langle (v, 11), (w, 33), (v, 44) \rangle \qquad \sigma \mathrel{\widehat{=}} \{(\mathrm{B}_p, B), (\mathrm{B}_q, BB), (u, 0), (v, 1), (w, 2)\}$$

Then we have $\mathit{cover}.B = \{(w, 33), (v, 44)\}$, i.e., for each variable in B its last corresponding value in B is picked. Hence, we have

$$\sigma \oplus \mathit{cover}.B = \{(\mathrm{B}_p, B), (\mathrm{B}_q, BB), (u, 0), (v, 44), (w, 33)\}$$

which replaces each mapping in σ by those in $\mathit{cover}.B$. □

We lift buffer effects to state predicates using $(\mathit{mask}.b.B).\sigma \mathrel{\widehat{=}} b.(\sigma \oplus \mathit{cover}.B)$, which states that b holds in a state σ covered by B. For the definitions in Example 5, both $(\mathit{mask}.(u = 0).B).\sigma$ and $(\mathit{mask}.(w < v).B).\sigma$ hold, but $(w < v).\sigma$ does not.

Processes evaluate state predicates (e.g., as part of a guard), however, in the presence of permissions and local buffers, evaluation is non-trivial. Firstly, one must ensure that a process p evaluating state predicate b is able to obtain read permission to each variable of b whenever the variable's value is fetched from memory. Note that this is only potentially problematic if the variable in question is shared (i.e., not a local variable of p) and not in p's buffer (p may can always access its local buffer). Secondly, the value of a variable v read by p must be the last value of v in p's buffer if it exists, and the value of v in memory, otherwise. Assuming $\mathit{vars}.B \mathrel{\widehat{=}} \{(B.i) \upharpoonright 1 \mid i \in \mathrm{dom}.B\}$ is the set of all variables in $B \in \mathrm{seq}.(\mathit{Var} \times \mathit{Val})$, we define:

$$\boxdot_p\, b \mathrel{\widehat{=}} \boxdot(\mathit{mask}.b.\mathrm{B}_p \wedge \forall v{:}\, \mathit{vars}.b \backslash \mathit{vars}.\mathrm{B}_p \bullet \mathcal{R}.p.v)$$

Thus, $(\boxdot_p\, b).\Delta.s$ models the evaluation of state predicate b by the process p, by either reading variables from p's buffer (if possible) or from main memory. Here $(\boxdot_p\, b).\Delta.s$ holds iff (i) b masked by B_p holds in s throughout Δ, and (ii) p has read permission to the free variables of b not in $\mathit{vars}.\mathrm{B}_p$ throughout Δ in s.[5] In Section 4.4, $(\boxdot_p\, b)$ is used to define expression evaluation, which in turn is used to model guards and the right hand side of assignments.

[5] In general, if b contains multiple shared variables, $\boxdot_p\, b$ is not an accurate model of evaluation because the variables in b may be read at different instants [17,10,12]. However, here, we assume that each expression/guard of each program under consideration contains at most one shared variable, in which case $\boxdot_p\, b$ is accurate (see [17,10,12]).

4.4 Commands under TSO

As already mentioned, processes that execute under TSO write only to their local buffers. The effects of these writes are not seen by other processes until a buffer is *flushed*, which moves the pending write from a buffer to shared memory. TSO buffers operate in a FIFO order, and hence, we define the following commands, where Φ models a single flush, $\widetilde{\Phi}$ models a flush or a non-empty idle, and Φ_{all} models a complete buffer flush.

$$\Phi.p \mathrel{\widehat{=}} \exists k \bullet \{B_p = k \wedge k \neq \langle\rangle\} \qquad \widetilde{\Phi} \mathrel{\widehat{=}} \Phi \vee \mathsf{nid}$$
$$B_p, (k.0 \restriction 1) \Leftarrow tail.k, (k.0 \restriction 2) \qquad \Phi_{all}.p \mathrel{\widehat{=}} \Phi^{+}.p\,\{B_p = \langle\rangle\}$$

Due to the fine-granularity of the concrete implementation, seemingly atomic statements become compound commands under TSO memory. Evaluation of a boolean expression b (e.g. a guard in an `if-then-else` block) is a compound statement that flushes or idles (zero or more times), evaluates b using the buffer-based evaluation semantics defined in Section 4.3, then flushes or idles again (zero or more times). A write of v with value k appends the pair (v,k) to the end of the local buffer. An assignment to a constant value k, potentially flushes or idles (zero or more times), appends the value to the buffer, then potentially flushes or idles (zero or more times). An assignment to a complex expression e, first evaluates the expression to a value k, then assigns k to v. Thus, we define:

$$[\![b]\!].p \mathrel{\widehat{=}} \widetilde{\Phi}^{*}.p\,;\,(\boxdot_p\, b \wedge \mathsf{nid}.p)\,;\,\widetilde{\Phi}^{*}.p$$
$$(v \leftarrow k).p \mathrel{\widehat{=}} B_p \Leftarrow B_p \frown \langle (v,k)\rangle$$
$$v := e \mathrel{\widehat{=}} \mathsf{if}\ e \in Val\ \mathsf{then}\ \widetilde{\Phi}^{*}\,;\,(v \leftarrow e)\,;\,\widetilde{\Phi}^{*}\ \mathsf{else}\ \exists k{:}\, Val \bullet [\![e = k]\!]\,;\, v := k$$

There are several TSO instructions that force the entire buffer to be flushed. These additionally may lock certain variables from being accessed while the flush all is being executed. We therefore define commands $pre_{\Phi}.\overline{v}.C.p$ and $post_{\Phi}.\overline{v}.C.p$, where $pre_{\Phi}.\overline{v}.C.p$ flushes the entire buffer (locking $\overline{v}$) before C is executed (and similarly $post_{\Phi}.\overline{v}.C.p$ flushes all after C).

$$pre_{\Phi}.\overline{v}.C.p \mathrel{\widehat{=}} \{B_p = \langle\rangle\}C.p \vee (\boxed{\overline{v} \bullet \Phi_{all}}\,;\,C).p$$
$$post_{\Phi}.\overline{v}.C.p \mathrel{\widehat{=}} C.p\{B_p = \langle\rangle\} \vee (C\,;\,\boxed{\overline{v} \bullet \Phi_{all}}).p$$

Some TSO instructions do not lock the memory while the buffer is being flushed. These may be modelled using $pre_{\Phi}.\varnothing.C$ and $post_{\Phi}.\varnothing.C$, which we abbreviate to $pre_{\Phi}.C$ and $post_{\Phi}.C$, respectively. A `lock` (e.g., `a2` in Fig. 2) acquires the memory lock then flushes the entire buffer; an `unlock` (e.g., `a5` in Fig. 2) flushes the entire buffer then releases the memory lock. Therefore, executing a command C within a `lock-unlock` block is modelled by

$$\boxed{\overline{v} \bullet C}_{\Phi} \mathrel{\widehat{=}} pre_{\Phi}.\overline{v}.(post_{\Phi}.\overline{v}.\boxed{\overline{v} \bullet C})$$

which executes C and ensures the buffer is empty before and after executing C. In addition it ensures that no reads and writes to $\overline{v}$ by other processes occur while C is being executed. Note however, that if a process p executes $\boxed{\overline{v} \bullet C}_{\Phi}$ and a process $q \neq p$ has a pending write to $\overline{v}$ in its local buffer, then q may read this value of $\overline{v}$ even while p is executing $\boxed{\overline{v} \bullet C}_{\Phi}$.

Example 6. Our modelling notation is used to formalising the behaviour of the concrete implementation as follows, where $Lck.p \mathrel{\widehat{=}} \boxed{x \bullet [\![x]\!] \;;\; (x := 0)}_\Phi$.

$$Acq.p \mathrel{\widehat{=}} \left(\boxed{x \bullet [\![\neg x]\!]}_\Phi \;;\; [\![\neg x]\!]^\omega \;;\; [\![x]\!]\right)^\omega ; Lck.p \qquad Rel.p \mathrel{\widehat{=}} x := 1$$

$$TryOK.p \mathrel{\widehat{=}} Lck.p \;;\; (r_p := true) \qquad TryFl.p \mathrel{\widehat{=}} \boxed{x \bullet [\![\neg x]\!]}_\Phi \;;\; (r_p := false)$$

$$Exec.p \mathrel{\widehat{=}} (Acq \;;\; \mathsf{id} \;;\; Rel \vee TryOK \;;\; \mathsf{id} \;;\; Rel \vee TryFl).p$$

$$Prog.P \mathrel{\widehat{=}} \{x\} \textstyle\bigwedge_{p:P} (\{B_p = \langle\rangle\}(\widetilde{\Phi}^+ \;;\; Exec)^\omega \;;\; post_\Phi.x.(\widetilde{\Phi}^+)).p$$

Within *Acq.p*, command $\boxed{x \bullet [\![\neg x]\!]}_\Phi$ models an execution consisting of the lock at `a2`, failed test at `a3`, unlock at `a7` (see Fig. 2). Command $[\![\neg x]\!]^\omega \;;\; [\![x]\!]$ models the while loop at `a8`. Therefore, the outermost $^\omega$ iteration in *Acq.p* models executions of the outermost loop of `acquire` that fail to acquire the lock. Command *Lck.p* models the lock at `a2`, successful test at `a3`, assignment at `a4`, and unlock at `a5` followed by the return at `a6`. The other operations are similar. □

5 Refinement and Local Refinement for TSO

5.1 Interval-Based Refinement

In this section, we develop a theory for proving that a command C *refines* another command A, providing a formal link between the behaviours of C and A. Here, A is an abstraction and therefore admits more behaviours than C, or conversely, any behaviour of C must also be a behaviour of A. In an interval-based setting, we use the following definition of refinement [11]. In the context of our example, if refinement holds, then whenever a spinlock client is able to enter its critical section, it must also be possible for the abstract client to enter the critical section.

Definition 1. *If C and A are commands, then C* refines *A with respect to $P \subseteq Proc$, (denoted $C \sqsupseteq_P A$) iff for any interval Δ and stream s, $C.P \Rrightarrow A.P$. We say C* is equivalent *to A with respect to P (denoted $C \sqsubseteq\!\!\sqsupseteq_P A$) iff both $C \sqsupseteq_P A$ and $A \sqsupseteq_P C$.*

Refinement is defined in terms of implication, and hence, relation $\sqsupseteq_P$ is both reflexive and transitive. In this paper, we use $\sqsupseteq_P$ as a basis for transforming the abstract specification *Spec* and the concrete program *Prog* individually. We use a notion of local buffer refinement (Definition 2) to relate concrete behaviours (with buffers) to abstract behaviours (without buffers).

Example 7. We transform the TSO implementation *Prog* into a form that is more amenable to verification. In particular, a difficulty encountered when verifying *Prog* in Example 6 directly is that for each process, p, command *Exec.p* is not guaranteed to end in a flush, and hence changes to x may not be globally visible until the start of the next iteration (which starts with a `lock` that performs a flush all). In particular, it is not immediately possible to match the behaviour of *Rel* with abstract *ARel* because *Rel* only places a pending write in the buffer, whereas *ARel* modifies the value of x in memory. Therefore, we aim to transform *Prog* to *Prog′* below (see Proposition 1), where the flush occurs at the end of execution. We use notation

$$\boxed{\overline{v} \bullet C}_\Phi \mathrel{\widehat{=}} post_\Phi.\overline{v}.\boxed{\overline{v} \bullet C}$$

Unlike $\boxed{\overline{v} \bullet C}_\Phi$, command $\dashbox{\overline{v} \bullet C}_\Phi$ only flushes the buffer at the end of execution. Below, we have used the property $g^\omega \equiv \varepsilon \vee g^{\omega+}$ to split *Acq* into two cases. Note that Acq'_1 is defined in terms of *Acq*.

$$\begin{array}{ll} Acq'_1.p \mathrel{\widehat=} \dashbox{x \bullet [\![\neg x]\!]}_\Phi \ ; [\![\neg x]\!]^* \ ; [\![x]\!] \ ; Acq.p & Acq'_2.p \mathrel{\widehat=} \dashbox{x \bullet [\![x]\!] \ ; (x := 0)}_\Phi \\ TryOK'.p \mathrel{\widehat=} \dashbox{x \bullet [\![x]\!] \ ; (x := 0)}_\Phi \ ; (r_p := true) & TryFl'.p \mathrel{\widehat=} \dashbox{x \bullet [\![\neg x]\!]}_\Phi \ ; (r_p := false) \\ Exec'.p \mathrel{\widehat=} ((Acq'_1 \ ; \mathsf{id} \ ; Rel) \vee (Acq'_2 \ ; \mathsf{id} \ ; Rel) \vee (TryOK' \ ; \mathsf{id} \ ; Rel) \vee TryFl').p & \\ Prog'.P \mathrel{\widehat=} \{x\} \bigwedge_{p:P} \left(\mathsf{id} \ ; \left(\{\mathrm{B}_p = \langle\rangle\}Exec' \ ; post_\Phi.x.(\widetilde{\Phi}^+)\right)^\omega\right).p & \end{array}$$ □

Clearly, transforming *Prog* to *Prog'* by reasoning at trace-based level of Definition 1 is infeasible. Therefore, we develop a number of refinement laws that are applied to our example. First, we have the following; the proof of each equivalence is straightforward.

Law 2. *If $p \in Proc$, C and D are commands, each C_i is a command and $\overline{v}$ is a vector of variables, then*

$$\boxed{\overline{v} \bullet C}_\Phi \sqsupseteq_p pre_\Phi.\overline{v}.\dashbox{\overline{v} \bullet C}_\Phi \tag{4}$$

$$pre_\Phi.C \sqsupseteq_p pre_\Phi.(\{\mathrm{B}_p = \langle\rangle\}C) \tag{5}$$

$$pre_\Phi.\overline{v}.\dashbox{\overline{v} \bullet C}_\Phi \ ; D \sqsupseteq_p pre_\Phi.\overline{v}.\left(\dashbox{\overline{v} \bullet C}_\Phi \ ; D\right) \qquad \text{provided } C.p \Rrightarrow \neg\varepsilon \tag{6}$$

$$\bigvee_i pre_\Phi.C_i \sqsupseteq_p pre_\Phi.\left(\bigvee_i C_i\right) \tag{7}$$

$$C \ ; pre_\Phi.D \sqsupseteq_p post_\Phi.C \ ; D \qquad \text{provided } C.p \vee D.p \Rrightarrow \neg\varepsilon \tag{8}$$

To transform *Prog* to *Prog'*, we develop a leapfrog theorem analogous to Law 1, whose proof uses the equivalences defined in Law 2 as well as $PF.\overline{v} \mathrel{\widehat=} post_\Phi.\overline{v}.\widetilde{\Phi}^+$.

Theorem 1 (Leapfrog Flush). *Suppose $p \in Proc$, each C_i and D_i is a command such that $C_i.p \Rrightarrow \neg\varepsilon$, and $\overline{v}$ is a vector of variables. Then*

$$\left(\widetilde{\Phi}^+ \ ; \left(\bigvee_i \boxed{\overline{v} \bullet C_i}_\Phi \ ; D_i\right)\right)^\omega ; PF.\overline{v} \ \sqsupseteq_p \ PF.\overline{v}; \left(\{\mathrm{B}_p = \langle\rangle\} \left(\bigvee_i \dashbox{\overline{v} \bullet C_i}_\Phi \ ; D_i\right); PF.\overline{v}\right)^\omega$$

The left hand side of Theorem 1 contains a disjunction that executes $\boxed{\overline{v} \bullet C_i}_\Phi$, which ensures the buffer is empty (via flushes if necessary) both before and after execution of C_i. After the end of the iteration, command $PF.\overline{v}$ is executed, which ensures the buffer is empty when the process terminates; flushes may be necessary due to the behaviour of D_i. On the right hand side, each iteration is guaranteed to start with an empty buffer and each disjunct starts with the weaker $\dashbox{\overline{v} \bullet C_i}_\Phi$, which only flushes the buffer at the end of execution. However, each iteration ends with $PF.\overline{v}$. Further note that on the right hand side, each iteration is guaranteed to begin in a state where the buffer of p is empty.

Proposition 1. *$Prog \sqsupseteq_P Prog'$*

Proof. Applying Theorem 1 to *Prog*, we obtain:

$$Prog''.P \mathrel{\widehat=} \{x\} \bigwedge_{p:P} \left(\{\mathrm{B}_p = \langle\rangle\}PF.x \ ; (\{\mathrm{B}_p = \langle\rangle\}Exec' \ ; PF.x)^\omega\right).p$$

Then, because $\{\mathrm{B}_p = \langle\rangle\}PF.x \sqsubseteq_p \mathsf{id}$, we have $Prog'' \sqsubseteq_P Prog'$. □

For the proof in Section 5.3, we find the following laws to be useful, each of which is proved in a straightforward manner. Note that for (11) and (14), the refinement only holds in one direction. Of these, (14) states that an assignment $\{\mathrm{B}_p = \langle\rangle\}v := k$ either ends with $\mathrm{B}_p = \langle(v,k)\rangle$, or the buffer B_p is flushed as part of the assignment semantics.

Law 3. *Suppose C and D are commands, p is a process, v is a variable and k a value. Then each of the following holds:*

$$\{\mathrm{B}_p = \langle\,\rangle\}\boxed{\overline{v} \bullet C}_{\Phi} \sqsupseteq_p \{\mathrm{B}_p = \langle\,\rangle\}\dbox{\overline{v} \bullet C}_{\Phi} \quad (9)$$

$$\dbox{\overline{v} \bullet C \mathbin{;} D}_{\Phi} \sqsupseteq_p \boxed{\overline{v} \bullet C} \mathbin{;} \dbox{\overline{v} \bullet D}_{\Phi} \quad \textit{provided } C.p \vee D.p \Rrightarrow \neg\varepsilon \quad (10)$$

$$\{\mathrm{B}_p = \langle\,\rangle\}\dbox{v \bullet v := k}_{\Phi} \sqsupseteq_p \boxed{v \bullet \mathsf{id} \mathbin{;} v \Lleftarrow k \mathbin{;} \mathsf{id}}\{\mathrm{B}_p = \langle\,\rangle\} \quad (11)$$

$$\boxed{\overline{v} \bullet C} \mathbin{;} \boxed{\overline{v} \bullet D} \sqsupseteq_p \boxed{\overline{v} \bullet C \mathbin{;} D} \quad (12)$$

$$\{\mathrm{B}_p = \langle\,\rangle\}\boxed{\overline{v} \bullet [\![b]\!]} \sqsupseteq_p \{\mathrm{B}_p = \langle\,\rangle\}\boxed{\overline{v} \bullet [b]} \quad (13)$$

$$\{\mathrm{B}_p = \langle\,\rangle\}v := k \sqsupseteq_p (\mathsf{id} \mathbin{;} v \mapsfrom k \mathbin{;} \mathsf{id})\{\mathrm{B}_p = \langle(v, k)\rangle\} \mathbin{;} (\varepsilon \vee (\Phi \mathbin{;} \mathsf{id})) \quad (14)$$

5.2 Local Buffer Refinement

In this section, we develop a novel method for proving refinement for TSO architectures, where buffer effects are taken into account. The method allows one to prove that a (concrete) process with buffer effects has the same behaviour as an (abstract) process without the buffer. In essence, one may pretend that the effect of local changes have already been flushed at the abstract level. This is allowed because in the context of the overall behaviour of a program, it makes no difference whether a variable local to process p has a pending write in p's local buffer, or in shared memory. This essentially removes the potential non-determinism that arises from reasoning about flushes for local variables. We let

$$LCover.P.(\sigma, \pi) \mathrel{\widehat{=}} \left(\sigma \oplus \textstyle\bigcup_{p:P} LVar.p \lhd cover.(\sigma.\mathrm{B}_p),\ \pi\right)$$
$$LBuffer.P.s \mathrel{\widehat{=}} \lambda\, t \colon \mathbb{Z} \bullet LCover.P.(s.t)$$

Here $cover.(\sigma.\mathrm{B}_p)$ generates a set of pairs from $\sigma.\mathrm{B}_p$ and $LVar.p \lhd cover.(\sigma.\mathrm{B}_p)$ restricts these to the local variables of p. Within $LCover.P.(\sigma, \pi)$ such a localised mapping is generated for each $p \in P$, and then, σ is overwritten by this mapping.

Definition 2. *If C and A are commands, and P is a non-empty set of processes, we say C* buffer refines *A* with respect to *P*, *(denoted $C \Supset_P A$) iff for any interval Δ and stream s, $(C.P).\Delta.s \Rightarrow (A.P).\Delta.(LBuffer.P.s)$.*

Thus, whenever C holds for a set of processes P over interval Δ in stream s, command A must hold for P in Δ for the masked stream $LBuffer.P.s$. In particular, this implies that A behaves as if the local buffer effects of the concrete command have already been applied. We write $\Supset_p$ for $\Supset_{\{p\}}$.

Clearly, reasoning at the level of Definition 2 is infeasible. Instead, we explore some higher level properties for the programming constructs we use. In general, buffer refinement is neither reflexive nor transitive[6], e.g., if C always reads from main memory regardless of the state of the buffer, then reflexivity does not hold. However, $\Supset_P$ may be combined with standard refinement as follows, which holds trivially after expanding the definitions of $\sqsupseteq_P$ and $\Supset_P$.

Theorem 2. *If $C' \sqsupseteq_P C$, $C \Supset_P A$ and $A \sqsupseteq_P A'$ then $C' \Supset_P A'$.*

[6] Fully exploring the properties of $\Supset_P$ lie outside the scope of this paper.

We immediately have the following laws, which help simplify verification of assignments and buffer flushes.

Law 4. *If* $v \in Var$, $p \in Proc$, $k \in Val$, $P \mathrel{\widehat{=}} \mathsf{B}_p \neq \langle\rangle \wedge (\mathsf{B}_p.0 \upharpoonright 1) \in LVar.p$, *and* $Q \mathrel{\widehat{=}} \mathsf{B}_p \neq \langle\rangle \wedge (\mathsf{B}_p.0 \upharpoonright 1) \notin LVar.p$, *then*

$$\{v \in LVar.p\} v \leftarrow\!\!\dashv k \mathrel{\Supset_p} v \Leftarrow\!\!\dashv k \quad (15) \qquad \{P\}\, \Phi \mathrel{\Supset_p} \mathsf{id} \quad (17)$$

$$\{v \notin LVar.p\} v \leftarrow\!\!\dashv k \mathrel{\Supset_p} \mathsf{id} \quad (16) \qquad \{Q\}\, \Phi \mathrel{\Supset_p} (\mathsf{B}_p.0 \upharpoonright 1) \Leftarrow\!\!\dashv (\mathsf{B}_p.0 \upharpoonright 2) \quad (18)$$

By conditions (15) and (16), adding a pending write (v, k) to p's buffer is a local buffer refinement of a global update to v whenever $v \in LVar.p$, and of id, otherwise. On the other hand, (17) states that flushing a local variable of process p from p's buffer has the same effect as executing id abstractly (because the effect of the variable has already occurred), and condition (18) states that flushing a global variable has the same effect as executing the corresponding write in memory at the abstract level.

The laws below allow local buffer refinement to be decomposed, and the pre/post assertions under local buffer refinement to strengthened.

Law 5. *If* $C \mathrel{\Supset_P} A$, $C_1 \mathrel{\Supset_P} A_1$ *and* $C_2 \mathrel{\Supset_P} A_2$ *for a set of processes* P, *and* b, c *are state predicates such that* $b \Rightarrow c$ *and* $(vars.b \cup vars.c) \cap (\bigcup_{p:P} LVar.p) = \varnothing$, *then*

$$C_1 \,;\, C_2 \mathrel{\Supset_P} A_1 \,;\, A_2 \quad (19) \qquad C_1 \vee C_2 \mathrel{\Supset_P} A_1 \vee A_2 \quad (22)$$

$$C^\omega \mathrel{\Supset_P} A^\omega \quad (20) \qquad \{b\}C \mathrel{\Supset_P} \{c\}A \quad (23)$$

$$C_1 \wedge C_2 \mathrel{\Supset_P} A_1 \wedge A_2 \quad (21) \qquad C\{b\} \mathrel{\Supset_P} A\{c\} \quad (24)$$

Law 6 (Parallel Composition). *If* $C'.P \mathrel{\widehat{=}} \bigwedge_{q:P} C.q$, $A'.P \mathrel{\widehat{=}} \bigwedge_{q:P} A.q$ *and* $C \mathrel{\Supset_q} A$ *holds for each for each* $q \in P$, *then* $C' \mathrel{\Supset_P} A'$.

5.3 Application: Spinlock Example

We now apply our rules to the running spinlock example (modelled by *Prog*), and prove it to be a refinement of the abstract program (modelled by *Spec*). Our notion of refinement is local buffer refinement (Definition 2), i.e., we show

$$Prog \mathrel{\Supset_P} Spec \tag{25}$$

for an arbitrarily chosen non-empty set of processes P. Various refinement rules may be introduced to generalise the theory as needed. By Theorem 2 and Proposition 1, (25) immediately reduces to a proof of $Prog' \mathrel{\Supset_P} Spec$. Using (23), followed by Law 6, proof of $Prog' \mathrel{\Supset_P} Spec$ again reduces to the following for some arbitrarily chosen $p \in P$.

$$\mathsf{id} \,;\, \left(\{\mathsf{B}_p = \langle\rangle\}Exec' \,;\, post_\Phi.x.(\widetilde{\Phi}^+)\right)^\omega \mathrel{\Supset_p} (\mathsf{id} \,;\, AExec)^\omega \,;\, \mathsf{id} \tag{26}$$

We use $\mathsf{id} \sqsupseteq_p \mathsf{id} \,;\, \mathsf{id}$ to split the first id on the right hand side of (26), then Law 1, to obtain $\mathsf{id} \,;\, (\mathsf{id} \,;\, AExec.p \,;\, \mathsf{id})^\omega$. Hence, using (19) followed by (20), the proof of (26) reduces again to a proof of

$$\{\mathsf{B}_p = \langle\rangle\}Exec' \,;\, post_\Phi.x.(\widetilde{\Phi}^+) \mathrel{\Supset_p} \mathsf{id} \,;\, AExec \,;\, \mathsf{id} \tag{27}$$

Then using (22), condition (2) and the fact that ; distributes over $\vee$, we are left with a number of proof obligations for each disjunct of $Exec'$. Of these, we present the most complex: the proof obligation for Acq_1'.

$$\{B_p = \langle\rangle\}Acq_1' ; \mathsf{id} ; Rel ; post_\Phi.x.(\widetilde{\Phi}^+) \ni_p \mathsf{id} ; AAcq ; \mathsf{id} ; ARel ; \mathsf{id} \qquad (28)$$

It is trivial to show $\{B_p = \langle\rangle\}\lceil x \bullet [\![\neg x]\!]\rceil_\Phi ; [\![\neg x]\!]^* ; [\![x]\!] ; Acq \sqsupseteq_p \mathsf{id} ; \{B_p = \langle\rangle\}Acq$, and using $\mathsf{id} \sqsubseteq_p \mathsf{id} ; \mathsf{id}$ the proof of (28) reduces as follows, where the initial part of Acq_1' is refined to id.

$$\{B_p = \langle\rangle\}Acq ; \mathsf{id} ; Rel ; post_\Phi.x.(\widetilde{\Phi}^+) \ni_p \mathsf{id} ; AAcq ; \mathsf{id} ; ARel ; \mathsf{id} \qquad (29)$$

We now focus on the initial part of the left hand side, where we distinguish between $ALoop \mathrel{\widehat{=}} \boxed{x \bullet [\![\neg x]\!]}_\Phi ; [\![\neg x]\!]^\omega ; [\![x]\!]$ and $ADo \mathrel{\widehat{=}} \boxed{x \bullet [\![x]\!] ; (x := 0)}_\Phi$.

$$\begin{array}{ll}
\{B_p = \langle\rangle\}Acq & \\
\sqsubseteq_p \{B_p = \langle\rangle\}ADo \vee \{B_p = \langle\rangle\}ALoop^{\omega+} ; ADo & \text{defn of } Acq\text{, then } g^\omega \equiv \varepsilon \vee g^{\omega+}
\end{array}$$

Using $\{B_p = \langle\rangle\}ALoop^{\omega+} \ni_p \mathsf{id}\{B_p = \langle\rangle\}$ (proof elided) condition (29) reduces to

$$\{B_p = \langle\rangle\}ADo ; \mathsf{id} ; Rel ; post_\Phi.x.(\widetilde{\Phi}^+) \ni_p AAcq ; \mathsf{id} ; ARel ; \mathsf{id} \qquad (30)$$

Again focusing on the initial part of the left hand side, we have:

$$\begin{array}{ll}
\{B_p = \langle\rangle\}\ ADo & \\
\sqsubseteq_p \{B_p = \langle\rangle\}\lceil x \bullet [\![x]\!] ; (x := 0)\rceil_\Phi & \text{defn of } ADo \text{ and by (9)} \\
\sqsubseteq_p \{B_p = \langle\rangle\}\boxed{x \bullet [\![x]\!]} ; \lceil x \bullet (x := 0)\rceil_\Phi & \text{by (10)} \\
\sqsupseteq_p \{B_p = \langle\rangle\}\boxed{x \bullet [\![x]\!]} ; \{B_p = \langle\rangle\}\lceil x \bullet (x := 0)\rceil_\Phi & \text{by logic} \\
\sqsupseteq_p \{B_p = \langle\rangle\}\boxed{x \bullet [\![x]\!]} ; \boxed{x \bullet \mathsf{id} ; (x \Leftarrow 0) ; \mathsf{id}}\{B_p = \langle\rangle\} & \text{by (11)} \\
\sqsubseteq_p \{B_p = \langle\rangle\}\boxed{x \bullet [\![x]\!] ; \mathsf{id}} ; \boxed{x \bullet (x \Leftarrow 0) ; \mathsf{id}}\{B_p = \langle\rangle\} & \text{using (12) twice} \\
\sqsupseteq_p \{B_p = \langle\rangle\}\boxed{x \bullet [\![x]\!]} ; \boxed{x \bullet (x \Leftarrow 0) ; \mathsf{id}}\{B_p = \langle\rangle\} & [\![x]\!] ; \mathsf{id} \sqsubseteq_p [\![x]\!] \\
\sqsupseteq_p \{B_p = \langle\rangle\}\boxed{x \bullet [x]} ; \boxed{x \bullet (x \Leftarrow 0) ; \mathsf{id}}\{B_p = \langle\rangle\} & \text{by (13)} \\
\sqsubseteq_p \boxed{x \bullet [x] ; (x \Leftarrow 0) ; \mathsf{id}}\{B_p = \langle\rangle\} & \text{by (12) and weakening}
\end{array}$$

It is straightforward to show $\boxed{x \bullet [x](x \Leftarrow 0) ; \mathsf{id}} \ni_p AAcq ; \mathsf{id}$, which expanding Rel and using (14) further reduces (30) to a proof of:

$$(\mathsf{id} ; x \leftarrow 1 ; \mathsf{id})\{B_p = \langle(x, 1)\rangle\} ; (\varepsilon \vee \Phi ; \mathsf{id}) ; post_\Phi.x.(\widetilde{\Phi}^+) \ni_p \mathsf{id} ; ARel ; \mathsf{id}$$

Using $\mathsf{id} \sqsubseteq_p \mathsf{id} ; \mathsf{id}$ the right hand side transforms to $\mathsf{id} ; \mathsf{id} ; ARel ; \mathsf{id}$. Then, using (16) and Theorem 2, we have $\mathsf{id} ; x \leftarrow 1 ; \mathsf{id} \ni_p \mathsf{id}$, and hence, using (19), we obtain $\{B_p = \langle(x, 1)\rangle\}(post_\Phi.x.(\widetilde{\Phi}^+) \vee (\Phi ; \mathsf{id} ; post_\Phi.x.(\widetilde{\Phi}^+)) \ni_p \mathsf{id} ; ARel ; \mathsf{id}$. Finally, because $\Phi ; \mathsf{id} ; post_\Phi.x.(\widetilde{\Phi}^+) \sqsupseteq_p post_\Phi.x.(\widetilde{\Phi}^+)$, we are left with

$$\{B_p = \langle(x, 1)\rangle\}post_\Phi.x.(\widetilde{\Phi}^+) \ni_p \mathsf{id} ; ARel ; \mathsf{id} \qquad (31)$$

Because there is exactly one item in the buffer, and because $post_\Phi.x.(\widetilde{\Phi}^+)$ guarantees that a flush occurs, the left hand side reduces to $\{B_p = \langle(x, 1)\rangle\}\mathsf{id} ; \Phi ; \mathsf{id}$. Finally, using the fact that B_p is a local variable of p and is not modified by id followed by (18), our proof is completed.

Notable in our verification is that concurrency aspects hardly need to be considered. The fact that the locking mechanisms guarantee safety is understood at the level of *Spec*. The refinement proof only requires consideration of the local buffer. This is in contrast to existing methods which require global conditions to be checked, e.g., [20] checks race conditions, [7,16,25] check linearizability, and [9] checks reduction. We conjecture that more complex examples will indeed require consideration of the behaviour of other processes. To this end, we will integrate compositional methods such as rely/guarantee into our framework [13,11].

6 Conclusions

Existing approaches to relaxed memory verification (e.g., [6,21,22,9,20,7,16]) focus on a low-level language (i.e., individual reads/writes), and hence, to perform a verification, programs need to be observed and understood in their (verbose) low-level representation. We are not aware of any approach that tries to lift memory model effects to a higher level of abstraction; our work here is hence unique in this sense [15].

The basic idea is to think of a statements as being executed over an interval of time or an execution window. Such execution windows can overlap if programs are executed concurrently and overlapping windows correspond to program instructions that can be executed in any order, representing the effect of concurrent executions and reorderings due to TSO. Overlapping execution windows may also interfere with each other and fixing the outcome of an execution within a window can influence the outcome within another. This paper presents several advances to the semantics in [15] by simplifying the interval logic, and program semantics, as well as developing buffer-specific rules for expression evaluation and refinement. The underlying rules are algebraic in nature, and hence, we provide generic transformation laws, which are in turn applied to our running example.

A difficulty when reasoning about TSO memory is that in addition to the normal non-determinism caused by concurrency, an additional level of non-determinism is introduced via use of local buffers. The methods in this paper allow one to reduce the non-determinism that must be considered when reasoning about local updates. In particular, we develop a notion of *local buffer refinement*, which allows one to proceed as if pending writes to local variables have already occurred in the abstract level. In particular, this means that local writes do not appear out of order. A similar observation is used for local transformation in the context of compilers for weak memory [8], however, these do not consider higher-level synchronisation instructions such as `lock`.

As part of future work, we aim to study the connections between local buffer refinement, and existing notions such as triangular race freedom [20] and reduction [9].

References

1. Adve, S.V., Gharachorloo, K.: Shared memory consistency models: A tutorial. IEEE Computer 29(12), 66–76 (1996)
2. Alglave, J.: A formal hierarchy of weak memory models. Formal Methods in System Design 41(2), 178–210 (2012)
3. Back, R.J.R., von Wright, J.: Reasoning algebraically about loops. Acta Informatica 36(4), 295–334 (1999)
4. Bovet, D., Cesati, M.: Understanding the Linux Kernel, 3rd edn. OReilly (2005)

5. Boyland, J.: Checking interference with fractional permissions. In: Cousot, R. (ed.) SAS 2003. LNCS, vol. 2694, pp. 55–72. Springer, Heidelberg (2003)
6. Burckhardt, S., Alur, R., Martin, M.M.K.: Checkfence: Checking consistency of concurrent data types on relaxed memory models. In: PLDI, pp. 12–21 (2007)
7. Burckhardt, S., Gotsman, A., Musuvathi, M., Yang, H.: Concurrent library correctness on the TSO memory model. In: Seidl, H. (ed.) Programming Languages and Systems. LNCS, vol. 7211, pp. 87–107. Springer, Heidelberg (2012)
8. Burckhardt, S., Musuvathi, M., Singh, V.: Verifying local transformations on relaxed memory models. In: Gupta, R. (ed.) CC 2010. LNCS, vol. 6011, pp. 104–123. Springer, Heidelberg (2010)
9. Cohen, E., Schirmer, B.: From total store order to sequential consistency: A practical reduction theorem. In: Kaufmann, M., Paulson, L.C. (eds.) ITP 2010. LNCS, vol. 6172, pp. 403–418. Springer, Heidelberg (2010)
10. Dongol, B., Derrick, J.: Data refinement for true concurrency. In: Derrick, J., Boiten, E.A., Reeves, S. (eds.) Refine. EPTCS, vol. 115, pp. 15–35 (2013)
11. Dongol, B., Derrick, J., Hayes, I.J.: Fractional permissions and non-deterministic evaluators in interval temporal logic. ECEASST 53 (2012)
12. Dongol, B., Hayes, I.J.: Deriving real-time action systems in a sampling logic. Sci. Comput. Program. 78(11), 2047–2063 (2013)
13. Dongol, B., Hayes, I.J., Derrick, J.: Deriving real-time action systems with multiple time bands using algebraic reasoning. Sci. of Comp. Prog. 85(Pt. B), 137–165 (2014)
14. Dongol, B., Hayes, I.J., Meinicke, L., Solin, K.: Towards an algebra for real-time programs. In: Kahl, W., Griffin, T.G. (eds.) RAMICS 2012. LNCS, vol. 7560, pp. 50–65. Springer, Heidelberg (2012)
15. Dongol, B., Travkin, O., Derrick, J., Wehrheim, H.: A high-level semantics for program execution under total store order memory. In: Liu, Z., Woodcock, J., Zhu, H. (eds.) ICTAC 2013. LNCS, vol. 8049, pp. 177–194. Springer, Heidelberg (2013)
16. Gotsman, A., Musuvathi, M., Yang, H.: Show no weakness: Sequentially consistent specifications of TSO libraries. In: Aguilera, M.K. (ed.) DISC 2012. LNCS, vol. 7611, pp. 31–45. Springer, Heidelberg (2012)
17. Hayes, I.J., Burns, A., Dongol, B., Jones, C.B.: Comparing degrees of non-determinism in expression evaluation. Comput. J. 56(6), 741–755 (2013)
18. Lamport, L.: How to make a multiprocessor computer that correctly executes multiprocess programs. IEEE Trans. Computers 28(9), 690–691 (1979)
19. Moszkowski, B.C.: A complete axiomatization of Interval Temporal Logic with infinite time. In: LICS, pp. 241–252 (2000)
20. Owens, S.: Reasoning about the implementation of concurrency abstractions on x86-TSO. In: D'Hondt, T. (ed.) ECOOP 2010. LNCS, vol. 6183, pp. 478–503. Springer, Heidelberg (2010)
21. Park, S., Dill, D.L.: An executable specification, analyzer and verifier for RMO (relaxed memory order). In: SPAA, pp. 34–41 (1995)
22. Sewell, P., Sarkar, S., Owens, S., Nardelli, F.Z., Myreen, M.O.: x86-TSO: a rigorous and usable programmer's model for x86 multiprocessors. Commun. ACM 53(7), 89–97 (2010)
23. Sorin, D.J., Hill, M.D., Wood, D.A.: A Primer on Memory Consistency and Cache Coherence. Synthesis Lectures on Computer Architecture. Morgan & Claypool (2011)
24. Spivey, J.M.: The Z Notation: A Reference Manual. Prentice Hall (1992)
25. Travkin, O., Mütze, A., Wehrheim, H.: SPIN as a linearizability checker under weak memory models. In: Bertacco, V., Legay, A. (eds.) HVC 2013. LNCS, vol. 8244, pp. 311–326. Springer, Heidelberg (2013)

Structural Refinement for the Modal nu-Calculus

Uli Fahrenberg, Axel Legay, and Louis-Marie Traonouez

Inria / IRISA, Campus de Beaulieu, 35042 Rennes CEDEX, France

Abstract. We introduce a new notion of structural refinement, a sound abstraction of logical implication, for the modal nu-calculus. Using new translations between the modal nu-calculus and disjunctive modal transition systems, we show that these two specification formalisms are structurally equivalent.

Using our translations, we also transfer the structural operations of composition and quotient from disjunctive modal transition systems to the modal nu-calculus. This shows that the modal nu-calculus supports composition and decomposition of specifications.

1 Introduction

There are two conceptually different approaches for the specification and verification of properties of formal models. *Logical* approaches make use of logical formulae for expressing properties and then rely on efficient model checking algorithms for verifying whether or not a model satisfies a formula. *Automata*-based approaches, on the other hand, exploit equivalence or refinement checking for verifying properties, given that models and properties are specified using the same (or a closely related) formalism.

The logical approaches have been quite successful, with a plethora of logical formalisms available and a number of successful model checking tools. One particularly interesting such formalism is the modal μ-calculus [21], which is universal in the sense that it generalizes most other temporal logics, yet mathematically simple and amenable to analysis.

One central problem in the verification of formal properties is *state space explosion*: when a model is composed of many components, the state space of the combined system quickly grows too big to be analyzed. To combat this problem, one approach is to employ *compositionality*. When a model consists of several components, each component would be model checked by itself, and then the components' properties would be composed to yield a property which automatically is satisfied by the combined model.

Similarly, given a global property of a model and a component of the model that is already known to satisfy a local property, one would be able to *decompose* automatically, from the global property and the local property, a new property which the rest of the model must satisfy. We refer to [23] for a good account of these and other features which one would wish specifications to have.

As an alternative to logical specification formalisms and with an eye to compositionality and decomposition, automata-based *behavioral* specifications were

G. Ciobanu and D. Méry (Eds.): ICTAC 2014, LNCS 8687, pp. 169–187, 2014.

$$\mathsf{AG}(\mathsf{req} \Rightarrow \mathsf{AX}(\mathsf{work}\ \mathsf{AW}\ \mathsf{grant}))$$

$$\nu X.([\mathsf{grant}, \mathsf{idle}, \mathsf{work}]X \wedge [\mathsf{req}]\nu Y.(\langle \mathsf{work}\rangle Y \vee \langle \mathsf{grant}\rangle X) \wedge [\mathsf{idle}, \mathsf{req}]\mathbf{ff})$$

$$X \overset{\nu}{=} [\mathsf{grant}, \mathsf{idle}, \mathsf{work}]X \wedge [\mathsf{req}]Y$$

$$Y \overset{\nu}{=} (\langle \mathsf{work}\rangle Y \vee \langle \mathsf{grant}\rangle X) \wedge [\mathsf{idle}, \mathsf{req}]\mathbf{ff}$$

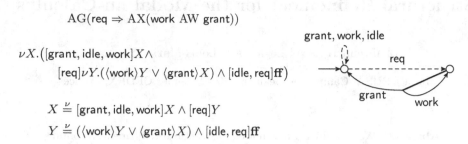

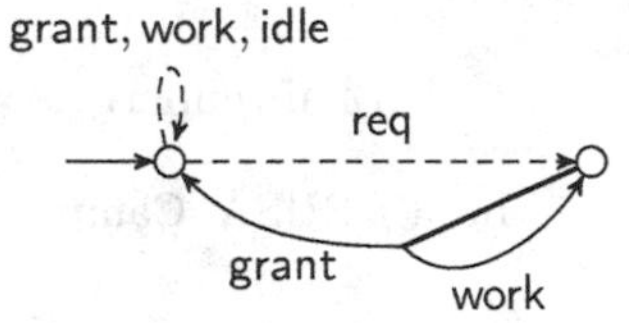

Fig. 1. An example property specified in CTL (top left), in the modal μ-calculus (below left), as a modal equation system (third left), and as a DMTS (right)

introduced in [22]. Here the specification formalism is a generalization of the modeling formalism, and the satisfaction relation between models and specifications is generalized to a refinement relation between specifications, which resembles simulation and bisimulation and can be checked with similar algorithms.

For an example, we refer to Fig. 1 which shows the property informally specified as "after a req(uest), no idle(ing) is allowed, but only work, until grant is executed" using the logical formalisms of CTL [14] and the modal μ-calculus [21] and the behavioral formalism of disjunctive modal transition systems [26].

The precise relationship between logical and behavioral specification formalisms has been subject to some investigation. In [22], Larsen shows that any *modal transition system* can be translated to a formula in Hennessy-Milner logic which is equivalent in the sense of admitting the same models. Conversely, Boudol and Larsen show in [11] that any formula in Hennessy-Milner logic is equivalent to a finite disjunction of modal transition systems.

We have picked up this work in [6], where we show that any *disjunctive modal transition system* (DMTS) is equivalent to a formula in the *modal ν-calculus*, the safety fragment of the modal μ-calculus which uses only maximal fixed points, and vice versa. (Note that the modal ν-calculus is equivalent to Hennessy-Milner logic with recursion and maximal fixed points.) Moreover, we show in [6] that DMTS are as expressive as (non-deterministic) *acceptance automata* [30,31]. Together with the inclusions of [7], this settles the expressivity question for behavioral specifications: they are at most as expressive as the modal ν-calculus.

In this paper, we show that not only are DMTS as expressive as the modal ν-calculus, but the two formalisms are *structurally equivalent.* Introducing a new notion of structural refinement for the modal ν-calculus (a sound abstraction of logical implication), we show that one can freely translate between the modal ν-calculus and DMTS, while preserving structural refinement.

DMTS form a *complete specification theory* [2] in that they both admit logical operations of conjunction and disjunction and structural operations of composition and quotient [6]. Hence they support full compositionality and decomposition in the sense of [23]. Using our translations, we can transport these notions to the modal ν-calculus, thus also turning the modal ν-calculus into a complete specification theory.

In order to arrive at our translations, we first recall DMTS and (non-deterministic) *acceptance automata* in Section 2. We also introduce a new hybrid modal logic, which can serve as compact representation for acceptance automata and should be of interest in itself. Afterwards we show, using the translations introduced in [6], that these formalisms are structurally equivalent.

In Section 3 we recall the modal ν-calculus and review the translations between DMTS and the modal ν-calculus which were introduced in [6]. These in turn are based on work by Boudol and Larsen in [11,22], hence fairly standard. We show that, though semantically correct, the two translations are structurally *mismatched* in that they relate DMTS refinement to two different notions of ν-calculus refinement. To fix the mismatch, we introduce a new translation from the modal ν-calculus to DMTS and show that using this translation, the two formalisms are structurally equivalent.

In Section 4, we use our translations to turn the modal ν-calculus into a complete specification theory. We remark that all our translations and constructions are based on a new *normal form* for ν-calculus expressions, and that turning a ν-calculus expression into normal form may incur an exponential blow-up. However, the translations and constructions preserve the normal form, so that this translation only need be applied once in the beginning.

We also note that composition and quotient operators are used in other logics such as *e.g.* spatial [13] or separation logics [32,28]. However, in these logics they are treated as *first-class* operators, *i.e.* as part of the formal syntax. In our approach, on the other hand, they are defined as operations on logical expressions which as results again yield logical expressions (without compositions or quotients).

Note that some proofs had to be omitted from this paper; these are available in its long version [17].

2 Structural Specification Formalisms

Let Σ be a finite set of labels. A *labeled transition system* (LTS) is a structure $\mathcal{I} = (S, S^0, \longrightarrow)$ consisting of a finite set of *states* S, a subset $S^0 \subseteq S$ of *initial states* and a *transition relation* $\longrightarrow \subseteq S \times \Sigma \times S$.

Disjunctive Modal Transition Systems. A *disjunctive modal transition system* (DMTS) is a structure $\mathcal{D} = (S, S^0, \dashrightarrow, \longrightarrow)$ consisting of finite sets $S \supseteq S^0$ of states and initial states, a *may*-transition relation $\dashrightarrow \subseteq S \times \Sigma \times S$, and a *disjunctive must*-transition relation $\longrightarrow \subseteq S \times 2^{\Sigma \times S}$. It is assumed that for all $(s, N) \in \longrightarrow$ and all $(a, t) \in N$, $(s, a, t) \in \dashrightarrow$.

As customary, we write $s \overset{a}{\dashrightarrow} t$ instead of $(s, a, t) \in \dashrightarrow$, $s \longrightarrow N$ instead of $(s, N) \in \longrightarrow$, $s \overset{a}{\dashrightarrow}$ if there exists t for which $s \overset{a}{\dashrightarrow} t$, and $s \overset{a}{\not\dashrightarrow}$ if there does not.

The intuition is that may-transitions $s \overset{a}{\dashrightarrow} t$ specify which transitions are permitted in an implementation, whereas a must-transitions $s \longrightarrow N$ stipulates a disjunctive requirement: at least one of the choices $(a, t) \in N$ must be implemented. A DMTS $(S, S^0, \dashrightarrow, \longrightarrow)$ is an *implementation* if $\longrightarrow = \{(s, \{(a, t)\}) \mid s \overset{a}{\dashrightarrow} t\}$; DMTS implementations are precisely LTS.

DMTS were introduced in [26] in the context of equation solving, or *quotient*, for specifications and are used *e.g.* in [5] for LTL model checking. They are a natural closure of *modal transition systems* (MTS) [22] in which all disjunctive must-transitions $s \longrightarrow N$ lead to singletons $N = \{(a,t)\}$.

Let $\mathcal{D}_1 = (S_1, S_1^0, \dashrightarrow_1, \longrightarrow_1)$, $\mathcal{D}_2 = (S_2, S_2^0, \dashrightarrow_2, \longrightarrow_2)$ be DMTS. A relation $R \subseteq S_1 \times S_2$ is a *modal refinement* if it holds for all $(s_1, s_2) \in R$ that

- for all $s_1 \overset{a}{\dashrightarrow} t_1$ there is $t_2 \in S_2$ with $s_2 \overset{a}{\dashrightarrow} t_2$ and $(t_1, t_2) \in R$, and
- for all $s_2 \longrightarrow N_2$ there is $s_1 \longrightarrow N_1$ such that for each $(a, t_1) \in N_1$ there is $(a, t_2) \in N_2$ with $(t_1, t_2) \in R$.

We say that $\mathcal{D}_1$ *modally refines* $\mathcal{D}_2$, denoted $\mathcal{D}_1 \leq_{\mathsf{m}} \mathcal{D}_2$, whenever there exists a modal refinement R such that for all $s_1^0 \in S_1^0$, there exists $s_2^0 \in S_2^0$ for which $(s_1^0, s_2^0) \in R$. We write $\mathcal{D}_1 \equiv_{\mathsf{m}} \mathcal{D}_2$ if $\mathcal{D}_1 \leq_{\mathsf{m}} \mathcal{D}_2$ and $\mathcal{D}_2 \leq_{\mathsf{m}} \mathcal{D}_1$. For states $s_1 \in S_1$, $s_2 \in S_2$, we write $s_1 \leq_{\mathsf{m}} s_2$ if the DMTS $(S_1, \{s_1\}, \dashrightarrow_1, \longrightarrow_1) \leq_{\mathsf{m}} (S_2, \{s_2\}, \dashrightarrow_2, \longrightarrow_2)$.

Note that modal refinement is reflexive and transitive, *i.e.* a preorder on DMTS. Also, the relation on states $\leq_{\mathsf{m}} \subseteq S_1 \times S_2$ defined above is itself a modal refinement, indeed the maximal modal refinement under the subset ordering.

The *set of implementations* of an DMTS $\mathcal{D}$ is $[\![\mathcal{D}]\!] = \{\mathcal{I} \leq_{\mathsf{m}} \mathcal{D} \mid \mathcal{I} \text{ implementation}\}$. This is, thus, the set of all LTS which satisfy the specification given by the DMTS $\mathcal{D}$. We say that $\mathcal{D}_1$ *thoroughly refines* $\mathcal{D}_2$, and write $\mathcal{D}_1 \leq_{\mathsf{th}} \mathcal{D}_2$, if $[\![\mathcal{D}_1]\!] \subseteq [\![\mathcal{D}_2]\!]$. We write $\mathcal{D}_1 \equiv_{\mathsf{th}} \mathcal{D}_2$ if $\mathcal{D}_1 \leq_{\mathsf{th}} \mathcal{D}_2$ and $\mathcal{D}_2 \leq_{\mathsf{th}} \mathcal{D}_1$. For states $s_1 \in S_1$, $s_2 \in S_2$, we write $[\![s_1]\!] = [\![(S_1, \{s_1\}, \dashrightarrow_1, \longrightarrow_1)]\!]$ and $s_1 \leq_{\mathsf{th}} s_2$ if $[\![s_1]\!] \subseteq [\![s_2]\!]$.

The below proposition, which follows directly from transitivity of modal refinement, shows that modal refinement is *sound* with respect to thorough refinement; in the context of specification theories, this is what one would expect, and we only include it for completeness of presentation. It can be shown that modal refinement is also *complete* for *deterministic* DMTS [8], but we will not need this here.

Proposition 1. *For all DMTS $\mathcal{D}_1$, $\mathcal{D}_2$, $\mathcal{D}_1 \leq_{\mathsf{m}} \mathcal{D}_2$ implies $\mathcal{D}_1 \leq_{\mathsf{th}} \mathcal{D}_2$.* □

We introduce a new construction on DMTS which will be of interest for us; intuitively, it adds all possible may-transitions without changing the implementation semantics. The *may-completion* of a DMTS $\mathcal{D} = (S, S^0, \dashrightarrow, \longrightarrow)$ is $\mathsf{mc}(\mathcal{D}) = (S, S^0, \dashrightarrow_{\mathsf{mc}}, \longrightarrow)$ with

$$\dashrightarrow_{\mathsf{mc}} = \{(s, a, t') \subseteq S \times \Sigma \times S \mid \exists (s, a, t) \in \dashrightarrow : t' \leq_{\mathsf{th}} t\}.$$

Note that to compute the may-completion of a DMTS, one has to decide thorough refinements, hence this computation (or, more precisely, deciding whether a given DMTS is may-complete) is EXPTIME-complete [9]. We show an example of a may-completion in Fig. 2.

Proposition 2. *For any DMTS $\mathcal{D}$, $\mathcal{D} \leq_{\mathsf{m}} \mathsf{mc}(\mathcal{D})$ and $\mathcal{D} \equiv_{\mathsf{th}} \mathsf{mc}(\mathcal{D})$.*

Proof. It is always the case that $\mathcal{D} \leq_{\mathsf{m}} \mathcal{D}$, and adding may transitions on the right side preserves modal refinement. Therefore it is immediate that $\mathcal{D} \leq_{\mathsf{m}} \mathsf{mc}(\mathcal{D})$, hence also $\mathcal{D} \leq_{\mathsf{th}} \mathsf{mc}(\mathcal{D})$.

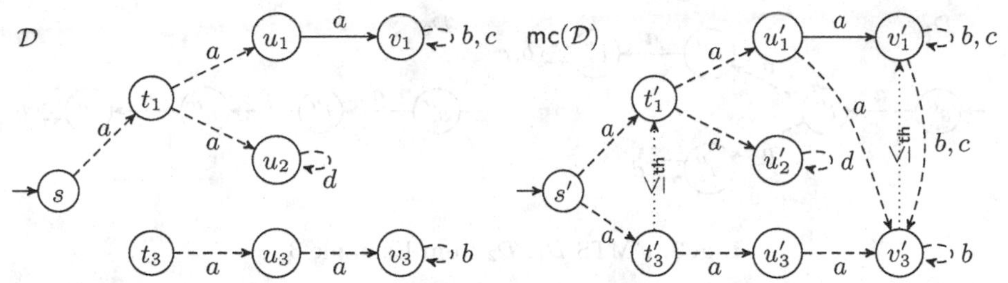

Fig. 2. A MTS $\mathcal{D}$ (left) and its may-completion $\mathsf{mc}(\mathcal{D})$ (right). In $\mathsf{mc}(\mathcal{D})$, the semantic inclusions which lead to extra may-transitions are depicted with dotted arrows.

To prove that $\mathsf{mc}(\mathcal{D}) \leq_{\mathsf{th}} \mathcal{D}$, we consider an implementation $\mathcal{I} \leq_{\mathsf{m}} \mathsf{mc}(\mathcal{D})$; we must prove that $\mathcal{I} \leq_{\mathsf{m}} \mathcal{D}$. Write $\mathcal{D} = (S, S^0, \dashrightarrow, \longrightarrow)$, $\mathcal{I} = (I, I^0, \dashrightarrow_I, \longrightarrow_I)$ and $\mathsf{mc}(\mathcal{D}) = (S, S^0, \dashrightarrow_{\mathsf{mc}}, \longrightarrow)$. Let $R \subseteq I \times S$ be the largest modal refinement between $\mathcal{I}$ and $\mathsf{mc}(\mathcal{D})$. We now prove that R is also a modal refinement between $\mathcal{I}$ and $\mathcal{D}$. For all $(i, d) \in R$:

- For all $i \overset{a}{\dashrightarrow}_I i'$, there exists $d' \in S$ such that $d \overset{a}{\dashrightarrow}_{\mathsf{mc}} d'$ and $(i', d') \in R$. Then by definition of $\dashrightarrow_{\mathsf{mc}}$, there exists $d'' \in S$ such that $d \overset{a}{\dashrightarrow} d''$ and $[\![d']\!] \subseteq [\![d'']\!]$. $(i', d') \in R$ implies $i' \in [\![d']\!]$, which implies $i' \in [\![d'']\!]$. This means that $i' \leq_{\mathsf{m}} d''$, and since R is the largest refinement relation in $I \times S$ it must be the case that $(i', d'') \in R$.
- The case of must transitions follows immediately, since must transitions are exactly the same in $\mathcal{D}$ and $\mathsf{mc}(\mathcal{D})$. $\square$

Example 3. The example in Fig. 2 shows that generally, $\mathsf{mc}(\mathcal{D}) \not\leq_{\mathsf{m}} \mathcal{D}$. First, $t_3 \leq_{\mathsf{th}} t_1$: For an implementation $\mathcal{I} = (I, I^0, \longrightarrow) \in [\![t_3]\!]$ with modal refinement $R \subseteq I \times \{t_3, u_3, v_3\}$, define $R' \subseteq I \times \{t_1, u_1, u_2, v_1\}$ by

$$\begin{aligned} R' = \{(i, t_1) \mid (i, t_3) \in R\} &\cup \{(i, v_1) \mid (i, v_3) \in R\} \\ &\cup \{(i, u_1) \mid (i, u_3) \in R, i \overset{a}{\longrightarrow}\} \\ &\cup \{(i, u_2) \mid (i, u_3) \in R, i \overset{a}{\not\longrightarrow}\}, \end{aligned}$$

then R' is a modal refinement $\mathcal{I} \leq_{\mathsf{m}} t_1$. Similarly, $t'_3 \leq_{\mathsf{th}} t'_1$ in $\mathsf{mc}(\mathcal{D})$.

On the other hand, $t_3 \not\leq_{\mathsf{m}} t_1$ (and similarly, $t'_3 \not\leq_{\mathsf{m}} t'_1$), because neither $u_3 \leq_{\mathsf{m}} u_1$ nor $u_3 \leq_{\mathsf{m}} u_2$. Now in the modal refinement game between $\mathsf{mc}(\mathcal{D})$ and $\mathcal{D}$, the may-transition $s' \overset{a}{\dashrightarrow} t'_3$ has to be matched by $s \overset{a}{\dashrightarrow} t_1$, but then $t'_3 \not\leq_{\mathsf{m}} t_1$, hence $\mathsf{mc}(\mathcal{D}) \not\leq_{\mathsf{m}} \mathcal{D}$.

Also, the may-completion does not necessarily preserve modal refinement: Consider the DMTS $\mathcal{D}$ from Fig. 2 and $\mathcal{D}_1$ from Fig. 3, and note first that $\mathsf{mc}(\mathcal{D}_1) = \mathcal{D}_1$. It is easy to see that $\mathcal{D} \leq_{\mathsf{m}} \mathcal{D}_1$ (just match states in $\mathcal{D}$ with their double-prime cousins in $\mathcal{D}_1$), but $\mathsf{mc}(\mathcal{D}) \not\leq_{\mathsf{m}} \mathsf{mc}(\mathcal{D}_1) = \mathcal{D}_1$: the may-transition $s' \overset{a}{\dashrightarrow} t'_3$ has to be matched by $s'' \overset{a}{\dashrightarrow} t''_1$ and $t'_3 \not\leq_{\mathsf{m}} t''_1$.

Lastly, the may-completion can also create modal refinement: Considering the DMTS $\mathcal{D}_2$ from Fig. 3, we see that $\mathcal{D}_2 \not\leq_{\mathsf{m}} \mathcal{D}$, but $\mathsf{mc}(\mathcal{D}_2) = \mathcal{D}_2 \leq_{\mathsf{m}} \mathsf{mc}(\mathcal{D})$.

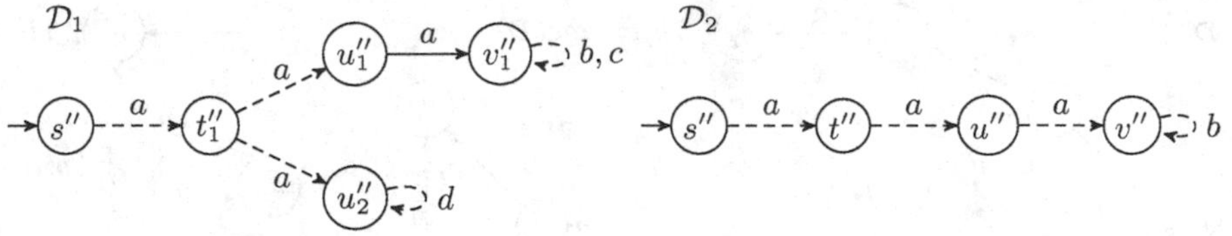

Fig. 3. DMTS $\mathcal{D}_1$, $\mathcal{D}_2$ from Example 3.

Acceptance Automata. A (non-deterministic) *acceptance automaton* (AA) is a structure $\mathcal{A} = (S, S^0, \text{Tran})$, with $S \supseteq S^0$ finite sets of states and initial states and $\text{Tran} : S \to 2^{2^{\Sigma \times S}}$ an assignment of *transition constraints*. We assume that for all $s^0 \in S^0$, $\text{Tran}(s^0) \neq \emptyset$.

An AA is an *implementation* if it holds for all $s \in S$ that $\text{Tran}(s) = \{M\}$ is a singleton; hence also AA implementations are precisely LTS. Acceptance automata were first introduced in [30] (see also [31], where a slightly different language-based approach is taken), based on the notion of acceptance trees in [20]; however, there they are restricted to be *deterministic*. We employ no such restriction here. The following notion of modal refinement for AA was also introduced in [30].

Let $\mathcal{A}_1 = (S_1, S_1^0, \text{Tran}_1)$ and $\mathcal{A}_2 = (S_2, S_2^0, \text{Tran}_2)$ be AA. A relation $R \subseteq S_1 \times S_2$ is a *modal refinement* if it holds for all $(s_1, s_2) \in R$ and all $M_1 \in \text{Tran}_1(s_1)$ that there exists $M_2 \in \text{Tran}_2(s_2)$ such that

- $\forall (a, t_1) \in M_1 : \exists (a, t_2) \in M_2 : (t_1, t_2) \in R$,
- $\forall (a, t_2) \in M_2 : \exists (a, t_1) \in M_1 : (t_1, t_2) \in R$.

As for DMTS, we write $\mathcal{A}_1 \leq_{\mathsf{m}} \mathcal{A}_2$ whenever there exists a modal refinement R such that for all $s_1^0 \in S_1^0$, there exists $s_2^0 \in S_2^0$ for which $(s_1^0, s_2^0) \in R$. Sets of implementations and thorough refinement are defined as for DMTS. Note that as both AA and DMTS implementations are LTS, it makes sense to use thorough refinement and equivalence *across* formalisms, writing *e.g.* $\mathcal{A} \equiv_{\mathsf{th}} \mathcal{D}$ for an AA $\mathcal{A}$ and a DMTS $\mathcal{D}$.

Hybrid Modal Logic. We introduce a hybrid modal logic which can serve as compact representation of AA. This logic is closely related to the Boolean modal transition systems of [7] and hybrid in the sense of [29,10]: it contains nominals, and the semantics of a nominal is given as all sets which contain the nominal.

For a finite set X of nominals, let $\mathcal{L}(X)$ be the set of formulae generated by the abstract syntax $\mathcal{L}(X) \ni \phi := \mathbf{tt} \mid \mathbf{ff} \mid \langle a \rangle x \mid \neg \phi \mid \phi \wedge \phi$, for $a \in \Sigma$ and $x \in X$. The semantics of a formula is a set of subsets of $\Sigma \times X$, given as follows: $(\!|\mathbf{tt}|\!) = 2^{\Sigma \times X}$, $(\!|\mathbf{ff}|\!) = \emptyset$, $(\!|\neg\phi|\!) = 2^{\Sigma \times X} \setminus (\!|\phi|\!)$, $(\!|\langle a \rangle x|\!) = \{M \subseteq \Sigma \times X \mid (a, x) \in M\}$, and $(\!|\phi \wedge \psi|\!) = (\!|\phi|\!) \cap (\!|\psi|\!)$. We also define disjunction $\phi_1 \vee \phi_2 = \neg(\phi_1 \wedge \phi_2)$.

An $\mathcal{L}$-*expression* is a structure $\mathcal{E} = (X, X^0, \Phi)$ consisting of finite sets $X^0 \subseteq X$ of variables and a mapping $\Phi : X \to \mathcal{L}(X)$. Such an expression is an *implementation* if $(\!|\Phi(x)|\!) = \{M\}$ is a singleton for each $x \in X$. It can easily be shown that $\mathcal{L}$-implementations precisely correspond to LTS.

Let $\mathcal{E}_1 = (X_1, X_1^0, \Phi_1)$ and $\mathcal{E}_2 = (X_2, X_2^0, \Phi_2)$ be $\mathcal{L}$-expressions. A relation $R \subseteq X_1 \times X_2$ is a *modal refinement* if it holds for all $(x_1, x_2) \in R$ and all $M_1 \in (\!|\Phi_1(x_1)|\!)$ that there exists $M_2 \in (\!|\Phi_2(x_2)|\!)$ such that

- $\forall (a, t_1) \in M_1 : \exists (a, t_2) \in M_2 : (t_1, t_2) \in R$,
- $\forall (a, t_2) \in M_2 : \exists (a, t_1) \in M_1 : (t_1, t_2) \in R$.

Again, we write $\mathcal{E}_1 \leq_{\mathsf{m}} \mathcal{E}_2$ whenever there exists such a modal refinement R such that for all $x_1^0 \in X_1^0$, there exists $x_2^0 \in X_2^0$ for which $(x_1^0, x_2^0) \in R$. Sets of implementations and thorough refinement are defined as for DMTS.

Structural Equivalence. We proceed to show that the three formalisms introduced in this section are structurally equivalent. Using the translations between AA and DMTS discovered in [6] and new translations between AA and hybrid logic, we show that these respect modal refinement.

The translations *al*, *la* between AA and our hybrid logic are straightforward: For an AA $\mathcal{A} = (S, S^0, \mathrm{Tran})$ and all $s \in S$, let

$$\Phi(s) = \bigvee_{M \in \mathrm{Tran}(s)} \Big(\bigwedge_{(a,t) \in M} \langle a \rangle t \wedge \bigwedge_{(b,u) \notin M} \neg \langle b \rangle u \Big)$$

and define the $\mathcal{L}$-expression $al(\mathcal{A}) = (S, S^0, \Phi)$.

For an $\mathcal{L}$-expression $\mathcal{E} = (X, X^0, \Phi)$ and all $x \in X$, let $\mathrm{Tran}(x) = (\!|\Phi(x)|\!)$ and define the AA $la(\mathcal{E}) = (X, X^0, \mathrm{Tran})$.

The translations *da*, *ad* between DMTS and AA were discovered in [6]. For a DMTS $\mathcal{D} = (S, S^0, \dashrightarrow, \longrightarrow)$ and all $s \in S$, let

$$\mathrm{Tran}(s) = \{ M \subseteq \Sigma \times S \mid \forall (a, t) \in M : s \overset{a}{\dashrightarrow} t, \forall s \longrightarrow N : N \cap M \neq \emptyset \}$$

and define the AA $da(\mathcal{D}) = (S, S^0, \mathrm{Tran})$.[1]

For an AA $\mathcal{A} = (S, S^0, \mathrm{Tran})$, define the DMTS $ad(\mathcal{A}) = (D, D^0, \dashrightarrow, \longrightarrow)$ as follows:

$$\begin{aligned} D &= \{ M \in \mathrm{Tran}(s) \mid s \in S \} \\ D^0 &= \{ M^0 \in \mathrm{Tran}(s^0) \mid s^0 \in S^0 \} \\ \longrightarrow &= \big\{ \big(M, \{(a, M') \mid M' \in \mathrm{Tran}(t)\}\big) \bigm| (a, t) \in M \big\} \\ \dashrightarrow &= \{ (M, a, M') \mid \exists M \longrightarrow N : (a, M') \in N \} \end{aligned}$$

Note that the state spaces of $\mathcal{A}$ and $ad(\mathcal{A})$ are not the same; the one of $ad(\mathcal{A})$ may be exponentially larger. The following lemma shows that this explosion is unavoidable:

Lemma 4. *There exists a one-state AA $\mathcal{A}$ for which any DMTS $\mathcal{D} \equiv_{\mathsf{th}} \mathcal{A}$ has at least 2^{n-1} states, where n is the size of the alphabet Σ.*

We notice that LTS are preserved by all translations: for any LTS $\mathcal{I}$, $al(\mathcal{I}) = la(\mathcal{I}) = da(\mathcal{I}) = ad(\mathcal{I}) = \mathcal{I}$. In [6] it is shown that the translations between

[1] Note that there is an error in the corresponding formula in [6].

AA and DMTS respect sets of implementations, *i.e.* that $da(\mathcal{D}) \equiv_{\mathsf{th}} \mathcal{D}$ and $ad(\mathcal{A}) \equiv_{\mathsf{th}} \mathcal{A}$ for all DMTS $\mathcal{D}$ and all AA $\mathcal{A}$. The next theorem shows that these and the other presented translations respect modal refinement, hence these formalisms are not only semantically equivalent, but *structurally equivalent.*

Theorem 5. *For all AA $\mathcal{A}_1$, $\mathcal{A}_2$, DMTS $\mathcal{D}_1, \mathcal{D}_2$ and $\mathcal{L}$-expressions $\mathcal{E}_1$, $\mathcal{E}_2$:*

1. $\mathcal{A}_1 \leq_{\mathsf{m}} \mathcal{A}_2$ *iff* $al(\mathcal{A}_1) \leq_{\mathsf{m}} al(\mathcal{A}_2)$,
2. $\mathcal{E}_1 \leq_{\mathsf{m}} \mathcal{E}_2$ *iff* $la(\mathcal{E}_1) \leq_{\mathsf{m}} la(\mathcal{E}_2)$,
3. $\mathcal{D}_1 \leq_{\mathsf{m}} \mathcal{D}_2$ *iff* $da(\mathcal{D}_1) \leq_{\mathsf{m}} da(\mathcal{D}_2)$, *and*
4. $\mathcal{A}_1 \leq_{\mathsf{m}} \mathcal{A}_2$ *iff* $ad(\mathcal{A}_1) \leq_{\mathsf{m}} ad(\mathcal{A}_2)$.

Proof (sketch). We give a few hints about the proofs of the equivalences; the details can be found in [17]. The first two equivalences follow easily from the definitions, once one notices that for both translations, $(\!|\Phi(x)|\!) = \text{Tran}(x)$ for all $x \in X$. For the third equivalence, we can show that a DMTS modal refinement $\mathcal{D}_1 \leq_{\mathsf{m}} \mathcal{D}_2$ is also an AA modal refinement $da(\mathcal{D}_1) \leq_{\mathsf{m}} da(\mathcal{D}_2)$ and vice versa.

The fourth equivalence is slightly more tricky, as the state space changes. If $R \subseteq S_1 \times S_2$ is an AA modal refinement relation witnessing $\mathcal{A}_1 \leq_{\mathsf{m}} \mathcal{A}_2$, then we can construct a DMTS modal refinement $R' \subseteq D_1 \times D_2$, which witnesses $ad(\mathcal{A}_1) \leq_{\mathsf{m}} ad(\mathcal{A}_2)$, by

$$\begin{aligned} R' = \{(M_1, M_2) \mid \exists (s_1, s_2) \in R : M_1 \in \text{Tran}_1(s_1), M_2 \in \text{Tran}(s_2), \\ \forall (a, t_1) \in M_1 : \exists (a, t_2) \in M_2 : (t_1, t_2) \in R, \\ \forall (a, t_2) \in M_2 : \exists (a, t_1) \in M_1 : (t_1, t_2) \in R\}. \end{aligned}$$

Conversely, if $R \subseteq D_1 \times D_2$ is a DMTS modal refinement witnessing $ad(\mathcal{A}_1) \leq_{\mathsf{m}} ad(\mathcal{A}_2)$, then $R' \subseteq S_1 \times S_2$ given by

$$R' = \{(s_1, s_2) \mid \forall M_1 \in \text{Tran}_1(s_1) : \exists M_2 \in \text{Tran}_2(s_2) : (M_1, M_2) \in R\}$$

is an AA modal refinement. □

The result on thorough equivalence from [6] now easily follows:

Corollary 6. *For all AA $\mathcal{A}$, DMTS $\mathcal{D}$ and $\mathcal{L}$-expressions $\mathcal{E}$, $al(\mathcal{A}) \equiv_{\mathsf{th}} \mathcal{A}$, $la(\mathcal{E}) \equiv_{\mathsf{th}} \mathcal{E}$, $da(\mathcal{D}) \equiv_{\mathsf{th}} \mathcal{D}$, and $ad(\mathcal{A}) \equiv_{\mathsf{th}} \mathcal{A}$.* □

Also soundness of modal refinement for AA and hybrid logic follows directly from Theorem 5:

Corollary 7. *For all AA $\mathcal{A}_1$ and $\mathcal{A}_2$, $\mathcal{A}_1 \leq_{\mathsf{m}} \mathcal{A}_2$ implies $\mathcal{A}_1 \leq_{\mathsf{th}} \mathcal{A}_2$. For all $\mathcal{L}$-expressions $\mathcal{E}_1$ and $\mathcal{E}_2$, $\mathcal{E}_1 \leq_{\mathsf{m}} \mathcal{E}_2$ implies $\mathcal{E}_1 \leq_{\mathsf{th}} \mathcal{E}_2$.* □

3 The Modal ν-Calculus

We wish to extend the structural equivalences of the previous section to the modal ν-calculus. Using translations between AA, DMTS and ν-calculus based on work in [22,11], it has been shown in [6] that ν-calculus and DMTS/AA are *semantically* equivalent. We will see below that there is a *mismatch* between the translations from [6] (and hence between the translations in [22,11]) which precludes structural equivalence and then proceed to propose a new translation which fixes the mismatch.

Syntax and Semantics. We first recall the syntax and semantics of the modal ν-calculus, the fragment of the modal μ-calculus [33,21] with only maximal fixed points. Instead of an explicit maximal fixed point operator, we use the representation by equation systems in Hennessy-Milner logic developed in [24].

For a finite set X of variables, let $\mathcal{H}(X)$ be the set of *Hennessy-Milner formulae*, generated by the abstract syntax $\mathcal{H}(X) \ni \phi ::= \mathbf{tt} \mid \mathbf{ff} \mid x \mid \langle a\rangle\phi \mid [a]\phi \mid \phi \wedge \phi \mid \phi \vee \phi$, for $a \in \Sigma$ and $x \in X$.

A *declaration* is a mapping $\Delta : X \to \mathcal{H}(X)$; we recall the maximal fixed point semantics of declarations from [24]. Let $(S, S^0, \longrightarrow)$ be an LTS, then an *assignment* is a mapping $\sigma : X \to 2^S$. The set of assignments forms a complete lattice with order $\sigma_1 \sqsubseteq \sigma_2$ iff $\sigma_1(x) \subseteq \sigma_2(x)$ for all $x \in X$ and lowest upper bound $\big(\bigsqcup_{i\in I} \sigma_i\big)(x) = \bigcup_{i\in I} \sigma_i(x)$.

The semantics of a formula is a subset of S, given relative to an assignment σ, defined as follows: $(\!|\mathbf{tt}|\!)\sigma = S$, $(\!|\mathbf{ff}|\!)\sigma = \emptyset$, $(\!|x|\!)\sigma = \sigma(x)$, $(\!|\phi \wedge \psi|\!)\sigma = (\!|\phi|\!)\sigma \cap (\!|\psi|\!)\sigma$, $(\!|\phi \vee \psi|\!)\sigma = (\!|\phi|\!)\sigma \cup (\!|\psi|\!)\sigma$, and

$$(\!|\langle a\rangle\phi|\!)\sigma = \{s \in S \mid \exists s \xrightarrow{a} s' : s' \in (\!|\phi|\!)\sigma\},$$
$$(\!|[a]\phi|\!)\sigma = \{s \in S \mid \forall s \xrightarrow{a} s' : s' \in (\!|\phi|\!)\sigma\}.$$

The semantics of a declaration Δ is then the assignment defined by

$$(\!|\Delta|\!) = \bigsqcup\{\sigma : X \to 2^S \mid \forall x \in X : \sigma(x) \subseteq (\!|\Delta(x)|\!)\sigma\};$$

the maximal (pre)fixed point of Δ.

A *ν-calculus expression* is a structure $\mathcal{N} = (X, X^0, \Delta)$, with $X^0 \subseteq X$ sets of variables and $\Delta : X \to \mathcal{H}(X)$ a declaration. We say that an LTS $\mathcal{I} = (S, S^0, \longrightarrow)$ *implements* (or models) the expression, and write $\mathcal{I} \models \mathcal{N}$, if it holds that for all $s^0 \in S^0$, there is $x^0 \in X^0$ such that $s^0 \in (\!|\Delta|\!)(x^0)$. We write $[\![\mathcal{N}]\!]$ for the set of implementations (models) of a ν-calculus expression $\mathcal{N}$. As for DMTS, we write $[\![x]\!] = [\![(X, \{x\}, \Delta)]\!]$ for $x \in X$, and thorough refinement of expressions and states is defined accordingly.

The following lemma introduces a *normal form* for ν-calculus expressions:

Lemma 8. *For any ν-calculus expression $\mathcal{N}_1 = (X_1, X_1^0, \Delta_1)$, there exists another expression $\mathcal{N}_2 = (X_2, X_2^0, \Delta_2)$ with $[\![\mathcal{N}_1]\!] = [\![\mathcal{N}_2]\!]$ and such that for any $x \in X$, $\Delta_2(x)$ is of the form*

$$\Delta_2(x) = \bigwedge_{i\in I}\Big(\bigvee_{j\in J_i} \langle a_{ij}\rangle x_{ij}\Big) \wedge \bigwedge_{a\in\Sigma} [a]\Big(\bigvee_{j\in J_a} y_{a,j}\Big) \tag{1}$$

for finite (possibly empty) index sets I, J_i, J_a, for $i \in I$ and $a \in \Sigma$, and all $x_{ij}, y_{a,j} \in X_2$. Additionally, for all $i \in I$ and $j \in J_i$, there exists $j' \in J_{a_{ij}}$ for which $x_{ij} \leq_{\mathsf{th}} y_{a_{ij},j'}$.

As this is a type of *conjunctive normal form*, it is clear that translating a ν-calculus expression into normal form may incur an exponential blow-up.

We introduce some notation for ν-calculus expressions in normal form which will make our life easier later. Let $\mathcal{N} = (X, X^0, \Delta)$ be such an expression and $x \in X$, with $\Delta(x) = \bigwedge_{i\in I}\big(\bigvee_{j\in J_i}\langle a_{ij}\rangle x_{ij}\big) \wedge \bigwedge_{a\in\Sigma}[a]\big(\bigvee_{j\in J_a} y_{a,j}\big)$ as in the lemma. Define $\Diamond(x) = \{\{(a_{ij}, x_{ij}) \mid j \in J_i\} \mid i \in I\}$ and, for each $a \in \Sigma$, $\Box^a(x) = \{y_{a,j} \mid j \in J_a\}$. Note that now $\Delta(x) = \bigwedge_{N\in\Diamond(x)}\big(\bigvee_{(a,y)\in N}\langle a\rangle y\big) \wedge \bigwedge_{a\in\Sigma}[a]\big(\bigvee_{y\in\Box^a(x)} y\big)$.

Refinement. In order to expose our structural equivalence, we need to introduce a notion of modal refinement for the modal ν-calculus. For reasons which will become apparent later, we define two different such notions:

Let $\mathcal{N}_1 = (X_1, X_1^0, \Delta_1)$, $\mathcal{N}_2 = (X_2, X_2^0, \Delta_2)$ be ν-calculus expressions in normal form and $R \subseteq X_1 \times X_2$. The relation R is a *modal refinement* if it holds for all $(x_1, x_2) \in R$ that

1. for all $a \in \Sigma$ and every $y_1 \in \Box_1^a(x_1)$, there is $y_2 \in \Box_2^a(x_2)$ for which $(y_1, y_2) \in R$, and
2. for all $N_2 \in \Diamond_2(x_2)$ there is $N_1 \in \Diamond_1(x_1)$ such that for each $(a, y_1) \in N_1$, there exists $(a, y_2) \in N_2$ with $(y_1, y_2) \in R$.

R is a *modal-thorough* refinement if, instead of 1., it holds that

1′. for all $a \in \Sigma$, all $y_1 \in \Box_1^a(x_1)$ and every $y_1' \in X_1$ with $y_1' \leq_{\mathsf{th}} y_1$, there is $y_2 \in \Box_2^a(x_2)$ and $y_2' \in X_2$ such that $y_2' \leq_{\mathsf{th}} y_2$ and $(y_1', y_2') \in R$.

We say that $\mathcal{N}_1$ *refines* $\mathcal{N}_2$ whenever there exists such a refinement R such that for every $x_1^0 \in X_1^0$ there exists $x_2^0 \in X_2^0$ for which $(x_1^0, x_2^0) \in R$. We write $\mathcal{N}_1 \leq_{\mathsf{m}} \mathcal{N}_2$ in case of modal and $\mathcal{N}_1 \leq_{\mathsf{mt}} \mathcal{N}_2$ in case of modal-thorough refinement.

We remark that whereas modal refinement for ν-calculus expressions is a simple and entirely syntactic notion, modal-thorough refinement involves semantic inclusions of states. Using results in [9], this implies that modal refinement can be decided in time polynomial in the size of the (normal-form) expressions, whereas deciding modal-thorough refinement is EXPTIME-complete.

Translation from DMTS to ν-calculus. Our translation from DMTS to ν-calculus is new, but similar to the translation from AA to ν-calculus given in [6]. This in turn is based on the *characteristic formulae* of [22] (see also [1]).

For a DMTS $\mathcal{D} = (S, S^0, \dashrightarrow, \longrightarrow)$ and all $s \in S$, we define $\Diamond(s) = \{N \mid s \longrightarrow N\}$ and, for each $a \in \Sigma$, $\Box^a(s) = \{t \mid s \overset{a}{\dashrightarrow} t\}$. Then, let

$$\Delta(s) = \bigwedge_{N\in\Diamond(s)}\Big(\bigvee_{(a,t)\in N}\langle a\rangle t\Big) \wedge \bigwedge_{a\in\Sigma}[a]\Big(\bigvee_{t\in\Box^a(s)} t\Big)$$

and define the (normal-form) ν-calculus expression $dh(\mathcal{D}) = (S, S^0, \Delta)$.

Note how the formula precisely expresses that we demand at least one of every choice of disjunctive must-transitions (first part) and permit all may-transitions (second part); this is also the intuition of the characteristic formulae of [22]. Using results of [6] (which introduces a very similar translation from AA to ν-calculus expressions), we see that $dh(\mathcal{D}) \equiv_{\mathsf{th}} \mathcal{D}$ for all DMTS $\mathcal{D}$.

Theorem 9. *For all DMTS $\mathcal{D}_1$ and $\mathcal{D}_2$, $\mathcal{D}_1 \leq_{\mathsf{m}} \mathcal{D}_2$ iff $dh(\mathcal{D}_1) \leq_{\mathsf{m}} dh(\mathcal{D}_2)$.*

Proof. For the forward direction, let $R \subseteq S_1 \times S_2$ be a modal refinement between $\mathcal{D}_1 = (S_1, S_1^0, \dashrightarrow_1, \longrightarrow_1)$ and $\mathcal{D}_2 = (S_2, S_2^0, \dashrightarrow_2, \longrightarrow_2)$; we show that R is also a modal refinement between $dh(\mathcal{D}_1) = (S_1, S_1^0, \Delta_1)$ and $dh(\mathcal{D}_2) = (S_2, S_2^0, \Delta_2)$. Let $(s_1, s_2) \in R$.

- Let $a \in \Sigma$ and $t_1 \in \Box_1^a(s_1)$, then $s_1 \overset{a}{\dashrightarrow}_1 t_1$, which implies that there is $t_2 \in S_2$ for which $s_2 \overset{a}{\dashrightarrow}_2 t_2$ and $(t_1, t_2) \in R$. By definition of $\Box_2^a$, $t_2 \in \Box_2^a(s_2)$.
- Let $N_2 \in \Diamond_2(s_2)$, then $s_2 \longrightarrow_2 N_2$, which implies that there exists $s_1 \longrightarrow_1 N_1$ such that $\forall (a, t_1) \in N_1 : \exists (a, t_2) \in N_2 : (t_1, t_2) \in R$. By definition of $\Box_1^a$, $N_1 \in \Box_1^a(s_1)$.

For the other direction, let $R \subseteq S_1 \times S_2$ be a modal refinement between $dh(\mathcal{D}_1)$ and $dh(\mathcal{D}_2)$, we show that R is also a modal refinement between $\mathcal{D}_1$ and $\mathcal{D}_2$. Let $(s_1, s_2) \in R$.

- For all $s_1 \overset{a}{\dashrightarrow}_1 t_1$, $t_1 \in \Box_1^a(s_1)$, which implies that there is $t_2 \in \Box_2^a(s_2)$ with $(t_1, t_2) \in R$, and by definition of $\Box_2^a$, $s_2 \overset{a}{\dashrightarrow}_2 t_2$.
- For all $s_2 \longrightarrow_2 N_2$, $N_2 \in \Diamond_2(s_2)$, which implies that there is $N_1 \in \Diamond_1(s_1)$ such that $\forall (a, t_1) \in N_1 : \exists (a, t_2) \in N_2 : (t_1, t_2) \in R$, and by definition of $\Box_1^a$, $s_1 \longrightarrow_1 N_1$. □

Old Translation from ν-calculus to DMTS. We recall the translation from ν-calculus to DMTS given in [6], which is based on a translation from Hennessy-Milner formulae (without recursion and fixed points) to sets of acyclic MTS in [11]. For a ν-calculus expression $\mathcal{N} = (X, X^0, \Delta)$ in normal form, let

$$\dashrightarrow = \{(x, a, y') \in X \times \Sigma \times X \mid \exists y \in \Box^a(x) : y' \leq_{\mathsf{th}} y\},$$
$$\longrightarrow = \{(x, N) \mid x \in X, N \in \Diamond(x)\}.$$

and define the DMTS $hd_{\mathsf{t}}(\mathcal{N}) = (X, X^0, \dashrightarrow, \longrightarrow)$.

Note how this translates diamonds to disjunctive must-transitions directly, but for boxes takes semantic inclusions into account: for a subformula $[a]y$, may-transitions are created to all variables which are semantically below y. This is consistent with the interpretation of formulae-as-properties: $[a]y$ means "for any a-transition, $\Delta(y)$ must hold"; but $\Delta(y)$ holds for all variables which are semantically below y.

It follows from results in [6] (which uses a slightly different normal form for ν-calculus expressions) that $hd_{\mathsf{t}}(\mathcal{N}) \equiv_{\mathsf{th}} \mathcal{N}$ for all ν-calculus expressions $\mathcal{N}$.

Theorem 10. *For all ν-calculus expressions, $\mathcal{N}_1 \leq_{\mathsf{mt}} \mathcal{N}_2$ iff $hd_{\mathsf{t}}(\mathcal{N}_1) \leq_{\mathsf{m}} hd_{\mathsf{t}}(\mathcal{N}_2)$.*

Proof. For the forward direction, let $R \subseteq X_1 \times X_2$ be a modal-thorough refinement between $\mathcal{N}_1 = (X_1, X_1^0, \Delta_1)$ and $\mathcal{N}_2 = (X_2, X_2^0, \Delta_2)$. We show that R is also a modal refinement between $hd_{\mathsf{t}}(\mathcal{N}_1) = (X_1, X_1^0, \dashrightarrow_1, \longrightarrow_2)$ and $hd_{\mathsf{t}}(\mathcal{N}_2) = (X_2, X_2^0, \dashrightarrow_2, \longrightarrow_2)$. Let $(x_1, x_2) \in R$.

- Let $x_1 \overset{a}{\dashrightarrow}_1 y_1'$. By definition of $\dashrightarrow_1$, there is $y_1 \in \Box_1^a(x_1)$ for which $y_1' \leq_{\mathsf{th}} y_1$. Then by modal-thorough refinement, this implies that there exists $y_2 \in$

$\Box_2^a(x_2)$ and $y_2' \in X_2$ such that $y_2' \leq_{\mathsf{th}} y_2$ and $(y_1', y_2') \in R$. By definition of $\dashrightarrow_2$ we have $x_2 \overset{a}{\dashrightarrow}_2 y_2'$.
- Let $x_2 \longrightarrow_2 N_2$, then we have $N_2 \in \Diamond_2(x_2)$. By modal-thorough refinement, this implies that there is $N_1 \in \Diamond_1(x_1)$ such that $\forall (a, y_1) \in N_1 : \exists (a, y_2) \in N_2 : (y_1, y_2) \in R$. By definition of $\longrightarrow_1$, $x_1 \longrightarrow_1 N_1$.

Now to the proof that $hd_t(\mathcal{N}_1) \leq_{\mathsf{m}} hd_t(\mathcal{N}_2)$ implies $\mathcal{N}_1 \leq_{\mathsf{mt}} \mathcal{N}_2$. We have a modal refinement (in the DMTS sense) $R \subseteq X_1 \times X_2$. We must show that R is also a modal-thorough refinement. Let $(x_1, x_2) \in R$.
- Let $a \in \Sigma$, $y_1 \in \Box_1^a(x_1)$ and $y_1' \in X_1$ such that $y_1' \leq_{\mathsf{th}} y_1$. Then by definition of $\dashrightarrow_1$, $x_1 \overset{a}{\dashrightarrow}_1 y_1'$. By modal refinement, this implies that there exists $x_2 \overset{a}{\dashrightarrow}_2 y'2$ with $(y_1', y_2') \in R$. Finally, by definition of $\dashrightarrow_2$, there exists $y_2 \in \Box_2^a(x_2)$ such that $y_2' \leq_{\mathsf{th}} y_2$.
- Let $N_2 \in \Diamond_2(x_2)$, then by definition of $\longrightarrow_2$, $x_2 \longrightarrow_2 N_2$. Then, by modal refinement, this implies that there exists $x_1 \longrightarrow_1 N_1$ such that $\forall (a, y_1) \in N_1 : \exists (a, y_2) \in N_2 : (y_1, y_2) \in R$. By definition of $\longrightarrow_1$, $N_1 \in \Box_1^a(x_1)$. $\Box$

Discussion. Notice how Theorems 9 and 10 expose a *mismatch* between the translations: *dh* relates DMTS refinement to ν-calculus *modal* refinement, whereas hd_t relates it to *modal-thorough* refinement. Both translations are well-grounded in the literature and well-understood, *cf.* [6,11,22], but this mismatch has not been discovered up to now. Given that the above theorems can be understood as universal properties of the translations, it means that there is no notion of refinement for ν-calculus which is consistent with them both.

The following lemma, easily shown by inspection, shows that this discrepancy is related to the may-completion for DMTS:

Lemma 11. *For any DMTS $\mathcal{D}$, $\mathsf{mc}(\mathcal{D}) = hd_t(dh(\mathcal{D}))$.* $\Box$

As a corollary, we see that modal refinement and modal-thorough refinement for ν-calculus are incomparable: Referring back to Example 3, we have $\mathcal{D} \leq_{\mathsf{m}} \mathcal{D}_1$, hence by Theorem 9, $dh(\mathcal{D}) \leq_{\mathsf{m}} dh(\mathcal{D}_1)$. On the other hand, we know that $\mathsf{mc}(\mathcal{D}) \not\leq_{\mathsf{m}} \mathsf{mc}(\mathcal{D}_1)$, *i.e.* by Lemma 11, $hd_t(dh(\mathcal{D})) \not\leq_{\mathsf{m}} hd_t(dh(\mathcal{D}_1))$, and then by Theorem 10, $dh(\mathcal{D}) \not\leq_{\mathsf{mt}} dh(\mathcal{D}_1)$.

To expose an example where modal-thorough refinement holds, but modal refinement does not, we note that $\mathsf{mc}(\mathcal{D}_2) \leq_{\mathsf{m}} \mathsf{mc}(\mathcal{D})$ implies, again using Lemma 11 and Theorem 10, that $dh(\mathcal{D}_2) \leq_{\mathsf{mt}} dh(\mathcal{D})$. On the other hand, we know that $\mathcal{D}_2 \not\leq_{\mathsf{m}} \mathcal{D}$, so by Theorem 9, $dh(\mathcal{D}_2) \not\leq_{\mathsf{m}} dh(\mathcal{D})$.

New Translation from ν-calculus to DMTS. We now show that the mismatch between DMTS and ν-calculus expressions can be fixed by introducing a new, simpler translation from ν-calculus to DMTS.

For a ν-calculus expression $\mathcal{N} = (X, X^0, \Delta)$ in normal form, let

$$\dashrightarrow = \{(x, a, y) \in X \times \Sigma \times X \mid y \in \Box^a(x)\},$$
$$\longrightarrow = \{(x, N) \mid x \in X, N \in \Diamond(x)\}.$$

and define the DMTS $hd(\mathcal{N}) = (X, X^0, \dashrightarrow, \longrightarrow)$. This is a simple syntactic translation: boxes are translated to disjunctive must-transitions and diamonds to may-transitions.

Theorem 12. *For all ν-calculus expressions, $\mathcal{N}_1 \leq_{\mathsf{m}} \mathcal{N}_2$ iff $hd(\mathcal{N}_1) \leq_{\mathsf{m}} hd(\mathcal{N}_2)$.*

Proof. Let $R \subseteq X_1 \times X_2$ be a modal refinement between $\mathcal{N}_1 = (X_1, X_1^0, \Delta_1)$ and $\mathcal{N}_2 = (X_2, X_2^0, \Delta_2)$; we show that R is also a modal refinement between $hd(\mathcal{N}_1) = (S_1, S_1^0, \dashrightarrow_1, \longrightarrow_1)$ and $hd(\mathcal{N}_2) = (S_2, S_2^0, \dashrightarrow_2, \longrightarrow_2)$. Let $(x_1, x_2) \in R$.

- Let $x_1 \overset{a}{\dashrightarrow}_1 y_1$, then $y_1 \in \Box_1^a(x_1)$, which implies that there exists $y_2 \in \Box_2^a(x_2)$ for which $(y_1, y_2) \in R$, and by definition of $\dashrightarrow_2$, $x_2 \overset{a}{\dashrightarrow}_2 y_2$.
- Let $x_2 \longrightarrow_2 N_2$, then $N_2 \in \Diamond_2(x_2)$, hence there is $N_1 \in \Diamond_1(x_1)$ such that $\forall (a, y_1) \in N_1 : \exists (a, y_2) \in N_2 : (y_1, y_2) \in R$, and by definition of $\longrightarrow_1$, $x_1 \longrightarrow_1 N_1$.

Now let $R \subseteq X_1 \times X_2$ be a modal refinement between $hd(\mathcal{N}_1)$ and $hd(\mathcal{N}_2)$, we show that R is also a modal refinement between $\mathcal{N}_1$ and $\mathcal{N}_2$. Let $(x_1, x_2) \in R$,

- Let $a \in \Sigma$ and $y_1 \in \Box_1^a(x_1)$. Then $x_1 \overset{a}{\dashrightarrow}_1 y_1$, which implies that there is $y_2 \in X_2$ for which $x_2 \overset{a}{\dashrightarrow}_2 y_2$ and $(y_1, y_2) \in R$, and by definition of $\dashrightarrow_2$, $t_2 \in \Box_2^a(s_2)$.
- Let $N_2 \in \Diamond_2(x_2)$, then $x_2 \longrightarrow_2 N_2$, so there is $x_1 \longrightarrow_1 N_1$ such that $\forall (a, y_1) \in N_1 : \exists (a, y_2) \in N_2 : (y_1, y_2) \in R$. By definition of $\longrightarrow_1$, $N_1 \in \Box_1^a(x_1)$. ☐

We finish the section by proving that also for the syntactic translation $hd(\mathcal{N}) \equiv_{\mathsf{th}} \mathcal{N}$ for all ν-calculus expressions; this shows that our translation can serve as a replacement for the partly-semantic hd_t translation from [6,11]. First we remark that dh and hd are inverses to each other:

Proposition 13. *For any ν-calculus expression $\mathcal{N}$, $dh(hd(\mathcal{N})) = \mathcal{N}$; for any DMTS $\mathcal{D}$, $hd(dh(\mathcal{D})) = \mathcal{D}$.* ☐

Corollary 14. *For all ν-calculus expressions $\mathcal{N}$, $hd(\mathcal{N}) \equiv_{\mathsf{th}} \mathcal{N}$.* ☐

4 The Modal ν-Calculus as a Specification Theory

Now that we have exposed a close structural correspondence between the modal ν-calculus and DMTS, we can transfer the operations which make DMTS a complete specification theory to the ν-calculus.

Refinement and Implementations. As for DMTS and AA, we can define an embedding of LTS into the modal ν-calculus so that implementation $\models$ and refinement $\leq_{\mathsf{m}}$ coincide. We say that a ν-calculus expression (X, X^0, Δ) in normal form is an *implementation* if $\Diamond(x) = \{\{(a, y)\} \mid y \in \Box^a(x), a \in \Sigma\}$ for all $x \in X$.

The ν-calculus translation of a LTS $(S, S^0, \longrightarrow)$ is the expression (S, S^0, Δ) in normal form with $\Diamond(s) = \{\{(a, t)\} \mid s \overset{a}{\longrightarrow} t\}$ and $\Box^a(s) = \{t \mid s \overset{a}{\longrightarrow} t\}$. This defines a bijection between LTS and ν-calculus implementations.

Theorem 15. *For any LTS $\mathcal{I}$ and any ν-calculus expression $\mathcal{N}$, $\mathcal{I} \models \mathcal{N}$ iff $\mathcal{I} \leq_{\mathsf{m}} \mathcal{N}$.*

Proof. $\mathcal{I} \models \mathcal{N}$ is the same as $\mathcal{I} \in [\![\mathcal{N}]\!]$, which by Corollary 14 is equivalent to $\mathcal{I} \in [\![hd(\mathcal{N})]\!]$. By definition, this is the same as $\mathcal{I} \leq_{\mathsf{m}} hd(\mathcal{N})$, which using Theorem 12 is equivalent to $\mathcal{I} \leq_{\mathsf{m}} \mathcal{N}$. □

Using transitivity, this implies that modal refinement for ν-calculus is sound:

Corollary 16. *For all ν-calculus expressions, $\mathcal{N}_1 \leq_{\mathsf{m}} \mathcal{N}_2$ implies $\mathcal{N}_1 \leq_{\mathsf{th}} \mathcal{N}_2$.* □

Disjunction and Conjunction. As for DMTS, disjunction of ν-calculus expressions is straight-forward. Given ν-calculus expressions $\mathcal{N}_1 = (X_1, X_1^0, \Delta_1)$, $\mathcal{N}_2 = (X_2, X_2^0, \Delta_2)$ in normal form, their *disjunction* is $\mathcal{N}_1 \vee \mathcal{N}_2 = (X_1 \cup X_2, X_1^0 \cup X_2^0, \Delta)$ with $\Delta(x_1) = \Delta_1(x_1)$ for $x_1 \in X_1$ and $\Delta(x_2) = \Delta_2(x_2)$ for $x_2 \in X_2$.

The *conjunction* of ν-calculus expressions like above is $\mathcal{N}_1 \wedge \mathcal{N}_2 = (X, X^0, \Delta)$ defined by $X = X_1 \times X_2$, $X^0 = X_1^0 \times X_2^0$, $\Box^a(x_1, x_2) = \Box_1^a(x_1) \times \Box_2^a(x_2)$ for each $(x_1, x_2) \in X$, $a \in \Sigma$, and for each $(x_1, x_2) \in X$,

$$\begin{aligned}\Diamond(x_1, x_2) = \big\{\{(a, (y_1, y_2)) \mid (a, y_1) \in N_1, (y_1, y_2) \in \Box^a(x_1, x_2)\} \bigm| N_1 \in \Diamond_1(x_1)\big\} \\ \cup \big\{\{(a, (y_1, y_2)) \mid (a, y_2) \in N_2, (y_1, y_2) \in \Box^a(x_1, x_2)\} \bigm| N_2 \in \Diamond_2(x_2)\big\}.\end{aligned}$$

Note that both $\mathcal{N}_1 \vee \mathcal{N}_2$ and $\mathcal{N}_1 \wedge \mathcal{N}_2$ are again ν-calculus expressions in normal form.

Theorem 17. *For all ν-calculus expressions $\mathcal{N}_1$, $\mathcal{N}_2$, $\mathcal{N}_3$ in normal form,*
- *$\mathcal{N}_1 \vee \mathcal{N}_2 \leq_{\mathsf{m}} \mathcal{N}_3$ iff $\mathcal{N}_1 \leq_{\mathsf{m}} \mathcal{N}_3$ and $\mathcal{N}_2 \leq_{\mathsf{m}} \mathcal{N}_3$,*
- *$\mathcal{N}_1 \leq_{\mathsf{m}} \mathcal{N}_2 \wedge \mathcal{N}_3$ iff $\mathcal{N}_1 \leq_{\mathsf{m}} \mathcal{N}_2$ and $\mathcal{N}_1 \leq_{\mathsf{m}} \mathcal{N}_3$,*
- *$[\![\mathcal{N}_1 \vee \mathcal{N}_2]\!] = [\![\mathcal{N}_1]\!] \cup [\![\mathcal{N}_2]\!]$, and $[\![\mathcal{N}_1 \wedge \mathcal{N}_2]\!] = [\![\mathcal{N}_1]\!] \cap [\![\mathcal{N}_2]\!]$.*

Theorem 18. *With operations $\vee$ and $\wedge$, the class of ν-calculus expressions forms a bounded distributive lattice up to $\equiv_{\mathsf{m}}$.*

The bottom element (up to $\equiv_{\mathsf{m}}$) in the lattice is the empty ν-calculus expression $\bot = (\emptyset, \emptyset, \emptyset)$, and the top element (up to $\equiv_{\mathsf{m}}$) is $\top = (\{s\}, \{s\}, \Delta)$ with $\Delta(s) = \mathbf{tt}$.

Structural Composition. The structural composition operator for a specification theory is to mimic, at specification level, the structural composition of implementations. That is to say, if $\|$ is a composition operator for implementations (LTS), then the goal is to extend $\|$ to specifications such that for all specifications $\mathcal{S}_1$, $\mathcal{S}_2$,

$$[\![\mathcal{S}_1 \| \mathcal{S}_2]\!] = \{\mathcal{I}_1 \| \mathcal{I}_2 \mid \mathcal{I}_1 \in [\![\mathcal{S}_1]\!], \mathcal{I}_2 \in [\![\mathcal{S}_2]\!]\}. \tag{2}$$

For simplicity, we use CSP-style synchronization for structural composition of LTS, however, our results readily carry over to other types of composition. Analogously to the situation for MTS [8], we have the following negative result:

Theorem 19. *There is no operator $\|$ for the ν-calculus which satisfies* (2).

Proof. We first note that due to Theorem 17, it is the case that implementation sets of ν-calculus expressions are closed under disjunction: for any ν-calculus expression $\mathcal{N}$ and $\mathcal{I}_1, \mathcal{I}_2 \in [\![\mathcal{N}]\!]$, also $\mathcal{I}_1 \vee \mathcal{I}_2 \in [\![\mathcal{N}]\!]$.

Now assume there were an operator as in the theorem, then because of the translations, (2) would also hold for DMTS. Hence for all DMTS $\mathcal{D}_1, \mathcal{D}_2$, $\{\mathcal{I}_1 \| \mathcal{I}_2 \mid \mathcal{I}_1 \in [\![\mathcal{D}_1]\!], \mathcal{I}_2 \in [\![\mathcal{D}_2]\!]\}$ would be closed under disjunction. But Example 7.8 in [8] exhibits two DMTS (actually, MTS) for which this is not the case, a contradiction. □

Given that we cannot have (2), the revised goal is to have a *sound* composition operator for which the right-to-left inclusion holds in (2). We can obtain one such from the structural composition of AA introduced in [6]. We hence define, for ν-calculus expressions $\mathcal{N}_1 = (X_1, X_1^0, \Delta_1)$, $\mathcal{N}_2 = (X_2, X_2^0, \Delta_2)$ in normal form, $\mathcal{N}_1 \parallel \mathcal{N}_2 = ah(ha(\mathcal{N}_1) \parallel_{\mathsf{A}} ha(\mathcal{N}_2))$, where $\parallel_{\mathsf{A}}$ is AA composition and we write $ah = dh \circ ad$ and $ha = da \circ hd$ for the composed translations.

Notice that the involved translation from AA to DMTS may lead to an exponential blow-up. Unraveling the definition gives us the following explicit expression for $\mathcal{N}_1 \parallel \mathcal{N}_2 = (X, X^0, \Delta)$:

- $X = \{\{(a,(y_1,y_2)) \mid \forall i \in \{1,2\} : (a, y_i) \in M_i\} \mid \forall i \in \{1,2\} : M_i \subseteq \Sigma \times X_i, \exists x_i \in X_i : \forall (a, y_i') \in M_i : y_i' \in \Box_i^a(x_i), \forall N_i \in \Diamond_i(x_i) : N_i \cap M_i \neq \emptyset\}$,
- $X^0 = \{\{(a,(y_1,y_2)) \mid \forall i \in \{1,2\} : (a, y_i) \in M_i\} \mid \forall i \in \{1,2\} : M_i \subseteq \Sigma \times X_i, \exists x_i \in X_i^0 : \forall (a, y_i') \in M_i : y_i' \in \Box_i^a(x_i), \forall N_i \in \Diamond_i(x_i) : N_i \cap M_i \neq \emptyset\}$,
- $\Diamond(x) = \{\{(a,\{(b,(z_1,z_2)) \mid \forall i \in \{1,2\} : (b, z_i) \in M_i\} \mid \forall i \in \{1,2\} : M_i \subseteq \Sigma \times X_i, \forall (a, z_i') \in M_i : z_i' \in \Box_i^b(y_i), \forall N_i \in \Diamond_i(y_i) : N_i \cap M_i \neq \emptyset\} \mid (a,(y_1,y_2)) \in x\}$ for each $x \in X$, and
- $\Box^a(x) = \{y \mid \exists N \in \Diamond(x) : (a, y) \in N\}$.

Theorem 20. *For all ν-calculus expressions $\mathcal{N}_1$, $\mathcal{N}_2$, $\mathcal{N}_3$, $\mathcal{N}_4$ in normal form, $\mathcal{N}_1 \leq_{\mathsf{m}} \mathcal{N}_3$ and $\mathcal{N}_2 \leq_{\mathsf{m}} \mathcal{N}_4$ imply $\mathcal{N}_1 \parallel \mathcal{N}_2 \leq_{\mathsf{m}} \mathcal{N}_3 \parallel \mathcal{N}_4$.*

Proof. This follows directly from the analogous property for AA [6] and the translation theorems 5, 9 and 12. □

This implies the right-to-left inclusion in (2), *i.e.* $\{\mathcal{I}_1 \parallel \mathcal{I}_2 \mid \mathcal{I}_1 \in [\![\mathcal{N}_1]\!], \mathcal{I}_2 \in [\![\mathcal{N}_2]\!]\} \subseteq [\![\mathcal{N}_1 \parallel \mathcal{N}_2]\!]$. It also entails *independent implementability*, in that the structural composition of the two refined specifications $\mathcal{N}_1, \mathcal{N}_2$ is a refinement of the composition of the original specifications $\mathcal{N}_3, \mathcal{N}_4$. Fig. 4 shows an example of the DMTS analogue of this structural composition.

Quotient. The quotient operator / for a specification theory is used to synthesize specifications for components of a structural composition. Hence it is to have the property, for all specifications $\mathcal{S}$, $\mathcal{S}_1$ and all implementations $\mathcal{I}_1$, $\mathcal{I}_2$, that

$$\mathcal{I}_1 \in [\![\mathcal{S}_1]\!] \text{ and } \mathcal{I}_2 \in [\![\mathcal{S} / \mathcal{S}_1]\!] \text{ imply } \mathcal{I}_1 \parallel \mathcal{I}_2 \in [\![\mathcal{S}]\!]. \qquad (3)$$

Furthermore, $\mathcal{S} / \mathcal{S}_1$ is to be as permissive as possible.

We can again obtain such a quotient operator for ν-calculus from the one for AA introduced in [6]. Hence we define, for ν-calculus expressions $\mathcal{N}_1$, $\mathcal{N}_2$ in

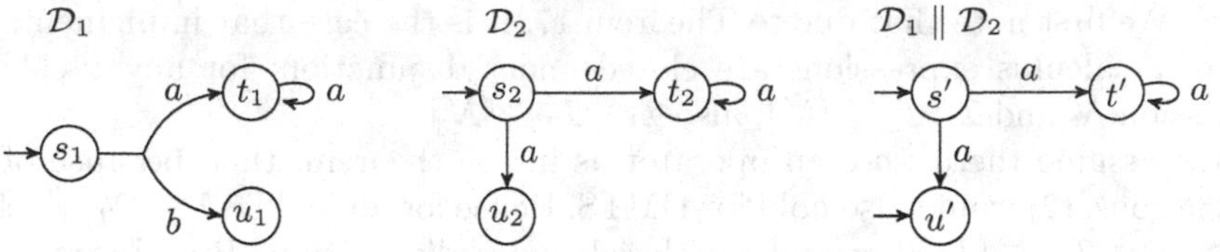

Fig. 4. DMTS $\mathcal{D}_1$, $\mathcal{D}_2$ and the reachable parts of their structural composition $\mathcal{D}_1 \parallel \mathcal{D}_2$. Here, $s' = \{(a,(t_1,t_2)),(a,(t_1,u_2))\}$, $t' = \{(a,(t_1,t_2))\}$ and $u' = \emptyset$. Note that $\mathcal{D}_1 \parallel \mathcal{D}_2$ has two initial states.

normal form, $\mathcal{N}_1 \,/\, \mathcal{N}_2 = ah(ha(\mathcal{N}_1) \,/_{\mathsf{A}}\, ha(\mathcal{N}_2))$, where $/_{\mathsf{A}}$ is AA quotient. We recall the construction of $/_{\mathsf{A}}$ from [6]:

Let $\mathcal{A}_1 = (S_1, S_1^0, \mathrm{Tran}_1)$, $\mathcal{A}_2 = (S_2, S_2^0, \mathrm{Tran}_2)$ be AA and define $\mathcal{A}_1 \,/_{\mathsf{A}}\, \mathcal{A}_2 = (S, \{s^0\}, \mathrm{Tran})$, with $S = 2^{S_1 \times S_2}$, $s^0 = \{(s_1^0, s_2^0) \mid s_1^0 \in S_1^0, s_2^0 \in S_2^0\}$, and Tran given as follows:

Let $\mathrm{Tran}(\emptyset) = 2^{\Sigma \times \{\emptyset\}}$. For $s = \{(s_1^1, s_2^1), \dots, (s_1^n, s_2^n)\} \in S$, say that $a \in \Sigma$ is *permissible from* s if it holds for all $i = 1, \dots, n$ that there is $M_1 \in \mathrm{Tran}_1(s_1^i)$ and $t_1 \in S_1$ for which $(a, t_1) \in M_1$, or else there is no $M_2 \in \mathrm{Tran}_2(s_2^i)$ and no $t_2 \in S_2$ for which $(a, t_2) \in M_2$.

For a permissible from s and $i \in \{1, \dots, n\}$, let $\{t_2^{i,1}, \dots, t_2^{i,m_i}\} = \{t_2 \in S_2 \mid \exists M_2 \in \mathrm{Tran}_2(s_2^i) : (a, t_2) \in M_2\}$ be an enumeration of the possible states in S_2 after an a-transition and define $pt_a(s) = \big\{\{(t_1^{i,j}, t_2^{i,j}) \mid i = 1, \dots, n, j = 1, \dots, m_i\} \bigm| \forall i : \forall j : \exists M_1 \in \mathrm{Tran}_1(s_1^i) : (a, t_1^{i,j}) \in M_1\big\}$, the set of all sets of possible assignments of next-a states from s_1^i to next-a states from s_2^i.

Now let $pt(s) = \{(a, t) \mid t \in pt_a(s), a \text{ admissible from } s\}$ and define $\mathrm{Tran}(s) = \{M \subseteq pt(s) \mid \forall i = 1, \dots, n : \forall M_2 \in \mathrm{Tran}_2(s_2^i) : M \rhd M_2 \in \mathrm{Tran}_1(s_1^i)\}$. Here $\rhd$ is the composition-projection operator defined by $M \rhd M_2 = \{(a, t \rhd t_2) \mid (a, t) \in M, (a, t_2) \in M_2\}$ and $t \rhd t_2 = \{(t_1^1, t_2^1), \dots, (t_1^k, t_2^k)\} \rhd t_2^i = t_1^i$ (note that by construction, there is precisely one pair in t whose second component is t_2^i).

Theorem 21. *For all ν-calculus expressions $\mathcal{N}, \mathcal{N}_1, \mathcal{N}_2$ in normal form, $\mathcal{N}_2 \le_{\mathsf{m}} \mathcal{N} \,/\, \mathcal{N}_1$ iff $\mathcal{N}_1 \parallel \mathcal{N}_2 \le_{\mathsf{m}} \mathcal{N}$.*

Proof. From the analogous property for AA [6] and Theorems 5, 9 and 12. □

As a corollary, we get (3): If $\mathcal{I}_2 \in [\![\mathcal{N} / \mathcal{N}_1]\!]$, *i.e.* $\mathcal{I}_2 \le_{\mathsf{m}} \mathcal{N} / \mathcal{N}_1$, then $\mathcal{N}_1 \| \mathcal{I}_2 \le_{\mathsf{m}} \mathcal{N}$, which using $\mathcal{I}_1 \le_{\mathsf{m}} \mathcal{N}_1$ and Theorem 20 implies $\mathcal{I}_1 \parallel \mathcal{I}_2 \le_{\mathsf{m}} \mathcal{N}_1 \parallel \mathcal{I}_2 \le_{\mathsf{m}} \mathcal{N}$. The reverse implication in Theorem 21 implies that $\mathcal{N} \,/\, \mathcal{N}_1$ is as permissive as possible.

Theorem 22. *With operations $\wedge$, $\vee$, $\parallel$ and $/$, the class of ν-calculus expressions forms a commutative residuated lattice up to $\equiv_{\mathsf{m}}$.*

The unit of $\parallel$ (up to $\equiv_{\mathsf{m}}$) is the ν-calculus expression corresponding to the LTS $\mathsf{U} = (\{u\}, \{u\}, \{(u, a, u) \mid a \in \Sigma\})$. We refer to [19] for a good reference on commutative residuated lattices.

5 Conclusion and Further Work

Using new translations between the modal ν-calculus and DMTS, we have exposed a structural equivalence between these two specification formalisms. This means that both types of specifications can be freely mixed; there is no more any need to decide, whether due to personal preference or for technical reasons, between one and the other. Of course, the modal ν-calculus can only express safety properties; for more expressivity, one has to turn to more expressive logics, and no behavioral analogue to these stronger logics is known (neither is it likely to exist, we believe).

Our constructions of composition and quotient for the modal ν-calculus expect (and return) ν-calculus expressions in normal form, and it is an interesting question whether they can be defined for general ν-calculus expressions. (For disjunction and conjunction this is of course trivial.) Larsen's [23] has composition and quotient operators for Hennessy-Milner logic (restricted to "deterministic context systems"), but we know of no extension (other than ours) to more general logics.

We also note that our hybrid modal logic appears related to the *Boolean equation systems* [27,25] which are used in some μ-calculus model checking algorithms. The precise relation between the modal ν-calculus, our $\mathcal{L}$-expressions and Boolean equation systems should be worked out. Similarly, acceptance automata bear some similarity to the *modal automata* of [12].

Lastly, we should note that we have in [4,3] introduced *quantitative* specification theories for weighted modal transition systems. These are well-suited for specification and analysis of systems with quantitative information, in that they replace the standard Boolean notion of refinement with a robust distance-based notion. We are working on an extension of these quantitative formalisms to DMTS, and hence to the modal ν-calculus, which should relate our work to other approaches at quantitative model checking such as *e.g.* [16,15,18].

References

1. Aceto, L., Ingólfsdóttir, A., Larsen, K.G., Srba, J.: Reactive Systems. Cambridge Univ. Press (2007)
2. Bauer, S.S., David, A., Hennicker, R., Guldstrand Larsen, K., Legay, A., Nyman, U., Wąsowski, A.: Moving from specifications to contracts in component-based design. In: de Lara, J., Zisman, A. (eds.) Fundamental Approaches to Software Engineering. LNCS, vol. 7212, pp. 43–58. Springer, Heidelberg (2012)
3. Bauer, S.S., Fahrenberg, U., Juhl, L., Larsen, K.G., Legay, A., Thrane, C.: Quantitative refinement for weighted modal transition systems. In: Murlak, F., Sankowski, P. (eds.) MFCS 2011. LNCS, vol. 6907, pp. 60–71. Springer, Heidelberg (2011)
4. Bauer, S.S., Fahrenberg, U., Legay, A., Thrane, C.: General quantitative specification theories with modalities. In: Hirsch, E.A., Karhumäki, J., Lepistö, A., Prilutskii, M. (eds.) CSR 2012. LNCS, vol. 7353, pp. 18–30. Springer, Heidelberg (2012)
5. Beneš, N., Černá, I., Křetínský, J.: Modal transition systems: Composition and LTL model checking. In: Bultan, T., Hsiung, P.-A. (eds.) ATVA 2011. LNCS, vol. 6996, pp. 228–242. Springer, Heidelberg (2011)

6. Beneš, N., Delahaye, B., Fahrenberg, U., Křetínský, J., Legay, A.: Hennessy-milner logic with greatest fixed points as a complete behavioural specification theory. In: D'Argenio, P.R., Melgratti, H. (eds.) CONCUR 2013 – Concurrency Theory. LNCS, vol. 8052, pp. 76–90. Springer, Heidelberg (2013)
7. Beneš, N., Křetínský, J., Larsen, K.G., Møller, M.H., Srba, J.: Parametric modal transition systems. In: Bultan, T., Hsiung, P.-A. (eds.) ATVA 2011. LNCS, vol. 6996, pp. 275–289. Springer, Heidelberg (2011)
8. Beneš, N., Křetínský, J., Larsen, K.G., Srba, J.: On determinism in modal transition systems. Th. Comp. Sci. 410(41), 4026–4043 (2009)
9. Beneš, N., Křetínský, J., Larsen, K.G., Srba, J.: EXPTIME-completeness of thorough refinement on modal transition systems. Inf. Comp. 218, 54–68 (2012)
10. Blackburn, P.: Representation, reasoning, and relational structures: a hybrid logic manifesto. Log. J. IGPL 8(3), 339–365 (2000)
11. Boudol, G., Larsen, K.G.: Graphical versus logical specifications. Th. Comp. Sci. 106(1), 3–20 (1992)
12. Bradfield, J., Stirling, C.: Modal mu-calculi. In: The Handbook of Modal Logic. Elsevier (2006)
13. Caires, L., Cardelli, L.: A spatial logic for concurrency. Inf. Comp. 186(2) (2003)
14. Clarke, E.M., Emerson, E.A.: Design and synthesis of synchronization skeletons using branching-time temporal logic. In: Kozen, D. (ed.) Logic of Programs 1981. LNCS, vol. 131, pp. 52–71. Springer, Heidelberg (1982)
15. de Alfaro, L.: Quantitative verification and control via the mu-calculus. In: Amadio, R.M., Lugiez, D. (eds.) CONCUR 2003. LNCS, vol. 2761, pp. 103–127. Springer, Heidelberg (2003)
16. de Alfaro, L., Henzinger, T.A., Majumdar, R.: Discounting the future in systems theory. In: Baeten, J.C.M., Lenstra, J.K., Parrow, J., Woeginger, G.J. (eds.) ICALP 2003. LNCS, vol. 2719, pp. 1022–1037. Springer, Heidelberg (2003)
17. Fahrenberg, U., Legay, A., Traonouez, L.-M.: Structural refinement for the modal nu-calculus. CoRR, 1402.2143 (2014), `http://arxiv.org/abs/1402.2143`
18. Gebler, D., Fokkink, W.: Compositionality of probabilistic Hennessy-Milner logic through structural operational semantics. In: Koutny, M., Ulidowski, I. (eds.) CONCUR 2012. LNCS, vol. 7454, pp. 395–409. Springer, Heidelberg (2012)
19. Hart, J.B., Rafter, L., Tsinakis, C.: The structure of commutative residuated lattices. Internat. J. Algebra Comput. 12(4), 509–524 (2002)
20. Hennessy, M.: Acceptance trees. J. ACM 32(4), 896–928 (1985)
21. Kozen, D.: Results on the propositional μ-calculus. Th. Comp. Sci. 27 (1983)
22. Larsen, K.G.: Modal specifications. In: Sifakis, J. (ed.) CAV 1989. LNCS, vol. 407, pp. 232–246. Springer, Heidelberg (1990)
23. Larsen, K.G.: Ideal specification formalism = expressivity + compositionality + decidability + testability +... In: Baeten, J.C.M., Klop, J.W. (eds.) CONCUR 1990. LNCS, vol. 458. Springer, Heidelberg (1990)
24. Larsen, K.G.: Proof systems for satisfiability in Hennessy-Milner logic with recursion. Th. Comp. Sci. 72(2&3), 265–288 (1990)
25. Larsen, K.G.: Efficient local correctness checking. In: Probst, D.K., von Bochmann, G. (eds.) CAV 1992. LNCS, vol. 663, pp. 30–43. Springer, Heidelberg (1993)
26. Larsen, K.G., Xinxin, L.: Equation solving using modal transition systems. In: LICS. IEEE Computer Society (1990)
27. Mader, A.: Verification of Modal Properties Using Boolean Equation Systems. PhD thesis, Technische Universität München (1997)

28. O'Hearn, P.W., Reynolds, J.C., Yang, H.: Local reasoning about programs that alter data structures. In: Fribourg, L. (ed.) CSL 2001 and EACSL 2001. LNCS, vol. 2142, pp. 1–19. Springer, Heidelberg (2001)
29. Prior, A.N.: Papers on Time and Tense. Clarendon Press, Oxford (1968)
30. Raclet, J.-B.: Residual for component specifications. Publication interne 1843, IRISA, Rennes (2007)
31. Raclet, J.-B.: Residual for component specifications. Electr. Notes Theor. Comput. Sci. 215, 93–110 (2008)
32. Reynolds, J.C.: Separation logic: A logic for shared mutable data structures. In: LICS. IEEE Computer Society (2002)
33. Scott, D., de Bakker, J.W.: A theory of programs. IBM, Vienna (1969) (unpublished manuscript)

Precise Interprocedural Side-Effect Analysis

Manuel Geffken, Hannes Saffrich, and Peter Thiemann

Universität Freiburg, Germany
{geffken,saffriha,thiemann}@informatik.uni-freiburg.de

Abstract. A side-effect analysis computes for each program phrase a set of memory locations that may be read or written to when executing this phrase. Our analysis expresses abstract objects, points-to and aliasing information, escape information, and side effects all in terms of a single novel abstract domain, generalized access graphs. This abstract domain represents sets of access paths precisely and compactly. It is suitable for intraprocedural analysis as well as for constructing method summaries for interprocedural analysis.

We implement the side-effect analysis for Java on top of the SOOT framework and report on its application to selected examples.

1 Introduction

A side-effect analysis computes for each program phrase a set of memory locations that may be read or written to when executing this phrase. The results of such an analysis have many uses in practice including the identification of *pure* methods (that have no side effects), of read-only parameters, and of objects that escape from a method. A compiler may perform aggressive code motion on a call to a pure method. Such a method may also be used in a specification [2]. In a concurrent program, methods with disjoint side effects may run in parallel without interfering [1]. Several program analyses require information on side effects of method calls to correctly transfer local analysis results across call sites [7,9,16]. Furthermore, the search space of a software model checker may be reduced by ignoring interleavings of methods with disjoint side effects [5].

Our analysis expresses abstract objects, points-to and aliasing information, escape information, and side effects all in terms of *generalized access graphs*, a novel abstract domain inspired by Deutsch's *symbolic access paths* (SAPs) [8] and Khedker and coworkers' access graphs [15]. A value in this domain represents information about heap-allocated objects using a regular language of access paths in the pre-state of a method. This condensation of information in one domain facilitates an elegant and economic description of the analysis that completely fits into this paper and it simplifies its implementation.

The analysis computes context insensitive may-information for method summaries. From the method summary, it is straightforward to determine whether parameters are read-only, whether a method is pure, and whether any heap-allocated objects may escape from the method.

G. Ciobanu and D. Méry (Eds.): ICTAC 2014, LNCS 8687, pp. 188–205, 2014.

The intraprocedural analysis that computes the method summaries is flow-sensitive and performs strong updates for local and global variables. The interprocedural analysis is an instance of a bottom-up analysis.

We implement the side-effect analysis for Java on top of the SOOT [23] framework and report on preliminary results of its application to selected examples.

1.1 Contributions

- We define the abstract domain of generalized access graphs (GAGs) as an extension of Khedker and coworkers' access graphs [15]. In comparison, our analysis requires fewer and simpler operations on the domain and can make do with one domain to maintain all kinds of information.
- We specify the intraprocedural GAG-based analysis for a CFG representation of a method in an object-based language (without pointer arithmetic). It is integrated with a context-insensitive bottom-up interprocedural analysis.
- We present preliminary results from applying our implementation (see `https://github.com/saffriha/ictac2014`) to a range of benchmarks.

1.2 Outline

Section 2 informally presents the GAG-based side-effect analysis and its supporting analyses, in particular, points-to analysis. Section 3 formally defines the domain of generalized access graphs and its operations and establishes its basic properties. Section 4 specifies the intraprocedural points-to analysis and Section 6 extends it to a side-effect analysis. Section 7 sketches the interprocedural analysis. Section 8 reports on our experiments with the implementation. Section 9 discusses related work and Section 10 concludes.

2 Side-Effect Analysis

This section presents motivational examples that demonstrate various aspects of our side-effect analysis and in particular uses of GAGs in method summaries.

Our running example concerns some methods written for objects of class `List`, which represent a node in a linked list with integer elements.

```
class List {
  int v;
  List n;
  List(int v, List n) {
    this.v = v;
    this.n = n;
  }
}
```

2.1 Simple Method

The method `foo` takes a list as its argument. It modifies the list node and returns the rest of the argument list.

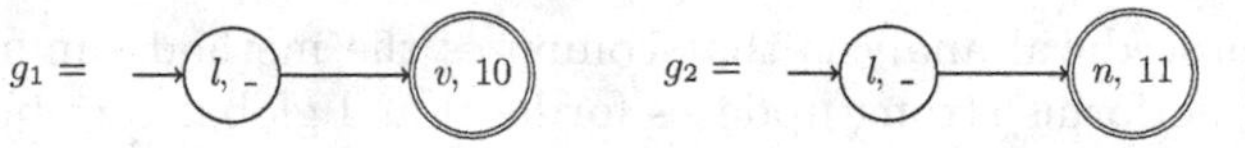

Fig. 1. Write effect and abstract return object of `foo`

```
 9 List foo(List l) {
10   l.v = 0;
11   return l.n;
12 }
```

The GAG g_1 in Fig. 1 describes the write effect of the method whereas g_2 describes the returned value (and also the read effect). Both specify the outcome in isolation, that is, without regard for the calling context of the method.

Each path through a GAG from a root node to an accepting node corresponds to a potential access path that starts from any object that may be supplied for the root node later on. Each identifier on this path is interpreted as a component of the access path. In general, a GAG may represent any number of paths.

In this case, the root node is l in both graphs. It stands for the formal parameter of the method. Further, each graph represents exactly one path, g_1 represents the path $l.v$ whereas g_2 represents $l.n$.

The GAG also includes the program point of each field access in its nodes. The example uses line numbers to indicate program points. If the program points are important, then we write access paths with program points as superscripts as in $l.v^{10}$. These program point annotations play a crucial role in effectively finding a fixpoint during the analysis.

Thus, the write effect of `foo` consists of the potential modification of $l.v$ whereas the returned object is described by $l.n$. That is, the GAG describes the set of parameter rooted SAPs written by the method. The GAG computed by the analysis yields access paths that are valid in the pre-state of the method, that is, in the program state at invocation time of the method.

The return value of `foo` is described by g_2 as the abstract object that can be reached via the path $l.n^{11}$. Again, the superscript 11 indicates the program point containing the access to the field n.

2.2 Loops

So far we have seen how GAGs represent summaries of the side effects and the return value of a simple method. Next, we put the body of `foo` in a loop to observe summarization at work. The comments in the listing contain the abstract object bound to r after analyzing the line in regular expression notation.

```
13 List loop(List l) {        // round 1 // round 2  // round 3
14   List r = l;              // r ↦ l
15   while (r.v != 42) {  //           // r ↦ l.n? // r ↦ l.n*
16     r.v = 0;
17     r = r.n;               // r ↦ l.n // r ↦ l.n+ // r ↦ l.n+
18   }                                             // r ↦ l.n*
19   return r;
20 }
```

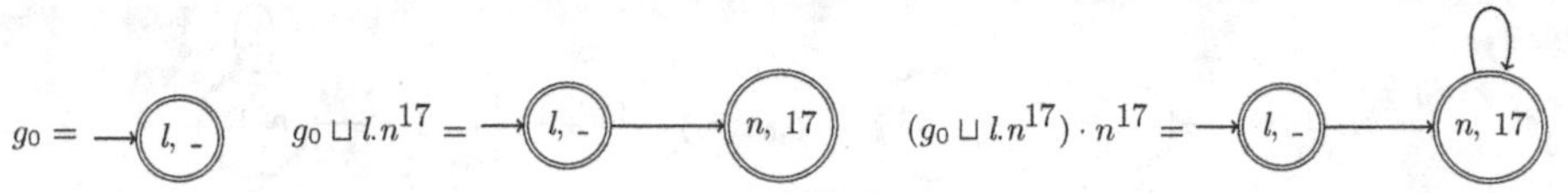

Fig. 2. GAGs to analyze `loop`

Before entering line 15, our analysis finds that r points to l. This fact is represented by an environment that maps r to the abstract object represented by the GAG g_0 (Fig. 2). After line 17, unsurprisingly, r points to the object represented by $l.n^{17}$. Then the body of the while loop is reanalyzed with r bound to the join of g_0 and $l.n^{17}$. By the end of the loop, r is updated to $(g_0 \sqcup l.n^{17}) \cdot n^{17}$, which does not change anymore. This fixpoint is reached because each access at a particular program point is represented at most once in an access graph. See Section 3 for the formal definition of the operators. Thus, the return value corresponds to the path set generated by the regular expression $l.n^*$. Similarly, the side effect may be computed as $l.n^*.v$ (see Fig. 4).

2.3 Method Calls

Next, we change the program to call the method `foo` in the loop.

```
21 List loopCall(List l) {
22   List r = l;
23   while (r != null)
24     r = foo(r);
25   return r;
26 }
```

Up to line 24, r is bound to g_0 as before. Analyzing the method call just fetches the return value g_2 from `foo`'s method summary. In general, this value is phrased in terms of the formal parameters of the callee, so that it must be translated to the caller. In this case, the translation replaces l by l, so the body of the while loop is reanalyzed with r bound to $g_0 \sqcup g_2$ (Fig. 3).

Reanalyzing the call yields the same return value g_2, but now the parameter value $g_0 \sqcup g_2$ must be substituted for l in g_2, which yields $(g_0 \sqcup g_2) \cdot n^{11}$ (Fig. 3) representing the same access paths as $(g_0 \sqcup l.n^{17}) \cdot n^{17}$ in the analysis of *loop*.

The write effect is computed in a similar way. In the first analysis of the method call, we replace l with l in `foo`'s summary resulting in a write effect of $l.v^{10}$. In the next iteration, $g_0 \sqcup g_2$ is substituted for the formal parameter in `foo`'s summary. The resulting write effect is $l.(n^{11})?.v^{10}$. The abstract object resulting from the return value is $l.(n^{11})^+$. As seen in Fig. 2, the concatenation of a field access at the same program point results in a loop on the GAG node representing this field access. The new environment entry for r is $r \mapsto l.(n^{11})^*$ with the resulting write effect $l.(n^{11})^*.v^{10}$.

The next iteration reaches the fixpoint. The method summary for *loopCall* consists of the abstract object from r's environment entry $l.(n^{11})^*$ and the write effect $l.(n^{11})^*.v^{10}$ (see Fig. 4).

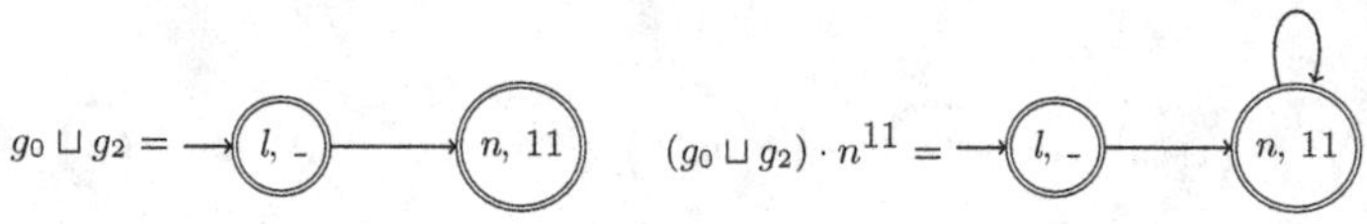

Fig. 3. GAGs to analyze `loopCall`

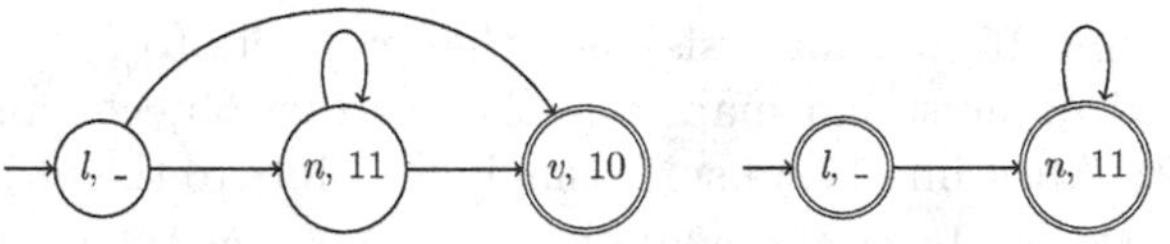

Fig. 4. `loopCall`'s write effects (left) and abstract return object (right)

2.4 New Object Abstraction

The next examples deal with allocation and aliasing. They unveil some further information that is computed by the analysis: points-to and escape information. To simplify the presentation, we ignore side effects in the following examples. Thus, for the rest of this section we only regard two components of a method summary:

1. may-points-to information that describes aliases created by the method and
2. the abstract object returned by the method.

First, we turn to the representation of new objects. Here is a method that constructs and attaches a new element to a list.

```
27 List newList(int v, List n) {
28   assert (v < 10);
29   List l = new List(v, n);
30   return l;
31 }
```

The constructor call on line 29 creates a points-to entry $\texttt{new}^{29}.n^6 \rightarrow n$ (with the 6 coming from the `List` constructor) and the assignment on line 29 creates an environment entry for l of $\texttt{new}^{29}$. Thus, the method summary consists of the points-to relation $\{\texttt{new}^{29}.n^6 \rightarrow n\}$ and the abstract return object $\texttt{new}^{29}$, which indicates that this object may escape.

Given the method summary, we consider a method that calls `newList` in a loop.

```
List newNList() {
  int i = 0;
  List r = null;
  while (i < 10) {
    r = newList(i, r);
    i++;
  }
  return r;
}
```

The points-to relation in the method summary of `newNList` is $\texttt{new}^{29}.n^{6} \rightarrow \texttt{new}^{29}$ and the abstract return object is $\texttt{new}^{29}$. We do not explicitly represent `null` in the abstract domain as we only consider may-points-to information.

2.5 Aliasing

Another group of examples demonstrates how aliasing is described in terms of the pre-state of the method. We omit program points and use the symbol • as the rightmost field selection operator in SAPs describing references. The following method swaps the first two entries of the passed list and returns the resulting list.

```
List swap(List l) {
  List ln = l.n;
  List lnn = ln.n;
  l.n = lnn;
  ln.n = l;
  return ln;
}
```

The method's points-to summary is $l \bullet n \rightarrow l.n.n,\ l.n \bullet n \rightarrow l$. The GAG $l.n$ describes the returned value. This result demonstrates that the summary consists of paths that refer to the pre-state of the method.

A final aliasing example illustrates the interaction between the information from the method summary and the points-to set at the call site. An auxiliary method `depth2` overwrites the n field two elements down the list. Method `depth1` overwrites the n field of the first element and then calls `depth2`.

```
48 void depth2 (List x, List y) {
49   List v = y.n;          // env: v ↦ y.n
50   v.n = x;               // points-to: {(y.n) • n → x}
51 }
52 void depth1 (List x, List y, List z) {
53   y.n = z;               // points-to: {y • n → z}
54   depth2 (x, y);         // points-to: {y • n → z, (y.n) • n → x, z • n → x}
55 }
```

The annotations in the listing state the intermediate results of the analysis after executing the statement on the line. The annotation `env` states the binding of v and `points-to` states the accumulated points-to set up to that point. Most annotations are straightforward, except the points-to set on line 54 after the call to `depth2`. The first entry, $y \bullet n \rightarrow z$, is carried through from the previous statement. The second entry, $(y.n) \bullet n \rightarrow x$, is obtained by substituting the abstraction of the formal parameters in the points-to summary of `depth2` from line 50. In this case, the substitution is the identity. The last entry, $z \bullet n \rightarrow x$, is generated by our resolution algorithm that integrates the points-to information at the call site with the method summary. The abstract object $y.n$ in the summary interacts with the first points-to entry that states that $y \bullet n$ may also point to z. Thus, the last entry results from contracting $y.n$ to z in the summary. No entry can be removed because the points-to information is not definitive (may-points-to information).

2.6 Global Variables

A final example demonstrates side effects on global variables.

```
void setGlobal(List l) {
  Global.g = l;
}
```

The points-to relation in `setGlobal`'s summary is $Global \bullet g \to l$.

3 Abstract Domain: Generalized Access Graphs

A generalized access graph represents a possibly infinite set of access paths relative to a set of roots. These roots are usually abstract objects like method parameters or allocation points.

Definition 1. *An* occurrence *of an identifier or an allocation in a program P is specified with an element from* $Occ_P = (ID_P \uplus \{\texttt{new}\}) \times PP_P$ *where* ID_P *is the set of identifiers occurring in P and* PP_P *is the set of program points of P. We write* $\texttt{C}.x \in ID_P$ *to refer to static global variable x of class* C*.*

In the rest of this section, we take for granted that all definitions are relative to an arbitrary, fixed program P and thus drop the $_P$ subscript.

Definition 2. *A* generalized access graph *for program P is a tuple* $\langle N, E, A, R\rangle$

- $N \subseteq Occ$ *is a set of identifiers or allocation occurrences of P,*
- $E \subseteq N \times N$ *is a set of directed edges,*
- $A \subseteq N$ *is a set of* accepting nodes, *and*
- $R \subseteq N$ *is a set of* root nodes.

If g is a generalized access graph, then we sometimes write $N(g)$, $E(g)$, $A(g)$, and $R(g)$ for its components.

As Occ is finite, the set GAG of all generalized access graphs for P is also finite. We write $id(n)$ and $pp(n)$ to extract the identifier and program point of a node $n \in N$.

Khedker et al. proposed a closely related notion of access graphs [15], which we generalize in several respects: we allow non-accepting nodes, we allow several root nodes instead of just a single root node, and we do not distinguish between "normal" access graphs and "remainder" graphs as they do.

We use two notations, a graphical one and another based on regular expressions, to concisely write generalized access graphs. Both are inspired by the close relationship to nondeterministic finite automata. Consider the access graph with $N = \{\langle x, 1\rangle, \langle f, 2\rangle, \langle f, 3\rangle\}$, $E = \{\langle\langle x, 1\rangle, \langle f, 2\rangle\rangle, \langle\langle f, 2\rangle, \langle f, 2\rangle\rangle, \langle\langle f, 2\rangle, \langle f, 3\rangle\rangle\}$, $A = \{\langle f, 3\rangle\}$, and $R = \{\langle x, 1\rangle\}$. Its regular expression notation is $x^1.f^2{+}.f^3$ and Fig. 5 shows its graphical representation.

Definition 3. *Let* $g \in GAG$ *be an access graph. The* indexed path language $L^p(g) \subseteq Occ^*$ *of g is defined as the language of the nondeterministic finite automaton* $\langle N \uplus \{q_0\}, Occ, \delta, \{q_0\}, A\rangle$ *where* δ *is the smallest relation such that*

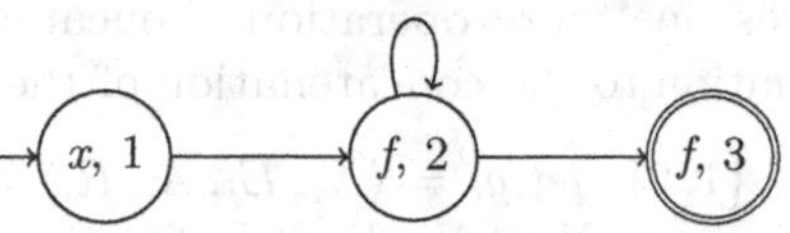

Fig. 5. Example access graph

- $(q_0, \langle x, p\rangle, \langle x, p\rangle) \in \delta$, *for each* $\langle x, p\rangle \in R(g)$,
- $(n, \langle x, p\rangle, \langle x, p\rangle) \in \delta$, *for each* $\langle n, \langle x, p\rangle\rangle \in E(g)$.

The path language *of* g *is* $L(g) = \{x_1 \dots x_n \mid \langle x_1, p_1\rangle \dots \langle x_n, p_n\rangle \in L^p(g)\}$.

Here are two examples:

$L($ 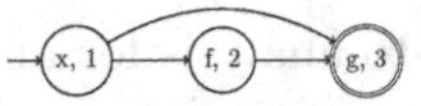 $) = \{x.g,\ x.f.g\}$ $L($ 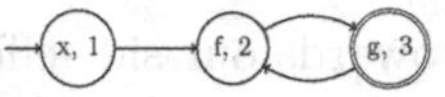 $) = x.(f.g)+$

Lemma 1. *Let g be an access graph. The path language of g, $L(g)$, is regular.*

Next, we define a join operation on access graphs that approximates the union of their path languages.

Definition 4. *For $i \in \{1, 2\}$, let $g_i = \langle N_i, E_i, A_i, R_i\rangle$ be access graphs. Define their* join $g_1 \sqcup g_2 = \langle N_1 \cup N_2, E_1 \cup E_2, A_1 \cup A_2, R_1 \cup R_2\rangle$.

Lemma 2. $L(g_1 \sqcup g_2) \supseteq L(g_1) \cup L(g_2)$.

In general, $L(g_1 \sqcup g_2)$ may contain words that are neither in $L(g_1)$ nor in $L(g_2)$ because their underlying node sets need not be disjoint. For example, consider

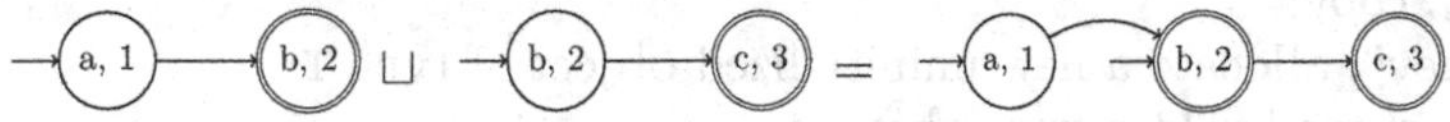

where the language of the joined graph contains the word $a.b.c$ which is not in the language of either argument graph: $\{a.b\}$ and $\{b.c\}$, respectively.

Theorem 1. *For each program P, the structure $\langle GAG, \sqcup, \sqcap, \perp, \top\rangle$ is a finite, complete lattice. The meet operation $\sqcap$ is componentwise intersection, $\perp = \langle \emptyset, \emptyset, \emptyset, \emptyset\rangle$, and $\top = \langle N, N \times N, N, N\rangle$ where $N = Occ \times PP$.*

The lattice ordering, which corresponds to the componentwise subset ordering, is defined by $g_1 \sqsubseteq g_2$ iff $g_1 \sqcup g_2 = g_2$. Furthermore, $g_1 \sqsubseteq g_2$ implies $L(g_1) \subseteq L(g_2)$.

Remark 1. Our analysis does not make use of the meet operation, but it is easy to see that $L(g_1 \sqcap g_2) \subseteq L(g_1) \cap L(g_2)$: Suppose there is a path in $g_1 \sqcap g_2$ from a root node $r \in R(g_1) \cap R(g_2)$ to an accepting node $a \in A(g_1) \cap A(g_2)$. Then this path exists in g_1 and g_2, too, as it exists in $E(g_1) \cap E(g_2)$.

As an example that the inclusion is proper consider $g_1 = a^1$ and $g_2 = a^2$. It holds that $L(g_1 \sqcap g_2) = \emptyset$ but $L(g_1) \cap L(g_2) \neq \emptyset$.

The analysis requires one more operation. Concatenation of access graphs computes an approximation to the concatenation of their languages.

Definition 5. *For* $i \in \{1, 2\}$, *let* $g_i = \langle N_i, E_i, A_i, R_i \rangle$ *be access graphs. Define their* concatenation $g_1 \cdot g_2 = \langle N_1 \cup N_2, E_1 \cup E_2 \cup (A_1 \times R_2), A_2, R_1 \rangle$.

Lemma 3. $L(g_1 \cdot g_2) \supseteq L(g_1) \cdot L(g_2)$.

As an example that the inclusion may be proper, consider $L(a^1) = \{a\}$ and $L(a^1 \cdot a^1) = a^+ \supsetneq \{a.a\}$.

Lemma 4. *Concatenation is monotone in both arguments.*

4 Intraprocedural Points-to Analysis

The first step towards our side-effect analysis is a points-to analysis for an imperative core language with objects, which is the essence of an intermediate representation for Java like Jimple. We consider a program one method at a time and we assume that each method is given in the form of a control-flow graph $CFG = (V, F)$ where the nodes $V = PP$ correspond to program points and the directed edges $F \subseteq V \times V$ correspond to potential control transfers between program points. The function $pred : V \to \mathcal{P}(V)$ maps a node to its set of predecessors: $pred(v) = \{v' \mid (v', v) \in F\}$. There are two distinct nodes that determine the entry point and the exit from the method.

Each node v in the CFG is associated with a statement $stm(v)$ of one of the following forms, where $x, y, \ldots$ range over local variables.

- $x = y \oplus z$, primitive operation;
- $x = c$, constant including `null`;
- $x = y$, copy;
- $x = \mathtt{new}^p$, allocate a new uninitialized object of type `T`;
- $x = y.a$, read field a from object y;
- $x.a = y$, write field a in object x;
- $x = \mathtt{call}\ f(y_1, \ldots, y_n)$, call method f.

4.1 Memory Abstraction

An abstract object is described by a generalized access graph. We use a different symbol to emphasize the interpretation of the graph as an abstract object.

$$o \in Obj = GAG$$

The object o represents the set of objects that can be reached in the pre-state of a method call via the access paths in $L(o)$. The graphs are anchored either in the formal parameters of the method, in global variables that can be accessed inside the method, or in objects newly allocated during the method call. Without loss of generality, we pretend that all such allocations take place before the method starts, so that they are representable in the pre-state already.

To represent points-to information, we need references to fields of abstract objects. Such a reference is a pair of an abstract object and a field name.

$$Ref \ni r ::= o \bullet a$$

Points-to information itself is represented by a points-to set of the form $P \in \mathcal{P}(Ref \times Obj)$. An element $\langle r, o\rangle \in P$ states that the reference r *may* point to abstracted object o. In addition, each reference $o \bullet a$ points to an implicit natural object, namely $o \cdot a$, which implements the may-nature of the analysis.

In principle, we might also represent points-to information with a partial mapping $Ref \hookrightarrow Obj$ by joining multiple target objects for the same reference. By keeping a set of target objects, we retain some more precision. We have yet to investigate whether it makes a difference in practice.

4.2 Method Summaries

We assume that, for each method, there is a method summary that describes the result of a method call and the potential side effects of the method. Specifically, for a method f

- $returns(f) \in Obj$ describes the return value of the method as an abstract object in terms of f's formal parameters;
- $exitSet(f) \in \mathcal{P}(Ref \times Obj)$ describes the potential modification of the points-to information by calling f;
- $reads(f) \in GAG$ describes the set of objects that may be read during execution of f;
- $writes(f) \in GAG$ describes the set of objects that may be written to during execution of f.

These functions do not take into account aliasing that is present at a call site of the method f. Thus, the method summary needs to be adapted to the circumstances at each call site. On the positive side, it means that our analysis is modular, because after generating the method summary for f, all further analysis can rely on the summary.

4.3 Dataflow Equations

The domain DP of the dataflow analysis consists of a local variable environment that maps a variable name to an abstract memory location and a points-to relation P as described above. We model the environment ρ as a partial map that we consider as a set of pairs when convenient: $\rho \in Env = Var \hookrightarrow Obj$.

$$DP = Env \times \mathcal{P}(Ref \times Obj)$$

The dataflow equations for the points-to analysis are typical for a forward analysis. For each node v in the CFG, they determine values $inP(v), outP(v) \in DP$ that accumulate the analysis result and there are functions $genP(v), killP(v) \in DP$

that compute information to add to or remove from an intermediate result. As a slight difference to the standard framework, the value of *genP*(v) often depends on *inP*(v).

$$inP(v) = \bigsqcup_{p \in pred(v)} outP(p) \qquad outP(v) = (inP(v) - killP(v)) \sqcup genP(v) \tag{1}$$

The join operation on the environment is the pointwise join of the abstract objects in the range. It is set union on the points-to sets. The "−" operation computes the set difference on the underlying sets. The initial state is given by $inP(v) = outP(v) = (\emptyset, \emptyset)$, with one exception:

$$outP(entry) = ([x \mapsto x^p \mid x \text{ formal parameter defined at program point } p], \emptyset)$$

The result of the analysis is the least fixpoint of the equations (1).

The *genP* and *killP* functions applied to node v are defined by case analysis on the statement at node v. If $stm(v)$ has the form $x = \ldots$, then $killP(v) = (\{(x, o) \mid o \in Obj\}, \emptyset)$, that is, the previous assignment to x is removed. For other forms of statements, we specify the kill set explicitly.

Generally, let $(\rho_{in}, P_{in}) = inP(v)$ in the following definition of *genP* and *killP*.

- If $stm(v)$ is $x = y \oplus z$ or $x = c$, then $genP(v) = (\emptyset, \emptyset)$.
- If $stm(v)$ is $x = y$, then $genP(v) = ([x \mapsto \rho_{in}(y)], \emptyset)$.
- If $stm(v)$ is $x = \mathtt{new}^p$, then $genP(v) = ([x \mapsto \ell^p], \emptyset)$, so that x points to an abstract object allocated at program point p.
- If $stm(v)$ is $x = y.a^p$, then $genP(v) = ([x \mapsto \bigsqcup objs(inP(v), y.a^p)], \emptyset)$. The function $objs : DP \times Var \times Field \times PP \to \mathcal{P}(Obj)$ resolves a field access under a given points-to set.

$$objs((\rho, P), y.a^p) = \{\rho(y) \cdot a^p\} \cup \{o' \mid (o \bullet b, o') \in P, mayAlias(o, \rho(y)), a = b\}$$

The first part concerns the direct effect of the field access. It concatenates the abstract object that is currently stored in the variable with the trivial access graph to field a^p. The second part is the indirect effect. If the points-to set contains evidence that $y.a$ may also point to some object o', then that object is also a potential result.

Checking the last part is more involved than comparing o and $\rho(y)$ for equality. As both are represented by method-local access graphs, it may be the case that they are not equal but nevertheless have some access paths in common. The function *mayAlias* checks the absence of such common access paths by checking disjointness of the path languages:

$$mayAlias(o_1, o_2) = L(o_1) \cap L(o_2) \neq \emptyset$$

As the path languages are regular (cf. Lemma 1), this check is effective. If we consider the booleans ordered by `false` $\sqsubseteq$ `true`, then *mayAlias* is monotone in both arguments and thus *objs* is also monotone in ρ and P.

– If $stm(v)$ is $x.a^p = y$, then $killP(v) = (\emptyset, \emptyset)$ because no local (or global) variable is overwritten but after the assignment $x.a$ may point to the object stored in y which must be reflected in the points-to information. Hence, $genP(v) = (\emptyset, refs(\rho_{in}, x.a^p) \times \{\rho_{in}(y)\})$. The function $refs : Env \times Var \times Field \times PP \rightarrow Ref$ resolves a field access to a reference, a symbolic left value.

$$refs(\rho, x.a^p) = \rho(x) \bullet a$$

– If $stm(v)$ is $x = \texttt{call}\ f(y_1, \ldots, y_k)$, then we first need to consult the call graph for the set of possible call targets $f_1, \ldots, f_t$. The gen-information of the method call is joined from the individual call targets: $genP(v) = \bigsqcup_j d_j$. For each target f_j with formal parameters $x_1, \ldots, x_k$, define $d_j = ([x \mapsto o'_j], P'_j)$ where o'_j describes the abstract object returned by the method call and P'_j describes the potential side-effect of the call on the parameters and global variables.

We obtain this data from the method summary of f_j, but as this summary contains information that is local to f_j, it needs to be translated to the calling context. In particular, the access graphs in f_j's summary refer to the x_i, the names of f_j's formal parameters. They need to be replaced by the abstract objects $\rho_{in}(y_i)$, for $1 \leq i \leq k$, representing the parameters of the call site.

However, this replacement alone is not sufficient, because the access paths in the result ignore aliasing (points-to information) that is present at the call site: a method is always analyzed under the assumption that its arguments are not aliased. This discrepancy has to be corrected by retracing the access paths in $returns(f_j)$ using the points-to information at the call site, which is represented by P_{in}. Thus, if $o = returns(f_j)$ then

$$o'_j = trans_o(\rho_{in}, P_{in}, o) = \bigsqcup_{n \in A(o)} Q(o, n)$$

where $Q : Obj \times Occ \rightarrow Obj$ is the smallest function such that

$$Q(o, \langle \texttt{new}, p \rangle) \sqsupseteq \ell^p \quad \text{if } \langle \texttt{new}, p \rangle \in R(o)$$

$$Q(o, \langle x_i, p \rangle) \sqsupseteq \rho_{in}(y_i) \quad \text{if } \langle x_i, p \rangle \in R(o) \qquad \text{substituting formal parameter}$$

$$Q(o, \langle a, p \rangle) \sqsupseteq Q(o, n) \cdot a^p \sqcup \bigsqcup \{t \mid \langle o' \bullet a, t \rangle \in P_{in}, mayAlias(o', Q(o, n))\}$$

$$\text{for all } n \text{ such that } \langle n, \langle a, p \rangle \rangle \in E(o)$$

A similar transformation has to be applied to the points-to set returned by the method.

$$P'_j = \{\langle trans_r(\rho_{in}, P_{in}, r), trans_o(\rho_{in}, P_{in}, o) \rangle \mid \langle r, o \rangle \in exitSet(f)\}$$

where

$$trans_r(\rho, P, o \bullet a) = trans_o(\rho, P, o) \bullet a$$

5 Global Variables

Global variables are straightforward to integrate into the analysis. The environment ρ also maintains information about the abstract objects contained in the global variables. That is, the initial environment in *outP(entry)* also contains bindings $[\mathtt{C}.a \mapsto \mathtt{C}.a^p]$, where p is the program point defining a in C.

There are two new cases for *genP*(v) where $(\rho_{in}, P_{in}) = inP(v)$ and the method call needs to be extended.

- If $stm(v)$ is $x = \mathtt{C}.a$, then $genP(v) = ([x \mapsto \rho_{in}(\mathtt{C}.a)], \emptyset)$.
- If $stm(v)$ is $\mathtt{C}.a^p = y$, then $killP(v) = (\{(\mathtt{C}.a, o) \mid o \in Obj\}, \emptyset)$ and $genP(v) = ([\mathtt{C}.a \mapsto \rho_{in}(y)], \emptyset)$.
- If $stm(v)$ is $x = \mathtt{call}\ f(y_1, \ldots, y_k)$, then we extend the previous treatment. Let $\rho_{out} = globals(f)$ be the environment at the exit node of f restricted to the bindings of the global variables (also part of the method summary). Then the environment part of $genP(v)$ needs to be extended with $[\mathtt{C}.a \mapsto trans_o(\rho, P, \rho_{out}(\mathtt{C}.a))]$ for each global variable C.a.[1]

  To transfer these entries successfully, we need to extend the Q function in the definition of $trans_o$ by

$$Q(o, \langle \mathtt{C}.a, p\rangle) \sqsupseteq \rho_{in}(\mathtt{C}.a) \quad \text{if } \langle \mathtt{C}.a, p\rangle \in R(o)$$

The treatment of function calls could be improved by additionally keeping track of which global variables are *definitely* overwritten by the call. The entries for these variables could be killed from the environment and replaced by the information from the method summary.

6 Intraprocedural Side-Effect Analysis

To perform the side-effect analysis, we assume that the results of the points-to analysis are available in *inP*(v) and *outP*(v), for each CFG node v. The domain for this analysis is the product lattice of two access graphs, the first one summarizing read accesses, the second one write accesses.

$$DS = GAG \times GAG$$

The analysis is again a forward analysis, but in this case there are no kill sets.

$$inS(v) = \bigsqcup_{p \in pred(v)} outS(p) \qquad outS(v) = inS(v) \sqcup genS(v) \qquad (2)$$

All values are initialized to the bottom of the lattice $inS(v) = outS(v) = (\bot, \bot)$. The result of the analysis is the least fixpoint of the equations (2).

Again, we define *genS*(v) by cases on the statement at node v. Let $(\rho_{in}, P_{in}) = inP(v)$ be the result of the points-to analysis at node v.

[1] In an implementation, it is sufficient to only keep entries for those variables that are actually used inside f.

```
INTERPROCEDURALANALYSIS()
Compute call graph and its SCC tree.
for each method f
    summaries[f] = (⊥, ∅, ⊥, ⊥)
while an unprocessed SCC exists
    Choose an unprocessed SCC S where all predecessors are processed.
    repeat
        done = true
        for each method f in S
            newSummary = INTRAPROCEDURALANALYSIS(f, summaries)
            if summaries[f] ≠ newSummary
                summaries[f] = newSummary
                done = false
    until done
    Mark S as processed.
```

Fig. 6. Algorithm for interprocedural analysis

- If $stm(v)$ is $x = y \oplus z$ or $x = c$ or $x = y$ or $x = \mathbf{new}^p$, then $genS(v) = (\emptyset, \emptyset)$.
- If $stm(v)$ is $x = y.a^p$, then $genS(v) = (\rho_{in}(y) \cdot a^p, \emptyset)$.
- If $stm(v)$ is $x.a^p = y$, then $genS(v) = (\emptyset, \rho_{in}(x) \cdot a^p)$.
- If $stm(v)$ is $x = \texttt{call}\ f(y_1, \ldots, y_n)$, then $genS(v) = (g_r, g_w)$ where

$$g_r = trans_o(\rho_{in}, P_{in}, reads(f)) \qquad g_w = trans_o(\rho_{in}, P_{in}, writes(f))$$

As in the points-to analysis, the method summary needs to be translated into the current environment and aliasing context, and we need to join the information of the possible call targets for f.

Allocations are not registered as write effects because they do not modify existing data structures. However, reads and writes to newly allocated data appear as side-effects. No special treatment is needed to cater for global variables.

7 Interprocedural Analysis

The interprocedural analysis computes the method summaries for the whole program by repeatedly applying the intraprocedural analysis to the program's functions until a fixpoint is reached. We sketch the algorithm in Fig. 6.

At first, the program's call graph and its strongly connected components (SCCs) are computed. Next, the SCCs are traversed bottom-up and for each SCC the fixpoints of its methods' summaries are computed starting from the bottom values of the respective domains.

The fixpoint computation recomputes the method summaries for all methods contained in the current SCC. This computation is repeated until all summaries stabilize. If any method summary changes, then all summaries in the current SCC have to be recomputed because they mutually depend on each other.

8 Experience

We implemented both the points-to analysis and side-effect analysis on top of the SOOT Java bytecode analysis and transformation framework in version 2.5.0. To increase the scalability of our analysis points-to pairs with structurally equal left hand sides are joined into a single pair.

Evaluation. The evaluation concentrates on analysis time and precision of our analyses. We focus on relatively small benchmark programs from the JOlden [3] benchmark suite. The suite consists of ten benchmark programs.

We stripped the benchmarks of the time measurement, statistics and printing functionality that is common to all programs in the JOlden suite to avoid analyzing large parts of the JDK. For example, the unstripped version of the Bisort benchmark had a call graph containing more than 8000 methods although Bisort's functionality is implemented in 11 user methods. All benchmarks were executed on a machine with AMD Phenom II X6 (2.8 GHz, 6 Cores) processor and 8 GB RAM on top of Archlinux 64bit, Kernel 3.13.8-1 and OpenJDK 7.0.

We present the results of running our points-to and side-effect analysis on nine of the JOlden benchmarks in Table 1. In each case, the analysis processed all methods, user and library, that are reachable from the main method. We excluded static initializers and external methods (methods without an active body) from the analysis. For each application, we present the total number of methods (including library methods) and the number of user methods. We measured the total run time and the run time of the call graph construction including the calculation of the SCCs. As a quality measure that is independent from our abstract domain we count the methods that do not introduce new aliases and the pure methods. Here, "pure" means that a method does not have any write effects on heap-allocated memory locations in the prestate of the method. Following the JML convention, we consider constructors that only mutate fields of "this" as pure.

For the benchmarks marked with * we excluded the JDK methods from the call graph, as our prototype apparently does not scale to the thousands of library methods that can be transitively invoked by these benchmarks.

Discussion. The run time of the analysis is within reasonable bounds for small programs, generally running in a fraction of the time taken for the call graph construction and finding the SCCs. We still need to gain experience with larger programs.

For the benchmark programs that we have analyzed our analysis gives useful results with a precision that is roughly comparable to that of others [21]. In the best case `TreeAdd` we can identify roughly 89% of the methods as pure. The other extreme is `MST` where we identify 21% of the methods as pure.

Table 1. Analysis results for the Java Olden benchmarks

Application	Methods		Run time			Method summaries	
	User	All	CG+SCC	Analysis	Total	No new aliases	Pure
BH	53	68	57.26s	1.17s	58.43s	76.47%	66.18%
BiSort	11	13	5.40s	1.05s	6.45s	61.54%	38.46%
Em3d*	16	16	54.50s	0.90s	55.40s	50.00%	37.50%
MST	29	34	55.45s	0.83s	56.28s	85.29%	20.59%
Perimeter	34	36	5.75s	0.95s	6.70s	94.44%	80.56%
Power	26	34	55.65s	2.81s	58.46s	88.24%	50.00%
TreeAdd	3	9	52.67s	0.45s	53.12s	88.89%	88.89%
TSP*	12	12	52.05s	1.03s	53.08s	66.67%	33.34%
Voronoi*	55	55	58.15s	1.69s	59.84s	89.09%	76.36%

9 Related Work

There is a plethora of literature on effect and points-to analysis for heap-allocated objects. We therefore focus on the distinguishing features of our abstract domain and compare our work to selected bottom-up points-to, shape and effect analyses.

Regarding our abstract domain, the most closely related work is by Khedker et al. [15]. They have introduced access graphs, which include program points in their nodes to deal with unboundedly large data structures. We extend access graphs to GAGs, which facilitate a compact representation of the points-to relation that cannot be achieved with (sets of) simple access graphs.

While Khedker et al. rely on access graphs for a number of analyses including alias analysis, they do not employ them for points-to analysis, as we do. In contrast to our abstract domain, they use partly "unresolved" access graphs to represent abstract references in their alias sets. That is, their access graphs can be rooted in a reference that requires further resolution to obtain the abstract object it stands for, whereas we use "resolved" (up to the unknown context) abstract objects being rooted in parameters or allocation points. We use such resolved GAGs, as they avoid repeated resolution of the same access graphs while preserving precision on updates to references, which we consider as an advantage in our flow-sensitive analysis.

Most of the following proposals share the property with ours that their abstract domains are based on Larus' *access paths* [18] or Deutsch's SAPs [8].

All data-flow algorithms must deal with the unbounded nature of recursive data structures. Many proposals [17,4,6] follow the *k-limiting* approach [12], which limits access paths by truncation.

While our proposal uses a *storeless model* (originally proposed by Jonkers [13]) other proposals [18,21] use a *store-based model* model employing some form of (rooted) directed graph representation or a compact representation thereof [11] for points-to or alias information. A store-based model can enable the description of regular patterns across references and the objects these references refer to.

In their side-effect analysis for Java, Sălcianu and Rinard [21] represent the points-to relation as a *points-to graph*, a rooted directed graph representation.

Their points-to graphs allow multiple root nodes (as our GAGs do), but do not include program points in their points-to graphs. Larus and Hilfinger's *alias graphs* [18] are similar to Sălcianu's points-to graphs.

Several authors [14,24] propose scalable bottom-up pointer analyses for C programs, but ignore heap-allocated data. Matosevic et al. [19] use a SAP-based abstract domain in their bottom-up side-effect analysis, but their loop abstraction mechanism can only detect three patterns of iteration. Moreover, they do not formally describe how they handle method calls. In contrast to our points-to relation, which can be viewed as a *total transfer function*, their abstract domain serves as a *partial transfer function* [20] that assumes the context to have certain properties. Partial transfer functions can be considered as an optimization that is also applicable to our analysis. Gulavani et al. [10] propose a bottom-up shape-analysis based on separation logic. Their *Logic of Iterated Separation Formulae* allows the computation of a loop summary from a loop body summary.

Another widely-used and highly scalable proposal for points-to analysis is Steensgaard's [22] type-based analysis using unification, which is most suitable for statically-typed languages. Our analysis can be combined with type-based information to improve both precision and scalability.

10 Conclusion

Side-effect analysis is an important tool in the programmer's toolbox. It aids program understanding, supports other program analyses, and it enables advanced program optimizations and safe parallel execution. Our analysis is based on a single comprehensive and precise abstract domain of generalized access graphs, which serve to express abstract objects, points-to information, escape information, as well as read and write effects. In our experience, the single abstract domain simplifies the implementation. The preliminary data gathered from our implementation shows that our approach is practically feasible, but we believe that further algorithmic tuning is possible.

References

1. Bocchino Jr., R.L., Adve, V.S., Dig, D., Adve, S.V., Heumann, S., Komuravelli, R., Overbey, J., Simmons, P., Sung, H., Vakilian, M.: A type and effect system for deterministic parallel Java. In: Arora, S., Leavens, G.T. (eds.) OOPSLA, pp. 97–116. ACM (2009)
2. Burdy, L., Cheon, Y., Cok, D.R., Ernst, M.D., Kiniry, J.R., Leavens, G.T., Leino, K.R.M., Poll, E.: An overview of JML tools and applications. Int. J. Softw. Tools Technol. Transf. 7(3), 212–232 (2005)
3. Cahoon, B., McKinley, K.S.: Data flow analysis for software prefetching linked data structures in java. In: IEEE PACT, pp. 280–291. IEEE Computer Society (2001)
4. Cherem, S., Rugina, R.: A practical escape and effect analysis for building lightweight method summaries. In: Adsul, B., Odersky, M. (eds.) CC 2007. LNCS, vol. 4420, pp. 172–186. Springer, Heidelberg (2007)

5. Corbett, J.C., Dwyer, M.B., Hatcliff, J., Laubach, S., Păsăreanu, C.S., Zheng, H.: Bandera: Extracting finite-state models from Java source code. In: Ghezzi, C., Jazayeri, M., Wolf, A.L. (eds.) ICSE, Limerick, Ireland, pp. 439–448. ACM (June 2000)
6. Dasgupta, S., Karkare, A., Reddy, V.K.: Precise shape analysis using field sensitivity. ISSE 9(2), 79–93 (2013)
7. DeLine, R., Fähndrich, M.: Typestates for objects. In: Odersky, M. (ed.) ECOOP 2004. LNCS, vol. 3086, pp. 465–490. Springer, Heidelberg (2004)
8. Deutsch, A.: Interprocedural alias analysis for pointers: Beyond k-limiting. In: Sarkar, V., Ryder, B.G., Soffa, M.L. (eds.) PLDI, pp. 230–241. ACM (1994)
9. Flanagan, C., Leino, K.R.M., Lillibridge, M., Nelson, G., Saxe, J.B., Stata, R.: Extended static checking for Java. In: Knoop, J., Hendren, L.J. (eds.) PLDI, Berlin, Germany, pp. 234–245. ACM Press (2002)
10. Gulavani, B.S., Chakraborty, S., Ramalingam, G., Nori, A.V.: Bottom-up shape analysis using lisf. ACM Trans. Program. Lang. Syst. 33(5), 17 (2011)
11. Hind, M., Burke, M.G., Carini, P.R., Choi, J.-D.: Interprocedural pointer alias analysis. ACM Trans. Program. Lang. Syst. 21(4), 848–894 (1999)
12. Jones, N.D., Muchnick, S.S.: A flexible approach to interprocedural data flow analysis and programs with recursive data structures. In: Proc. of the 9th ACM Symp. POPL, Albuquerque, New Mexico, USA, pp. 66–74. ACM Press (1982)
13. Jonkers, H.B.M.: Abstract storage structures. In: de Bakker, van Vllet (eds.) Algorithmic Languages. IFIP, pp. 321–343 (1981)
14. Kang, H.-G., Han, T.: A bottom-up pointer analysis using the update history. Information & Software Technology 51(4), 691–707 (2009)
15. Khedker, U.P., Sanyal, A., Karkare, A.: Heap reference analysis using access graphs. ACM TOPLAS 30(1) (2007)
16. Kuncak, V., Lam, P., Rinard, M.C.: Role analysis. In: Launchbury, J., Mitchell, J.C. (eds.) POPL, pp. 17–32. ACM (2002)
17. Landi, W., Ryder, B.G., Zhang, S.: Interprocedural modification side effect analysis with pointer aliasing. In: Cartwright, R. (ed.) PLDI, pp. 56–67. ACM (1993)
18. Larus, J.R., Hilfinger, P.N.: Detecting conflicts between structure accesses. In: Wexelblat, R.L. (ed.) PLDI, pp. 21–34. ACM (1988)
19. Matosevic, I., Abdelrahman, T.S.: Efficient bottom-up heap analysis for symbolic path-based data access summaries. In: Eidt, C., Holler, A.M., Srinivasan, U., Amarasinghe, S.P. (eds.) CGO, San Jose, CA, USA, pp. 252–263 (March 2012)
20. Murphy, B.R., Lam, M.S.: Program analysis with partial transfer functions. In: Lawall, J.L. (ed.) PEPM, pp. 94–103. ACM (2000)
21. Sălcianu, A., Rinard, M.: Purity and side effect analysis for Java programs. In: Cousot, R. (ed.) VMCAI 2005. LNCS, vol. 3385, pp. 199–215. Springer, Heidelberg (2005)
22. Steensgaard, B.: Points-to analysis in almost linear time. In: Proc. 1996 ACM Symp. POPL, St. Petersburg, FL, USA, pp. 32–41. ACM Press (January1996)
23. Vallée-Rai, R., Hendren, L., Sundaresan, V., Lam, P., Gagnon, E., Co, P.: Soot - a Java optimization framework. In: Proc. CASCON 1999, pp. 125–135 (1999)
24. Yu, H., Xue, J., Huo, W., Feng, X., Zhang, Z.: Level by level: making flow- and context-sensitive pointer analysis scalable for millions of lines of code. In: Moshovos, A., Steffan, J.G., Hazelwood, K.M., Kaeli, D.R. (eds.) CGO, pp. 218–229. ACM (2010)

Expressiveness via Intensionality and Concurrency

Thomas Given-Wilson*

INRIA, Paris, France
thomas.given-wilson@inria.fr

Abstract. Computation can be considered by taking into account two dimensions: extensional versus intensional, and sequential versus concurrent. Traditionally sequential extensional computation can be captured by the λ-calculus. However, recent work shows that there are more expressive intensional calculi such as SF-calculus. Traditionally process calculi capture computation by encoding the λ-calculus, such as in the π-calculus. Following this increased expressiveness via intensionality, other recent work has shown that concurrent pattern calculus is more expressive than π-calculus. This paper formalises the relative expressiveness of all four of these calculi by placing them on a square whose edges are irreversible encodings. This square is representative of a more general result: that expressiveness increases with both intensionality and concurrency.

1 Introduction

Computation can be characterised in two dimensions: *extensional* versus *intensional*; and *sequential* versus *concurrent*. Extensional sequential computation models are those whose *functions* cannot distinguish the internal structure of their *arguments*, here characterised by the λ-calculus [3]. However, Jay & Given-Wilson show that λ-calculus does not support all sequential computation [20]. In particular, there are intensional Turing-computable functions, characterised by *pattern-matching*, that can be represented within SF-calculus but not within λ-calculus [20]. Of course λ-calculus can encode Turing computation, but this is a weaker claim. Ever since Milner et al. showed that the π-calculus generalises λ-calculus [24, 26], concurrency theorists expect process calculi to subsume sequential computation as represented by λ-calculus [24, 26, 25]. Following from this, here extensional concurrent computation is characterised by process calculi that do not communicate terms with internal structure, and, at least, support λ-calculus. Intensional concurrent computation is represented by process calculi whose communication includes terms with internal structure, and reductions that depend upon the internal structure of terms. Here intensional concurrent computation is demonstrated by *concurrent pattern calculus* (CPC) that not only generalises intensional pattern-matching from sequential computation to *pattern-unification* in a process calculus, but also increases the *symmetry* of interaction [14, 15].

These four calculi form the corners of a *computation square*

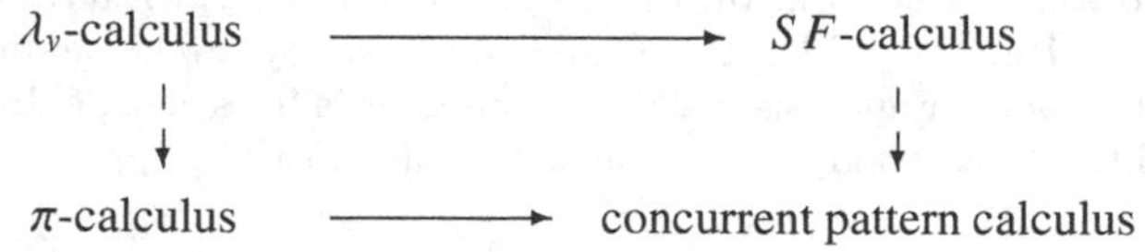

* This work has been partially supported by the project ANR-12-IS02-001 PACE.

G. Ciobanu and D. Méry (Eds.): ICTAC 2014, LNCS 8687, pp. 206–223, 2014.

where the left side is merely extensional and the right side also intensional; the top edge is sequential and the bottom edge concurrent. All arrows are defined via valid encodings [18]. The horizontal (solid) arrows are *homomorphisms* in that they also preserve *application* or *parallel composition*. The vertical (dashed) arrows are *parallel encodings* in that application is mapped to a parallel composition (with some machinery). Each arrow represents increased expressive power with CPC completing the square.

This paper presents the formalisation of these expressiveness results for the four calculi above. This involves adapting some popular definitions of encodings [16–18] and then building upon various prior results [8, 24, 26, 14, 20, 11]. These can be combined to yield the new expressiveness results here captured by the computation square.

The organisation of the paper is as follows. Section 2 reviews prior definitions of encodings and defines the ones used in this paper. Section 3 reviews λ-calculus and combinatory logic while introducing common definitions. Section 4 summarises intensionality in the sequential setting and formalises the arrow across the top of the square. Section 5 begins concurrency through π-calculus and its parallel encoding of λ_v-calculus. Section 6 recalls concurrent pattern calculus and completes the results of the computation square. Section 7 draws conclusions, considers related work, and discusses future work.

2 Encodings

This section recalls valid encodings [18] for formally relating process calculi and adapts the definition to define homomorphisms and parallel encodings. The validity of valid encodings in developing expressiveness studies emerges from the various works [16–18], that have also recently inspired similar works [22, 23, 31]. Here the adaptations are precise definitions of homomorphisms that give stronger positive results (the negative results are not required to be as strong). Also, parallel encodings are defined to account for the mixture of sequential and concurrent languages considered.

An *encoding* of a language $\mathcal{L}_1$ into another language $\mathcal{L}_2$ is a pair $([\![\cdot]\!], \varphi_{[\![\,]\!]})$ where $[\![\cdot]\!]$ translates every $\mathcal{L}_1$-term into an $\mathcal{L}_2$-term and $\varphi_{[\![\,]\!]}$ maps every name (of the source language) into a tuple of k names (of the target language), for $k > 0$. The translation $[\![\cdot]\!]$ turns every term of the source language into a term of the target; in doing this, the translation may fix some names to play a precise rôle or may translate a single name into a tuple of names. This can be obtained by exploiting $\varphi_{[\![\,]\!]}$.

Now consider only encodings that satisfy the following properties. Let a *k-ary context* $C(\_{1};\ldots;\_{k})$ be a term with k holes $\{\_{1};\ldots;\_{k}\}$ that appear exactly once each. Moreover, denote with $\longmapsto_i$ and $\Longmapsto_i$ the relations $\longmapsto$ (reduction relation) and $\Longmapsto$ (the reflexive transitive closure of $\longmapsto$) in language $\mathcal{L}_i$; denote with $\longmapsto_i^{\omega}$ an infinite sequence of reductions in $\mathcal{L}_i$. Moreover, let $\equiv_i$ denote the structural equivalence relation for a language $\mathcal{L}_i$, and $\sim_i$ denote the reference behavioural equivalence for language $\mathcal{L}_i$. For simplicity the notation $T \longmapsto_i\equiv_i T'$ denotes that there exists T'' such that $T \longmapsto_i T''$ and $T'' \equiv_i T'$, and may also be used with $\Longmapsto_i$ or $\sim_i$. Also, let $P \Downarrow_i$ mean that there exists P' such that $P \Longmapsto_i P'$ and $P' \equiv_i P'' \mid \surd$, for some P'' where $\surd$ is a specific process to indicate success. Finally, to simplify reading, let S range over terms of the source language (viz., $\mathcal{L}_1$) and T range over terms of the target language (viz., $\mathcal{L}_2$).

Definition 1 (Valid Encoding (from [18])). *An encoding* $(\llbracket \cdot \rrbracket, \varphi_{\llbracket\,\rrbracket})$ *of* $\mathcal{L}_1$ *into* $\mathcal{L}_2$ *is* valid *if it satisfies the following five properties:*

1. Compositionality: *for every k-ary operator* op *of* $\mathcal{L}_1$ *and for every subset of names* N*, there exists a k-ary context* $C^N_{\mathsf{op}}(\_1;\ldots;\_k)$ *of* $\mathcal{L}_2$ *such that, for all* $S_1,\ldots,S_k$ *with* $\mathsf{fn}(S_1,\ldots,S_k)=N$*, it holds that* $\llbracket \mathsf{op}(S_1,\ldots,S_k) \rrbracket = C^N_{\mathsf{op}}(\llbracket S_1 \rrbracket;\ldots;\llbracket S_k \rrbracket)$.
2. Name invariance: *for every* S *and name substitution* σ*, it holds that*
$$\llbracket \sigma S \rrbracket \begin{cases} = \sigma'\llbracket S \rrbracket & \text{if } \sigma \text{ is injective} \\ \asymp_2 \sigma'\llbracket S \rrbracket & \text{otherwise} \end{cases}$$
where σ' *is such that* $\varphi_{\llbracket\,\rrbracket}(\sigma(a)) = \sigma'(\varphi_{\llbracket\,\rrbracket}(a))$ *for every name* a.
3. Operational correspondence:
   - *for all* $S \Longmapsto_1 S'$*, it holds that* $\llbracket S \rrbracket \Longmapsto_2 \asymp_2 \llbracket S' \rrbracket$;
   - *for all* $\llbracket S \rrbracket \Longmapsto_2 T$*, there exists* S' *such that* $S \Longmapsto_1 S'$ *and* $T \Longmapsto_2 \asymp_2 \llbracket S' \rrbracket$.
4. Divergence reflection: *for every* S *such that* $\llbracket S \rrbracket \longmapsto^{\omega}_2$*, it holds that* $S \longmapsto^{\omega}_1$.
5. Success sensitiveness: *for every* S*, it holds that* $S \Downarrow_1$ *if and only if* $\llbracket S \rrbracket \Downarrow_2$.

Observe that the definition of valid encoding is very general and, with the exception of success sensitiveness, can apply to sequential languages such as λ-calculus as well as process calculi. (On the understanding that a name substitution for sequential calculi is a mapping from names/variables to names/variables *not* terms.) However, the relations presented in this work bring together a variety of prior results and account for them in a stronger and more uniform manner. To this end, the following definitions support the results. The first two define homomorphism in the sequential and concurrent settings.

Definition 2 (Homomorphism (Sequential)). *A* (sequential) homomorphism *is a translation* $\llbracket \cdot \rrbracket$ *from one language to another that satisfies: compositionality, name invariance, operational correspondence, and divergence reflection; and that preserves application, i.e. where* $\llbracket S_1\, S_2 \rrbracket = \llbracket S_1 \rrbracket\,\llbracket S_2 \rrbracket$.

Definition 3 (Homomorphism (Concurrent)). *A* (concurrent) homomorphism *is a valid encoding whose translation preserves parallel composition, i.e.* $\llbracket P_1 \mid P_2 \rrbracket = \llbracket P_1 \rrbracket \mid \llbracket P_2 \rrbracket$.

The next is for encoding sequential languages into concurrent languages and exploits that $\llbracket \cdot \rrbracket_c$ indicates an encoding from source terms to target terms that is parametrised by a name c.

Definition 4 (Parallel Encoding). *An encoding* $(\llbracket \cdot \rrbracket_c, \varphi_{\llbracket\,\rrbracket})$ *of* $\mathcal{L}_1$ *into* $\mathcal{L}_2$ *is a* parallel encoding *if it satisfies the first four properties of a valid encoding (compositionality, name invariance, operational correspondence, and divergence reflection) and the following additional property.*

5. Parallelisation: *The translation of the application* MN *is of the form* $\llbracket MN \rrbracket_c \stackrel{\text{def}}{=} (\nu n_1)(\nu n_2)(\mathcal{A}(c,n1,n2) \mid \llbracket M \rrbracket_{n1} \mid \llbracket N \rrbracket_{n2})$ *where* $\mathcal{A}$ *is a process parametrised by* c *and* $n1$ *and* $n2$.

Parallelisation is a restriction on the more general compositionality criteria. Here this ensures that in addition to compositionality, the translation must allow for independent reduction of the components of an application. As the shift from sequential to concurrent computation can exploit this to support parallel reductions, the definition of parallel encoding encourages more flexibility in reduction since components can be reduced independently.

The removal of the success sensitiveness property is for simplicity when using prior results. It is not difficult to include success sensitiveness, this involves adding the success primitive to the sequential languages and defining $S \Downarrow$, e.g. $S \Downarrow$ means that $S \longmapsto^* \surd$. Additionally, this requires adding a test process Q_c to the definition of parallel encoding with success sensitiveness defined by: "for every S, it holds that $S \Downarrow_1$ if and only if $[\![S]\!]_c \mid Q_c \Downarrow_2$. However, since adding the success state $\surd$ to λ-calculus and combinatory logics[1] would require redoing many existing results, it is easier to avoid the added complexity since no clarity or gain in significance is made by adding it.

Encodings from concurrent languages into sequential ones have not been defined specifically here since they prove impossible. The proof of these results relies merely on the requirement of operational correspondence, and so shall be done on a case-by-case basis.

3 Sequential Extensional Computation

Both λ-calculus and traditional combinatory logic base reduction rules upon the application of a function to one or more arguments. Functions in both models are extensional in nature, that is a function does not have direct access to the internal structure of its arguments. Thus, functions that are extensionally equal are indistinguishable within either model even though they may have different normal forms.

The relationship between the λ-calculus and traditional combinatory logic is closer than sharing application-based reduction and extensionality. There is a homomorphism from call-by-value λ_v-calculus into any combinatory logic that supports the combinators S and K [8, 3]. There is also a homomorphism from traditional combinatory logic to a λ-calculus with more generous operational semantics [8, 3].

3.1 λ-calculus

The *term* syntax of the λ-calculus is given by

$$t ::= x \mid t\,t \mid \lambda x.t\ .$$

The *free variables* of a term are defined in the usual manner. A *substitution* σ is defined as a partial function from variables to terms. The *domain* of σ is denoted $\mathsf{dom}(\sigma)$; the free variables of σ, written $\mathsf{fv}(\sigma)$, is given by the union of the sets $\mathsf{fv}(\sigma x)$ where $x \in \mathsf{dom}(\sigma)$. The *variables* of σ, written $\mathsf{vars}(\sigma)$, are $\mathsf{dom}(\sigma) \cup \mathsf{fv}(\sigma)$. A substitution σ *avoids* a variable x (or collection of variables μ) if $x \notin \mathsf{vars}(\sigma)$ (respectively $\mu \cap$

[1] The results for intensional combinatory logics require that success behaves as a *constructor* as discussed for various combinatory logics in [20].

$\mathsf{vars}(\sigma) = \{\}$). Note that all substitutions considered in this paper have finite domain. The application of a substitution σ to a term t is defined as usual, as is α-conversion $=_\alpha$.

There are several variations of the λ-calculus with different operational semantics. For construction of the computation square by exploiting the results of Milner et al. [24], it is necessary to choose an operation semantics, such as *call-by-value* λ_v-calculus or *lazy* λ_l-calculus. The choice here is to use call-by-value λ_v-calculus, although the results can be reproduced for lazy λ_l-calculus as well. In addition a more generous operation semantics for λ-calculus will be presented for later discussion and relations.

To formalise the reduction of call-by-value λ_v-calculus requires a notion of *value* v. These are defined in the usual way, by

$$v ::= x \mid \lambda x.t$$

consisting of variables and λ-abstractions.

Computation in the λ_v-calculus is through the β_v-reduction rule

$$(\lambda x.t)v \quad \longmapsto_v \quad \{v/x\}t\ .$$

When an abstraction $\lambda x.t$ is applied to a value v then substitute v for x in the body t. The *reduction relation* (also denoted $\longmapsto_v$) is the smallest that satisfies the following rules

$$\frac{}{(\lambda x.t)v \longmapsto_v \{v/x\}t} \qquad \frac{s \longmapsto_v s'}{s\,t \longmapsto_v s'\,t} \qquad \frac{t \longmapsto_v t'}{s\,t \longmapsto_v s\,t'}\ .$$

The transitive closure of the reduction relation is denoted $\longmapsto_v^*$ though the star may be elided if it is obvious from the context.

The more generous operational semantics for the λ-calculus allows any term to be the argument when defining β-reduction. Thus the more generous β-reduction rule is

$$(\lambda x.s)t \quad \longmapsto \quad \{t/x\}s$$

where t is any term of the λ-calculus. The reduction relation $\longmapsto$ and the transitive closure thereof $\longmapsto^*$ are obvious adaptations from those for the λ_v-calculus. Observe that any reduction $\longmapsto_v$ of λ_v-calculus is also a reduction $\longmapsto$ of λ-calculus.

3.2 Traditional Combinatory Logic

A *combinatory calculus* is given by a finite collection $\mathcal{O}$ of *operators* (meta-variable O) that are used to define the $\mathcal{O}$-*combinators* (meta-variables M, N, X, Y, Z) built from these by application

$$M, N ::= O \mid MN\ .$$

The $\mathcal{O}$-*combinatory calculus* or $\mathcal{O}$-*calculus* is given by the combinators plus their reduction rules.

Traditional combinatory logic can be represented by two combinators S and K [8] so the SK-calculus has *reduction rules*

$$\begin{aligned} SMNX &\longmapsto MX(NX) \\ KXY &\longmapsto X\ . \end{aligned}$$

The combinator $SMNX$ duplicates X as the argument to both M and N. The combinator KXY eliminates Y and returns X. The *reduction relation* $\longmapsto$ is as for λ-calculus.

Although this is sufficient to provide a direct account of functions in the style of λ-calculus, an alternative is to consider the representation of arbitrary computable functions that act upon combinators.

A *symbolic function* is defined to be an n-ary partial function $\mathcal{G}$ of some combinatory logic, i.e. a function of the combinators that preserves their equality, as determined by the reduction rules. That is, if $X_i = Y_i$ for $1 \leq i \leq n$ then $\mathcal{G}(X_1, X_2, \ldots, X_n) = \mathcal{G}(Y_1, Y_2, \ldots, Y_n)$ if both sides are defined. A symbolic function is *restricted* to a set of combinators, e.g. the normal forms, if its domain is within the given set.

A combinator G in a calculus *represents* $\mathcal{G}$ if

$$GX_1 \ldots X_n = \mathcal{G}(X_1, \ldots, X_n)$$

whenever the right-hand side is defined. For example, the symbolic functions $\mathcal{S}(X_1, X_2, X_3) = X_1X_3(X_2X_3)$ and $\mathcal{K}(X_1, X_2) = X_1$ are represented by S and K, respectively, in SK-calculus. Consider the symbolic function $\mathcal{I}(X) = X$. In SKI-calculus where I has the rule $IY \longmapsto Y$ then $\mathcal{I}$ is represented by I. In both SKI-calculus and SK-calculus, $\mathcal{I}$ is represented by any combinator of the form SKX since

$$SKXY \longmapsto KY(XY) \longmapsto Y\ .$$

For convenience define the *identity combinator* I in SK-calculus to be SKK.

3.3 Relations

One of the goals of combinatory logic is to give an equational account of variable binding and substitution, particularly as it appears in λ-calculus. In order to represent λ-abstraction, it is necessary to have some variables to work with. Given O as before, define the *O-terms* by

$$M, N ::= x \mid O \mid MN$$

where x is as in λ-calculus. Free variables, substitutions, and symbolic computations are defined just as for O-calculus.

Given a variable x and term M define a symbolic function $\mathcal{G}$ on terms by

$$\mathcal{G}(X) = \{X/x\}M\ .$$

Note that if M has no free variables other than x then $\mathcal{G}$ is also a symbolic computation of the combinatory logic. If every such function $\mathcal{G}$ on O-combinators is representable then the O-combinatory logic is *combinatorially complete* in the sense of Curry [8, p. 5].

Given S and K then $\mathcal{G}$ above can be represented by a term $\lambda^*x.M$ given by

$$\begin{aligned} \lambda^*x.x &= I & \lambda^*x.O &= KO \\ \lambda^*x.y &= Ky \quad \text{if } y \neq x & \lambda^*x.MN &= S\,(\lambda^*x.M)(\lambda^*x.N)\ . \end{aligned}$$

The following lemmas are central results of combinatory logic [8] and Theorem 2.3 of [20]. This is sufficient to show there is a homomorphism from λ_v-calculus to any combinatory calculus that represents S and K.

Lemma 1. *For all terms M and N and variables x there is a reduction* $(\lambda^* x.M)\, N \longmapsto^* \{N/x\}M$.

Lemma 2. *Any combinatory calculus that is able to represent S and K is combinatorially complete.*

Theorem 1. *There is a homomorphism (Definition 2) from λ-calculus into SK-calculus.*

Proof. Compositionality, name invariance, and preservation of application hold by construction. Operational correspondence and divergence reflection can by proved via Lemma 2.

Below is a standard translation from SK-calculus into λ-calculus that preserves reduction and supports the following lemma [8, 3].

$$[\![S]\!] = \lambda g.\lambda f.\lambda x.g\, x\, (f\, x) \qquad [\![K]\!] = \lambda x.\lambda y.x \qquad [\![MN]\!] = [\![M]\!]\, [\![N]\!]$$

Lemma 3 (Theorem 2.3.3 of [11]). *Translation from SK-calculus to λ-calculus preserves the reduction relation.*

Theorem 2. *There is a homomorphism (Definition 2) from SK-calculus into λ-calculus.*

Proof. Compositionality and preservation of application hold by construction. Name invariance is trivial. Operational correspondence and divergence reflection are proved via Lemma 3.

Although the top left corner of the computation square is populated by λ_v-calculus, the arrows out allow for either λ_v-calculus or SK-calculus to be used. Indeed, the homomorphisms in both directions between λ-calculus and SK-calculus allow these two calculi to be considered equivalent.

4 Sequential Intensional Computation

Intuitively intensional functions are more expressive than merely extensional functions, however populating the top right corner of the computation square requires more formality than intuition. The cleanest account of this is by considering combinatory logic.

Even in SK-calculus there are Turing-computable functions defined upon the combinators that cannot be represented within SK-calculus. For example, consider the function that reduces any combinator of the form SKX to X. Such a function cannot be represented in SK-calculus, or λ-calculus, as all combinators of the form SKX represent the identity function. However, such a function is Turing-computable and definable upon the combinators. This is an example of a more general problem of *factorising* combinators that are both applications and stable under reduction.

Exploiting this factorisation is SF-calculus [20] that is able to support intensional functions on combinators including a structural equality of normal forms. Thus SF-calculus sits at the top right hand corner of the computation square. The arrow across the top of the square is formalised by showing a homomorphism from SK-calculus into SF-calculus. The lack of a converse has been proven by showing that the intensionality of SF-calculus cannot be represented within SK-calculus, or λ-calculus [20].

4.1 Symbolic Functions

Symbolic functions need not be merely extensional, indeed it is possible to define symbolic functions that consider the structure of their arguments. Observe that each operator O has an *arity* given by the minimum number of arguments it requires to instantiate a rule. Thus, K has arity 2 while S has arity 3. A *partially applied operator* is a combinator of the form $OX_1 \ldots X_k$ where k is less than the arity of O. An operator with a positive arity is an *atom* (meta-variable A). A partially applied operator that is an application is a *compound*. Hence, the partially applied operators of SK-calculus are the atoms S and K, and the compounds SM, SMN and KM for any M and N.

Now define a *factorisation function* $\mathcal{F}$ on combinators by

$$\begin{array}{rll} \mathcal{F}(A,M,N) & \longmapsto M & \text{if } A \text{ is an atom} \\ \mathcal{F}(XY,M,N) & \longmapsto NXY & \text{if } XY \text{ is a compound.} \end{array}$$

Lemma 4 (Theorem 3.2 of [20]). *Factorisation of SK-combinators is a symbolic computation that is not representable within SK-calculus.*

Proof. Suppose that there is an SK-combinator F that represents $\mathcal{F}$. Then, for any combinator X it follows that $F(SKX)S(KI) \longmapsto KI(SK)X \longmapsto X$. Translating this to λ-calculus as in Lemma 1 yields $[\![F(SKX)S(KI)]\!] \longmapsto [\![X]\!]$ and also $[\![F(SKX)S(KI)]\!] = [\![F]\!]\ [\![(SKX)]\!]\ [\![S]\!]\ [\![KI]\!] \longmapsto [\![F]\!]\ (\lambda x.x)\ [\![S]\!]\ [\![KI]\!]$. Hence, by confluence of reduction in λ-calculus, all $[\![X]\!]$ share a reduct with $[\![F]\!]\ (\lambda x.x)\ [\![S]\!]\ [\![KI]\!]$ but this is impossible since $[\![S]\!]$ and $[\![K]\!]$ are distinct normal forms. Hence $\mathcal{F}$ cannot be represented by an SK-combinator.

4.2 *SF*-calculus

When considering intensionality in a combinatory logic it is tempting to specify a factorisation combinator F as a representative for $\mathcal{F}$. However, $\mathcal{F}$ is defined using partially applied operators, which cannot be known until all reduction rules are given, including those for F. This circularity of definition is broken by beginning with a syntactic characterisation of the combinators that are to be factorable.

The SF-calculus [20] has *factorable forms* given by $S \mid SM \mid SMN \mid F \mid FM \mid FMN$ and *reduction rules*

$$\begin{array}{rll} SMNX & \longmapsto MX(NX) & \\ FOMN & \longmapsto M & \text{if } O \text{ is } S \text{ or } F \\ F(XY)MN & \longmapsto NXY & \text{if } XY \text{ is a factorable form.} \end{array}$$

The expressive power of SF-calculus subsumes that of SK-calculus since K is here defined to be FF and I is defined to be SKK as before.

Lemma 5. *There is a homomorphism (Definition 2) from SK-calculus into SF-calculus.*

Theorem 3. *There is a homomorphism (Definition 2) from λ_v-calculus to SF-calculus.*

Proof. By Theorem 1 and Lemma 5.

Lemma 6. *There is no reduction preserving translation* $[\![\cdot]\!]$ *from* SF*-calculus to* λ_v*-calculus.*

Proof. By Lemma 4.

Theorem 4. *There is no homomorphism (Definition 2) from* SF*-calculus to* λ_v*-calculus.*

Proof. Lemma 6 shows that operational correspondence is impossible.

This completes the top edge of the computation square by showing that SF-calculus subsumes λ_v-calculus and that the subsumption is irreversible. Indeed, these results hold for λ-calculus [11, Theorem 5.2.6] and SK-calculus (by Lemma 4) as well.

5 Concurrent Extensional Computation

The bottom left corner of the computation square considers extensional concurrent computation, here defined to be extensional process calculi that subsume λ-calculus. The π-calculus [26] holds a pivotal rôle amongst process calculi due to popularity, being the first to represent topological changes, and subsuming λ_v-calculus [24]. Note that although there are many π-calculi, the one here is that used by Milner so as to more easily exploit previous results [24] (and here augmented with a success process $\surd$).

The processes for the π-calculus are given as follows and exploit a class of names (denoted $m, n, x, y, z, \ldots$ similar to variables in the λ-calculus):

$$P \quad ::= \quad \mathbf{0} \mid P|P \mid !P \mid (\nu a)P \mid a(b).P \mid \overline{a}\langle b\rangle.P \mid \surd\,.$$

The names of the π-calculus are used for channels of communication and for information being communicated. The *free names* of a process $\mathsf{fn}(P)$ are as usual. *Substitutions* in the π-calculus are partial functions that map names to names, with domain, range, free names, names, and avoidance, all straightforward adaptations from substitutions of the λ-calculus. The application of a substitution to a process is defined in the usual manner. Issues where substitutions must avoid restricted or input names are handled by α-conversion $=_\alpha$ that is the congruence relation defined in the usual manner. The general *structural equivalence relation* $\equiv$ is defined by:

$$\begin{array}{ccc} P \mid \mathbf{0} \equiv P & P \mid Q \equiv Q \mid P & P \mid (Q \mid R) \equiv (P \mid Q) \mid R \\ !P \equiv P \mid !P & (\nu n)\mathbf{0} \equiv \mathbf{0} & (\nu n)(\nu m)P \equiv (\nu m)(\nu n)P \end{array}$$
$$P \mid (\nu n)Q \equiv (\nu n)(P \mid Q) \text{ if } n \notin \mathsf{fn}(P)$$

The π-calculus has one *reduction rule* given by

$$a(b).P \mid \overline{a}\langle c\rangle.Q \quad \longmapsto \quad \{c/b\}P \mid Q\,.$$

The reduction rule is then closed under parallel composition, restriction and structural equivalence to yield the reduction relation $\longmapsto$ as follows:

$$\frac{P \longmapsto P'}{P \mid Q \longmapsto P' \mid Q} \qquad \frac{P \longmapsto P'}{(\nu n)P \longmapsto (\nu n)P'} \qquad \frac{P \equiv Q \quad Q \longmapsto Q' \quad Q' \equiv P'}{P \longmapsto P'}\,.$$

Now that the π-calculus and process calculus concepts are recalled, it remains to demonstrate that Milner's encoding [24] can meet the criteria for a parallel encoding. As the β_v-reduction rule depends upon the argument being a value the translation into π-calculus must be able to recognise values. Thus, Milner defines the following

$$[\![y := \lambda x.t]\!] \stackrel{\text{def}}{=} !y(w).w(x).w(c).[\![t]\!]_c \qquad [\![y := x]\!] \stackrel{\text{def}}{=} !y(w).\overline{x}\langle w\rangle\,.$$

Also the following translation of λ_v-terms

$$\begin{aligned} [\![v]\!]_c &\stackrel{\text{def}}{=} (\nu y)\overline{c}\langle y\rangle.[\![y := v]\!] && y \text{ not free in } v \\ [\![s\ t]\!]_c &\stackrel{\text{def}}{=} (\nu q)(\nu r)(\mathsf{ap}(c,q,r) \mid [\![s]\!]_q \mid [\![t]\!]_r) \\ \mathsf{ap}(p,q,r) &\stackrel{\text{def}}{=} q(y).(\nu v)\overline{y}\langle v\rangle.r(z).\overline{v}\langle z\rangle.\overline{v}\langle p\rangle\,. \end{aligned}$$

Lemma 7. *The translation $[\![\cdot]\!]_c$ preserves and reflects reduction. That is:*

1. *If $s \longmapsto_v t$ then $[\![s]\!]_c \Longmapsto\sim [\![t]\!]_c$;*
2. *if $[\![s]\!]_c \longmapsto Q$ then there exists Q' and s' such that $Q \Longmapsto Q'$ and $Q' \sim [\![s']\!]_c$ and either $s \longmapsto_v s'$ or $s = s'$.*

Proof. The first part can be proved by exploiting Milner's Theorem 7.7 [24]. The second is by considering the reduction $[\![s]\!]_c \longmapsto Q$ which must arise from the encoding of an application. It is then straightforward to show that either: the reductions $Q \Longmapsto Q'$ correspond only to translated applications and thus $Q' \sim [\![s]\!]_c$; or the reductions are due to a λ_v-abstraction and thus $Q' \sim [\![s']\!]_c$ and $s \longmapsto_v s'$.

Theorem 5. *The translation $[\![\cdot]\!]_c$ is a parallel encoding (Definition 4) from λ_v-calculus to π-calculus.*

Proof. Compositionality, parallelisation, and name invariance hold by construction. Operational correspondence follows from Lemma 7. Divergence reflection can be proved by observing that the only reductions introduced in the translation that do not correspond to reductions in the source language are from translated applications, and these are bounded by the size of the source term.

There is some difficulty in attempting to define the analogue of a parallel encoding or homomorphism from a language with a parallel composition operator into a language without. However, this difficulty can be avoided by observing that any valid encoding, parallel encoding, or homomorphism must preserve reduction. Reduction preservation can then be exploited to show when an encoding is impossible. Here this is by exploiting Theorem 14.4.12 of Barendregt [3], showing that λ-calculus is unable to render concurrency or support concurrent computations.

Theorem 6. *There is no reduction preserving encoding of π-calculus into λ-calculus.*

Proof. Define the parallel-or function and show that it can be represented in π-calculus but not λ-calculus. The parallel-or function is a function $g(x, y)$ that satisfies the following three rules $g(\bot, \bot) \longmapsto^* \bot$ and $g(\mathrm{T}, \bot) \longmapsto^* \mathrm{T}$ and $g(\bot, \mathrm{T}) \longmapsto^* \mathrm{T}$ where $\bot$ represents non-termination and T represents true. Such a function is trivial to encode in π-calculus by $g(n_1, n_2) = G = n_1(x).\overline{m}\langle x\rangle.\mathbf{0} \mid n_2(x).\overline{m}\langle x\rangle.\mathbf{0}$. Consider G in parallel with two processes P_1 and P_2 that output their result on n_1 and n_2, respectively. If either P_1 or P_2 outputs T then G will also output T along m. Clearly π-calculus can represent the parallel-or function, and since Barendregt's Theorem 14.4.12 shows that λ-calculus cannot, there cannot be any reduction preserving encoding of π-calculus into λ-calculus.

6 Concurrent Intensional Computation

Intensionality in sequential computation yields greater expressive power so it is natural to consider intensional concurrent computation. Intensionality in CPC is supported by a generalisation of pattern-matching to symmetric *pattern-unification* that provides the basis for defining interaction.

6.1 Concurrent Pattern Calculus

The *patterns* (meta-variables $p, p', p_1, q, q', q_1, \ldots$) are built using a class of *names* familiar from π-calculus and have the following forms

$$p ::= \lambda x \mid x \mid \ulcorner x \urcorner \mid p \bullet p$$

Binding names λx denote an input sought by the pattern. Variable names x may be output or tested for equality. Protected names $\ulcorner x \urcorner$ can only be tested for equality. A compound combines two patterns p and q, its *components*, into a pattern $p \bullet q$ and is left associative. The *atoms* are patterns that are not compounds and the atoms x and $\ulcorner x \urcorner$ are defined to *know* x. The binding names of a pattern must be pairwise distinct.

A *communicable* pattern contains no binding or protected names. Given a pattern p, the binding names $\mathsf{bn}(p)$, variable names $\mathsf{vn}(p)$, and protected names $\mathsf{pn}(p)$, are as expected, with the free names $\mathsf{fn}(p)$ being the union of variable and protected names.

A *substitution* σ (also denoted $\sigma_1, \rho, \rho_1, \theta, \theta_1, \ldots$) is a partial function from names to communicable patterns. Otherwise substitutions and their properties are familiar from earlier sections and are applied to patterns in the obvious manner. (Observe that protection can be extended to a communicable pattern by $\ulcorner p \bullet q \urcorner = \ulcorner p \urcorner \bullet \ulcorner q \urcorner$ in the application of a substitution to a protected name.)

The *symmetric matching* or *unification* $\{p \parallel q\}$ of two patterns p and q attempts to unify p and q by generating substitutions upon their binding names. When defined, the result is some pair of substitutions whose domains are the binding names of p and of q, respectively. The rules to generate the substitutions are:

$$\begin{array}{rcll}
\{x \parallel x\} = \{x \parallel \ulcorner x \urcorner\} & = & \{\ulcorner x \urcorner \parallel x\} = \{\ulcorner x \urcorner \parallel \ulcorner x \urcorner\} \stackrel{\text{def}}{=} (\{\}, \{\}) & \\
\{\lambda x \parallel q\} & \stackrel{\text{def}}{=} & (\{q/x\}, \{\}) & \text{if } q \text{ is communicable} \\
\{p \parallel \lambda x\} & \stackrel{\text{def}}{=} & (\{\}, \{p/x\}) & \text{if } p \text{ is communicable} \\
\{p_1 \bullet p_2 \parallel q_1 \bullet q_2\} & \stackrel{\text{def}}{=} & (\sigma_1 \cup \sigma_2 \,,\, \rho_1 \cup \rho_2) & \text{if } \{p_i \parallel q_i\} = (\sigma_i, \rho_i) \text{ for } i \in \{1, 2\}
\end{array}$$

Two atoms unify if they know the same name. A binding name unifies with any communicable pattern to produce a binding for its underlying name. Two compounds unify if their corresponding components do; the resulting substitutions are given by taking unions of those produced by unifying the components. Otherwise the patterns cannot be unified and the unification is undefined.

The processes of CPC are the same as π-calculus except the input and output are replaced by the *case* $p \rightarrow P$ with pattern p and body P. A case with the null process as the body $p \rightarrow \mathbf{0}$ may also be written p when no ambiguity may occur.

The free names of processes, denoted $\mathsf{fn}(P)$, are defined as usual for all the traditional primitives and $\mathsf{fn}(p \rightarrow P) = \mathsf{fn}(p) \cup (\mathsf{fn}(P) \backslash \mathsf{bn}(p))$ for the case. As expected the binding names of the pattern bind their free occurrences in the body. The application σP of a substitution σ to a process P is defined in the usual manner to avoid name capture. For cases this ensures that substitution avoids the binding names in the pattern: $\sigma(p \rightarrow P) = (\sigma p) \rightarrow (\sigma P)$ if σ avoids $\mathsf{bn}(p)$. Renaming via α-conversion is defined in the usual manner [14, 11, 15]. The general *structural equivalence relation* $\equiv$ is defined just as in π-calculus.

CPC has one *interaction axiom* given by

$$(p \rightarrow P) \mid (q \rightarrow Q) \quad \longmapsto \quad (\sigma P) \mid (\rho Q) \qquad \text{if } \{p \parallel q\} = (\sigma, \rho)\ .$$

It states that if the unification of two patterns p and q is defined and generates (σ, ρ), then apply the substitutions σ and ρ to the bodies P and Q, respectively. If the matching of p and q is undefined then no interaction occurs. The interaction rule is then closed under parallel composition, restriction and structural equivalence in the usual manner. The reflexive and transitive closure of $\longmapsto$ is denoted $\Longmapsto$. Finally, the reference behavioural equivalence relation $\sim$ for CPC is already well detailed [11, 13, 15].

6.2 Completing the Square

Support for both intensionality and concurrency places CPC at the bottom right corner of the computation square. This section shows how SF-calculus and π-calculus can both be subsumed by CPC, and thus completes the computation square.

Down the right side of the square there is a parallel encoding from SF-calculus into CPC that also maps the combinators S and F to reserved names S and F, respectively. The impossibility of finding a parallel encoding of CPC into SF-calculus is proved in the same manner as the relation between λ_v-calculus and π-calculus. Interestingly, in contrast with the parallel encoding of λ-calculus into π-calculus, the parallel encoding of SF-calculus into CPC does *not* fix a reduction strategy for SF-calculus. This is achieved by exploiting the intensionality of CPC to directly encode the reduction rules for SF-calculus into an SF-reducing process, or SF-machine. In turn, this process can then operate on translated combinators and so support reduction and rewriting.

The square is completed by showing a homomorphism from π-calculus into CPC, and by showing that there cannot be any homomorphism (or indeed a more general valid encoding) from CPC into π-calculus.

***SF*-calculus.** The SF-calculus combinators can be easily encoded into patterns by defining the *construction* $(\!| \cdot |\!)$, exploiting reserved names S and F, as follows

$$(\!| S |\!) \stackrel{\text{def}}{=} S \qquad (\!| F |\!) \stackrel{\text{def}}{=} F \qquad (\!| MN |\!) \stackrel{\text{def}}{=} (\!| M |\!) \bullet (\!| N |\!)\ .$$

$$
\begin{array}{l}
!\lambda c \bullet (S \bullet \lambda m \bullet \lambda n \bullet \lambda x) \to c \bullet (m \bullet x \bullet (n \bullet x)) \\
\mid !\lambda c \bullet (F \bullet S \bullet \lambda m \bullet \lambda n) \to c \bullet m \\
\mid !\lambda c \bullet (F \bullet F \bullet \lambda m \bullet \lambda n) \to c \bullet m \\
\mid !\lambda c \bullet (F \bullet (S \bullet \lambda q) \bullet \lambda m \bullet \lambda n) \to c \bullet (n \bullet S \bullet q) \\
\mid !\lambda c \bullet (F \bullet (F \bullet \lambda q) \bullet \lambda m \bullet \lambda n) \to c \bullet (n \bullet F \bullet q) \\
\mid !\lambda c \bullet (F \bullet (S \bullet \lambda p \bullet \lambda q) \bullet \lambda m \bullet \lambda n) \to c \bullet (n \bullet (S \bullet p) \bullet q) \\
\mid !\lambda c \bullet (F \bullet (F \bullet \lambda p \bullet \lambda q) \bullet \lambda m \bullet \lambda n) \to c \bullet (n \bullet (F \bullet p) \bullet q) \\
\mid !\lambda c \bullet (\lambda u \bullet \lambda v \bullet \lambda w \bullet \lambda x \bullet \lambda y) \\
\quad \to (\nu d) d \bullet (u \bullet v \bullet w \bullet x) \to d \bullet \lambda z \to c \bullet (z \bullet y) \\
\mid !\lambda c \bullet (\lambda m \bullet \lambda n \bullet \lambda o \bullet (\lambda u \bullet \lambda v \bullet \lambda w \bullet \lambda x)) \\
\quad \to (\nu d) d \bullet (u \bullet v \bullet w \bullet x) \to d \bullet \lambda z \to c \bullet (m \bullet n \bullet o \bullet z) \\
\mid !\lambda c \bullet (\lambda m \bullet \lambda n \bullet (\lambda u \bullet \lambda v \bullet \lambda w \bullet \lambda x) \bullet \lambda p) \\
\quad \to (\nu d) d \bullet (u \bullet v \bullet w \bullet x) \to d \bullet \lambda z \to c \bullet (m \bullet n \bullet z \bullet p) \\
\mid !\lambda c \bullet (\lambda m \bullet (\lambda u \bullet \lambda v \bullet \lambda w \bullet \lambda x) \bullet \lambda o \bullet \lambda p) \\
\quad \to (\nu d) d \bullet (u \bullet v \bullet w \bullet x) \to d \bullet \lambda z \to c \bullet (m \bullet z \bullet o \bullet p)
\end{array}
$$

Fig. 1. The SF-reducing process $\mathcal{R}$

Observe that the first two rules map the operators to the same names. The third rule maps application to a compound of the components $(\!|M|\!)$ and $(\!|N|\!)$.

By representing SF-calculus combinators in the pattern of a CPC case, the reduction is driven by cases that recognise a reducible structure and perform the appropriate operations. The reduction rules can be captured by matching on the structure of the left hand side of the rule and reducing to the structure on the right. So (considering each possible instance for the F reduction rules) they can be encoded by cases as follows

$$
\begin{array}{c}
S \bullet \lambda m \bullet \lambda n \bullet \lambda x \to m \bullet x \bullet (n \bullet x) \\
F \bullet S \bullet \lambda m \bullet \lambda n \to m \\
F \bullet F \bullet \lambda m \bullet \lambda n \to m \\
\cdots \\
F \bullet (F \bullet \lambda p \bullet \lambda q) \bullet \lambda m \bullet \lambda n \to n \bullet (F \bullet p) \bullet q\,.
\end{array}
$$

These processes capture the reduction rules, matching the pattern for the left hand side and transforming it to the structure on the right hand side. Of course these process do not capture the possibility of reduction of a sub-combinator, so further rules are required. Rather than detail them all, consider the example of a reduction $MNOP \longmapsto MN'OP$ that can be captured by

$$\lambda m \bullet (\lambda u \bullet \lambda v \bullet \lambda w \bullet \lambda x) \bullet \lambda o \bullet \lambda p \to u \bullet v \bullet w \bullet x \to \lambda z \to m \bullet z \bullet o \bullet p$$

This process unifies with a combinator $MXOP$ where X is reducible (observable from the structure), here binding the components of X to four names u, v, w and x. These four names are then shared as a pattern, which can then be unified with another process that can perform the reduction. The result will then (eventually) unify with λz and be substituted back into $m \bullet z \bullet o \bullet p$ to complete the reduction.

To exploit these processes in constructing a parallel encoding requires the addition of a name, used like a channel, to control application. Thus, prefix each pattern that matches the structure of an SF-combinator with a binding name λc and add this to the result, e.g. $\lambda c \bullet (F \bullet S \bullet \lambda m \bullet \lambda n) \rightarrow c \bullet m$. Now the processes that handle each possible reduction rule can be placed under a replication and in parallel composition with each other. This yields the SF-reducing process $\mathcal{R}$ as shown in Figure 1 where the last four replications capture reduction of sub-combinators.

The translation $[\![\cdot]\!]_c$ from SF-combinators into CPC processes is here parametrised by a name c and combines application with a process $\mathsf{ap}(c,m,n)$. This is similar to Milner's encoding from λ_v-calculus into π-calculus and allows the parallel encoding to exploit compositional encoding of sub-terms as processes and thus parallel reduction, while preventing confusion of application.

The translation $[\![\cdot]\!]_c$ of SF-combinators into CPC, exploiting the SF-reducing process $\mathcal{R}$ and reserved names S and F, is defined as follows:

$$[\![S]\!]_c \stackrel{\text{def}}{=} c \bullet S \mid \mathcal{R} \qquad [\![F]\!]_c \stackrel{\text{def}}{=} c \bullet F \mid \mathcal{R}$$
$$[\![MN]\!]_c \stackrel{\text{def}}{=} (\nu m)(\nu n)(\mathsf{ap}(c,m,n) \mid [\![M]\!]_m \mid [\![N]\!]_n)$$
$$\mathsf{ap}(c,m,n) \stackrel{\text{def}}{=} m \bullet \lambda x \rightarrow n \bullet \lambda y \rightarrow c \bullet (x \bullet y) \mid \mathcal{R}\,.$$

The following lemmas are at the core of the operational correspondence and divergence reflection components of the proof of valid encoding, similar to Milner's Theorem 7.7 [24]. Further, it provides a general sense of how to capture the reduction of combinatory logics or similar rewrite systems. (Note that the results exploit that $\mathcal{R} \mid \mathcal{R} \sim \mathcal{R}$ to remove redundant copies of $\mathcal{R}$ [11, Theorem 8.7.2].)

Lemma 8. *Given an SF-combinator M the translation $[\![M]\!]_c$ has a reduction sequence to a process of the form $c \bullet (\!|M|\!) \mid \mathcal{R}$.*

Proof. The proof is by induction on the structure of M.

Lemma 9 (Theorem 7.1.2 of [11]). *Given an SF-combinator M the translation $[\![M]\!]_c$ preserves reduction.*

Proof. The proof is routine by considering each reduction rule and Lemma 8.

Lemma 10. *The translation $[\![\cdot]\!]_c$ preserves and reflects reduction. That is:*

1. *If $M \longmapsto N$ then $[\![M]\!]_c \Longmapsto\sim [\![N]\!]_c$;*
2. *if $[\![M]\!]_c \longmapsto Q$ then there exists Q' and N such that $Q \Longmapsto Q'$ and $Q' \sim [\![N]\!]_c$ and either $M \longmapsto N$ or $M = N$.*

Proof. The first part can be proved by exploiting Lemmas 8 and 9. The second is by considering the reduction $[\![M]\!]_c \longmapsto Q$ which must arise from the encoding of an application. It is then straightforward to show that either: the reductions $Q \Longmapsto Q'$ correspond only to rebuilding the structure as in Lemma 8; or the reductions correspond to a reduction $M \longmapsto N$ and $Q' \sim [\![N]\!]_c$.

Theorem 7. *The translation $[\![\cdot]\!]_c$ is a parallel encoding from SF-calculus to CPC.*

Proof. Compositionality, parallelisation, and name invariance hold by construction. Operational correspondence follows from Lemma 9. Divergence reflection can be proved by observing that the only reductions introduced in the translation that do not correspond to reductions in the source language are from translated applications, and these are bounded by the size of the source term.

The lack of an encoding of CPC (or even π-calculus) into SF-calculus can be proved in the same manner as Theorem 6 for showing no encoding of π-calculus into λ-calculus.

Theorem 8. *There is no reduction preserving encoding from CPC into SF-calculus.*

It may appear that the factorisation operator F adds some expressiveness that could be used to capture the parallel-or function g. Perhaps use F to switch on the result of the first function so that (assuming true is some operator T then) $g(x,y)$ is represented by $Fx\mathrm{T}(K(Ky))$ that reduces to T when $x = \mathrm{T}$ and to $K(Ky)MN \Longmapsto y$ when $x = MN$ that somehow is factorable but not terminating. However, this kind of attempt is equivalent to exploiting factorisation to detect termination and turns out to be paradoxical as demonstrated in the proof of Theorem 5.1 of [20].

This completes the arrow down the right side of the computation square. The rest of this section discusses some properties of translations and the diagonal from the top left to the bottom right corner of the square.

Observe that the parallel encoding from SF-calculus into CPC does not require the choice of a reduction strategy, unlike Milner's encodings from λ-calculus into π-calculus. The structure of patterns and peculiarities of pattern-unification allow the reduction relation to be directly rendered by CPC. In a sense this is similar to the approach in [12] of encoding the SF-combinators as the tape of a Turing Machine, the pattern $(\!|\cdot|\!)$, and providing another process that reads the tape and performs operations upon it, the SF-reducing process $\mathcal{R}$. This approach can also be adapted in a straightforward manner to support a parallel encoding of SK-calculus into CPC, that like the encoding of SF-calculus does not fix a reduction strategy.

Theorem 9. *There is a translation $[\![\cdot]\!]_c$ that is a parallel encoding from SK-calculus into CPC.*

The translation from SF-calculus to CPC presented here is designed to map application to parallel composition (with some restriction and process R) so as to meet the compositionality and parallelisation criteria for a parallel encoding. However, the construction $(\!|\cdot|\!)$ can be used to provide a cleaner translation if these are not required (while still supporting the other criteria). Consider an alternative translation $[\![\cdot]\!]^c$ parametrised by a name c as usual and defined by $[\![M]\!]^c \stackrel{\text{def}}{=} c \bullet (\!|M|\!) \mid \mathcal{R}$.

π-calculus. Across the bottom of the computation square there is a homomorphism from π-calculus into CPC. The converse separation result can be proved multiple ways [14, 11, 15].

The translation $[\![\cdot]\!]$ from π-calculus into CPC is homomorphic on all process forms except for the input and output which are translated as follows:

$$[\![a(b).P]\!] \stackrel{\text{def}}{=} a \bullet \lambda b \bullet \mathsf{in} \rightarrow [\![P]\!] \qquad [\![\overline{a}\langle b\rangle.P]\!] \stackrel{\text{def}}{=} a \bullet b \bullet \lambda\mathsf{in} \rightarrow [\![P]\!]$$

Here in is a fresh name (due to the renaming policy to avoid all other names in the translation) that prevents the introduction of new reductions due to CPC's unification.

Lemma 11 (Corollary 7.2.3 of [11]). *The translation $[\![\cdot]\!]$ from π-calculus into CPC is a valid encoding.*

Theorem 10. *There is a homomorphism (Definition 3) from π-calculus into CPC.*

Thus the translation provided above is a homomorphism from π-calculus into CPC. Now consider the converse separation result.

Lemma 12 (Theorem 7.2.5 of [11]). *There is no valid encoding of CPC into π-calculus.*

Proof (Sketch). Define the *self-reducing* CPC process $P = n \to \surd$. Observe that $P \not\Downarrow$ and $P \mid P \Downarrow$. However, for every π-calculus process T such that $T \mid T \Downarrow$ it holds that $T \Downarrow$. This is sufficient to show contradiction of any possible valid encoding.

Theorem 11. *There is no homomorphism (Definition 3) from CPC into π-calculus.*

7 Conclusions and Future Work

This work illustrates that there are increases in expressive power by shifting along two dimensions from: extensional to intensional, and sequential to concurrent. This is seen in the computation square relating λ_v-calculus, SF-calculus, π-calculus, and CPC

$$\begin{array}{ccc}
\lambda_v\text{-calculus} & \longrightarrow & SF\text{-calculus} \\
\downarrow & & \downarrow \\
\pi\text{-calculus} & \longrightarrow & \text{concurrent pattern calculus}
\end{array}$$

where the left side is extensional, the right side intensional, the top side sequential, and the bottom side concurrent. The horizontal arrows are homomorphisms that map application/parallel composition to itself. The vertical arrows are parallel encodings that map application to parallel composition (with some extra machinery). Further, there are no reverse arrows as each arrow signifies an increase in expressive power.

Such a square identifies relations that are more general than simply the choice of calculi here. The top left corner could be populated by λ_v-calculus or λ_l-calculus with minimal changes to the proofs. Alternatively, choosing λ-calculus or SK-calculus may also hold, although a parallel encoding into π-calculus requires some work. The top right corner could be populated by any of the structure complete combinatory logics [20, 11]. It may also be possible to place a pattern calculus [21, 19], at the top right. The bottom left corner is also open to many other calculi: monadic/polyadic synchronous/asynchronous π-calculus could replace π-calculus with no significant changes to the results [11, 15]. Similarly there are, and will be, other process calculi that can take the place of CPC at the bottom right. For Spi calculus [1] an encoding of SF-calculus is delicate due to correctly handling reduction and not introducing infinite reductions or blocking on Spi calculus primitives and reductions. For Psi calculi [4] the encoding can be achieved very similarly to CPC, although the implicit computation component of Psi calculi could simply allow for SF-calculus with the rest being moot. Although multiple process calculi may populate the bottom right hand corner, the elegance of CPC's intensionality is illustrated by the construction $(\!|\cdot|\!)$ for combinatory logics and [12].

Related Work. The choice of relations here is influenced by existing approaches. Homomorphisms in the sequential setting are typical [8, 3, 10]. Valid encodings are popular [16–18, 22, 23, 31] albeit not the only approach as other ways to relate process calculi are also used that vary on the choice to map parallel composition to parallel composition (i.e. homomorphism here) [28, 6, 9, 27, 31]. Since the choice here is to build on prior results, valid encodings are the obvious basis but no doubt this could be formalised under different criteria. Finally, the definition of parallel encodings here is to exploit the existing encodings in the literature. However, other approaches are possible [24, 29] and many more as encoding λ-calculus into process calculi is common [5, 26, 7, 25].

The separation results here build upon results already in the literature. For showing the inability to encoding concurrent languages into sequential, the work of Abramsky [2] and Plotkin [30] can also be considered. The impossibility of encoding CPC into π-calculus can be proved by using matching degree or symmetry [11, proofs for Theorem 7.2.5].

Future Work. Future work may proceed along several directions. The techniques used to encode SF-calculus (here) and Turing Machines [12] into CPC can be generalised for any combinatory logic, indeed perhaps a general result can be proved for all similar rewrite systems. Another path of exploration is to consider intensionality in concurrency with full results in a general manner, this could include formalising the intensionality (or lack of) of Spi calculus, Psi calculi, and other popular process calculi.

References

1. Abadi, M., Gordon, A.: A calculus for cryptographic protocols: The spi calculus. Information and Computation 148(1), 1–70 (1999)
2. Abramsky, S.: The lazy lambda calculus. In: Research Topics in Functional Programming, pp. 65–116. Addison-Wesley (1990)
3. Barendregt, H.P.: The Lambda Calculus. Its Syntax and Semantics. Studies in Logic and the Foundations of Mathematics. Elsevier Science Publishers B.V. (1985)
4. Bengtson, J., Johansson, M., Parrow, J., Victor, B.: Psi-calculi: a framework for mobile processes with nominal data and logic. Logical Methods in Computer Science 7(1) (2011)
5. Berry, G., Boudol, G.: The chemical abstract machine. In: POPL 1990: Proceedings of the 17th ACM SIGPLAN-SIGACT Symposium on Principles of Programming Languages, pp. 81–94. ACM, New York (1990)
6. Busi, N., Gorrieri, R., Zavattaro, G.: On the expressiveness of linda coordination primitives. Inf. Comput. 156(1-2), 90–121 (2000)
7. Cardelli, L., Gordon, A.D.: Mobile ambients. In: Nivat, M. (ed.) FOSSACS 1998. LNCS, vol. 1378, pp. 140–155. Springer, Heidelberg (1998)
8. Curry, H.B., Feys, R.: Combinatory Logic, vol. I. North-Holland, Amsterdam (1958)
9. De Nicola, R., Gorla, D., Pugliese, R.: On the expressive power of klaim-based calculi. Theor. Comput. Sci. 356(3), 387–421 (2006)
10. Felleisen, M.: On the expressive power of programming languages. Science of Computer Programming 17(1-3), 35–75 (1991)
11. Given-Wilson, T.: Concurrent Pattern Unification. PhD thesis, University of Technology, Sydney, Australia (2012)

12. Given-Wilson, T.: An Intensional Concurrent Faithful Encoding of Turing Machines. In: 7th Interaction and Concurrency Experience (ICE 2014), Berlin, Germany (June 2014)
13. Given-Wilson, T., Gorla, D.: Pattern matching and bisimulation. In: De Nicola, R., Julien, C. (eds.) COORDINATION 2013. LNCS, vol. 7890, pp. 60–74. Springer, Heidelberg (2013)
14. Given-Wilson, T., Gorla, D., Jay, B.: Concurrent pattern calculus. In: Calude, C.S., Sassone, V. (eds.) TCS 2010. IFIP AICT, vol. 323, pp. 244–258. Springer, Heidelberg (2010)
15. Given-Wilson, T., Gorla, D., Jay, B.: A Concurrent Pattern Calculus. To appear in: Logical Methods in Computer Science (2014)
16. Gorla, D.: Comparing communication primitives via their relative expressive power. Information and Computation 206(8), 931–952 (2008)
17. Gorla, D.: A taxonomy of process calculi for distribution and mobility. Distributed Computing 23(4), 273–299 (2010)
18. Gorla, D.: Towards a unified approach to encodability and separation results for process calculi. Information and Computation 208(9), 1031–1053 (2010)
19. Jay, B.: Pattern Calculus: Computing with Functions and Data Structures. Springer (2009)
20. Jay, B., Given-Wilson, T.: A combinatory account of internal structure. Journal of Symbolic Logic 76(3), 807–826 (2011)
21. Jay, B., Kesner, D.: First-class patterns. Journal of Functional Programming 19(2), 191–225 (2009)
22. Lanese, I., Pérez, J.A., Sangiorgi, D., Schmitt, A.: On the expressiveness of polyadic and synchronous communication in higher-order process calculi. In: Abramsky, S., Gavoille, C., Kirchner, C., Meyer auf der Heide, F., Spirakis, P.G. (eds.) ICALP 2010. LNCS, vol. 6199, pp. 442–453. Springer, Heidelberg (2010)
23. Lanese, I., Vaz, C., Ferreira, C.: On the expressive power of primitives for compensation handling. In: Gordon, A.D. (ed.) ESOP 2010. LNCS, vol. 6012, pp. 366–386. Springer, Heidelberg (2010)
24. Milner, R.: Functions as processes. In: Proceedings of the Seventeenth International Colloquium on Automata, Languages and Programming, pp. 167–180. Springer-Verlag New York, Inc., New York (1990)
25. Milner, R.: Communicating and Mobile Systems: the Pi-Calculus. Cambridge University Press (June 1999)
26. Milner, R., Parrow, J., Walker, D.: A calculus of mobile processes, part I/II. Information and Computation 100, 1–77 (1992)
27. Nielsen, L., Yoshida, N., Honda, K.: Multiparty symmetric sum types. In: EXPRESS, pp. 121–135 (2010)
28. Palamidessi, C.: Comparing the expressive power of the synchronous and the asynchronous pi-calculus. CoRR, cs.PL/9809008 (1998)
29. Parrow, J., Victor, B.: The fusion calculus: Expressiveness and symmetry in mobile processes. In: Proc. of LICS, pp. 176–185. IEEE Computer Society (1998)
30. Plotkin, G.: Full abstraction, totality and pcf. Math. Structures Comput. Sci. 9 (1997)
31. van Glabbeek, R.J.: Musings on encodings and expressiveness. In: Proc. of EXPRESS/SOS. EPTCS, vol. 89, pp. 81–98 (2012)

Optimally Streaming Greedy Regular Expression Parsing*

Niels Bjørn Bugge Grathwohl, Fritz Henglein, and Ulrik Terp Rasmussen

Department of Computer Science, University of Copenhagen (DIKU), Denmark

Abstract. We study the problem of *streaming* regular expression parsing: Given a regular expression and an input stream of symbols, how to output a serialized syntax tree representation as an output stream *during* input stream processing.

We show that *optimally streaming* regular expression parsing, outputting bits of the output as early as is semantically possible for any regular expression of size m and any input string of length n, can be performed in time $O(2^{m \log m} + mn)$ on a unit-cost random-access machine. This is for the wide-spread *greedy* disambiguation strategy for choosing parse trees of grammatically ambiguous regular expressions. In particular, for a fixed regular expression, the algorithm's run-time scales linearly with the input string length. The exponential is due to the need for preprocessing the regular expression to analyze state coverage of its associated NFA, a PSPACE-hard problem, and tabulating all reachable *ordered* sets of NFA-states.

Previous regular expression parsing algorithms operate in multiple phases, *always* requiring processing or storing the whole input string before outputting the first bit of output, not only for those regular expressions and input prefixes where reading to the end of the input is strictly *necessary*.

1 Introduction

In programming, regular expressions are often used to extract information from an input, which requires an intensional interpretation of regular expressions as denoting parse trees, and not just their ordinary language-theoretic interpretation as denoting strings.

This is a nontrivial change of perspective. We need to deal with grammatical ambiguity—*which* parse tree to return, not just that it has one—and memory requirements become a critical factor: Deciding whether a string belongs to the language denoted by $(\mathtt{ab})^\star + (\mathtt{a} + \mathtt{b})^\star$ can be done in constant space, but outputting the first bit, whether the string matches the first alternative or only the second, may require buffering the whole input string. This is an instructive

* This work has been partially supported by The Danish Council for Independent Research under Project 11-106278, "Kleene Meets Church: Regular Expressions and Types". The order of authors is insignificant.

G. Ciobanu and D. Méry (Eds.): ICTAC 2014, LNCS 8687, pp. 224–240, 2014.

case of deliberate grammatical ambiguity to be resolved by the prefer-the-left-alternative policy of greedy disambiguation: Try to match the left alternative; if that fails, return a match according to the right alternative as a fallback. Straight-forward application of automata-theoretic techniques does not help: $(\mathtt{ab})^\star + (\mathtt{a} + \mathtt{b})^\star$ denotes the same *language* as $(\mathtt{a} + \mathtt{b})^\star$, which is unambiguous and corresponds to a small DFA, but is also useless: it doesn't represent any more when a string consists of a sequence of *ab*-groups.

Previous parsing algorithms [9,3,5,10,13,6] require at least one full pass over the input string before outputting any output bits representing the parse tree. This is the case even for regular expressions requiring only bounded lookahead such as one-unambiguous regular expressions [1].

In this paper we study the problem of *optimally streaming* parsing. Consider $(\mathtt{ab})^\star + (\mathtt{a} + \mathtt{b})^\star$, which is ambiguous and in general requires unbounded input buffering, and consider the particular input string $\mathtt{ab} \ldots \mathtt{abaabababababab} \ldots$. An *optimally* streaming parsing algorithm needs to buffer the prefix $\mathtt{ab} \ldots \mathtt{ab}$ in some form because the complete parse might match either of the two alternatives in the regular expression, but once encountering $\mathtt{aa}$, only the right alternative is possible. At this point it outputs this information and the output representation for the buffered string as parsed by the second alternative. After this, it outputs a bit for each input symbol read, with no internal buffering: input symbols are discarded before reading the next symbol. Optimality means that output bits representing the eventual parse tree must be produced *earliest possible*: as soon as they are semantically determined by the input processed so far under the assumption that the parse will succeed.

Outline. In Section 2 we recall the *type interpretation* of regular expressions, where a regular expression denotes parse trees, along with the *bit-coding* of parse trees.

In Section 3 we introduce a class of Thompson-style augmented nondeterministic finite automata (aNFAs). Paths in such an aNFA naturally represent *complete* parse trees, and paths to intermediate states represent *partial* parse trees for prefixes of an input string.

We recall the greedy disambiguation strategy in Section 4, which specifies a deterministic mapping of accepted strings to NFA-paths.

Section 5 contains a definition of what it means to be an optimally streaming implementation of a parsing function.

We define what it means for a set of aNFA-states to *cover* another state in Section 6, which constitutes the computationally hardest part needed in our algorithm.

Section 7 contains the main results. We present *path trees* as a way of organizing partial parse trees, and based on these we present our algorithm for an optimally streaming parsing function and analyze its asymptotic run-time complexity.

Finally, in Section 8, the algorithm is demonstrated by illustrative examples alluding to its expressive power and practical utility.

2 Preliminaries

In the following section, we recall definitions of regular expressions and their interpretation as types [10].

Definition 1 (Regular Expression). *A regular expression (RE) over a finite alphabet Σ is an expression E generated by the grammar*

$$E ::= \mathbf{0} \mid \mathbf{1} \mid a \mid E_1 E_2 \mid E_1 + E_2 \mid E_1^\star$$

where $a \in \Sigma$.

Concatenation (juxtaposition) and alternation (+) associates to the right; parentheses may be inserted to override associativity. Kleene star ($\star$) binds tightest, followed by concatenation and alternation.

The standard interpretation of regular expressions is as descriptions of regular languages.

Definition 2 (Language Interpretation). *Every RE E denotes a language $\mathcal{L}[\![E]\!] \subseteq \Sigma^\star$ given as follows:*

$$\begin{array}{lll} \mathcal{L}[\![\mathbf{0}]\!] = \emptyset & \mathcal{L}[\![E_1 E_2]\!] = \mathcal{L}[\![E_1]\!]\mathcal{L}[\![E_2]\!] & \mathcal{L}[\![a]\!] = \{a\} \\ \mathcal{L}[\![\mathbf{1}]\!] = \{\epsilon\} & \mathcal{L}[\![E_1 + E_2]\!] = \mathcal{L}[\![E_1]\!] \cup \mathcal{L}[\![E_2]\!] & \mathcal{L}[\![E_1^\star]\!] = \bigcup_{n \geq 0} \mathcal{L}[\![E_1]\!]^n \end{array}$$

where we have $A_1 A_2 = \{w_1 w_2 \mid w_1 \in A_1, w_2 \in A_2\}$, and $A^0 = \{\epsilon\}$ and $A^{n+1} = AA^n$.

Proviso: Henceforth we shall restrict ourselves to REs E such that $\mathcal{L}[\![E]\!] \neq \emptyset$.

For regular expression parsing, we consider an alternative interpretation of regular expressions as types.

Definition 3 (Type Interpretation). *Let the syntax of* values *be given by*

$$v ::= () \mid \mathsf{inl}\ v_1 \mid \mathsf{inr}\ v_1 \mid \langle v_1, v_2 \rangle \mid [v_1, v_2, ..., v_n]$$

Every RE E can be seen as a type *describing a set $\mathcal{T}[\![E]\!]$ of well-typed values:*

$$\begin{array}{ll} \mathcal{T}[\![\mathbf{0}]\!] = \emptyset & \mathcal{T}[\![E_1 E_2]\!] = \{\langle v_1, v_2 \rangle \mid v_1 \in \mathcal{T}[\![E_1]\!], v_2 \in \mathcal{T}[\![E_2]\!]\} \\ \mathcal{T}[\![\mathbf{1}]\!] = \{()\} & \mathcal{T}[\![E_1 + E_2]\!] = \{\mathsf{inl}\ v \mid v \in \mathcal{T}[\![E_1]\!]\} \cup \{\mathsf{inr}\ v \mid v \in \mathcal{T}[\![E_2]\!]\} \\ \mathcal{T}[\![a]\!] = \{a\} & \mathcal{T}[\![E_1^\star]\!] = \{[v_1, \ldots, v_n] \mid n \geq 0 \wedge \forall 1 \leq i \leq n. v_i \in \mathcal{T}[\![E_1]\!]\} \end{array}$$

We write $|v|$ for the *flattening* of a value, defined as the word obtained by doing an in-order traversal of v and writing down all the symbols in the order they are visited. We write $\mathcal{T}_w[\![E]\!]$ for the restricted set $\{v \in \mathcal{T}[\![E]\!] \mid |v| = w\}$. Regular expression *parsing* is a generalization of the acceptance problem of determining whether a word w belongs to the language of some RE E, where additionally we produce a parse tree from $\mathcal{T}_w[\![E]\!]$. We say that an RE E is *ambiguous* iff there exists a w such that $|\mathcal{T}_w[\![E]\!]| > 1$.

Any well-typed value can be serialized into a sequence of bits.

Definition 4 (Bit-Coding). *Given a value* $v \in \mathcal{T}[\![E]\!]$, *we denote its* bit-code *by* $\ulcorner v \urcorner \subseteq \{0,1\}^\star$, *defined as follows:*

$$\ulcorner () \urcorner = \epsilon \qquad \ulcorner a \urcorner = \epsilon \qquad \ulcorner \mathsf{inl}\ v \urcorner = 0 \ulcorner v \urcorner$$
$$\ulcorner \langle v_1, v_2 \rangle \urcorner = \ulcorner v_1 \urcorner \ulcorner v_2 \urcorner \qquad \ulcorner [v_1, ..., v_n] \urcorner = 0 \ulcorner v_1 \urcorner ... 0 \ulcorner v_n \urcorner 1 \qquad \ulcorner \mathsf{inr}\ v \urcorner = 1 \ulcorner v \urcorner$$

We write $\mathcal{B}[\![E]\!]$ for the set $\{\ulcorner v \urcorner \mid v \in \mathcal{T}[\![E]\!]\}$ and $\mathcal{B}_w[\![E]\!]$ for the set restricted to bit-codes for values with a flattening w. Note that for any RE E, bit-coding is an isomorphism when seen as a function $\ulcorner \cdot \urcorner_E : \mathcal{T}[\![E]\!] \to \mathcal{B}[\![E]\!]$.

3 Augmented Automata

In this section we recall from [6] the construction of finite automata from regular expressions. Our construction is similar to that of Thompson [15], but augmented with extra annotations on non-deterministic ϵ-transitions. The resulting automata can be seen as non-deterministic transducers which for each accepted input string in the language of the underlying regular expression outputs the bit-codes for the corresponding parse trees.

Definition 5 (Augmented Non-deterministic Finite Automaton). *An* augmented non-deterministic finite automaton *(aNFA) is a tuple* $(\mathsf{State}, \delta, q^{\mathsf{in}}, q^{\mathsf{fin}})$, *where* State *is a finite set of* states, $q^{\mathsf{in}}, q^{\mathsf{fin}} \in \mathsf{State}$ *are* initial *and* final *states, respectively, and* $\delta \subseteq \mathsf{State} \times \Gamma \times \mathsf{State}$ *is a labeled transition relation with labels* $\Gamma = \Sigma \uplus \{0, 1, \epsilon\}$.

Transition labels are divided into the disjoint sets Σ (symbol labels); $\{0,1\}$ (bit-labels); and $\{\epsilon\}$ (ϵ-labels). Σ-transitions can be seen as input actions, and bit-transitions as output actions.

Definition 6 (aNFA construction). *Let* E *be an RE and define an aNFA* $M_E = (\mathsf{State}_E, \delta_E, q^{\mathsf{in}}_E, q^{\mathsf{fin}}_E)$ *by induction on* E. *We give the definition diagrammatically by cases:*

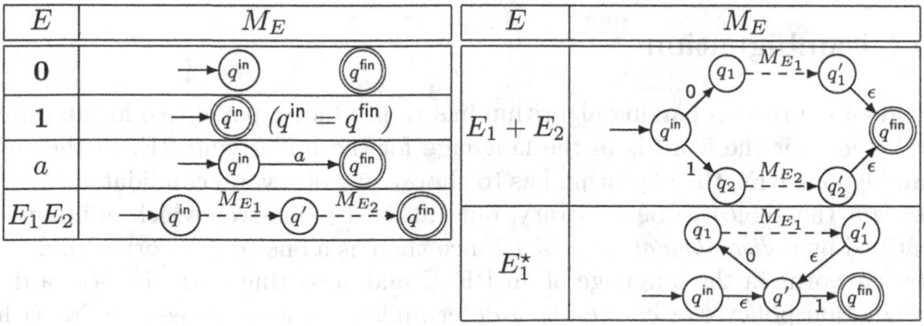

In the above, the notation $(q_1) \overset{M}{\dashrightarrow} (q_2)$ *means that* q_1, q_2 *are initial and final states, respectively, in some (sub-)automaton* M.

See Figure 1 for an example.

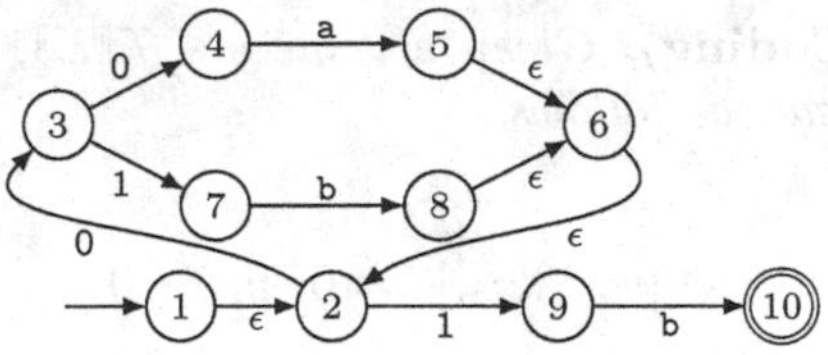

Fig. 1. Example automaton for the RE $(\mathtt{a}+\mathtt{b})^{\star}\mathtt{b}$

Definition 7 (Path). *A* path *in an aNFA is a finite non-empty sequence $\alpha \in \mathsf{State}^{\star}$ of the form $\alpha = p_0\, p_1 \ldots p_{n-1}$ such that for each $i < n$, we have $(p_i, \gamma_i, p_{i+1}) \in \delta_E$ for some γ_i. As a shorthand for this fact we might write $p_0 \overset{\alpha}{\rightsquigarrow} p_{n-1}$ (note that a single state is a path to itself).*

Each path α is associated with a (possibly empty) sequence of labels $\mathsf{lab}(\alpha)$: we let $\mathsf{read}(\alpha)$ and $\mathsf{write}(\alpha)$ refer to the corresponding subsequences of $\mathsf{lab}(\alpha)$ filtered by Σ and $\{0,1\}$, respectively. An automaton *accepts* a word w iff $q^{\mathsf{in}} \overset{\alpha}{\rightsquigarrow} q^{\mathsf{fin}}$ for some α where $\mathsf{read}(\alpha) = w$. There is a one-to-one correspondence between bit-codes and accepting paths:

Proposition 1. *For any RE E with aNFA M_E, we have for each $w \in \mathcal{L}[\![E]\!]$ that*

$$\{\mathsf{write}(\alpha) \mid q^{\mathsf{in}} \overset{\alpha}{\rightsquigarrow} q^{\mathsf{fin}}, \mathsf{read}(\alpha) = w\} = \mathcal{B}_w[\![E]\!].$$

Determinization. Given a state set Q, define its *closure* as the set $\mathsf{closure}(Q) = \{q' \mid q \in Q \wedge \exists \alpha.\mathsf{read}(\alpha) = \epsilon \wedge q \overset{\alpha}{\rightsquigarrow} q'\}$. For any aNFA $M = (\mathsf{State}, \delta, q^{\mathsf{in}}, q^{\mathsf{fin}})$, let $D(M) = (\mathsf{DState}_M, I_M, F_M, \Delta_M)$ be the deterministic automaton obtained by applying the standard subset construction: Here, $I_M = \mathsf{closure}(\{q^{\mathsf{in}}\})$ is the *initial state*, and $\mathsf{DState}_M \subseteq \mathbf{2}^{\mathsf{State}}$ is the set of states, defined to be the smallest set containing I_M and closed under the transition function $\Delta_M(Q, a) = \mathsf{closure}(\{q' \mid (q, a, q') \in \delta, q \in Q\})$. The set of *final states* F_M is the set $\{Q \in \mathsf{DState}_M \mid q^{\mathsf{fin}} \in Q\}$.

4 Disambiguation

A regular expression parsing algorithm has to produce a parse tree for an input word whenever the word is in the language for the underlying RE. In the case of ambiguous REs, the algorithm has to choose one of several candidates. We do not want the choice to be arbitrary, but rather a parse tree which is uniquely identified by a *disambiguation policy*. Since there is a one-to-one correspondence between words in the language of an RE E and accepting paths in M_E, a disambiguation policy can be seen as a deterministic choice between aNFA paths recognizing the same string.

We will focus on greedy disambiguation, which corresponds to choosing the first result that would have been found by a backtracking regular expression parsing algorithm such as the one found in the Perl programming language [16].

The greedy strategy has successfully been implemented in previous work [5,6], and is simpler to define and implement than other strategies such as POSIX [8,4] whose known parsing algorithms are technically more complicated [11,13,14].

Greedy disambiguation can be seen as picking the accepting path with the lexicographically least bitcode. A well-known problem with backtracking parsing is non-termination in the case of regular expressions with nullable subexpressions under Kleene star, which means that the lexicographically least path is not always well-defined. This problem can easily be solved by not considering paths with non-productive loops, as in [5].

5 Optimal Streaming

In this section we specify what it means to be an *optimally streaming* implementation of a function from sequences to sequences.

We write $w \sqsubseteq w''$ if w is a *prefix* of w'', that is $ww' = w''$ for some w'. Note that $\sqsubseteq$ is a partial order with greatest lower bounds for nonempty sets: $\bigsqcap L = w$ if $w \sqsubseteq w''$ for all $w'' \in L$ and $\forall w'.(\forall w'' \in S.w' \sqsubseteq w'') \Rightarrow w' \sqsubseteq w$. $\bigsqcap L$ is the longest common prefix of all words in L.

Definition 8 (Completions). *The set of* completions $C_E(w)$ *of* w *in* E *is the set of all words in* $\mathcal{L}[\![E]\!]$ *that have* w *as a prefix:*

$$C_E(w) = \{w'' \mid w \sqsubseteq w'' \wedge w'' \in \mathcal{L}[\![E]\!]\}.$$

Note that $C_E(w)$ may be empty.

Definition 9 (Extension). *For nonempty* $C_E(w)$ *the unique* extension $\hat{w}_E$ *of* w *under* E *is the longest extension of* w *with a suffix such that all successful extensions of* w *to an element of* $\mathcal{L}[\![E]\!]$ *are also extensions of* $\hat{w}$:

$$\hat{w}_E = \bigsqcap C_E(w).$$

Word w *is* extended *under* E *if* $w = \hat{w}$; *otherwise it is unextended.*

Extension is a closure operation: $\hat{\hat{w}} = \hat{w}$; in particular, extensions are extended.

Definition 10 (Reduction). *For empty* $C_E(w)$ *the unique* reduction $\bar{w}_E$ *of* w *under* E *is the longest prefix* w' *of* w *such that* $C_E(w') \neq \emptyset$.

Given parse function $\mathsf{P}_E(\cdot) : \mathcal{L}[\![E]\!] \to \mathcal{B}[\![E]\!]$ for complete input strings, we can now define what it means for an implementation of it to be optimally streaming:

Definition 11 (Optimally Streaming). *The* optimally streaming *function corresponding to* $\mathsf{P}_E(\cdot)$ *is*

$$O_E(w) = \begin{cases} \bigsqcap\{\mathsf{P}_E(w'') \mid w'' \in C_E(w)\} & \text{if } C_E(w) \neq \emptyset \\ (\bigsqcap O_E(\bar{w}))\sharp & \text{if } C_E(w) = \emptyset. \end{cases}$$

The first condition expresses that after seeing prefix w the function must output *all* bits that are a common prefix of all bit-coded parse trees of words in $\mathcal{L}[\![E]\!]$ that w can be extended to. The second condition expresses that as soon as it is clear that a prefix has no extension to an element of $\mathcal{L}[\![E]\!]$, an indicator $\sharp$ of failure must be emitted, with no further output after that. In this sense O_E is *optimally* streaming: It produces output bits at the semantically earliest possible time during input processing.

It is easy to check that O_E is a streaming function:

$$w \sqsubseteq w' \Rightarrow O_E(w) \sqsubseteq O_E(w')$$

The definition has the, at first glance, surprising consequence that O_E may output bits for parts of the input it has not even read yet:

Proposition 2. $O_E(w) = O_E(\hat{w})$

E.g. for $E = (\mathtt{a} + \mathtt{a})(\mathtt{a} + \mathtt{a})$ we have $O_E(\epsilon) = \mathtt{00}$; that is, O_E outputs 00 off the bat, before reading any input symbols, in anticipation of aa being the only possible successful extension. Assume the input is ab. After reading a it does not output anything, and after reading b it outputs $\sharp$ to indicate a failed parse, the total output being $\mathtt{00}\sharp$.

6 Coverage

Our algorithm is based on simulating aNFAs in lock-step, maintaining a set of partial paths reading the prefix w of the input that has been consumed so far. In order to be optimally streaming, we have to identify partial paths which are guaranteed not to be a prefixes of a greedy parse for a word in $C_E(w)$.

In this section, we define a *coverage relation* which our parsing algorithm relies on in order to detect the aforementioned situation. In the following, fix an RE E and its aNFA $M_E = (\mathsf{State}_E, \delta_E, q_E^{\mathsf{in}}, q_E^{\mathsf{fin}})$.

Definition 12 (Coverage). *Let $p \in \mathsf{State}_E$ be a state and $Q \subseteq \mathsf{State}_E$ a state set. We say that Q* covers *p, written $Q \sqsupseteq p$, iff*

$$\{\mathsf{read}(\alpha) \mid q \overset{\alpha}{\rightsquigarrow} q^{\mathsf{fin}}, q \in Q\} \supseteq \{\mathsf{read}(\beta) \mid p \overset{\beta}{\rightsquigarrow} q^{\mathsf{fin}}\} \quad (1)$$

Coverage can be seen as a slight generalization of language inclusion. That is, if $Q \sqsupseteq p$, then every word suffix read by a path from p to the final state can also be read by a path from one of the states in Q to the final state.

Let $\overline{M_e}$ refer to the automaton obtained by reversing the direction of all transitions and swapping the initial and final states. It can easily be verified that if (1) holds for some Q, p, then the following property also holds in the *reverse* automaton $\overline{M_E}$:

$$\{\mathsf{read}(\alpha) \mid q^{\mathsf{in}} \overset{\alpha}{\rightsquigarrow} q, q \in Q\} \supseteq \{\mathsf{read}(\beta) \mid q^{\mathsf{in}} \overset{\alpha}{\rightsquigarrow} p\} \quad (2)$$

If we consider $D(\overline{M_E})$, the deterministic automaton generated from $\overline{M_E}$, then we see that (2) is satisfied iff

$$\forall S \in \mathsf{DState}_{\overline{M_E}}.\ p \in S \Rightarrow Q \cap S \neq \emptyset \qquad (3)$$

This is true since a DFA state S is reachable by reading a word w in $D(\overline{M_E})$ iff every $q \in S$ is reachable by reading w in $\overline{M_E}$. Since a DFA accepts the same language as the underlying aNFA, this implies that condition (2) must hold iff Q has a non-empty intersection with *all* DFA states containing p.

The equivalence of (1) and (3) gives us a method to decide $\sqsupseteq$ in an aNFA M, provided that we have computed $D(\overline{M})$ beforehand. Checking (3) for a particular Q and p can be done by intersecting all states of $\mathsf{DState}_{\overline{M_E}}$ with Q, using time $O(|Q||\mathsf{DState}_{\overline{M_E}}|) = O(|Q|2^{O(m)})$, where m is the size of the RE E.

The exponential cost appears to be unavoidable – the problem of deciding coverage is inherently hard to compute:

Proposition 3. *The problem of deciding coverage, that is the set* $\{(E, Q, p) \mid Q \subseteq \mathsf{State}_E \wedge Q \sqsupseteq p\}$, *is PSPACE-hard.*

Proof. We can reduce regular expression equivalence to coverage: Given regular expressions E and F, produce an aNFA M_{E+F} for $E + F$ and observe that M_E and M_F are subautomata. Now observe that there is a path $q^{\mathsf{in}}_{E+F} \overset{\alpha}{\rightsquigarrow} q^{\mathsf{fin}}_E$ (respectively $q^{\mathsf{in}}_{E+F} \overset{\beta}{\rightsquigarrow} q^{\mathsf{fin}}_F$) in M_{E+F} iff there is a path $q^{\mathsf{in}}_E \overset{\alpha'}{\rightsquigarrow} q^{\mathsf{fin}}_E$ with $\mathsf{read}(\alpha) = \mathsf{read}(\alpha')$ in M_E (respectively $q^{\mathsf{in}}_F \overset{\beta'}{\rightsquigarrow} q^{\mathsf{fin}}_F$ with $\mathsf{read}(\beta) = \mathsf{read}(\beta')$ in M_F). Hence, we have $\{q^{\mathsf{in}}_F\} \sqsupseteq q^{\mathsf{in}}_E$ in M_{E+F} iff $\mathcal{L}[\![E]\!] \subseteq \mathcal{L}[\![F]\!]$. Since regular expression containment is PSPACE-complete [12] this shows that coverage is PSPACE-hard. □

Even after having computed a determinized automaton, the decision version of the coverage problem is still NP-complete, which we show by reduction to and from MIN-COVER, a well-known NP-complete problem. Let STATE-COVER refer to the problem of deciding membership for the language $\{(M, D(M), p, k) \mid \exists Q.\ |Q| = k \wedge p \notin Q \wedge Q \sqsupseteq p \text{ in } M\}$. Recall that MIN-COVER is the problem of deciding membership for the language $\{(X, \mathcal{F}, k) \mid \exists \mathcal{C} \subseteq \mathcal{F}. |\mathcal{C}| = k \wedge X = \bigcup \mathcal{C}\}$.

Proposition 4. STATE-COVER *is NP-complete.*

Proof. STATE-COVER $\Rightarrow$ MIN-COVER: Let $(M, D(M), p, k)$ be given. Define $X = \{S \in \mathsf{DState}_M \mid p \in S\}$ and $\mathcal{F} = \{R_q \mid q \in \bigcup X\}$ where $R_q = \{S \in X \mid q \in S\}$. Then any k-sized set cover $\mathcal{C} = \{R_{q_1}, \dots, R_{q_k}\}$ gives a state cover $Q = \{q_1, \dots, q_k\}$ and vice-versa.

MIN-COVER $\Rightarrow$ STATE-COVER: Let $(X, \mathcal{F}, k)$ be given, where $|X| = m$ and $|\mathcal{F}| = n$. Construct an aNFA $M_{X,\mathcal{F}}$ over the alphabet $\Sigma = X \uplus \{\$\}$. Define its states to be the set $\{q^{\mathsf{in}}, q^{\mathsf{fin}}, p\} \cup \{F_1, \dots, F_n\}$, and for each F_i, add transitions $F_i \overset{\$}{\rightarrow} q^{\mathsf{fin}}$ and $q^{\mathsf{in}} \overset{x_{ij}}{\rightarrow} F_i$ for each $x_{ij} \in F_i$. Finally add transitions $p \overset{\$}{\rightarrow} q^{\mathsf{fin}}$ and $q^{\mathsf{in}} \overset{x}{\rightarrow} p$ for each $x \in X$.

Observe that $D(M_{X,\mathcal{F}})$ will have states $\{\{q^{\mathsf{in}}\}, \{q^{\mathsf{fin}}\}\} \cup \{S_x \mid x \in X\}$ where $S_x = \{F \in \mathcal{F} \mid x \in F\} \cup \{p\}$, and $\Delta(\{q^{\mathsf{in}}\}, x) = S_x$. Also, the time to compute $D(M_{X,\mathcal{F}})$ is bounded by $O(|X||\mathcal{F}|)$. Then any k-sized state cover $Q = \{F_1, ..., F_k\}$ is also a set cover. □

7 Algorithm

Our parsing algorithm produces a bit-coded parse tree from an input string w for a given RE E. We will simulate M_E in lock-step, reading a symbol from w in each step. The simulation maintains a set of all partial paths that read the prefix of w that has been consumed so far; there are always only finitely many paths to consider, since we restrict ourselves to paths without non-productive loops. When a path reaches a non-deterministic choice, it will "fork" into two paths with the same prefix. Thus, the path set can be represented as a tree of states, where the root is the initial state, the edges are transitions between states, and the leaves are the reachable states.

Definition 13 (Path Trees). *A* path tree *is a rooted, ordered, binary tree with* internal nodes *of outdegrees* 1 *or* 2. *Nodes are labeled by aNFA-states and edges by $\Gamma = \Sigma \cup \{0, 1\} \cup \{\epsilon\}$. Binary nodes have a pair of* 0*- and* 1*-labeled edges (in this order only), respectively.*

We use the following notation:

- $\mathsf{root}(T)$ is the root node of path tree T.
- $\mathsf{path}(n, c)$ is the path from n to c, where c is a descendant of n.
- $\mathsf{init}(T)$ is the path from the root to the first binary node reachable or to the unique leaf of T if it has no binary node.
- $\mathsf{leaves}(T)$ is the *ordered list* of leaf nodes.
- $\mathsf{Tr}_{\mathsf{empty}}$ is the empty tree.

As a notational convenience, the tree with a root node labeled q and no children is written $q\langle\cdot\rangle$, where q is an aNFA-state. Similarly, a tree with a root labeled q with children l and r is written $q\langle 0 : l, 1 : r\rangle$, where q is an aNFA-state and l and r are path trees and the edges from q to l and r are labeled 0 and 1, respectively. Unary nodes are written $q\langle \ell : c\rangle$, denoting a tree rooted at q with only one ℓ-labelled child c.

In the following we shall use T_w to refer to a path tree created after processing input word w and T to refer to path trees in general, where the input string giving rise to the tree is irrelevant.

Definition 14 (Path Tree Invariant). *Let T_w be a path tree and w a word. Define $I(T_w)$ as the proposition that* all *of the following hold:*

(i) The $\mathsf{leaves}(T_w)$ *have pairwise distinct node labels; all labels are* symbol sources, *that is states with a single symbol transition, or the accept state.*
(ii) All paths from the root to a leaf read w:
$\forall n \in \mathsf{leaves}(T_w).\ \mathsf{read}(\mathsf{path}(\mathsf{root}(T_w), n)) = w$.
(iii) For each leaf $n \in \mathsf{leaves}(T_w)$ there exists $w'' \in C_E(w)$ such that the bit-coded parse of w'' starts with $\mathsf{write}(\mathsf{path}(\mathsf{root}(T_w), n))$.

Algorithm 1. Optimally streaming greedy regular expression parsing algorithm

Require: An aNFA M, a coverage relation $\sqsupseteq$, and an input stream S.
Ensure: The greedy leftmost parse tree, emitted in an optimally-streaming fashion.

function STREAM-PARSE(M, $\sqsupseteq$, S)
  $w \leftarrow \epsilon$
  $(T_\epsilon, \_\_) \leftarrow$ CLOSURE$(M, \emptyset, q^{\mathsf{in}})$ ▷ Initialize path tree as the output of CLOSURE
  while S has another input symbol a **do**
    if $C_E(wa) = \emptyset$ **then**
      return $\mathsf{write}(\mathsf{init}(T_w))$ followed by $\sharp$ and exit.
    $T_{wa} \leftarrow$ ESTABLISH-INVARIANT$(T_w, a, \sqsupseteq)$
    Output new bits on the path to the first binary node in T_{wa}, if any.
    $w \leftarrow wa$
  if $q^{\mathsf{fin}} \in \mathsf{leaves}(T_w)$ **then**
    return $\mathsf{write}(\mathsf{path}(\mathsf{root}(T_w), q^{\mathsf{fin}}))$
  else
    return $\mathsf{write}(\mathsf{init}(T_w))$ followed by $\sharp$

Algorithm 2. Establishing invariant $I(T_{wa})$

Require: A path tree T_w satisfying invariant $I(T_w)$, a character a, and a coverage relation $\sqsupseteq$.
Ensure: A path tree T_{wa} satisfying invariant $I(T_{wa})$.

1: **function** ESTABLISH-INVARIANT(T_w, a, $\sqsupseteq$)
2:   Remove leaves from T_w that do not have a transition on a.
3:   Extend T_w to T_{wa} by following all a-transitions.
4:   **for** each leaf n in T_{wa} **do**
5:     $(T', \_\_) \leftarrow$ CLOSURE$(M, \emptyset, n)$.
6:     Replace the leaf n with the tree T' in T_{wa}.
7:   **return** PRUNE$(T_{wa}, \sqsupseteq)$

(iv) For each $w'' \in C_E(w)$ there exists $n \in \mathsf{leaves}(T_w)$ such that the bit-coded parse of w'' starts with $\mathsf{write}(\mathsf{path}(\mathsf{root}(T_w), n))$.

The path tree invariant is maintained by Algorithm 2: line 2 establishes part i; line 3 establishes part ii; and lines 4–7 establish part iii and iv.

Theorem 1 (Optimal Streaming Property). *Assume extended w, $C_E(w) \neq \emptyset$. Consider the path tree T_w after reading w upon entry into the while-loop of the algorithm in Algorithm 1. Then $\mathsf{write}(\mathsf{init}(T_w)) = O_E(w)$.*

In other words, the initial path from the root of T_w to the first binary node in T_w is the longest common prefix of all paths accepting an extension of w. Operationally, whenever that path gets longer by pruning branches, we output the bits on the extension.

Proof. Assume w extended, that is $w = \hat{w}$; assume $C_E(w) \neq \emptyset$, that is there exists w'' such that $w \sqsubseteq w''$ and $w'' \in \mathcal{L}[\![E]\!]$.

Algorithm 3. Pruning algorithm

Require: A path tree T and a covering relation $\sqsupseteq$.
Ensure: A pruned path tree T' where all leaves are alive.

```
function PRUNE(T, ⊒)
    for each l in reverse(leaves(T)) do
        S ← {n | n comes before l in leaves(T)}
        if S ⊒ l then
            p ← parent(l)
            Delete l from T
            T ← CUT(T, p)
    return T
function CUT(T, n)                              ▷ Cuts a chain of 1-ary nodes.
    if |children(n)| = 0 then
        p ← parent(n)
        T' ← T with n removed
        return CUT(T', p)
    else
        return T
```

Claim: $|\mathsf{leaves}(T_w)| \geq 2$ or the unique node in $\mathsf{leaves}(T_w)$ is labeled by the accept state. Proof of claim: Assume otherwise, that is $|\mathsf{leaves}(T_w)| = 1$, but its node is not the accept state. By i of $I(T_w)$, this means the node must have a symbol transition on some symbol a. In this case, all accepting paths $C_E(wa) = C_E(w)$ and thus $\hat{w} = \widehat{wa}$; in particular $\hat{w} \neq w$, which, however, is a contradiction to the assumption that w is extended.

This means we have two cases. The case $|\mathsf{leaves}(T_w)| = 1$ with the sole node being labeled by the accept state is easy: It spells a single path from initial to accept state. By ii and iii of $I(T_w)$ we have that that path is correct for w. By iv and since the accept state has no outgoing transitions, we have $C_E(w) = \{w\}$, and the theorem follows for this case.

Let us consider the case $|\mathsf{leaves}(T_w)| \geq 2$ then. Recall that $C_E(w) \neq \emptyset$ by assumption. By iv of $I(T_w)$ the accepting path of every $w'' \in C_E(w)$ starts with $\mathsf{path}(\mathsf{root}(T_w), n)$ for some $n \in \mathsf{leaves}(T_w)$, and by iii each path from the root to a leaf is the start of some accept path. Since $|\mathsf{leaves}(T_w)| \geq 2$ we know that there exists a binary node in T_w. Consider the first on the path from the root to a leaf. It has both 0- and 1-labeled out-edges. Thus the longest common prefix of $\{\mathsf{write}(p) \mid n \in \mathsf{leaves}(T_w), p \in \mathsf{path}(\mathsf{root}(T_w), n)\}$ is $\mathsf{write}(\mathsf{init}(T_w))$, the bits on the initial path from the root of T_w to its first binary node. □

The algorithm, as given, is only optimally streaming for extended prefixes. It can be made to work for all prefixes by enclosing it in an outer loop that for each prefix w computes $\hat{w}$ and calls the given algorithm with $\hat{w}$. The outer loop then checks that subsequent symbols match until $\hat{w}$ is reached. By Proposition 2 the resulting algorithm gives the right result for all input prefixes, not only extended ones.

Algorithm 4. ϵ-closure with path tree construction.

Require: An aNFA M, a set of visited states V, and a state q
Ensure: A path tree T and a set of visited states V'

```
function CLOSURE(M, V, q)
    if q -0-> q_l and q -1-> q_r then
        (T^l, V_l) ← CLOSURE(M, V ∪ {q}, q_l)          ▷ Try left option first.
        (T^r, V_lr) ← CLOSURE(M, V_l, q_r)     ▷ Use V_l to skip already-visited nodes.
        return (q⟨T^l : T^r⟩, V_lr)
    if q -ε-> p then
        if p ∈ V then                                           ▷ Stop loops.
            return (Tr_empty, V)
        else
            (T', V') ← CLOSURE(M, V ∪ {q}, p)
            return (q⟨ε : T'⟩, V')
    else                                 ▷ q is a symbol source or the final state.
        return (q⟨·⟩, V)
```

Theorem 2. *The optimally streaming algorithm can be implemented to run in time $O(2^{m \log m} + mn)$, where $m = |E|$ and $n = |w|$.*

Proof (Sketch). As shown in Section 6, we can decide coverage in time $O(m2^{O(m)})$. The set of ordered lists $\mathsf{leaves}(T)$ for any T reachable from the initial state can be precomputed and covered states marked in it. (This requires unit-cost random access since there are $O(2^{m \log m})$ such lists.) The ϵ-closure can be computed in time $O(m)$ for each input symbol, and pruning can be amortized over ϵ-closure computation by charging each edge removed to its addition to a tree path. □

For fixed regular expression E this is linear time in n and thus asymptotically optimal. An exponential in m as an additive preprocessing cost appears practically unavoidable since we require the coverage relation, which is inherently hard to compute (Proposition 3).

8 Example

Consider the RE $(\mathtt{aaa} + \mathtt{aa})^\star$. A simplified version of its aNFA is shown in Figure 2. The following two observations are requirements for an earliest parse of this expression:

- After one a has been read, the algorithm *must* output a 0 to indicate that one iteration of the Kleene star has been made, but:
- *five* consecutive as determine that the leftmost possibility in the Kleene star choice was taken, meaning that the first *three* as are consumed in that branch.

The first point can be seen by noting that any parse of a non-zero number of as must follow a path through the Kleene star. This guarantees that *if* a successful parse is eventually performed, it must be the case that at least one iteration was made.

The second point can be seen by considering the situation where only four input as have been read: It is not known whether these are the only four or more input symbols in the stream. In the former case, the correct (and only) parse is two iterations with the right alternative, but in the latter case, the first three symbols are consumed in the left branch instead.

These observations correspond intuitively to what "earliest" parsing is; as soon as it is impossible that an iteration was *not* made, a bit indicating this fact is emitted, and as soon as the first three symbols must have been parsed in the left alternative, this fact is output. Furthermore, a 0-bit is emitted to indicate that (at least) another iteration is performed.

Figure 2 shows the evolution of the path tree during execution with the RE $(\mathtt{aaa} + \mathtt{aa})^\star$ on the input aaaaa.

By similar reasoning as above, after five as it is safe to commit to the left alternative after every third a. Hence, for the inputs $\mathtt{aaaaa(aaa)}^n$, $\mathtt{aaaaa(aaa)}^n\mathtt{a}$, and $\mathtt{aaaaa(aaa)}^n\mathtt{aa}$ the "commit points" are placed as follows ($\cdot$ indicate end-of-input):

$$\underset{0}{\mathtt{a}} \mid \underset{00}{\mathtt{aaaa}} \mid \underbrace{\left(\underset{00}{\mathtt{aaa}} \mid \cdots \mid \underset{00}{\mathtt{aaa}}\right)}_{n \text{ times}} \mid \underset{11}{\cdot} \qquad \underset{0}{\mathtt{a}} \mid \underset{00}{\mathtt{aaaa}} \mid \underbrace{\left(\underset{00}{\mathtt{aaa}} \mid \cdots \mid \underset{00}{\mathtt{aaa}}\right)}_{n \text{ times}} \mid \underset{01}{\mathtt{a}\cdot}$$

$$\underset{0}{\mathtt{a}} \mid \underset{00}{\mathtt{aaaa}} \mid \underbrace{\left(\underset{00}{\mathtt{aaa}} \mid \cdots \mid \underset{00}{\mathtt{aaa}}\right)}_{n \text{ times}} \mid \underset{1011}{\mathtt{aa}\cdot}$$

Complex coverage. The previous example does not exhibit any non-trivial coverage, i.e., situations where a state n is covered by $k > 1$ other states. One can construct an expression that contains non-trivial coverage relations by observing that if each symbol source s in the aNFA is associated with the RE representing the language recognized from s, coverage can be expressed as a set of (in)equations in Kleene algebra. Thus, the coverage $\{n_0, n_1\} \sqsupseteq n$ becomes $RE(n_0) + RE(n_1) \geq RE(n)$ in KA, where $RE(\cdot)$ is the function that yields the RE from a symbol source in an aNFA.

Any expression of the form $x_1 z y_1 + x_2 z y_2 + x_3 z(y_1 + y_2)$ satisfies the property that two subterms cover a third. If the coverage is to play a role in the algorithm, however, the languages denoted by x_1 and x_2 must not subsume that of x_3, otherwise the part starting with x_3 would never play a role due to greedy leftmost disambiguation.

Choose $x_1 = x_2 = (\mathtt{aa})^\star$, $x_3 = \mathtt{a}^\star$, $y_1 = \mathtt{a}$, and $y_2 = \mathtt{b}$. Figure 3 shows the expression

$$(\mathtt{aa})^\star(\mathtt{za} + \mathtt{zb}) + \mathtt{a}^\star\mathtt{z}(\mathtt{a} + \mathtt{b}).$$

The earliest point where any bits can be output is when the z is reached. Then it becomes known whether there was an even or odd number of as. Due to the coverage $\{8, 13\} \sqsupseteq 20$ state 20 is pruned away on the input aazb, thereby causing the path tree to have a large trunk that can be output.

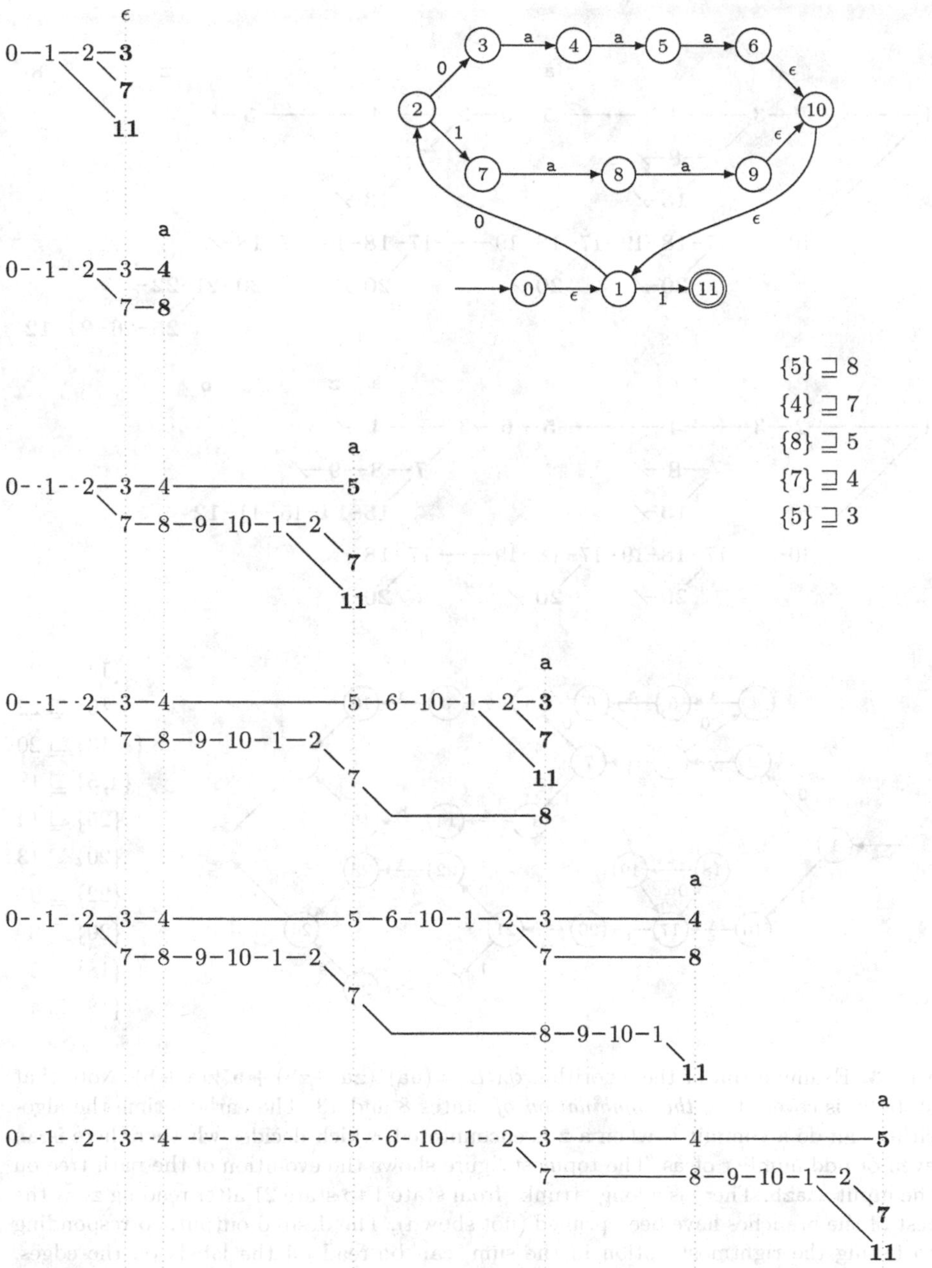

Fig. 2. Example run of the algorithm on the regular expression $E = (\mathtt{aaa} + \mathtt{aa})^\star$ and the input string aaaaa. The dashed edges represent the partial parse trees that can be emitted: thus, after one a we can emit a 0, and after five as we can emit 00 because the bottom "leg" of the tree has been removed in the pruning step. The automaton for E and its associated minimal covering relation are shown in the inset.

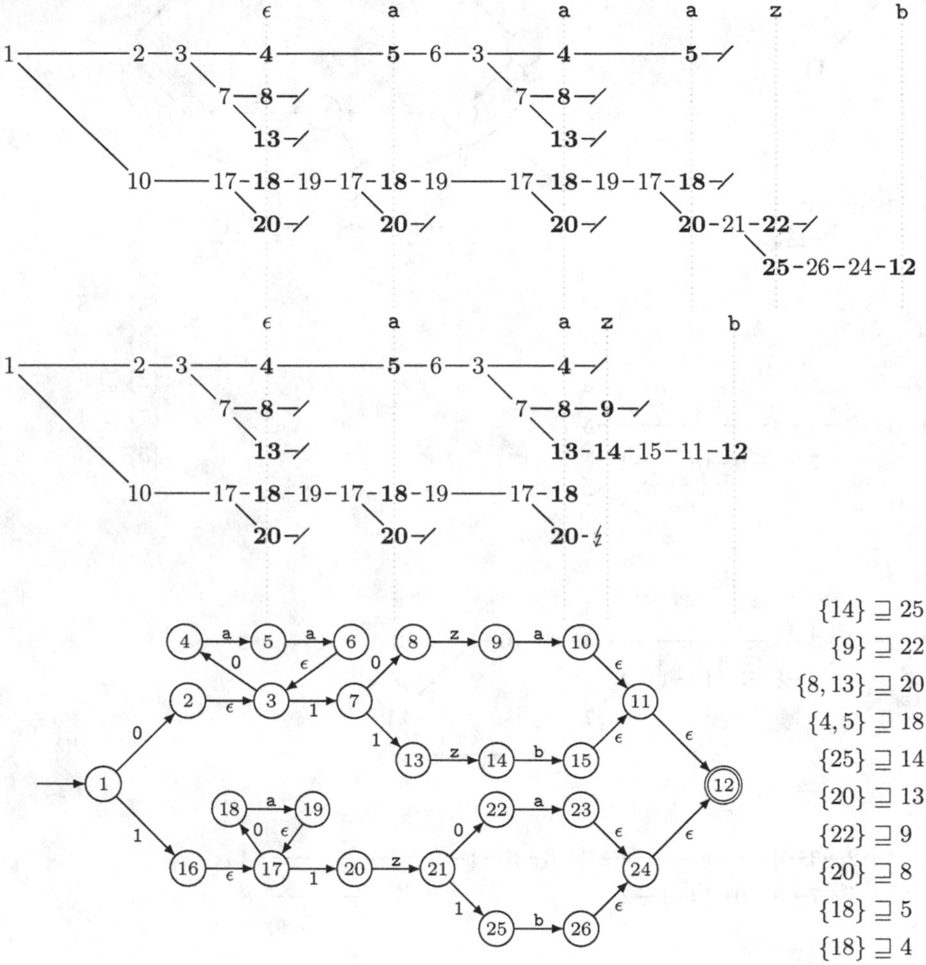

Fig. 3. Example run of the algorithm on $E = (\mathtt{aa})^\star(\mathtt{za} + \mathtt{zb}) + \mathtt{a}^\star\mathtt{z}(\mathtt{a} + \mathtt{b})$. Note that state 20 is covered by *the combination of* states 8 and 13. The earliest time the algorithm can do a commit is when a **z** is encountered, which decides whether there is an even or odd number of **a**s. The topmost figure shows the evolution of the path tree on the input **aaazb**. There is a long "trunk" from state 1 to state 21 after reading **z**, as the rest of the branches have been pruned (not shown). The desired output, corresponding to taking the rightmost option in the sum, can be read off the labels on the edges. Likewise in the second figure, we see that if the **z** comes after an even number of **a**s, a binary-node-free path from 1 to 7 emerges. Due to the cover $\{8, 13\} \sqsupseteq 20$, the branch starting from 20 is not expanded further, even though there could be a **z**-transition on it. This is indicated with $\lightning$. Overall, the resulting parse tree corresponds to the leftmost option in the sum.

CSV files. The expression $((\mathtt{a}+\mathtt{b})^\star(;(\mathtt{a}+\mathtt{b})^\star)^\star\mathtt{n})^\star$ defines the format of a simple semicolon-delimited data format, with data consisting of words over $\{\mathtt{a},\mathtt{b}\}$ and rows separated by the newline character, n. Our algorithm emits the partial parse trees after each letter has been parsed, as illustrated on the example input below:

```
a;ba;a
b;;b
```

a	;	b	a	;	a	n	b	;	;	a	n	·
000	10	01	00	10	00	11	001	10	10	00	11	1

Due to the star-height of three, many widespread implementations would not be able to meaningfully handle this expression using only the RE engine. Capturing groups under Kleene stars return either the first or last match, but not a *list* of matches—and certainly not a list of lists of matches! Hence, if using an implementation like Perl's [16], one is forced to rewrite the expression by removing the iteration in the outer Kleene star and reintroduce it as a looping construct in Perl.

9 Related and Future Work

Parsing regular expressions is not new [6,5,3,10,14], and streaming parsing of XML documents has been investigated for more than a decade in the context of XQuery and XPath—see, e.g., [2,7,17]. However, *streaming regular expression* parsing appears to be new.

In earlier work [6] we described a compact "lean log" format for storing intermediate information required for two-phase regular expression parsing. The algorithm presented here may degenerate to two passes, but requires often just one pass in the sense being effectively streaming, using only $O(m)$ work space, independent of n. The preprocessing of the regular expression and the intermediate data structure durig input string processing are more complex, however. It may be possible to merge the two approaches using a tree of lean log frames with associated counters, observing that edges in the path tree that are *not* labeled 0 or 1 are redundant. This is future work.

Acknowledgements. This work is supported by the Danish Independent Research Council under Project "Kleene Meets Church: Regular Expressions and Types". We would like to thank the anonymous referees for their helpful criticisms.

References

1. Brüggemann-Klein, A., Wood, D.: One-unambiguous regular languages. Information and computation 140(2), 229–253 (1998), `http://dx.doi.org/10.1006/inco.1997.2688`
2. Debarbieux, D., Gauwin, O., Niehren, J., Sebastian, T., Zergaoui, M.: Early nested word automata for XPath query answering on XML streams. In: Konstantinidis, S. (ed.) CIAA 2013. LNCS, vol. 7982, pp. 292–305. Springer, Heidelberg (2013)

3. Dubé, D., Feeley, M.: Efficiently building a parse tree from a regular expression. Acta Informatica 37(2), 121–144 (2000), http://dx.doi.org/10.1007/s002360000037
4. Fowler, G.: An interpretation of the POSIX regex standard (January 2003), http://www2.research.att.com/~astopen/testregex/re-interpretation.html
5. Frisch, A., Cardelli, L.: Greedy regular expression matching. In: Díaz, J., Karhumäki, J., Lepistö, A., Sannella, D. (eds.) ICALP 2004. LNCS, vol. 3142, pp. 618–629. Springer, Heidelberg (2004), http://dx.doi.org/10.1007/978-3-540-27836-8_53
6. Grathwohl, N.B.B., Henglein, F., Nielsen, L., Rasmussen, U.T.: Two-pass greedy regular expression parsing. In: Konstantinidis, S. (ed.) CIAA 2013. LNCS, vol. 7982, pp. 60–71. Springer, Heidelberg (2013), http://dx.doi.org/10.1007/978-3-642-39274-0_7
7. Gupta, A.K., Suciu, D.: Stream processing of XPath queries with predicates. In: Proc. 2003 ACM SIGMOD International Conference on Management of Data, SIGMOD 2003, pp. 419–430. ACM, New York (2003), http://dx.doi.org/10.1145/872757.872809
8. IEEE Computer Society: Standard for Information Technology - Portable Operating System Interface (POSIX), Base Specifications, Issue 7. IEEE (2008), http://dx.doi.org/10.1109/IEEESTD.2008.4694976, IEEE Std 1003.1
9. Kearns, S.: Extending regular expressions with context operators and parse extraction. Software - Practice and Experience 21(8), 787–804 (1991), http://dx.doi.org/10.1002/spe.4380210803
10. Nielsen, L., Henglein, F.: Bit-coded regular expression parsing. In: Dediu, A.-H., Inenaga, S., Martín-Vide, C. (eds.) LATA 2011. LNCS, vol. 6638, pp. 402–413. Springer, Heidelberg (2011), http://dx.doi.org/10.1007/978-3-642-21254-3_32
11. Okui, S., Suzuki, T.: Disambiguation in regular expression matching via position automata with augmented transitions. In: Domaratzki, M., Salomaa, K. (eds.) CIAA 2010. LNCS, vol. 6482, pp. 231–240. Springer, Heidelberg (2011), http://dx.doi.org/10.1007/978-3-642-18098-9_25
12. Stockmeyer, L.J., Meyer, A.R.: Word problems requiring exponential time (preliminary report). In: Proc. Fifth Annual ACM Symposium on Theory of Computing, pp. 1–9. ACM (1973)
13. Sulzmann, M., Lu, K.Z.M.: Regular expression sub-matching using partial derivatives. In: Proc. 14th Symposium on Principles and Practice of Declarative Programming, PPDP 2012, pp. 79–90. ACM, New York (2012), http://dx.doi.org/10.1145/2370776.2370788
14. Sulzmann, M., Lu, K.Z.M.: POSIX regular expression parsing with derivatives. In: Codish, M., Sumii, E. (eds.) FLOPS 2014. LNCS, vol. 8475, pp. 203–220. Springer, Heidelberg (2014), http://dx.doi.org/10.1007/978-3-319-07151-0_13
15. Thompson, K.: Programming techniques: Regular expression search algorithm. Commun. ACM 11(6), 419–422 (1968), http://dx.doi.org/10.1145/363347.363387
16. Wall, L., Christiansen, T., Orwant, J.: Programming Perl. O'Reilly Media, Incorporated (2000)
17. Wu, X., Theodoratos, D.: A survey on XML streaming evaluation techniques. The VLDB Journal 22(2), 177–202 (2013), http://dx.doi.org/10.1007/s00778-012-0281-y

Learning Cover Context-Free Grammars from Structural Data

Mircea Marin[1,*] and Gabriel Istrate[1,2]

[1] Department of Computer Science, West University of Timişoara, Timişoara RO-300223, Romania
mmarin@info.uvt.ro
[2] e-Austria Research Institute, Timişoara RO-300223, Romania

Abstract. We consider the problem of learning an unknown context-free grammar when the only knowledge available and of interest to the learner is about its structural descriptions with depth at most ℓ. The goal is to learn a *cover context-free grammar* (CCFG) with respect to ℓ, that is, a CFG whose structural descriptions with depth at most ℓ agree with those of the unknown CFG. We propose an algorithm, called LA^ℓ, that efficiently learns a CCFG using two types of queries: structural equivalence and structural membership. We show that LA^ℓ runs in time polynomial in the number of states of a minimal deterministic finite cover tree automaton (DCTA) with respect to ℓ. This number is often much smaller than the number of states of a minimum deterministic finite tree automaton for the structural descriptions of the unknown grammar.

Keywords: automata theory and formal languages, structural descriptions, grammatical inference.

1 Introduction

Angluin's approach to grammatical inference [1] is an important contribution to computational learning, with extensions to problems, such as compositional verification and synthesis [4,11], that go beyond the usual applications to natural language processing and computational biology [5].

Practical concerns, e.g. [9], seem to require going beyond regular languages to classes of languages with regular tree nature. However, Angluin and Kharitonov have shown that learning CFGs from membership and equivalence queries is intractable under plausible cryptographic assumptions [2]. A way out is to learn structural descriptions of context free languages. Sakakibara has shown that Angluin's algorithm extends to this setting [12]. His approach has applications in learning the structural descriptions of natural languages, which describe the shape of the parse trees of well chosen CFGs. Often, these structural descriptions are subject to additional restrictions arising from modelling considerations. For

* Corresponding author.

G. Ciobanu and D. Méry (Eds.): ICTAC 2014, LNCS 8687, pp. 241–258, 2014.

instance, in natural language understanding, the bounded memory restriction on human comprehension seems to limit the recursion depth of such a parse tree to a constant. A natural example with a similar flavour is the limitation imposed by the LaTeX system, that limits the number of nestings of itemised environments to a small constant.

Imposing such a restriction leads to the idea of learning cover languages, that is, languages that are accurate up to an equivalence. For regular languages modulo a finite prefix such an approach has been pursued by Ipate [8] (see also [6]).

In this paper, we extend this approach to context-free languages with structural descriptions. We propose an algorithm called LA^ℓ which asks two types of queries: structural equivalence and structural membership queries, both restricted to structural descriptions with depth at most ℓ, where ℓ is a constant. LA^ℓ stores the answers retrieved from the teacher in an *observation table* which is used to guide the learning protocol and to construct a minimal DCTA of the unknown context-free grammar with respect to ℓ. Our main result shows that LA^ℓ runs in time polynomial in n and m, where n is the number of states of a minimal DCTA of the unknown CFG with respect to ℓ, and m is the maximum size of a counterexample returned by a failed structural membership query.

The paper is structured as follows. Section 2 introduces the basic notions and results to be used later in the paper. It also describes algorithm LA. In Sect. 4 we introduce the main concepts related to the specification and analysis of our learning algorithm LA^ℓ. They are natural generalisations to languages of structural descriptions of the concepts proposed by Ipate [8] in the design and study of his algorithm L^ℓ. In Sect. 5 we analyse the space and time complexity of LA^ℓ and show that its time complexity is a polynomial in n and m, where n is the number of states of a minimal deterministic finite cover automaton w.r.t. ℓ of the language of structural descriptions of interest, and m is an upper bound to the size of counterexamples returned by failed structural equivalence queries.

2 Preliminaries

We write $\mathbb{N}$ for the set of nonnegative integers, A^* for the set of finite strings over a set A, and ϵ for the empty string. If $v, w \in A^*$, we write $v \leq w'$ if there exists $w' \in A^*$ such that $vw' = w$; $v < v'$ if $v \leq v'$ and $v \neq v'$; and $v \perp w$ if neither $v \leq w$ nor $w \leq v$.

Trees, Terms, Contexts, and Context-Free Grammars

A *ranked alphabet* is a finite set $\mathcal{F}$ of function symbols together with a finite *rank* relation $rk(\mathcal{F}) \subseteq \mathcal{F} \times \mathbb{N}$. We denote the subset $\{f \in \mathcal{F} \mid (f, m) \in rk(\mathcal{F})\}$ by $\mathcal{F}_m$, the set $\{m \mid (f, m) \in rk(\mathcal{F})\}$ by $ar(f)$, and $\bigcup_{f \in \mathcal{F}} ar(f)$ by $ar(\mathcal{F})$. The *terms* of the set $\mathcal{T}(\mathcal{F})$ are the strings of symbols defined recursively by the grammar $t ::= a \mid f(t_1, \ldots, t_m)$ where $a \in \mathcal{F}_0$ and $f \in \mathcal{F}_m$ with $m > 0$. The *yield* of a term $t \in \mathcal{T}(\mathcal{F})$ is the finite string $yield(t) \in \mathcal{F}_0^*$ defined as follows: $yield(a) := a$ if $a \in \mathcal{F}_0$, and $yield(f(t_1, \ldots, t_m)) := w_1 \ldots w_m$ where $w_i = yield(t_i)$ for $1 \leq i \leq m$.

A *finite ordered tree* over a set of labels $\mathcal{F}$ is a mapping t from a nonempty and prefix closed set $Pos(t) \subseteq (\mathbb{N} \setminus \{0\})^*$ into $\mathcal{F}$. Each element in $Pos(t)$ is called a *position*. The tree t is *ranked* if $\mathcal{F}$ is a ranked alphabet, and t satisfies the following additional property: For all $p \in Pos(t)$, there exists $m \in \mathbb{N}$ such that $\{i \in \mathbb{N} \mid pi \in Pos(t)\} = \{1, \ldots, m\}$ and $t(p) \in \mathcal{F}_m$.

Thus, any term $t \in \mathcal{T}(\mathcal{F})$ may be viewed as a finite ordered ranked tree, and we will refer to it by "tree" when we mean the finite ordered tree with the additional property mentioned above. The *depth* of t is $\mathtt{d}(t) := \max\{\|p\| \mid p \in Pos(t)\}$ where $\|p\|$ denotes the length of p as sequence of numbers. The *size* $\mathtt{sz}(t)$ of t is the number of elements of the set $\{p \in Pos(t) \mid \|p\| \neq \mathtt{d}(t)\}$, that is, the number of internal nodes of t.

The *subterm* $t|_p$ of a term t at position $p \in Pos(t)$ is defined by the following: $Pos(t|_p) := \{i \mid pi \in Pos(t)\}$, and $t|_p(p') := t(pp')$ for all $p' \in Pos(t|_p)$. We denote by $t[u]_p$ the term obtained by replacing in t the subterm $t|_p$ with u, that is: $Pos(t[u]_p) = (Pos(t) - \{pp' \mid p' \in Pos(t|_p)\}) \cup \{pp'' \mid p'' \in Pos(u)\}$, and

$$t[u]_p(p') := \begin{cases} u(p'') & \text{if } p' = pp'' \text{ with } p'' \in Pos(u), \\ t(p') & \text{otherwise.} \end{cases}$$

The set $\mathcal{C}(\mathcal{F})$ of *contexts* over $\mathcal{F}$ is the set of terms over $\mathcal{F} \cup \{\bullet\}$, where:

- $\bullet$ is a distinguished fresh symbol with $ar(\bullet) = \{0\}$, called *hole*,
- $rk(\mathcal{F} \cup \{\bullet\}) = rk(\mathcal{F}) \cup \{(\bullet, 0)\}$, and
- every element $C \in \mathcal{C}(\mathcal{F})$ contains only one occurrence of $\bullet$. This is the same as saying that $\{p \in Pos(C) \mid C(p) = \bullet\}$ is a singleton set.

If $C \in \mathcal{C}(\mathcal{F})$ and $u \in \mathcal{C}(\mathcal{F}) \cup \mathcal{T}(\mathcal{F})$ then $C[u]$ stands for the context or term $C[u]_p$, where $C(p) = \bullet$. The *hole depth* of a context $C \in \mathcal{C}(\mathcal{F})$ is $\mathtt{d}_\bullet(C) := \|p\|$ where p is the unique position of C such that $C(p) = \bullet$. From now on, whenever M is a set of terms, P is a set of contexts, and m is a non-negative integer, we define the sets $M_{[m]} := \{t \in M \mid \mathtt{d}(t) \leq m\}$ and $P_{\langle m \rangle} := \{C \in P \mid \mathtt{d}_\bullet(C) \leq m\}$.

We assume that the reader is acquainted with the notions of CFG and the context-free language $\mathcal{L}(G)$ generated by a CFG G, see, e.g., [13]. A CFG is ϵ-free if it has no productions of the form $X \to \epsilon$. It is well known [7] that every ϵ-free context-free language L (that is, $\epsilon \notin L$) is generated by an ϵ-free CFG. The derivation trees of an ϵ-free CFG $G = (N, \Sigma, P, S)$ correspond to terms from $\mathcal{T}(N \cup \Sigma)$ with $ar(a) = \{0\}$ for al $a \in \Sigma$ and $ar(X) = \{m \mid \exists (X \to \alpha) \in P$ with $\|\alpha\| = m\}$ for all $X \in N$. The sets $D_G(U)$ of derivation trees issued from $U \in N \cup \Sigma$, and $D(G)$ of derivation trees of G, are defined recursively as follows:

$$D_G(a) := \{a\} \text{ if } a \in \Sigma,$$

$$D_G(X) := \bigcup_{(X \to U_1 \ldots U_m) \in P} \{X(t_1, \ldots, t_m) \mid t_1 \in D_G(U_1) \wedge \ldots \wedge t_m \in D_G(U_m)\},$$

$D(G) := D_G(S)$. Note that $\mathcal{L}(G) = \{yield(t) \mid t \in D(G)\}$.

Structural Descriptions and Cover Context-Free Grammars

A *skeletal alphabet* is a ranked alphabet $Sk = \{\sigma\}$, where σ is a special symbol with $ar(\sigma)$ a finite subset of $\mathbb{N} \setminus \{0\}$, and a *skeletal set* is a ranked alphabet $Sk \cup A$ where $Sk \cap A = \emptyset$ and $ar(a) = \{0\}$ for all $a \in A$. Skeletal alphabets are intended to describe the structures of the derivation trees of ϵ-free CFGs. For an ϵ-free CFG $G = (N, \Sigma, P, S)$ we consider the skeletal alphabet Sk with $ar(\sigma) := \{\|\alpha\| \mid (X \to \alpha) \in P\}$, and the skeletal set $Sk \cup \Sigma$. The *skeletal* (or *structural*) *description* of a derivation tree $t \in D_G(U)$ is the term $\mathtt{sk}(t) \in \mathcal{T}(Sk \cup \Sigma)$ where

$$\mathtt{sk}(t) := \begin{cases} a & \text{if } t = a \in \Sigma, \\ \sigma(\mathtt{sk}(t_1), \ldots, \mathtt{sk}(t_m)) & \text{if } t = X(t_1, \ldots, t_m) \text{ with } m > 0. \end{cases}$$

For example, if G is the grammar $(\{\mathtt{S}, \mathtt{A}\}, \{a, b\}, \{\mathtt{S} \to \mathtt{A}, \mathtt{A} \to a\mathtt{A}b, \mathtt{A} \to ab\}, \mathtt{S})$ then $t = \mathtt{S}(\mathtt{A}(a, \mathtt{A}(a, b), b)) \in D_G(\mathtt{S})$ and $\mathtt{sk}(t) = \sigma(\sigma(a, \sigma(a, b), b)) \in \mathcal{T}(\{\sigma, a, b\})$, where $ar(\sigma) = \{1, 2, 3\}$ and $ar(a) = ar(b) = \{0\}$. Graphically, we have

$$t = \begin{array}{c} S \\ | \\ A \\ a \diagup \overset{|}{A} \diagdown b \\ a \diagup \diagdown b \end{array} \qquad \Rightarrow \qquad \mathtt{sk}(t) = \begin{array}{c} \sigma \\ | \\ \sigma \\ a \diagup \overset{|}{\sigma} \diagdown b \\ a \diagup \diagdown b \end{array}$$

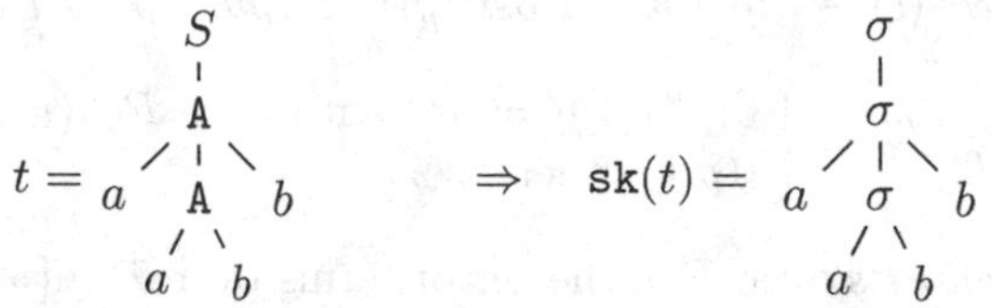

If M is a set of ranked trees, then the set of its structural descriptions is $K(M) := \{\mathtt{sk}(t) \mid t \in M\}$. Two context-free grammars G_1 and G_2 over the same alphabet of terminals are *structurally equivalent* if $K(D(G)) = K(D(G'))$.

Definition 1 (cover CFG). *Let ℓ be a positive integer and G_U be an ϵ-free CFG of a language $U \subseteq \Sigma^*$. A cover context-free grammar of G_U with respect to ℓ is an ϵ-free CFG G' such that $K(D(G'))_{[\ell]} = K(D(G_U))_{[\ell]}$.*

Tree Automata

The definition of tree automaton presented here is equivalent with that given in [12]. It is non-standard in the sense that it cannot accept any tree of depth 0.

Definition 2. *A nondeterministic (bottom-up) finite tree automaton (NFTA) over $\mathcal{F}$ is a quadruple $\mathcal{A} = (\mathcal{Q}, \mathcal{F}, \mathcal{Q}_f, \Delta)$ where $\mathcal{Q}$ is a finite set of states, $\mathcal{Q}_f \subseteq \mathcal{Q}$ is the set of final states, and Δ is a set of transition rules of the form $f(q_1, \ldots, q_m) \to q$ where $m \geq 1$, $f \in \mathcal{F}_m$, $q_1, \ldots, q_m \in \mathcal{F}_0 \cup \mathcal{Q}$, and $q \in \mathcal{Q}$.*

Such an automaton $\mathcal{A}$ induces a *move* relation $\to_{\mathcal{A}}$ on the set of terms $\mathcal{T}(\mathcal{F} \cup \mathcal{Q})$ where $ar(q) = \{0\}$ for all $q \in \mathcal{Q}$, as follows:

$t \to_{\mathcal{A}} t'$ if there exist $C \in \mathcal{C}(\mathcal{F} \cup \mathcal{Q})$ and $f(q_1, \ldots, q_m) \to q \in \Delta$ such that $t = C[f(q_1, \ldots, q_m)]$ and $t' = C[q]$.

The *language accepted by* $\mathcal{A}$ is $\mathcal{L}(\mathcal{A}) := \{t \in \mathcal{T}(\mathcal{F}) \mid t \rightarrow^*_{\mathcal{A}} q \text{ for some } q \in \mathcal{Q}_{\mathtt{f}}\}$ where $\rightarrow^*_{\mathcal{A}}$ is the reflexive-transitive closure of $\rightarrow_{\mathcal{A}}$. In this paper, a *regular tree language* is a language accepted by such an NFTA. Two NFTAs are *equivalent* if they accept the same language.

$\mathcal{A} = (\mathcal{Q}, \mathcal{F}, \mathcal{Q}_{\mathtt{f}}, \Delta)$ is *deterministic* (DFTA) if the transition rules of Δ describe a mapping δ which assigns to every $m \in ar(\mathcal{F})$ a function δ_m such that $\delta_0 : \mathcal{F}_0 \rightarrow \mathcal{F}_0$, $\delta_0(a) = a$ for all $a \in \mathcal{F}_0$, and $\delta_m : \mathcal{F}_m \rightarrow (\mathcal{F}_0 \cup \mathcal{Q})^m \rightarrow \mathcal{Q}$ if $m > 0$. This implies that $f(q_1, \ldots, q_m) \rightarrow q \in \Delta$ if and only if $\delta_m(f)(q_1, \ldots, q_m) = q$. The extension δ^* of $\{\delta_m \mid m \in ar(\mathcal{F})\}$ to $\mathcal{T}(\mathcal{F})$ is defined as expected: $\delta^*(a) = a$ if $a \in \mathcal{F}_0$, and $\delta^*(f(t_1, \ldots, t_m)) := \delta_m(f)(\delta^*(t_1), \ldots, \delta^*(t_m))$ otherwise. Note that, if $\mathcal{A}$ is a DFTA then $\mathcal{L}(\mathcal{A}) = \{t \in \mathcal{T}(\mathcal{F}) \mid \delta^*(t) \in \mathcal{Q}_{\mathtt{f}}\}$.

Two DFTAs $\mathcal{A}_1 = (\mathcal{Q}, \mathcal{F}, \mathcal{Q}_{\mathtt{f}}, \delta)$ and $\mathcal{A}_2 = (\mathcal{Q}', \mathcal{F}, \mathcal{Q}'_{\mathtt{f}}, \delta')$ are *isomorphic* if there exists a bijection $\varphi : \mathcal{Q} \rightarrow \mathcal{Q}'$ such that $\varphi(\mathcal{Q}_{\mathtt{f}}) = \mathcal{Q}'_{\mathtt{f}}$ and for every $f \in \mathcal{F}_m$, $q_1, \ldots, q_m \in \mathcal{F}_0 \cup \mathcal{Q}$, $\varphi(\delta_m(f)(q_1, \ldots, q_m)) = \delta'_m(f)(\varphi(q_1), \ldots, \varphi(q_m))$. A *minimum DFTA* of a regular tree language $L \subseteq \mathcal{T}(\mathcal{F}) \setminus \mathcal{F}_0$ is a DFTA $\mathcal{A}$ with minimum number of states such that $\mathcal{L}(A) = L$.

There is a strong correspondence between tree automata and ϵ-free CFGs. The NFTA corresponding to an ϵ-free CFG $G = (N, \Sigma, P, S)$ is $NA(G) = (N, Sk \cup \Sigma, \{S\}, \Delta)$ with $\Delta := \{\sigma(U_1, \ldots, U_m) \rightarrow X \mid (X \rightarrow U_1 \ldots U_m) \in P\}$. Conversely, the ϵ-free CFG corresponding to an NFTA $\mathcal{A} = (\mathcal{Q}, Sk \cup \Sigma, \mathcal{Q}_{\mathtt{f}}, \Delta)$ over the skeletal set $Sk \cup \Sigma$ is $G(\mathcal{A}) = (\mathcal{Q} \cup \{S\}, \Sigma, P, S)$ where S is a fresh symbol and $P := \{q \rightarrow q_1 \ldots q_m \mid (\sigma(q_1, \ldots, q_m) \rightarrow q) \in \Delta\} \cup \{S \rightarrow q_1 \ldots q_m \mid (\sigma(q_1, \ldots, q_m) \rightarrow q) \in \Delta \text{ with } q \in \mathcal{Q}_{\mathtt{f}}\}$. These constructs are dual to each other, in the following sense:

(A_1) If G is an ϵ-free CFG then $\mathcal{L}(NA(G)) = K(D(G))$. [12, Prop. 3.4]

(A_2) If $\mathcal{A} = (\mathcal{Q}, Sk \cup \Sigma, \mathcal{Q}_{\mathtt{f}}, \Delta)$ is an NFTA for the skeletal set $Sk \cup \Sigma$ then $K(D(G(\mathcal{A}))) = \mathcal{L}(\mathcal{A})$. That is, the set of structural descriptions of $G(\mathcal{A})$ coincides with the set of trees accepted by $\mathcal{A}$. [12, Prop. 3.6]

We recall the following well-known results: every NFTA is equivalent to an DFTA [10], and every two minimal DFTAs are isomorphic [3].

Cover Tree Automata

Definition 3 (DCTA). *Let* $\ell \in \mathbb{N}^+$ *and* A *be a tree language over the ranked alphabet* $\mathcal{F}$. *A* deterministic cover tree automaton *(DCTA) of* A *with respect to* ℓ *is a DFTA* $\mathcal{A}$ *over a skeletal set* $Sk \cup \mathcal{F}_0$ *such that* $\mathcal{L}(\mathcal{A})_{[\ell]} = K(A)_{[\ell]}$.

The correspondence between tree automata and ϵ-free CFGs is carried over to a correspondence between cover tree automata and cover CFGs. More precisely, it can be shown that if G_U is an ϵ-free CFG, then a DFTA $\mathcal{A}$ is a DCTA of $K(D(G_U))$ w.r.t. ℓ if and only if $G(\mathcal{A})$ is a cover CFG of G_U w.r.t. ℓ.

3 Learning Context-Free Grammars

In [12], Sakakibara assumes a *learner* eager to learn a CFG which is structurally equivalent with the CFG G_U of an unknown context-free language $U \subseteq \Sigma^*$ by asking questions to a *teacher*. We assume that the learner and the teacher share the skeletal set $Sk \cup \Sigma$ for the structural descriptions in $K(D(G_U))$. The learner can pose the following types of queries:

1. *Structural membership queries*: the learner asks if some $s \in \mathcal{T}(Sk \cup \Sigma)$ is in $K(D(G_U))$. The answer is *yes* if so, and *no* otherwise.
2. *Structural equivalence queries*: The learner proposes a CFG G' and asks whether G' is structurally equivalent to G_U. If the answer is *yes*, the process stops with the learned answer G. Otherwise, the teacher provides a counterexample s from the symmetric set difference $K(D(G')) \triangle K(D(G_U))$.

This learning protocol is based on what is called *minimal adequate teacher* in [1]. Ultimately, the learner constructs a minimal DFTA $\mathcal{A}$ of $K(D(G_U))$ from which it can infer immediately the CFG $G' = G(\mathcal{A})$ which is structurally equivalent to G_U, that is, $K(D(G')) = K(D(G_U))$. In order to understand how $\mathcal{A}$ gets constructed, we shall introduce a few auxiliary notions.

For any subset S of $\mathcal{T}(Sk \cup \Sigma)$, we define the sets

$$\sigma_\bullet\langle S\rangle := \bigcup_{m\in ar(\sigma)} \bigcup_{i=1}^{m} \{\sigma(s_1,\ldots,s_m)[\bullet]_i \mid s_1,\ldots,s_m \in S\cup\Sigma\},$$

$$X(S) := \{C_1[s] \mid C_1 \in \sigma_\bullet\langle S\rangle, s \in S\cup\Sigma\} \setminus S.$$

Note that $\sigma_\bullet\langle S\rangle = \{C \in \mathcal{C}(Sk \cup \Sigma) \setminus \{\bullet\} \mid C|_p \in S \cup \Sigma \cup \{\bullet\}$ for all $p \in Pos(C) \cap \mathbb{N}\}$.

Definition 4. *A subset E of $\mathcal{C}(Sk \cup \Sigma)$ is* **•-prefix closed** *with respect to a set $S \subseteq \mathcal{T}(Sk \cup \Sigma)$ if $C \in E \setminus \{\bullet\}$ implies the existence of $C' \in E$ and $C_1 \in \sigma_\bullet\langle S\rangle$ such that $C = C'[C_1]$. If $E \subseteq \mathcal{C}(Sk \cup \Sigma)$ and $S \subseteq \mathcal{T}(Sk \cup \Sigma)$ then $E[S]$ denotes the set of structural descriptions defined by $E[S] = \{C[s] \mid C \in E, s \in S\}$.*

We say that $S \subseteq \mathcal{T}(Sk \cup \Sigma)$ is **subterm closed** *if $\mathtt{d}(s) \geq 1$ for all $s \in S$, and $s' \in S$ whenever s' is a subterm of some $s \in S$ with $\mathtt{d}(s') \geq 1$.*

An *observation table* for $K(D(G_U))$, denoted by (S, E, T), is a tabular representation of the finitary function $T : E[S \cup X(S)] \to \{0, 1\}$ defined by $T(t) := 1$ if $t \in K(D(G_U))$, and 0 otherwise, where S is a finite nonempty subterm closed subset S of $\mathcal{T}(Sk \cup \Sigma)$, and E is a finite nonempty subset of $\mathcal{C}(Sk \cup \Sigma)$ which is •-prefix closed with respect to S. Such an observation table is visualised as a matrix with rows labeled by elements from $S \cup X(S)$, columns labeled by elements from E, and the entry for row of s and column of C equal to $T(C[s])$. If we fix a listing $\langle C_1, \ldots, C_r\rangle$ of all elements of E, then the row of values of some $s \in S \cup X(S)$ corresponds to the vector $row(s) = \langle T(C_1[s]), \ldots, T(C_r[s])\rangle$. In fact, for every such s, $row(s)$ is a finitary representation of the function $f_s : E \to \{0, 1\}$ defined by $f_s(C) = T(C[s])$.

The observation table (S, E, T) is *closed* if every $row(x)$ with $x \in X(S)$ is identical to some $row(s)$ of $s \in S$. It is *consistent* if whenever $s_1, s_2 \in S$ such that $row(s_1) = row(s_2)$, we have $row(C_1[s_1]) = row(C_1[s_2])$ for all $C_1 \in \sigma_\bullet\langle S\rangle$.

The DFTA *corresponding to a closed and consistent observation table* (S, E, T) is $\mathcal{A}(S, E, T) = (\mathcal{Q}, Sk \cup \Sigma, \mathcal{Q}_{\mathfrak{f}}, \delta)$ where $\mathcal{Q} := \{row(s) \mid s \in S\}$, $\mathcal{Q}_{\mathfrak{f}} := \{row(s) \mid s \in S \text{ and } T(s) = 1\}$, and δ is uniquely defined by

$$\delta_m(\sigma)(q_1, \ldots, q_m) := row(\sigma(r_1, \ldots, r_m)) \quad \text{for all } m \in ar(\sigma),$$

where $r_i := a$ if $q_i = a \in \Sigma$, and $r_i := s_i$ if $q_i = row(s_i) \in \mathcal{Q}$.

It is easy to check that, under these assumptions, $\mathcal{A}(S, E, T)$ is well-defined, and that $\delta^*(s) = row(s)$. Furthermore, Sakakibara proved that the following properties hold whenever (S, E, T) is a closed and consistent observation table:

1. $\mathcal{A}(S, E, T)$ is consistent with T, that is, for all $s \in S \cup X(S)$ and $C \in E$ we have $\delta^*(C[s]) \in \mathcal{Q}_{\mathfrak{f}}$ iff $T(C[s]) = 1$. [12, Lemma 4.2]
2. If $\mathcal{A}(S, E, T) = (\mathcal{Q}, Sk \cup \Sigma, \delta, \mathcal{Q}_{\mathfrak{f}})$ has n states, and $\mathcal{A}' = (\mathcal{Q}', Sk \cup \Sigma, \delta', \mathcal{Q}'_{\mathfrak{f}})$ is any DFTA consistent with T that has n or fewer states, then $\mathcal{A}'$ is isomorphic to $\mathcal{A}(S, E, T)$. [12, Lemma 4.3]

The *LA* Algorithm

In this subsection we briefly recall Sakakibara's algorithm LA. *LA* extends the observation table whenever one of the following situations occurs: the table is not consistent, the table is not closed, or the table is both consistent and closed but the CFG corresponding to the resulting automaton $\mathcal{A}(S, E, T)$ is not structurally equivalent to G_U (in which case a counterexample is produced). The first two situations trigger an extension of the observation table with one distinct row. From properties (A_1) and (A_2), if n is the number of states of the minimum bottom-up tree automaton for the structural descriptions of G_U, then the number of unsuccessful consistency and closedness checks during the whole run of this algorithm is at most $n - 1$. For each counterexample of size at most m returned by a structural equivalence query, at most m subtrees are added to S. Since the algorithm encounters at most n counterexamples, the total number of elements in S cannot exceed $n + m \cdot n$, thus *LA* must terminate. It also follows that the number of elements of the domain $E[S \cup X(S)]$ of the function T is at most $(n + m \cdot n + (l + m \cdot n + k)^d) \cdot n = O(m^d \cdot n^{d+1})$, where l is the number of distinct ranks of $\sigma \in Sk$, and d is the maximum rank of a symbol in Sk. A careful analysis of *LA* reveals that its time complexity is indeed bounded by a polynomial in m and n [12, Thm. 5,3].

4 Learning Cover Context-Free Grammars

We assume we are given a teacher who knows an ϵ-free CFG G_U for a language $U \subseteq \Sigma^*$, and a learner who knows the skeletal set $Sk \cup \Sigma$ for $K(D(G_U))$. The teacher and learner both know a positive integer ℓ, and the learner is interested to learn a cover CFG G' of G_U w.r.t. ℓ or, equivalently, a cover DCTA of $K(D(G_U))$ w.r.t. ℓ. The learner is allowed to pose the following types of questions:

1. *Structural membership queries*: the learner asks if some $s \in \mathcal{T}(Sk \cup \Sigma)_{[\ell]}$ is in $K(D(G_U))$. The answer is *yes* if so, and *no* otherwise.
2. *Structural equivalence queries*: The learner proposes a CFG G', and asks if G' is a cover CFG of G_U w.r.t. ℓ. If the answer is yes, the process stops with the learned answer G'. Otherwise, the teacher provides a counterexample from the set $(K(D(G_U))_{[\ell]} - K(D(G'))) \cup (K(D(G'))_{[\ell]} - K(D(G_U)))$.

We will describe an algorithm LA^ℓ that learns a cover CFG of G_U with respect to ℓ in time that is polynomial in the number of states of a minimal DCTA of the rational tree language $K(D(G_U))$.

4.1 The Observation Table

LA^ℓ is a generalisation of the learning algorithm L^ℓ proposed by Ipate [8]. Ipate's algorithm is designed to learn a minimal finite cover automaton of an unknown finite language of words in polynomial time, using membership queries and language equivalence queries that refer to words and languages of words with length at most ℓ. Similarly, LA^ℓ is designed to learn a minimal DCTA $\mathcal{A}'$ for $K(D(G_U))$ with respect to ℓ by maintaining an observation table (S, E, T, ℓ) for $K(D(G_U))$ which differs from the observation table of LA in the following respects:

1. S is a finite nonempty subterm closed subset of $\mathcal{T}(Sk \cup \Sigma)_{[\ell]}$.
2. E is a finite nonempty subset of $\mathcal{C}(Sk \cup \Sigma)_{\langle \ell-1\rangle} \cap \mathcal{C}(Sk \cup \Sigma)_{[\ell]}$ which is $\bullet$-prefix closed with respect to S.
3. $T : E[S \cup X(S)_{[\ell]}] \to \{1, 0, -1\}$ is defined by

$$T(t) := \begin{cases} 1 & \text{if } t \in K(D(G_U))_{[\ell]}, \\ 0 & \text{if } t \in \mathcal{T}(Sk \cup \Sigma)_{[\ell]} \setminus K(D(G_U)), \\ -1 & \text{if } t \notin \mathcal{T}(Sk \cup \Sigma)_{[\ell]}. \end{cases}$$

In a tabular representation, the observation table (S, E, T, ℓ) is a two-dimensional matrix with rows labeled by elements from $S \cup X(S)_{[\ell]}$, columns labeled by elements from E, and the entry corresponding to the row of t and column of C equal to $T(C[t])$. If we fix a listing $\langle C_1, \ldots, C_k\rangle$ of all elements from E, then the row of t in the observation table is described by the vector $\langle T(C_1[t]), \ldots, T(C_k[t])\rangle$ of values from $\{-1, 0, 1\}$. The rows of an observation table are used to identify the states a a minimal DCTA for $K(D(G_U))$ with respect to ℓ. But, like Ipate [8], we do not compare rows by equality but by a similarity relation.

4.2 The Similarity Relation

This time, the rows in the observation table correspond to terms from $S \cup X(S)_{[\ell]}$, and the comparison of rows should take into account only terms of depth at most ℓ. For this purpose, we define a relation $\sim_k$ of *k-similarity*, which is a generalisation to terms of Ipate's relation of k-similarity on strings [8].

Definition 5 (k-similarity). *For $1 \le k \le \ell$ we define the relation $\sim_k$ on the elements of the set $S \cup X(S)$ of an observation table (S, E, T, ℓ) as follows:*

$$s \sim_k t \text{ if, for every } C \in E_{\langle k-\max\{\mathtt{d}(s),\mathtt{d}(t)\}\rangle},\ T(C[s]) = T(C[t]).$$

*When the relation $\sim_k$ does not hold between two terms $s, t \in S \cup X(S)$, we write $s \not\sim_k t$ and say that s and t are k-*dissimilar*. When $k = \ell$ we simply say that s and t are* similar *or* dissimilar *and write $s \sim t$ or $s \not\sim t$, respectively.*

We say that a context C **ℓ-distinguishes** *s_1 and s_2, where $s_1, s_2 \in S$, if $C \in E_{\langle \ell-\max\{\mathtt{d}(s_1),\mathtt{d}(s_2)\}\rangle}$ and $T(C[s_1]) \ne T(C[s_2])$.*

Note that only the contexts $C \in E_{\langle k-\max\{\mathtt{d}(s),\mathtt{d}(t)\}\rangle}$ with $\mathtt{d}(C) \le \ell$ are relevant to check whether $s \sim_k t$, because if $\mathtt{d}(C) > \ell$ then $\mathtt{d}(C[s]) > \ell$ and $\mathtt{d}(C[t]) > \ell$, and therefore $T(C[s]) = -1 = T(C[t])$. Also, if $t \in S \cup X(S)$ with $\mathtt{d}(t) > \ell$ then it must be the case that $t \in X(S)$, and then $t \sim_k s$ for all $s \in S \cup X(S)$ and $1 \le k \le \ell$ because $E_{\langle k-\max \mathtt{d}(t),\mathtt{d}(s)\}\rangle} = \emptyset$.

The relation of k-similarity is obviously reflexive and symmetric, but not transitive. The following example illustrates this fact.

Example 1. Let $\Sigma = \{a, b\}$, $k = 1$, $\ell = 2$, $S = \{\sigma(a), \sigma(b), \sigma(\sigma(a), b)\}$, $E = \{\bullet, \sigma(\bullet, b)\}$, $t_1 = \sigma(a)$, $t_2 = \sigma(\sigma(a), b)$, $t_3 = \sigma(b)$, and

$$G_U = (\{\mathtt{S}, \mathtt{A}\}, \{a, b\}, \{\mathtt{S} \to a, \mathtt{S} \to b, \mathtt{S} \to \mathtt{A}b, \mathtt{A} \to a, \mathtt{A} \to \mathtt{A}b\}, \mathtt{S}).$$

S is a nonempty subterm closed subset of $\mathcal{T}(Sk \cup \Sigma)_{[\ell]}$, and E is a nonempty subset of $\mathcal{C}(Sk \cup \Sigma)_{\langle \ell-1\rangle}$ which is $\bullet$-prefix closed with respect to S. We have $K(D(G_U))_{[\ell]} = \{t_1, t_2, t_3\}$, $t_1 \sim_\ell t_2$ because $E_{\langle \ell-\max\{\mathtt{d}(t_1),\mathtt{d}(t_2)\}\rangle} = \{\bullet\}$ and $T(\bullet[t_1]) = 1 = T(\bullet[t_2])$, and $t_2 \sim_\ell t_3$ because $E_{\langle \ell-\max\{\mathtt{d}(t_2),\mathtt{d}(t_3)\}\rangle} = \{\bullet\}$ and $T(\bullet[t_2]) = 1 = T(\bullet[t_3])$, However, $t_1 \not\sim_\ell t_3$ because $C = \sigma(\bullet, b) \in E_{\langle 1\rangle} = E_{\langle \ell-\max\{\mathtt{d}(t_1),\mathtt{d}(t_3)\}\rangle}$ and $T(C[t_1]) = T(\sigma(\sigma(a), b)) = T(t_2) = 1$, but $T(C[t_3]) = T(\sigma(\sigma(b), b)) = 0$. □

Still, k-similarity has a useful property, captured in the following lemma.

Lemma 1. *Let (S, E, T, ℓ) be an observation table. If $s, t, x \in S \cup X(S)$ such that $\mathtt{d}(x) \le \max\{\mathtt{d}(s), \mathtt{d}(t)\}$, then $s \sim_k t$ whenever $s \sim_k x$ and $x \sim_k t$.*

In addition, we will also assume a total order $\prec$ on the alphabet Σ, and the following total orders induced by $\prec$ on $\mathcal{T}(Sk \cup \Sigma)$ and $\mathcal{C}(Sk \cup \Sigma)$.

Definition 6. *The total order $\prec_{\mathrm{T}}$ on $\mathcal{T}(Sk \cup \Sigma)$ induced by a total order $\prec$ on Σ is defined as follows: $s \prec_{\mathrm{T}} t$ if either (a) $\mathtt{d}(s) < \mathtt{d}(t)$, or (b) $\mathtt{d}(s) = \mathtt{d}(t)$ and*

1. *$s, t \in \Sigma$ and $s \prec t$, or else*
2. *$s \in \Sigma$ and $t \notin \Sigma$, or else*
3. *$s = \sigma(s_1, \ldots, s_m)$, $t = \sigma(t_1, \ldots, t_n)$ and there exists $1 \le k \le \min(m, n)$ such that $s_k \prec_{\mathrm{T}} t_k$ and $s_i = t_i$ for all $1 \le i < k$, or else*
4. *$s = \sigma(s_1, \ldots, s_m)$ and $t = \sigma(t_1, \ldots, t_n)$, $m < n$, and $s_i = t_i$ for $1 \le i \le m$.*

The total order $\prec_{\mathrm{C}}$ on $\mathcal{C}(Sk \cup \Sigma)$ induced by a total order $\prec$ on Σ is defined as follows: $C_1 \prec_{\mathrm{C}} C_2$ if either (a) $\mathtt{d}_\bullet(C_1) < \mathtt{d}_\bullet(C_2)$, or (b) $\mathtt{d}_\bullet(C_1) = \mathtt{d}_\bullet(C_2)$ and $C_1 \prec_{\mathrm{T}} C_2$ where C_1, C_2 are interpreted as terms over the signature with Σ extended with the constant $\bullet$ such that $\bullet \prec a$ for all $a \in \Sigma$.

Definition 7 (Representative). *Let* (S, E, T, ℓ) *be an observation table and* $x \in S \cup X(S)$*. We say* x has a representative *in* S *if* $\{s \in S \mid s \sim x\} \neq \emptyset$*. If so, the* representative *of* x *is* $\mathtt{r}(x) := \min_{\prec_T}\{s \in S \mid x \sim s\}$*.*

We will show later that the construction of an observation table (S, E, T, ℓ) is instrumental to the construction of a cover tree automaton, and the states of the automaton correspond to representatives of the elements from $S \cup X(S)$. Note that, if (S, E, T, ℓ) is an observation table and $x \in S \cup X(S)$ has $\mathtt{d}(x) > \ell$ then $x \in X(S)$ and $x \sim s$ for all $s \in S$. Then $s \prec_T x$ because $\mathtt{d}(s) \leq \ell < \mathtt{d}(x)$ for all $s \in S$. Thus x has a representative in S, and $\mathtt{r}(x) = \min_{\prec_T} S$. For this reason, only the rows for elements $x \in S \cup X(S)_{[\ell]}$ are kept in an observation table.

4.3 Consistency and Closedness

The consistency and closedness of an observation table are defined as follows.

Definition 8 (Consistency). *An observation table* (S, E, T, ℓ) *is consistent if, for every* $k \in \{1, \ldots, \ell\}$*,* $s_1, s_2 \in S$*, and* $C_1 \in \sigma_\bullet\langle S\rangle$*, the following implication holds: If* $s_1 \sim_k s_2$ *then* $C_1[s_1] \sim_k C_1[s_2]$*.*

The following lemma captures a useful property of consistent observation tables.

Lemma 2. *Let* (S, E, T, ℓ) *be a consistent observation table. Let* $m \in ar(\sigma)$*,* $1 \leq k \leq \ell$*, and* $s_1, \ldots, s_m, t_1, \ldots, t_m \in S \cup \Sigma$ *such that, for all* $1 \leq i \leq m$*, either* $s_i = t_i \in \Sigma$*, or* $s_i, t_i \in S$*,* $s_i \sim_k t_i$*, and* $\mathtt{d}(s_i) \leq \mathtt{d}(t_i)$*, and* $s = \sigma(s_1, \ldots, s_m)$*,* $t = \sigma(t_1, \ldots, t_m)$*. Then* $s \sim_k t$*.*

Definition 9 (Closedness). *An observation table* (S, E, T, ℓ) *is closed if, for all* $x \in X(S)$*, there exists* $s \in S$ *with* $\mathtt{d}(s) \leq \mathtt{d}(x)$ *such that* $x \sim s$*.*

The next five lemmata capture important properties of closed observation tables, which will be used to justify the correctness of the learning algorithm we are about to introduce.

Lemma 3. *If* (S, E, T, ℓ) *is closed then every* $x \in S \cup X(S)$ *has a representative, and* $\mathtt{d}(\mathtt{r}(x)) \leq \mathtt{d}(x)$*.*

Lemma 4. *If* (S, E, T, ℓ) *is closed,* $r_1, r_2 \in \{\mathtt{r}(x) \mid x \in S \cup X(S)\}$*, and* $r_1 \sim r_2$ *then* $r_1 = r_2$*.*

Lemma 5. *If* (S, E, T, ℓ) *is closed and* $r \in \{\mathtt{r}(x) \mid x \in S \cup X(S)\}$*, then* $\mathtt{r}(r) = r$*.*

Proof. Let $r_1 = \mathtt{r}(r)$. Then $r_1 \sim r$ and $r_1, r \in \{\mathtt{r}(x) \mid x \in S \cup X(S)\}$. By Lemma 4, $r = r_1$. □

Lemma 6. *If* (S, E, T, ℓ) *is closed, then for every* $x \in S \cup X(S)$ *and* $C_1 \in \sigma_\bullet\langle S\rangle$*, there exists* $s \in S$ *such that* $\mathtt{r}(C_1[\mathtt{r}(x)]) = \mathtt{r}(s)$*.*

Lemma 7. *Let* (S, E, T, ℓ) *be closed,* $r \in \{\mathtt{r}(x) \mid x \in S \cup X(S)\}$*,* $C_1 \in \sigma_\bullet\langle S\rangle$*, and* $s \in S$*. If* $C_1[s] \sim r$ *then* $\mathtt{d}(r) \leq \mathtt{d}(C_1[s])$*.*

The Automaton $\mathcal{A}(\mathbb{T})$

Like L^ℓ, our algorithm relies on the construction of a consistent and closed observation table of the unknown context-free grammar. The table is used to build an automaton which, in the end, turns out to be a minimal DCTA for the structural descriptions of the unknown grammar.

Definition 10. *Suppose $\mathbb{T} = (S, E, T, \ell)$ is a closed and consistent observation table. The automaton corresponding to this table, denoted by $\mathcal{A}(\mathbb{T})$, is the DFTA $(\mathcal{Q}, Sk \cup \Sigma, \mathcal{Q}_\mathtt{f}, \delta)$ where $\mathcal{Q} := \{\mathtt{r}(s) \mid s \in S\}$, $\mathcal{Q}_\mathtt{f} = \{q \in \mathcal{Q} \mid T(q) = 1\}$, and δ is uniquely defined by $\delta_m(\sigma)(q_1, \ldots, q_m) := \mathtt{r}(\sigma(q_1, \ldots, q_m))$ for all $m \in ar(\sigma)$.*

The transition function δ is well defined because, for all $m \in ar(\sigma)$ and $q_1, \ldots, q_m$ from $\mathcal{Q}$, $C_1 := \sigma(\bullet, q_2, \ldots, q_m) \in \sigma_\bullet\langle S\rangle$, thus $\sigma(q_1, \ldots, q_m) = C_1[q_1] \in S \cup X(S)$ and $\mathtt{r}(C_1[q_1]) = \mathtt{r}(s)$ for some $s \in S$, by Lemma 6. Hence, $\mathtt{r}(\sigma(q_1, \ldots, q_m)) \in \mathcal{Q}$. Also, the set $\mathcal{Q}_\mathtt{f}$ can be read off directly from the observation table because $\bullet \in E$ (since E is $\bullet$-prefix closed), thus $q = \bullet[q] \in E[(S \cup X(S)_{[\ell]}]$ for all $q \in \mathcal{Q}$, and we can read off from the observation table all $q \in \mathcal{Q}$ with $T(q) = 1$.

In the rest of this subsection we assume that $\mathbb{T} = (S, E, T, \ell)$ is closed and consistent, and δ is the transition function of the corresponding DFTA $\mathcal{A}(\mathbb{T})$.

Lemma 8. *$\delta^*(x) \sim x$ and $\mathtt{d}(\delta^*(x)) \leq \mathtt{d}(x)$ for every $x \in S \cup X(S)$.*

Corollary 1. *$\delta^*(x) = x$ for all $x \in \{\mathtt{r}(s) \mid s \in S \cup X(S)\}$.*

Proof. By Lemma 8, $x \sim \delta^*(x)$. Since both $\delta^*(x)$ and x belong to the set of representatives $\{\mathtt{r}(s) \mid s \in X \cup X(S)\}$, $x = \delta^*(x)$ by Lemma 4. □

The following theorem shows that the DFTA of a closed and consistent observation table is consistent with the function T on terms with depth at most ℓ.

Theorem 1. *Let $\mathbb{T} = (S, E, T, \ell)$ be a closed and consistent observation table. For every $s \in S \cup X(S)$ and $C \in E$ such that $\mathtt{d}(C[s]) \leq \ell$ we have $\delta^*(C[s]) \in \mathcal{Q}_\mathtt{f}$ if and only if $T(C[s]) = 1$.*

Theorem 2. *Let $\mathbb{T} = (S, E, T, \ell)$ be a closed and consistent observation table, and N be the number of states of $\mathcal{A}(\mathbb{T})$. If $\mathcal{A}'$ is any other DFTA with N or fewer states, that is consistent with T on terms with depth at most ℓ, then $\mathcal{A}'$ has exactly N states and $\mathcal{L}(\mathcal{A}(\mathbb{T}))_{[\ell]} = \mathcal{L}(\mathcal{A}')_{[\ell]}$.*

Corollary 2. *Let $\mathcal{A}$ be the automaton corresponding to a closed and consistent observation table (S, E, T, ℓ) of the skeletons of a CFG G_U of an unknown language U, and N be its number of states. Let n be the number of states of a minimal DCTA of $K(D(G_U))$ with respect to ℓ. If $N \geq n$ then $N = n$ and $\mathcal{A}$ is a minimal DCTA of $K(D(G_U))$ with respect to ℓ.*

The LA^ℓ Algorithm

The algorithm LA^ℓ extends the observation table $\mathbb{T} = (S, E, T, \ell)$ whenever one of the following situations occurs: the table is not consistent, the table is not closed, or the table is both consistent and closed but the resulting automaton $\mathcal{A}(\mathbb{T})$ is not a cover tree automaton of $K(D(G_U))$ with respect to ℓ.

The pseudocode of the algorithm is shown below.

ask if $(\{\mathtt{S}\}, \Sigma, \emptyset, \mathtt{S})$ is a cover CFG of G_U w.r.t. ℓ
if answer is *yes* **then** halt and output the CFG $(\{\mathtt{S}\}, \Sigma, \emptyset, \mathtt{S})$
if answer is *no* with counterexample t **then**
  set $S := \{s \mid s$ is a subterm of t with depth at least $1\}$ and $E = \{\bullet\}$
  construct the table $\mathbb{T} = (S, E, T, \ell)$ using structural membership queries
  repeat
    repeat
      /* check consistency */
      for every $C \in E$, in increasing order of $i = \mathtt{d}_\bullet(C)$ **do**
        search for $s_1, s_2 \in S$ with $\mathtt{d}(s_1), \mathtt{d}(s_2) \leq \ell - i - 1$ and $C_1 \in \sigma_\bullet\langle S\rangle$
          such that $C[C_1[s_1]]), C[C_1[s_2]] \in \mathcal{T}(Sk \cup \Sigma)_{[\ell]}$,
            $s_1 \sim_k s_2$ where $k = \max\{\mathtt{d}(s_1), \mathtt{d}(s_2)\} + i + 1$,
            and $T(C[C_1[s_1]]) \neq T(C[C_1[s_2]])$
        if found **then**
          add $C[C_1]$ to E
          extend T to $E[S \cup X(S)_{[\ell]}]$ using structural membership queries
      /* check closedness */
      new_row_added := `false`
      repeat for every $s \in S$, in increasing order of $\mathtt{d}(s)$
        search for $C_1 \in \sigma_\bullet\langle S\rangle$ such that $C_1[s] \nsim t$ for all $t \in S_{[\mathtt{d}(C_1[s])]}$
        if found **then**
          add $C_1[s]$ to S
          extend T to $E[S \cup X(S)_{[\ell]}]$ using structural membership queries
          new_row_added := `true`
      until *new_row_added* = `true` or all elements of S have been processed
    until *new_row_added* = `false`
    /* $\mathbb{T}$ is now closed and consistent */
    make the query whether $G(\mathcal{A}(\mathbb{T}))$ is a cover CFG of G_U w.r.t. ℓ
    if the reply is *no* with a counterexample t **then**
      add to S all subterms of t, including t, with depth at least 1,
          in the increasing order given by $\prec_{\mathrm{T}}$
      extend T to $E[S \cup X(S)_{[\ell]}]$ using structural membership queries
  until the reply is *yes* to the query if $G(\mathcal{A}(\mathbb{T}))$ is a cover CFG of G_U w.r.t. ℓ
  halt and output $G(\mathcal{A}(\mathbb{T}))$.

Consistency is checked by searching for $C \in E$ and $C_1 \in \sigma_\bullet\langle S\rangle$ such that $C[C_1]$ will ℓ-distinguish two terms $s_1, s_2 \in S$ not distinguished by any other context $C' \in E$ with $\mathtt{d}_\bullet(C') \leq \mathtt{d}_\bullet(C[C_1])$. Whenever such a pair of contexts (C, C_1) is found, $C[C_1]$ is added to E. Note that $C[C_1] \in \mathcal{C}(Sk \cup \Sigma)_{\langle \ell-1\rangle} \cap \mathcal{C}(Sk \cup \Sigma)_{[\ell]}$

because only such contexts can distinguish terms from S, and the addition of $C[C_1]$ to E yields a $\bullet$-prefix closed subset of $\mathcal{C}(Sk \cup \Sigma)_{\langle \ell-1 \rangle} \cap \mathcal{C}(Sk \cup \Sigma)_{[\ell]}$.

The search of such a pair of contexts (C, C_1) is repeated in increasing order of the hole depth of C, until all contexts from E have been processed. Therefore, any context $C[C_1]$ with $C \in E$ and $C_1 \in \sigma_\bullet\langle S \rangle$ that was added to E because of a failed consistency check will be processed itself in the same **for** loop.

The algorithm checks closedness by searching for $s \in S$ and $C_1 \in \sigma_\bullet\langle S \rangle$ such that $C_1[s] \not\sim t$ for all $t \in S$ for which $\mathtt{d}(t) \leq \mathtt{d}(C_1[s])$. The search is performed in increasing order of the depth of s. If s and C_1 are found, $C_1[s]$ is added to the S component of the observation table, and the algorithm checks again consistency. Note that adding $C_1[s]$ to S yields a subterm closed subset of $\mathcal{T}(Sk \cup \Sigma)_{[\ell]}$. Also, closedness checks are performed only on consistent observation tables.

When the observation table is both consistent and closed, the corresponding DFTA is constructed and it is checked whether the language accepted by the constructed automaton coincides with the set of skeletal descriptions of the unknown context-free grammar G_U (this is called a *structural equivalence query*). If this query fails, a counterexample from $\mathcal{L}(\mathcal{A}(\mathbb{T}))_{[\ell]} \triangle K(D(G_U))_{[\ell]}$ is produced, the component S of the observation table is expanded to include t and all its subterms with depth at least 1, and the consistency and closedness checks are performed once more. At the end of this step, the component S of the observation table is subterm closed, and E is unchanged, thus $\bullet$-prefix closed.

Thus, at any time during the execution of algorithm LA^ℓ, the defining properties of an observation table are preserved: the component S is a subterm closed subset of $\mathcal{T}(Sk \cup \Sigma)_{[\ell]}$, and the component E is a $\bullet$-prefix closed subset of $\mathcal{C}(Sk \cup \Sigma)_{\langle \ell-1 \rangle} \cap \mathcal{C}(Sk \cup \Sigma)_{[\ell]}$.

5 Algorithm Analysis

We notice that the number of states of the DFTA constructed by algorithm LA^ℓ will always increase between two successive structural equivalence queries. When this number of states reaches the number of states of a minimal DCTA of $K(D(G_U))$, the constructed DFTA is actually a minimal DCTA of $K(D(G_U))$ (Corollary 2) and the algorithm terminates.

From now on we assume implicitly that n is the number of states of a minimal DCTA of $K(D(G_U))$ with respect to ℓ, and that $\mathbb{T}(\mathtt{t})$ is the observation table $(S^\mathtt{t}, E^\mathtt{t}, T, \ell)$ before execution step $\mathtt{t}$ of the algorithm. By Corollary 2, $\mathcal{Q}^\mathtt{t}$ will always have between 1 and n elements. Note that the representative of an element $s \in S$ in $\mathcal{Q}^\mathtt{t}$ is a notion that depends on the observation table $\mathbb{T}(\mathtt{t})$. Therefore, we will use the notation $\mathtt{r}_\mathtt{t}(s)$ to refer to the representative of $s \in S^\mathtt{t}$ in the observation table $\mathbb{T}(\mathtt{t})$. With this notation, $\mathcal{Q}^\mathtt{t} = \{\mathtt{r}_\mathtt{t}(s) \mid s \in S^\mathtt{t}\}$.

Note that the execution of algorithm LA^ℓ is a sequence of steps characterised by the detection of three kinds of failure: closedness, consistency, and structural equivalence query. The $\mathtt{t}$-th execution step is

1. a failed closedness check when the algorithm finds $C_1 \in \sigma_\bullet\langle S^\mathtt{t} \rangle$ and $s \in S^\mathtt{t}$ such that $C_1[s] \not\sim t$ for all $t \in S^\mathtt{t}$ with $\mathtt{d}(t) \leq \mathtt{d}(C_1[s])$,

2. a failed consistency check when the algorithm finds $C \in E^{\mathtt{t}}$ with $\mathtt{d}_\bullet(C) = i$, $s_1, s_2 \in S^{\mathtt{t}}$ with $\mathtt{d}(s_1), \mathtt{d}(s_2) \leq \ell - i - 1$, and $C_1 \in \sigma_\bullet\langle S^{\mathtt{t}}\rangle$, such that $C[C_1[s_1]]$, $C[C_1[s_2]] \in \mathcal{T}(Sk \cup \Sigma)_{[\ell]}$, $s_1 \sim_k s_2$ where $k = \max\{\mathtt{d}(s_1), \mathtt{d}(s_2)\} + i + 1$, and $T(C[C_1[s_1]]) \neq T(C[C_1[s_2]])$,
3. a failed structural equivalence query when the observation table $\mathbb{T}(\mathtt{t})$ is closed and consistent, and the learning algorithm receives from the teacher a counterexample $t \in \mathcal{T}(Sk \cup \Sigma)_{[\ell]}$ as answer to the structural equivalence query with the grammar $G(\mathcal{A}(S^{\mathtt{t}}, E^{\mathtt{t}}, T, \ell))$.

In the following subsections we perform a complexity analysis of the algorithm by identifying upper bound estimates to the computations due to failed consistency checks, failed closeness checks, and failed structural equivalence queries.

5.1 Failed Closedness Checks

We recall that the t-th execution step is a failed closedness check if the algorithm finds a context $C_1 \in \sigma_\bullet\langle S\rangle$ and a term $s \in S^{\mathtt{t}}$ such that $C_1[s] \nsim t$ for all $t \in S^{\mathtt{t}}$ with $\mathtt{d}(t) \leq \mathtt{d}(C_1[s])$. We will show that the number of failed closedness checks performed by algorithm LA^ℓ has an upper bound which is a polynomial in n. To prove this fact, we will rely on the following auxiliary notions:

- For $r, r' \in \mathcal{Q}^{\mathtt{t}}$, we define $r \prec^{\mathtt{t}}_{\mathbb{T}} r'$ if either $\mathtt{d}(r) < \mathtt{d}(r')$ or $\mathtt{d}(r) = \mathtt{d}(r')$ and there exists $\mathtt{t}' < \mathtt{t}$ such that $r \in \mathcal{Q}^{\mathtt{t}'}$ but $r' \notin \mathcal{Q}^{\mathtt{t}'}$ (that is, r became a representative in the observation table before r').
- To every set of representatives $\mathcal{Q}^{\mathtt{t}} = \{r_1, \ldots, r_m\}$ with $r_1 \prec^{\mathtt{t}}_{\mathbb{T}} \ldots \prec^{\mathtt{t}}_{\mathbb{T}} r_m$ we associate the tuple $\mathtt{tpl}(\mathcal{Q}^{\mathtt{t}}) := (d_1, \ldots, d_n) \in \{1, \ldots, \ell+1\}^n$ where $d_i := \mathtt{d}(r_i)$ if $1 \leq i \leq m$, and $d_i := \ell + 1$ if $m < i \leq n$.
- We consider the following partial order on $\mathbb{N}^n$: $(x_1, \ldots, x_n) < (x'_1, \ldots, x'_n)$ iff there exists $i \in \{1, \ldots, n\}$ such that $x_i < x'_i$ and $x_j \leq x'_j$ for all $1 \leq j \leq n$.
- We denote by $\mathtt{st}_{\mathtt{t}}(i)$ the i-th component of $\mathcal{Q}^{\mathtt{t}}$ in the order given by $\prec^{\mathtt{t}}_{\mathbb{T}}$.

Lemma 9. *Suppose s has been introduced in $S^{\mathtt{t}+1}$ as a result of a failed closedness check. There exists $p \in Pos(s)$ such that $\|p\| = \mathtt{d}(s)$ and for every prefix p' of p different from p, $\mathtt{d}(\mathtt{r}_{\mathtt{t}+1}(s|_{p'})) = \mathtt{d}(s|_{p'})$.*

Corollary 3. *Whenever the t-th execution step is a failed closedness check, the term introduced in $S^{\mathtt{t}+1}$ is in $\mathcal{Q}^{\mathtt{t}+1} \setminus \mathcal{Q}^{\mathtt{t}}$ and its depth is at most j, where j is the position in $\mathcal{Q}^{\mathtt{t}+1}$ of the newly introduced element according to ordering $\prec^{\mathtt{t}}_{\mathbb{T}}$.*

Corollary 4. *$\mathtt{d}(s) \leq n$ for all $s \in S^{\mathtt{t}}$ which was introduced in the table by a failed closedness check.*

Proof. $\mathtt{d}(s) \leq j$ by Cor. 3, and $j \leq n$ because $|\mathcal{Q}^{\mathtt{t}}| \leq n$ for all t. Thus $\mathtt{d}(s) \leq n$.

Lemma 10. *Let j be the position of the element introduced in $\mathcal{Q}^{\mathtt{t}+1}$ by a failed closedness check. Then $\mathtt{tpl}(\mathcal{Q}^{\mathtt{t}+1}) < \mathtt{tpl}(\mathcal{Q}^{\mathtt{t}})$ and $\mathtt{d}(\mathtt{st}_{\mathtt{t}+1}(j)) < \mathtt{d}(\mathtt{st}_{\mathtt{t}}(j))$.*

Theorem 3. *The number of failed closedness checks performed during the entire run of LA^ℓ is at most $n(n+1)/2$.*

5.2 Failed Consistency Checks

The $\mathtt{t}$-th execution step is a failed consistency check if the algorithm finds $C \in E^{\mathtt{t}}$ with $\mathtt{d}_\bullet(C) = i$, $s_1, s_2 \in S^{\mathtt{t}}$ with $\mathtt{d}(s_1), \mathtt{d}(s_2) \leq \ell - i - 1$, and $C_1 \in \sigma_\bullet\langle S^{\mathtt{t}}\rangle$, such that $C[C_1[s_1]], C[C_1[s_2]] \in \mathcal{T}(Sk \cup \Sigma)_{[\ell]}$, $s_1 \sim_k s_2$ where $k = \max\{\mathtt{d}(s_1), \mathtt{d}(s_2)\} + i + 1$, and $T(C[C_1[s_1]]) \neq T(C[C_1[s_2]])$. In this case, the context $C[C_1]$ is newly introduced in the component $E^{\mathtt{t}+1}$ of the observation table $\mathbb{T}(\mathtt{t}+1)$.

We will show that the number of failed consistency checks performed by the learning algorithm LA^ℓ has an upper bound which is a polynomial in n. To prove this fact, we rely on the following auxiliary notions:

- For $C, C' \in E^{\mathtt{t}}$, we define $C \prec_{\mathtt{C}}^{\mathtt{t}} C'$ if either $\mathtt{d}_\bullet(C) < \mathtt{d}_\bullet(C')$ or $\mathtt{d}_\bullet(C) = \mathtt{d}_\bullet(C')$ and there exists $\mathtt{t}' < \mathtt{t}$ such that $C \in E^{\mathtt{t}'}$ but $C' \notin E^{\mathtt{t}'}$ (that is, C became an experiment in the observation table before C').
- We define $\delta_{\mathtt{t}}(s_1, s_2) := \min_{\prec_{\mathtt{C}}}\{C \in E^{\mathtt{t}} \mid C\ \ell\text{-distinguishes } s_1 \text{ and } s_2\}$ for every $s_1, s_2 \in S^{\mathtt{t}}$ such that $s_1 \nsim s_2$.
- A nonempty subset U of $E^{\mathtt{t}}$ induces a partition of a subset R of $S^{\mathtt{t}}$ into equivalence classes $Q_1, \ldots, Q_m$ if the following conditions are satisfied:
  1. $\bigcup_{j=1}^m Q_j = R$ and $Q_i \cap Q_j = \emptyset$ whenever $1 \leq i \neq j \leq m$,
  2. Whenever $1 \leq i \neq j \leq m$, $s_1 \in Q_i$, and $s_2 \in Q_j$, there exists $C \in U$ that ℓ-distinguishes s_1 and s_2.
  3. Whenever $s_1, s_2 \in Q_j$ for some $1 \leq j \leq m$, there is no $C \in U$ that ℓ-distinguishes s_1 and s_2.

Let $\mathcal{E}^{\mathtt{t}} := \{\delta_{\mathtt{t}}(s_1, s_2) \mid s_1, s_2 \in S^{\mathtt{t}}, s_1 \nsim s_2\}$. Since $\sim$ is not an equivalence, not every subset of $E^{\mathtt{t}}$ induces a partition of $S^{\mathtt{t}}$ into equivalence classes. However, the next lemma shows that $\mathcal{E}^{\mathtt{t}}$ induces a partition of $\mathcal{Q}^{\mathtt{t}}$ into at least $|\mathcal{E}^{\mathtt{t}}|$ classes.

Theorem 4. *If $\mathcal{E}^{\mathtt{t}} = \{C_1, \ldots, C_k\}$ with $C_1 \prec_{\mathtt{C}} \ldots \prec_{\mathtt{C}} C_k$ then, for every $1 \leq i \leq k$, $\{C_1, \ldots, C_i\}$ induces a partition of $\mathcal{Q}^{\mathtt{t}}$ into at least i classes.*

Corollary 5. *For any $\mathtt{t}$, $\mathcal{E}^{\mathtt{t}}$ has at most n elements.*

We will compute an upper bound on the number of failed consistency checks by examining the evolution of $\mathcal{E}^{\mathtt{t}}$ during the execution of LA^ℓ. Initially, $\mathcal{E}^0 = \{\bullet\}$.

Lemma 11. *At any time during the execution of the algorithm, if $\mathcal{Q}^{\mathtt{t}}$ has $i \geq 2$ elements, then the hole depth of any context in $E^{\mathtt{t}}$ is less than or equal to $i - 2$.*

Let $\mathcal{E}^{\mathtt{t}} = \{C_1', \ldots, C_k'\}$ before some execution step $\mathtt{t}$ of the algorithm LA^ℓ, where $C_1' \prec_{\mathtt{C}} \ldots \prec_{\mathtt{C}} C_k'$. Then $k \leq n$ by Cor. 5. We associate to every such $\mathcal{E}^{\mathtt{t}}$ the n-tuple $\mathtt{tpl}(\mathcal{Q}^{\mathtt{t}}) = (y_1, \ldots, y_n) \in \{0, 1, \ldots, n-1\}^n$, where, for every $1 \leq j \leq n$, y_j is defined as follows:

- If $\mathcal{Q}^{\mathtt{t}}$ has at least $j+1$ elements then, if i is the minimum integer such that $\{C_1', \ldots, C_i'\}$ partitions $\mathcal{Q}^{\mathtt{t}}$ into at least $j+1$ classes then $y_j = \mathtt{d}_\bullet(C_j')$. Since every $\{C_1', \ldots, C_i'\}$ partitions $\mathcal{Q}^{\mathtt{t}}$ into at least i classes (by Lemma 4) and we assume that $\mathcal{E}^{\mathtt{t}} = \{C_1', \ldots, C_k'\}$ partitions $\mathcal{Q}^{\mathtt{t}}$ into $|\mathcal{Q}^{\mathtt{t}}| \geq j+1$ classes, we conclude that such i exists.
- otherwise $y_j = n - 1$.

For $1 \leq j \leq n$ we denote the j-th component of $\mathtt{tpl}(\mathcal{E}^{\mathtt{t}})$ by $\mathtt{dh_t}(j)$. Note that, for all $1 \leq i \leq k$, $\mathtt{d}_\bullet(C_i') \leq |\mathcal{Q}^{\mathtt{t}}| - 2$ by Theorem 11, and $|\mathcal{Q}^{\mathtt{t}}| \leq n$, hence $\mathtt{d}_\bullet(C_i') \leq n-2$. Therefore, we can always distinguish the components y_i of $\mathtt{tpl}(\mathcal{Q}^{\mathtt{t}})$ that correspond to the defining case (1) from those in case (2).

Lemma 12. $\mathtt{dh_t}(j) \leq j-1$ *whenever* $2 \leq j \leq n$ *and* $\mathtt{dh_t}(j) \neq n-1$.

Theorem 5. *If* $\mathcal{Q}^{\mathtt{t}}$ *has at least 2 elements then the number of failed consistency checks over the entire run of* LA^ℓ *is at most* $n(n-1)/2$.

5.3 Failed Structural Equivalence Queries

Every failed structural equivalence query yields a counterexample which increases the number of representatives in $\mathcal{Q}^{\mathtt{t}}$. Thus

Theorem 6. *The number of failed structural equivalence queries is at most* n.

5.4 Space and Time Complexity

We are ready now to express the space and time complexity of LA^ℓ in terms of the following parameters:

- $n =$ the number of states of a minimal DFCA for the language of structural descriptions of the unknown grammar with respect to ℓ,
- $m =$ the maximum size of a counterexample returned by a failed structural equivalence query,
- $p =$ the cardinality of the alphabet Σ of terminal symbols, and
- $d =$ the maximum rank (or arity) of the symbol $\sigma \in Sk$.

First, we determine the space needed by the observation table. The number of elements in $S^{\mathtt{t}}$ is initially 0 (i.e., $|S^0| = 0$) and is increased either by a failed closedness check or by a failed structural membership query. By Theorem 3, the number of failed closedness checks is at most $n(n+1)/2$, and each of them adds one element to S. By Theorem 6, the number of failed structural equivalence queries is at most n. A failed structural equivalence query which produces a counterexample t with $\mathtt{sz}(t) \leq m$, adds at most m terms to $S^{\mathtt{t}}$. Thus, $|S^{\mathtt{t}}| \leq n(n+1)/2 + n\,m = O(m\,n + n^2)$ and $|S^{\mathtt{t}} \cup \Sigma| = O(m\,n + n^2 + p)$, therefore $|\sigma_\bullet\langle S^{\mathtt{t}}\rangle| \leq \sum_{j=0}^{d-1}(j+1)\,|S^{\mathtt{t}} \cup \Sigma|^j = O((d+1)\,(m\,n + n^2 + p)^d)$ and $|X(S)| \leq \sum_{j=1}^{d} |S^{\mathtt{t}} \cup \Sigma|^j = O(d\,(m\,n + n^2 + p)^d)$. Thus $S^{\mathtt{t}} \cup X(S^{\mathtt{t}})_{[\ell]}$ has $O(d\,(m\,n + n^2 + p)^d)$ elements. By Theorem 5, there may be at most $n(n-1)/2$ failed consistency checks, and each of them adds a context to $E^{\mathtt{t}}$. Thus $E^{\mathtt{t}}$ has $O(n^2)$ elements and $E^{\mathtt{t}}[S^{\mathtt{t}} \cup X(S^{\mathtt{t}})_{[\ell]}]$ has $O(n^2 d\,(m\,n + n^2 + p)^d)$ elements. By Lemma 12, $\mathtt{d}_\bullet(C) \leq n-1$ for all $C \in E^{\mathtt{t}}$. We also know that, if $s \in S^{\mathtt{t}}$, then $\mathtt{d}(s) \leq m$ if it originates from a failed structural equivalence query, and $\mathtt{d}(s) \leq n$ if it originates from a failed closedness check (by Cor. 4). Therefore $\mathtt{d}(s) \leq \max(m,n)$ for all $s \in S^{\mathtt{t}}$, and thus $\mathtt{d}(x) \leq 1 + \max(m,n) \leq 1 + m + n$ for all $x \in S^{\mathtt{t}} \cup X(S^{\mathtt{t}})$ and $\mathtt{d}(t) \leq m + 2\,n$ for all $t \in E^{\mathtt{t}}[S^{\mathtt{t}} \cup X(S^{\mathtt{t}})_{[\ell]}]$. Since the number of positions of such a term t is $\sum_{j=0}^{m+2\,n} d^j = O((m+2\,n+1)d^{m+2\,n})$, we conclude that the total space occupied by an observation table at any time is $O\left(n^2(m\,n+n^2+p)^d(m+2\,n+1)d^{m+2\,n+1}\right)$.

Next, we examine the time complexity of the algorithm by looking at the time needed to perform each kind of operation.

Since the consistency checks of the observation table are performed in a **for** loop which checks the result produced by $s_1 \sim_k s_2$ (where $s_1, s_2 \in S^{\mathtt{t}}$) in increasing order of k, the result produced by $s_1 \sim_k s_2$ can be reused in checking $s_1 \sim_{k+1} s_2$ and so the corresponding elements in the rows of s_1 and s_2 are compared only once. Thus, the total time needed to check if the observation table is consistent involves at most $(|S^{\mathtt{t}}| \cdot (|S^{\mathtt{t}}|-1)/2) \cdot |E^{\mathtt{t}}| \cdot (1+|\sigma_{\bullet}\langle S^{\mathtt{t}}\rangle|)$, comparisons. As $\sigma_{\bullet}\langle S^{\mathtt{t}}\rangle$ has $O(d\,(m\,n+n^2+p)^d)$ elements, a consistency check of the table takes $O((m\,n+n^2)^2 n^2 d\,(m\,n+n^2+p)^d) = O(n^2 d\,(m\,n+n^2+p)^{d+2})$ time. As there are at most $(n\,(n+1)/2+1)\,(n+1) = O(n^3)$ consistency checks, the total time needed to check if the table is consistent is $O(n^5 d\,(m\,n+n^2+p)^{d+2})$.

Checking if the observation table is closed takes at most $|S^{\mathtt{t}}|^2 \cdot |\sigma_{\bullet}\langle S^{\mathtt{t}}\rangle| \cdot |E^{\mathtt{t}}|$ time, which is $O((m\,n+n^2)^2 d\,(m\,n+n^2+p)^d n^2) = O(n^2 d\,(m\,n+n^2+p)^{d+2})$.

Extending an observation table $\mathbb{T}(\mathtt{t})$ with a new element in $S^{\mathtt{t}+1}$ requires the addition of $\sum_{k=2}^{d}(2^{k-1}-1) = 2^d - d - 1$ contexts to $\sigma_{\bullet}\langle S^{\mathtt{t}+1}\rangle \setminus \sigma_{\bullet}\langle S^{\mathtt{t}}\rangle$, thus the addition of at most $2^d - d$ new rows for the new elements of $S^{\mathtt{t}+1} \cup X(S^{\mathtt{t}+1})$ in the observation table $\mathbb{T}(\mathtt{t}+1)$. This extension requires at most $(2^d - d) \cdot |E^{\mathtt{t}}| \cdot (1+|\sigma_{\bullet}\langle S^{\mathtt{t}}\rangle|) = O(n^2 d\,(2^d - d)\,(m\,n+n^2+p)^d)$ membership queries. The number of elements added to $S^{\mathtt{t}}$ as a result of a failed structural equivalence query is at most m. As there will be at most n failed structural equivalence queries and at most $n(n+1)/2$ failed closedness checks, the maximum number of elements added to $S^{\mathtt{t}}$ is $n(n+1)/2+m\,n = O(m\,n+n^2)$. Thus the total time spent on inserting new elements in the S-component of the observation table is $O(n^2 d\,(2^d-d)\,(m\,n+n^2)(m\,n+n^2+p)^d)$. Adding a context to $E^{\mathtt{t}}$ requires at most $|S^{\mathtt{t}}+X(S^{\mathtt{t}})_{[\ell]}| = O(d\,(m\,n+n^2+p)^d)$ membership queries. These additions are performed only by failed consistency checks, and there are at most $n(n-1)/2$ of them. Thus, the total time spent to insert new contexts in the E-component of the observation table is $O(n^2 d\,(m\,n+n^2+p)^d)$. We conclude that the total time spent to add elements to the components S and E of the observation table is $O(n^2 d\,(2^d-d)\,(m\,n+n^2)(m\,n+n^2+p)^d)$, which is polynomial.

The identification of the representative $\mathtt{r}_{\mathtt{t}}(s)$ for every $s \in S^{\mathtt{t}}$ can be done by performing $((|S^{\mathtt{t}}|)(S^{\mathtt{t}}-1)/2)\,|E^{\mathtt{t}}| = O((m\,n+n^2)^2 n^2)$ comparisons.

Thus, all DFCAs $\mathcal{A}(\mathbb{T}(\mathtt{t}))$ corresponding to consistent and closed observation tables $\mathbb{T}(\mathtt{t})$ can be constructed in time polynomial in m and n. Since the algorithm encounters at most n consistent and closed observation tables, the total running time of the algorithm is polynomial in m and n.

6 Conclusions and Acknowledgments

We have presented an algorithm, called LA^{ℓ}, for learning cover context-free grammars from structural descriptions of languages of interest. LA^{ℓ} is an adaptation of Sakakibara's algorithm LA for learning context-free grammars from structural descriptions, by following a methodology similar to the design of Ipate's algorithm L^ℓ as a nontrivial adaptation of Angluin's algorithm L^*. Like L^*, our algorithm synthesizes a minimal deterministic cover automaton consistent with an observation table maintained via a learning protocol based on what is called in the literature a "minimally adequate teacher" [1]. And again, like algorithm L^*, our algorithm is guaranteed to synthesize the desired automaton in time polynomial in n and m, where n is its number of states and m is the maximum size of a counterexample to a structural equivalence query. As the size of a minimal finite cover automaton is usually much smaller than that of a minimal automaton that accepts that language, the algorithm LA^ℓ is a better choice than algorithm LA for applications where we are interested only in an accurate characterisation of the structural descriptions with depth at most ℓ.

This work has been supported by CNCS IDEI Grant PN-II-ID-PCE-2011-3-0981 "Structure and computational difficulty in combinatorial optimization: an interdisciplinary approach."

References

1. Angluin, D.: Learning regular sets from queries and counterexamples. Information and Computation 75, 87–106 (1987)
2. Angluin, D., Kharitonov, M.: When won't membership queries help? Journal of Computer and System Sciences 50(2), 336–355 (1995)
3. Brainerd, W.S.: The minimalization of tree automata. Information and Control 13(5), 484–491 (1968)
4. Farzan, A., Chen, Y.-F., Clarke, E.M., Tsay, Y.-K., Wang, B.-Y.: Extending automated compositional verification to the full class of omega-regular languages. In: Ramakrishnan, C.R., Rehof, J. (eds.) TACAS 2008. LNCS, vol. 4963, pp. 2–17. Springer, Heidelberg (2008)
5. De la Higuera, C.: Grammatical inference: learning automata and grammars. Cambridge University Press (2010)
6. Holzer, M., Jakobi, S.: From equivalence to almost-equivalence, and beyond—minimizing automata with errors. In: Yen, H.-C., Ibarra, O.H. (eds.) DLT 2012. LNCS, vol. 7410, pp. 190–201. Springer, Heidelberg (2012)
7. Hopcroft, J.E., Motwani, R., Ullman, J.D.: Introduction to Automata Theory, Languages, and Computation, 2nd edn. Pearson Addison Wesley (2003)
8. Ipate, F.: Learning finite cover automata from queries. Journal of Computer and System Sciences 78(1), 221–244 (2012)
9. Kumar, V., Madhusudan, P., Viswanathan, M.: Minimization, learning, and conformance testing of boolean programs. In: Baier, C., Hermanns, H. (eds.) CONCUR 2006. LNCS, vol. 4137, pp. 203–217. Springer, Heidelberg (2006)
10. Levy, L.S., Joshi, A.K.: Skeletal structural descriptions. Information and Control 39(3), 192–211 (1978)
11. Maler, O., Pnueli, A.: On the learnability of infinitary regular sets. Information and Computation 118(2), 316–326 (1995)
12. Sakakibara, Y.: Learning context-free grammars from structural data in polynomial time. Theoretical Computer Science 76, 223–242 (1990)
13. Sipser, M.: Introduction to the Theory of Computation, 2nd edn. Thomson (2006)

Context-Free Sequences

Didier Caucal[1] and Marion Le Gonidec[2]

[1] CNRS, LIGM-Université Paris-Est
caucal@univ-mlv.fr
[2] LIM, Université de la Réunion
marion.le-gonidec@univ-reunion.fr

Abstract. A sequence w over a finite alphabet A is generated by a uniform automaton if there exists an automaton labelled on $\{0, \ldots, k-1\}$ for some $k > 1$ and recognizing for each output a in A the set of positions of a in w expressed in base k. Automatic sequences are generated by finite automata. By considering pushdown automata instead of finite ones, we generate exactly the context-free sequences. We distinguish the subfamilies of unambiguous, deterministic, real-time deterministic context-free sequences associated with the corresponding families of pushdown automata. We study the closure under shift, product, morphisms, inverse substitutions and various extractions of these four families of context-free sequences. Additionally, we show that only using multiplicatively dependent bases yields the same set of context-free sequences.

1 Introduction

Automatic sequences are well known objects appearing in many area of theoretical computer science and mathematics [Co 72, AS 03]. To generate these sequences, one can use finite k-automata. Fixing an integer k greater than 1, a k-automaton G has states labelled over an alphabet A and edges labelled on $[\![k]\!] = \{0, \ldots, k-1\}$ so that the family of languages $L(G, a)$ recognized by G from an initial state to a state labelled a in A forms a partition of the set of words over $[\![k]\!]$ labelling paths in G. A k-automaton G generates the sequence w over A when $L(G, a)$ is the set of positions of a in w properly expressed in base k.

When G is finite, the sequence w is k-automatic. Finite automata are also well known devices which recognize regular languages [Kl 56].

As automatic sequences can be characterized in several other ways (using regular languages, uniform morphisms or their kernel finiteness property), there are different ways to extend this notion: context-free sequences, morphic sequences, regular sequences (see [AS 03] for example).

One can also extend this notion using a larger class of automata, the most natural one being, following Chomsky's hierarchy [Ch 56], k-pushdown automata (pushdown automata over $[\![k]\!]$). It turns out that this generalization of automatic sequences, also yields to context-free sequences defined in [Ha 98], [Mo 08] and [AS 03] (open problem 6.3, p 208): a sequence w is k-context-free if and only if

G. Ciobanu and D. Méry (Eds.): ICTAC 2014, LNCS 8687, pp. 259–276, 2014.

for each letter a, the set of occurrences in base k of a in w is a context-free language.

Contrary to finite automata, we obtain distinct sub-families of sequences by considering only unambiguous, deterministic, or real-time deterministic k-pushdown automata. These (proper) sub-families are respectively named unambiguous, deterministic, real-time deterministic k-context-free sequences.

The closure properties of these four families of sequences are also different. This fact underlying the different consequences of the considered transformations (synchronization product, shift, uniform morphism, inverse injective k-uniform substitution, regular extractions) over the unambiguity and the determinism of involved languages and automata. They sometimes share the same closure properties (Propositions 12 and 14) or not (Propositions 13 and 15).

Moreover, tools used to prove these properties depend on the considered family: while, for context-free sequences and unambiguous context-free sequences, closure properties follow from easy considerations over languages, for deterministic context-free sequences and real-time deterministic context-free sequences one has to use tools from infinite graph theory. Indeed, for these two last subfamilies, classes of languages associated to involved pushdown automata no longer have suitable closure properties and the usual internal representation of k-pushdown automata is not really suitable to compute automaton transformations induced by the transformations of sequences we consider. Then, we choose to favor their external representation (by their transition graphs) using the equivalent notion of regular k-automata [MS 85].

Finally, we investigate base dependence properties of context-free sequences.

2 Generating Sequences with Automata

We recall the general notion of automaton, and the particular cases of unambiguous and deterministic automata. The languages recognized by automata over the alphabet $\{0,\ldots,k-1\}$ of $k>1$ digits, called k-automata, can be seen as sets of natural numbers expressed in base k. This in turn allows such automata to define sequences.

2.1 Automata and Languages Terminology

Let A be an *alphabet i.e.*, a finite set of symbols called *letters*. A *word* u over A of *length* $n \geq 0$ is a mapping from the set $[\![n]\!] = \{0,\ldots,n-1\}$ into A which is denoted by the n-tuple $(u(0),\ldots,u(n-1)) \in A^n$ or simply by $u = u_0 \cdots u_{n-1}$ where $u_i = u(i)$ for each $i \in [\![n]\!]$; we write $|u|$ the length n of u. The word of length 0, *i.e.*, the 0-tuple () is the *empty word* denoted by ε. The set A^* of words over A is the free monoid $(A^*,\cdot,\varepsilon)$ generated by A with the *concatenation* operator $\cdot$ which can be omitted: $u_0 \cdots u_{m-1} \cdot v_0 \cdots v_{n-1} = u_0 \cdots u_{m-1} v_0 \cdots v_{n-1}$.

Any subset of A^* is a *language* over A. The family 2^{A^*} of languages over A is a semiring $(2^{A^*},\cup,\cdot,\emptyset,\{\varepsilon\})$ for the *language concatenation* defined by $L \cdot M = \{\, u \cdot v \mid u \in L \wedge v \in M \,\}$ for any $L, M \subseteq A^*$. The closure under concatenation of a language L is the language $L^* = \bigcup_{n\geq 0} L^n$ for $L^n = \{\, u_1 \cdot \ldots \cdot u_n \mid u_1,\ldots,u_n \in L \,\}$ the concatenation n-times of L. Recall that a language is a *regular language* if it can be obtained from $\emptyset$ and the elementary languages $\{a\}$ with $a \in A$ by finite application of the operations $\cup, \cdot, ^*$. The regular languages are also the languages recognized by finite automata.

We fix a symbol ι. An *automaton* over A and an alphabet C of *colours* (output alphabet) with $\iota \notin C$, is a directed graph G whose each edge is labelled by a letter in A and whose vertices can be labelled by letters in $C \cup \{\iota\}$:

$$G \subseteq V{\times}A{\times}V \;\cup\; (C \cup \{\iota\}){\times}V \quad \text{for some (possibly infinite) set } V.$$

A triple $(s,a,t) \in G$ is an *edge* labelled by a from *source* s to *target* t; it is identified with the transition $s \xrightarrow{a}_G t$ or directly $s \xrightarrow{a} t$ if G is understood. A pair $(c,s) \in G$ or directly $cs \in G$ means that s is labelled by c. A vertex can be uncoloured. The set of *vertices* of G is

$$V_G \;=\; \{\, s \mid \exists\, c\,(cs \in G) \,\} \;\cup\; \{\, s \mid \exists\, a,t\;(s \xrightarrow{a}_G t \;\vee\; t \xrightarrow{a}_G s) \,\}.$$

So G is a finite automaton if and only if its vertex set V_G is finite. An *input vertex* is a vertex s labelled by ι: $\iota s \in G$.

The *induced automaton* $G_{|P}$ of an automaton G to a set P is the restriction of G to its vertices in P:

$$G_{|P} \;=\; \{\, (s,a,t) \in G \mid s,t \in P \,\} \;\cup\; \{\, (c,s) \in G \mid s \in P \,\}.$$

The *inverse* G^{-1} of an automaton G is the automaton

$$G^{-1} \;=\; \{\, (t,a,s) \mid (s,a,t) \in G \,\} \;\cup\; \{\, (c,s) \mid (c,s) \in G \,\}.$$

The *out-degree* $d^+_G(s) = |G \cap \{s\}{\times}A{\times}V_G|$ and the *in-degree* $d^-_G(s) = d^+_{G^{-1}}(s)$ of a vertex s are the number of edges of respectively source and target s; the *degree* of s is $d(s) = d^+(s) + d^-(s)$. An automaton is of *finite (resp. in-, out-) degree* if any vertex is of finite (resp. in-, out-) degree. A finite degree automaton with finitely many vertex degrees is of *bounded degree.*

Any tuple $(s_0,a_1,s_1,\ldots,a_n,s_n) \in (VA)^*V$ for $n \geq 0$ with $s_0 \xrightarrow{a_1}_G s_1,\ldots,$ $s_{n-1} \xrightarrow{a_n}_G s_n$ is a *path* from s_0 to s_n labelled by $u = a_1 \cdots a_n$; we write $s_0 \xrightarrow{u}_G s_n$ or directly $s_0 \xrightarrow{u} s_n$ if G is understood; we say that s_n is *accessible* from s_0 and also write $s_0 \longrightarrow^* s_n$ if we do not want to specify a path label. An *accessible automaton* G means that any vertex is accessible from an input vertex: $\forall\, s \in V_G\; \exists\, r \in V_G\; (\iota r \in G \wedge r \longrightarrow^* s)$.

We say that G is *deterministic* if it has a unique input vertex and there are no two edges with the same source and the same label: $(r \xrightarrow{a} s \wedge r \xrightarrow{a} t) \implies s = t$. We also say that G is *unambiguous* if there is no couple of distinct paths labelled with the same word from the set of input vertices to the set of coloured vertices. We abbreviate unambiguous by 'una.' and deterministic by 'det.'.

The *language recognized* by an automaton G (from ι) to $c \in C$

$$\mathrm{L}(G,c) \;=\; \{\, u \in A^* \mid \exists\, s,t \in V_G\; (s \xrightarrow{u}_G t \;\wedge\; \iota s,\, ct \in G) \,\}$$

is the set of words labelling the paths from an input vertex to a vertex coloured by c. Kleene's theorem [Kl 56] states that a language is regular if and only if it is recognized by a finite (resp. and det.) automaton.

A *colouring* over (A, C) is a mapping $\Lambda : C \longrightarrow 2^{A^*}$. To any automaton G, we associate the colouring Λ_G defined for any $c \in C$ by $\Lambda_G(c) = \mathrm{L}(G, c)$. We say that a colouring Λ is a *partition* of $P \subseteq A^*$ if $\bigcup_{c \in C} \Lambda(c) = P$ and $\Lambda(c) \cap \Lambda(d) = \emptyset$ for any $c \neq d$ in C.

2.2 Uniform Automata and Sequences

A *sequence* w over A is a mapping from the set $\mathbb{N}$ of nonnegative integers into A denoted by $w = (w_n)_{n \geq 0}$. Let A^ω be the set of sequences over A.

We fix an integer $k > 1$. Recall that $[\![k]\!] = \{0, \ldots, k-1\}$ and we denote

$$[\![k]\!]^\diamond = [\![k]\!]^* - 0.[\![k]\!]^*$$

the set of words over $[\![k]\!]$ not beginning with 0.

The (proper) *representation in base* k of any $n \in \mathbb{N}$ is the word

$$(n)_k \;=\; n_0 \cdots n_r \in [\![k]\!]^\diamond \quad \text{such that} \quad \textstyle\sum_{i=0}^{r} n_i k^{r-i} = n$$

Conversely to any word $u \in [\![k]\!]^*$, we associate the integer

$$[u]_k \;=\; \textstyle\sum_{i=0}^{|u|-1} u_i k^{|u|-i-1}.$$

So for any $n \geq 0$ and $u \in [\![k]\!]^*$, we have $[(n)_k]_k = n$ and $([u]_k)_k$ is the greatest suffix of u in $[\![k]\!]^\diamond$. For any $w \in A^\omega$, we define the k-colouring Λ_w^k associating with any $a \in A$ the language

$$\Lambda_w^k(a) \;=\; \{\, u \in [\![k]\!]^\diamond \mid w_{[u]_k} = a \,\} \;=\; \{\, (n)_k \mid w_n = a \,\}$$

of the representations in base k of the positions of a in w.

So the colouring Λ_w^k is a partition of $[\![k]\!]^\diamond$.

Definition 1. *A* k-automaton *over* A *is an accessible automaton* G *on the set* $[\![k]\!]$ *of edge labels and on the set* A *of colours such that* Λ_G *is a partition of* $[\![k]\!]^\diamond$. *We talk about* uniform automaton *when* k *and* A *are not specified.*

Any vertex of a uniform automaton has at most one colour. For a det. uniform automaton, each vertex has a unique colour and its input vertex has no ingoing edge. A non-det. uniform automaton can have several paths with same labels starting from distinct input vertices and leading to vertices possibly coloured but with the same colour.

Definition 2. *A* k*-automaton* G generates *the sequence* $\mathrm{Seq}(G) \in A^\omega$ *if for any integer* n, *the* n*-th letter* $\mathrm{Seq}(G)_n$ *is the colour* a *such that* $(n)_k \in \Lambda_G(a)$, *i.e.,* $\Lambda^k_{\mathrm{Seq}(G)} = \Lambda_G$.

Example 3. Let G be the following det. 2-automaton over $\{a, b\}$:

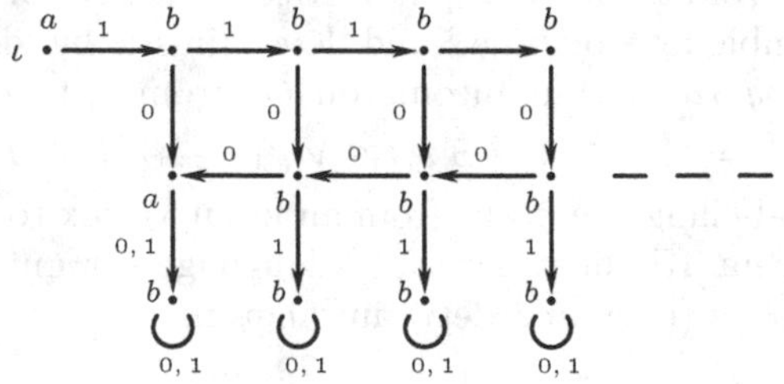

The automaton G generates the sequence $w = \mathrm{Seq}(G)$ characterized by:

$$\Lambda_w(a) = \{\, 1^n 0^n \mid n \geq 0 \,\} \quad \text{and} \quad \Lambda_w(b) = \{0,1\}^\diamond - \Lambda_w(a),$$

that is $w_n = a$ if and only if there exits p in $\mathbb{N}$ such that $n = 2^p(2^{p+1} - 1)$.

3 Context-Free Sequences

The k-automatic sequences can be characterized using languages. Precisely, a sequence w is a k-automatic sequence if the colouring Λ_w^k only contains regular languages. In [Ha 98] and [AS 03] (open problem 6.3, p 208), authors start considering the sequences for which the colouring Λ_w^k is made of context-free languages [AU 79].

Definition 4. *A sequence w over A is a k-*context-free sequence *if $\Lambda_w^k(a)$ is context-free for all $a \in A$.*

Moreover, a sequence w is a k-automatic sequence if and only if it is generated by a finite k-automaton G: $w = \mathrm{Seq}(G)$ [Co 72]. It is natural to expect that k-context-free sequences are sequences generated by pushdown automata over $[\![k]\!]$ (with no 0-edges starting from input vertices), called k-pushdown automata, as these machines recognize context-free languages. As the usual internal representation of k-pushdown automata is not really convenient to play with automata transformations involved by transformations of sequences we will consider. So we use the equivalent notion (via their transition graphs) of regular k-automata [MS 85] and we show that k-context-free sequences are sequences generated by regular k-automata (See Proposition 6).

3.1 Regular Automata

Regular automata, which form a family of infinite automata (including the finite automata), is the set of finitely decomposable automata.

Let us make precise the notion of decomposition of an automaton G according to a *graduation* which is a mapping $g : V_G \longrightarrow \mathbb{N}$ such that only finitely many vertices have the same value by g: $g^{-1}(n)$ is finite for every $n \geq 0$.

Let $\mathrm{Con}(G, g, n)$ be the set of connected components of $G_{|\{\, s \mid g(s) \geq n \,\}}$ for any $n \geq 0$. In particular $\mathrm{Con}(G, g, 0) = \{G\}$ for G connected.

We complete elements in $\mathrm{Con}(G, g, n)$ to obtain a partition $\mathrm{Dec}(G, g, n)$ of $G_{|\{\, s \mid g(s) \geq n \,\}} \cup \{\, s \xrightarrow{a}_G t \mid g(s) > n \ \vee\ g(t) > n \,\}$ as follows:

$$\mathrm{Dec}(G, g, n) = \bigcup_{H \in \mathrm{Con}(G,g,n)} H \cup \left\{ s \xrightarrow{a}_G t \;\middle|\; \vee \begin{array}{l} (s \in V_H \wedge g(s) > n) \\ (t \in V_H \wedge g(t) > n) \end{array} \right\}$$

The information added in $\mathrm{Dec}(G, g, n)$ consists in specifying the way(s) a graph of $\mathrm{Con}(G, g, n)$ can be linked with others connected components in G. This additionnal information is useful when G contains at least a vertex of infinite degree.

The finiteness of the *decomposition* $\mathrm{Dec}(G, g) = \bigcup_{n \geq 0} \mathrm{Dec}(G, g, n)$ is expressed by isomorphism respecting the frontiers. The *frontier* $\mathrm{Fr}_G(H)$ of $H \subseteq G$

is the set of vertices common to H and $G - H$ *i.e.*,

$$\mathrm{Fr}_G(H) \;=\; V_H \cap V_{G-H}\,.$$

We say that $H, K \subseteq G$ are *strongly isomorphic* if there exists a bijection $h : V_H \longrightarrow V_K$ preserving edges, colouring and frontiers:

$$h(H) = K \;\text{ and }\; h(\mathrm{Fr}_G(H)) \;=\; \mathrm{Fr}_G(K).$$

An automaton G is *finitely decomposable* by a graduation g if

$\mathrm{Dec}(G, g)$ has finitely many non-strongly-isomorphic automata.

Finally G is a *regular automaton* if G is finitely decomposable by some graduation.

In particular, a regular automaton has finitely many non-isomorphic connected components but notice that there exists a non-regular automaton G with a graduation g giving a decomposition $\mathrm{Dec}(G, g)$ with finitely many non-isomorphic components. Moreover, a regular automaton of finite degree is of bounded degree.

Example 5. Let us consider the following 4-automaton of vertex set $\mathbb{Z}$:

$$G = \{\, n \xrightarrow{0} -n \mid n \in \mathbb{Z} - \{0\} \,\} \cup \{\, n \xrightarrow{i} n+i \mid n \in \mathbb{Z}\,, i \in \{1,2\}\}$$
$$\cup \{\, n \xrightarrow{3} n-1 \mid n \in \mathbb{Z} \,\} \cup \{\iota 0\} \cup \{\, a\,0 \,\} \cup \{\, b\,n \mid n \neq 0 \,\}$$

This is a 4-automaton which is represented (with vertices in bold faces, with $-n$ denoted by $\overline{n}$) as follows:

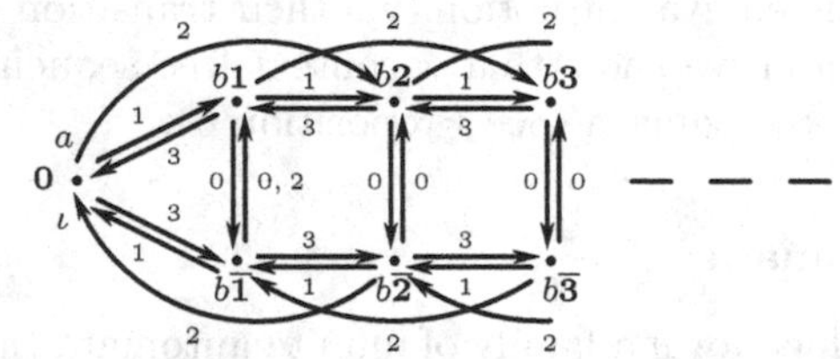

Such an automaton G is regular because it is finitely decomposable by the graduation $g(n) = |n|$ of the absolute value. In fact, $\mathrm{Dec}(G, g)$ has only three non-strongly-isomorphic automata: G and the following two automata where the vertices of the frontier are circled:

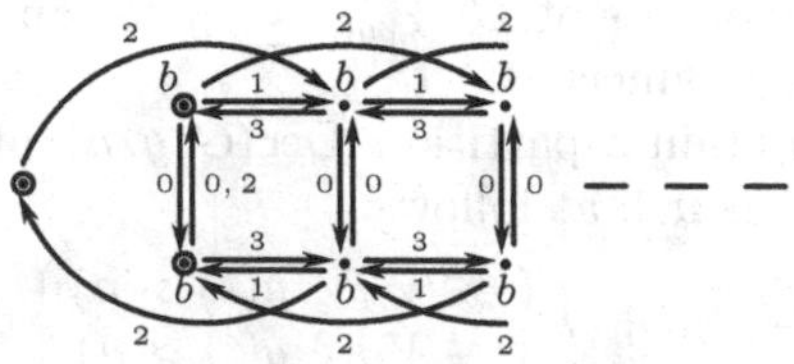

and

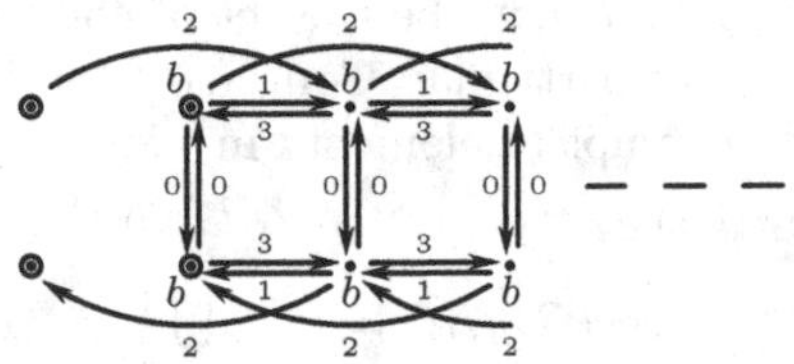

For any connected automaton G of finite degree and with finitely many input vertices *i.e.*, $\{\, s \mid \iota\, s \in G \,\}$ is finite, a standard graduation is the *distance* from the input vertices: for any vertex $s \in V_G$:

$$\mathrm{Dist}_G(s) \;=\; \min\{\, |u| \mid \exists\, r\; (\iota r \in G \;\wedge\; r \xrightarrow{u}_{G \,\cup\, G^{-1}} s) \,\}$$

The decomposition $\mathrm{Dec}(G, \mathrm{Dist}, n)$ of any automaton G at distance $n \geq 0$ is the set of connected components obtained by removing in G all the vertices at distance from ι less than n. For the automaton presented in Example 5, we have $\mathrm{Dist}_G(n) = \left\lceil \frac{|n|}{2} \right\rceil$ for any $n \in \mathbb{Z}$.

The regular k-automata are isomorphic to the transition graphs of k-pushdown automata (ε-transitions in their transition graphs being removed by gluing vertices, and with regular set of configurations labelled by each colour) [MS 85]. So the languages recognized by the family of regular automata are the context-free languages. Moreover, det. (resp. det. and of finite degree, resp. una.) regular k-automata are isomorphic to transition graphs of det. (resp. real-time det., resp. una.) k-pushdown automata, so these classes of regular k-automata recognizing respectively the det. context-free languages, the real-time det. context-free languages, and una. context-free languages of $[\![k]\!]^{\diamond}$ [Ca 07]. We abbreviate real-time by 'rt.'.

3.2 Regular Automata and Context-Free Sequences

We extend the characterization of automatic sequences by finite automata to context-free sequences using regular k-automata.

Proposition 6. *A sequence is a k-context-free sequence if and only if it is generated by a regular k-automaton G: $w = \mathrm{Seq}(G)$.*

Proof. Let w be a k-context-free sequence. As context-free languages over $[\![k]\!]$ are recognized by pushdown automata over $[\![k]\!]$, for any letter $a \in A$, $\Lambda^k_w(a) = \mathrm{L}(G_a, a)$ for some regular k-automaton G_a having a unique colour a. Notice that in general, the usual product of two regular automata is not regular. As automata G_a can be choosen with distinct vextex sets, w is generated by the disjoint finite union $\cup_{a\in A} G_a$ which remains a regular k-automaton.

For a sequence generated by a regular k-automaton G, the language $\Lambda_w(a)$ with $a \in A$ is recognized by the regular k-automaton obtained from G by removing colours $c \neq a$, so $\Lambda_w(a)$ is context-free. □

It is natural to introduce the sub-families of k-context-free sequences associated with the different classes of regular automata/pushdown automata presented at the end of Section 3.1.

Definition 7. *A sequence is a* una. *(resp.* det., rt. det.*)* k-context-free sequence *if it is generated by a una. (resp. det. , det. and of finite degree) regular k-automaton.*

Let us note $\mathrm{RtDet}Cf_k(A^\omega)$, $\mathrm{DetCf}_k(A^\omega)$, $\mathrm{UnaCf}_k(A^\omega)$, $\mathrm{Cf}_k(A^\omega)$ *for the respective four families of rt. det., det., una., k-context-free sequences over A.*

For any $k > 1$, we have

$$\mathrm{RtDetCf}_k(A^\omega) \subsetneq \mathrm{DetCf}_k(A^\omega) \subsetneq \mathrm{UnaCf}_k(A^\omega) \subsetneq \mathrm{Cf}_k(A^\omega)$$

The strict inclusions follow from these on context-free languages and we can refine Proposition 6.

Proposition 8. *For any* $w \in A^\omega$ *and* $k > 1$,

$$w \text{ is a (resp. una.) } k\text{-context-free sequence}$$
$$\Longleftrightarrow \Lambda_w^k(a) \text{ is (resp. una.) context-free for any } a \in A.$$

Furthermore

$$w \text{ is a (resp. rt.) det. } k\text{-context-free sequence}$$
$$\Longrightarrow \Lambda_w^k(a) \text{ is (resp. rt.) det. context-free for any } a \in A.$$

For the unambiguous case, the proof of Proposition 6 holds, as if all automata G_a are una., the disjoint union $\cup_{a \in A} G_a$ remains una. The second implication comes from the fact that families of det. and rt. det. regular automata recognize respectively det. and rt. det. families of context-free languages [Ca 07]. The converse of the implication is true for $|A| = 2$ (because the complement of a det. context-free language is also context-free) but false in general (see Proposition 16).

Notice also that the characterization of k-automatic sequences as morphic sequences obtained with k-uniform morphism [Co 72] can be extended to the smallest subfamily $\mathrm{RtDet}_k(A^\omega)$ of k-context-free sequences [LeG 12] using a notion of context-free morphic sequence which extend the usual notion of morphic sequence [Lo 05].

4 Closure Properties

The set of k-automatic sequences is stable by various transformations: regular modifications of letters, shift, application of a uniform morphism, inverse substitution, various extractions. It underlies the robustness of this concept, and one can ask whereas these properties remain true or not for the four families of k-context-free sequences.

In this section, we present closure properties of $\mathrm{Cf}_k(A^\omega)$ and $\mathrm{DetCf}_k(A^\omega)$ but these properties and their proofs remain valid by replacing $\mathrm{Cf}_k(A^\omega)$ by $\mathrm{UnaCf}_k(A^\omega)$ and by replacing $\mathrm{DetCf}_k(A^\omega)$ by $\mathrm{RtDetCf}_k(A^\omega)$.

By Proposition 6, we get closure properties of $\mathrm{Cf}_k(A^\omega)$ and $\mathrm{UnaCf}_k(A^\omega)$ from the closure properties of context-free languages. We get closure properties of $\mathrm{DetCf}_k(A^\omega)$ and $\mathrm{RtDetCf}_k(A^\omega)$) from the preservation of deterministic regular automata by inverse regular path functions.

4.1 An Important Tool: Regular Path Functions

Let us recall the notion of a path function.

We define the set Exp of *regular expressions* as the smallest language over

$$C \cup A \cup \{\varepsilon, (,), {}^{-1}, \neg, \vee, \wedge, \cdot, {}^{+}\}$$

such that $C \cup A \cup \{\varepsilon\} \subseteq \mathrm{Exp}$ and for any u, v in Exp, the expressions (u^{-1}), $(\neg u)$, $(u \vee v)$, $(u \wedge v)$, $(u \cdot v)$, (u^+) are in Exp.

We can remove parentheses using the associativity of $\vee$, $\wedge$, $\cdot$ and by assigning priorities to operators as usual. Finally $\cdot$ can be omitted. The expression u^* corresponds to $\varepsilon \vee u^+$ and we will use A instead of $\bigvee_{a \in A} a$.

A *finite expression* is a regular expression without the operator $^+$ and we denote by FinExp the set of finite expressions.

The label $a \in A$ of an edge $s \xrightarrow{a}_G t$ from s to t of an automaton G is extended to a regular expression $u \in \mathrm{Exp}$ by induction on the length of u: for any $c \in C$ and $u, v \in \mathrm{Exp}$,

$$s \xrightarrow{c} t \quad \text{if} \quad s = t \wedge cs \qquad ; \qquad s \xrightarrow{\varepsilon} t \quad \text{if} \quad s = t$$

$$s \xrightarrow{u^{-1}} t \quad \text{if} \quad t \xrightarrow{u} s \qquad ; \qquad s \xrightarrow{\neg u} t \quad \text{if} \quad \neg\,(s \xrightarrow{u} t)$$

$$s \xrightarrow{u \vee v} t \quad \text{if} \quad s \xrightarrow{u} t \vee s \xrightarrow{v} t \qquad ; \qquad s \xrightarrow{u \wedge v} t \quad \text{if} \quad s \xrightarrow{u} t \wedge s \xrightarrow{v} t$$

$$s \xrightarrow{uv} t \quad \text{if} \quad \exists\, r\,(s \xrightarrow{u} r \wedge r \xrightarrow{v} t) \quad ; \quad s \xrightarrow{u^+} t \quad \text{if} \quad s\,(\xrightarrow{u})^+\, t.$$

A regular expression formalizes a path pattern in a graph.

Example 9. For instance, if u and v are words over A and c in C,

$s \xrightarrow{\varepsilon \wedge u} t$ means that there is a cycle on vertex s labelled by u;

$s \xrightarrow{\varepsilon \wedge uu^{-1}} t$ means that $s = t$ and there is a path labelled by u starting from s;

$s \xrightarrow{c} t$ or $s \xrightarrow{c} s$ means that the vertex $s = t$ is coloured by c;

$s \xrightarrow{(\varepsilon \wedge ucu^{-1})v} t$ means that there is a path labelled by u starting from s and ending to a vertex coloured with c and a path from s to t labelled by v.

$s \xrightarrow{A^*cA^*} t$ means that there is a path labelled from s to t which goes through a vertex colored by c.

A function $h : C \cup A \longrightarrow \mathrm{Exp}$ (resp. FinExp) is called a *regular* (resp. *finite*) *path function* and is applied by inverse to any automaton G to get the automaton:

$$h^{-1}(G) \;=\; \{\, s \xrightarrow{a} t \mid a \in A \,\wedge\, s \xrightarrow{h(a)}_G t \,\} \,\cup\, \{\, cs \mid c \in C \,\wedge\, s \xrightarrow{h(c)}_G s \,\}.$$

That is, the graph $h^{-1}(G)$ is obtained by replacing any path of type $h(a)$ from s to t by an edge $s \xrightarrow{a} t$ and colouring of a vertex s by c if a path $h(c)$ loops on s.

Let us give an example to illustrate the notion of inverse regular path function.

Example 10. For instance we take the following automaton $G = \{\, n \xrightarrow{a} n{+}1 \mid n \geq 0 \,\}$ depicted as follows:

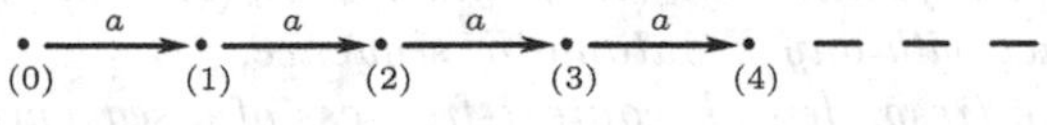

and the finite path function h defined by $h(a) = a$ and $h(\iota) = \neg(a^{-1}a)$. So $h^{-1}(G)$ is the following automaton:

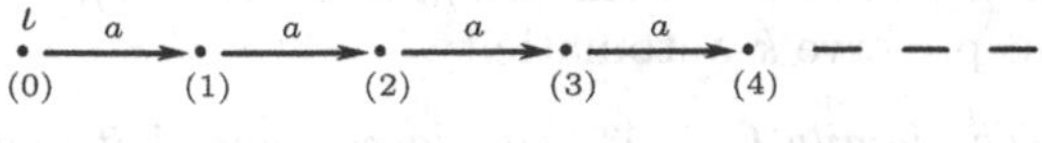

By applying to this automaton by inverse the following regular path function g:

$$g(\iota) = \iota \quad ; \quad g(o) = \iota \vee a^{-1}\iota a \quad ;$$
$$g(a) = (\varepsilon \wedge (a^{-1})^* \iota (aa)^*)\, a\, a \quad ;$$
$$g(b) = (\varepsilon \wedge (a^{-1})^* \iota (aa)^*)\, a^{-1} \quad \vee \quad (\varepsilon \wedge (a^{-1})^* \iota\, a(aa)^*)\, a^{-1} a^{-1} \, .$$

we get the following automaton $g^{-1}(h^{-1}(G))$ depicted as follows:

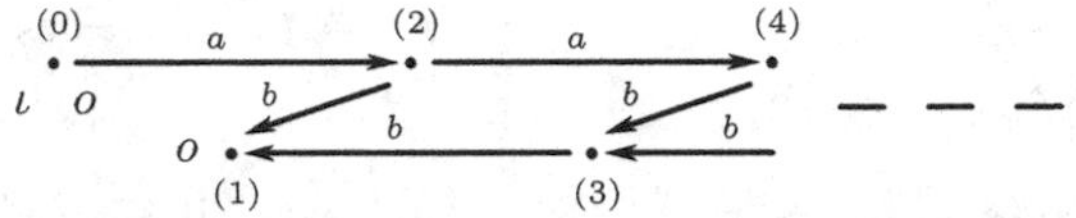

The regularity of automata and the finiteness of its degree are preserved by inverse of regular path function under conditions [CK 01].

Proposition 11. *Let G be a regular automaton and h a regular path function.*

1. *If $h^{-1}(G)$ is deterministic or of finite degree, then $h^{-1}(G)$ is regular.*
2. *If the degree of G is finite and h is a finite path function, then $h^{-1}(G)$ is regular and of finite degree.*

Notice that $h^{-1}(G)$ is deterministic if it has a unique input vertex and for all a in A and s in V_G, $s \xrightarrow{h(a)}_G t \wedge s \xrightarrow{h(a)}_G t' \implies t = t'$.

4.2 Synchronization Product and Shift

Let us start with a preservation result by regularly modifying letters. Taking a mapping $* : A \times A \longrightarrow A$, the *synchronization product* $w * w'$ of sequences $w, w' \in A^\omega$ is the sequence $w * w' = (w_n * w'_n)_{n \geq 0}$.

As the intersection of two rt. det. context-free languages can be context-sensitive but not context-free, the (resp. det.) k-context-freeness of sequences is not preserved by synchronization product. We have to restrict one sequence to be automatic and use the fact that the synchronization product of a (resp. det.) regular automaton with a finite automaton remains a (resp. det.) regular automaton [Ca 07].

Proposition 12. *The families* $\mathrm{Cf}_k(A^\omega)$ *and* $\mathrm{DetCf}_k(A^\omega)$ *are closed under synchronization product with any k-automatic sequence.*

In particular, the (resp. det.) k-context-freeness of a sequence is preserved by modifying finitely many letters.

The *(left) shift* of a sequence $w = (w_n)_{n\geq 0}$ is the sequence $\mathrm{S}(w) = (w_n)_{n>0}$ obtained from w by removing its first letter. The *right shift* of a sequence w by a letter a is the sequence aw, with $(aw)_0 = a$ and $(aw)_n = w_{n-1}$ for $n > 0$. The shift operations preserve k-automaticity.

Proposition 13. *The family* $\mathrm{Cf}_k(A^\omega)$ *is closed under left and right shifts.*

The family $\mathrm{DetCf}_k(A^\omega)$ *is not closed under left and right shifts.*

Proof. We check this proposition for the left shift operation S. Let $w \in \mathrm{Cf}_k(A^\omega)$. For each $a \in A$, $\Lambda^k_w(a)$ is a context-free language.

The transformation $R : (n)_k \longrightarrow (n-1)_k$ is realized by the following finite transducer:

$$\{\, p \overset{1/\varepsilon}{\longrightarrow} r \,\} \cup \{\, p \overset{i/i}{\longrightarrow} q \mid 0 \le i < k \,\} \;\cup\; \{\, q \overset{i/i}{\longrightarrow} q \mid 0 \le i < k \,\}$$
$$\cup\; \{\, q \overset{i/i-1}{\longrightarrow} r \mid 0 < i < k \,\} \;\cup\; \{ r \overset{0/k-1}{\longrightarrow} r \}$$

of input state p and of final state r. Hence, the language $\Lambda^k_{\mathrm{S}(w)}(a) = R(\Lambda^k_w(a))$ is context-free for each $a \in A$, that is $\mathrm{S}(w) \in \mathrm{Cf}_k(A^\omega)$.

Note that this transducer also preserves unambiguous context-freeness.

Furthermore let us consider the following deterministic regular 2-automaton G:

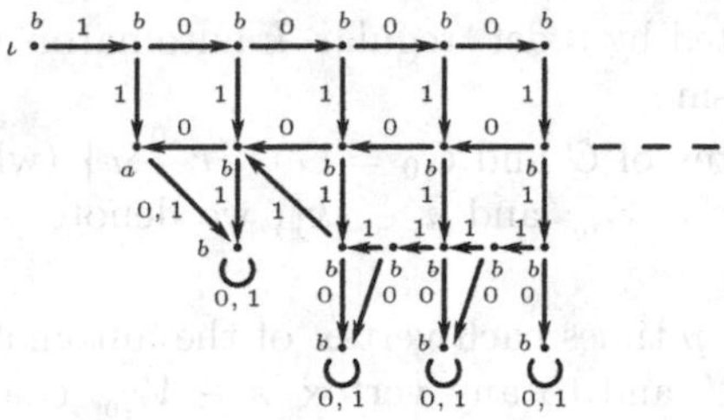

So $w = \mathrm{Seq}(G) \in \mathrm{RtDetCf}_2(\{a,b\}^\omega)$ and the language

$$\Lambda^2_w(a) \;=\; \{11\} \cup \{\, 10^{m+n+1}10^n1^{2m}0 \mid m,n \ge 0 \,\}$$

is det. context-free. For the shifted sequence $S(w)$, we have:

$$\Lambda^2_{\mathrm{S}(w)}(a) \;=\; \{10\} \cup \{\, 10^{n+2}1^{n+1} \mid n \ge 0 \,\}$$
$$\cup\; \{\, 10^{m+n+1}10^n1^{2m-1}01 \mid m > 0, n \ge 0 \,\}$$

which is not a deterministic context-free language since the language

$\Lambda^2_{\mathrm{S}(w)}(a) \cap (10^*1^+ + 10^*1^+01) \;=\; \{\, 10^{n+1}1^n \mid n > 0 \,\} \cup \{\, 10^{n+1}1^{2n}01 \mid n > 0 \,\}$

is not deterministic context-free [Yu 89]. So $\mathrm{S}(w)$ is not in $\mathrm{DetCf}_k(A^\omega)$. □

4.3 Morphisms and Inverse Substitutions

A p-uniform morphism over an alphabet A is a function $\sigma : A \longrightarrow A^p$, extended to A^* and A^ω by concatenation of images.

Proposition 14. *The families* $\mathrm{Cf}_k(A^\omega)$ *and* $\mathrm{DetCf}_k(A^\omega)$ *are closed under any finite p-uniform morphism with* $p > 0$.

Proof. Let $w \in \mathrm{Cf}_k(A^\omega)$ and $\sigma : A \longrightarrow A^p$. Let us show that $\sigma(w) \in \mathrm{Cf}_k(A^\omega)$. For each $a \in A$, $\Lambda^k_w(a)$ is a context-free language and we have to check that $\Lambda^k_{\sigma(w)}$ remains a context-free colouring. By denoting $\widetilde{u}$ the mirror of a (finite) word u, the relation $R = \{\, (u,v) \in [\![k]\!]^* \times [\![k]\!]^* \mid [\widetilde{v}]_k = p \times [\widetilde{u}]_k \,\}$ is recognized by the following finite transducer [Be 72] :

$$\{\, r \overset{q/t}{\longrightarrow} s \mid r,s \in [\![p]\!],\ q,t \in [\![k]\!],\ pq + r = ks + t \,\}$$
$$\cup\; \{\, r \overset{\varepsilon/[r]_k}{\longrightarrow} p \mid r \in [\![p]\!] \,\} \;\cup\; \{ p \overset{\varepsilon/0}{\longrightarrow} p \}$$

of input state 0 and of final state p. So for any context-free language $L \subseteq [\![k]\!]^*$, $p \times L = \{ (p \times [u]_k)_k \mid u \in L \} = \widetilde{R(\widetilde{L})} \cap [\![k]\!]^\diamond$ is a context-free language.

By applying $r \geq 0$ right shifts, $p \times L + r = \{ (p \times [u]_k + r)_k \mid u \in L \}$ remains context-free (see proof of Proposition 13). Also, note that this transformation $L \mapsto p \times L + r$ preserves unambiguous context-freeness.

To conclude the proof for $\mathrm{Cf}_k(A^\omega)$, it remains to see that for any $a \in A$,

$$\Lambda^k_{\sigma(w)}(a) = \bigcup_{b \in A\,,\,\sigma(b)_r = a} p \times \Lambda^k_w(b) + r.$$

For the family $\mathrm{DetCf}_k(A^\omega)$, we cannot use the same argument as the family of det. context-free languages is not stable by mirror image. However, this transformation of context-free sequences is done by inverse of a finite path function on associated det. regular automata.

Let $w \in A^\omega$ generated by a det. regular k-automaton G. Let $\sigma : A \longrightarrow A^p$ be a p-uniform morphism.

Let r the initial vertex of G and $G_0 = G \cup \{r \xrightarrow{0} r\}$ (which is also a regular automaton). For any $s \in V_{G_0}$ and $i \in [\![k]\!]$, we denote si the i-th *successor* of s: $s \xrightarrow{i}_{G_0} si$.

We start by copying p times each vertex of the automaton G_0: we take new symbols $0', \ldots, (p-1)'$ and to any vertex $s \in V_{G_0}$ coloured by $a \in A$, we associate new vertices $s_{|0'}, s_{|1'}, \ldots, s_{|(p-1)'}$ respectively colored by letters $\sigma(a)_0$, $\sigma(a)_1, \ldots, \sigma(a)_{p-1}$ and linked to s by edges $s \xrightarrow{i'} s_{|i'}$ in order to complete G_0 into the following automaton:

$$G' = G_0 \cup \{ s \xrightarrow{i'} s_{|i'} \mid s \in V_{G_0} \wedge i \in [\![p]\!] \} \cup \{ \sigma(a)_i\, s_{|i'} \mid a\, s \in G_0 \wedge i \in [\![p]\!] \}.$$

So G' remains a deterministic regular automaton.

By linking the vertices $s_{|i'}$ for $s \in V_G$ and $i \in [\![p]\!]$ with a finite path function h, we will construct a new k-automaton $h^{-1}(G')$ generating $\sigma(w)$.

Let r be the input vertex of G.

In $h^{-1}(G')$, for any $s \in V_G$, the kp successors of $s_{|0'}, \ldots, s_{|(p-1)'}$ are

$$s0_{|0'}, \ldots, s0_{|(p-1)'}, s1_{|0'}, \ldots, s1_{|(p-1)'}, \ldots, s(k-1)_{|0'}, \ldots, s(k-1)_{|(p-1)'},$$

where the k first are successors of $s_{|0'}$ ordered by increasing index of edges $i \in [\![k]\!]$, the following k are successors of $s_{|1'}$ ordered by increasing index of edges $i \in [\![k]\!]$ and so on.

Formally, we can define the finite path function h as follows. We denote

$$[i,j] = ki + j \text{ for any } i \in [\![p]\!] \text{ and } j \in [\![k]\!]$$

and the Euclidian division of any integer $n \geq 0$ by p is denoted by

$$n = pq_n + r_n \text{ with } 0 \leq r_n < p.$$

We define h by

for any $a \in A$, h(a)=a and $h(\iota) = (0')^{-1}\iota 0'$,

for any $j \in [\![k]\!]$, $j \neq 0$, $h(j) = \bigvee_{i \in [\![p]\!]} (i')^{-1} \cdot q_{[i,j]} \cdot (r_{[i,j]})'$.

and $h(0) = (0')^{-1} \cdot (\neg\iota)0 \cdot (0')' \vee \bigvee_{i \in [\![p]\!], i \neq 0} (i')^{-1} \cdot q_{[i,0]} \cdot (r_{[i,0]})'$.

For instance for $p = 3$ and $k = 2$, the table of $q_{[i,j]}, r_{[i,j]}$ for $i \in [\![3]\!]$ and $j \in [\![2]\!]$ is given by:

$i \setminus j$	0	1
0	0, 0	0, 1
1	0, 2	1, 0
2	1, 1	1, 2

hence

$$h(0) \equiv (0')^{-1} \cdot (\neg\iota) \cdot 0 \cdot 0' \vee (1')^{-1} \cdot 0 \cdot 2' \vee (2')^{-1} \cdot 1 \cdot 1'$$
$$h(1) \equiv (0')^{-1} \cdot 0 \cdot 1' \quad\quad \vee (1')^{-1} \cdot 1 \cdot 0' \vee (2')^{-1} \cdot 1 \cdot 2'$$

By Propositions 11, $h^{-1}(G')$ is a det. regular k-automaton. The det. regular k-automaton $h^{-1}(G')$ generates $\sigma(w)$. □

Let us apply p-uniform morphism by inverse on a sequence: starting from position 0 each sequence of length p is replaced by a letter. A *substitution* σ on a set V is a mapping $V \longrightarrow 2^{V^*}$ and it is a p-uniform substitution for $p \geq 0$ if $\sigma(s) \subseteq V^p$ for every $s \in V$. A p-uniform substitution σ is *total* if $\bigcup_{s \in V} \sigma(s) = A^p$. A substitution σ is *injective* if $\sigma(s) \cap \sigma(t) = \emptyset$ for any $s, t \in V$ with $s \neq t$. We apply by inverse an injective and total p-uniform substitution σ on any sequence $w \in A^\omega$ to get the sequence $\sigma^{-1}(w)$ defined by:

$$\sigma^{-1}(w) = \left(\sigma^{-1}(w_{np} \ldots w_{(n+1)p-1})\right)_{n \geq 0}.$$

Proposition 15. *The family* $\mathrm{DetCf}_k(A^\omega)$ *is closed under inverse of any injective and total* k^p*-uniform substitution with* $p \geq 0$.
The family $\mathrm{Cf}_k(A^\omega)$ *is not closed under inverse of injective total* k*-uniform substitutions.*

Proof. Let $w = \mathrm{Seq}(G)$ for some det. k-automaton G.
Let $p \geq 0$ and σ be an injective and total k^p-uniform substitution.
Let $G' = G \cup \{r \xrightarrow{0} r\}$ for $\iota r \in G$. So G' remains deterministic and regular.
Let h be the finite path function on G' which renames colours as follows: the initial vertex does not change $h(\iota) = \iota$ and for any $a \in A$, a vertex s is relabelled by a if there exists a word u in $\sigma(a)$ such that for any $v \in [\![k]\!]^p$, the v-path from s ends to a vertex coloured by the $[v]_k$-th letter of u, that is, formally: $h(a) = \bigvee_{u \in \sigma(a)} \bigwedge_{v \in [\![k]\!]^p} (\varepsilon \wedge v\, u_{[v]_k}\, v^{-1})$ where $(v)^{-1}$ stands for $v_p^{-1} \cdots v_2^{-1} v_1^{-1}$
Finally, the 0-loop on the inital vertex is removed and labels on edges are unchanged: $h(0) = (\neg\iota)0$ and $h(i) = i$ for any $0 < i < k$.

Since h is a regular path function, $h^{-1}(G')$ is a det. regular k-automaton and we have $\mathrm{Seq}(h^{-1}(G')) = \sigma^{-1}(w)$.

Let us check the non-closure of $\mathrm{Cf}_k(A^\omega)$ under the inverse of an injective k-substitution.

We consider the following real-time deterministic context-free languages:

$$L = 1^+ \cdot \{\, 0^n 1^n \mid n > 0 \,\} \quad \text{and} \quad M = \{\, 1^n 0^n \mid n > 0 \,\} \cdot 1^+$$

The language $L \cdot 0 \cup M \cdot 1$ is context-free but not deterministic [Yu 89].

Let $a, b \in A$. We define the 2-context-free sequence $w \in \{a,b\}^\omega$ by

$$\Lambda_w^k(a) = L{\cdot}0 \ \cup\ M{\cdot}1 \quad \text{and } \Lambda_w^k(b) = [\![k]\!]^\diamond - \Lambda_w^k(a).$$

The letter a appears in w in positions $[1^m0^n1^n0]_k$ or $[1^n0^n1^m]_k$ for $m, n > 0$.
Moreover, the pattern aa appears in w in positions $[1^n0^n1^n0]_k$ for $n > 0$.
Let σ be the k-uniform substitution

$$\sigma(a) \ = \ aab^{k-2} \quad \text{and} \quad \sigma(b) \ = \ A^k - \sigma(a)$$

which is total and injective. As $\Lambda_{\sigma^{-1}(w)}^k(a) \ = \ L \cap M \ = \ \{\, 1^n0^n1^n \mid n > 0 \,\}$ is not context-free, $\sigma^{-1}(w)$ is not a k-context-free sequence. □

A consequence of this last proposition is that the converse of the implication of Proposition 8 is false:

Proposition 16. *There exists a sequence w such that $\Lambda_w^k(a)$ is a rt. det. context-free language for all a, and w is not a deterministic k-context-free sequence.*

Proof. We use the same languages L, M as in the proof of Proposition 15. Let $A = \{a, \overline{a}, b, \overline{b}, c\}$ be a five letters alphabet. We define $w \in A^\omega$ by
$\Lambda_w^k(a) = L{\cdot}0$; $\Lambda_w^k(\overline{a}) = ([\![k]\!]^\diamond - (L \cup \{\varepsilon\})){\cdot}0$;
$\Lambda_w^k(b) = M{\cdot}1$; $\Lambda_w^k(\overline{b}) = ([\![k]\!]^\diamond - M){\cdot}1$; $\Lambda_w^k(c) \ = \ [\![k]\!]^\diamond\{2, \ldots, k-1\} \cup \{\varepsilon\}$.
For any $a \in A$, the language $\Lambda_w^k(a)$ is a rt. det. context-free language and the pattern ab only appears in w in positions $[1^n0^n1^n0]_k$ for $n > 0$.
Let σ be the k-substitution defined by $\sigma(a) \ = \ abc^{k-2}$ and $\sigma(b) \ = \ A^k - \sigma(a)$.
As $\Lambda_{\sigma^{-1}(w)}^k(a) \ = \ L \cap M$, the sequence $\sigma^{-1}(w)$ is not k-context-free.
By Proposition 15, w cannot be a deterministic k-context-free sequence. □

4.4 1-Context-Free Sequences and Extractions of Ultimately Periodic Subsequences

The k-automatic sequences are characterized by the finiteness of their k-kernels. The *k-kernel* $K_k(w)$ of a sequence $w \in A^\omega$ is the set of subsequences $k_u(w)$ for $u \in [\![k]\!]^*$ obtained by only picking up letters in positions of type $[vu]_k$, for $v \in [\![k]\!]^\diamond$, that is, $k_u(w) \ = \ \left(w_{k^{|u|}n+[u]_k}\right)_{n \geq 0}$.

Proposition 17. *The k-kernel of any (resp. det.) k-context-free sequence only contains (resp. det.) k-context-free sequence.*

Proof. Let $i \in [\![k]\!]$ and $w \in \mathrm{Cf}_k(A^\omega)$. As we have $k_\varepsilon(w) = w$ and $k_{uv}(w) = k_v(k_u(w))$ for any $u, v \in [\![k]\!]^*$, we just need to check that $k_i(w) \in \mathrm{Cf}_k(A^\omega)$.
As $k_i(w) \ = \ (w_{kn+i})_{n \geq 0}$, the language $\Lambda_{k_i(w)}^k(a) \ = \ \Lambda_w^k(a)i^{-1}$ is the right residual by i of $\Lambda_w^k(a)$, for any $a \in A$, thus context-free. It follows that $k_i(w)$ is a k-context-free sequence.
Now assume that $w \in \mathrm{DetCf}_k(A^\omega)$. We define the injective and total k-substitution h by $h(a) \ = \ A^{i-1}aA^{k-i}$ for any $a \in A$. Hence by Proposition 15, $k_i(w) = h^{-1}(w)$ so $k_i(w) \in \mathrm{DetCf}_k(A^\omega)$. □

To generate the ultimately periodic sequences, we extend the definition of a k-automaton to the case $k = 1$.

Definition 18. *A* 1-automaton *over A is an accessible automaton G on the set $[\![1]\!] = \{0\}$ of edge labels and on the set A of colours such that Λ_G is a partition of $\{0\}^*$.*
A 1-automaton G generates the sequence $\mathrm{Seq}(G) \in A^\omega$ such that for any $n \geq 0$, its n-th letter $Seq(G)_n$ is the colour a such that $0^n \in \Lambda_G(a)$.

The 1-*colouring* Λ^1_w of any $w \in A^\omega$ is the mapping associating with any $a \in A$ the language $\Lambda^1_w(a) = \{ 0^n \mid w_n = a \}$ of the representations in base 1 of the positions of a in w.

Definition 19. *A sequence w over A is 1-automatic (resp. 1-context-free) if $\Lambda^1_w(a)$ is a regular (resp. context-free) for all a in A.*

Lemma 20. *For any $w \in A^\omega$, the following four statements are equivalent:*
- a) *w is a ultimately periodic sequence,*
- b) *w is a 1-automatic sequence,*
- c) *w is a 1-context-free sequence,*
- d) *$\Lambda^1_w(a)$ is a regular language for any $a \in A$.*

Proof. The equivalences a) $\Longleftrightarrow$ b) $\Longleftrightarrow$ d) are well known (see [AS 03] for example) and b) $\Longrightarrow$ c) because any finite automaton is regular.
To show c) $\Longrightarrow$ d), Let $w = \mathrm{Seq}(G)$ for some regular 1-automaton G. For any $a \in A$, $\Lambda^1_w(a) = \Lambda_G(a)$ is a context-free language over $\{0\}$. By Parikh's lemma, every context-free language over a unique letter is regular. □

We can extract ultimately periodic sequences in any k-context-free sequence by picking k-regularly letters.

Proposition 21. *Let w be a k-context-free sequence and $u, v_1, \ldots, v_p \in [\![k]\!]^*$ with $p \geq 1$, $uv_1 \in [\![k]\!]^\diamond$ and $v_1, \ldots, v_p \neq \varepsilon$. For any $q \geq 0$ and $i \in \{1, \ldots, p\}$, we denote $v^{q+\frac{i}{p}} = (v_1 \ldots v_p)^q v_1 \ldots v_i$.*
The sequence $\left(w_{[uv^{\frac{n}{p}}]_k} \right)_{n \geq 0}$ is ultimately periodic.

Proof. We have $w = \mathrm{Seq}(G)$ for some regular k-automaton G.
Let h be the finite path function on G defined as follows:

$h(\iota) = u^{-1}\,\iota\, u \quad ; \quad h(a) = a$ for any $a \in A \quad ; \quad h(i) = v_i$ for any $1 \leq i \leq p$.

So $h^{-1}(G)$ is a prefix-recognizable automaton (got from the complete binary tree by inverse regular path functions).
We take the following finite deterministic automaton:

$$H = \{ i \xrightarrow{i} i{+}1 \mid 1 \leq i < p \} \cup \{ p \xrightarrow{p} 1 \} \cup \{ \iota 1 \} \cup \{ a i \mid a \in A \wedge 1 \leq i \leq p \}$$

The following synchronisation product of $h^{-1}(G)$ and H

$$\begin{aligned} h^{-1}(G) \times H &= \{ (s,p) \xrightarrow{0} (t,q) \mid \exists\, i\ (s \xrightarrow{i}_{h^{-1}(G)} t \wedge p \xrightarrow{i}_H q) \} \\ &\cup \{ a(s,p) \mid a\,s \in h^{-1}(G) \wedge a\,p \in H \} \end{aligned}$$

is a deterministic prefix-recognizable automaton, hence by [CK 01] is a regular automaton.

Let K be the restriction by accessibility from ι of $h^{-1}(G) \times H$.

So K is a regular 1-automaton generating $\mathrm{Seq}(K) = \left(w_{[uv^{\frac{n}{p}}]_k} \right)_{n \geq 0}$.

By Lemma 20, $\mathrm{Seq}(K)$ is an ultimately periodic sequence. □

4.5 About Base Dependence

A famous theorem of Cobham states that sequences which are automatic in two multiplicatively independent bases are the ultimately periodic ones [Co 69]. This section presents two results in the direction of a possible extension of this statement for context-free sequences. First, the context-freeness is preserved for any non-null power of the base (but not the deterministic context-freeness). Second, we have the same set of context-free sequences only for multiplicatively dependent bases.

Proposition 22. *For every $k, p > 1$, we have*

$$\mathrm{Cf}_k(A^\omega) = \mathrm{Cf}_{k^p}(A^\omega) \quad \textit{and} \quad \mathrm{DetCf}_k(A^\omega) \subsetneq \mathrm{DetCf}_{k^p}(A^\omega).$$

Proof. Let $h : [\![k^p]\!] \longrightarrow [\![k]\!]^p$ be the bijective mapping associating with any $n \in [\![k^p]\!]$ its $(n+1)$-th word of $[\![k]\!]^p$ by (length) lexicographic order:

$$h(n) = 0^{p - \lceil \log_k (n+1) \rceil} (n)_k \quad \text{for every } n \in [\![k^p]\!]$$

which is extended by morphism on $[\![k^p]\!]^*$. For any $w \in A^\omega$ and $a \in A$, we have

$$\Lambda^k_w(a) = (0^*)^{-1} h(\Lambda^{k^p}_w(a)) \cap [\![k]\!]^\diamond \quad \text{and} \quad \Lambda^{k^p}_w(a) = h^{-1}(0^* \Lambda^k_w(a)) \cap [\![k^p]\!]^\diamond$$

We deduce that $\mathrm{Cf}_k(A^\omega) = \mathrm{Cf}_{k^p}(A^\omega)$.

Let $w = \mathrm{Seq}(G)$ for some regular deterministic k-automaton G. We complete G by adding a 0-loop to its input vertex: $G' = G \cup \{ r \xrightarrow{0} r \}$ for $\iota r \in G$. The mapping h is extended to a finite path function by adding its behaviours on colours $h(\iota) = \iota$ and $h(a) = a$ for any $a \in A$. Let $H = h^{-1}(G') - \{ r \xrightarrow{0} r \}$. We have $\mathrm{Seq}(H) = \mathrm{Seq}(G)$ and by Proposition 11, H is a regular deterministic k^p-automaton. Thus $\mathrm{DetCf}_k(A^\omega) \subseteq \mathrm{DetCf}_{k^p}(A^\omega)$.

Let us check that this inclusion is strict. As $k, p > 1$, we have $k^p \geq 4$ and

$$h(0) = 0^p \quad ; \quad h(1) = 0^{p-1}1 \quad ; \quad h(k) = 0^{p-2}10 \quad ; \quad h(k^{p-1}) = 10^{p-1}$$

Let $a, b \in A$. We define the sequence $w \in \{a,b\}^\omega$ by

$$\begin{aligned} \Lambda^{k^p}_w(a) &= 1\{ 0^{n+1}(k^{p-1})^{2n} \mid n \geq 0 \} \cup k\{ 0^n 1^{n+1} \mid n \geq 0 \}, \\ \Lambda^{k^p}_w(b) &= [\![k^p]\!]^\diamond - \Lambda^{k^p}_w(a). \end{aligned}$$

So $w \in \mathrm{DetCf}_{k^p}(A^\omega)$ and

$$\begin{aligned} \Lambda^k_w(a) &= 1\{ (0^p)^{n+1}(10^{p-1})^{2n} \mid n \geq 0 \} \cup 10\{ (0^p)^n (0^{p-1}1)^{n+1} \mid n \geq 0 \} \\ &= \{ 10^{pn+p}(10^{p-1})^{2n} \mid n \geq 0 \} \cup \{ 10^{pn+p}(10^{p-1})^n 1 \mid n \geq 0 \} \end{aligned}$$

which is not a deterministic context-free language, so $w \notin \mathrm{DetCf}_k(A^\omega)$. □

Proposition 23. *For any* $p, q > 1$,
$$\mathrm{Cf}_p(A^\omega) = \mathrm{Cf}_q(A^\omega) \iff \exists\, i, j > 0,\ p^i = q^j.$$

Proof. $\Longleftarrow$: This implication is straightforward from Proposition 22.
$\Longrightarrow$: We just transpose the construction given in [Be 72] for the context-free integer sets.
Let $a, b \in A$. We define the sequence $w \in \{a,b\}^\omega$ by
$$\Lambda^p_w(a) = 10^* \quad \text{and} \quad \Lambda^p_w(b) = [\![p]\!]^\diamond - 10^*.$$
So $w \in \mathrm{Cf}_p(A^\omega) = \mathrm{Cf}_q(A^\omega)$ thus $\Lambda^q_w(a)$ is an infinite context-free language. As a corollary of the pumping lemma on context-free languages, there exists u, v, x, y, z in $[\![q]\!]^*$ such that $u \neq \varepsilon$ and for every $n \geq 0$, $xu^n yv^n z \in \Lambda^q_w(a)$. The integer mapping f defined for every $n \geq 0$ by $[xu^n yv^n z]_q = p^{f(n)}$ is increasing.
Note that for any $s, t \in [\![q]\!]^*$, $[st]_q = [s]_q q^{|t|} + [t]_q$, hence for any $n \geq 0$,
$$\begin{aligned}[s^n t]_q &= [s]_q q^{|s^{n-1}t|} + \ldots + [s]_q q^{|st|} + [s]_q q^{|t|} + [t]_q \\ &= [s]_q q^{|t|} \left(1 + q^{|s|} + [t]_q + \ldots + (q^{|s|})^{n-1}\right) \\ &= [s]_q q^{|t|} \tfrac{q^{n|s|}-1}{q-1} + [t]_q\end{aligned}$$
Thus for any $n \geq 0$, $p^{f(n)} = [xu^n yv^n z]_q = A\, q^{n|uv|} + B\, q^{n|v|} + C$ with
$$A = q^{|yz|}\left([x]_q + \tfrac{[u]_q}{q-1}\right) \;;\; B = q^{|z|}\left([y]_q + \tfrac{[v]_q - q^{|y|}[u]_q}{q-1}\right) \;;\; C = [z]_q - \tfrac{q^{|z|}[v]_q}{q-1}$$
hence
$$\frac{p^{f(n+1)}}{p^{f(n)}} \sim \frac{A\, q^{(n+1)|uv|}}{A\, q^{n|uv|}} = q^{|uv|} \quad i.e., \quad \lim_{n\to\infty} p^{f(n+1)-f(n)} = q^{|uv|}.$$
This last equality on integers implies that there exists n_0 such that
$$p^{f(n_0+1)-f(n_0)} = q^{|uv|},$$
meaning that p and q are multiplicatively dependent. □

5 Conclusion and Open Problems

Let us mention again that the results of Section 4 remain valid when substituting UnaCf_k for Cf_k, and $\mathrm{RtDetCf}_k$ for DetCf_k.

The difference of behaviours of these families of context-free sequences under transformations and the difference of involved tools (from languages or from gaph theory) also allows to deeper understand from where come the strong robustness of automatic sequences, for which concepts of unambiguity of languages and determinism of automata are totally erased.

Some properties of k-context-free sequences have to be further studied, for instance the structure of their k-kernels, properties of symbolic dynamical systems associated with these sequences, their degenerated cases (how to decide whenever a k-context-free sequence is k-automatic, periodic, etc.). Moreover, results of Section 4.5 are encouraging for a possible extension of the Cobham's theorem on base dependence.

We expect that the most of presented closure properties extend to similar constructions of sequences using indexed languages [Ah 68] and higher order indexed languages following Maslov's hierarchy of languages [Ma 74]. As regular and context-free languages are the first two level of this hierarchy, this paper is a second step towards a theory of the infinite hierarchy of higher order indexed automatic sequences following Maslov's hierarchy, the automatic sequences and context-free automatic sequences being the first two levels.

References

[Ah 68] Aho, A.: Indexed grammars - an extension of context-free grammars. Journal of the Association for Computing Machinery 15(4), 647–671 (1968)

[AU 79] Aho, A., Ullman, J.: Introduction to automata theory, languages and computation, p. 418. Addison-Wesley (1979), doi:A. Aho

[AS 03] Allouche, J.-P., Shallit, J.: Automatic sequences: theory, applications, generalizations. Cambridge University Press (2003)

[Be 72] Berstel, J.: On sets of numbers recognized by push-down automata. In: 13th Symposium on Switching and Automata Theory (SWAT - FOCS), pp. 200–206. IEEE (1972)

[Ca 90] Caucal, D.: On the regular structure of prefix rewriting. In: Arnold, A. (ed.) HCC11 2014. LNCS, pp. 87–102 (1990); Theoretical Computer Science 106, 61–86 (1992)

[Ca 07] Caucal, D.: Deterministic graph grammars. In: Flum, J., Grädel, E., Wilke, T. (eds.) Texts in Logic and Games 2, pp. 169–250. Amsterdam University Press (2007)

[CK 01] Caucal, D., Knapik, T.: On internal presentation of regular graphs by prefix-recognizable graphs. Theory of Computing Systems 34(4), 299–336 (2001)

[Ch 56] Chomsky, N.: Three models for the description of language. IRE Transactions on Information Theory 2, 113–124 (1956)

[Co 72] Cobham, A.: Uniform-tag sequences. Mathematical Systems Theory 6, 164–192 (1972)

[Co 69] Cobham, A.: On the base-dependence of sets of numbers recognizable by finite automata. Mathematical Systems Theory 3, 186–192 (1969)

[Ha 98] Hamm, D.: Contributions to formal language theory: Fixed points, complexity, and context-free sequences, M.Sc. Thesis, University of Waterloo (1998)

[Kl 56] Kleene, S.: Representation of events in nerve nets and finite automata. In: Automata studies. Annals of Mathematics Studies, pp. 3–41. Princeton University Press (1956)

[LeG 08] Le Gonidec, M.: Drunken man infinite words complexity. RAIRO - Theoretical Informatics and Applications 42, 599–613 (2008)

[LeG 12] Le Gonidec, M.: On the complexity of a family of k-context-free sequences. Theoretical Computer Science 44(1), 47–54 (2012)

[Lo 05] Lothaire, M.: Applied Combinatorics on Words. Cambridge University Press (2005)

[Ma 74] Maslov, A.: The hierarchy of indexed languages of an arbitrary level. Doklady Akademii Nauk SSSR 217, 1013–1016 (1974)

[Mo 08] Moshe, Y.: On some questions regarding k-regular and k-context-free sequences. Theoretical Computer Science 400, 62–69 (2008)

[MS 85] Muller, D., Schupp, P.: The theory of ends, pushdown automata, and second-order logic. Theoretical Computer Science 37, 51–75 (1985)

[Yu 89] Yu, S.: A pumping lemma for deterministic context-free languages. Information Processing Letters 31(1), 47–51 (1989)

Modular Reasoning for Message-Passing Programs*

Jinjiang Lei and Zongyan Qiu

LMAM and Department of Informatics, School of Math, Peking University, P.R. China
{jinjiang.lei,qzy}@math.pku.edu.cn

Abstract. Verification of concurrent systems is difficult because of the inherent nondeterminism. Modern verification requires better locality and modularity. Reasoning of shared memory systems has gained much progress in these aspects. However, modular verification of distributed systems is still in demand. In this paper, we propose a new reasoning system for message-passing programs. It is a novel logic that supports Hoare style triples to specify and verify distributed programs modularly. We concretize the concept of event traces to represent interactions among distributed agents, and specify behaviors of agents by their local traces with regard to environmental assumptions. Based on trace semantics, the verification is compositional in both temporal and spatial dimensions. As an example, we show how to modularly verify an implementation of merging network.

1 Introduction

With the blossom of multi-core processors, concurrency has become a crucial element in software systems. In general, concurrency can be roughly categorized into shared memory model and message-passing model. They are both notoriously difficult to be verified because of non-deterministic interleaves of memory accesses or message passing.

Verification of shared memory models has gained great progress since the emergence of Separation Logic (SL) [16], e.g., Concurrent Separation Logic (CSL) [2], and lots of other separation-based reasoning [17,18]. On the other hand, although message-passing programs have been extensively studied using various process calculi [5,11,13], fewer Hoare type reasoning systems are developed, especially modular reasoning systems. In this paper, based on a classic graphical semantics (see Lamport [8]) of message-passing models, we propose a novel modular reasoning system for distributed programs.

Lamport introduced *event graphs* (or *event traces*) as a representation for the semantics of message-passing programs. Event graphs are essentially Directed Acyclic Graphs (DAGs) composed by nodes and directed arrows, where nodes represent atomic actions (e.g., send/receive events), and directed arrows represent inter-agent communications. Each event graph is associated with a partial order — *happens-before* $\prec$ — among nodes, which is defined as the transitive closure of agent local order and directed arrows, to reveal the *causality relation* among events. This will be formally defined later.

In terms of formal verification, for any semantics based on event graphs, the crux is to modularly specify the structures of graphs. As a graph represents a collection of

* Supported by NNSF of China, Grant No. 61272160, 61100061, and 61202069.

G. Ciobanu and D. Méry (Eds.): ICTAC 2014, LNCS 8687, pp. 277–294, 2014.

ordered and correlated events (nodes), the modularity could only be achieved when we can carve out irrelevant events in reasoning a local behavior. To make this possible, we view the structure of an event graph from two dimensions: the spatial dimension and the temporal dimension. We introduce a *separating conjunction* operator, $*$, to depict the spatial dimension by specifying separated traces; and an operator, $\circ$, to represent *sequential conjunction*, which defines the temporal dimension based on happens-before relation, and takes a stronger condition than the spatial one.

Our logic adopts the Hoare style triples to specify message-passing programs, and makes local reasoning of message-passing programs a reality. Generally, the semantics of a set of agents D is specified by a triple as follows:

$$\{r, p\}\, D\, \{r', q\}$$

where r and r' specify D's expectations (or assumptions) about its *environment* (the behaviors of other agents), and p and q specify the local states (changes) of D. The reasoning of D relies on its environmental assumptions, that is, the local behaviors of D, which are specified by p and q, are correlated with its environmental expectations. Local agents are able to calculate and strengthen its environmental assumption in r' in order to fulfill certain local function that specified by q. Other agents should satisfy the expectation of D in order to parallel composite with D.

Now we give a tiny example to show the reasoning in our system and how the spatial and temporal modularity is achieved. We prove a simple program as follows:

```
send (2, pt);  ||  x := recv (pt);
send (3, pt);  ||  y := recv (pt);
```

The left agent sends messages 2 and 3 to the port *pt* sequentially, and the right agent is the owner of *pt*, who withdraws the messages and stores them into local variables.

The proof given in Fig. 1 is modular, because the system is proved agent by agent, and an agent is separately proved command by command.

Lines 1 – 5 are the proof for the first receive command: it starts from {emp, emp}, where no assumption for the environment is made (the first emp) and no local action is taken (the second emp); then in line 2, we assume the environment sends message X to *pt* ($pt!X$), where X is an implicitly existential-quantified logical variable and its scope is confined within the environmental part; after executing the receive command, there is a receive event in the local state ($pt?Y \wedge x = Y$), where Y is existential-quantified to represent the message received and its scope is the local state; line 4 is deduced from line 3 by strengthening the environmental assumption ($pt!2 \Rightarrow \exists X \cdot pt!X$); and line 5 comes because a receive should match with its sender ($pt!2 * pt?Y \Rightarrow Y = 2$)[1].

The other receive is proved separately (lines 6 – 7) which is similar as lines 1 – 5. Lines 8 – 9 are obtained from lines 6 – 7 by adding a "frame" — $pt!2 * pt?2$ — ahead of current event graph, where $pt!2$ is added ahead of the environment, and $pt?2$ is ahead of the local state. Note that the frame does not affect existing proofs and program state. The "ahead of" relation is formally called "happens-before", which is served by the

[1] This form of writing is just for easy understanding, precisely it should be $pt?Y \overset{pt!2}{\Longrightarrow} pt?2$, which will be formally discussed in Section 5.

1. $\{\mathsf{emp}, \mathsf{emp}\}$
2. $\{pt!X, \mathsf{emp}\}$
   `x := recv (pt);`
3. $\{pt!X, pt?Y \wedge x = Y\}$
4. $\{pt!2, pt?Y \wedge x = Y\}$
5. $\{pt!2, pt?2 \wedge x = 2\}$

6. $\{\mathsf{emp}, \mathsf{emp}\}$
   `y := recv (pt);`
7. $\{pt!3, pt?3 \wedge y = 3\}$

8. $\{pt!2, pt?2\}$
   `y := recv (pt);`
9. $\left\{\begin{array}{l} pt!2 \circ pt!3, \\ pt?2 \circ pt?3 \wedge y = 3 \end{array}\right\}$

10. $\{\mathsf{emp}, \mathsf{emp}\}$
    `x := recv (pt);`
    `y := recv (pt);`
11. $\left\{\begin{array}{l} pt!2 \circ pt!3, \\ pt?2 \circ pt?3 \wedge x = 2 \wedge y = 3 \end{array}\right\}$

12. $\{\mathsf{emp}, \mathsf{emp}\}$
13. $\{\mathsf{emp}, \mathsf{emp}\}$ `send (2, pt); send (3, pt);` || $\{\mathsf{emp}, \mathsf{emp}\}$ `x := recv (pt); y := recv (pt);`
14. $\{\mathsf{emp}, pt!2 \circ pt!3\}$ {see line 11}
15. $\left\{\begin{array}{l} \mathsf{emp}, \\ x = 2 \wedge y = 3 \wedge (pt!2 \circ pt!3) * (pt?2 \circ pt?3) \end{array}\right\}$

Fig. 1. Modular Proof of a Tiny Example

operator "$\circ$". We can add another frame "$x = 2$" to the local state in line 8 and 9 in order to ensure line 8 is the same as line 5. This step is trivial and thus omitted.

Having the proofs for the two receives, the whole specification of the first agent is obtained by sequentially combining the two proofs (lines 10 – 11).

The specification of the sending agent shown in lines 13 – 14 is trivial that need no explanation. The composition for the two agents is shown in lines 12 – 15. In line 15, the environmental assumption of the receiver (line 11) is satisfied by the local state of the sender (line 14), therefore, the environmental part of line 15 is emp. The local part of line 15 is just the composition of local state of both agents.

In summary, the logic we developed makes the following contributions:

- It concretizes the concept of *event graph* [8] to represent the interactions among agents, and proposes a set of trace predicates to specify the properties of traces;
- It supports two-dimensional modularity: temporal modularity, as the separated proofs of the receives; and spatial modularity, as the separated proofs of the agents;
- Each agent can be proved locally with explicitly calculated assumptions about its environment, and proofs of separate agents can be combined as long as their local behaviors could mutually satisfy the environmental assumptions of other agents.

In the rest of the paper, we give a formal definition of event trace, and present a trace algebra for separating and sequential conjunctions in Section 2; and a formal presentation for the model and operational semantics in Section 3. The assertion language for specifying trace structures and the reasoning logic are presented in Section 4 and 5. Section 6 gives a case study, and Section 7 discusses the related work and concludes.

2 Event Trace

A distributed system is composed by a set of *agents*, each of which represents a computational process that can own several *ports* for receiving messages. Each port belongs to one agent, while an agent can own multiple ports. We adopt asynchronous message passing: send commands will not be blocked, while receives will be blocked if there is no message in the designated ports. We assume the state of a port is a queue, and messages transmitted following the FIFO-principle [3].

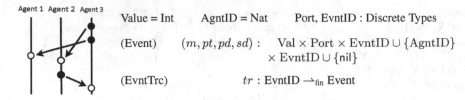

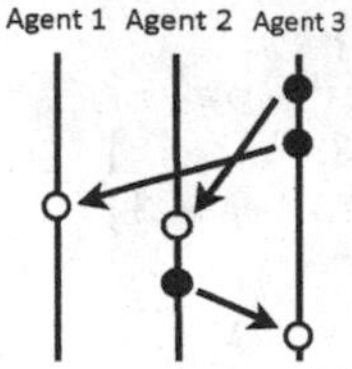

Value = Int AgntID = Nat Port, EvntID : Discrete Types

$$\text{(Event)} \quad (m, pt, pd, sd) : \quad \text{Val} \times \text{Port} \times \text{EvntID} \cup \{\text{AgntID}\} \times \text{EvntID} \cup \{\mathsf{nil}\}$$

$$\text{(EvntTrc)} \quad tr : \text{EvntID} \rightharpoonup_{\text{fin}} \text{Event}$$

Fig. 2. Trace State

2.1 Traces

We use *event graphs* [8] to depict the semantics of distributed programs. The left part of Fig. 2 shows the trace of a program execution, where solid nodes represent send events and hollow nodes are receives. The picture also reveals that event traces are time-space graphs, where space is measured by agents, and time is measured by agent local orders and inter-agent arrows.

The state of traces is defined formally in Fig. 2 (right part). Each event is a quad referred by a unique reference, $tr(e) = (m, pt, pd, sd)$, where: m (referred by $e.\mathsf{val}$) and pt (referred by $e.\mathsf{port}$) are the value and port for the message respectively; pd (referred by $e.\mathsf{pred}$) is e's local direct predecessor; and sd (referred by $e.\mathsf{send}$) refers to e's corresponding send event when e is a receive. If e is the first event of an agent, $e.\mathsf{pred}$ is the agent ID where e lives in; and $e.\mathsf{send} = \mathsf{nil}$ if e is a send event. We use $\mathsf{isSend}(e)$ and $\mathsf{isRecv}(e)$ to specify the type of e:

$$\mathsf{isSend}(e) \stackrel{\text{def}}{=} e.\mathsf{send} = \mathsf{nil} \qquad\qquad \mathsf{isRecv}(e) \stackrel{\text{def}}{=} e.\mathsf{send} \neq \mathsf{nil}$$

We recursively define a function $\mathsf{agent}(e)$ to return the agent ID of the event referred by e:

$$\mathsf{agent}(e) \stackrel{\text{def}}{=} \begin{cases} e.\mathsf{pred} & e.\mathsf{pred} \in \text{AgntID} \\ \mathsf{agent}(e.\mathsf{pred}) & \text{otherwise} \end{cases}$$

2.2 Well-Formed Traces

Event traces are specific structures to record the communicating history among agents. In this section we present an axiomatic definition of traces. As Lamport postulated, events are partially ordered by *happens-before* relation $\prec$.

Definition 1 (Happens-Before). *The* happens-before *relation for* tr, $\prec$, *is defined as:*

$$e \prec e' \stackrel{\text{def}}{=} e = e'.\mathsf{pred} \vee e = e'.\mathsf{send} \vee (\exists e'' \cdot e \prec e'' \wedge (e'' = e'.\mathsf{pred} \vee e'' = e'.\mathsf{send}))$$
$$\textit{where } \{e, e'\} \subseteq \mathsf{dom}(tr)$$

Based on happens-before relation, we define six axioms to specify *well-formed traces*. Axioms 1 – 4 are general axioms which are proposed originally in [1]; axioms 5 – 6 are specifically proposed for our model. We use tr to denote an event trace.

Axiom 1. *tr is self-closed:*

$$\forall e \in \mathsf{dom}(tr) \cdot e.\mathsf{pred} \notin \text{AgntID} \Rightarrow e.\mathsf{pred} \in \mathsf{dom}(tr), \textit{ and}$$
$$\forall e \in \mathsf{dom}(tr) \cdot e.\mathsf{send} \neq \mathsf{nil} \Rightarrow e.\mathsf{send} \in \mathsf{dom}(tr).$$

Axiom 2. *Relation $\prec$ is strongly well founded. There exists $f : \mathsf{dom}(tr) \rightarrow \text{Nat}$ that:*

$$\forall e, e' \in \mathsf{dom}(tr) \cdot e \prec e' \Rightarrow f(e) < f(e').$$

Axiom 3. *Maps $\bullet.\mathsf{pred}$ and $\bullet.\mathsf{send}$ are injective:*

$$\forall e, e' \in \mathsf{dom}(tr) \cdot e.\mathsf{pred} = e'.\mathsf{pred} \Rightarrow e = e', \textit{ and}$$
$$\forall e, e' \in \mathsf{dom}(tr) \cdot e.\mathsf{send} = e'.\mathsf{send} \wedge e.\mathsf{send} \neq \mathsf{nil} \Rightarrow e = e'.$$

Axiom 4. *The* send *field of a receive event refers to its corresponding send event:*

$$\forall e \in \mathsf{dom}(tr) \cdot \mathsf{isRecv}(e) \Rightarrow \exists e' \cdot e.\mathsf{send} = e' \wedge \mathsf{isSend}(e') \wedge e.\mathsf{val} = e'.\mathsf{val} \wedge e.\mathsf{port} = e'.\mathsf{port}.$$

Axiom 5. *Communications are robust that there is no lost message. Let e_1 and e_2 be two send events:*

$$e_1 \prec e_2 \wedge e_1.\mathsf{port} = e_2.\mathsf{port} \wedge \exists e_2' \cdot e_2'.\mathsf{send} = e_2 \Rightarrow \exists e_1' \cdot e_1'.\mathsf{send} = e_1.$$

That is, if e_2 is received, all send events that happen before e_2 on the same channel must have been received.

Axiom 6. *Messages are sent and received by the FIFO principle. Let e_1 and e_2 be two send events:*

$$e_1 \prec e_2 \wedge e_1.\mathsf{port} = e_2.\mathsf{port} \wedge e_1'.\mathsf{send} = e_1 \wedge e_2'.\mathsf{send} = e_2 \Rightarrow \neg(e_2' \prec e_1').$$

We use $\mathcal{A}_1, \ldots, \mathcal{A}_6$ to represent the above axioms, and $\mathcal{A}$ to denote their conjunction: $\mathcal{A} \stackrel{\text{def}}{=} \mathcal{A}_1 \wedge \ldots \wedge \mathcal{A}_6$.

Theorem 1. *Let Prop be the type of propositions over tr, then for all P : EvntID $\rightarrow$ Prop:*

$$(\forall e' \cdot (\forall e \cdot e \prec e' \wedge P(e) \rightarrow P(e'))) \Rightarrow \forall e \cdot P(e)$$

Theorem 1 is the induction over event traces: take any event e' in the trace, if all events that happen before e' satisfy P leads the truth of P at e', then P holds all over the trace.

We use $f \uplus g$ as the union of f and g but require f and g have disjointed domains.

Definition 2 (Well-formed Trace). *Trace tr is well-formed, $WF(tr)$, iff there exist tr' and tr'' such that:*

$$tr'' = tr \uplus tr' \wedge tr'' \models \mathcal{A}$$

That is, any well-formed trace, tr, is a sub-trace of some "complete" trace tr'' such that tr'' entails $\mathcal{A}$.

$$tr_1 = tr_2 \Rightarrow tr_1 * tr_3 = tr_2 * tr_3 \qquad tr_1 * tr_2 = tr_2 * tr_1$$
$$tr_1 * tr_2 = tr_1 * tr_3 \Rightarrow tr_2 = tr_3 \qquad tr = tr_1 \circ tr_2 \Rightarrow tr = tr_1 * tr_2$$
$$tr_1 \circ (tr_2 \circ tr_3) = (tr_1 \circ tr_2) \circ tr_3 \qquad tr = tr_1 \circ (tr_2 * tr_3) \Rightarrow tr = (tr_1 \circ tr_2) * tr_3$$
$$tr_1 * (tr_2 * tr_3) = (tr_1 * tr_2) * tr_3 \qquad tr = (tr_1 * tr_2) \circ tr_3 \Rightarrow tr = (tr_1 \circ tr_3) * tr_2$$

Fig. 3. Selected Properties for Traces

2.3 Trace Separation and Algebra

To structurally specify event traces, we introduce two operators, separating conjunction $*$ and sequential conjunction $\circ$, where:

$tr_1 * tr_2$ is the union of all the events in tr_1 and tr_2 as long as tr_1 and tr_2 contain disjointed set of events.

$tr_1 \circ tr_2$ returns $tr_1 * tr_2$ if three additional conditions hold: (1) no event in tr_2 happens before any event in tr_1; (2) if e_1 ($\in tr_1$) and e_2 ($\in tr_2$) send messages to a same port, then e_1 happens before e_2; and (3) if e_1 ($\in tr_1$) and e_2 ($\in tr_2$) receive messages from the same port, then e_1 happens before e_2.

We give the formal definitions of $*$ and $\circ$ as follows:

$$tr * tr' \stackrel{\text{def}}{=} tr \uplus tr' \text{ iff } WF(tr \uplus tr')$$
$$tr \circ tr' \stackrel{\text{def}}{=} tr * tr' \text{ iff } \forall e \in \mathsf{dom}(tr), e' \in \mathsf{dom}(tr')\cdot$$
$$\neg(e' \prec e) \wedge (\mathsf{isSend}(e) \wedge \mathsf{isSend}(e') \wedge e.\mathsf{port} = e'.\mathsf{port} \Rightarrow e \prec e')$$
$$\wedge (\mathsf{isRecv}(e) \wedge \mathsf{isRecv}(e') \wedge e.\mathsf{port} = e'.\mathsf{port} \Rightarrow e \prec e')$$

In [19], Wehrman *et al.* defined another semantics for $tr \circ tr'$, which only requires events in tr' do not happen before events in tr. Our semantic definition of *sequential composition* is stronger. Take the trace $(pt!88 \circ pt!14) * (pt?x \circ pt?y)$ for instance, we can deduce $x = 88 \wedge y = 14$ with the additional conditions, otherwise the $x = 14 \wedge y = 88$ is permitted as well. Fig. 3 lists some selected properties for trace structures, which are sound based on the semantics.

3 Programming Language

In this section, we define a programming language for constructing the distributed programs (system models) and its operational semantics.

Fig. 4 gives the programming language. We use E and B to denote numerical and boolean expressions. Command **send** (E, pt) sends message E to port pt; and $x :=$ **recv** (pt) withdraws a message from pt and stores it into the local variable x. A distributed program is a parallel composition of agents C_i, where each agent is tagged with a unique agent ID ($i_1, \ldots, i_k$ in Fig. 4). For simplicity, we don't consider memory management in our model.

The program state is defined in Fig. 5. A state, $\sigma = (s, tr)$ is composed by a store and a trace, where s maps variable names to values, and tr is already defined in Fig. 2.

(Expr) $E ::= x \mid X \mid n \mid E + E \mid E - E \mid \ldots$
(BExp) $B ::= \textbf{true} \mid \textbf{false} \mid E = E \mid E \neq E \mid \ldots$
(Comd) $c ::= x := E \mid \textbf{skip} \mid$
$\textbf{send}\,(E, pt) \mid x := \textbf{recv}\,(pt)$
(Stmts) $C ::= c \mid C_1; C_2 \mid \textbf{while}\ B\ \textbf{do}\ C \mid$
$\textbf{if}\ B\ \textbf{then}\ C_1\ \textbf{else}\ C_2$
(Prog) $D ::= i_1 : C_1 || \ldots || i_k : C_k$

Fig. 4. The Language

Loc = Int Var : Discrete Type

(*Store*) s : Var $\rightharpoonup_{\text{fin}}$ Value
(*EvntTrc*) tr : EvntID $\rightharpoonup_{\text{fin}}$ Event
(*State*) $\sigma ::= (s, tr)$

Fig. 5. State Definition

We define the predicate $\mathsf{fstUnMchd}(e, tr, pt)$ to state that e is the first pending send event on port pt, that is, e is a send event on pt which has not matched with a receive yet, and no other unmatched send event on pt happens before e. Formally:

$$\begin{aligned}\mathsf{fstUnMchd}(e, tr, pt) \stackrel{\text{def}}{=}\ & \mathsf{isSend}(e) \wedge e.\mathsf{port} = pt \wedge \\ & \neg\exists e' \in \mathsf{dom}(tr) \cdot e'.\mathsf{send} = e \wedge \\ & \neg\exists e' \cdot \mathsf{isSend}(e') \wedge e'.\mathsf{port} = pt \wedge e' \prec e.\end{aligned}$$

Function $\mathsf{last}(tr, i)$ returns the last event of agent i in tr. If there is no event at agent i, then the function returns "i" since the $\bullet$.pred field of the first event is an agent ID.

$$\mathsf{last}(tr, i) \stackrel{\text{def}}{=} \begin{cases} i & \{e \mid e \in \mathsf{dom}(tr) \wedge \mathsf{agent}(e) = i\} = \varnothing \\ \max\limits_{\prec}\left\{ e \;\middle|\; \begin{array}{c} e \in \mathsf{dom}(tr) \\ \wedge\ \mathsf{agent}(e) = i \end{array} \right\} & \text{otherwise} \end{cases}$$

The operational semantics is defined by a set of rules which describe configuration transitions caused by program execution. These rules take the following form:

$$(D, s, tr) \rightsquigarrow (D', s', tr')$$

If there is only one agent i ($D = i : C$), the transition can take the form of:

$$(C, s, tr) \rightsquigarrow_i (C', s', tr')$$

Fig. 6 gives the operational semantics, where $\{i_1 : v_1; \ldots; i_n : v_n\}$ denotes a function f with $\mathsf{dom}(f) = \{i_1, \ldots, i_n\}$ and $f(i_j) = v_j$; $f\{x \nrightarrow v\}$ remaps x of f to v; $f \uplus g$ is function union when $\mathsf{dom}(f) \cap \mathsf{dom}(g) = \varnothing$; $[\![E]\!]_s$ and $[\![B]\!]_s$ evaluate the numerical and boolean expressions based on store s.

Semantics of local primitives, e.g., assignment, control flow commands are regular. Rules for send and receive primitives are added, which create new events in the trace. For receive, it should find the first unmatched send event on the port in tr according to $\mathsf{fstUnMchd}(e, tr, pt)$ and then create a corresponding receive event.

Theorem 2. *Let (D, σ) be the initial state, and σ_{tr} be the trace of σ, then the execution traces of any distributed program would entail $\mathcal{A}$:*

$$((D, \sigma) \rightsquigarrow^* (D', \sigma')) \wedge \sigma_{tr} \models \mathcal{A} \Rightarrow \sigma'_{tr} \models \mathcal{A}.$$

$$\frac{[\![E]\!]_s = n}{(x := E, s, tr) \rightsquigarrow_i (\textbf{skip}, s\{x \leftrightarrow n\}, tr)} \quad \frac{[\![E]\!]_s \text{ undefined}}{(x := E, s, tr) \rightsquigarrow_i \textbf{abort}}$$

$$\frac{\mathsf{fstUnMchd}(e, tr, pt) \quad e.\mathsf{val} = n \quad e' = \mathsf{last}(tr, i) \quad e'' \notin \mathsf{dom}(tr)}{(x := \textbf{recv}\,(pt), s, tr) \rightsquigarrow_i (\textbf{skip}, s\{x \leftrightarrow n\}, tr \uplus \{e'' : (n, pt, e', e)\})}$$

$$\frac{[\![E]\!]_s = n \quad e = \mathsf{last}(tr, i) \quad e' \notin \mathsf{dom}(tr)}{(\textbf{send}\,(E, pt), s, tr) \rightsquigarrow_i (\textbf{skip}, (s, tr \uplus \{e' : (n, pt, e, \mathsf{nil})\}))} \quad \frac{[\![E]\!]_s \text{ undefined}}{(\textbf{send}\,(E, pt), s, tr) \rightsquigarrow_i \textbf{abort}}$$

$$\frac{[\![B]\!]_{\sigma_s} = \textbf{true}}{(\textbf{if } B \textbf{ then } C_1 \textbf{ else } C_2, s, tr) \rightsquigarrow_i (C_1, s, tr)} \quad \frac{[\![B]\!]_s = \textbf{false}}{(\textbf{if } B \textbf{ then } C_1 \textbf{ else } C_2, s, tr) \rightsquigarrow_i (C_2, s, tr)}$$

$$\frac{[\![B]\!]_s = \textbf{true}}{(\textbf{while } B \textbf{ do } C, s, tr) \rightsquigarrow_i (C; \textbf{while } B \textbf{ do } C, s, tr)} \quad \frac{[\![B]\!]_s = \textbf{false}}{(\textbf{while } B \textbf{ do } C, s, tr) \rightsquigarrow_i (\textbf{skip}, s, tr)}$$

$$\frac{[\![B]\!]_s \text{ undefined}}{(\textbf{if } B \textbf{ then } C_1 \textbf{ else } C_2, s, tr) \rightsquigarrow_i \textbf{abort}} \quad \frac{[\![B]\!]_s \text{ undefined}}{(\textbf{while } B \textbf{ do } C, s, tr) \rightsquigarrow_i \textbf{abort}}$$

$$\frac{(C_1, \sigma) \rightsquigarrow_i (C_1', \sigma')}{(C_1; C_2, \sigma) \rightsquigarrow_i (C_1'; C_2, \sigma')} \quad \frac{(C_1, \sigma) \rightsquigarrow_i \textbf{abort}}{(C_1; C_2, \sigma') \rightsquigarrow_i \textbf{abort}} \quad \frac{}{(\textbf{skip}; C, \sigma) \rightsquigarrow_i (C, \sigma)}$$

$$\frac{(D_1, \sigma) \rightsquigarrow (D_1', \sigma')}{(D_1 \parallel D_2, \sigma) \rightsquigarrow (D_1' \parallel D_2, \sigma')} \quad \frac{(D_2, \sigma) \rightsquigarrow (D_2', \sigma')}{(D_1 \parallel D_2, \sigma) \rightsquigarrow (D_1 \parallel D_2', \sigma')}$$

$$\frac{(D_1, \sigma) \rightsquigarrow \textbf{abort} \text{ or } (D_2, \sigma) \rightsquigarrow \textbf{abort}}{(D_1 \parallel D_2, \sigma) \rightsquigarrow \textbf{abort}}$$

Fig. 6. Operational Semantics

4 Assertion Language

This section defines an assertion language for specifying traces. We assume an infinite set of logical variables $LVar = \{X, Y, \ldots\}$. The assertion language is a mixture of store predicates, and trace predicates with the following syntax:

$$\begin{array}{lll} p, q ::= & E = E \mid E > E \mid \ldots & \text{(store predicates)} \\ & \mid \mathsf{emp} \mid \mathsf{true} \mid pt!E \mid pt?E & \text{(trace predicates)} \\ & \mid \neg p \mid p \wedge q \mid \exists X \cdot p \mid p * q \mid p \circ q \mid \ldots & \text{(connectives)} \end{array}$$

The semantics of assertions is defined in Fig. 7, where emp and true specify empty trace and any well-formed trace respectively; $pt!E$ and $pt?E$ specify singleton events, where $pt!E$ represents sending message E to pt and $pt?E$ says receiving E from pt; boolean expression holds only if it is true over the state; $\sigma_1 \uplus \sigma_2$ is the conjunction of separated states, σ_1 and σ_2, which have separated traces; $\sigma_1 \odot \sigma_2$ is the conjunction of sequential states which have sequentially connected traces. $p * q$ says p and q holds over separated states; $p \circ q$ holds over sequential states.

Definition 3 (Pure Assertion). *Assertion p is* pure, $\mathsf{Pure}(p)$, iff *the validity of p does not rely on the state of trace, i.e.,*

$$\textit{if } (s, tr) \models p, \textit{ then for all } tr', \ (s, tr') \models p.$$

Syntactically, *pure assertions* do not contain trace predicates.

Fig. 8 lists some selected proof rules, which are sound based on the semantics.

$$
\begin{array}{lll}
(s,tr) \models \mathsf{emp} & \text{iff} & \mathsf{dom}(tr) = \varnothing \qquad (s,tr) \models \mathsf{true} \;\; \text{iff} \;\; WF(tr) \\
(s,tr) \models B & \text{iff} & [\![B]\!]_s = \mathbf{true} \\
(s,tr) \models pt!E & \text{iff} & \exists l,n,e \cdot [\![E]\!]_s = n \land \mathsf{dom}(tr) = \{e\} \land tr(e) = (n, pt, \_, \mathsf{nil}) \\
(s,tr) \models pt?E & \text{iff} & \exists l,n,e \cdot [\![E]\!]_s = n \land \mathsf{dom}(tr) = \{e\} \land tr(e) = (n, pt, \_, \neg\mathsf{nil})
\end{array}
$$

$$
(s,tr) \uplus (s',tr') \stackrel{\text{def}}{=} \begin{cases} (s, tr * tr') & \text{if } s = s' \land tr * tr' \text{ defined} \\ \text{undefined} & \text{otherwise} \end{cases}
$$

$$
(s,tr) \odot (s',tr') \stackrel{\text{def}}{=} \begin{cases} (s, tr \circ tr') & \text{if } s = s' \land tr \circ tr' \text{ defined} \\ \text{undefined} & \text{otherwise} \end{cases}
$$

$$
tr^* \stackrel{\text{def}}{=} \varnothing \cup tr \cup tr \circ tr \cup \ldots \qquad \sigma^* \stackrel{\text{def}}{=} (s, tr^*) \quad \text{where } \sigma = (s, tr)
$$

$$
\begin{array}{lll}
\sigma \models p_1 * p_2 & \text{iff} & \exists \sigma_1, \sigma_2 \cdot \sigma_1 \uplus \sigma_2 = \sigma \land \sigma_1 \models p_1 \land \sigma_2 \models p_2 \\
\sigma \models p_1 \circ p_2 & \text{iff} & \exists \sigma_1, \sigma_2 \cdot \sigma_1 \odot \sigma_2 = \sigma \land \sigma_1 \models p_1 \land \sigma_2 \models p_2 \\
\sigma \models p^* & \text{iff} & \exists \sigma' \cdot \sigma' \models p \land \sigma_s = \sigma'_s \land \sigma_{tr} = {\sigma'_{tr}}^* \\
\sigma \models \neg p & \text{iff} & \sigma \not\models p \qquad \sigma \models p \land q \;\; \text{iff} \;\; \sigma \models p \land \sigma \models q \\
\sigma \models p \Rightarrow q & \text{iff} & \text{if } \sigma \models p, \text{then } \sigma \models q \qquad \sigma \models \exists X \cdot p \;\; \text{iff} \;\; \exists n \in \mathsf{Val} \cdot \sigma \models p[n/X]
\end{array}
$$

Fig. 7. Semantics of Assertions

$$
\frac{}{p * q \Leftrightarrow q * p} \quad \frac{}{p \circ (q \circ r) \Leftrightarrow (p \circ q) \circ r} \quad \frac{}{(p * q) \circ r \Rightarrow (p \circ r) * q} \quad \frac{}{p \circ q \Rightarrow p * q}
$$

$$
\frac{}{p \circ (q * r) \Rightarrow (p \circ q) * r} \quad \frac{}{(r_1 * p_1) \circ (r_2 * p_2) \Rightarrow (r_1 \circ r_2) * (p_1 \circ p_2)}
$$

$$
\frac{\mathsf{Pure}(p) \text{ or } \mathsf{Pure}(q)}{p \circ q \Leftrightarrow p \land q} \quad \frac{\mathsf{Pure}(p) \text{ or } \mathsf{Pure}(q)}{p * q \Leftrightarrow p \land q}
$$

$$
\frac{\mathsf{Pure}(p)}{p \land (q \circ r) \Rightarrow (p \land q) \circ (p \land r)} \quad \frac{\mathsf{Pure}(r)}{p \circ (q \land r) \Leftrightarrow (p \circ q) \land r} \quad \frac{\mathsf{Pure}(q)}{(p \land q) \circ r \Leftrightarrow (p \circ r) \land q}
$$

Fig. 8. Selected Proof Rules

5 Inference System

In this section, we introduce our inference system, which is a separation-based system for reasoning distributed programs.

The inference rules are given in Fig. 9 and 10. In order to avoid tedious side conditions, Syntactic Control of Interference (SCI) [15] is adopted here. There are two syntactic context: $\mathcal{O}_{\mathsf{var}}$ for variable context, and $\mathcal{O}_{\mathsf{port}}$ for port context.

$$
\mathcal{O}_{\mathsf{var}} ::= x_1, x_2, x_3, \ldots \qquad \mathcal{O}_{\mathsf{port}} ::= pt_1, pt_2, pt_3, \ldots
$$

$\mathcal{O}_{\mathsf{var}}$ denotes the ownership of a set of variables, and $\mathcal{O}_{\mathsf{var}}, \mathcal{O}'_{\mathsf{var}}$ is the conjunctive ownership of two separated sets of variables. $\mathcal{O}_{\mathsf{port}}$ is a set of port names, which specifies the access permissions of ports. If $pt \in \mathcal{O}_{\mathsf{port}}$, the current agent can withdraw messages out of pt, and other agents can only send messages to pt. For simplicity, we consider full permissions. It is possible to extend this definition with fractional permissions [14] as well. SCI specifies the well-formedness of variables, expressions, assertions, and programs. The formal definitions are given in the extended technical report [9] .

$$\frac{x \in \mathcal{O}_{\mathsf{var}} \quad pt \in \mathcal{O}_{\mathsf{port}} \quad x \notin \mathsf{freeVar}(r) \cup \mathsf{freeVar}(p)}{\mathcal{O}_{\mathsf{var}}, X; \mathcal{O}_{\mathsf{port}} \vdash_i \{r, p\}\, x := \mathbf{recv}\,(pt)\, \{r, p \circ pt?X \wedge x = X\}} \text{ (RECV)}$$

$$\frac{x \in \mathcal{O}_{\mathsf{var}} \quad pt \notin \mathcal{O}_{\mathsf{port}}}{\mathcal{O}_{\mathsf{var}}, \mathcal{O}_{\mathsf{port}} \vdash_i \{r, p\}\, \mathbf{send}\,(x, pt)\, \{r, p \circ pt!x\}} \text{ (SEND)}$$

$$\frac{\mathcal{O}_{\mathsf{var}}; \mathcal{O}_{\mathsf{port}} \vdash_i \{r, p\}\, C_1\, \{r', p'\} \quad \mathcal{O}_{\mathsf{var}}; \mathcal{O}_{\mathsf{port}} \vdash_i \{r', p'\}\, C_2\, \{r'', p''\}}{\mathcal{O}_{\mathsf{var}}; \mathcal{O}_{\mathsf{port}} \vdash_i \{r, p\}\, C_1; C_2\, \{r'', p''\}} \text{ (SEQ)}$$

$$\frac{\mathcal{O}_{\mathsf{var}}; \mathcal{O}_{\mathsf{port}} \vdash_i \{r, (p \wedge B)\}\, C_1\, \{r', q\} \quad \mathcal{O}_{\mathsf{var}}; \mathcal{O}_{\mathsf{port}} \vdash_i \{r, (p \wedge \neg B)\}\, C_2\, \{r', q\}}{\mathcal{O}_{\mathsf{var}}; \mathcal{O}_{\mathsf{port}} \vdash_i \{r, p\}\, \mathbf{if}\ B\ \mathbf{then}\ C_1\ \mathbf{else}\ C_2\, \{r', q\}} \text{ (IF)}$$

$$\frac{\mathcal{O}_{\mathsf{var}}; \mathcal{O}_{\mathsf{port}} \vdash_i \{r^*, p^* \wedge B\}\, C\, \{r^*, p^*\}}{\mathcal{O}_{\mathsf{var}}; \mathcal{O}_{\mathsf{port}} \vdash_i \{r^*, p^*\}\, \mathbf{while}\ B\ \mathbf{do}\ C\, \{r^*, p^* \wedge \neg B)\}} \text{ (WHILE)}$$

Fig. 9. Selected Inference Rules — Basics

The specification of a message-passing program takes the form as:

$$\mathcal{O}_{\mathsf{var}}; \mathcal{O}_{\mathsf{port}} \vdash \{r, p\}\, D\, \{r', q\}$$

It specifies the partial correctness of D: if D starts from a pre-state satisfying $r * p$, where r for the environmental state and p for the local state, then D will not abort, and when D terminates, if the environmental state satisfies r', then the local state satisfies q. If D contains only one agent, e.g., $i : C$, the specification can be written as:

$$\mathcal{O}_{\mathsf{var}}; \mathcal{O}_{\mathsf{port}} \vdash_i \{r, p\}\, C\, \{r', q\}$$

It is innovative that we syntactically separate the pre- and post-conditions into two parts: one assumption for the environment, and one specification for the local state. In the precondition, r is the environmental assumption before the execution, and p specifies the local pre-state. During execution, D may receive (send) messages from (to) the environment, so we can calculate the post-assumption of environment from D's local requirements. When D terminates, its environmental assumption becomes r' and local state becomes q. Clearly, the trace specified by r' and q will not be shorter than r and p.

Note that we always consider well-formed triples in this paper. Take $\mathcal{O}_{\mathsf{var}}; \mathcal{O}_{\mathsf{port}} \vdash \{r, p\}\, D\, \{r', q\}$ for instance, the environmental trace does not receive messages from $\mathcal{O}_{\mathsf{port}}$, which is the set of ports owned by D, and the local trace does not send messages to any port in $\mathcal{O}_{\mathsf{port}}$:

$$r \vee r' \Rightarrow (\mathsf{true} * pt?\_ \Rightarrow pt \notin \mathcal{O}_{\mathsf{port}}) \quad \text{and} \quad p \vee q \Rightarrow (\mathsf{true} * pt!\_ \Rightarrow pt \notin \mathcal{O}_{\mathsf{port}})$$

In Fig. 9, the rule (RECV) for receiving commands is straightforward. In this rule, variable X is a fresh logical variable to represent the message received by the current agent. No environmental assumption should be made at this stage. The rule for send event is similar. Note that no agent can send messages to itself ($pt \notin \mathcal{O}_{\mathsf{port}}$). Both (SEQ) for sequential composition and (IF) for conditional are trivial. (WHILE) is normal too, where each iteration should maintain the validity of loop invariant $\{r^*, p^*\}$.

$$\frac{p \Rightarrow p' \quad q' \Rightarrow q \quad \mathcal{O}_{\mathsf{var}}; \mathcal{O}_{\mathsf{port}} \vdash \{r, p'\}\, D\, \{r', q'\}}{\mathcal{O}_{\mathsf{var}}; \mathcal{O}_{\mathsf{port}} \vdash \{r, p\}\, D\, \{r', q\}} \text{ (CONSEQ-A)}$$

$$\frac{r_1 \Rightarrow r \quad r_2 \Rightarrow r' \quad \mathcal{O}_{\mathsf{var}}; \mathcal{O}_{\mathsf{port}} \vdash \{r, p\}\, D\, \{r', q\}}{\mathcal{O}_{\mathsf{var}}; \mathcal{O}_{\mathsf{port}} \vdash \{r_1, p\}\, D\, \{r_2, q\}} \text{ (CONSEQ-B)}$$

$$\frac{\mathcal{O}_{\mathsf{var}}; \mathcal{O}_{\mathsf{port}} \vdash \{r, p\}\, D\, \{r_1, q_1\} \; \mathcal{O}_{\mathsf{var}}; \mathcal{O}_{\mathsf{port}} \vdash \{r, p\}\, D\, \{r_2, q_2\}}{\mathcal{O}_{\mathsf{var}}; \mathcal{O}_{\mathsf{port}} \vdash \{r, p\}\, D\, \{r_1 \vee r_2, q_1 \vee q_2\}} \text{ (DISJ)}$$

$$\frac{\mathcal{O}_{\mathsf{var}}; \mathcal{O}_{\mathsf{port}} \vdash \{r, p\}\, D\, \{r_1, q_1\} \; \mathcal{O}_{\mathsf{var}}; \mathcal{O}_{\mathsf{port}} \vdash \{r, p\}\, D\, \{r_2, q_2\}}{\mathcal{O}_{\mathsf{var}}; \mathcal{O}_{\mathsf{port}} \vdash \{r, p\}\, D\, \{r_1 \wedge r_2, q_1 \wedge q_2\}} \text{ (CONJ)}$$

$$\frac{\mathcal{O}_{\mathsf{var}}; \mathcal{O}_{\mathsf{port}} \vdash \{r, p\}\, D\, \{r', q\} \quad \mathcal{O}'_{\mathsf{var}} \vdash r'' \textbf{ Assert} \quad \mathsf{notInterfere}(r'', \mathcal{O}_{\mathsf{port}})}{\mathcal{O}_{\mathsf{var}}, \mathcal{O}'_{\mathsf{var}}; \mathcal{O}_{\mathsf{port}} \vdash \{r * r'', p\}\, D\, \{r' * r'', q\}} \text{ (FRM-ENV)}$$

$$\frac{\mathcal{O}_{\mathsf{var}}; \mathcal{O}_{\mathsf{port}} \vdash \{r, p\}\, D\, \{r', q\} \quad \mathcal{O}'_{\mathsf{var}} \vdash p' \textbf{ Assert} \quad \mathsf{notInterfere}(p', \mathcal{O}_{\mathsf{port}})}{\mathcal{O}_{\mathsf{var}}, \mathcal{O}'_{\mathsf{var}}; \mathcal{O}_{\mathsf{port}} \vdash \{r, p * p'\}\, D\, \{r', q * p'\}} \text{ (FRM-LOC)}$$

$$\frac{\mathcal{O}_{\mathsf{var}}; \mathcal{O}_{\mathsf{port}} \vdash \{r, p\}\, D\, \{r', q\} \quad \mathcal{O}'_{\mathsf{var}} \vdash r'' \textbf{ Assert} \quad r' \bowtie_{\mathcal{O}_{\mathsf{port}}} q}{\mathcal{O}_{\mathsf{var}}, \mathcal{O}'_{\mathsf{var}}; \mathcal{O}_{\mathsf{port}} \vdash \{r, p\}\, D\, \{r' \circ r'', q\}} \text{ (FRM-BHD)}$$

$$\frac{\mathcal{O}_{\mathsf{var}}; \mathcal{O}_{\mathsf{port}} \vdash \{r, p\}\, D\, \{r', q\} \quad \mathcal{O}'_{\mathsf{var}} \vdash r'', p' \textbf{ Assert} \quad r'' \bowtie_{\mathcal{O}_{\mathsf{port}}} p'}{\mathcal{O}_{\mathsf{var}}, \mathcal{O}'_{\mathsf{var}}; \mathcal{O}_{\mathsf{port}} \vdash \{r'' \circ r, p' \circ p\}\, D\, \{r'' \circ r', p' \circ q\}} \text{ (FRM-AHD)}$$

$$\frac{\begin{array}{l}\mathcal{O}_{\mathsf{var}1}; \mathcal{O}_{\mathsf{port}_1} \vdash \{\mathsf{emp}, p_1\}\, D_1\, \{r_1, q_1\} \quad q_1 * r \Rightarrow r_2 * r'_2 \quad \mathsf{notInterfere}(r'_2, \mathcal{O}_{\mathsf{port}_2})\\ \mathcal{O}_{\mathsf{var}2}; \mathcal{O}_{\mathsf{port}_2} \vdash \{\mathsf{emp}, p_2\}\, D_2\, \{r_2, q_2\} \quad q_2 * r \Rightarrow r_1 * r'_1 \quad \mathsf{notInterfere}(r'_1, \mathcal{O}_{\mathsf{port}_1})\end{array}}{\mathcal{O}_{\mathsf{var}1}, \mathcal{O}_{\mathsf{var}2}; \mathcal{O}_{\mathsf{port}_1}, \mathcal{O}_{\mathsf{port}_2} \vdash \{\mathsf{emp}, p_1 * p_2\}\, D_1 \| D_2\, \{r, q_1 * q_2\}} \text{ (PAR)}$$

Fig. 10. Selected Inference Rules — Others

Definition 4 (Hooked Assertions). *Assertion p is* hooked with *q by $\mathcal{O}_{\mathsf{port}}$, $p \bowtie_{\mathcal{O}_{\mathsf{port}}} q$, iff for any trace tr such that $tr \models p * q$, any event $e \in \mathsf{dom}(tr)$, and any port $pt \in \mathcal{O}_{\mathsf{port}}$, the following conditions hold:*

$$\mathsf{isSend}(e) \wedge e.\mathsf{port} = pt \Rightarrow \exists e' \in \mathsf{dom}(tr) \cdot e'.\mathsf{sender} = e, \text{ and}$$
$$\mathsf{isRecv}(e) \wedge e.\mathsf{port} = pt \Rightarrow \exists e' \in \mathsf{dom}(tr) \cdot e.\mathsf{sender} = e'.$$

Intuitively, $p \bowtie_{\mathcal{O}_{\mathsf{port}}} q$ says that for any trace tr which satisfies $p * q$, there is no pending send or receive event that accesses ports within $\mathcal{O}_{\mathsf{port}}$.

Example 1. Let $\mathcal{O}_{\mathsf{port}} = \{pt\}$, $posiSend = (\exists X \cdot pt!X \wedge X > 0)^*$, $posiRecv = (\exists X \cdot pt?X \wedge X > 0)^*$, we will have

- $posiSend \bowtie_{\mathcal{O}_{\mathsf{port}}} posiRecv$ does not hold, and
- $(posiSend \circ pt!0) \bowtie_{\mathcal{O}_{\mathsf{port}}} (posiRecv \circ pt?0)$ holds.

Here $posiSend$ is a sequence of events sending positive numbers, and $posiRecv$ is a sequence of events receiving positive numbers. These two assertions are not hooked, because there may exist pending events in $posiSend$. However, the second pair above is hooked, since each sequence is appended with a sentinel 0 at the end, which enforces each send to be paired with a receive and vice versa. □

There are some rules for hooked assertions, which are useful for program reasoning.

$$\frac{p_1 \ltimes\!\!\!\rtimes_{\mathcal{O}_{\mathsf{port}}} q_1 \quad p_2 \ltimes\!\!\!\rtimes_{\mathcal{O}_{\mathsf{port}}} q_2}{p_1 \circ p_2 \ltimes\!\!\!\rtimes_{\mathcal{O}_{\mathsf{port}}} q_1 \circ q_2} \qquad \frac{p_1 \circ p_2 \ltimes\!\!\!\rtimes_{\mathcal{O}_{\mathsf{port}}} q_1 \circ q_2 \quad p_1 \ltimes\!\!\!\rtimes_{\mathcal{O}_{\mathsf{port}}} q_1}{p_2 \ltimes\!\!\!\rtimes_{\mathcal{O}_{\mathsf{port}}} q_2}$$

$$\frac{p_1 \ltimes\!\!\!\rtimes_{\mathcal{O}_{\mathsf{port}}} q_1 \quad p_2 \ltimes\!\!\!\rtimes_{\mathcal{O}_{\mathsf{port}}} q_2}{p_1 * p_2 \ltimes\!\!\!\rtimes_{\mathcal{O}_{\mathsf{port}}} q_1 * q_2} \qquad \frac{p_1 * p_2 \ltimes\!\!\!\rtimes_{\mathcal{O}_{\mathsf{port}}} q_1 * q_2 \quad p_1 \ltimes\!\!\!\rtimes_{\mathcal{O}_{\mathsf{port}}} q_1}{p_2 \ltimes\!\!\!\rtimes_{\mathcal{O}_{\mathsf{port}}} q_2}$$

Definition 5 (Environmental-Aided Implication). *Environmental-aided implication, written as* $p \overset{r}{\Rightarrow} p'$, *implies* p' *from* p *with an extra coupled trace* r. *It is true when:*

$$\forall \sigma_1 = (s, tr_1), \sigma_2 = (s, tr_2) \cdot \begin{array}{c} \sigma_1 * \sigma_2 \models r * p \wedge \sigma_1 \models r \wedge \sigma_2 \models p \\ \wedge \forall e \in tr_2 \cdot \mathsf{isRecv}(e) \Rightarrow e.\mathsf{send} \in \mathsf{dom}(tr_1) \end{array} \Rightarrow \sigma_2 \models p'$$

Environmental-aided implication enables local deduction with the extra knowledge about environmental state. In this definition, we require all receives in the local state are matched with some sends in the environment. For instance, we have $pt?X \overset{pt!2}{\Longrightarrow} X = 2$, while $pt?X * pt!2 \Rightarrow X = 2$ is not true since $*$ does not enforce send-receive matches in the trace.

In Fig. 10, there are two rules of consequences. (CONSEQ-A) is for local deduction. To weaken a specification, we can either strengthen its local precondition, or weaken its local postcondition. (CONSEQ-B) is for the environmental state. Different from (CONSEQ-A), it only allows strengthening the assumption of environment, either at precondition or postcondition. These two rules are usually applied together with rule (RECV). (RECV) makes no assumption about the message received. Using (CONSEQ-B), the current agent can make assumption for the received value by strengthening the predicate in the assumption part. Then, by using (CONSEQ-A), we can deduce the local state with the aid of a stronger environmental assumption.

Definition 6 (Non-Interference). *For an assertion* r, *we say that* r *does not interfere with* $\mathcal{O}_{\mathsf{port}}$, *written as* $\mathsf{notInterfere}(r, \mathcal{O}_{\mathsf{port}})$, *when:*

$$r \Rightarrow (\mathsf{true} * (pt!\_ \vee pt?\_) \Rightarrow pt \notin \mathcal{O}_{\mathsf{port}})$$

Predicate $\mathsf{notInterfere}(r, \mathcal{O}_{\mathsf{port}})$ says that the trace specified by r does not interfere with $\mathcal{O}_{\mathsf{port}}$, that is, it does not send or receive messages via any port in $\mathcal{O}_{\mathsf{port}}$.

Our system supports spatial modularity, because it allows the proof of a local agent to be extended with a frame by $*$, as long as the frame does not interfere with existing proofs. This is described by rules (FRM-ENV) and (FRM-LOC) in Fig. 10. Rule (FRM-ENV) is the frame rule for the environment. It allows the environment to be extended with frame r'', as long as r'' does not interfere with D, i.e., r'' must not race with r' by sending messages to D; and not race with q by receiving messages form r'. Rule (FRM-LOC) is the frame rule for local state. If p' does not contain any message passing predicates, this rule is reduced to the standard frame rule in SL. If p' contains some message passing events, it must not interfere the existing communication between r' and q. Note that $\mathcal{O}'_{\mathsf{var}} \vdash r''$ **Assert** and $\mathcal{O}'_{\mathsf{var}} \vdash p'$ **Assert** require all free variables of r'' and p' are within $\mathcal{O}'_{\mathsf{var}}$.

On the other hand, we support temporal modularity as well, that is, we allow a trace to be connected ahead or behind of current trace. Temporal modularity is supported by rules (FRM-BHD) and (FRM-AHD). Rule (FRM-BHD) allows appending an extra r'' to the end of environmental assumption. To ensure soundness, r' should be hooked with q' so that the extra trace r'' in the environment should not affect the behavior of existing trace q. Rule (FRM-AHD) allows appending the frame $r'' * p'$ ahead of the current trace, where r'' is added to the environment, and p' is added to the local trace. The two assertions must be hooked together so that they do not affect the communication of later traces. This rule can be applied when proving sequential programs.

Informally, in order to prove $C_1; C_2$, we can prove C_1 and C_2 independently in the first to get, e.g., $\{r_1, p_1\}\, C_1\, \{r_1', p_1'\}$ and $\{\mathsf{emp}, \mathsf{emp}\}\, C_2\, \{r_2', p_2'\}$. By applying (FRM-AHD), C_2 satisfies $\{r_1', p_1'\}\, C_2\, \{r_1' \circ r_2', p_1' \circ p_2'\}$. Therefore by rule (SEQ), the conjunct program can be specified by $\{r_1, p_1\}\, C_1; C_2\, \{r_1' \circ r_2', p_1' \circ p_2'\}$.

Rule (PAR) is for parallel composition of separated agents. For $D_1 || D_2$, the local trace of D_1 becomes the environment of D_2, and vice versa. In the rule, r is the environment of D_1 and D_2. Therefore, D_1's environment is $r * \mathsf{traceOf}(D_2)$, and D_2's environment is $r * \mathsf{traceOf}(D_1)$. $q_1 * r \Rightarrow r_2 * r_2'$ ensures D_2's environment is satisfied; and $q_2 * r \Rightarrow r_1 * r_1'$ ensures D_1's environment is satisfied.

Due to page limit, the semantics of all these rules and their soundness proofs are given in technical report [9] .

6 Example: Filters

Filters form a common class of distributed systems. A filter is an agent that receives messages from one or more ports and send messages to some other ports. In this section, we prove a filter example — Merging Network.

Fig. 11 (upper part) shows the architecture of a merging network. Each agent in the network is a filter that merges two monotonic positive streams into one monotonic stream, and 0 marks the end of streams. Fig. 11 (lower part) shows an implementation of agent 5. Here each port takes a unique ID, and $k@i$ denotes port k of agent i.

Agent 5 owns two variables and two ports, and is sequentially composed by three while loops. As the proof of the trivial example in Section 1, we prove these loops separately and then sequentially compose these independent proofs together.

For clarity, we define the following predicates to simplify descriptions:

$$\begin{aligned}
\mathsf{mono}(pt) &\stackrel{\text{def}}{=} (\mathsf{true} * pt!X) \circ (\mathsf{true} * pt!Y) \Rightarrow 0 < X \leq Y \\
\mathsf{monoEnd}(pt) &\stackrel{\text{def}}{=} \mathsf{mono}(pt) \circ pt!0 \\
\mathsf{large}(var) &\stackrel{\text{def}}{=} \forall n \cdot (\mathsf{true} * 2@6!n \Rightarrow var = 0 \vee var \geq n) \\
\mathsf{eqLast}(pt, var) &\stackrel{\text{def}}{=} \mathsf{true} \circ pt?X \Rightarrow var = X
\end{aligned}$$

$\mathsf{mono}(pt)$ says the environment send positive monotonic messages to pt, and $\mathsf{monoEnd}(pt)$ additionally says the stream has ended; $\mathsf{large}(var)$ says var is either larger than any message that previously sent to 2@6 or equal to 0; $\mathsf{eqlast}(pt, var)$ says var equals to the last message that received from pt.

Fig. 12 lists the proof for the first loop. Line 1 is the loop invariant. It assumes that the environment send monotonic streams to 1@5 and 2@5, and its local trace is a composition of some receive events of that two ports and a set of monotonic sends to

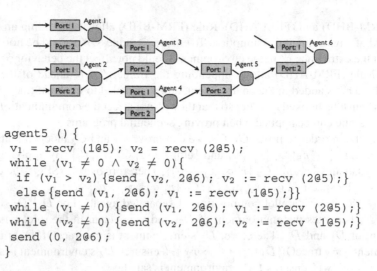

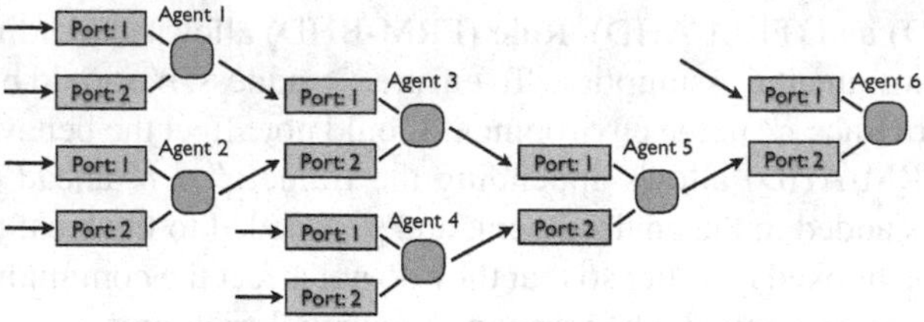


```
agent5 () {
  v1 = recv (1@5); v2 = recv (2@5);
  while (v1 ≠ 0 ∧ v2 ≠ 0){
   if (v1 > v2){send (v2, 2@6); v2 := recv (2@5);}
   else{send (v1, 2@6); v1 := recv (1@5);}}
  while (v1 ≠ 0){send (v1, 2@6); v1 := recv (2@5);}
  while (v2 ≠ 0){send (v2, 2@6); v2 := recv (1@5);}
  send (0, 2@6);
}
```

Fig. 11. Merge Sort

1 $$\left\{\mathsf{monoEnd}(1@5) * \mathsf{monoEnd}(2@5),\ \begin{array}{l}(1@5?\_)^* * (2@5?\_)^* * \mathsf{mono}(2@6)\\ \wedge\ \mathsf{large}(v_1) \wedge \mathsf{large}(v_2)\\ \wedge\ \mathsf{eqLast}(1@5, v_1) \wedge \mathsf{eqLast}(2@5, v_2)\end{array}\right\}$$

```
while (v1 ≠ 0 ∧ v2 ≠ 0){
```

2 $$\left\{\mathsf{monoEnd}(1@5) * \mathsf{monoEnd}(2@5),\ \begin{array}{l}(1@5?\_)^* * (2@5?\_)^* * \mathsf{mono}(2@6) \wedge v_1 \neq 0\\ \wedge\ v_2 \neq 0 \wedge \mathsf{large}(v_1) \wedge \mathsf{large}(v_2)\\ \wedge\ \mathsf{eqLast}(1@5, v_1) \wedge \mathsf{eqLast}(2@5, v_2)\end{array}\right\}$$

```
 if (v1 > v2){
```

3 $$\left\{\mathsf{monoEnd}(2@5),\ \begin{array}{l}(2@5?\_)^* * \mathsf{mono}(2@6) \wedge \mathsf{large}(v_1) \wedge \mathsf{large}(v_2)\\ \wedge\ \mathsf{eqLast}(2@5, v_2) \wedge v_1 > v_2 > 0\end{array}\right\}$$

```
  send (v2, 2@6);
```

4 $$\left\{\mathsf{monoEnd}(2@5),\ \begin{array}{l}(2@5?\_)^* * \mathsf{mono}(2@6) \wedge \mathsf{large}(v_1) \wedge \mathsf{large}(v_2)\\ \wedge\ \mathsf{eqLast}(2@5, v_2) \wedge v_1 > v_2 > 0\end{array}\right\}$$

```
  v2 := recv (2@5);}
```

5 $$\{\mathsf{monoEnd}(2@5), (2@5?\_)^* * \mathsf{mono}(2@6) \wedge \mathsf{large}(v_1) \wedge \mathsf{large}(v_2) \wedge \mathsf{eqLast}(2@5, v_2)\}$$

```
 else{
  send (v1, 2@6);
  v1 := recv (1@5);}
}
```

6 $$\left\{\mathsf{monoEnd}(1@5) * \mathsf{monoEnd}(2@5),\ \begin{array}{l}(1@5?\_)^* * (2@5?\_)^* * \mathsf{mono}(2@6)\\ \wedge\ (v_1 = 0 \vee v_2 = 0) \wedge \mathsf{large}(v_1) \wedge \mathsf{large}(v_2)\\ \wedge\ \mathsf{eqLast}(1@5, v_1) \wedge \mathsf{eqLast}(2@5, v_2)\end{array}\right\}$$

Fig. 12. Proof of the First While-loop

port 2@6. Line 2 is obtained from line 1 by conjoining the boolean condition of the while. Line 3 falls into a branch of the if statement. In this branch, the agent sends v_2 and receives messages from 2@5. Therefore, we treat some assertions about v_1 and 1@5 in line 2 as frame, and line 3 is deduced by framing out those irrelevant traces. Line 4 is the same as line 3, because according to $\mathsf{large}(v_2) \wedge v_2 \neq 0$, v_2 is larger than any message that previously sent to 2@6, so $\mathsf{mono}(2@6)$ holds in line 4; also since $v_1 > v_2$,

```
while (v1 ≠ 0){
```

7 $$\left\{\mathsf{monoEnd}(1@5), \begin{array}{l}(1@5?\_)^* * \mathsf{mono}(2@6) \wedge v_1 \neq 0 \wedge v_2 = 0 \\ \wedge\ \mathsf{large}(v_1) \wedge \mathsf{eqLast}(1@5, v_1)\end{array}\right\}$$

```
  send (v1, 2@6);
  v1 := recv (2@5);
```

8 $$\left\{\mathsf{monoEnd}(1@5), \begin{array}{l}(1@5?\_)^* * \mathsf{mono}(2@6) \wedge v_2 = 0 \wedge \mathsf{large}(v_1) \\ \wedge\ \mathsf{eqLast}(1@5, v_1)\end{array}\right\}$$

```
}
```

9 $$\left\{\mathsf{monoEnd}(1@5) * \mathsf{monoEnd}(2@5), \begin{array}{l}(1@5?\_)^* * (2@5?\_)^* * \mathsf{mono}(2@6) \\ \wedge\ v_1 = 0 \wedge v_2 = 0\end{array}\right\}$$

Fig. 13. Proof of the Second While-loop

$\mathsf{large}(v_1)$ remains to be true in line 4. The deduction from line 4 to 5 is also the standard local deduction. Deductions for the other branch are symmetric and thus omitted. Line 6 is obtained from line 5 by conjoining with the frame that was put aside in line 3.

Proof of the second loop is given in Fig 13. Line 7 is obtained from line 6 by framing out irrelavent assertions about 2@5 and adjoining the boolean predicate that guarded by the loop. The proof from line 7 to line 8 is just local reasoning as the proof from line 3 to 5. Line 9 is obtained by conjoining the frame of line 7 back to the post-condition. The third loop is symmetric with the second, and therefore we omit its proof.

The sketch of the overall proof is given in Fig. 14, where we only present the assertions at the critical places, e.g., the position where a framework is put aside or token back. As we have already discussed, the overall proof is obtained by sequentially conjoining several separated proofs of some locally connected commands. The post condition says if the environment send monotonic streams to the two ports of agent 5, then the agent will send monotonic streams to 2@6.

Note that the specification of agent 5 can be joined with the specification of other agents, just like the proof of the tiny example in Section 1. It is feasible if other agents take different algorithms as long as the assumption of agent 5 is satisfied.

7 Related Work and Conclusions

Now we summarize some related work, and then give a conclusion. Program verification has been studied from various standpoints for decades. Verification of concurrent systems is especially interesting because of their inherent non-determinism.

Process Calculus. For the message-passing models, there exist many famous process calculi, e.g., CSP (Communicating Sequential Processes) [5], CCS (Calculus of Communicating System) [10], π-calculus [12], and KPN (Kahn Process Network) [7]. However, those algebraic systems focus mainly on agent behavior deductions and equivalence, e.g., bi-simulation. It is unclear how to apply those calculi to modularly specify and reason the properties of local agents, which are written in real code and defined with stated-based semantics, that is our focus in this work.

Separation Logic. Recently there is a clear trend that concurrency verification should support better modularity and locality. Modular verification of shared memory models

```
agent5 (){
 {emp, emp}
 v1 = recv (1@5);
 v2 = recv (2@5);
 {1@5!X * 2@5!Y, 1@5?X ∘ 2@5?Y ∧ v1 = X ∧ v2 = Y}
 {monoEnd(1@5) * monoEnd(2@5), 1@5?X ∘ 2@5?Y ∧ v1 = X ∧ v2 = Y}
 {monoEnd(1@5) * monoEnd(2@5), (1@5?_)* * (2@5?_)* * mono(2@6)
                               ∧ large(v1) ∧ large(v2)
                               ∧ eqLast(1@5, v1) ∧ eqLast(2@5, v2)}
 while (v1 ≠ 0 ∧ v2 ≠ 0){
  if (v1 > v2){send (v2, 2@6); v2 := recv (2@5);}
  else{send (v1, 2@6); v1 := recv (1@5);}
 }
 {monoEnd(1@5) * monoEnd(2@5), (1@5?_)* * (2@5?_)* * mono(2@6)
                               ∧ (v1 = 0 ∨ v2 = 0) ∧ large(v1) ∧ large(v2)
                               ∧ eqLast(1@5, v1) ∧ eqLast(2@5, v2)}
 while (v1 ≠ 0){
  {monoEnd(1@5), (1@5?_)* * mono(2@6) ∧ v1 ≠ 0 ∧ v2 = 0
                 ∧ large(v1) ∧ eqLast(1@5, v1)}
  send (v1, 2@6); v1 := recv (2@5);
  {monoEnd(1@5), (1@5?_)* * mono(2@6) ∧ v2 = 0 ∧ large(v1) ∧ eqLast(1@5, v1)}
 }
 while (v2 ≠ 0){send (v2, 2@6); v2 := recv (1@5);}
 {monoEnd(1@5) * monoEnd(2@5), (1@5?_)* * (2@5?_)* * mono(2@6) ∧ v1=0 ∧ v2=0}
 send (0, 2@6);
 {monoEnd(1@5) * monoEnd(2@5), (1@5?_)* * (2@5?_)* * monoEnd(2@6)}
}
```

Fig. 14. Proof Sketch of Merge Sort

has gained much process since the development of Separation Logic (SL) [16]. SL treats program state as *resource*, and modularity is achieved by curving irrelevant resource (*frame*) out of the current state and conjoining the frame back when merging the local state into the environment. A typical shared memory model is defined based on the state of *heap*, a mapping from locations to values.

However, in message-passing models, event traces are, unlike heap, well-organized structures that associate with many add-on restrictions, e.g., acyclic, send-receive match, *etc*. The complexity of trace structures impedes state-based Hoare type reasoning. The challenge (and also a shining spot of our paper) is to structurally specify event traces so that local reasoning could be achieved by curving out irrelevant events, as in other SL-related works. Our framework solves this problem by introducing two operators to depict the separation of traces, so that traces could be either separately connected or temporally connected; and introducing four frame rules, so that frames could be added in four ways: adding in environmental or local trace, or adding ahead or behind of the current trace. These make our system flexible and powerful.

Rely-Guarantee Based Reasoning. Rely-guarantee (RG) reasoning [6] has been extended with SL to verify concurrent systems, e.g., Vafeiadis *et al.* [17] and Wehrman *et al.* [19].

In RG, rely condition is agents' local assumption of the environmental interferences, and guarantee condition is agents' interference upon other agents. Our system gets clearly some ideas from RG. However, comparing with the regular RG reasoning, ours has some innovations: (1) The rely condition of RG should be pre-defined and fixed, here we can dynamically calculate and alter environmental assumption. (2) Our environmental assumption is hidden once it is satisfied by other agents, rather than remained permanently as RG reasoning. (3) In RG reasoning, pre- and post-conditions are required to be *stable*, i.e., the assertions should remain valid no matter how environmental interferes. Stability is a rather strong requirement that requires thoughtful assertion definitions. In our system, this requirement is eliminated, that makes the reasoning process easier.

Other Works. W.de Roever *et al.* [4] published a book which made an excellent summarization with good coverage of previous works on the state-based verification of concurrent programs. The leading theme of the book is compositional techniques for concurrency verification. The book makes a comprehensive discussion about verification of both shared memory and message-passing models, and clearly, our work can be viewed as a new development on the same theme. However, there are some fundamental differences and contributions that distinguish our logic from W.de Roever *et al.*'s and many others: (1) by our limited knowledge, although Lamport's trace semantics has been proposed for decades, there are no state-based reasoning system defined based on this semantics; (2) the two-dimensional (temporal and spatial) modularity of our work, which is a benefit extracted from Lamport's semantics, has not been clearly touched by others; (3) our logic directly reason about imperative programming language, rather than high level mathematical descriptions.

There are also many other work aiming at specifying and reasoning message-passing programs based on trace semantics. For instance, Bickford *et al.* [1] formally defined the event trace structures, and gave a minimal set of axioms for trace reasoning. Comparing with other trace-based reasoning, ours supports better modularity, and allows directly reasoning over existing code modules and conjoining separated proofs based on several explicit conditions.

Villard *et al.* [18] proposed a separation-based logic for copyless message-passing models. One feature of their work is the support of ownership transfer. It is possible to extend our logic for ownership transfer, e.g., by adding the notion of "resource" for each agent and specifying those transmitted resource inside environmental assumptions. However, this solution is similar with CSL, and not be able to provide many theoretical innovations. In another aspect, Villard's method is still defined based on shared memory models, while ours is for pure message-passing systems.

Conclusion and Future Work. We propose a compositional reasoning system for verifying distributed programs with asynchronous message passing. The work inherits and integrates some ideas from SL, RG reasoning, etc., and archives very good modularity in both specification and reasoning. This reasoning framework exhibits two major contributions: first, we embody the concept of event graphs for distributed systems, and supports modular specification at both temporal and spatial dimensions; second, we propose an innovative Hoare triple, which syntactically separates environmental and local assertions, to better specify and reason interactions between agents and the environment. We have applied this method to reason about some programs, while a proof for a filter network is presented in this paper. In the future, we will further test its applicability with more applications. It is also interesting to explore the possibility of building tools to automate the verification process.

References

1. Bickford, M., Constable, R.L.: A causal logic of events in formalized computational type theory (2005)
2. Brookes, S.D.: A semantics for concurrent separation logic. In: Gardner, P., Yoshida, N. (eds.) CONCUR 2004. LNCS, vol. 3170, pp. 16–34. Springer, Heidelberg (2004)
3. Charron-Bost, B., Mattern, F., Tel, G.: Synchronous, asynchronous, and causally ordered communication. Distributed Computing 9(4), 173–191 (1996)
4. de Roever, W.-P., de Boer, F., Hanneman, U., Hooman, J., Lakhnech, Y., Poel, M., Zwiers, J.: Concurrency Verification: Introduction to Compositional and Noncompositional Proof Methods. Cambridge University Press (January 2012)
5. Hoare, C.A.R.: Communicating sequential processes. Commun. ACM 21(8), 666–677 (1978)
6. Jones, C.B.: Tentative steps toward a development method for interfering programs. ACM Trans. Program. Lang. Syst. 5(4), 596–619 (1983)
7. Kahn, G.: The semantics of simple language for parallel programming. In: IFIP Congress, pp. 471–475 (1974)
8. Lamport, L.: Time, clocks, and the ordering of events in a distributed system. Commun. ACM 21(7), 558–565 (1978)
9. Lei, J., Qiu, Z.: Modular reasoning for message-passing programs. Technical report, School of Mathematical Sciences, Peking University (June 2014), `https://sites.google.com/site/jinjianglei/publications/rgdsep`
10. Milner, R.: A Calculus of Communication Systems. LNCS, vol. 92. Springer, Heidelberg (1980)
11. Milner, R.: Communication and concurrency. PHI Series in computer science. Prentice Hall (1989)
12. Milner, R.: The polyadic pi-calculus (abstract). In: Cleaveland, W.R. (ed.) CONCUR 1992. LNCS, vol. 630, p. 1. Springer, Heidelberg (1992)
13. Milner, R.: Communicating and mobile systems - the Pi-calculus. Cambridge University Press (1999)
14. Parkinson, M.J., Bornat, R., Calcagno, C.: Variables as resource in hoare logics. In: LICS, pp. 137–146 (2006)
15. Reddy, U.S., Reynolds, J.C.: Syntactic control of interference for separation logic. In: POPL, pp. 323–336 (2012)
16. Reynolds, J.C.: Separation logic: A logic for shared mutable data structures. In: LICS, pp. 55–74 (2002)
17. Vafeiadis, V., Parkinson, M.J.: A marriage of rely/guarantee and separation logic. In: Caires, L., Vasconcelos, V.T. (eds.) CONCUR 2007. LNCS, vol. 4703, pp. 256–271. Springer, Heidelberg (2007)
18. Villard, J., Lozes, É., Calcagno, C.: Proving copyless message passing. In: Hu, Z. (ed.) APLAS 2009. LNCS, vol. 5904, pp. 194–209. Springer, Heidelberg (2009)
19. Wehrman, I., Hoare, C.A.R., O'Hearn, P.W.: Graphical models of separation logic. Inf. Process. Lett. 109(17), 1001–1004 (2009)

Symbolic Analysis Tools for CSP

Liyi Li, Elsa Gunter, and William Mansky

Department of Computer Science,
University of Illinois at Urbana-Champaign
{liyili2,egunter,mansky1}@illinois.edu

Abstract. Communicating Sequential Processes (CSP) is a well-known formal language for describing concurrent systems, where transition semantics for it has been given by Brookes, Hoare and Roscoe [1]. In this paper, we present trace refinement model analysis tools based on a generalized transition semantics of CSP, which we call HCSP, that merges the original transition system with ideas from Floyd-Hoare Logic and symbolic computation. This generalized semantics is shown to be sound and complete with respect to the original trace semantics. Traces in our system are symbolic representations of families of traces as given by the original semantics. This more compact representation allows us to expand the original CSP systems to effectively and efficiently model check some CSP programs that are difficult or impossible for other CSP systems to analyze. In particular, our system can handle certain classes of non-deterministic choices as a single transition, while the original semantics would treat each choice separately, possibly leading to large or unbounded case analyses. All the work described in this paper has been carried out in the theorem prover Isabelle [2]. This then provides us with a framework for automated and interactive analysis of CSP processes. It also gives us the ability to extract Ocaml code for an HCSP-based simulator directly from Isabelle. Based on the HCSP semantics and traditional trace refinement, we develop an idea of symbolic trace refinement and build a model checker based on it. The model checker was transcribed by hand into Maude [3] as automatic extraction of Maude code is not yet supported by the Isabelle system.

1 Introduction

Communicating Sequential Processes (CSP) is a process algebra to describe the behavior and interactions of concurrent systems. Due to the expressive features of external and internal choice together with the parallel composition in CSP, it has been used practically in industry for specifying and verifying concurrent features of various systems, especially ones combining human operators and automations, such as the medical mediator system in Gunter *et al.* [4], the airline ticket reservation system in Wong and Gibbons' paper [5] and interactive systems with human error tolerance in Wright *et al.* [6].

In the traditional semantics of CSP, processes are given semantics via the set of traces they may generate, the set of sequences of individual actions the processes may execute. For example, the CSP process $c?x : B \rightarrow P$ is generally

G. Ciobanu and D. Méry (Eds.): ICTAC 2014, LNCS 8687, pp. 295–313, 2014.

modeled as receiving a single value across a channel c from the set $\{x|B\}$ and proceeding as P with that value. The set of possible traces will depend on the size of that set. Previous CSP simulators and model checkers have followed this semantics by enumerating all traces individually. In practice, if the set $\{x|B\}$ is an infinite set, current CSP simulators and model checkers actually create an endless number of similar processes and wait for other parts of the program to stop these processes. This affects the efficiency and decreases the scope of analyzable problems for these tools, particularly for model checking.

In this paper, we present a simulator HSIM to effectively generate the behaviors of CSP programs, and a model checking tool HMC to check trace refinement properties of CSP programs based on Holistic CSP (HCSP) semantics, a new semantics for CSP processes that uses a symbolic representation of actions to capture a group of properties simultaneously instead of considering only a single element with a single property. The approach we take in this work is to represent families of transitions in CSP by a single transition in HCSP. This allows us to view a set of actions as a whole in some contexts, but also divide it based on various properties in other contexts.

$$A.\ (\bigsqcap_{x:x>0\wedge x<10} \mathsf{A}.x \rightarrow \mathsf{SKIP})[\![\{k.x|k = A \wedge x > 0 \wedge x < 100\}]\!](\mathsf{A}?x : x > 0 \wedge x < 150 \rightarrow \mathsf{SKIP})$$

$$B.\ (\bigsqcap_{x:x>0\wedge x<10^5} \mathsf{A}.x \rightarrow \mathsf{SKIP})[\![\{k.x|k = A \wedge x > 0 \wedge x < 10^6\}]\!](\mathsf{A}?x : x > 0 \wedge x < 15 * 10^5 \rightarrow \mathsf{SKIP})$$

$$C.\ (\bigsqcap_{x:x>0\wedge x<10} \mathsf{A}.x \rightarrow \mathsf{SKIP})[\![\{k.x|k = A \wedge x > 0 \wedge x < 100\}]\!](\mathop{\Box}_{y:y=1-1} \mathsf{A}?x : x > y \rightarrow \mathsf{SKIP})$$

$$D.\ (\bigsqcap_{x:x>0\wedge x<10} \mathsf{A}.x \rightarrow \mathsf{SKIP})[\![\{k.x|k = A \wedge x > 0 \wedge x < 100\}]\!]\begin{array}{r}(\mathsf{A}?x : x > 0 \wedge x < 100 \rightarrow \mathsf{SKIP})\\ \Box(\mathsf{A}?x : x > 99 \rightarrow \mathsf{SKIP})\end{array}$$

Fig. 1. Example

The differences between the original CSP semantics based tools and the HCSP semantics based tools can be demonstrated by some very simple examples. Four such examples are shown in Figure 1. Each process is the parallel composition of a process selecting a value from a range with a process receiving a value restricted to be in another range, with synchronization requiring the shared value to be in a third range. For each process, the problem was posed, does the process refine itself. Logically, the are at most three cases to be considered: is the value chosen within the range of synchronization, and if so is it in the range to be received. The model checker FDR2 [7] can handle case A easily, but it fails to terminate on cases B because of the large data sets for each restricting set, and C because of the infinite restricting set on the receiving process. On process B it begins to run, but eventually generates a stack overflow. When process C is directly input into the CSPM-based simulator ProBE, the whole program crashes. However, HMC, the HCSP model checker we have derived from the semantics we discuss in this paper, easily verifies the trace refinement properties of the processes A - D with respect to themselves in the same amount of time. We will show more details of the experiments in Section 5, but from these examples, we can clearly see that the running time of FDR2 depends on the size of the sets bounding choice and parallel composition in each process.

These facts reflect, in part, that the original CSP and Machine-Readable CSP (CSPM) semantics view replicated operators (Replicated Internal Choice, Replicated External Choice, etc) as macros of their binary versions over sets. This means that the original CSP and CSPM semantics cannot express a replicated operator if the set of the replicated operator is infinite, such as the second Replicated Internal Choice operator in process C. Even if the set is finite, the cost is very expensive for CSP-semantics-based tools to run a small replicated process in a large macro, such as in the process B. On the other hand, HCSP-semantics-based tools can overcome this problem and run CSP processes regardless of the size of sets bounding replicated operators. By using HCSP semantics, the three processes in the example will have the same number of possible next moves. This property allows the HCSP-semantics-based tools to run faster than the traditional CSP-semantics-based tools in some cases. In addition, HCSP-semantics-based tools can expand the set of possible CSP processes to analyze; processes B and C above are examples of this fact. We will see that this fact is useful in some real applications such as the medical mediator system in Gunter *et al.* [4].

This paper's contribution is a general methodology for the translation of traditional transition semantics of a process algebra to its symbolic semantics, and from that symbolic semantics the derivation of tools for simulation and model checking, combined with the direct application of this methodology to the process algebra CSP and a demonstration of the advantages acquired by the derived tools. Our methodology replaces transitions between individual processes with transitions between configurations of parametrized processes and propositions describing constraints on the process parameters. This generalized framework allows us to expand the processes given semantics to include those ranging over infinite or dynamically calculated sets of data (such as actions), including infinite choice and parallel composition operators. The translated symbolic semantics for CSP, HCSP is proved to be sound and complete (for the common subset) with respect to the traditional transition semantics in the theorem prover Isabelle. We directly extracted the simulator HSim is from the HCSP semantics in Isabelle. In a similar manner, we define a general symbolic trace refinement relation that is compatible with the symbolic transition relation, and prove it equivalent to traditional trace refinement. This symbolic trace refinement property together with our symbolic transition relation in turn directly translates to the procedures of our model checker HMC. The symbolic nature of HMC allows it to handle a wider class of processes and to decide trace refinement more efficiently for processes within scope of existing model checkers, but where the size of the data causes them to be handle extremely inefficiently or not at all.

2 Syntax and Semantics

The syntax of HCSP and informal meaning is given in Figure 2. For the remainder of this paper, the following name conventions will be used. We will use P and Q for processes. Lower case p refers to an HCSP process name. The letter c represents an HCSP channel, while the letter a represents an HCSP action. The letter B is a proposition describing the property of a set. In HCSP, we include both *variables* and *parameters*, which are distinct types. We use k for variables ranging over HCSP channels and x for variables over actions. We use U and V to refer to parameters ranging over channels and actions. Variables and parameters serve similar functions, but differ as follows: variables may occur free or bound in HCSP processes and may be replaced by actions or channels by substitution, while parameters occur essentially as local constants not subject to binding or substitution. In the rest of the paper, we will use *freeParams* to refer to a function returning all free parameters in an expression of arbitrary type. We will use l to represent a transition label. Finally, the Greek letter ρ refers to an assignment function that assigns values to parameters. To facilitate the application of HCSP to specific examples, it is parameterized by four user-defined types: a type of expressions for actions and channels (acts), a type of propositions, a type of process names and a type of values to be assigned to acts. One remark must be made here concerning the scope of variables. In the processes $c?x : B \rightarrow P$, $\bigsqcap_{x:B} P$ and $\square_{x:B} P$, the scope of variable x is both the proposition B and the process P, while the scope of the variables k and x is only the proposition B in the processes $P[\![\{k.x|B\}]\!]Q$ and $P \setminus \{k.x|B\}$.

$P =$

Ω	Successful termination	| SKIP	Awaiting successful termination
| STOP	Unexpected termination	| $c.a \rightarrow P$	Prefix by action a on channel c
| if b then P else Q	If statement	| $\$p$	Process name p as prcoess
| $P;Q$	Sequential execution	| $P \square Q$	Binary external choice
| $P \sqcap Q$	Binary internal choice	| $\bigsqcap_{x:B} P$	Replicated internal choice
| $c?x : B \rightarrow P$	External set prefix		
| $P[\![\{k.x\|B\}]\!]Q$	Parallel composition	| $\square_{x:B} P$	Replicated external choice
| let $p = P$ in Q	Local process name binding		
		| $P \setminus \{k.x\|B\}$	Hiding over a set of actions

Fig. 2. HCSP Syntax

The syntax of HCSP differs from that of CSPM by Bryan Scattergood [8] in three ways. Firstly, the actions of CSP are explicitly divided into channels and actions (written $c.a$) in HCSP syntax. Secondly, for the sets used in constructs such as the parallel composition of two processes or replicated internal choice, we use a set comprehension notation. This decomposition of sets into variables and predicates will facilitate the statement of the transition rules of HCSP semantics. Finally, HCSP currently lacks the CSP Renaming operator.

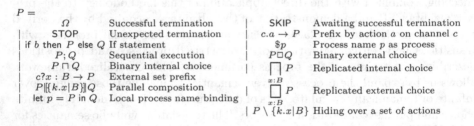

Representative rules for the semantics for HCSP is given in Figures 3 - 4. The semantics is a merge of the original CSP transition semantics given by Brookes *et al.* [10] with ideas from Floyd-Hoare Logic [11] and symbolic computation. In order to describe the HCSP semantics, there are some functions that need to be supplied for the evaluations of user-defined types. A family of substitution functions $T[a/x]$ is needed for the replacement of variables by acts in each of acts, propositions, and process names. Using these, we define the substitution function for processes. There also needs to be a family of user-defined evaluation functions for acts and a "models" function, $\models$, for checking whether a proposition is true under a given assignment function. We define the functions $sem(\rho, P)$ and $sem(\rho, l)$ as interpretation functions to interpret a given HCSP process or label as a CSP process or label with respect to valuation ρ. The labels of the HCSP semantics will be ranged over by l as follows:

$$l = \surd \mid \tau \mid (U.V)$$

The label $\surd$ represents process completion, the label τ represents a process performing an invisible action, and the label $(U.V)$ represents a pair of parameters, one for a channel and one for a real action. In any execution of a process in accordance with this semantics, the sequence of transitions is labeled with mutually distinct pairs of parameters $(U.V)$, when not labeled by $\surd$ or τ.

Rule Replacement	HCSP Corresponding Rule
$P \sqcap Q \xrightarrow{\tau} P$	$(\alpha, \gamma, S, P \sqcap Q) \xrightarrow{\tau} (\alpha, \gamma, S, P)$ Int_choice1
$\dfrac{P \xrightarrow{(c.a)} P' \quad (c.a) \notin \{k.x \mid B\}}{P [\![\{k.x \mid B\}]\!] Q \xrightarrow{(c.a)} P' [\![\{k.x \mid B\}]\!] Q}$	$\dfrac{(\alpha, \gamma, S, P) \xrightarrow{(U.V)} (\alpha', \gamma', S', P') \quad \exists \rho \,.\, \rho \models (\neg B[U/k][V/x] \wedge \gamma')}{(\alpha, \gamma, S, P [\![\{k.x \mid B\}]\!] Q) \xrightarrow{(U.V)} (\alpha', \neg B[U/k][V/x] \wedge \gamma', S', P' [\![\{k.x \mid B\}]\!] Q)}$ Par_out1
$\dfrac{(c.a) \in \{k.x \mid B\} \quad P \xrightarrow{(c.a)} P' \quad Q \xrightarrow{(c.a)} Q'}{P [\![\{k.x \mid B\}]\!] Q \xrightarrow{(c.a)} P' [\![\{k.x \mid B\}]\!] Q'}$	$\dfrac{\exists \rho \,.\, \rho \models (U = U' \wedge V = V' \wedge B[U/k][V/x] \wedge \gamma'') \quad (\alpha, \gamma, S, P) \xrightarrow{(U.V)} (\alpha', \gamma', S', P') \quad (\alpha', \gamma', S', Q) \xrightarrow{(U'.V')} (\alpha'', \gamma'', S'', Q')}{(\alpha, \gamma, S, P [\![\{k.x \mid B\}]\!] Q) \xrightarrow{(U.V)} (\alpha'', U = U' \wedge V = V' \wedge B[U/k][V/x] \wedge \gamma'', S'', P' [\![\{k.x \mid B\}]\!] Q')}$ Par_in

Fig. 3. HCSP semantics (part of category One and category Two)

We present a labeled transition system for HCSP over quadruples of the form (α, γ, S, P) or (β, ϕ, T, Q), which are called configurations in this paper, where P and Q are HCSP processes, γ and ϕ are environment condition propositions in HCSP that are intended to state the current requirements for parameters "in scope", including those occurring free in P, and α and β are sets of parameters large enough to contain all parameters occurring free in P or γ. The tuples $(l, \alpha, \gamma, S, P)$ or (l, β, ϕ, T, Q) are called moves in this paper. We carry α (or β) with us to allow for the choice of fresh parameter names guaranteed not to clash with a potentially bigger scope than the one locally presented by P (or Q) and γ (or ϕ). S and T are the interpretation functions for interpreting a process name p in a given HCSP process. The environment conditions γ (or ϕ) play the role of providing the pre- and post-condition for each transition. The values potentially represented by labels of the form $(U.V)$ are progressively restricted by the conditions in each of the subsequent quadruples resulting from each transition in the execution. In this way, a single execution in the transition semantics of HCSP potentially represents a parameterized family of executions from the original CSP semantics.

Rule Replacement	HCSP Corresponding Rule
$c.a \to P \stackrel{(c.a)}{\longrightarrow} P$	$\dfrac{U \notin \alpha \quad V \notin \{U\} \cup \alpha \qquad \exists \rho\,.\,\rho \models (U = c \wedge V = a \wedge \gamma)}{(\alpha, \gamma, S, c.a \to P) \stackrel{(U.V)}{\longrightarrow} (\{U, V\} \cup \alpha, U = c \wedge V = a \wedge \gamma, S, P)}$ Act_prefix
Macro Replacement	**HCSP Corresponding Rule**
$c?x : B \to P = \Box_{x:B}\, c.x \to P$	$\dfrac{U \notin \alpha \quad V \notin \{U\} \cup \alpha \qquad \exists \rho\,.\,\rho \models (U = c \wedge B[V/x] \wedge \gamma)}{(\alpha, \gamma, S, c?x : B \to P) \stackrel{(U.V)}{\longrightarrow} (\{U, V\} \cup \alpha, U = c \wedge B[V/x] \wedge \gamma, S, P[V/x])}$ Ext_prefix
$\prod_{x:B} P = P[a_1/x] \sqcap \ldots \sqcap P[a_n/x]$ $a_1 \ldots a_n \in \{x \mid B\}$	$\dfrac{U \notin \alpha \qquad \exists \rho\,.\,\rho \models (B[U/x] \wedge \gamma)}{(\alpha, \gamma, S, \prod_{x:B} P) \stackrel{\tau}{\longrightarrow} (\{u\} \cup \alpha, B[U/x] \wedge \gamma, S, P[U/x])}$ Rep_int_choice
$\Box_{x:B} P = P[a_1/x] \Box \ldots \Box P[a_n/x]$ $a_1 \ldots a_n \in \{x \mid B\}$	$\dfrac{U \notin \alpha \quad \exists \rho\,.\,\rho \models \gamma' \qquad (\{u\} \cup \alpha, B[U/x] \wedge \gamma, S, P[U/x]) \stackrel{\tau}{\longrightarrow} (\alpha', \gamma', S', P')}{(\alpha, \gamma, S, \Box_{x:B} P) \stackrel{\tau}{\longrightarrow} (\alpha', \gamma', S', (\Box_{x:B \wedge x \neq U} P) \Box P')}$ Rep_ext_tau
$\Box_{x:B} P = P[a_1/x] \Box \ldots \Box P[a_n/x]$ $a_1 \ldots a_n \in \{x \mid B\}$	$\dfrac{U \notin \alpha \quad l \neq \tau \quad \exists \rho\,.\,\rho \models \gamma' \qquad (\{u\} \cup \alpha, B[U/x] \wedge \gamma, S, P[U/x]) \stackrel{l}{\longrightarrow} (\alpha', \gamma', S', P')}{(\alpha, \gamma, S, \Box_{x:B} P) \stackrel{l}{\longrightarrow} (\alpha', \gamma', S', P')}$ Rep_ext_nor

Fig. 4. HCSP semantics (category Three)

We describe our translation process by dividing the CSP semantics into three categories. The first category contains rules for basic operators having no side conditions other than $\surd$-τ label constraints. The second category contains rules with side conditions that need to be translated into the HCSP framework, i.e., the operators with set or boolean guard information. The third category includes rules for operators that are treated as macros in CSP, including all replicated operators. When translating the original CSP semantics into HCSP semantics, the main task is to merge information about actions and channels into the environment condition γ. Basically, we do the translation by a two step processes. Step One is to solve a problem that often arises when translating a substitution-based semantics to an environment-based semantics. The problem is misinterpreting free variables via free-variable capture. We divide the identifiers into *variables* and *parameters* to solve the problem. Parameters are used in each rule that involves passing through a variable binding. Those rules will be altered to always replace existing *variables* with new fresh *parameters* when evaluating a given process.

Step Two is to simultaneously translate a substitution-based semantics of CSP to an environment-based transition semantics of CSP, and then transform the environment-based transition semantics into a symbolic semantics by making the global environment into a boolean predicate. By this strategy, we can represent the state as the pre- and post-conditions that appear in Hoare Logic. This can be seen as a generalization of the environment by treating each assignment of a variable to an expression (or value) as an equation stating the variable is equal to that expression. In addition, we can treat every side-condition in the transition semantics as a constraint and conjoin it to the boolean predicate constraining the environment.

The HCSP semantics has been proved sound and complete with respect to the original CSP semantics. We will state only the soundness and relative completeness theorems below.

Theorem 1 (Soundness). *For all HCSP processes P, P', assignments ρ, environment conditions γ, γ' and process environments S, S' such that $\rho \models \gamma$ and $\rho \models \gamma'$, and parameter set α such that $freeParams(P) \cup freeParams(\gamma) \subseteq \alpha$ and $(\forall p.p \in \mathtt{dom}(S) \Rightarrow freeParams(S(p)) = \emptyset)$, if $(\alpha, \gamma, S, P) \xrightarrow{l} (\alpha', \gamma', S', P')$, then $sem(\rho, P) \xrightarrow{sem(\rho,l)} sem(\rho, P')$.*

Theorem 2 (Relative Completeness). *Let P be an HCSP process, ρ be an assignment, γ be an environment condition, S be a process environment such that $\rho \models \gamma$, and α be a parameter set such that $freeParams(P) \cup freeParams(\gamma) \subseteq \alpha$ and $(\forall p.p \in \mathtt{dom}(S) \Rightarrow freeParams(S(p)) = \emptyset)$, such that $sem(\rho, P) \xrightarrow{i} T$ in CSP semantics. Then there exist an HCSP process P', an assignment ρ', an environment condition γ', a parameter set α', a process environment S', and a label l such that $i = sem(\rho', l)$, $\rho'|_\alpha = \rho|_\alpha$, $T = sem(\rho', P')$, $\rho' \models \gamma'$ and $(\alpha, \gamma, S, P) \xrightarrow{l} (\alpha', \gamma', S', P')$.*

All the work was done in the interactive theorem prover Isabelle/HOL [2] and can be found at `http://www.cs.illinois.edu/ egunter/fms/HCSP/hcsp.tar.gz`. All details of semantics can be found in in the technical report [9].

3 Symbolic Trace Refinement

In order to describe trace refinement model checking in Section 4, we define trace refinement in Definition 1, which is given by Roscoe in [12]. In implementations of CSP trace refinement checker, such as FDR2, they actually check the refinement property by using some relation which is similar to trace simulation. We list it in Definition 2. This definition is more similar to the trace simulation definition. Current trace refinement checkers, such as FDR2, use this approach with memoization to prove the trace refinement between two programs. In Back and Wright's paper [13], they actually prove that this relation implies the trace refinement relation described in Definition 1.

Definition 1 (Trace Refinement). $Q \sqsubseteq_T P = trace(P) \subseteq trace(Q)$.

Definition 2 (Simulation Trace Refinement). *The relation P* trace refines *Q, written $Q \sqsubseteq_S P$, is the smallest relation satisfying the following:*

- $\Omega \sqsubseteq_S \Omega$
- $Q \sqsubseteq_S \mathit{STOP}$
- *If there exists Q' such that $Q \xrightarrow{\tau} Q'$ and $Q' \sqsubseteq_S P$, then $Q \sqsubseteq_S P$*
- *If for all P' we have $P \xrightarrow{l} P'$ implies*
  - *either $l = \tau$ and $Q \sqsubseteq_S P'$,*
  - *or $l \neq \tau$ and there exists Q' such that $Q \xrightarrow{l} Q'$ and $Q' \sqsubseteq_S P'$,*

  then $Q \sqsubseteq_S P$

Definition 2 only works in traditional CSP semantics. In order to describe the symbolic semantics, we need to be able to connect environment predicates with processes. Hence, we define symbolic trace refinement in Definition 3. This definition is actually more similar to the definition of trace simulation, but we call it symbolic trace refinement here because it is equivalent to the simulation trace refinement definition above. In this definition, the function $\eta : (\phi, \beta, \alpha) \to (\alpha', \phi')$ is a renaming function to rename all parameters of a given environment predicate ϕ that are in the set β to new parameters that are not in the set $\alpha \cup \beta$, and return the resulting environment condition ϕ' and the new parameter set α' containing all parameters in the set α and the environment condition ϕ'. This function is useful in the final rule of symbolic trace refinement. In each inductive step $(\beta_i, \phi_i, T_i, Q_i) \sqsubseteq_T (\alpha'_i, \gamma' \wedge \phi'_i, S', P')$ for $i \in [1..n]$ in the final rule, we want to bind the environment condition of the implementation configuration γ' by an extra constraint ϕ'_i.

Definition 3 (Symbolic Trace Refinement). *The relation $(\beta, \phi, T, Q) \sqsubseteq_{ST} (\alpha, \gamma, S, P)$ is the smallest relation satisfying the following:*

- $(\beta, \phi, T, \Omega) \sqsubseteq_{ST} (\alpha, \gamma, S, \Omega)$
- $(\beta, \phi, T, Q) \sqsubseteq_{ST} (\alpha, \gamma, S, \mathit{STOP})$
- *If there exists (β', ϕ', T', Q') such that $(\beta, \phi, T, Q) \xrightarrow{\tau} (\beta', \phi', T', Q')$ and $(\beta', \phi', T', Q') \sqsubseteq_{ST} (\alpha, \gamma, S, P)$, then $(\beta, \phi, T, Q) \sqsubseteq_{ST} (\alpha, \gamma, S, P)$*
- *If for all $(\alpha', \gamma', S', P')$ we have $(\alpha, \gamma, S, P) \xrightarrow{l} (\alpha', \gamma', S', P')$ implies*
  - *either $l = \tau$ and $(\beta, \phi, T, Q) \sqsubseteq_{ST} (\alpha', \gamma', S', P')$,*
  - *or $l = \surd$ and there exists (β', ϕ', T', Q') such that $(\beta, \phi, T, Q) \xrightarrow{\surd} (\beta', \phi', T', Q')$ and $(\beta', \phi', T', Q') \sqsubseteq_{ST} (\alpha', \gamma', S', P')$,*
  - *or $l = (U.V)$ and there exists a natural number n, configurations $(\beta_i, \phi_i, T_i, Q_i)$, parameters U' and V', parameter sets α'_i and environment conditions ϕ'_i such that,*
    - ▷ $(\alpha \cup \alpha' \cup \bigcup_{i=1}^{n} \alpha'_i) \cap (\beta \cup \bigcup_{i=1}^{n} \beta_i) = \emptyset$
    - ▷ $\bigwedge_{i=1}^{n} ((\beta, \phi, T, Q) \xrightarrow{(U_i.V_i)} (\beta_i, \phi_i, T_i, Q_i))$

$\triangleright \bigwedge_{i=1}^{n}(\exists \rho.\rho \models \gamma' \wedge \phi_i[U/U_i][V/V_i])$

$\triangleright (\exists \rho'.\forall c\ a\ .\rho'[U \mapsto c, V \mapsto a] \models \gamma' \Rightarrow \bigvee_{i=1}^{n} \rho'[U_i \mapsto c, V_i \mapsto a] \models \phi_i)$

$\triangleright \bigwedge_{i=1}^{n}((\alpha_i', \phi_i') = \eta(\phi_i[U/U_i][V/V_i], \beta_i, \alpha'))$

$\triangleright \bigwedge_{i=1}^{n}((\beta_i, \phi_i, T_i, Q_i) \sqsubseteq_{ST} (\alpha_i', \gamma' \wedge \phi_i', S', P'))$

then $(\beta, \phi, T, Q) \sqsubseteq_{ST} (\alpha, \gamma, S, P)$

The symbolic trace refinement definition has the following relation with the original trace refinement definition.

Theorem 3 (Symbolic Trace Refinement Relation). *For all HCSP configurations* (α, γ, S, P) *and* (β, ϕ, T, Q)*, assignments* ρ *and* δ *such that* $\beta \cap \alpha = \emptyset$*, freeParams*$(P) \cup$ *freeParams*$(\gamma) \subseteq \alpha$*,* $(\forall p.p \in \mathtt{dom}(S) \Rightarrow \mathit{freeParams}(S(p)) = \emptyset)$*,* $\rho \models \gamma$ *and* $\delta \models \phi$*, we have* $(\beta, \phi, T, Q) \sqsubseteq_S T(\alpha, \gamma, S, P)$ *if, and only if* $sem(\delta, Q) \sqsubseteq_T sem(\rho, P)$.

Proof. (*Sketch*) We first do an induction on rules of trace refinement to prove the "only if" side of the theorem. We prove this direction by using relatively completeness described in Theorem 2. We then prove the "if" side of the theorem by induction on rules of symbolic trace refinement with the soundness theorem described in Definition 1 □

4 HCSP Simulator and Model Checker

To put the theory of HCSP into practice, we have specified an HCSP simulator with a rich mutually recursive datatype for actions and propositions in Isabelle. OCaml code for the simulator, which we call HSIM, is then extracted from the Isabelle specification directly. The core of HSIM is included with the package for the soundness and completeness theorems. In HSIM, we have limited the propositions to quantifier-free first order logic with Presburger arithmetic in order to maintain decidability. In doing so, we render the single-step transition relation computable as a function generating a finite set of possibilities. We then represent all possible traces with a lazy stream data structure supporting back-tracking. This enables us to incrementally compute the requirements for a given trace, which can be inspected at each step, and can be back-tracked when the requirements are proven to be unsatisfiable. Using the HCSP semantics, we can indefinitely delay the calculation of a specific trace using trace patterns and pre- and post-conditions, until one trace pattern / condition is selected. At this point, satisfiability analysis can be used to generate an instance trace, if such is desired.

In the case of the medical mediator example, we were able to use the simulator specification in Isabelle to enumerate the possible trace patterns for the System, and to verify that all traces satisfying each pattern-condition so enumerated satisfy a pattern-condition of the Safety process.

In addition to HSIM, we have implemented in the rewriting logic engine Maude [3] a model checker HMC to check the trace refinement property of HCSP programs. This implementation is a hand transcription of the Isabelle specification of HMC. Previous model checkers for CSP, including FDR2 and PAT, check trace refinement between two processes by explicit trace enumeration with circularity checking. In contrast, as we did with the simulator HSIM, the model checker HMC is based on the symbolic semantics and symbolic trace refinement described in Section 2 and 3.

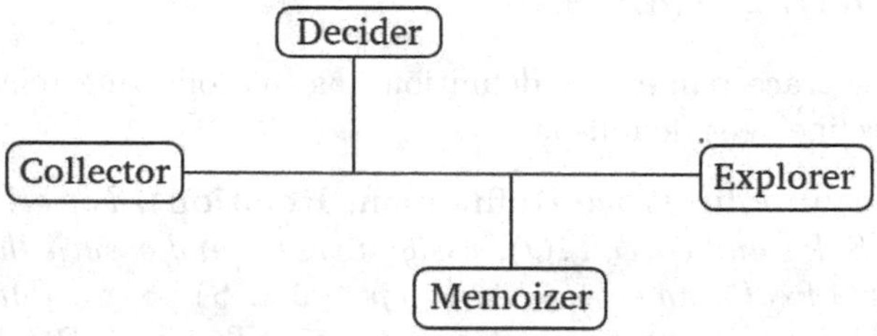

Fig. 5. Algorithm Components

We describe the algorithm of HMC by breaking it into four components: the explorer, the decider, the memoizer and the collector. The explorer is the core of the trace refinement graph searching algorithm. It controls how we do the trace refinement at the meta level regardless of the details of the HCSP semantics. The decider is an auxiliary SMT solver to determine whether a next possible move of a given HCSP configuration is valid based on the satisfiability of the constraint in the next possible move. The memoizer controls how we determine the one configuration is an instance of another configuration that arises during circularity checking, while the collector controls how we collect the next possible moves of a given HCSP configuration.

At the explorer level, our algorithm can be generalized as a four-step procedure based on the definition of symbolic trace refinement described in Section 3. We start with a pair of configurations of the implementation and specification, two τ-labeled memoizing sets (one for the implementation and one for the specification) and a configuration-pair memoizing set for checking a pair of configurations. Step One is to get the next possible moves of a given implementation configuration and a given specification configuration by the collector. Based on these sets of moves, we remove the inconsequential and invalid moves. A move is inconsequential if it is a τ-labeled move that is an instance to one in the implementation or specification τ-memoizing set. An invalid move is one with an unsatisfiable environment condition. In Maude, Step One is implemented by the Maude function verify.

Step Two is to replace all τ-labeled moves of the implementation from the set collected in Step One by the set of non-τ-labeled moves that are reachable by a possibly empty sequence of τ-labeled moves from the τ-labeled move being replaced. Both the τ-labeled move being replaced and all subsequent τ-labeled moves transitioned over in a sequence leading to a replacing non-τ-labeled move are added to the implementation τ-memoizing set, if it is not an instance to one already present. If it is an instance to one already present, we cut that search branch and backtrack to look for other τ-labeled sequences. The result of Step

Two is that the all moves collected for the implementation have the label form $U.V$. In Maude, the function verifyTau will handle this step.

Step Three is to check for each move of the implementation if there exists a move of the specification that refines the given move of the implementation. It has three operations. The first operation processes the next-moves set for the specification generated in Step One. This operation preforms the same reduction/replacement action on τ-labeled moves of the specification as was done in Step Two for the implementation, except that the replacement is only of a single reachable non-τ move, rather than the full set. All τ-labeled moves in the sequence for the one in the given next-moves set to its replacement are added to the specification τ-memoizing set. This operation is only performed when there do not exist any more $U.V$-labeled moves in the specification next-moves set. We are repeatedly applying this operation until there is a move of the specification with the label form $U.V$, or there are no more moves of the specification. In the latter case, we answer false.

In the second operation, a move of the specification with the label $U'.V'$ and environment condition γ' is selected. For the given move of the implementation with label $U.V$ and environment condition γ, we check whether or not the formula $\forall U.V.\gamma \Rightarrow \gamma'[U/U'][V/V']$ is satisfiable, using the decider also implemented in Maude. Here, we are taking advantage of our symbolic semantics for HSCP. If one such specification move can be found, we can move onto Step Four directly. If not, we will go to the third operation. In the third operation, we assume that we have the given move of the implementation $(U.V, \alpha, \gamma, S, P)$. We check whether there is a set of next moves of the specification $(U_i'.V_i', \beta_i, \phi_i, T_i, Q_i)$ for $i = 1, \ldots, n$ such that for each i the formula $\gamma \wedge \phi_i[U/U_i'][V/V_i']$ is satisfiable, and the formula $\forall U\, V.\gamma \Rightarrow \bigvee_{i=1}^{n} \phi_i[U/U_i'][V/V_i']$ is satisfiable. If there are such moves of the specification, we construct the new parameter sets and new environment conditions as $(\alpha_i', \phi_i') = \eta(\phi_i[U/U_i'][V/V_i'], \beta_i, \alpha)$ for $i = 1, \ldots, n$, where η is the renaming function of Section 2. We also construct the new pairs of configurations $((\alpha_i', \gamma \wedge \phi_i', S, P), (\beta_i, \phi_i, T_i, Q_i))$ for $i = 1, \ldots, n$. We then allocate Step Four checks for each of these pairs of configurations. If there do not exist such moves of the specification, we will answer false. Step Three is implemented as functions verifyAction and verifyActionAux in Maude.

Step Four checks if the new input pair of implementation-specification configurations from Step Three is an instance of one in the current configuration-pair memoizing set. If it is, then we answer true; otherwise, we go back to Step One to check the new implementation-specification configuration pair with adding the implementation-specification configuration pair into the current configuration-pair memoizing set. In Maude, we implement this step by the function preverify.

The four-step procedure is a general framework and specification for the symbolic graph searching algorithm. There are many ways to implement this procedure. We implement one such transcription in Maude. It is a combination of a breadth-first search algorithm, a depth-first search algorithm and circularity checking.

The memoizer checks when an element is an instance of an element in a given memoizing set. The memoizer can be easily implemented by checking whether an element is an instance of a specific element in a set except that we need to define instance for configurations and extend to configuration pair with covariance in the first component and contravariance in the second.

Definition 4 (Configuration Instance). *We say that* $((\alpha, \gamma, S, P)$ *is an* instance *of* $(\alpha', \gamma', S', P'))$ *and write* $(\alpha, \gamma, S, P) \rightarrow (\alpha', \gamma', S', P')$ *if there exists renaming functions* ζ *for parameters and* ξ *for process variables, such that* $P = \zeta(\xi(Q))$*, and for all* p *as process variable,* $S(p) = \xi(S'(\xi^{-1}(p)))$ *and* $\gamma \Rightarrow \zeta(\gamma')$*. A configuration pair* $((\alpha, \gamma, S, P),\ (\beta, \phi, T, Q))$ *is an* instance *of* $((\alpha', \gamma', S', P'), (\beta', \phi', T', Q'))$ *if* $(\alpha, \gamma, S, P) \rightarrow (\alpha', \gamma', S', P')$ *and* $(\beta', \phi', T', Q') \rightarrow (\beta, \phi, T, Q)$*.*

In the implementation in Maude, rather than using a SMT solver to check the formula $(\gamma \Rightarrow \zeta(\gamma'))$ in Definition 4, we do a weaker syntactic check on the formula. This implementation reduces the heavy use of the SMT solver and might lead to a searching on unnecessary branches, but it will not lead to a false result.

At the collector level, we first implement our HCSP rules in Maude, then we collect all next possible moves of an HCSP configuration by unifying the HCSP configuration with the set of symbolic semantic rules and applying all possible rules on the configuration and put the results in a set. There is a central problem in the implementation of the collector: the structures of an environment condition in an HCSP configuration. we revisit them by borrowing an idea from the Union Find algorithm and Binary decision diagrams [14] in the implementation of HMC. In the HCSP semantics, we observe that at each move we only add a new constraint conjunctively to the current pre-condition. As a result, when we evaluate an HCSP process long enough, the condition of the environment gets quite large. Since SMT solvers operate in time exponential in the size o fthe input problem, it is critical to keep the size of the problems passed to them as small as possible. The collector serves to accomplish this.

Observation of executing CSP programs indicates that any given conjunct of the environment condition typically shares parameters with only a very few other individual conjuncts. We say two predicates satisfy the *parameter relation* if they share one or more parameters in common. The collector represents the environment condition as a set of sets of individual conjuncts, where each set of individual conjuncts is a connected component of the parameter relation. When a next move is calculated, the new conjunct it contributes to the environment condition is merged with the sets of conjuncts with which it satisfies the parameter relation, forming a new connected component. Since each environment condition is checked for satisfiability, checking the new environment condition can be done by checking satisfiability of just the connected component of the new conjunct contributed by the move. This typically is a much smaller formula to pass to the SMT solver than the entirety of the new environment condition.

After we have confirmed that the connected component of the new conjunct is satisfiable, the collector is also used to reduce the environment condition itself.

This is done by calculating the parameters still present in the process of the next move, and removing from the new environment condition all those connected components whose parameters are disjoint from the process parameters. Subsequent moves of a process cannot constraint any previously existing parameters not present in the process. Therefore, given that the current connected components of the environment condition are satisfiable, the satisfiability of subsequent environment conditions is not impacted by the connected components whose parameters are disjoint from the process parameters.

5 Examples and Experiment

$$\mathsf{Clicker}(c,r) = \bigsqcap_{\substack{s:s>0\wedge\\ s\leq N}} K.r.c.s \rightarrow \mathsf{Clicker}(c,r)$$

$$\mathsf{Broadcast}(r) = \Box_{c:true}\; K.r.c?s : s > 0 \wedge s \leq N \rightarrow \mathsf{Out}.r.s \rightarrow \mathsf{Broadcast}(r)$$

$$\mathsf{Room}(r) = (\mathsf{Clicker}(c1_r, r) [\![\{\}]\!] \mathsf{Clicker}(c2_r, r)) [\![\{k.x | \exists c\ s.\ \mathsf{hd}(k) = K\}]\!] \mathsf{Broadcast}(r)$$

$$\mathsf{Center} = \mathsf{Room}(1) [\![\{\}]\!] \mathsf{Room}(2)$$

$$\mathsf{ATM1} = \mathsf{Incard}?c : (M < c < N) \rightarrow \mathsf{PIN}.c \rightarrow \mathsf{Req}?n : (99 < n) \rightarrow \bigsqcap_{x:x=n\wedge bx<2000} \mathsf{Dispense}.x \rightarrow \mathsf{Outcard}.c \rightarrow \mathsf{ATM1}$$

$$\mathsf{ATM2} = \mathsf{Incard}?c : (M < c < N) \rightarrow \mathsf{PIN}.c \rightarrow \mathsf{Req}?n : (99 < n) \rightarrow \bigsqcap_{x:x=n\wedge bx<2000} \mathsf{Dispense}.x \rightarrow \mathsf{Outcard}.c \rightarrow \mathsf{ATM2} \sqcap (\mathsf{Refuse}.1 \rightarrow \mathsf{ATM2} \sqcap \mathsf{Outcard}.c \rightarrow \mathsf{ATM2})$$

Fig. 6. Examples

HSIM keeps track of the constraints of data in an HCSP process, while traditional CSP simulator, such as ProBE, lists values of data in a CSP process. The difference between two kinds of simulators is large enough without doing experiment. In this section, we focus on the experiment between HMC and traditional trace refinement model checker.

Besides the small examples in Fig. 6 and the medical mediator example from Gunter *et al.* [4], there are many other real implementations that can benefit from modeling in the HCSP system. Generally speaking, every real model with several users trying to access one or more copies of a very large database can benefit from the HCSP system. A song broadcasting system and an ATM are two such small examples, which are used to show some systems which cannot be model checked in current CSP model checkers and can be benefited from HMC.

Song broadcasting systems are used in entertainment businesses such as discos and karaokes to allow people to select songs from a large database. Such systems typically have a large collection of songs; a collection in excess of 200,000 would not be uncommon. A typical karaoke bar has more than twenty rooms for separate entertainment. Typically, each room has two remote clickers for selecting the next song to be played. After a user selects a song, the remote clicker will send the song selection to the song broadcasting system, which will play it in the room. Since only one song can be broadcast at a time, if two people send selections simultaneously, only one signal will be honored immediately, while the other one will be delayed for later action. We model the karaoke center in CSP in Figure 6. For simplicity, we assume the Karaoke center only has two rooms.

In Fig. 6, the capital letter N refers to an arbitrary number to represent the size of the database that contains all songs in the song broadcasting system. Typically, we know that the number N is a large number, but we do not know exactly how large it is. In order to verify properties in the system, such as safety and deadlock-freedom, it is better to leave the number N to be unspecified. We will set the number N to be 500 and 500,000 as test cases in the experiment.

Likewise, we implement two ATMs in Fig. 6. The two ATM processes are to describe the procedures of a machine that is receiving commands from humans and responding to them. One can easily see that ATM1 can refine ATM2 but not vice versa. We will test positive trace refinement cases of ATM1 to ATM2 as well as negative trace refinement cases of ATM2 to ATM1. In these ATMs, the numbers N and M specify the range of the debit or credit cards that can be read. Typically, a debit or credit card will have sixteen digits. We test the cases when the numbers N and M are one, four and sixteen digits.

Programs	Specifications	FDR2	HMC
Process A	Process A	< 2 secs	< 2 secs
Process B	Process B	N/A	< 2 secs
Process C	Process C	N/A	< 2 secs
Process C	Process D	N/A	< 2 secs
ATM1 one digit	ATM2 one digit	45 secs	< 2 secs
ATM1 four digits	ATM2 four digits	> 12 hours	< 2 secs
ATM1 16 digits	ATM2 16 digits	N/A	< 2 secs
ATM2 one digit	ATM1 one digit	< 2 secs	< 2 sec
ATM2 four digits	ATM1 four digits	> 12 hours	
ATM2 16 digits	ATM1 16 digits	N/A	< 2 secs
Karaoke 5	Karaoke 5	< 2 secs	37 secs
Karaoke 500	Karaoke 500	> 12 hours	37 secs
Karaoke 500,000	Karaoke 500,000	N/A	37 secs
Medical one nurse	Safety one nurse	55 mins	3.6 hours
Medical two nurses	Safety two nurses	N/A	3.8 hours

Fig. 7. Experiment Results

We have compared the efficiencies of FDR2 and HMC in some programs. The experiment was run on an Intel core i7 machine with eight gigabytes of memory and a Ubuntu 13.04 system. The testing programs are processes A, B, C and D from Section 1 and the ATMs with N and M being one, four and sixteen digits. We have tested positive trace refinement cases of ATM2 to ATM1 and negative ones of ATM1 to ATM2. In addition, we have tested the Karaoke center examples when N is equal to 500 and 500,000; and the medical mediator examples in which there are two mediators, one nurse, three patients and three devices and when there are two mediators, two nurses, three patients and three devices. The results can be found in Figure 7.

From the table, we can see that HMC can finish all the jobs, while FDR2 fails to execute some programs. In most cases, HMC is more efficient at verifying the trace refinement property of programs than is FDR2. In addition, FDR2 is very sensitive to the size of the input data, and it cannot recognize the similarity of different programs. It succeeds in model checking some programs, but fails when we change them a little bit. For example, FDR2 can execute process A completely, but fails to even read processes C and D. The medical mediator is a more representative example. Because of the sensitivity of FDR2 with respect to the input data, it can finish the job when there is only one nurse, but not when there are two. On the other hand, even though HMC needs a longer time to finish a job when the input data is small, it can successfully model check the trace refinement property no matter how big the input data gets replicated choice operators.

6 Related Work

Currently, there are several existing CSP simulators and model checkers based on the original CSP transition semantics. CSPM [8] gives a standard CSP syntax and semantics in machine readable form, introduced by Bryan Scattergood, which is based on the transition semantics introduced by Brookes and Roscoe [10]. It provides a standard for many CSP tools, including FDR2 by Formal Systems (Europe) Ltd., the industry standard for CSP model checkers [7]. ProBE [15] is a simulator created by the same group, which simulates a CSP process by listing all the actions and states one by one as a tree structure [15]. Jun Sun *et al.* [16] merged partial order reduction with the trace refinement model checking in the tool called PAT. CSP-Prover is a theorem proving tool built on top of Isabelle [17]. It provides a denotational semantics of CSP in Higher-Order Logic. CSPsim [18] is another simulator based on the CSPM standard. Its major innovation is the use of "lazy evaluation". The basic idea of CSPsim is to keep track of all the current actions, then compare them with the actions of the outside world and only select the possible executable actions for the very next step [18]. The phrase "lazy evaluation" refers to a pre-processing step in which CSPsim selects some processes that contain fewer actions and generates some conditions in advance. After that, CSPsim evaluates the whole program based on these conditions.

These tools use the traditional view of actions as single elements, and tend to generate a large number of states when comparing multiple possibilities for actions. Additionally they treat some operators, especially replicated operators, as macros, and hence, even though it is possible for some tools (CSP-Prover) to analyze some complicated programs, such as medical mediator, by the theorem proving setting, it is impossible for these tools to generate traces when the replicated set is infinite. The medical mediator project by Gunter *et al.* [4] provides an example of the advantages of HCSP over CSP-semantics based tools. The main goal of the medical mediator project is to prove that the set of traces of the process $System|[Vis]|Given$ is a subset of the traces of $Safety|[Vis]|Given$, where *Vis* is a set defined as $\{y.(\exists n\ d\ m\ x.\ y = RFIDChan_d^{n,m}.x) \vee (\exists m\ z.\ y = EHRBECh_m.z)\}$. This requires exploring all possible traces generated by the System process. Tools based on the original CSP semantics, such as FDR2, fail when dealing with large or unbounded sets. For example, the Med process as given does not put any restrictions on the sets of values that may be received over various channels. The simulator and model checker we have built based on HCSP semantics benefits from being able to handle such large or unbounded sets uniformly as single actions, thus avoiding state explosion problems.

Along the way we are writting our paper, FDR3 comes out [19]. FDR3 develops a parallelized algorithm for model checking trace refinement property over FDR2. When we use FDR3 to test our programs, the performence is better than the performence of FDR2. However, it still cannot catch the behavior of infinite sets in replicated choice operators. For example, when we test the process C and D in Figure 1, FDR3 fails to terminate.

In terms of symbolic semantics, there are several existing symbolic semantics for process algebras, mainly, serving process algebras having similar structure to the Π-calculus. Early work of Hennessy and Lin [20] provides a framework of symbolic semantics for value-passing process algebras. Later, Sangiorgi applied this symbolic semantics idea to the Π-calculus [21]. Bonchi and Montanari revisited the symbolic semantics of the Π-calculus [22], providing a symbolic transition semantics for the Π-calculus by including the predicate environment condition as a part of a label in a transition system. In their symbolic transition semantics, they only discovered the relation in the parallel operator in the Π-calculus.

LOTOS is a kind of process algebras that contains features from both the Π-calculus and CSP. Its parallel operator is similar to that of CSP. The parallel operator contains a middle set to restrict the communication actions between the left and right processes. Calder and Shankland provided a symbolic semantics for LOTOS [23]. As in the work done by Bonchi and Montanari, Calder and Shankland both represented their condition in the label position instead of dividing it into pre- and post-conditions. Pugliese, Tiezzi and Yoshida proposed a symbolic semantics for service-oriented computing COWS that is similar to the Π-calculus [24]. For the works above providing symbolic transition semantics, the condition is placed in the label instead of dividing it into pre- and post-conditions, which makes their symbolic semantics fail to answer differently for different input environments for a same program. In many cases, it is necessary to consider different initial environment conditions and these different conditions lead to different results in CSP programs.

mCRL2 is a process algebra designed to execute symbolically [25]. It is a well-known π-calculus like generic language with symbolic transition semantics to model point-to-point communication. They claimed that people can catch the behavior of infinite set in their choice summation operator. However, the set needs to be determined statically. It means that it cannot catch the behavior similar to process C in Figure 1. In addition, even though mCRL2 claimed the language is generic and we can translate other process algebra into mCRL2, it is very hard for mCRL2 to model a broadcast communication system with point-to-group communication, because they need to know the total number of processes in the universe and send enough messages to each individual process in a group. However, knowing total number of processes is very hard in some cases. For example, if we want to model a group of people who are in a conference reach a consensus at the same time. It is almost impossible for mCRL2 to model this procedure. On the other hand, we can model this procedure easily by using a replicated choice operator to select the total number of people in the conference, then using a replicated parallel operator with consensus value in between to model the fact that all people communicate with each other by the consensus value.

7 Conclusion and Future Work

In this paper we have presented a new semantics for CSP, the HCSP semantics. HSCP provides an alternative way to model CSP processes by viewing transitions as bundles of the original transitions, where all transitions in the bundle can be described by a uniform property derived from the process. By this translation, we can allow HCSP-based tools to run some CSP programs which are not able to run in the original CSP-based tools. We have shown the HCSP semantics to be equivalent to the original CSP transition semantics. We have also presented an HCSP-based simulator, which is extracted directly from the Isabelle code for the HCSP semantics. We show an HCSP-based model checker to check the trace refinement of CSP programs and show that the model checker is very efficient to deal with some CSP programs by experiment. By using several examples in the experiment, we show that the HCSP semantics based trace refinement model checker can overcome some difficulties that traditional CSP-semantics-based tools cannot handle.

For further study, we are interested in adding semantics to deal with the replicated parallel operators in HCSP and use it in the model checker and extending our trace refinement model checker to check trace failure and failure-divergence refinement properties of CSP programs. We believe that it will significantly increase the efficiency in the model checker to answer trace refinement problem. We also want to generalize our framework to deal with other kinds of transition semantics.

Acknowledgments. This material is based upon work supported in part by NASA Contract NNA10DE79C and NSF Grant 0917218. Any opinions, findings, and conclusions or recommendations expressed in this material are those of the authors and do not necessarily reflect the views of NASA or NSF.

References

1. Brookes, S.D., Hoare, C.A.R., Roscoe, A.W.: A theory of communicating sequential processes. J. ACM 31(3), 560–599 (1984)
2. Paulson, L.C.: Isabelle: The next 700 theorem provers. In: Odifreddi, P. (ed.) Logic and Computer Science, pp. 361–386. Academic Press (1990)
3. Clavel, M., Durán, F., Eker, S., Lincoln, P., Martí-Oliet, N., Meseguer, J., Quesada, J.F.: Maude: Specification and programming in rewriting logic (2001)
4. Gunter, E.L., Yasmeen, A., Gunter, C.A., Nguyen, A.: Specifying and analyzing workflows for automated identification and data capture. In: HICSS, pp. 1–11. IEEE Computer Society (2009)
5. Wong, P.Y.H., Gibbons, J.: A process-algebraic approach to workflow specification and refinement. In: Lumpe, M., Vanderperren, W. (eds.) SC 2007. LNCS, vol. 4829, pp. 51–65. Springer, Heidelberg (2007)
6. Wright, P., Fields, B., Harrison, M.: Deriving human-error tolerance requirements from tasks. In: Proc. of ICRE 1994 IEEE Intl. Conf. on Requirements Engineering, pp. 135–142. IEEE (1994)
7. Ltd., F.S.E.: Failures-divergence refinement. FDR2 user manual. In: FDR2 User Manual (2010)
8. Scattergood, J.: The Semantics and Implementation of Machine-Readable CSP. D.phil., Oxford University Computing Laboratory (1998)
9. Li, L., Gunter, E.L., Mansky, W.: Symbolic semantics for csp. Technical report, Department of Computer Science, University of Illinois at Urbana-Champaign (2013)
10. Brookes, S.D., Roscoe, A.W., Walker, D.J.: An operational semantics for CSP. Technical report, Oxford University Computing Laboratory (1986)
11. Hoare, C.A.R.: An axiomatic basis for computer programming. Commun. ACM 12(10), 576–580 (1969)
12. Roscoe, A.W.: The Theory and Practice of Concurrency. Prentice Hall PTR, Upper Saddle River (1997)
13. Back, R.J.R., Wright, J.V.: Trace refinement of action systems. In: Jonsson, B., Parrow, J. (eds.) CONCUR 1994. LNCS, vol. 836, pp. 367–384. Springer, Heidelberg (1994)
14. Groote, J.F.: Binary decision diagrams for first order predicate logic
15. Ltd, F.S.E.: Process behaviour explorer. ProBE user manual. In: ProBE User Manual (2003)
16. Sun, J., Liu, Y., Dong, J.S.: Model checking csp revisited: Introducing a process analysis toolkit. In: Margaria, T., Steffen, B. (eds.) ISoLA 2008, Springer. CCIS, vol. 17, pp. 307–322. Springer, Heidelberg (2008)
17. Isobe, Y., Roggenbach, M.: A complete axiomatic semantics for the CSP stable-failures model. In: Baier, C., Hermanns, H. (eds.) CONCUR 2006. LNCS, vol. 4137, pp. 158–172. Springer, Heidelberg (2006)
18. Brooke, P.J., Paige, R.F.: Lazy exploration and checking of CSP models with CSPsim. In: McEwan, A.A., Schneider, S.A., Ifill, W., Welch, P.H. (eds.) CPA. Concurrent Systems Engineering Series, vol. 65, pp. 33–49. IOS Press (2007)
19. Gibson-Robinson, T., Armstrong, P., Boulgakov, A., Roscoe, A.W.: Fdr3 - a modern refinement checker for csp. In: Ábrahám, E., Havelund, K. (eds.) TACAS 2014 (ETAPS). LNCS, vol. 8413, pp. 187–201. Springer, Heidelberg (2014)
20. Hennessy, M., Lin, H.: Symbolic bisimulations. Theor. Comput. Sci. 138(2), 353–389 (1995)

21. Sangiorgi, D.: A theory of bisimulation for the pi-calculus. In: Best, E. (ed.) CONCUR 1993. LNCS, vol. 715, pp. 127–142. Springer, Heidelberg (1993)
22. Bonchi, F., Montanari, U.: Symbolic semantics revisited. In: Amadio, R.M. (ed.) FOSSACS 2008. LNCS, vol. 4962, pp. 395–412. Springer, Heidelberg (2008)
23. Calder, M., Shankland, C.: A symbolic semantics and bisimulation for full lotos. In: Proc. Formal Techniques for Networked and Distributed Systems, pp. 184–200. Kluwer Academic Publishers, FORTE xXIV (2001)
24. Pugliese, R., Tiezzi, F., Yoshida, N.: A symbolic semantics for a calculus for service-oriented computing. Electron. Notes Theor. Comput. Sci. 241, 135–164 (2009)
25. Groote, J.F., Mathijssen, A., Reniers, M., Usenko, Y., Weerdenburg, M.V.: The formal specification language mcrl2. In: Proceedings of the Dagstuhl Seminar. MIT Press (2007)

A Heterogeneous Characterisation of Component-Based System Design in a Categorical Setting

Carlos Gustavo Lopez Pombo[1,3], Pablo F. Castro[2,3], Nazareno Aguirre[2,3], and Tomas S.E. Maibaum[4]

[1] Department of Computing, FCEyN, Universidad de Buenos Aires, Argentina
[2] Department of Computing, FCEFQyN, Universidad Nacional de Río Cuarto, Argentina
[3] Consejo Nacional de Investigaciones Científicas y Tecnológicas (CONICET), Argentina
[4] Department of Computing & Software, McMaster University, Canada

Abstract. In component-based design, components and communication mechanisms have a different nature; while the former represent the agents that cooperate to fulfill a certain goal, the latter formalise the communication mechanism through which these agents interact. A proper formalisation of the heterogeneity that arises from this difference requires one to employ the most adequate formalism for each of the parts of a specification and then proceed to merge the parts of the system specification characterised in different languages. The approach we propose in this paper is based on the notion of *institution*, and makes extensive use of *institution representations* in order to relate the specifications of components and communication mechanisms, each of which might be expressed in different formalisms. The contribution focuses on providing tools needed to engineer heterogeneous languages arising from particular choices for the specification of components and communication devices.

1 Introduction

Nowadays, software artefacts are ubiquitous in our lives, being an essential part of home appliances, cars, cell phones, and even in more critical activities like aeronautics and health sciences. In this context, software failures may produce enormous losses, either economical or, in the worst case, in human lives. In order to provide better guarantees for the correct functioning of software, various elements are necessary, among which *formal foundations*, that enable a precise reasoning about software, and *modularity*, that helps in dealing with software complexity, are crucial.

The importance of modularity has been acknowledged since the foundational work of Parnas, which promoted building software artefacts (and more specifically software specifications) modularly, enhancing reusability of modules and contributing to a better separation of concerns, and leading to an improved quality in specification and development. Modularisation is generally understood as the process of dividing a system specification, or implementation, into

G. Ciobanu and D. Méry (Eds.): ICTAC 2014, LNCS 8687, pp. 314–332, 2014.

manageable parts: the *modules* or *components*. It leads to a structural view of systems, called *architecture*, as described in [16], in which the relevance of *component interaction* is brought out. Aside from its crucial relevance for managing the complexity of systems, a system's architectural structure also plays an important role in its functional and non-functional characteristics.

Given the relevance of software architecture, its formal foundations are essential to guarantee the correct functioning of component based systems. There exist various approaches to formally capture component based systems, which are either language-specific (e.g., formalisations of schema operators in Z or structuring mechanisms in B), making its results difficult to generalise to other component-based settings, or target specific ways of communicating components (e.g., formalising particular communication mechanisms, such as synchronisation in process algebra based approaches). Moreover, these approaches support communication mechanisms that are influenced (or defined) by the nature of the components they communicate (e.g., synchronisation in event based models, or shared memory in state based models).

In this work, we tackle the above described limitations of existing formalisations of component based systems by introducing an *abstract* and *heterogeneous* categorical characterisation of component-based systems. Our characterisation is presented in a logic or language independent setting, by making use of the notion of *institution*. Moreover, the approach is *heterogeneous*, favouring a more genuine separation of concerns in the specification of components, and that of the communication mechanisms. Finally, although there exist other abstract and heterogeneous approaches, most notably the work related to The Heterogeneous Tool Set HETS [26,27], our approach differs from these in that it focuses on providing a formal characterisation of the elements of the domain of component-based software architecture (to be further discussed in Sec. 5).

The practical usefulness of heterogeneous specification formalisms in the context of component-based systems is acknowledged by the existence of languages such as Acme [17] (and others), designed with the aim of putting together specifications originating in different formalisms. Generally, heterogeneity arises from two different angles: it arises from the fact that the description of each component or module could be given in a different specification language; and as a consequence of components and communication mechanisms being of a different nature. The existing literature concentrates on the first kind of heterogeneity; we will devote this work to providing formal foundations for the second one.

The formal tools we use in this paper are those coming from the field of category theory, more specifically from the domain of algebraic specifications [6]. They have been shown to be useful for enabling the formal characterisation of different kinds of specification structuring mechanisms and refinement in different settings; a few examples are: [7,9,22,33]. We employ a well established abstract definition of logical systems, known as *institutions* [18], to achieve generality (in the sense of the approach being language independent). In order to appropriately combine different formalisms, we make extensive use of *institution representations* [30]. These serve the purpose of relating and, consequently, combining different (abstract) logics used for different purposes in a given specification, e.g., those used for component and connector specifications.

In summary, the main contributions of this work are: *a*) providing a formal, and language independent, interpretation of the concepts arising from the field of software architecture, and *b*) providing formal foundations for the heterogeneity observed when components interact through communication channels.

2 Preliminaries

From now on, we assume that the reader has a nodding acquaintance with the basic definitions of category theory, including the concepts of category, functor, natural transformation, etc. We mainly follow the notation of [23]: given a category $\mathbf{C}$, $|\mathbf{C}|$ denotes its collection of objects, while $||\mathbf{C}||$ denotes its collection of arrows. $g \circ f : A \to C$ denotes the composition of arrows $f : A \to B$ and $g : B \to C$. Natural transformations will be indicated with the arrow $\dot{\to}$.

The theory of institutions [18] provides a formal definition of logical system, and of how specifications in a logical system can be put together. They also serve as a suitable framework for addressing the problem of heterogeneity [27,32].

Definition 1. *An* institution *is a structure of the form* $\langle \mathsf{Sign}, \mathbf{Sen}, \mathbf{Mod}, \{\models^{\Sigma}\}_{\Sigma \in |\mathsf{Sign}|}\rangle$ *satisfying the following conditions: a)* Sign *is a category of signatures, b)* $\mathbf{Sen} : \mathsf{Sign} \to \mathsf{Set}$ *is a functor (let* $\Sigma \in |\mathsf{Sign}|$*, then* $\mathbf{Sen}(\Sigma)$ *corresponds to the set of* Σ*-formulae), c)* $\mathbf{Mod} : \mathsf{Sign}^{\mathsf{op}} \to \mathsf{Cat}$ *is a functor (let* $\Sigma \in |\mathsf{Sign}|$*, then* $\mathbf{Mod}(\Sigma)$ *corresponds to the category of* Σ*-models), d)* $\{\models^{\Sigma}\}_{\Sigma \in |\mathsf{Sign}|}$*, where* $\models^{\Sigma} \subseteq |\mathbf{Mod}(\Sigma)| \times \mathbf{Sen}(\Sigma)$*, is a family of binary relations. Such that, for every signature morphism* $\sigma : \Sigma \to \Sigma' \in ||\mathsf{Sign}||$*,* $\phi \in \mathbf{Sen}(\Sigma)$ *and* $\mathcal{M}' \in |\mathbf{Mod}(\Sigma)|$ *the following* $\models$*-invariance condition must hold:* $\mathcal{M}' \models^{\Sigma'} \mathbf{Sen}(\sigma)(\phi)$ *iff* $\mathbf{Mod}(\sigma^{\mathsf{op}})(\mathcal{M}') \models^{\Sigma} \phi$.

Institutions are an abstract formulation of the notion of logical system, more specifically, of its model theory, where the concepts of languages, models and truth are characterised in a category theoretic way. Examples of institutions are: propositional logic, equational logic, first-order logic, first-order logic with equality, dynamic logics and temporal logics (a detailed list is given in [18]).

Let $\langle \mathsf{Sign}, \mathbf{Sen}, \mathbf{Mod}, \{\models^{\Sigma}\}_{\Sigma \in |\mathsf{Sign}|}\rangle$ be an institution, $\Sigma \in |\mathsf{Sign}|$ and $\Gamma \subseteq \mathbf{Sen}(\Sigma)$ then, we define the functor $\mathbf{Mod}(\Sigma, \Gamma)$ as the full subcategory of $\mathbf{Mod}(\Sigma)$ determined by those models $\mathcal{M} \in |\mathbf{Mod}(\Sigma)|$ such that for all $\gamma \in \Gamma$, $\mathcal{M} \models^{\Sigma} \gamma$. We also overload the symbol $\models^{\Sigma}$, to define a relation between sets of formulae and formulae, as follows: $\Gamma \models^{\Sigma} \alpha$ if and only if $\mathcal{M} \models^{\Sigma} \alpha$ for all $\mathcal{M} \in |\mathbf{Mod}(\Sigma, \Gamma)|$. where $\alpha \in \mathbf{Sen}(\Sigma)$.

Definition 2. *We define the category of theory presentations as the pair* $\langle \mathcal{O}, \mathcal{A}\rangle$ *(usually denoted as* Th*) where:* $\mathcal{O} = \{\, \langle \Sigma, \Gamma\rangle \mid \Sigma \in |\mathsf{Sign}|$ *and* $\Gamma \subseteq \mathbf{Sen}(\Sigma) \,\}$*, and*

$$\mathcal{A} = \left\{ \sigma : \langle \Sigma, \Gamma\rangle \to \langle \Sigma', \Gamma'\rangle \;\middle|\; \begin{array}{l} \langle \Sigma, \Gamma\rangle, \langle \Sigma', \Gamma'\rangle \in \mathcal{O}, \sigma : \Sigma \to \Sigma' \in ||\mathsf{Sign}|| \\ \text{and } \Gamma' \models^{\Sigma'} \mathbf{Sen}(\sigma)(\Gamma) \end{array} \right\}.$$

In addition, if a morphism $\sigma : \langle \Sigma, \Gamma \rangle \to \langle \Sigma', \Gamma' \rangle$ satisfies $\mathbf{Sen}(\sigma)(\Gamma) \subseteq \Gamma'$, it is called *axiom preserving*. By keeping only those morphisms of Th which are axiom preserving we obtain the category Th_0.

As we mentioned before, an institution is a structure $\langle \mathsf{Sign}, \mathbf{Sen}, \mathbf{Mod}, \{\models^{\Sigma}\}_{\Sigma \in |\mathsf{Sign}|} \rangle$ so from now on, whenever we make reference to a given institution I we will be referring to the structure $\mathsf{I} = \langle \mathsf{Sign}^{\mathsf{I}}, \mathbf{Sen}^{\mathsf{I}}, \mathbf{Mod}^{\mathsf{I}}, \{\models^{\mathsf{I}}_{\Sigma}\}_{\Sigma \in |\mathsf{Sign}^{\mathsf{I}}|} \rangle$.

Next, we define the notion of *institution representation* (also find in the literature under the name of co-morphism)[30, Def. 12, Sec. 5].

Definition 3. *Let I and I' be institutions then, the structure $\langle \gamma^{Sign}, \gamma^{Sen}, \gamma^{Mod} \rangle : I \to I'$ is an* institution representation *if and only if: a) $\gamma^{Sign} : \mathbf{Sign} \to \mathbf{Sign}'$ is a functor, b) $\gamma^{Sen} : \mathbf{Sen} \dot{\to} \mathbf{Sen}' \circ \gamma^{Sign}$, is a natural transformation such that for every $\Sigma_1, \Sigma_2 \in |\mathbf{Sign}|$ and $\sigma : \Sigma_1 \to \Sigma_2 \in ||\mathbf{Sign}||$, $\gamma^{Sen}_{\Sigma_2} \circ \mathbf{Sen}(\sigma) = \mathbf{Sen}'(\gamma^{Sign}(\sigma)) \circ \gamma^{Sen}_{\Sigma_1}$, c) $\gamma^{Mod} : \mathbf{Mod}' \circ (\gamma^{Sign})^{\mathsf{op}} \dot{\to} \mathbf{Mod}$, is a natural transformation such that for every $\Sigma_1, \Sigma_2 \in |\mathbf{Sign}|$ and $\sigma : \Sigma_1 \to \Sigma_2 \in ||\mathbf{Sign}||$, $\mathbf{Mod}(\sigma^{\mathsf{op}}) \circ \gamma^{Mod}_{\Sigma_2} = \gamma^{Mod}_{\Sigma_1} \circ \mathbf{Mod}'((\gamma^{Sign})^{\mathsf{op}}(\sigma^{\mathsf{op}}))$, such that, for any $\Sigma \in |\mathbf{Sign}|$, $\gamma^{Sen}_{\Sigma} : \mathbf{Sen}(\Sigma) \to \mathbf{Sen}'(\gamma^{Sign}(\Sigma))$ and $\gamma^{Mod}_{\Sigma} : \mathbf{Mod}'(\gamma^{Sign}(\Sigma)) \to \mathbf{Mod}(\Sigma)$ preserve the following satisfaction condition: for any $\alpha \in \mathbf{Sen}(\Sigma)$ and $\mathcal{M}' \in |\mathbf{Mod}(\gamma^{Sign}(\Sigma))|$, $\mathcal{M}' \models_{\gamma^{Sign}(\Sigma)} \gamma^{Sen}_{\Sigma}(\alpha)$ iff $\gamma^{Mod}_{\Sigma}(\mathcal{M}') \models_{\Sigma} \alpha$.*

An institution representation intuitively corresponds to the relationship between a given institution, and how it is interpreted into another one, in a semantics preserving way. It may be regarded as a realisation for institutions of the established concept of property preserving translation between two theories in two different logics.

Property 1. Let $\mathbf{I}$ and $\mathbf{I}'$ be institutions, and $\rho : \mathbf{I} \to \mathbf{I}'$ an institution representation. For every signature $\Sigma \in |\mathbf{Sign}|$, set $\Phi \subseteq \mathbf{Sen}(\Sigma)$ of Σ-sentences, and Σ-sentence $\varphi \in \mathbf{Sen}(\Sigma)$, if $\Phi \models_{\Sigma} \varphi$, then $\rho^{Sen}_{\Sigma}(\Phi) \models'_{\rho^{Sign}(\Sigma)} \rho^{Sen}_{\Sigma}(\varphi)$.

We will also use a few other categorical concepts, such as that of a *bicategory* and a *lax functor*; the interested reader is referred to [23].

3 A Characterisation of Component-Based Design

Two main goals of our approach to component-based specification are *generality* and *abstraction*, in the sense that we have designed our characterisation of component-based systems to be specification language independent. We start by providing a category theoretic characterisation of the concepts used in the field of software architecture. There, the building blocks for software design are *components*, *glues*, *ports*, *roles* and *adaptors*.

Components. A component describes an independent computational unit of a system. They are formally characterised by theories in a given institution. Note that the approach presented here does not prevent the introduction of another level of heterogeneity where each component is specified in a different formalism,

Component: *Producer*
Attributes: *p-current*: Bit, *p-waiting*: Bool,
ready-in: Bool
Actions: *produce-0*, *produce-1*, *send-0*, *send-1*,
p-init
Axioms:
1. $\Box(p\text{-}init \rightarrow \bigcirc(p\text{-}current = 0 \wedge \neg p\text{-}waiting))$
2. $\Box(produce\text{-}0 \vee produce\text{-}1 \rightarrow \neg p\text{-}waiting \wedge \bigcirc p\text{-}waiting)$
3. $\Box(produce\text{-}0 \rightarrow \bigcirc(p\text{-}current = 0))$
4. $\Box(produce\text{-}1 \rightarrow \bigcirc(p\text{-}current = 1))$
5. $\Box((send\text{-}0 \rightarrow p\text{-}current = 0) \wedge (send\text{-}1 \rightarrow p\text{-}current = 1))$
6. $\Box(send\text{-}0 \vee send\text{-}1 \rightarrow p\text{-}waiting \wedge \bigcirc\neg p\text{-}waiting)$
7. $\Box(send\text{-}0 \vee send\text{-}1 \rightarrow ready\text{-}in \wedge p\text{-}current = \bigcirc p\text{-}current)$
8. $\Box(send\text{-}0 \vee send\text{-}1 \vee produce\text{-}0 \vee produce\text{-}1 \vee p\text{-}init \vee (p\text{-}current = \bigcirc p\text{-}current \wedge p\text{-}waiting = \bigcirc p\text{-}waiting))$

Component: Consumer
Attributes: *c-current*: Bit, *c-waiting*: Bool,
ready-ext: Bool
Actions: *consume*, *extract-0*, *extract-1*, *c-init*
Axioms:
1. $\Box(c\text{-}init \rightarrow \bigcirc(c\text{-}current = 0 \wedge c\text{-}waiting))$
2. $\Box(extract\text{-}0 \vee extract\text{-}1 \rightarrow c\text{-}waiting \wedge \bigcirc\neg c\text{-}waiting \wedge ready\text{-}ext)$
3. $\Box(extract\text{-}0 \rightarrow \bigcirc(c\text{-}current = 0))$
4. $\Box(extract\text{-}1 \rightarrow \bigcirc(c\text{-}current = 1))$
5. $\Box(consume \rightarrow \neg c\text{-}waiting \wedge \bigcirc c\text{-}waiting)$
6. $\Box(consume \rightarrow c\text{-}current = \bigcirc c\text{-}current)$
7. $\Box(consume \vee extract\text{-}0 \vee extract\text{-}1 \vee c\text{-}init \vee (c\text{-}current = \bigcirc c\text{-}current \wedge c\text{-}waiting = \bigcirc c\text{-}waiting))$

Fig. 1. A Producer-Consumer specification

in which case the language must deal with this internally, e.g., by instantiating the institution for describing components with a Grothendieck institution formed by all the specification languages needed for the task, as proposed in [5,27]. Without loss of generality we will assume a single language for component specification formalised as an institution, which will be referred to as $\mathsf{I}^{\mathsf{Comp}}$.

Example 1. (A simple producer and consumer specification) In Fig. 1 we present a formalisation of a producer and a consumer in *propositional temporal logic* [13], a specification language based on *linear temporal logic* [24]; as a consequence, our specifications are state-based. For simplicity, we assume that messages are bits identified by the type *Bit*. The producer's state is defined by a bit-typed attribute *p-current* to store a produced element, a *boolean* attribute *p-waiting* to indicate whether an item is already produced and ready to be sent (so that null values for items are not necessary), and a boolean attribute *ready-in*, so that a producer is informed by the environment when this is ready to receive a product. This specification consists of a set of sorts (*Bit* and *Bool*, in this case), a set of attributes (i.e., flexible variables), some of which are supposed to be controlled by the environment, and a set of action symbols (they are flexible boolean variables indicating the occurrence of an action). The axioms of the specification are linear temporal logic formulae characterising the behaviour of the component, in a rather obvious way. The consumer component can be specified in a similar way.

Notice that these components are coherent with the notion of component in software architecture [16]. The theory associated with a component represents the computational aspects of it, in our case indicating via axioms (and their corresponding consequences) the behaviour of the actions of the component and their effect on its state. It is worth noticing that components described below do not formalise any aspect of the communication between them.

Ports. Ports constitute the communication *interfaces* of a component. As in [13], ports can be captured by using *channels*, which consist of theories with no axioms. Given $A \in |\mathsf{Th_0}^{\mathsf{I}^{\mathsf{Comp}}}|$, a port for A is a morphism $\sigma : \mathbf{Th}(\Sigma) \to A \in ||\mathsf{Th_0}^{\mathsf{I}^{\mathsf{Comp}}}||$ such that $\Sigma \in |\mathsf{Sign}^{\mathsf{I}^{\mathsf{Comp}}}|$.[1]

In software architecture, emphasis is put on the explicit description of communication aspects of a system, separated from the computational, component related, aspects. Below we define the elements relevant in our formalisation, various of which are inspired by the formal approach to interaction characterisation put forward in [12].

Glues. In a communication mechanism between two components, a glue captures the way in which these components interact, that is, computational aspects of the interaction, e.g., a protocol. In our setting, glues are also required to be organised in an institution, thus being theories in a given specification language. From now on, the institution used to specify glues will be denoted as $\mathsf{I}^{\mathsf{Glue}}$. Also in this case, similarly to components, the use of more than one specification language can be considered, so different glues can be described by means of different languages.

Roles. Roles constitute the interfaces of glues. Thus, given $G \in |\mathsf{Th_0}^{\mathsf{I}^{\mathsf{Glue}}}|$, a role for G is a morphism $\pi : \mathbf{Th}(\Sigma) \to G \in ||\mathsf{Th_0}^{\mathsf{I}^{\mathsf{Glue}}}||$ such that $\Sigma \in |\mathsf{Sign}^{\mathsf{I}^{\mathsf{Glue}}}|$.

Connector. A connector represents a mechanism for interconnecting two components, and establishes: *a*) the roles of the glue, and *b*) the glue itself. Given a glue $G \in |\mathsf{Th_0}^{\mathsf{I}^{\mathsf{Glue}}}|$, and the roles $\sigma_1 : \mathbf{Th}(\Sigma_1) \to G, \sigma_2 : \mathbf{Th}(\Sigma_2) \to G \in ||\mathsf{Th_0}^{\mathsf{I}^{\mathsf{Glue}}}||$ for G, a connector with behaviour G and roles σ_1 and σ_2, is a structure of the form $\langle \sigma_1, G, \sigma_2 \rangle$. If we restrict ourselves to binary connectors, they are required to be organised as a subcategory of the bicategory $\mathbf{co-spans}(\mathsf{Th}_0^{\mathsf{I}^{\mathsf{Glue}}})$, denoted by $\mathbf{Connector}(\mathsf{I}^{\mathsf{Glue}})$. The generalisation to n-ary connectors is straightforward but their category theoretic characterisation is no longer a bicategory but a generalised version of it.

As mentioned above, heterogeneity arises, in architecture description languages, as a reflection of the different nature of components and communication mechanisms. This led us to the use of separate institutions for formalising components (and their ports) and the glues (and their roles). So, we need a way of establishing the relationship between component and connector specifications, or more generally, between their corresponding institutions. There exist various mechanisms for relating institutions, each with a particular meaning when interpreted in the context of software design. The interested reader is referred to [30,19], where the authors make a thorough study of these mechanisms. We can

[1] $\mathbf{Th} : \mathsf{Sign}^{\mathsf{I}^{\mathsf{Comp}}} \to \mathsf{Th_0}^{\mathsf{I}^{\mathsf{Comp}}}$ is the right adjoint of the forgetful functor $\mathbf{Sign} : \mathsf{Th}^{\mathsf{I}^{\mathsf{Comp}}} \to \mathsf{Sign}^{\mathsf{I}^{\mathsf{Comp}}}$.

use Property 1 to draw the relationship between the institutions $\mathsf{I}^{\mathsf{Comp}}$ and $\mathsf{I}^{\mathsf{Glue}}$ needed to be able to obtain a complete description of a system as communicating components. Such a relationship is captured as follows. Let $\mathsf{I}^{\mathsf{Sys}}$ be an institution, and let $\gamma^{Comp} : \mathsf{I}^{\mathsf{Comp}} \to \mathsf{I}^{\mathsf{Sys}}$ and $\gamma^{Glue} : \mathsf{I}^{\mathsf{Glue}} \to \mathsf{I}^{\mathsf{Sys}}$ be institution representations. Thus, $\mathsf{I}^{\mathsf{Sys}}$ serves as a common formal language, in which the components and connectors of a system can be interpreted and put together.

Example 2. (Connecting components and connectors directly) A straightforward way of establishing links between components and connectors is by requiring the roles and the ports one wants to connect to be equal when they are translated to the $\mathsf{I}^{\mathsf{Sys}}$ institution. This situation is illustrated in Fig. 2, a) for a component denoted as **A** and a glue denoted as **G**; connections between **G** and a different component **B** are analogous. Observing the upper part of the diagram, component **A** communicates using ports $\pi_A \to \mathbf{A}$, using a medium characterised by the connector formed by the glue **G** and the role $\rho_l \to \mathbf{G}$, to be attached to the port $\pi_A \to \mathbf{A}$. Dashed arrows express the application of the corresponding institution representation to the theories and morphisms appearing in the upper part of the diagram in order to provide a homogeneous description of the whole system in the institution $\mathsf{I}^{\mathsf{Sys}}$. The bottom part of the diagram shows how things are put together in $\mathsf{I}^{\mathsf{Sys}}$, thus obtaining a diagram, in the usual sense of category theory, consisting of the behaviour associated with component **A** and glue **G**.

This simple way of connecting the components, though correct, has some limitations. The differences between $\mathsf{I}^{\mathsf{Comp}}$ and $\mathsf{I}^{\mathsf{Glue}}$ might not be merely syntactical, but sometimes their semantics also need to be "*harmonized*". Assume, for instance, that we use Propositional Dynamic Logic (PDL) [20] in order to describe the components of a system, whereas the glues are formalised in Linear Temporal Logic (LTL). The models of these two logics have different structures, since LTL models are interpret formulas along traces, while PDL models have state-based semantics. Even when a more expressive logic might be capable of interpreting both PDL and LTL theories, the coordination of the semantic objects cannot always be obtained merely by a syntactic identification in the more expressive logic. Following the principles of software architecture, we deal with this problem using so called *adaptors.*

Adaptor. An adaptor is a connector in $\mathsf{I}^{\mathsf{Sys}}$. The intuition behind the inclusion of adaptors is that roles will interact with ports, not only at a syntactic level as shown in Ex. 2, but mediated by a semantic synchronisation of models, induced by the axioms of the theory characterising the connector. Adaptors in software architecture serve the purpose of solving or alleviating architectural mismatches. In our case, the (potential) mismatch is related to the difference between the logics used for the specification of the components and the connectors.

Example 3. Adaptors help in establishing the links between roles and ports. This situation is illustrated in Fig. 2 b) for a component denoted as **A**, a glue denoted as **G** and an adaptor Γ_{GA}; connections between **G** and a different component **B** are analogous.

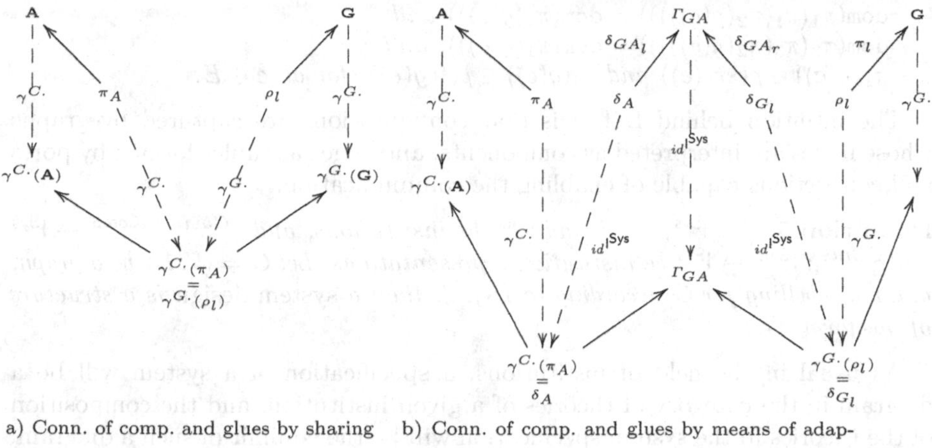

a) Conn. of comp. and glues by sharing ports and roles.

b) Conn. of comp. and glues by means of adaptors.

Fig. 2. Connections between components, glues, adaptors, roles and ports. **Note:** $\gamma^{C\cdot} \to \gamma^{Comp^{\mathsf{Th}_0}}$, $\gamma^{G\cdot} \to \gamma^{Glue^{\mathsf{Th}_0}}$.

Connection. A *connection* is formed by a connector together with a pair of adaptors linking the ports of the components participating in the communication. Let $\mathsf{I}^{\mathsf{Sys}}$, $\mathsf{I}^{\mathsf{Comp}}$ and $\mathsf{I}^{\mathsf{Glue}}$ be institutions, $\gamma^{Comp} : \mathsf{I}^{\mathsf{Comp}} \to \mathsf{I}^{\mathsf{Sys}}$ and $\gamma^{Glue} : \mathsf{I}^{\mathsf{Glue}} \to \mathsf{I}^{\mathsf{Sys}}$ be institution representations. Let $\pi = \langle \pi_l : \rho_l \to G, G, \pi_r : \rho_r \to G \rangle \in |\mathbf{Connector}(\mathsf{I}^{\mathsf{Glue}})|$ and $\delta_{GA} = \langle \delta_{GAl} : \delta_A \to \Gamma_{GA}, \Gamma_{GA}, \delta_{GAr} : \delta_{G_l} \to \Gamma_{GA} \rangle, \delta_{GB} = \langle \delta_{GB_l} : \delta_{G_r} \to \Gamma_{GB}, \Gamma_{GB}, \delta_{GB_r} : \delta_B \to \Gamma_{GB} \rangle \in |\mathbf{Connector}(\mathsf{I}^{\mathsf{Sys}})|$. Then a *connection* is a structure of the form $\langle \delta_{GA}, \pi, \delta_{GB} \rangle$ such that $\delta_{G_l} = \gamma^{Glue^{\mathsf{Th}_0}}(\rho_l)$ and $\delta_{G_r} = \gamma^{Glue^{\mathsf{Th}_0}}(\rho_r)$. Given the institutions $\mathsf{I}^{\mathsf{Sys}}$ and $\mathsf{I}^{\mathsf{Glue}}$ such that $\gamma^{Glue} : \mathsf{I}^{\mathsf{Glue}} \to \mathsf{I}^{\mathsf{Sys}}$ is an institution representation, the connections definable over these two institutions will be the complete subcategory of $\mathbf{Connector}(\mathsf{I}^{\mathsf{Sys}}) \times \mathbf{Connector}(\mathsf{I}^{\mathsf{Glue}}) \times \mathbf{Connector}(\mathsf{I}^{\mathsf{Sys}})$ whose objects are those triples satisfying the conditions stated above and will be denoted as $\mathbf{Connection}(\mathsf{I}^{\mathsf{Glue}}, \mathsf{I}^{\mathsf{Sys}})$.

The previous definitions allow us to formalise in a categorical setting the main notions involved in component-based designs as a labeled graph. The next definition formalises graph labelings.

Definition 4. *Let $\mathsf{I}^{\mathsf{Sys}}$, $\mathsf{I}^{\mathsf{Comp}}$ and $\mathsf{I}^{\mathsf{Glue}}$ be institutions, and $\gamma^{Comp} : \mathsf{I}^{\mathsf{Comp}} \to \mathsf{I}^{\mathsf{Sys}}$ and $\gamma^{Glue} : \mathsf{I}^{\mathsf{Glue}} \to \mathsf{I}^{\mathsf{Sys}}$ be institution representations. Let $G = \langle V, E \rangle$ be a graph; then a* labeling *ι for G is a structure of the form $\langle f : V \to |\mathsf{Th}_0{}^{\mathsf{I}^{\mathsf{Comp}}}|, p : V \to 2^{||\mathsf{Th}_0{}^{\mathsf{I}^{\mathsf{Comp}}}||}, g : E \to |\mathsf{Th}_0{}^{\mathsf{I}^{\mathsf{Comp}}}| \times \mathbf{Connection}(\mathsf{I}^{\mathsf{Glue}}, \mathsf{I}^{\mathsf{Sys}}) \times |\mathsf{Th}_0{}^{\mathsf{I}^{\mathsf{Comp}}}| \rangle$ such that:*

- $p(v) \subseteq \left\{ \mathbf{Th}(\sigma) \,\middle|\, \sigma : \Sigma \to \mathbf{Sign}(f(v)) \in ||\mathsf{Sign}^{\mathsf{I}^{\mathsf{Comp}}}|| \right\}$*, for all $v \in V$,*
- *let π_1, π_2 and π_3 are the first, second and third projections of a tuple, respectively, and* `dom` *retrieves the domain of a morphism, then for all $e \in E$,*

$\mathtt{dom}(\pi_1(\pi_1(\pi_2(g(e))))) = \mathtt{dom}(\pi_1(g(e)))$ *and*
$\mathtt{dom}(\pi_3(\pi_3(\pi_2(g(e))))) = \mathtt{dom}(\pi_1(g(e)))$, *and*

– $\pi_1(g(e)) \in p(src(e))$ *and* $\pi_3(g(e)) \in p(trg(e))$, *for all* $e \in E$.

The intuition behind Def. 4 is that configurations are captured by graphs whose nodes are interpreted as components, and edges as tuples formed by ports and connections capable of enabling the communication.

Definition 5. *Let* $\mathsf{I}^{\mathsf{Sys}}$, $\mathsf{I}^{\mathsf{Comp}}$ *and* $\mathsf{I}^{\mathsf{Glue}}$ *be institutions, and* $\gamma^{Comp} : \mathsf{I}^{\mathsf{Comp}} \to \mathsf{I}^{\mathsf{Sys}}$ *and* $\gamma^{Glue} : \mathsf{I}^{\mathsf{Glue}} \to \mathsf{I}^{\mathsf{Sys}}$ *be institution representations. Let* $G = \langle V, E \rangle$ *be a graph, and* ι *a labelling for* G *according to Def. 4, then a* system design *is a structure of the form* $\langle G, \iota \rangle$.

As usual in the field of institutions, a specification of a system will be a diagram in the category of theories of a given institution, and the composition of the theories in the system specification will be the co-limit of such a diagram. This requires the category of theories to be finitely co-complete which, by [18, Thm. 11], follows directly when the category of signatures is finitely co-complete. In our case, the diagram is obtained by using the fact that the graph is expressed in terms of two institutions, $\mathsf{I}^{\mathsf{Comp}}$ and $\mathsf{I}^{\mathsf{Glue}}$, for which there exists an institution $\mathsf{I}^{\mathsf{Sys}}$ and institution representations $\gamma^{Comp} : \mathsf{I}^{\mathsf{Comp}} \to \mathsf{I}^{\mathsf{Sys}}$ and $\gamma^{Glue} : \mathsf{I}^{\mathsf{Glue}} \to \mathsf{I}^{\mathsf{Sys}}$, guaranteeing that a common interpretation is feasible.

The following theorem will be an important tool. Intuitively, this theorem tells us that whenever a connector (a co-span in $\mathsf{Th_0}^{\mathsf{I}^{\mathsf{Glue}}}$) is translated from $\mathsf{I}^{\mathsf{Glue}}$ to $\mathsf{I}^{\mathsf{Sys}}$, using an institution representation, it yields a co-span in $\mathsf{Th_0}^{\mathsf{I}^{\mathsf{Sys}}}$, thus complying with the restrictions associated with composition in the bicategory **co-span**$(\mathsf{Th_0}^{\mathsf{I}^{\mathsf{Sys}}})$.

Theorem 1. *Let* I *and* I' *be institutions such that* $\mathsf{Sign}^{\mathsf{I}}$ *and* $\mathsf{Sign}^{\mathsf{I}'}$ *are co-complete and have pushouts, and let* $\gamma : \mathsf{I} \to \mathsf{I}'$ *be an institution representation. Then, the pointwise extension of* $\gamma^{\mathsf{Th_0}^{\mathsf{I}}} : \mathsf{Th_0}^{\mathsf{I}} \to \mathsf{Th_0}^{\mathsf{I}'}$, $\widehat{\gamma^{\mathsf{Th_0}^{\mathsf{I}}}} :$ ***co-span***$(\mathsf{Th_0}^{\mathsf{I}}) \to$ ***co-span***$(\mathsf{Th_0}^{\mathsf{I}'})$*, is a lax functor.*

The following definition is based on the previous result, and enables us to integrate the computational parts of the glue and the adaptors in a communication mechanism. As the reader will notice, the connections, which are triples involving a connector and two adaptors, are translated into a single connector in the richer institution used to integrate components and connectors.

Definition 6. *Let* $\mathsf{I}^{\mathsf{Sys}}$, $\mathsf{I}^{\mathsf{Comp}}$ *and* $\mathsf{I}^{\mathsf{Glue}}$ *be institutions, and* $\gamma^{Comp} : \mathsf{I}^{\mathsf{Comp}} \to \mathsf{I}^{\mathsf{Sys}}$ *and* $\gamma^{Glue} : \mathsf{I}^{\mathsf{Glue}} \to \mathsf{I}^{\mathsf{Sys}}$ *be institution representations. Let* $G = \langle V, E \rangle$ *be a graph and* $\iota = \langle f, p, g \rangle$ *an interpretation for* G*. We define* $F(\langle G, \iota \rangle) = \langle \delta_0, \delta_1 \rangle : G_\iota \to$ **graph**$(\mathsf{Th_0}^{\mathsf{I}^{\mathsf{Sys}}})$ *as follows:*

$$G_\iota = \langle V \cup \bigcup_{e \in E} \{r_e^1, g_e, r_e^2\}, \bigcup_{e \in E} \{e_1, e_1', e_2', e_2\} \rangle \text{ such that:}$$
$$src(e_1) = r_e^1 \text{ and } trg(e_1) = src(e),$$
$$src(e_1') = r_e^1 \text{ and } trg(e_1') = g_e,$$
$$src(e_2') = r_e^2 \text{ and } trg(e_2') = g_e, \text{ and}$$
$$src(e_2) = r_e^2 \text{ and } trg(e_2) = trg(e),$$

$$\delta_0(v) = \begin{cases} \gamma^{Comp^{\mathsf{Th}_0}}(f(v)) & , \textit{if } v \in V. \\ \mathtt{dom}(\pi_1(\pi_1(\pi_2(g(e))))) & , \textit{if } v = r_e^1. \\ \pi_2(\pi_1(\pi_2(g(e)))\ ;\widehat{\gamma^{Glue}(\pi_2(g(e)));\ \pi_3(\pi_2(g(e)))}) & , \textit{if } v = g_e. \\ \mathtt{dom}(\pi_3(\pi_3(\pi_2(g(e))))) & , \textit{if } v = r_e^2. \end{cases}$$

$$\delta_1(e) = \begin{cases} \gamma^{Comp^{\mathsf{Th}_0}}(\pi_1(g(e))) & , \textit{if } src(e) = r_e^1 \textit{ and } trg(e) = a. \\ \pi_1(\pi_1(g(e))\ ;\widehat{\gamma^{Glue}(\pi_2(g(e)));\ \pi_3(g(e))}) & , \textit{if } src(e) = r_e^1 \textit{ and } trg(e) = g_e. \\ \pi_3(\pi_1(g(e))\ ;\widehat{\gamma^{Glue}(\pi_2(g(e)));\ \pi_3(g(e))}) & , \textit{if } src(e) = r_e^2 \textit{ and } trg(e) = g_e. \\ \gamma^{Comp^{\mathsf{Th}_0}}(\pi_3(g(e))) & , \textit{if } src(e) = r_e^2 \textit{ and } trg(e) = b. \end{cases}$$

In order to make the previous construction clearer, we illustrate how this applies to our previously introduced example of the producer and consumer, when we interconnect the parts and form a system design.

Example 4. (Putting the producer and the consumer together in a synchronous way) Putting the Producer and Consumer component specifications together in a synchronous way requires just *coordinating* them. As put forward in [9], this can be achieved by indicating how attributes are "connected" or identified with attributes of other components, and by synchronising actions.

This is a straightforward way of connecting two components, which simply expresses a correlation between the symbols of the components. In our example, we may want to make the components interact by synchronising the `send-i` and `extract-i` actions, of the producer and consumer, respectively, and by identifying `ready-in` and `p-waiting`, in the producer, with `c-waiting` and `ready-ext` in the consumer, respectively. This situation requires the system design to be over a single institution, so components and glues are expressed in a common language, as theories in $\mathsf{I}^{\mathsf{LTL}}$. To make it clearer, $\Sigma = [\textbf{Attributes} : \textbf{x}, \textbf{y} : \textbf{Bool}; \textbf{Actions} : \textbf{a}, \textbf{b}]$, and $\gamma^{Comp^{\mathsf{Th}_0}} = \gamma^{Glue^{\mathsf{Th}_0}} = id_{\mathsf{Th}_0{}^{\mathsf{I}^{\mathsf{LTL}}}}$.

Putting together **Producer** and **Consumer** in a synchronous way can be done in a homogeneous setting. Of course, the machinery we have defined will actually demonstrate its potential when dealing with heterogeneous specifications. Example 5 generalises the previous one, in which the components need to be connected asynchronously, and the communication mechanism is specified in a formalism different from that used for components.

Example 5. (Putting the producer and the consumer together in an asynchronous way) Consider a more complex communicating scenario for the producer and the consumer, in which these components need to interact via an asynchronous communication channel. The idea is to maintain the specifications for producer and consumer, which have already been appropriately characterised, and model the asynchronous nature of the channel within the communication specification, i.e., in the connector. This cannot be captured simply by identification of symbols in the interconnected parts. We will assume the state of the glue is characterised just by a queue whose functional behaviour is described in equational logic (Fig. 3). Now, we put these specifications together, so that the producer and the consumer communicate via a buffer of bit messages specified by the above queue.

Component: BitQueue
Sorts: Queue
Ops:
empty : Queue,
isEmpty? : Queue $\rightarrow$ Bool,
enqueue : Queue $\times$ Bit $\rightarrow$ Queue,
dequeue : Queue $\rightarrow$ Queue,
front : $\{ q : \text{Queue} \mid \neg isEmpty?(q) \} \rightarrow$ Bit.

Axioms: vars : q : Queue, b, b' : Bit
1. $isEmpty?(empty) = true$
2. $isEmpty?(enqueue(b, q)) = false$
3. $front(enqueue(b, empty)) = b$
4. $front(enqueue(b', enqueue(b, q))) = front(enqueue(b, q))$
5. $dequeue(enqueue(b, empty)) = empty$
6. $dequeue(enqueue(b, enqueue(b', q))) = enqueue(b, dequeue(enqueue(b', q)))$

Vars: q: Queue

Fig. 3. Producer and Consumer with Asynchronous communication

As opposed to the previous example, now we have a different formalism for the communication specification, i.e., I^{Glue} is **Eq** (equational logic), and the problem of putting together the three components cannot be syntactically solved.

We need to find an appropriate institution $\mathsf{I}^{\mathsf{Sys}}$, expressive enough to interpret, in a semantics preserving way, both linear temporal logic and equational logic. We will use first-order linear temporal logic [24]. The institution representation γ^{Comp} is the standard embedding of propositional temporal logic into first-order temporal logic. The institution representation γ^{Glue} is the embedding of equational logic into first-order logic with equality.

The reader should notice that, since the components and the glue are specified in different logics, we need suitable adaptors to put them together, which have to be specified in the richer institution $\mathsf{I}^{\mathsf{Sys}}$. Figs. 4 a) and 4 b) correspond to the adaptors in first-order LTL. Note that in the axioms q is a flexible variable, and q' is a rigid or logical (specification) variable. The reader should notice that even when the adaptors presented in Figs. 4 a) and 4 b) look complex in relation to the components being connected, they would remain the same, independently of the complexity of the components; this means that one could consider a more complex specification of the producer and the consumer, including the formalisation of the internal processes by which the information is produced and consumed, which could be highly complex. Objects originating ports and roles are the axiomless theories with signatures: 1. $\{send\text{-}i_{i=1,2}, ready\text{-}in, p\text{-}init\}$ for π_A, 2. $\{\text{Bool}, q, isEmpty? : \text{Queue} \rightarrow \text{Bool}, enqueue : \text{Queue} \times \text{Bit} \rightarrow \text{Queue}\}$ for ρ_l, 3. $\{extract\text{-}i_{i=1,2}, ready\text{-}ext, c\text{-}init\}$ for π_B, and 4. $\{\text{Bool}, q, isEmpty? : \text{Queue} \rightarrow \text{Bool}, dequeue : \text{Queue} \times \text{Bit} \rightarrow \text{Queue}, front : \text{Queue} \rightarrow \text{Bit}\}$ for ρ_r. The morphisms relating π_A, ρ_l, π_B and ρ_r with the corresponding theories associated with components, adaptors and glues, are inclusions in the corresponding category of signatures.

The reader should notice that the way in which the definitions and methodology we provided above interpret the elements of an architecture, allowed us to go from a model of a producer and a consumer connected in a synchronous way (see Ex. 4) to a model of a producer and a consumer connected in an asynchronous way (see Ex. 5), just by replacing the connection without modifying the components involved in the architectural design.

Component: Adapt
Sorts: Queue
Attributes: q: Queue
Actions: *p-init*, *send-0*, *send-1*, *ready-in*
Functions:
$enqueue$: Queue × Bit → Queue,
$isEmpty?$: Queue → Bool
Axioms:
1. $(\forall q' : \text{Queue}) q = q' \wedge send\text{-}0 \rightarrow \bigcirc(q = enqueue(0, q'))$
2. $(\forall q' : \text{Queue}) q = q' \wedge send\text{-}1 \rightarrow \bigcirc(q = enqueue(1, q'))$
3. $(\forall q' : \text{Queue}) q = q' \wedge \neg send\text{-}1 \rightarrow \neg send\text{-}0 \rightarrow \bigcirc(q = q')$
4. $p\text{-}init \rightarrow isEmpty?(q)$
5. $ready\text{-}in \leftrightarrow true$

a) A first-order LTL specification of **Adapt**.

Component: Adapt'
Sorts: Queue
Attributes: q: Queue
Actions: *c-init*, *extract-0*, *extract-1* , *ready-ext*
Functions:
$isEmpty?$: Queue → Bool,
$dequeue$: Queue → Queue,
$front : \{ q : \text{Queue} \mid \neg isEmpty?(q) \} \rightarrow \text{Bit}$
Axioms:
1. $isEmpty?(q) \rightarrow \neg extract\text{-}0 \wedge \neg extract\text{-}1$
2. $(\forall q' : Queue)(q = q' \wedge extract\text{-}0) \rightarrow (front(q) = 0 \wedge \bigcirc(q' = dequeue(q)))$
3. $(\forall q' : Queue)(q = q' \wedge extract\text{-}1) \rightarrow (front(q) = 1 \wedge \bigcirc(q' = dequeue(q)))$
4. $(\forall q' : Queue)(q = q' \wedge \neg extract\text{-}1 \wedge \neg extract\text{-}0) \rightarrow \bigcirc(q = q')$
5. $c\text{-}init \rightarrow isEmpty?(q)$
6. $ready\text{-}ext \leftrightarrow \neg isEmpty?(q)$

b) A first-order LTL specification of **Adapt'**.

Fig. 4. Specification of an adaptor

4 On the Institutions for Systems

In this section we introduce some results that allow us to obtain an institution of systems in a systematic way based on suitable specification languages for describing components and communications. A property that we want for such an institution is that both components and communications interpreted in the system institution can be mapped back to their original languages. This requirement emerges from the fact that it is often useful to be able to move back and forth from the (perhaps less expressive) specification languages used for components and communications to the formalism used to build the complete descriptions of the system. Moving from the component (resp. communication) specification language to the system specification language enables one to promote properties; moving from the system back to the components (resp. communications) allows us, for instance, to identify problems in the specifications of our "building blocks" when a counterexample of a property of the (whole) system is found.

Glueing two institutions together in a general way We provide a simple and general way of glueing two institutions into a new one. The motivation for doing so is, as we mentioned before, to help the specifier in the development of a suitable logic in which to express the system description, when one does not have in hand such a formalism.

Once $\mathsf{I}^{\mathsf{Comp}}$ and $\mathsf{I}^{\mathsf{Glue}}$ are fixed, it is possible to characterise an institution $\mathsf{I}^{\#}$ in which $\mathsf{I}^{\mathsf{Comp}}$ and $\mathsf{I}^{\mathsf{Glue}}$ can be put together. Furthermore, $\mathsf{I}^{\mathsf{Sys}}$ can be obtained by extending $\mathsf{I}^{\#}$ with additional logical structure depending on the properties required to be expressed. This must be done in such a way that there exists an institution representation $\iota_C : \mathsf{I}^{\mathsf{Comp}} \rightarrow \mathsf{I}^{\#}$, $\iota_G : \mathsf{I}^{\mathsf{Glue}} \rightarrow \mathsf{I}^{\#}$ and $\epsilon : \mathsf{I}^{\#} \rightarrow \mathsf{I}^{\mathsf{Sys}}$.

Let I^C and I^G be institutions. The following definition provides an institution constructed out of I^C and I^G. It is inspired by the construction presented by Sannella and Tarlecki in [29, Sec. 4.1.2], but it is slightly different. In [29, Ex. 4.1.44] and [29, Ex. 4.1.45], Sannella and Tarlecki provide the definitions of co-product and product of institutions, respectively. In the first case, as explained by the authors, the construction corresponds to putting two institutions together with no interaction; in the second case, the construction provides a way of putting them together but synchronising formulae by means of pairs. In our case, we need formulae to remain independent (requiring a co-product), but composite models to be pairs, each model coming from the corresponding institution (requiring a product). This need will become clear in Ex. 6 where we will extend the institution of the next definition with boolean operators combining formulae coming from any of the two logical systems, thus requiring models to give semantics to them.

Definition 7. $\mathsf{I}^{\#}(\mathsf{I}^{Comp}, \mathsf{I}^{Glue})$ *is defined as follows:*

- $\mathsf{Sign}^{\#} = \mathsf{Sign}^{C} \times \mathsf{Sign}^{G}$,
- $\mathbf{Sen}^{\#} = (\mathbf{Sen}^{C} \circ \pi_{left}) + (\mathbf{Sen}^{G} \circ \pi_{right})$,
- $\mathbf{Mod}^{\#} = \left(\mathbf{Mod}^{C} \circ \pi_{left}\right) \times \left(\mathbf{Mod}^{G} \circ \pi_{right}\right)$,
- *Let* $\alpha \in \mathbf{Sen}^{\#}(\langle \Sigma^{C}, \Sigma^{G} \rangle)$ *and* $\langle \mathcal{M}^{C}, \mathcal{M}^{G} \rangle \in |\mathbf{Mod}^{\#}(\langle \Sigma^{C}, \Sigma^{G} \rangle)|$, *then we say that* $\langle \mathcal{M}^{C}, \mathcal{M}^{G} \rangle \models^{\#}_{\langle \Sigma^{C}, \Sigma^{G} \rangle} \alpha$ *if and only if: 1.* $\exists \alpha^{C} \in \mathbf{Sen}^{C}(\Sigma^{C}) | \alpha = in_{left}(\alpha^{C}) \wedge \mathcal{M}^{C} \models^{\mathsf{Comp}}_{\Sigma^{C}} \alpha^{C}$, *or 2.* $\exists \alpha^{G} \in \mathbf{Sen}^{G}(\Sigma^{G}) | \alpha = in_{right}(\alpha^{G}) \wedge \mathcal{M}^{G} \models^{\mathsf{Glue}}_{\Sigma^{G}} \alpha^{G}$.

Theorem 2. *Let* I^{C} *and* I^{G} *be institutions. Then,* $\mathsf{I}^{\#}(\mathsf{I}^{C}, \mathsf{I}^{G})$ *is an institution.*

Property 2. Let I^{C} and I^{G} be institutions such that Sign^{C} and Sign^{G} are finitely co-complete. Then $\mathsf{Sign}^{\#}$ is finitely co-complete.

Definition 8. $\iota_{C} = \langle \gamma_{C}^{Sign}, \gamma_{C}^{Sen}, \gamma_{C}^{Mod} \rangle : \mathsf{I}^{C} \to \mathsf{I}^{\#}(\mathsf{I}^{C}, \mathsf{I}^{G})$ *is defined as follows:*

- *for* $\Sigma \in |\mathsf{Sign}^{C}|$, $\gamma_{C}^{Sign}(\Sigma) = \langle \Sigma, \emptyset^{G} \rangle$, *where* $\emptyset^{G}$ *is the empty signature in* I^{G}, *and if* $\sigma \in ||\mathsf{Sign}^{C}||$, *then* $\gamma_{C}^{Sign}(\sigma) = \langle \sigma, id_{\emptyset^{G}} \rangle$,
- *for* $\Sigma \in |\mathsf{Sign}^{C}|$, *we define* ${\gamma^{Sen}_{Comp}}_{\Sigma} : \mathbf{Sen}^{Comp}(\Sigma) \to \mathbf{Sen}^{\#} \circ \gamma_{C}^{Sign}(\Sigma)$, *as* ${\gamma_{C}^{Sen}}_{\Sigma} = in_{left}$, *and*
- *for* $\Sigma \in |\mathsf{Sign}^{C}|$, ${\gamma_{C}^{Mod}}_{\Sigma} : \mathbf{Mod}^{\#} \circ (\gamma_{C}^{Sign})^{\mathsf{op}}(\Sigma) \to \mathbf{Mod}^{C}(\Sigma)$, *is defined as* ${\gamma_{C}^{Mod}}_{\Sigma} = \pi_{left}$.

$\iota_{G} = \langle \gamma_{G}^{Sign}, \gamma_{G}^{Sen}, \gamma_{G}^{Mod} \rangle : \mathsf{I}^{G} \to \mathsf{I}^{\#}(\mathsf{I}^{C}, \mathsf{I}^{G})$ *is defined in an analogous way.*

Theorem 3. *Let* I^{C} *and* I^{G} *be institutions. Then,* ι_{C} *and* ι_{G} *are institution representations.*

So far we have put together components and glues in a single language. However, it is obvious that we have not achieved any interaction between the languages as there is no actual "coordination" of their semantics. To deal with this issue, we can extend $\mathsf{I}^{\#}$ by adding logical behaviour that "coordinates" elements from I^{C} and I^{G}. The idea consists of extending $\mathsf{I}^{\#}(\mathsf{I}^{C}, \mathsf{I}^{G})$ to a new institution I^{Sys} where the additional logical behaviour is incorporated, but satisfying the following conditions: 1. $\mathsf{Sign}^{Sys} = \mathsf{Sign}^{\#}$, 2. for all $\Sigma \in |\mathsf{Sign}^{Sys}|$, $\mathbf{Sen}^{Sys}(\Sigma) \supseteq \mathbf{Sen}^{\#}(\Sigma)$, 3. for all $\Sigma \in |\mathsf{Sign}^{Sys}|$, $\mathbf{Mod}^{Sys}(\Sigma) = \mathbf{Mod}^{\#}(\Sigma)$, and 4. for all $\Sigma \in |\mathsf{Sign}^{Sys}|$, $\alpha \in \mathbf{Sen}^{\#}$ and $\mathcal{M} \in \mathbf{Mod}^{Sys}(\Sigma)$: $\mathcal{M} \models^{Sys}_{\Sigma} \alpha$ iff $\mathcal{M} \models^{\#}_{\Sigma} \alpha$.

Then, if $\mathsf{I}^{Sys} = \langle \mathsf{Sign}^{Sys}, \mathbf{Sen}^{Sys}, \mathbf{Mod}^{Sys}, \{\models^{Sys}_{\Sigma}\}_{\Sigma \in |\mathsf{Sign}^{Sys}|} \rangle$ is an institution, we have the guarantee that an institution representation ϵ exists simply by taking it to be the trivial inclusion institution representation, which of course satisfies the satisfaction invariance condition.

The following example shows how to extend an institution with boolean operators. Our construction, although very similar to the one presented by Sannella and Tarlecki in [29, Ex. 4.1.41], requires a slightly different treatment of formulae because their satisfaction, as we demonstrated before, must be evaluated in the corresponding model of the pair. Notice that as we are building composite formulae out of formulae coming from different institutions, the only way to assert their satisfaction by a model is by having a notion of model capable of interpreting every piece, thus justifying the need for a definition of institution whose formulae is the co-product of the sets of formulae of the two institutions and whose class of models is the product of the corresponding classes of models. Extending $\mathsf{I}^{\#}(\mathsf{I}^C, \mathsf{I}^G)$ with boolean operators provides the most basic coordination of behavior by synchronising models through formulae they must be satisfy. More complex extensions can be made by choosing other logics to build on top of $\mathsf{I}^{\#}(\mathsf{I}^C, \mathsf{I}^G)$; some of them also require a more complex class of models.

Example 6. Let I^C and I^G be institutions. Then, I^{Sys} is defined as the structure $\langle \mathsf{Sign}^{Sys}, \mathbf{Sen}^{Sys}, \mathbf{Mod}^{Sys}, \{\models^{Sys\,\Sigma}\}_{\Sigma \in |\mathsf{Sign}^{Sys}|} \rangle$ where:

- $\mathsf{Sign}^{Sys} = \mathsf{Sign}^{\#}$,
- for all $\Sigma^C \in |\mathsf{Sign}^{Comp}|$, $\Sigma^G \in |\mathsf{Sign}^{Glue}|$:
  - $in_{left}(\alpha) \in \mathbf{Sen}^{Sys}(\langle \Sigma^C, \Sigma^G \rangle)$, for all $\alpha \in \mathbf{Sen}^C(\Sigma^C)$,
  - $in_{right}(\alpha) \in \mathbf{Sen}^{Sys}(\langle \Sigma^C, \Sigma^G \rangle)$, for all $\alpha \in \mathbf{Sen}^G(\Sigma^G)$,
  - if $\alpha, \beta \in \mathbf{Sen}^{Sys}(\langle \Sigma^C, \Sigma^G \rangle)$, then $\{\neg\alpha, \alpha \vee \beta\} \in \mathbf{Sen}^{\mathsf{Sys}}(\langle \Sigma^C, \Sigma^G \rangle)$.
- $\mathbf{Mod}^{Sys} = \mathbf{Mod}^{\#}$, and
- for all $\langle \Sigma^C, \Sigma^G \rangle \in |\mathsf{Sign}^{Sys}|$, $\langle \mathcal{M}^C, \mathcal{M}^G \rangle \in |\mathbf{Mod}^{\#}(\langle \Sigma^C, \Sigma^G \rangle)|$:

$$\begin{aligned}
&\langle \mathcal{M}^C, \mathcal{M}^G \rangle \models^{\#}_{\langle \Sigma^C, \Sigma^G \rangle} in_{left}(\alpha^C) \text{ iff } \mathcal{M}^C \models^{C}_{\Sigma^C} \alpha^C \\
&\langle \mathcal{M}^C, \mathcal{M}^G \rangle \models^{\#}_{\langle \Sigma^C, \Sigma^G \rangle} in_{right}(\alpha^G) \text{ iff } \mathcal{M}^G \models^{C}_{\Sigma^G} \alpha^G \\
&\langle \mathcal{M}^C, \mathcal{M}^G \rangle \models^{\#}_{\langle \Sigma^C, \Sigma^G \rangle} \neg\alpha \text{ iff not } \langle \mathcal{M}^C, \mathcal{M}^G \rangle \models^{\#}_{\langle \Sigma^C, \Sigma^G \rangle} \alpha \\
&\langle \mathcal{M}^C, \mathcal{M}^G \rangle \models^{\#}_{\langle \Sigma^C, \Sigma^G \rangle} \alpha \vee \beta \text{ iff} \\
&\qquad \langle \mathcal{M}^C, \mathcal{M}^G \rangle \models^{\#}_{\langle \Sigma^C, \Sigma^G \rangle} \alpha \text{ or} \langle \mathcal{M}^C, \mathcal{M}^G \rangle \models^{\#}_{\langle \Sigma^C, \Sigma^G \rangle} \beta
\end{aligned}$$

Proving that I^{Sys} is an institution is simple because $\mathsf{I}^{\#}$ is an institution and the boolean addition constitutes no problem in the proof. Equally simple is the proof that there exists an institution representation $\epsilon : \mathsf{I}^{\#} \to \mathsf{I}^{Sys}$.

Glueing two institutions in a known logic When one has in hand a logical system I formalised as an institution $\langle \mathsf{Sign}, \mathbf{Sen}, \mathbf{Mod}, \{\models^{\Sigma}\}_{\Sigma \in |\mathsf{Sign}|}\rangle$ such that, given I^C and I^G as defined in the previous example, there exists institution representations $\gamma_C : \mathsf{I}^C \to \mathsf{I}$ and $\gamma_G : \mathsf{I}^G \to \mathsf{I}$, part of the problem is already solved. We however need a way of getting, from a whole system's specification, the parts that composed it in their original formalisms. The main technical difficulty at this point arises from the fact that symbols coming from components and glues may be identified as a single symbol in the system's language.

Definition 9. *Let* Sign^C *and* Sign^G *have pushouts of arbitrary co-spans and have initial objects* $\emptyset^C$ *and* $\emptyset^G$ *respectively; and suppose* $\gamma_C^{Sign}(\emptyset^C) = \gamma_G^{Sign}(\emptyset^G)$. *Then, we define* $\mathsf{I}^{Sys}(\mathsf{I})$ *in the following way:*

- $\mathsf{Sign}^{Sys} = \langle \mathcal{O}, \mathcal{A}\rangle$ *such that:*
  - $\mathcal{O} = \left\{\langle \sigma_c : \Sigma^C \to \Sigma', \sigma_g : \Sigma^G \to \Sigma'\rangle \text{ pushout in } \mathsf{Sign}\right\}$,
  - $\mathcal{A} = \left\{\left\langle \sigma_l : \Sigma^C \to \Sigma^{C'}, \sigma_s : \Sigma^S \to \Sigma^{S'}, \sigma_r : \Sigma^G \to \Sigma^{G'}\right\rangle \mid \right.$ $\left. \langle \sigma_c, \sigma_g\rangle, \langle \sigma_c', \sigma_g'\rangle \in \mathcal{O} \text{ and } \sigma_c' \circ \sigma_l = \sigma_s \circ \sigma_c, \sigma_s \circ \sigma_g = \sigma_g' \circ \sigma_r\right\}$

  Identities and composition are defined component-wise.
- *for all* $\langle \sigma_c : \Sigma^C \to \Sigma', \sigma_g : \Sigma^G \to \Sigma'\rangle \in |\mathsf{Sign}^{Sys}|$ *and* $\langle \sigma_l : \Sigma^C \to \Sigma^{C'}, \sigma_s : \Sigma^S \to \Sigma^{S'}, \sigma_r : \Sigma^G \to \Sigma^{G'}\rangle \in ||\mathsf{Sign}^{Sys}||$*:* $\mathbf{Sen}^{Sys}(\langle \sigma_c, \sigma_g\rangle) = \mathbf{Sen}(\Sigma')$, $\mathbf{Sen}^{Sys}(\langle \sigma_l, \sigma_s, \sigma_r\rangle) = \mathbf{Sen}(\sigma_s)$,
- *for all* $\langle \sigma_c : \Sigma^C \to \Sigma', \sigma_g : \Sigma^G \to \Sigma'\rangle \in |\mathsf{Sign}^{Sys}|$ *and* $\langle \sigma_l : \Sigma^C \to \Sigma^{C'}, \sigma_s : \Sigma^S \to \Sigma^{S'}, \sigma_r : \Sigma^G \to \Sigma^{G'}\rangle \in ||\mathsf{Sign}^{Sys}||$*:* $\mathbf{Mod}^{Sys}(\langle \sigma_c, \sigma_g\rangle) = \mathbf{Mod}(\Sigma')$ *and* $\mathbf{Mod}^{Sys}(\langle \sigma_l, \sigma_s, \sigma_r\rangle) = \mathbf{Mod}(\sigma_s)$,
- *for all* $\langle \sigma_c : \Sigma^C \to \Sigma', \sigma_g : \Sigma^G \to \Sigma'\rangle \in |\mathsf{Sign}^{Sys}|$, $\alpha \in \mathbf{Sen}^{Sys}(\langle \sigma_c, \sigma_g\rangle)$ *and* $\mathcal{M} \in \mathbf{Mod}^{Sys}(\langle \sigma_c, \sigma_g\rangle)$, $\mathcal{M} \models^{Sys}_{\langle \sigma_c, \sigma_g\rangle} \alpha$ *iff* $\mathcal{M} \models_{\Sigma'} \alpha$.

Notice that the definition of $\mathsf{I}^{Sys}(\mathsf{I})$ only differs from I in the category of signatures. This is because having pushouts as signatures opens up the possibility of tracing back the source of the objects we are dealing with. In the case of sentences and models, we only consider the signature that is in the target of the morphisms constituting the pushout. This construction is particularly useful in the cases there is a need to identify both the common part of the partial descriptions of the system (the fraction of the description on which the synchronisation of the languages takes place), and the elements that correspond to only one of the descriptions[2].

Theorem 4. *Let* I *be an institution. Then,* $\mathsf{I}^{Sys}(\mathsf{I})$ *is an institution.*

Theorem 5. *Let* I *be an institution such that* Sign *has an initial object* $\emptyset^I$ *and pushouts for arbitrary co-spans. Then,* Sign^{Sys} *is finitely co-complete.*

[2] In [3] we use institution representations to give semantics to schema promotion in Z notation. There, whenever a manager for the whole system is constructed, the only way to prove the commutativity of the diagrams (see [3, Sec. 4.1]) is by preserving the information revealing the language from which each of the elements originates.

Definition 10. *Let* Sign^C *and* Sign^G *have pushouts of arbitrary co-spans and have initial objects* $\emptyset^C$ *and* $\emptyset^G$ *respectively, and* $\gamma_C^{Sign}(\emptyset^C) = \gamma_G^{Sign}(\emptyset^G)$*. Then, we define* $\iota_C = \langle \iota_C^{Sign}, \iota_C^{Sen}, \iota_C^{Mod} \rangle : \mathsf{I}^C \to \mathsf{I}^{Sys}(\mathsf{I})$ *as follows:*

– *if* $\Sigma \in |\mathsf{Sign}^C|$*, then* $\iota_C^{Sign}(\Sigma) = \langle \sigma^C, \sigma^G \rangle$ *such that* $\langle \sigma^C, \sigma^G \rangle$ *is the pushout of* $\langle \gamma_C^{Sign}(\emptyset^C \to \Sigma), \gamma_G^{Sign}(id_{\emptyset^G}) \rangle$*; if* $\sigma \in ||\mathsf{Sign}^C||$*, then* $\iota_C^{Sign}(\sigma) = \langle \gamma_C^{Sign}(\sigma), \gamma_C^{Sign}(\sigma), \gamma_G^{Sign}(id_{\emptyset^G}) \rangle$*,*

– *if* $\Sigma \in |\mathsf{Sign}^C|$*, then we define* ${\iota_C^{Sen}}_\Sigma : \mathbf{Sen}^C(\Sigma) \to \mathbf{Sen}^{Sys} \circ \iota_C^{Sign}(\Sigma)$*, as* ${\iota_C^{Sen}}_\Sigma = {\gamma_C^{Sen}}_\Sigma$*,*

– *we define* ${\iota_C^{Mod}}_\Sigma : \mathbf{Mod}^{Sys} \circ (\iota_C^{Sign})^{\mathsf{op}}(\Sigma) \to \mathbf{Mod}^C(\Sigma)$*, as* ${\iota_C^{Mod}}_\Sigma = {\gamma_C^{Mod}}_\Sigma$*.*

$\iota_G = \langle \gamma_G^{Sign}, \gamma_G^{Sen}, \gamma_G^{Mod} \rangle : \mathsf{I}^G \to \mathsf{I}^{Sys}(\mathsf{I})$ *is defined analogously.*

Theorem 6. *Let* Sign^C *and* Sign^C *have pushouts of arbitrary co-spans and have initial objects* $\emptyset^C$ *and* $\emptyset^G$ *respectively, and* $\gamma_C^{Sign}(\emptyset^C) = \gamma_G^{Sign}(\emptyset^G)$*. Then,* ι_C *and* ι_G *are institution representations.*

5 Conclusions and Related Work

We have presented an abstract and heterogeneous categorical characterisation of component-based systems. Our characterisation is logic/language independent, based on the categorical notion of *institution*. The heterogeneity of the approach is aimed at favouring a more genuine separation of concerns in the specification of components, and how these communicate. Our characterisation is based on the view that the different elements of a software architecture, such as components, connectors, roles, ports and adaptors, may be more faithfully specified in different formalisms, which have then to be put together into a setting in which one can reason about these parts and the whole system in a coordinated way. While institutions are used to abstractly capture specification formalisms, we employ *institution representations* to relate the different formalisms. In particular, we show how to build a *system institution*, in which the various parts of the specification can be represented as identifiable pieces of the overall specification, and we can reason about system properties by performing relevant formal analyses over it. Our contribution involves then a the formal characterisation of the conditions to combine formalisms in a heterogeneous setting, heavily relying on the notions of institution and institution representation.

Our work is related to various formalisms for the specification of component-based systems, in particular those seeking heterogeneity and abstraction. A main source of inspiration is the categorical approach put forward by Fiadeiro et al [9,15,11] in relation to the architecture description language *CommUnity* [15,13]. *CommUnity* comprises a specific component-based design language, in which components and connectors are specified in a particular way. In our approach, components and connectors might be defined in any formalism, *CommUnity* being a particular case. *Acme* [17] seeks similar objectives; it is defined as an *interchange architecture description language*, a setting where different formalisms might be combined. However, Acme has no actual formal semantics, and the examples of translations from particular architecture description languages to Acme are defined in an *ad-hoc* manner, generally dealt with only at a syntactic level. Thus, questions such as the coherence of resulting Acme specifications, cannot be answered in Acme's context.

CASL [2] is an algebraic language for formal specification, which uses the notion of institutions to achieve a high degree of abstraction. Architectural specifications in CASL [1] are built by using basic relationships between modules like refinement or extension, i.e., the architectural structure of a system in terms of components and connections are not explicitly captured in CASL.

In [21] an heterogeneous approach for specifying service-oriented systems is presented. The basic idea is to use two different institutions to capture the different levels involved in a service-oriented system. One institution is used to specify the local behaviour of services, while the other institution is used as a global logic to describe the orchestration of services. The two levels are related via a co-morphism (or institution representation). Notice that the global logic is used for the description of the common behaviour of components, but it is not used for describing coordination mechanisms in an abstract way, as is done by glues in the present paper. Also notice that this approach is heterogeneous but not language independent. HETS [26,27] is a framework for integrating different institutions to support heterogeneous specifications of systems. We share the interest in heterogeneous frameworks of institutions, but our focus is on formalising the notions from software architecture and the corresponding kinds of entities, namely components, connectors, roles, ports, and adaptors, to structure heterogeneous system specifications. Nevertheless, we are committed to the HETS view that whenever the design requires the use of different languages for describing components (reap. glues) Grothendieck institutions provide the most suitable tool to deal with that level of heterogeneity. In [28], Mossakowski and Tarlecki presented a framework meant to be a tool for heterogeneous software specification. This framework exploits the use of morphisms and co-morphisms between institutions in a coordinated way, not only allowing moving a specification to a more expressive language, but also to project a part of the system into a less expressive one. Differences and similarities between their framework and ours are essentially the same as mentioned in the comparison above with HETS.

References

1. Bidoit, M., Sannella, D., Tarlecki, A.: Architectural Specifications in CASL. In: Haeberer, A.M. (ed.) AMAST 1998. LNCS, vol. 1548, pp. 341–357. Springer, Heidelberg (1998)
2. Mossakowski, T., Haxthausen, A., Sannella, D., Tarlecki, A.: CASL: The common algebraic specification language: Semantics and proof theory. Computing and Informatics 22 (2003)
3. Castro, P.F., Aguirre, N., López Pombo, C.G., Maibaum, T.: A Categorical Approach to Structuring and Promoting Z Specifications. In: Păsăreanu, C.S., Salaün, G. (eds.) FACS 2012. LNCS, vol. 7684, pp. 73–91. Springer, Heidelberg (2013)

4. Cengarle, M.V., Knapp, A., Tarlecki, A., Wirsing, M.: A Heterogeneous Approach To UML Semantics. In: Degano, P., De Nicola, R., Meseguer, J. (eds.) Concurrency, Graphs and Models. LNCS, vol. 5065, pp. 383–402. Springer, Heidelberg (2008)
5. Diaconescu, R., Futatsugi, K.: Logical foundations of CafeOBJ. Theor. Comp. Sc. 285(2) (2002)
6. Ehrig, H., Mahr, B.: Fundamentals of Algebraic Specification 2. Springer (1990)
7. Ehrig, H., Große-Rhode, M., Wolter, U.: On the Role of Category Theory in the Area of Algebraic Specification. In: Haveraaen, M., Owe, O., Dahl, O.-J. (eds.) Recent Trends in Data Type Specification. LNCS, pp. 17–48. Springer, Heidelberg (1996)
8. Allen Emerson, E.: Temporal and modal logic. Handbook of Theoretical Computer Science, vol. B. Elsevier (1990)
9. Fiadeiro, J., Maibaum, T.: Temporal Theories as Modularisation Units for Concurrent System Specification. Formal Asp. of Comp. 4(3) (1992)
10. Fiadeiro, J., Maibaum, T.: Describing, Structuring and Implementing Objects. In: de Bakker, J.W., de Roever, W.P., Rozenberg, G. (eds.) Proc. of the REX Workshop. LNCS, vol. 489, pp. 274–310. Springer, Heidelberg (1991)
11. Fiadeiro, J., Wermelinger, M.: A graph transformation approach to software architecture reconfiguration. Sc. of Comp. Prog. 44(2) (2002)
12. Fiadeiro, J.L., Schmitt, V.: Structured Co-spans: An Algebra of Interaction Protocols. In: Mossakowski, T., Montanari, U., Haveraaen, M. (eds.) CALCO 2007. LNCS, vol. 4624, pp. 194–208. Springer, Heidelberg (2007)
13. Fiadeiro, J.: Categories for Software Engineering. Springer (2004)
14. Fiadeiro, J., Maibaum, T.S.E.: A Mathematical Toolbox for the Software Architect. In: Proc. Workshop on Software Specification and Design. IEEE (1995)
15. Fiadeiro, J., Maibaum, T.S.E.: Categorical Semantics of Parallel Program Design. Sc. of Comp. Prog. 28 (1997)
16. Garlan, D.: Software Architecture: A Roadmap. ACM (2000)
17. Garlan, D., Monroe, R., Wile, D.: Acme: an architecture description interchange language. In: Proc. of CASCON 1997 (1997)
18. Goguen, J., Burstall, R.: Institutions: Abstract Model Theory for Specification and Programming. Journal of the ACM 39(1) (1992)
19. Goguen, J., Rosu, G.: Institution Morphisms. Formal Asp. of Comp. 13 (2002)
20. Harel, D., Kozen, D., Tiuryn, J.: Dynamic Logic. MIT Press (2000)
21. Knapp, A., Marczynski, G., Wirsing, M., Zawlocki, A.: A Heterogeneous Approach to Service-Oriented Systems Specification. In: Proc. of SAC 2010. ACM (2010)
22. Lopes, A., Fiadeiro, J.: Superposition: composition vs refinement of non-deterministic, action-based systems. Formal Asp. of Comp. 16(1) (2004)
23. McLane, S.: Categories for working mathematicians. Springer (1971)
24. Manna, Z., Pnueli, A.: The Temporal Logic of Reactive and Concurrent Systems. Springer (1991)
25. Meseguer, J.: General Logics. In: Logic Colloquium 1987. North-Holland (1989)
26. Mossakowski, T.: Heterogeneous Theories and the Heterogeneous Tool Set. In: Semantic Interoperability and Integration, Dagstuhl Seminar Proc. (2005)
27. Mossakowski, T., Maeder, C., Lüttich, K.: The Heterogeneous Tool Set, Hets. In: Grumberg, O., Huth, M. (eds.) TACAS 2007. LNCS, vol. 4424, pp. 519–522. Springer, Heidelberg (2007)
28. Mossakowski, T., Tarlecki, A.: Heterogeneous Logical Environments for Distributed Specifications. In: Corradini, A., Montanari, U. (eds.) WADT 2008. LNCS, vol. 5486, pp. 266–289. Springer, Heidelberg (2009)

29. Sannella, D., Tarlecki, A.: Foundations of Algebraic Specification and Formal Software Development. Springer (2012)
30. Tarlecki, A.: Moving Between Logical Systems. In: Haveraaen, M., Owe, O., Dahl, O.-J. (eds.) Proc. of COMPASS. LNCS, vol. 1130, pp. 478–502. Springer, Heidelberg (1996)
31. Tarlecki, A.: Toward Specifications for Reconfigurable Component Systems. In: Kleijn, J., Yakovlev, A. (eds.) ICATPN 2007. LNCS, vol. 4546, pp. 24–28. Springer, Heidelberg (2007)
32. Tarlecki, A.: Towards Heterogeneous Specifications. Frontiers of Combining Systems 2 (2000)
33. Wermelinger, M., Fiadeiro, J.: A graph transformation approach to software architecture reconfiguration. Sc. of Comp. Prog. 44(2) (2002)

On Unary Fragments of MTL and TPTL over Timed Words

Khushraj Madnani[1], Shankara Narayanan Krishna[1], and Paritosh K. Pandya[2]

[1] IIT Bombay, Powai, Mumbai, India
[2] Tata Institute of Fundamental Research, Colaba, Mumbai, India
{khushraj,krishnas}@cse.iitb.ac.in, pandya@tcs.tifr.res.in

Abstract. Real time logics such as Metric Temporal Logic, MTL and Timed Propositional Temporal Logic (TPTL) exhibit considerable diversity in expressiveness and decidability properties based on the permitted set of modalities, the nature of time interval constraints and restriction on models. We study the expressiveness and decidability properties of various *unary* fragments of MTL incorporating strict as well as non-strict modalities. We show that, from the point of view of expressive power, $\mathsf{MTL}[\Diamond_I] \subsetneq \mathsf{MTL}[\Diamond_I^s] \subsetneq \mathsf{MTL}[\Diamond_I, \mathsf{O}] \equiv \mathsf{MTL}[\Diamond_I^s, \mathsf{O}] \subsetneq \mathsf{MTL}[\mathsf{U}_I^s]$, in pointwise semantics. We also sharpen the decidability results by showing that, in the pointwise semantics, $\mathsf{MTL}[\Diamond_I]$ (which is the least expressive amongst the unary fragments considered) already has non-primitive-recursive complexity and is $\mathbf{F}_{\omega^\omega}$-hard for satisfiability checking over finite timed words, and that $\mathsf{MTL}[\Diamond_I, \diamondminus_I]$ is undecidable and Σ_1^0-hard. Next we explore, in the pointwise models, the decidability of $\mathsf{TPTL}[\Diamond_I]$ (unary TPTL) and show that 2-variables unary TPTL has undecidable satisfiability, while the single variable fragment $\mathsf{TPTL}[\mathsf{U}^s]$ incorporating even the most expressive operator U^s operator is decidable over finite timed words. We provide a comprehensive picture of the decidability and expressiveness properties of unary fragments of TPTL and MTL over pointwise time.

1 Introduction

Temporal Logics are a well established formalism for specifying ordering constraints on a sequence of events. Timed Temporal Logics extend this by allowing to specify quantitative constraints between events. Metric Temporal Logic (MTL) introduced by Koymans [12] and Timed Propositional Temporal Logic (TPTL) introduced by Alur and Henzinger [1] are two prominent linear time temporal logics. The logic $\mathsf{TPTL}[\mathsf{U},\mathsf{S}]$ makes use of freeze quantifiers along with untimed temporal modalities and explicit constraints on frozen time values; the logic $\mathsf{MTL}[\mathsf{U}_I,\mathsf{S}_I]$ uses time interval constrained modalities U_I and S_I. For example, the $\mathsf{TPTL}[\mathsf{U},\mathsf{S}]$ formula $x.(a\mathsf{U}(b \wedge 0 < x < 1))$ and the $\mathsf{MTL}[\mathsf{U}_I,\mathsf{S}_I]$ formula $a\mathsf{U}_{(0,1)}b$ both characterize the set of timed behaviours that have a symbol b at a time point < 1, such that the only letters preceding this b are a. Timed logics are defined over timed words (also called pointwise time models) or over signals

G. Ciobanu and D. Méry (Eds.): ICTAC 2014, LNCS 8687, pp. 333–350, 2014.

(also called continuous time models). A finite timed word is a finite sequence of letters each of which carries a time stamp giving the time of its occurrence. Weak monotonicity (as against strict monotonicity) of timed words allows a sequence of events to occur at the same time point. In this paper we confine ourselves to logics interpreted over finite timed words as models.

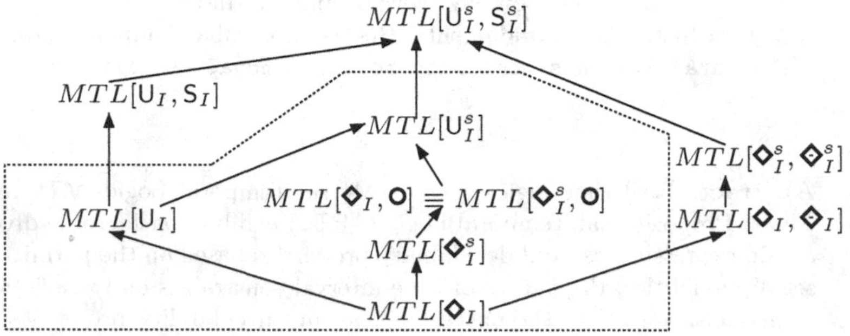

Fig. 1. Expressiveness without restriction of strict monotonicity on finite timed words

In their full generality, $\mathsf{MTL}[\mathsf{U}_I, \mathsf{S}_I]$ and $\mathsf{TPTL}[\mathsf{U}, \mathsf{S}]$ both have undecidable satisfiability even for finite timed words. Several restrictions have been proposed to get decidable subclasses (see Ouakinine and Worrel for a recent survey)[3]. In their seminal paper, Alur and Henzinger [1] proposed a subclass MITL of MTL having only non-punctual intervals, where the satisfiability is decidable with EXPSPACE complete complexity. The satisfiability of $\mathsf{MTL}[\mathsf{U}_I]$ was considered to be undecidable for a long time, until Ouaknine and Worrell proved that the satisfiability of $\mathsf{MTL}[\mathsf{U}_I]$ over finite timed words is decidable, albeit with a non-primitive recursive lower bound. Subsequently, in [16], it was shown that over infinite timed words, the satisfiability of $\mathsf{MTL}[\mathsf{U}_I]$ is undecidable. The satisfiability of $\mathsf{MTL}[\mathsf{U}_I]$ over continuous time models is also undecidable. In this paper, we sharpen the known undecidability results for MTL, by showing that over finite timed words, the full unary fragment $\mathsf{MTL}[\Diamond_I, \overleftarrow{\Diamond}_I]$ is undecidable. Further, we also show that checking satisfiability of the unary fragment $\mathsf{MTL}[\Diamond_I]$ over finite timed words has a non primitive recursive lower bound. Hence, restriction of unariness in modalities does not simplify the satisfiability problem of MTL. Our decidability and complexity results are established for the unary fragments of MTL with "non-strict" modalities, interpreted over weakly monotonic timed words. A strict modality is more expressive as it guarantees that it accesses a time point strictly in future (or past).

In order to study the expressive power of various unary fragments of MTL (with both strict and nonstrict modalities respectively), we use the tool of EF games for MTL introduced in [17].

We show that from the point of expressiveness, working on weakly monotonic timed words, $\mathsf{MTL}[\Diamond_I]$ is strictly contained in $\mathsf{MTL}[\Diamond^s_I]$ as it cannot specify strict monotonicity on the underlying models. Moreover, $\mathsf{MTL}[\Diamond^s_I]$ is strictly contained in $\mathsf{MTL}[\Diamond_I, \bigcirc]$ as unary operators cannot say anything about the next time point. Also, $\mathsf{MTL}[\Diamond_I, \bigcirc]$ is equivalent to $\mathsf{MTL}[\Diamond^s_I, \bigcirc]$, and $\mathsf{MTL}[\Diamond^s_I, \bigcirc]$ is a strict subset of $\mathsf{MTL}[\mathsf{U}^s_I]$. Note that over strictly monotonic timed words, the unary fragment of MTL with strict operators collapses to the unary fragment of MTL with non-strict operators, while this is not true of MTL with binary operators, until and since. Logic MTL with strict until (since) is more expressive than non strict until (since) due to its ability of encoding next (previous).

Figure 1 shows the expressiveness relationship between fragments of MTL over arbitrary timed words, where as Figure 2 shows these relationship when models are confined to strictly monotonic timed words. In these figures, $X \to Y$ means X is contained in Y, and if there does not exist a path between 2 classes then they are incomparable. We also indicate the subclasses which have decidable satisfiability: these are contained within dotted polygon. All these decidable logics have non primitive recursive decision complexity. Indeed, it is rare to find a timed logic with elementary decision complexity [18]. We believe that this paper gives a comprehensive characterization of the decidability as well as expressiveness of unary fragments of MTL.

Investigating the logic TPTL, we show that over finite timed words, the unary fragment $\mathsf{TPTL}[\Diamond]$ is undecidable with two freeze variables, while the one variable fragment of $\mathsf{TPTL}[\mathsf{U}^s_I]$ is decidable. The rest of the paper is organized as follows: In Section 2, we give all the preliminaries required for the further sections. Section 3 discusses the expressiveness of unary fragments of MTL. Section 4 discusses the decidability and complexity of unary MTL, as well as TPTL.

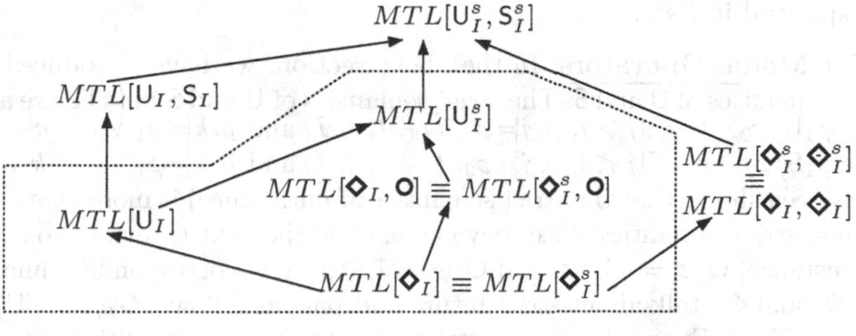

Fig. 2. Expressiveness assuming strict monotonicity on finite timed words

2 Preliminaries

Timed Word: Let Σ be a finite set of propositions. A finite timed word over Σ is of the form $\rho = (\sigma_1, t_1)(\sigma_2, t_2) \ldots (\sigma_n, t_n)$, where $\sigma_i \subseteq \Sigma$ and $t_i \in \mathbb{R}_{\geq 0}$. We also represent ρ by the tuple (σ, τ) where $\sigma = \sigma_1\sigma_2 \ldots \sigma_n$ and $\tau = t_1 t_2 \ldots t_n$. The positions $\{1, 2, \ldots, n\}$ in the word ρ is denoted by $dom(\rho)$. If $t_i < t_j$ for all $i, j \in dom(\rho)$ with $i < j$, the word ρ is said to be strictly monotonic; if $t_i \leq t_j$, ρ is weakly monotonic. Let $T\Sigma^*$ denote the set of all timed words over Σ.

Metric Temporal Logic: In this section, we describe the syntax and semantics of MTL in the *point-wise* sense. Given Σ, the formulae of MTL are built from Σ using boolean connectives and time constrained versions of the modalities U and S as follows:

$$\varphi ::= a(\in \Sigma)\ |true\ |\varphi \wedge \varphi \mid \neg\varphi \mid \varphi \mathsf{U}_I \varphi \mid \varphi \mathsf{S}_I \varphi \mid \bigcirc_I \varphi \mid \ominus_I \varphi$$

where I is an open, half-open or closed interval with end points in $\mathbb{N} \cup \{\infty\}$.

Semantics : Given a finite timed word ρ over Σ, and an MTL formula φ, in the point-wise semantics, the temporal connectives of φ quantify over a finite set of positions in ρ. For an alphabet Σ, a timed word $\rho = (\sigma, \tau)$, a position $i \in dom(\rho)$, and an MTL formula φ, the satisfaction of φ at a position i of ρ is denoted $(\rho, i) \models \varphi$, and is defined as follows:

$\rho, i \models a \quad \leftrightarrow \quad a \in \sigma_i,$
$\rho, i \models \neg\varphi \quad \leftrightarrow \quad \rho, i \not\models \varphi,$
$\rho, i \models \varphi_1 \wedge \varphi_2 \quad \leftrightarrow \quad \rho, i \models \varphi_1$ and $\rho, i \models \varphi_2,$
$\rho, i \models \bigcirc_I \varphi \quad \leftrightarrow \quad \rho, i+1 \models \varphi$ and $t_{i+1} - t_i \in I,$
$\rho, i \models \varphi_1 \mathsf{U}_I \varphi_2 \quad \leftrightarrow \exists j \geq i,\ \rho, j \models \varphi_2,\ t_j - t_i \in I,$ and $\rho, k \models \varphi_1\ \forall\, i \leq k < j,$
$\rho, i \models \ominus_I \varphi \quad \leftrightarrow \quad \rho, i-1 \models \varphi$ and $t_i - t_{i-1} \in I,$
$\rho, i \models \varphi_1 \mathsf{S}_I \varphi_2 \quad \leftrightarrow \quad \exists j \leq i,\ \rho, j \models \varphi_2,\ t_i - t_j \in I,$ and $\rho, k \models \varphi_1\ \forall\, j < k \leq i.$

We say that ρ satisfies φ denoted $\rho \models \varphi$ iff $\rho, 1 \models \varphi$. Let $L(\varphi) = \{\rho \mid \rho, 1 \models \varphi\}$. Additional temporal connectives are defined in the standard way: we have the constrained future and past eventuality operators $\Diamond_I a \equiv true \mathsf{U}_I a$ and $\overleftarrow{\Diamond}_I a \equiv true \mathsf{S}_I a$, and their duals $\Box_I a \equiv \neg\Diamond_I \neg a$, $\overleftarrow{\Box}_I a \equiv \neg\overleftarrow{\Diamond}_I \neg a$. We denote by $\mathsf{MTL}[\mathsf{U}_I, \mathsf{S}_I, \bigcirc_I, \ominus_I]$ the class of all MTL formulae in the pointwise sense, with modalities $\mathsf{U}_I, \mathsf{S}_I, \bigcirc_I, \ominus_I$. MTL[*list*] denotes the fragment of MTL using modalities specified in *list*.

Strict Modal Operators: In the above section, we have introduced the non-strict semantics of U and S. The strict semantics of U and S in MTL are as follows:

$\rho, i \models \varphi_1 \mathsf{U}_I^s \varphi_2 \quad \leftrightarrow \quad \exists j > i,\ \rho, j \models \varphi_2,\ t_j - t_i \in I,$ and $\rho, k \models \varphi_1\ \forall\, i < k < j.$
$\rho, i \models \varphi_1 \mathsf{S}_I^s \varphi_2 \quad \leftrightarrow \quad \exists j < i,\ \rho, j \models \varphi_2,\ t_i - t_j \in I,$ and $\rho, k \models \varphi_1\ \forall\, j < k < i.$

It has been shown that the strict semantics of until(since) is more expressive than the non-strict semantics since they can encode the next (previous) operators [5]. For instance, $\bigcirc_I \varphi = \bot \mathsf{U}_I^s \varphi$, and $\ominus_I \varphi = \bot \mathsf{S}_I^s \varphi$. The corresponding unary operators $\Diamond^s$ and $\overleftarrow{\Diamond}^s$ talk about strict future and past as follows: $\Diamond_I^s \varphi = \top \mathsf{U}_I^s \varphi$, while $\overleftarrow{\Diamond}_I^s \varphi = \top \mathsf{S}_I^s \varphi$. Expressiveness of strict and non-strict modalities are compared in section 3.

Timed Propositional Temporal Logic: In this section, we define the syntax and semantics of TPTL in the *point-wise* sense.

$$\varphi ::= a(\in \Sigma)\ |true\ |\varphi \wedge \varphi \mid \neg\varphi \mid \varphi \mathsf{U} \varphi \mid \varphi \mathsf{S} \varphi \mid \bigcirc \varphi \mid \ominus \varphi \mid y.\varphi \mid y \in I$$

where C is the set of clock variables progressing at same rate, $y \in \mathcal{C}$, and I is an open, half-open or closed interval with end points in $\mathbb{N} \cup \{\infty\}$.

Semantics: Given a finite timed word ρ over Σ, and an TPTL formula φ, in the point-wise semantics, the truth of a formula is interpreted at a position $i \in \mathbb{N}$ along the word. We define the satisfiability relation, $\rho, i, \nu \models \varphi$ saying that the formula φ is true at position i of the timed word ρ with valuation ν of all the clock variables.

$\rho, i, \nu \models a \quad \leftrightarrow \quad a \in \sigma_i,$
$\rho, i, \nu \models \neg\varphi \quad \leftrightarrow \quad \rho, i, \nu \not\models \varphi,$
$\rho, i, \nu \models \varphi_1 \wedge \varphi_2 \quad \leftrightarrow \quad \rho, i, \nu \models \varphi_1$ and $\rho, i, \nu \models \varphi_2,$
$\rho, i, \nu \models x.\varphi \quad \leftrightarrow \quad \rho, i, \nu[x \leftarrow t_i] \models \varphi,$
$\rho, i, \nu \models x \in I \quad \leftrightarrow \quad t_i - \nu(x) \in I,$
$\rho, i, \nu \models \mathsf{O}\varphi \quad \leftrightarrow \quad \rho, i+1, \nu \models \varphi,$
$\rho, i, \nu \models \varphi_1 \mathsf{U} \varphi_2 \quad \leftrightarrow \quad \exists j \geq i, \rho, j, \nu \models \varphi_2,$ and $\rho, k, \nu \models \varphi_1 \; \forall\, i \leq k < j,$
$\rho, i, \nu \models \ominus\varphi \quad \leftrightarrow \quad \rho, i-1, \nu \models \varphi,$
$\rho, i, \nu \models \varphi_1 \mathsf{S} \varphi_2 \quad \leftrightarrow \quad \exists\, j \leq i, \rho, j, \nu \models \varphi_2,$ and $\rho, k, \nu \models \varphi_1 \; \forall\, j < k \leq i.$

We say that ρ satisfies φ denoted $\rho \models \varphi$ iff $\rho, 1, \bar{0} \models \varphi$. Here $\bar{0}$ is the valuation obtained by setting all clock variables to 0. We denote by $\mathsf{TPTL}[\mathsf{U}, \mathsf{S}, \mathsf{O}, \ominus]$ the class of all TPTL formulae in the pointwise sense, with modalities $\mathsf{U}, \mathsf{S}, \mathsf{O}, \ominus$. TPTL[*list*] denotes the fragment of TPTL using modalities specified in *list*. Strict modalities U^s and S^s can be defined for TPTL in a way similar to that done for MTL.

Ehrenfeucht Fraïssé Games for MTL: In this section, we recall EF games [17], used in separating various fragments of MTL based on the expressive power. EF games for First-Order Logic defined by Ehrenfeucht and Fraïssé [22] is a very useful tool for characterizing FO-definable languages; Etessami and Wilke defined EF games for LTL [8]. These EF-games for LTL were extended to MTL in [17] over the pointwise and continuous semantics. We discuss EF Games for MTL on point-wise semantics only. Our expressiveness results in section 3 use these games.

A n-round MTL EF game is played between two players (*Spoiler* and *Duplicator*) on a pair of timed words (ρ_0, ρ_1). A configuration of the game is pair of points i_0, i_1 where $i_0 \in domain(\rho_0)$ and $i_1 \in domain(\rho_1)$. A configuration is called partially isomorphic, denoted $isop(i_0, i_1)$ iff $\sigma_{i_0} = \sigma_{i_1}$. From a starting configuration (i_0, i_1), the game is defined as follows:

- Either *Spoiler* or *Duplicator* eventually wins the game.
- A 0-round EF game is won by the *Duplicator* iff $isop(i_0, i_1)$.
- The n round game is played by first playing one round from the starting position. Either the *Spoiler* wins the round, and the game is terminated or the *Duplicator* wins the round, and now the second round is played from this new configuration and so on. The *Duplicator* wins the game only if it wins all the rounds. Following are the rules of game starting configuration being (i_0, i_1).
  - If $isop(i_0, i_1)$ is not true, then *Spoiler* wins the 0 round game.
  - The *Spoiler* chooses one of the words by choosing $\rho_x, x \in \{0, 1\}$. *Duplicator* has to play on the other word ρ_y, $x \neq y$. Then *Spoiler* chooses one of the U_I, S_I, U^s_I, S^s_I move, along with the interval I (such that the end points of the intervals are non-negative integers). Given the current configuration as (i_x, i_y), the rest of the round is played as follows:

  - If the chosen move of *Spoiler* is U_I, then *Spoiler* chooses a position $i'_x \in dom(\rho_x)$ such that $i_x \leq i'_x$ and $(t_{i'_x} - t_{i_x}) \in I$ (in case the chosen move is U^s_I, then $i_x < i'_x$)
  - The *Duplicator* responds to the U_I move by choosing $i'_y \in dom(\rho_y)$ in the other word such that $i_y \leq i'_y$ and $(t_{i'_y} - t_{i_y}) \in I$. (In case of U^s_I move, $i_y < i'_y$). If the *Duplicator* cannot find such a position, the *Spoiler* wins the round and the game. Otherwise, the game continues and *Spoiler* chooses one of the following options.
  - $\Diamond$ Part: The round ends with the configuration (i'_0, i'_1).
  - U Part: *Spoiler* chooses a position i''_y in ρ_y such that $i_y \leq i''_y < i'_y$ (in case of U^s move $i_y < i''_y < i'_y$). The *Duplicator* responds by choosing a position i''_x in ρ_x such that $i_x \leq i''_x < i'_x$ (in case of U^s move $i_x < i''_x < i'_x$). The round ends with the configuration (i''_0, i''_1). If the *Duplicator* cannot choose an i''_x , the game ends and the *Spoiler* wins.
- S_I (and the corresponding strict since) move is analogous to U_I move described above. The only difference is the *Spoiler* and the *Duplicator* choose points in the past of the present configuration. In this case, the $\Diamond$ or U parts are replaced with the $\diamondminus$ or S parts.
- We can restrict various moves according to the modalities provided by the logic. For example, if we restrict ourselves to playing an EF game for $\mathsf{MTL}[\Diamond_I, \diamondminus_I]$, then given a configuration (i_x, i_y), a round will simply consist of *Spoiler* choosing a $\Diamond_I$ or $\diamondminus_I$ move, and a position $i'_x \sim i_x$ in his word ρ_x, while *Duplicator* has to respond with a suitable choice of position $i'_y \sim i_y$ in his word ρ_y. $\sim=\leq$ for a $\diamondminus_I$ move, and is $<$ for a $\diamondminus^s_I$ move; we have $\sim\in \{\geq, >\}$ for a $\Diamond_I$ or $\Diamond^s_I$ move. The game proceeds to the configuration (i'_x, i'_y); in case *Duplicator* is unable to produce a i'_y, *Spoiler* wins.

- **Game equivalence:** $(\rho_0, i_0) \approx_k (\rho_1, i_1)$ iff for every k-round $\mathsf{MTL}[\mathsf{U}_I, \mathsf{S}_I]$ EF-game over the words ρ_0, ρ_1 starting from the configuration (i_0, i_1), the *Duplicator* always has a winning strategy.
- **Formula equivalence:** $(\rho_0, i_0) \equiv_k (\rho_1, i_1)$ iff for every $\mathsf{MTL}[\mathsf{U}_I, \mathsf{S}_I]$ formula ϕ of modal depth $\leq k$, $\rho_0, i_0 \models \phi \iff \rho_1, i_1 \models \phi$

Theorem 1. $(\rho_0, i_0) \approx_k (\rho_1, i_1)$ *iff* $(\rho_0, i_0) \equiv_k (\rho_1, i_1)$

Given 2 fragments M_1, M_2 of MTL, we use the above theorem to prove results of the form $M_1 - M_2 \neq \phi$. For instance, $M_1 = \mathsf{MTL}[\mathsf{U}_I]$, while $M_2 = \mathsf{MTL}[\Diamond_I]$. First, *Duplicator* chooses a formula $\varphi \in M_1$. In response, *Spoiler* choses a number n which indicates the maximum number of rounds that can be played before *Duplicator* can win. In response to n, *Duplicator* choses a pair of words ρ_0, ρ_1 such that one of the words satisfies φ while the other does not. Now we play n-round EF game using the modalities of M_2 (for the instance chosen above, we play a $\Diamond_I$ game) on these words as explained above. If we prove that for any $n \in \mathbb{N}$ *Duplicator* has a winning strategy, then according to the theorem 1, we cannot have a formula of modal depth n in logic M_2 which can distinguish ρ_0 and ρ_1. This shows that φ has no equivalent formula in M_2.

Counter Machines: Our undecidability results in section 4 are obtained by reduction of the halting problem of two counter machines. A deterministic 2-counter machine is a 3 tuple $\mathcal{M} = (P, C_1, C_2)$, where

1. C_1, C_2 are counters taking values in $\mathbb{N}$ (their initial values are set to zero);
2. P is a finite set of n instructions. Let $p_1, \ldots, p_{n-1}, p_n$ be the unique labels of these instructions. There is a unique instruction labeled HALT. For $E \in \{C_1, C_2\}$, the instructions P are of the following forms:
   (a) p_i: $Inc(E)$, goto p_j,
   (b) p_i: If $E = 0$, goto p_j, else go to p_k,
   (c) p_i: $Dec(E)$, goto p_j,
   (d) p_n: HALT.

A configuration $W = (i, c_1, c_2)$ of $\mathcal{M}$ is given by the value of the current program counter i and valuation c_1, c_2 of the counters C_1, C_2. A move of the counter machine $(l, c_1, c_2) \rightarrow (l', c_1', c_2')$ denotes that configuration (l', c_1', c_2') is obtained from (l, c_1, c_2) by executing the lth instruction p_l.

Theorem 2. *[14] The halting problem for 2-counter machines is undecidable.*

Insertion Channel machine With Emptiness Testing: In section 4, we show that the complexity of satisfiability checking $\mathsf{MTL}[\Diamond_I]$ in the pointwise sense has a non-primitive recursive lower bound and is $\mathbf{F}_{\omega^\omega}$-hard. This is shown by a reduction from the reachability problem for *Insertion Channel machine With Emptiness Testing* (ICMET), which is NPR and $\mathbf{F}_{\omega^\omega}$-complete [20].

A channel machine consists of a finite-state automaton acting on finite set of unbounded FIFO channels, or queues. More precisely, a channel machine is a tuple $\mathcal{C} = (S, M, \Delta, C)$, where S is a finite set of control states, C is a finite set of channels, M is a finite set of messages, and $\Delta \subseteq S \times \Sigma \times S$ is the transition relation over the label set $\Sigma = \{c!m, c?m, c = \emptyset \mid m \in M, c \in C\}$. A transition labelled $c!m$ writes message m to the tail of the channel c, and a transition labelled $c?m$ reads message m from the head of the channel c. The transition $c = \emptyset$ checks that channel c is empty. $c?m$ is only enabled when the channel c is non-empty, while the emptiness check $c = \phi$ is only enabled when the channel c is empty. A global state of the channel machine at any point of time is given by the contents of all the channels and the current state s_i. A global state is written as

$$\langle s_i, (c_1 = (m_{h_1} \ldots m_{t_1}), c_2 = (m_{h_2} \ldots m_{t_2}), \ldots, c_k = (m_{h_k} \ldots m_{t_k}))\rangle$$

where m_{h_i} refers to head of the ith channel and m_{t_i} refers to the tail of the ith channel. A transition of the channel machine is defined as :

1. Write to channel i:
   $\langle s, (c_1, \ldots, c_i = x, \ldots, c_k)\rangle \xrightarrow{c_i!m} \langle s', (c_1, \ldots, c_i = xm, \ldots, c_k)\rangle$
2. Read from channel i:
   $\langle s, (c_1, \ldots, c_i = m.x, \ldots, c_k)\rangle \xrightarrow{c_i?m} \langle s', (c_1, \ldots, c_i = x, \ldots, c_k)\rangle$
3. Emptiness check for channel i:
   $\langle s, (c_1, \ldots, c_i = \emptyset, \ldots, c_k)\rangle \xrightarrow{c_i=\phi} \langle s', (c_1, \ldots, c_i = \emptyset, \ldots, c_k)\rangle$

If we only allow the transitions indicated above, then we call C an error-free channel machine. A computation of such a machine is a finite sequence of transitions of the above kind. We also consider channel machines that are subject to insertion errors. Given $x, y \in M^*$, define an ordering $x \sqsubseteq y$ if x can be obtained from y by deletion of any number of characters. For example, PL $\sqsubseteq$ PROLOG. Insertion errors are modelled by extending the transitions as follows: If $(s, x_1) \xrightarrow{\alpha} (s', y_1)$, and $x_2 \sqsubseteq x_1$,$y_1 \sqsubseteq y_2$ then $(s, x_2) \xrightarrow{\alpha} (s', y_2)$. Transitions in an ICMET allow such transitions with insertion errors. A run in ICMET is a finite or infinite sequence of transitions between global states $(s_0, C_0) \xrightarrow{\alpha_0} (s_1, C_1) \xrightarrow{\alpha_1} (s_2, C_2) \xrightarrow{\alpha_2} \ldots$, with $s_0 = init$ and C_i is the of contents of all the channels at transition i.

The control-state reachability problem asks, given a channel machine $\mathcal{C} = (S, M, \Delta, C)$ and two distinct control states $s, s' \in S$, whether there is a finite computation of $\mathcal{C}$ starting in global state $\langle s, (\epsilon, \ldots, \epsilon)\rangle$ and ending in global state $\langle s', (x_1, \ldots, x_k)\rangle$ for some $x_i \in M^*$. This problem was proved to be decidable for ICMETs with NPR lower bound by Schnoebelen [21]. The recurrent-state problem, on the other hand, asks whether $\mathcal{C}$ has an infinite computation that visits some state infinitely often, irrespective of channel contents. This was shown to be undecidable in [16].

Theorem 3. *The control state reachability problem for ICMETs is decidable with non - primitive recursive complexity [21] and is $\mathbf{F}_{\omega^\omega}$-complete [20]. The recurrent state problem for ICMETs is undecidable and is Π_1^0-hard [16].*

3 Expressiveness

In this section we study the expressiveness of different classes of unary MTL in the pointwise sense. The unary fragment of MTL is the one which uses only the modalities $\Diamond_I, \overleftarrow{\Diamond}_I$; recall that the full MTL uses $\mathsf{U}_I, \mathsf{S}_I$.

Lemma 1. $\mathsf{MTL}[\Diamond_I^s]$ *is strictly more expressive than* $\mathsf{MTL}[\Diamond_I]$.

Proof. Any formula in $\mathsf{MTL}[\Diamond_I]$ can be expressed in $\mathsf{MTL}[\Diamond_I^s]$ as follows: For an interval of the form $I' = [0, l\rangle$, we have $\Diamond_{I'}(\varphi) = \varphi \vee \Diamond_{I'}^s(\varphi)$. For intervals I which are not of the form $[0, l\rangle$, we have $\Diamond_I(\varphi) = \Diamond_I^s(\varphi)$.

Next, we show that $\mathsf{MTL}[\Diamond^s] - \mathsf{MTL}[\Diamond_I, \overleftarrow{\Diamond}_I] \neq \phi$ by playing an EF game with modalities $\Diamond_I, \overleftarrow{\Diamond}_I$. Consider the candidate formula $\varphi = a \wedge \Diamond_{[0,0]}^s a$. The formula says that there are at least 2 a's at 0. This cannot be expressed with non strict unary modalities. Consider words $\rho_1 = (a, 0)$ and $\rho_2 = (a, 0)(a, 0)$. Clearly $\rho_1 \nvDash \varphi$ while $\rho_2 \models \varphi$.

If *Spoiler* starts on ρ_1, he will have to stay at the start position irrespective of $\Diamond_{[0,x\rangle}$ or $\overleftarrow{\Diamond}_{[0,x\rangle}$ move chosen, and the *Duplicator* can easily replicate the move on ρ_2. Assume now that the game begins by *Spoiler* choosing ρ_2. For any choice of moves $\Diamond_{[0,x\rangle}$ or $\overleftarrow{\Diamond}_{[0,x\rangle}$ of *Spoiler* in ρ_2, it is possible for *Duplicator* to stay at the ony position in ρ_1. Thus, *Duplicator* will always win a $\Diamond_I, \overleftarrow{\Diamond}_I$ game on these words. □

Lemma 2. $\mathsf{MTL}[\Diamond_I, \bigcirc]$ *is more expressive than* $\mathsf{MTL}[\Diamond_I^s, \overleftarrow{\Diamond}_I^s]$.

Proof. We first show that $\mathsf{MTL}[\Diamond_I, \mathsf{O}] - \mathsf{MTL}[\Diamond^s_I, \overleftarrow{\Diamond}^s_I] \neq \phi$. Consider the formula $\varphi = \Box(a \to \mathsf{O}b)$. Playing a $\Diamond^s_I, \overleftarrow{\Diamond}^s_I$ EF game for any $n > 0$ rounds, we show that φ has no equivalent formula in $\mathsf{MTL}[\Diamond^s_I, \overleftarrow{\Diamond}^s_I]$.

Choose some $m \in \mathbb{N}$ and δ such that $0 < 4m\delta < 1$ and $m >> n$. Consider words $\rho_1 = (a, \delta)(b, 2\delta) \ldots (a, (2m-1)\delta)(b, 2m\delta) \ldots (a, (4m-1)\delta)(b, 4m\delta)$, and $\rho_2 = (a, \delta)(b, 2\delta) \ldots (a, (2m-1)\delta)(a, \theta)(b, 2m\delta) \ldots (a, (4m-1)\delta)(b, 4m\delta)$. ρ_2 differs from ρ_1 only with respect to the extra (a, θ) inserted between $(a, (2m-1)\delta)$ and $(b, 2m\delta)$. Clearly $\rho_1 \models \varphi$ while ρ_2 does not.

Spoiler can choose to be on any word to begin with. If *Spoiler* is at a position other than (a, θ) on either word, *Duplicator* can play copy cat for any move chosen. Consider the case when *Spoiler* comes on (a, θ). In this case, *Duplicator* can come to $(a, (2m-1)\delta)$ or $(a, (2m+1)\delta)$ in the other word. If *Spoiler* moves to any previous/future a, b in ρ_2, *Duplicator* can also move to a corresponding a, b by choosing the same move. Thus, *Duplicator* wins the the n round game with $\Diamond^s_I, \overleftarrow{\Diamond}^s_I$ moves.

Next, we see that $\Diamond^s$ can be easily encoded using $\Diamond, \mathsf{O}$. For intervals I not of the form $[0, y\rangle$, $\Diamond^s_I$and $\Diamond_I$ are equivalent. Note that $\Diamond^s_{[0,y\rangle}(\varphi) = \mathsf{O}\Diamond_{[0,y\rangle}(\varphi)$. Note that this also implies that $\mathsf{MTL}[\Diamond_I, \mathsf{O}] = \mathsf{MTL}[\Diamond^s_I, \mathsf{O}]$. □

Lemma 3. $\mathsf{MTL}[\Diamond_I, \mathsf{O}_I] \equiv \mathsf{MTL}[\Diamond_I, \mathsf{O}]$

Proof. We need to encode O_I using $\Diamond_I$ and O. A formula $\mathsf{O}_{[l,u]}(\varphi)$ can be written as $\neg\mathsf{O}\Diamond_{[0,l)}(\top) \wedge \Diamond_{[l,u]}(\top) \wedge \mathsf{O}(\varphi)$. □

Lemma 4. $\mathsf{MTL}[\Diamond_I, \mathsf{O}_I]$ *and* $\mathsf{MTL}[\mathsf{U}_I]$ *are incomparable.*

Proof. We first show that $\mathsf{LTL}[\mathsf{U}] - \mathsf{MTL}[\Diamond_I, \mathsf{O}_I] \neq \phi$. $\mathsf{LTL}[\mathsf{U}]$ is the fragment of $\mathsf{MTL}[\mathsf{U}_I]$ where the intervals I are only of the form $[0, \infty)$.

By Lemma 3, we need to only consider the untimed O operator. Consider the formula $\varphi = a\mathsf{U}b \in \mathsf{LTL}[\mathsf{U}]$. We show that for any $n > 0$, *Duplicator* can win an n round $\Diamond_I, \mathsf{O}_I$ game. Pick $m >> n$ and δ, κ such that $0 < m\delta + 2m\kappa < 1$. Consider the words $\rho_1 = A(0)A'(\delta)A'(2\delta) \ldots A'(m\delta)$ and $\rho_2 = A'(0)A'(\delta)A'(2\delta) \ldots A'(m\delta)$. Here, $A(t) = (a, t)(a, t+\kappa) \ldots (b, t + 2m\kappa)$ and $A'(t) = (a, t)(a, t+\kappa) \ldots (a, t + m\kappa)(c, t+\theta)(a, t+(m+1)\kappa) \ldots (b, t+2m\kappa)$. Clearly, $\rho_1 \models \varphi$ and $\rho_2 \not\models \varphi$.

For any $\Diamond_I, \mathsf{O}_I$ move of *Spoiler* landing on an a or a b, *Duplicator* can play copy cat. When *Spoiler* visits the c at $t+\theta$ for $t \geq 0$, *Duplicator* can visit a c at $t + \theta + p\delta$, $p \geq 1$. *Spoiler* cannot achieve anything using the O move since the number of a's before the c is much more than n, the number of rounds. Any $\Diamond_I$ move of *Spoiler* can be duplicated with the gap of at most $(2m+1)\delta$ difference in time-stamp. Hence, *Duplicator* wins the n round $\Diamond_I, \mathsf{O}_I$ game.

We next show that $\mathsf{MTL}[\Diamond_I, \mathsf{O}_I] - \mathsf{MTL}[\mathsf{U}_I, \mathsf{S}_I] \neq \phi$. Consider the formula $\varphi = \Box(a \to \mathsf{O}b)$ and words ρ_1 and ρ_2 from Lemma 2. Consider an n round $\mathsf{U}_I, \mathsf{S}_I$ EF-game on those pair of words. An interesting move for *Spoiler* is to come to $(a, (2m-1)\delta)$ in ρ_2. *Duplicator* is on some $(a, (2k-1)\delta)$ in ρ_1. If *Spoiler* swaps words, and invokes an U_I move coming to $(b, 2k\delta)$ in ρ_1, then *Duplicator* comes to $(b, 2m\delta)$ in ρ_2. *Spoiler* can now pick the (a, θ) in duplicator's word in between the positions $(a, (2m-1)\delta)$ and $(b, 2m\delta)$. In response, *Duplicator* can always pick the position $(a, (2k-1)\delta)$ in ρ_1, since the until move is non-strict. Note that if the until move was a strict one, *Duplicator* would have to pick a position strictly in between $(2k-1)\delta$ and $2k\delta$, and would have lost. Similar argument will work for since moves. Hence, *Duplicator* wins the n round $\mathsf{U}_I, \mathsf{S}_I$ game. □

Lemma 5. $\mathsf{MTL}[\Diamond_I^s, \bigcirc_I]$ *is strictly contained in* $\mathsf{MTL}[\mathsf{U}_I^s]$.

Proof. By lemma 3 we can eliminate timed $\bigcirc$ and by lemma 4 we can say that $MTL[\Diamond_I^s, \bigcirc_I]$ cannot simulate U. By the equivalences $\bigcirc_I\varphi = \bot\mathsf{U}_I^s\varphi$ and $\Diamond^s\varphi = \top\mathsf{U}^s\varphi$, U^s can simulate both $\bigcirc, \Diamond$. Hence proved. □

Lemma 6. *$MTL[\Diamond_I^s, \Delta]$ is equivalent to $MTL[\Diamond_I, \Delta]$ where Δ is any list of operators if the models are restricted to be strictly monotonic words.*

Proof. For strictly monotonic words, any two points i, j with $i > j$ has $t_i > t_j$. We know that $\Diamond_I^s(\varphi) = \Diamond_I(\varphi)$ for any interval I not of the form of $[0, u\rangle$. Also, for intervals of type $I' = [0, u\rangle$, $\Diamond_{I'}\varphi = \varphi \vee \Diamond_{I'}^s(\varphi)$. Due to strict monotonicity of the underlying model, $\Diamond_{[0,u\rangle}^s(\varphi) = \Diamond_{(0,u\rangle}^s(\varphi) = \Diamond_{(0,u\rangle}(\varphi)$. Hence proved. □

4 Unary MTL and Undecidability

In this section, we explore the decidability and complexity of the satisfiability of the unary fragment of MTL in the pointwise sense. The undecidability of $\mathsf{MTL}[\Diamond_I, \Diamond\!\!\!\!-_I]$ follows by construction of an appropriate MTL formula φ simulating a deterministic 2-counter machine $\mathcal{M}$ such that φ is satisfiable iff $\mathcal{M}$ halts. Since the non-emptiness problem for two counter machines is Σ_1^0-complete, we obtain the Σ_1^0-hardness of the satisfiability for $\mathsf{MTL}[\Diamond_I, \Diamond\!\!\!\!-_I]$. We also show the non primitive recursive lower bound for satisfiability of $\mathsf{MTL}[\Diamond_I]$ by reduction of reachability problem for ICMETs. Since the reachability problem for ICMETs is $\mathbf{F}_{\omega^\omega}$-complete, we obtain the $\mathbf{F}_{\omega^\omega}$-hardness of the satisfiability of $\mathsf{MTL}[\Diamond_I]$.

Encoding Minsky Machines in $\mathsf{MTL}[\Diamond_I, \Diamond\!\!\!\!-_I]$

We encode each computation of a 2-counter machine $\mathcal{M}$ using timed words over the alphabet $\Sigma_{\mathcal{M}} = \{b_1, b_2, \ldots, b_n, a\}$. We then generate a formula $\varphi_{\mathcal{M}} \in \mathsf{MTL}[\Diamond_I, \Diamond\!\!\!\!-_I]$ such that $L_{\mathcal{M}} = L(\varphi_{\mathcal{M}})$. The encoding is done in the following way: A configuration $\langle i, c_1, c_2\rangle$ of the counter machine is represented by equivalence class of the words of the form $(b_i^+, t_0)(a^+, t_1)\ldots(a^+, t_{c_1})(a^+, t_1')\ldots(a^+, t_{c_2}')$, where for any time-stamp t, (x^+, t) is some element of $(X^+, t) = \bigcup_{i>0}\{(x, t)^i\}$ for any $x \in \Sigma$. Thus, (x^+, t) denotes a sequence of one or more symbols of the form (x, t). For brevity, we will denote (x^+, t) with x^+. Note that we are working on words that are *not* strictly monotonic : hence, we allow several symbols x at the same time stamp.

A computation of $\mathcal{M}$ is encoded by concatenating sequences of individual configurations. We encode the jth configuration of $\mathcal{M}$ in the time interval $[5j, 5(j+1))$ as follows: For $j \in \mathbb{N}$,

1. $b^+_{i_j}$ (representing instruction p_{i_j}) occurs at time $5j$;
2. The value of counter C_q, $q \in \{1,2\}$, in the jth configuration is given by the number of a^+'s with distinct time stamps in the interval $(5j+2q-1, 5j+2q)$;
3. The a's can appear only in the intervals $(5j+2q-1, 5j+2q)$, $q \in \{1,2\}$;
4. The intervals $(5j, 5j+1), (5j+2, 5j+3)$ and $(5j+4, 5j+5)$ have no events. Thus, after any unit interval encoding the value of a counter, the next unit interval has no events. The computation must start with initial configuration and the final configuration must be the $HALT$ instruction.

$\varphi_{\mathcal{M}}$ is obtained as a conjunction of several formulae. Let B be a shorthand for $\bigvee_{i \in \{1,\ldots,n\}} b_i$. The interesting part is to encode precise copy, increment and decrement of counters in the successive configurations. We define the macros $COPY_i, INC_i, DEC_i$ for copying, incrementing and decrementing the contents of counter C_i in successive configurations.

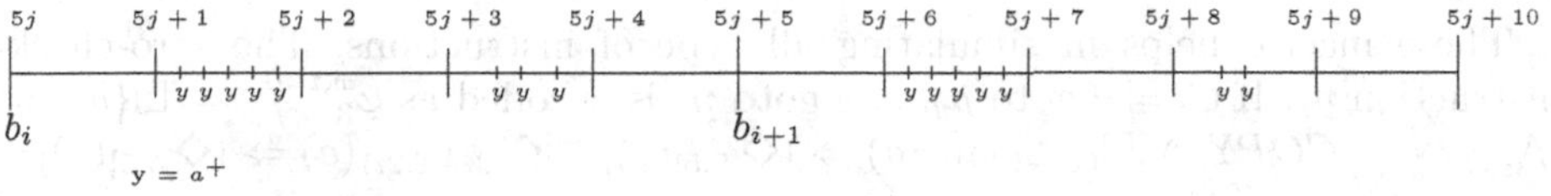

Fig. 3. b_i: Decrement counter C_2 and move next

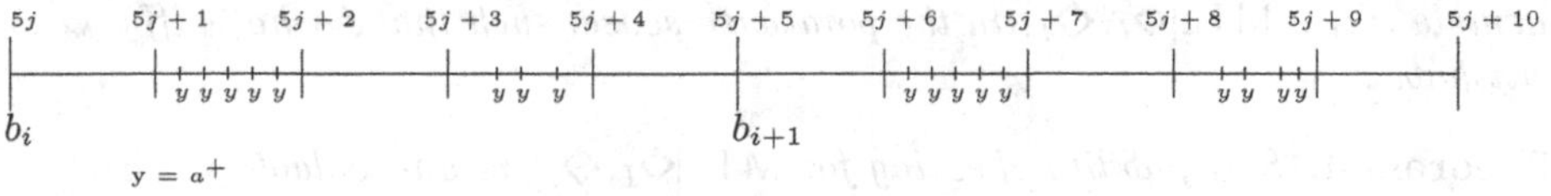

Fig. 4. b_i: Increment counter C_2 and move next

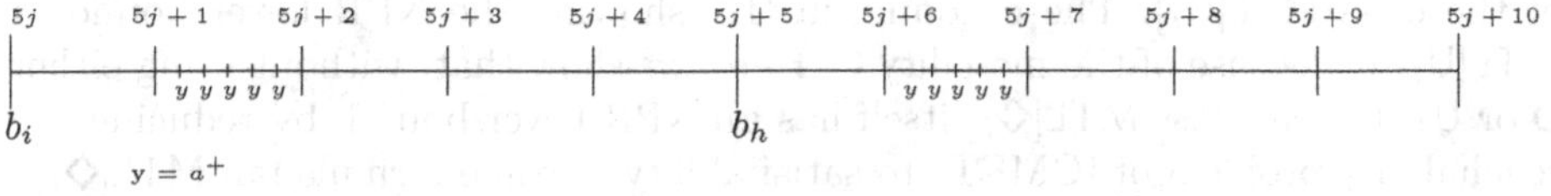

Fig. 5. b_i: If counter $C_2 = 0$ then move to b_h

– $COPY_i$: Every (a^+, t) occurring in the interval $(5j+2i-1, 5j+2i)$ has a copy at a future distance 5, and every (a^+, t) occurring in the next interval has an a^+ at a past distance 5. That is, corresponding to all the time stamps $t_1, \ldots, t_{c_i} \in (5j+2i-1, 5j+2i)$ where a^+ holds, we have precisely and only time stamps $t_1+5, \ldots, t_{c_i}+5 \in (5(j+1)+2i-1, 5(j+1)+2i)$ where a^+, for $i \in \{1,2\}$. This ensures the absence of insertion errors. $COPY_i = \Box_{(2i-1,2i)}[(a \Rightarrow \Diamond_{[5,5]} a)] \wedge \Box_{(5+2i-1,5+2i)}[(a \Rightarrow \Diamond\!\!\!\!\cdot_{[5,5]} a)]$.

- INC_i: All a^+'s in the current configuration are copied to the next, at a future distance 5; in the next configuration, every a^+ except the last, has an a^+ at past distance 5. This corresponds to propagating time stamps $t_1, \dots, t_{c_i}$ to $t_1 + 5, \dots, t_{c_i} + 5$. Also, there is a $t_{c_i+1} + 5$ which has no counterpart t_{c_i+1} in the previous configuration.
  $$INC_i = \Box_{(0,5)}\{ [a \Rightarrow \Diamond_{[5,5]}a] \wedge[(a \wedge \neg\Diamond_{(0,1)}a) \Rightarrow (\Diamond_{(5,6)}a \wedge \Box_{(5,6)}(a \Rightarrow \Box_{(0,1)}(false)))] \} \wedge \Box_{(5+2i-1,5+2i)}[(a \wedge \Diamond_{[0,1]}(a)) \Rightarrow \overleftarrow{\Diamond}_{[5,5]}a]$$
- DEC_i: All the a^+'s in the current configuration, except the last, have a copy at future distance 5. All the a^+'s in the next configuration have a copy at past distance 5. That is, all the time stamps $t_1, \dots, t_{c_i-1}$ in the interval $(5j + 2i - 1, 5j + 2i)$ have corresponding time stamps $t_1 + 5, \dots, t_{c_i-1} + 5$ in $(5(j + 1) + 2i - 1, 5(j + 1) + 2i)$. Moreover, $t_{c_i-1} + 5$ is the last time stamp in $(5(j + 1) + 2i - 1, 5(j + 1) + 2i)$.
  $DEC_i = \Box_{(2i-1,2i)}\{[(a \wedge \Diamond_{(0,1)}a) \Rightarrow \Diamond_{[5,5]}a] \wedge [(a \wedge \neg\Diamond_{[0,1]}a) \Rightarrow \neg\Diamond_{[5,5]}a]\} \wedge \Box_{(5+2i-1,5+2i)}[(a \Rightarrow \overleftarrow{\Diamond}_{[5,5]}a)]$.

These macros helps in simulating all type of instructions. The zero-check instruction p_x: If $C_i = 0$ goto p_y, else goto p_z is encoded as $\varphi_3^{x,i=0} = \Box\{b_x \Rightarrow (\bigwedge_{i\in\{1,\dots,n\}} COPY_i \wedge [\Box_{(2i-1,2i)}(\neg a) \Rightarrow (\Diamond_{[5,5]}b_y)] \wedge [\Diamond_{(2i-1,2i)}(a) \Rightarrow (\Diamond_{[5,5]}b_z)]\}$. The final formula we construct is $\varphi_{\mathcal{M}} = \bigwedge_{i=0}^{6} \varphi_i$, where φ_3 is the conjunction of formulae $\varphi_3^{x,inc_i}, \varphi_3^{x,dec_i}, \varphi_3^{x,i=0}$. We thus obtain:

Lemma 7. *Let $\mathcal{M}$ be a 2-counter Minsky machine. Then, we can synthesize a formula $\varphi_{\mathcal{M}} \in \mathsf{MTL}[\Diamond_I, \overleftarrow{\Diamond}_I]$ in the pointwise sense, such that $\mathcal{M}$ halts iff $\varphi_{\mathcal{M}}$ is satisfiable.*

Theorem 4. *Satisfiability checking for $\mathsf{MTL}[\Diamond_I, \overleftarrow{\Diamond}_I]$ is undecidable.*

We next look at the fragment $\mathsf{MTL}[\Diamond_I]$. The decidability of this fragment follows from the decidability of the more general class $\mathsf{MTL}[\mathsf{U}_I]$ [15]. Further, [16] established a non-primitive recursive (NPR) lower bound for satisfiability of the class $\mathsf{MTL}[\mathsf{U}_I]$. The encoding in [16] showing the NPR lower bound of $\mathsf{MTL}[\mathsf{U}_I]$ makes use of the modality O. Here, we show that, without using either O or U_I, the subclass $\mathsf{MTL}[\Diamond_I]$ itself has an NPR lower bound, by reducing the reachability problem of ICMETs to satisfiability of some formulae in $\mathsf{MTL}[\Diamond_I]$.

Encoding *ICMET* in $\mathsf{MTL}[\Diamond_I]$

Consider an ICMET $\mathcal{C} = (S, M, \Delta, C)$. For encoding configurations of ICMETs, let us work with the alphabet $\Sigma = M \cup \Delta \cup S \cup b$. Here is how we encode the jth configuration of the ICMET in a timed word:

1. A configuration j, $j \geq 1$ is encoded in the interval $[(2k+2)j, (2k+2)(j+1)]$ where k refers to number of channels.
2. At time $(2k + 2)j$, the current state of the ICMET at configuration j is encoded.

3. At intervals $(2i-1, 2i)$, $i \geq 1$, from the start of the jth configuration, the contents of ith channel are encoded as shown in the figure below.
4. The first string $m_{h_i}^{+}$ in the interval $(2i-1, 2i)$ is the head of the channel i and denotes that m_{h_i} is the message stored at the head of the channel. The last string $m_{t_i}^{+}$ in the interval is the tail of the channel, and denotes that message m_{t_i} is the message stored at the tail of the channel.
5. The Intervals $(2i-2, 2i-1)$ from the start of the jth configuration are "no action" intervals.
6. Exactly at $2k+1$ time unit after the start of the jth configuration, α_i^{+} holds. α_i encodes the transition from the state at j^{th} configuration($[(2k+2)j, (2k+2)(j+1))$ to the $(j+1)^{st}$ configuration($[(2k+2)j, (2k+2)(j+1))$. Note that α has the form $(s, c!m, s')$ or $(s, c?m, s')$ or $(s, c = \emptyset, s')$.
7. We introduce a special symbol b, which acts as separator between the head of the message and the remaining contents, for each channel.

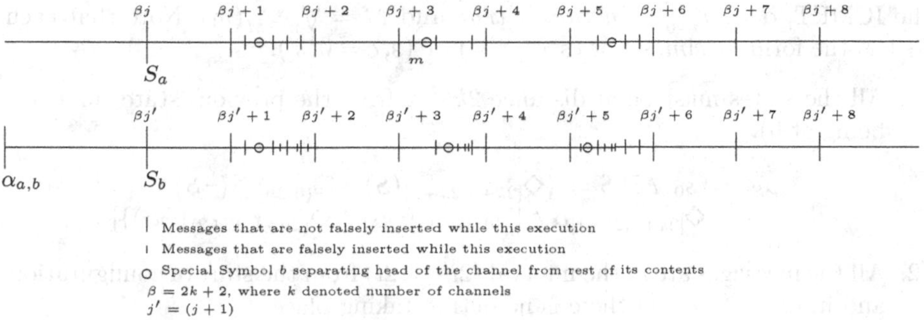

Fig. 6. S_i, $?m$, S_j

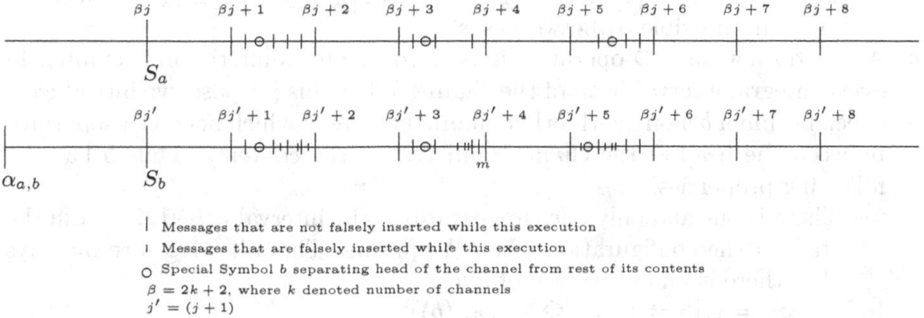

Fig. 7. S_i, $!m$, S_j

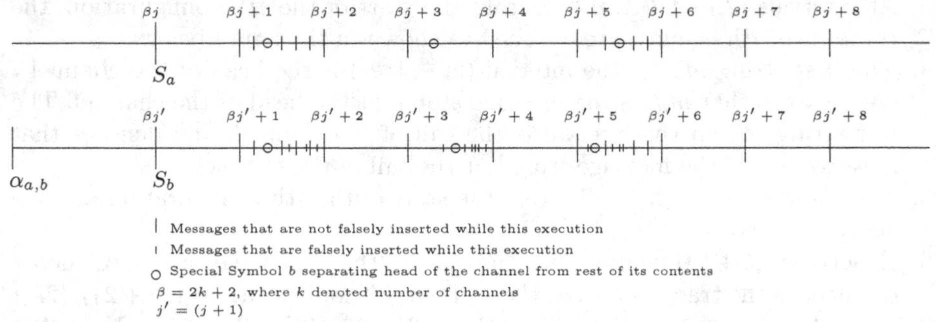

Fig. 8. $S_i,\ C_2 = \phi S_j$

Here we use $\mathsf{MTL}[\Diamond_I]$ formulae to specify the working of ICMET using timed words as discussed above. We use $S = \bigvee_{i=0}^{n} s_i$ to denote any of the states s_i of the ICMET, $\alpha = \bigvee_{i=0}^{m} \alpha_i$, $action = true$ and $M = \bigvee_{m \in M}(m)$. Note that each α_i has the form $(s, c!m, s')$ or $(s, c?m, s')$ or $(s, c = \emptyset, s')$.

1. All the states must be at distance $2k + 2$ from the previous state (first one being at 0).

$$\begin{aligned}\varphi_S \;&=\; s_0 \vee \Box[S \Rightarrow \{\Diamond_{[2k+2,2k+2]}(S) \wedge \Box_{(0,2k+2)}(\neg S)\wedge \\ &\Diamond_{[2k+1,2k+1]}\alpha \wedge \Box_{[0,2k+1)}(\neg\alpha) \wedge \Diamond_{(2k+1,2k+2)}(\neg\alpha)\}]\end{aligned}$$

2. All the messages are in the interval $(2i-1, 2i)$ from the start of configuration, and in $(2i-2, 2i-1)$ there is no action taking place.
   $\varphi_m = \Box\{S \Rightarrow \bigwedge_{i=1}^{k} \Box_{(2i-1,2i)}(M \vee b) \wedge \Box_{(2i-2,2i-1)}(\neg action)\}$
3. Consecutive source and target states must be in accordance with a transition α. For example, s_j appears consecutively after s_i reading α_i iff α_i is of the form $(s_i, y, s_j) \in \Delta$, with $y \in \{c_i!m, c_i?m, c_i = \emptyset\}$.
   $\varphi_\Delta = \bigwedge_{s,s' \in S} \Box\{(s \wedge \Diamond_{[2k+2,2k+2]} s') \Rightarrow (\Diamond_{[2k+1,2k+1]} \bigvee \Delta_{s,s'})\}$ where $\Delta_{s,s'}$ are possible α_i between s, s'.
4. As we do not have $\bigcirc$ operator, it is hard to pin point the first symbol in every integral interval(head of the channel). For this purpose, we introduce a special symbol b along with other channel contents which acts as a separator between the head of the channel and rest of the contents. Thus b has the following properties:
   - There is one and only one time-stamp in the interval $(2i-1, 2i)$ from the start of the configuration where b appears. The following formula says that there is an occurrence of a b^+:
     $\varphi_{b_1} = \Box[S \Rightarrow (\bigwedge_{i=1}^{k} \Diamond_{(2i-1,2i)}(b))]$
     The following formula says that there can be only one b^+:
     $\varphi_{b_2} = \Box(b \Rightarrow \neg\Diamond_{(0,1)} b)$
   - If the channel is not empty (there is at least one message m in the interval $(2i-1, 2i)$ corresponding to channel i contents) then there is one

and only one m before b. The following formula says that there can be at most one m before b.

$\varphi_{b_3} = \Box[\neg\{M \wedge \Diamond_{(0,1)}(M \wedge \Diamond_{(0,1)}(b)\}]$
The following formula says that there is a message m in the channel, if the channel is non-empty.

$\varphi_{b_4} = \Box[S \Rightarrow \{\bigwedge_{j=1}^{k}(\Diamond_{(2j-1,2j)}(M) \Rightarrow \Diamond_{(2j-1,2j)}(M \wedge \Diamond_{(0,1)}b))\}]$
Let $\varphi_b = \varphi_{b_1} \wedge \varphi_{b_2} \wedge \varphi_{b_3} \wedge \varphi_{b_4}$. The formula φ_b encodes the behaviour of b as described above.

5. Encoding transitions:
   - If the transition is of the form $c_i = \phi$. The following formulae checks that there is no event in the interval $(2i-1, 2i)$ corresponding to channel i, while all the other channel contents are copied. If there was some m^+ at time t, it is copied to $t+2k+2$, representing the channel contents in the next configuration.
   $\varphi_{c_i=\phi} = S \wedge \Box_{(2i-1,2i)}(\neg action) \wedge \Box_{(0,2k+2)}(\bigwedge_{m\in M}(m \Rightarrow \Diamond_{[2k+2,2k+2]}(m))$
   - If the transition is of the form $c_i!m$ where $m \in M$. An extra message is appended to the tail of channel i, and all the m^+'s are copied to the next configuration. $M \wedge \Box_{(0,1)}(\neg M))$ denotes the last message of channel i; if this occurs at time t, we know that this is copied at a distance $2k+2+t$, now we assert that from $2k+2+t$, we see a message in $\Diamond_{(0,1)}$.
   $\varphi_{c_i!m} = S \wedge \Box_{(0,2k+2)}\{\bigwedge_{m\in M}(m \Rightarrow \Diamond_{[2k+2,2k+2]}(m))\}$
   $\wedge \Diamond_{[2i-1,2i)}\{(M \wedge \Box_{(0,1)}(\neg M)) \Rightarrow \Diamond_{[2k+2,2k+2]}(\Diamond_{(0,1)}(m))\}$
   - If the transition is of the form $c_i?m$ where $m \in M$. The contents of all channels other than i are copied at a distance $2k+2$ corresponding to the next configuration. We check the existence of a first message in channel i; such a message has a b at distance $(0,1)$ from it. The rest of channel i contents are copied at distance $2k+2$.
   $\varphi_{c_i?m} = S \wedge \bigwedge_{j\neq i,j=1}^{k} \Box_{[2j-1,2j]}\{\bigwedge_{m\in M} m \Rightarrow \Diamond_{[2k+2,2k+2]}(m)\} \wedge \Diamond_{(2i-1,2i)}\{m \wedge \Diamond_{(0,1)}(b)\} \wedge \Box_{[2i-1,2i]}\{\bigwedge_{m\in M}(m \wedge \neg\Diamond_{(0,1)}b) \Rightarrow \Diamond_{[2k+2,2k+2]}(m)\}$
6. Channel contents must change in accordance to the relevant transition. Let $l \in L$ and α_l be a transition labeled l.
   $\varphi_C = \Box[S \Rightarrow \bigwedge_{l\in L}(\Diamond_{[2k+1,2k+1]}(\bigvee \alpha_l \Rightarrow \phi_l))]$ where ϕ_l are the formulae encoding transitions.
7. Let t be a state of the ICMET whose reachability we are interested in. Check t is reachable from s_0.
   $\phi_{reach} = \Diamond(t)$
8. Mutual Exclusion: There is only one type of event taking place at any particular time-stamp
   $\varphi_{mutex} = \bigwedge_{y\in\Sigma}(y \Rightarrow \neg\Diamond_{[0,0]}(\bigvee_{x\in\Sigma\setminus\{y\}}(x))$
   Thus the formula encoding ICMET is:
   $\varphi^3 = \varphi_S \wedge \varphi_\Delta \wedge \varphi_m \wedge \varphi_b \wedge \varphi_C \wedge \varphi_{reach} \wedge \varphi_{mutex}$

Lemma 8. *Let $\mathcal{M}$ be a ICMET. Then, we can synthesize a formula $\varphi_{\mathcal{M}} \in$ MTL[$\Diamond_I$] such that $\mathcal{M}$ reaches a state t starting with s_0 iff $\varphi_{\mathcal{M}}$ is satisfiable.*

Theorem 5. *Satisfiability checking for* $\mathsf{MTL}[\Diamond_I]$ *has non-primitive recursive complexity over finite words and is undecidable for infinite words.*

Proof. Lemma 8 and Theorem 3 together say that satisfiability of $\mathsf{MTL}[\Diamond_I]$ is non primitive recursive. The $\mathbf{F}_{\omega^\omega}$-hardness of $\mathsf{MTL}[\Diamond_I]$ over finite words follows from the $\mathbf{F}_{\omega^\omega}$-completeness of the reachability problem for ICMETs [20].

It follows from Lemma 8 that the recurrent state problem of $ICMET$ can also be encoded. Due to the Π_1^0-hardness of the recurrent-state problem for ICMETs, the satisfiability of $\mathsf{MTL}[\Diamond_I]$ over infinite words is Π_1^0-hard. Consider the formula $\varphi^4 = \varphi_S \wedge \varphi_\Delta \wedge \varphi_m \wedge \varphi_b \wedge \varphi_C \wedge \varphi_{rec} \wedge \varphi_{mutex}$ where all the RHS formula are from the section 4 (except φ_{rec}) and $\varphi_{rec} = \Box\Diamond(t)$. Thus φ^4 is satisfiable iff the state t occurs infinitely often. Hence $\mathsf{MTL}[\Diamond_I]$ is undecidable over infinite words. □

We next show the undecidability of the unary fragment of TPTL with only 2 variables. This is done by encoding two counter machines.

Encoding Minsky Machines in TPTL[◇] with 2 Clock Variables

The timed word used for encoding, as well as the formulae are similar to the one used for proving undecidability of $\mathsf{MTL}[\Diamond_I, \overleftarrow{\Diamond}_I]$. The only difference is that we explicitly write formulae to copy all the non-last a^+ to optimize the number of freeze variables. We only give the formula $COPY$ which tells about copying all the non-last a^+'s without any insertion errors. All other formulae are there in the appendix. Note that a non-last a^+ in an interval means the bunch of a's that have the same time-stamp t, and there is no other time-stamp in that interval greater than t where a holds. Let x, y be the freeze variables.
$COPY$: All the non-last a^+ should be copied precisely at distance 5. Also, if (a^+, t_1) and (a^+, t_2) denote two consecutive a blocks, that is, there is no time stamp t' such that $t_1 < t' < t_2$, then we must see a^+ at $t_1 + 5$ as well as $t_2 + 5$. Also, there must not be a time stamp $t'' + 5$ such that $t_1 + 5 < t'' + 5 < t_2 + 5$. We separate the copying of non-last symbols to optimize the number of clock variables (or freeze variables). Thus, the $COPY$ formulae is $COPY = COPY_1 \wedge COPY_2$. $COPY_1$ specifies that all the non-last a^+'s are copied precisely at distance 5, possibly with insertion errors. $COPY_2$ ensures that there is no insertion error. It says *freeze* x *at every non-last* a^+ *and freeze* y *at some* a^+ *in the future within* $(0,1)$ *of freezing* x*. Then assert that there is no* a^+ *in the region* $[T - x + 5, T - y + 5]$. This is only possible if the point where y is frozen is consecutive to the point where x was frozen (other wise $COPY_1$ will be violated). Thus $COPY_2$ ensures that there is no insertion in between the copied a^+'s.

- $COPY_1 = \Box x.[(a \wedge \Diamond(a \wedge x \in (0,1))) \Rightarrow \Diamond(a \wedge x \in [5,5])]$.
- $COPY_2 = \Box x.[(a \wedge \Diamond(a \wedge x \in (0,1))) \Rightarrow (\Diamond y.(a \wedge x \in (0,1) \wedge \neg\Diamond(a \wedge x \in (5,\infty) \wedge y \in (0,5))))]$.

Lemma 9. *Let* $\mathcal{M}$ *be a 2-counter Minsky machine. Then, we can synthesize a formula* $\varphi_{\mathcal{M}} \in \mathsf{TPTL}[\Diamond]$ *with 2 clock variables such that* $\mathcal{M}$ *halts iff* $\varphi_{\mathcal{M}}$ *is satisfiable.*

Theorem 6. *Satisfiability checking for* $\mathsf{TPTL}[\Diamond]$ *with 2 freeze variables is undecidable (Σ_1^0-hard), but with 1 freeze variable is decidable.*

Proof. Lemma 9 along with theorem 2 proves that $\mathsf{TPTL}[\Diamond]$ with 2 freeze variables is undecidable. The main idea of the proof for lemma 9 is shown in the encoding of *COPY* formula given above. Decidability of $\mathsf{TPTL}[\Diamond]$ with 1 freeze variable relies on the conversion of any given formula to 1-clock Alternating Timed Automata, shown to be decidable [15]. □

5 Discussion

In this paper, we sharpen some decidability and complexity results pertaining to MTL. We show that $\mathsf{MTL}[\Diamond_I]$ over finite weakly monotonic words has a NPR lower bound and $\mathsf{MTL}[\Diamond_I]$ over infinite weakly monotonic words is undecidable. An examination of the NPR lower bound of MTL over finite words of Ouaknine and Worrell show that they actually prove the NPR lower bound for $\mathsf{MTL}[\Diamond_I, \mathsf{O}]$ for finite strictly monotonic words. Their results carry over to weakly monotonic time as well, since strict monotonicity can be defined using $\Diamond_I, \mathsf{O}$. In our case, we show that $\mathsf{MTL}[\Diamond_I]$ is sufficient to encode lossy channel machines over weakly monotonic time although strict monotonicity cannot be defined in the logic. Our encoding is therefore, more complex. We can also show that over finite words, satisfiability of $\mathsf{MTL}[\Diamond_I]$ over continuous time, is undecidable. This is also an improvement of the known result that $\mathsf{MTL}[\mathsf{U}_I]$ is undecidable over finite continuous time.

Coming to expressiveness, Pandya and Shah [18] explored the expressiveness and decidability of unary MITL but the case of unary MTL was not addressed. [18] showed that the unary fragment with only lower bound constraints $\mathsf{MITL}[\Diamond_\infty, \overleftarrow{\Diamond}_\infty]$ has NP-complete satisfiability, while the bounded unary fragment $\mathsf{MITL}[\Diamond_b, \overleftarrow{\Diamond}_b]$ has NEXPTIME-complete satisfiability. It is henceforth, an interesting question to explore restrictions on unary MTL which still allows punctual time intervals, but has efficient satisfiability checking and model checking properties. Just as we show here that $\mathsf{MTL}[\Diamond_I, \overleftarrow{\Diamond}_I]$ cannot express the restriction of strict monotonicity, it would be interesting to characterize the limits of expressiveness of different fragments of MTL. Regarding logic TPTL, apart from the positive fragment [4] of TPTL, not much is known about decidable subclasses of TPTL with $n > 1$ clocks over finite and infinite words. Exploring the n-clock fragment of TPTL which is decidable is an interesting question and a future work.

References

1. Alur, R., Henzinger, T.A.: Logics and Models of Real Time: A Survey. In: Proceedings of REX Workshop, pp. 74–106 (1991)
2. Alur, R., Feder, T., Henzinger, T.A.: The Benefits of Relaxing Punctuality. Journal of the ACM 43(1), 116–146 (1996)

3. Ouaknine, J., Worrell, J.B.: Some Recent Results in Metric Temporal Logic. In: Cassez, F., Jard, C. (eds.) FORMATS 2008. LNCS, vol. 5215, pp. 1–13. Springer, Heidelberg (2008)
4. Bouyer, P., Chevalier, F., Markey, N.: On the expressiveness of TPTL and MTL. In: Sarukkai, S., Sen, S. (eds.) FSTTCS 2005. LNCS, vol. 3821, pp. 432–443. Springer, Heidelberg (2005)
5. D'Souza, D., Prabhakar, P.: On the expressiveness of MTL in the pointwise and continuous semantics. Technical Report IISc, India (2005)
6. D'Souza, D., Raj Mohan, M., Prabhakar, P.: Eliminating past operators in Metric Temporal Logic. In: Perspectives in Concurrency, pp. 86–106 (2008)
7. Demri, S., Lazic, R.: LTL with freeze quantifier and register automata. In: LICS 2006, pp. 17–26 (2006)
8. Etessami, K., Wilke, T.: An Until Hierarchy for Temporal Logic. In: Proceedings of LICS 1996, pp. 108–117 (1996)
9. Henzinger, T.A.: The Temporal Specification and Verification of Real-time Systems. Ph.D Thesis, Stanford University (1991)
10. Hirshfeld, Y., Rabinovich, A.: Logics for Real Time: Decidability and Complexity. Fundam. Inform. 62(1), 1–28 (2004)
11. Kini, D.R., Krishna, S.N., Pandya, P.K.: On Construction of Safety Signal Automata for $MITL[\mathsf{U}_I, \mathsf{S}_I]$ using Temporal Projections. In: Fahrenberg, U., Tripakis, S. (eds.) FORMATS 2011. LNCS, vol. 6919, pp. 225–239. Springer, Heidelberg (2011)
12. Koymans, R.: Specifying Real-Time Properties with Metric Temporal Logic. Real Time Systems 2(4), 255–299 (1990)
13. Maler, O., Nickovic, D., Pnueli, A.: Real Time Temporal Logic: Past, Present, Future. In: Pettersson, P., Yi, W. (eds.) FORMATS 2005. LNCS, vol. 3829, pp. 2–16. Springer, Heidelberg (2005)
14. Minsky, M.: Finite and infinite machines. Prentice Hall, New Jersey (1967)
15. Ouaknine, J., Worrell, J.: On the Decidability of Metric Temporal Logic. In: Proceedings of LICS 2005, pp. 188–197 (2005)
16. Ouaknine, J., Worrell, J.B.: On Metric Temporal Logic and Faulty Turing Machines. In: Aceto, L., Ingólfsdóttir, A. (eds.) FOSSACS 2006. LNCS, vol. 3921, pp. 217–230. Springer, Heidelberg (2006)
17. Pandya, P.K., Shah, S.: On Expressive Powers of Timed Logics: Comparing Boundedness, Non-punctuality, and Deterministic Freezing. In: Katoen, J.-P., König, B. (eds.) CONCUR 2011. LNCS, vol. 6901, pp. 60–75. Springer, Heidelberg (2011)
18. Pandya, P.K., Shah, S.: The Unary Fragments of Metric Interval Temporal Logic: Bounded versus Lower Bound Constraints. In: Chakraborty, S., Mukund, M. (eds.) ATVA 2012. LNCS, vol. 7561, pp. 77–91. Springer, Heidelberg (2012)
19. Prabhakar, P., D'Souza, D.: On the Expressiveness of MTL with Past Operators. In: Asarin, E., Bouyer, P. (eds.) FORMATS 2006. LNCS, vol. 4202, pp. 322–336. Springer, Heidelberg (2006)
20. Chambart, P., Schnobelen, P.: The ordinal recursive complexity of lossy channel systems. In: Proceedings of LICS, pp. 205–216 (2008)
21. Schnobelen, P.: Verifying lossy channel systems has nonprimitive recursive complexity. Info. Proc. Lett. 83(5), 251–261 (2002)
22. Straubing, H.: Finite Automata, Formal Logic and Circuit Complexity. Birkhauser, Boston (1994)

A Behavioral Congruence for Concurrent Constraint Programming with Nondeterministic Choice*

Luis F. Pino[1], Filippo Bonchi[2], and Frank D. Valencia[1]

[1] Comète, LIX, Laboratoire de l'École Polytechnique associé à l'INRIA
[2] CNRS - Laboratoire de l'Informatique du Parallélisme, ENS Lyon

Abstract. Concurrent constraint programming (ccp) is a well-established model of concurrency for reasoning about systems of multiple agents that interact with each other by posting and querying partial information on a shared space. (Weak) bisimilarity is one of the most representative notions of behavioral equivalence for models of concurrency. A notion of weak bisimilarity, called weak saturated bisimilarity ($\dot{\approx}_{sb}$), was recently proposed for ccp. This equivalence improves on previous bisimilarity notions for ccp that were too discriminating and it is a congruence for the choice-free fragment of ccp. In this paper, however, we show that $\dot{\approx}_{sb}$ is not a congruence for ccp with nondeterministic choice. We then introduce a new notion of bisimilarity, called weak full bisimilarity ($\approx_f$), and show that it is a congruence for the full language of ccp. We also show the adequacy of $\approx_f$ by establishing that it coincides with the congruence induced by closing $\dot{\approx}_{sb}$ under all contexts. The advantage of the new definition is that, unlike the congruence induced by $\dot{\approx}_{sb}$, it does not require quantifying over infinitely many contexts.

1 Introduction

The Context. Concurrency theory studies the description and the analysis of systems made of interacting *processes*. Processes are typically viewed as infinite objects, in the sense that they can produce arbitrary and possibly endless interactions with their environment. *Process calculi* treat these processes much like the λ-calculus treats computable functions. They provide a formal language in which processes are represented by terms, and a set of rewriting rules to represent process evolution (or transitions). For example, the term $P \parallel Q$ represents the process that results from the parallel composition of the processes P and Q. A (labeled) transition $P \xrightarrow{\alpha} P'$ represents the evolution of P into P' given an interaction α with the environment.

Concurrent Constraint Programming (ccp) [25,26] is a well-established formalism that combines the traditional algebraic and operational view of process calculi with a declarative one based upon first-order logic. Ccp processes can then be seen as computing agents as well as first-order logic formulae. In ccp, processes interact asynchronously by *posting* (or *telling*) and querying (or *asking*) information, traditionally referred to as *constraints*, in a shared-medium referred to as *the store*. Furthermore,

* This work has been partially supported by the project ANR 12IS02001 PACE, ANR-09-BLAN-0169-01 PANDA, and by the French Defence procurement agency (DGA) with a PhD grant.

G. Ciobanu and D. Méry (Eds.): ICTAC 2014, LNCS 8687, pp. 351–368, 2014.

ccp is parametric in a *constraint system* indicating interdependencies (entailment) between constraints and providing for the specification of data types and other rich structures. The above features have recently attracted a renewed attention as witnessed by the works [21,9,5,4] on calculi exhibiting data-types, logic assertions as well as tell and ask operations. More recently in [14] the authors proposed the post and ask interaction model of ccp as an abstraction of *social networks*.

In any computational model of processes, a central notion is that of *behavioral equivalences* [11]. These equivalences determine what processes are deemed indistinguishable and they are expected to be *congruences*. The congruence issue is of great importance for algebraic and *compositional* reasoning: If two processes are equivalent, one should be able to replace one with the other in any context and preserve the equivalence (see e.g, [13]). For example, if $\bowtie$ is a behavioral congruence, then $P \bowtie Q$ should imply $P \parallel R \bowtie Q \parallel R$.

Reasoning on processes and their equalities therefore means dealing with, and comparing, infinite structures. For this, a widely used mathematical tool is *coinduction* (see e.g. [1]). Coinduction is the dual of induction; while induction is a pervasive tool to reason about finite and stratified structures, coinduction offers similar strengths on structures that are circular or infinite. The most widely applied coinductive concept is *bisimulation*: *bisimilarity* is used to study behavioral equivalences, and the bisimulation proof method is used to prove such equivalences. In fact, most process calculi are equipped with a notion of bisimilarity.

The Problem. There have been few attempts to define notions of bisimilarity equivalence for ccp processes. These equivalences are, however, not completely satisfactory: As shown in [2], the one in [25] is too fine grained; i.e. it may tell apart processes whose logic interpretation is identical. The one in [16] quantifies over all possible inputs from the environment, and hence it is not clear whether it can lead to a feasible proof technique. The notion introduced in [2], called (weak) saturated barbed bisimilarity ($\dot{\approx}_{sb}$), solves the above-mentioned issues and it is a congruence for ccp without nondeterministic choice. Unfortunately, as we will show in this paper, it is not a congruence for the full language of ccp. In particular, in ccp with nondeterministic choice, $P \dot{\approx}_{sb} Q$ does not imply $P \parallel R \dot{\approx}_{sb} Q \parallel R$.

The goal of this paper is therefore to provide ccp with an adequate behavioral congruence based on the bisimulation proof method.

Our Approach. We build on a result of [2] showing that $\dot{\approx}_{sb}$ can be characterized by a novel bisimulation game (called, for simplicity, weak bisimulation) which relies at the same time on both *barbs* and *labeled transitions*. Barbs are basically predicates on the states, processes or configuration stating the observation we can make of them. This is rather peculiar with respect to the existing notions of bisimulations introduced for other process calculi where one usually exploits labeled transitions to avoid thinking about barbs and contexts. Indeed, labeled transitions usually capture barbs, in the sense that a state exposes a certain barb if and only if it performs a transition with a certain label. This is not the case of ccp, where barbs are observations on the store, while labeled transitions are determined by the processes. A more abstract understanding of this peculiarity of ccp can be given within the framework of [7] which is an extension of [15] featuring barbs and weak semantics.

As it is customary for weak barbed equivalences, in our weak bisimulation game whenever a player exposes a barb $\downarrow_e$, the opponent should expose the weak barb $\Downarrow_e$, i.e. it should be able to reach a state satisfying $\downarrow_e$, but then the game restarts from the original state ignoring the arriving state. One of our contributions is to show that for ccp the arriving state cannot be ignored.

Our Contributions. In this work, we prove that $\dot{\approx}_{sb}$ is a congruence for ccp without nondeterministic choice but not for the full language of ccp. We then propose a new notion of bisimilarity, called *(weak) full bisimilarity* ($\approx_f$). We show that $\approx_f$ is a congruence for the full language of ccp. We also show the adequacy of the new notion by establishing that it is the largest congruence included in $\dot{\approx}_{sb}$. In other words $\approx_f$ coincides with the congruence induced by closing $\dot{\approx}_{sb}$ under all contexts. Beyond being a congruence, the advantage of $\approx_f$ is that it does not require quantifying over infinitely many contexts. This is also important as it may simplify decision procedures for the equivalence. To the best of our knowledge, this is the first behavioral equivalence, which does not appeal to quantification over arbitrary process contexts in its definition, that is a congruence for ccp with nondeterministic choice.

A technical report with detailed proofs of this paper can be found in [22].

Structure of the paper. The paper is organized as follows: In Section 2 we recall the ccp formalism. In Section 3 we introduce the standard notion of observational equivalence ($\sim_o$) for ccp (from [26]), we then show its relation with the weak saturated barbed bisimilarity ($\dot{\approx}_{sb}$) (from [2]) for ccp with nondeterministic choice. We also prove that $\dot{\approx}_{sb}$ is not a congruence for the full ccp. In Section 4 we introduce our new notion $\approx_f$, and we prove that (i) $\approx_f$ coincides with $\dot{\approx}_{sb}$ in the choice-free fragment of ccp; (ii) $\approx_f$ is a congruence for ccp with summation; and (iii) $\approx_f$ coincides with the equivalence obtained after closing $\dot{\approx}_{sb}$ under any context. In Section 5 we present our conclusions and future work.

2 Background

We begin this section by recalling the notion of constraint system. We then present the concurrent constraint programming (ccp) formalism.

2.1 Constraint Systems

The ccp model is parametric in a *constraint system (cs)* specifying the structure and interdependencies of the information that processes can ask or and add to a *central shared store*. This information is represented as assertions traditionally called *constraints*.

Following [10,16] we regard a cs as a complete algebraic lattice in which the ordering $\sqsubseteq$ is the reverse of an entailment relation: $c \sqsubseteq d$ means d *entails* c, i.e., d contains "more information" than c. The top element *false* represents inconsistency, the bottom element *true* is the empty constraint, and the *least upper bound* $\sqcup$ is the join of information.

Definition 1 (Constraint Systems). *A constraint system (cs)* **C** *is a complete algebraic lattice* $(Con, Con_0, \sqsubseteq, \sqcup, true, false)$ *where* Con*, the set of constraints, is a partially ordered set w.r.t.* $\sqsubseteq$*,* Con_0 *is the subset of* compact *elements of* Con*,* $\sqcup$ *is the lub operation defined on all subsets, and* $true, false$ *are the least and greatest elements of* Con*, respectively.*

Recall that **C** is a *complete lattice* if every subset of Con has a least upper bound in Con. An element $c \in Con$ is *compact* if for any directed subset D of Con, $c \sqsubseteq \bigsqcup D$ implies $c \sqsubseteq d$ for some $d \in D$. **C** is *algebraic* if each element $c \in Con$ is the least upper bound of the compact elements below c.

In order to model *hiding* of local variables and *parameter passing*, in [25,26] the notion of constraint system is enriched with *cylindrification operators* and *diagonal elements*, concepts borrowed from the theory of cylindric algebras [20].

Let us consider a (denumerable) set of variables Var with typical elements $x, y, z, \ldots$ and let us define $\exists_{Var}$ as the family of operators $\exists_{Var} = \{\exists_x \mid x \in Var\}$ (*cylindric operators*) and D_{Var} as the set $D_{Var} = \{d_{xy} \mid x, y \in Var\}$ (*diagonal elements*).

A *cylindric constraint system* over a set of variables Var is a constraint system whose underlying support set $Con \supseteq D_{Var}$ is closed under the cylindric operators $\exists_{Var}$ and quotiented by Axioms C1 – C4, and whose ordering $\sqsubseteq$ satisfies Axioms C5 – C7:

C1. $\exists_x \exists_y c = \exists_y \exists_x c$	C2. $d_{xx} = true$
C3. if $z \neq x, y$ then $d_{xy} = \exists_z (d_{xz} \sqcup d_{zy})$	C4. $\exists_x (c \sqcup \exists_x d) = \exists_x c \sqcup \exists_x d$
C5. $\exists_x c \sqsubseteq c$	C6. *if* $c \sqsubseteq d$ *then* $\exists_x c \sqsubseteq \exists_x d$
C7. if $x \neq y$ then $c \sqsubseteq d_{xy} \sqcup \exists_x (c \sqcup d_{xy})$	

where c and d indicate compact constraints, and $\exists_x c \sqcup d$ stands for $(\exists_x c) \sqcup d$. For our purposes, it is enough to think the operator $\exists_x$ as *existential quantifier* and the constraint d_{xy} as the equality $x = y$.

Cylindrification and diagonal elements allow us to model the variable renaming of a formula ϕ; in fact, by the aforementioned axioms, we have that the formula $\exists_x (d_{xy} \sqcup \phi)$ can be depicted as the formula $\phi[y/x]$, i.e., the formula obtained from ϕ by replacing all free occurrences of x by y.

We assume notions of *free variable* and of *substitution* that satisfy the following conditions, where $c[y/x]$ is the constraint obtained by substituting x by y in c and $fv(c)$ is the set of free variables of c: (1) if $y \notin fv(c)$ then $(c[y/x])[x/y] = c$; (2) $(c \sqcup d)[y/x] = c[y/x] \sqcup d[y/x]$; (3) $x \notin fv(c[y/x])$; (4) $fv(c \sqcup d) = fv(c) \cup fv(d)$.

We now illustrate a constraint system for linear-order arithmetic.

Example 1 (A Constraint System of Linear Order Arithmetic). Consider the following syntax:

$$\phi, \psi \ldots := t = t' \mid t > t' \mid \phi \vee \psi \mid \neg \phi$$

where the terms t, t' can be elements of a set of variables Var, or constant symbols $0, 1, \ldots$. Assume an underlying first-order structure of linear-order arithmetic with the obvious interpretation in the natural numbers ω of $=, >$ and the constant symbols.

A variable assignment is a function $\mu : Var \longrightarrow \omega$. We use $\mathcal{A}$ to denote the set of all assignments; $\mathcal{P}(X)$ to denote the powerset of a set X, $\emptyset$ the empty set and $\cap$ the intersection of sets. We use $\mathcal{M}(\phi)$ to denote the set of all assignments that *satisfy* the formula ϕ, where the definition of *satisfaction* is as expected.

We can now introduce a *constraint system* as follows: the set of constraints is $\mathcal{P}(\mathcal{A})$, and define $c \sqsubseteq d$ iff $c \supseteq d$. The constraint *false* is $\emptyset$, while *true* is $\mathcal{A}$. Given two constraints c and d, $c \sqcup d$ is the intersection $c \cap d$. By abusing the notation, we will often use a formula ϕ to denote the corresponding constraint, i.e., the set of all assignments satisfying ϕ. E.g. we use $x > 1 \sqsubseteq x > 5$ to mean $\mathcal{M}(x > 1) \sqsubseteq \mathcal{M}(x > 5)$. For this constraint system one can show that e is a compact constraint (i.e., e is in Con_0) iff e is a co-finite set in $\mathcal{A}$ (i.e., iff the complement of e in $\mathcal{A}$ is a finite set). For example, $x > 10 \wedge y > 42$ is a compact constraint for $Var = \{x, y\}$.

From this structure, let us now define the *cylindric constraint system* $\mathcal{S}$ as follows. We say that an assignment μ' is *an x-variant* of μ if $\forall y \neq x$, $\mu(y) = \mu'(y)$. Given $x \in Var$ and $c \in \mathcal{P}(\mathcal{A})$, the constraint $\exists_x c$ is the set of assignments μ such that exists $\mu' \in c$ that is an x-variant of μ. The diagonal element d_{xy} is $x = y$. □

Assumption 1. *We shall assume that the constraint system is well-founded and, for practical reasons, that its ordering $\sqsubseteq$ is decidable. Well-foundedness is needed for technical reasons in the definition of the labeled transition semantics in Section 3.2.*

2.2 Syntax of CCP

Let $\mathbf{C} = (Con, Con_0, \sqsubseteq, \sqcup, true, false)$ be a constraint system. The ccp processes are given by the following syntax:

$$P, Q, \ldots ::= \mathbf{tell}(c) \mid \sum_{i \in I} \mathbf{ask}\ (c_i) \rightarrow P_i \mid P \parallel Q \mid \exists_x P \mid p(z)$$

where I is a finite set of indexes and $c, c_i \in Con_0$. We use *Proc* to denote the set of all processes.

Finite processes. Intuitively, the tell process $\mathbf{tell}(c)$ adds c to the global store. The addition is performed regardless the generation of inconsistent information. The process $P \parallel Q$ stands for the *parallel execution* of P and Q.

The guarded-choice $\sum_{i \in I} \mathbf{ask}\ (c_i) \rightarrow P_i$ where I is a finite set of indexes, represents a process that can *nondeterministically* choose one of the P_j (with $j \in I$) whose corresponding guard constraint c_j is entailed by the store. The chosen alternative, if any, precludes the others. We shall often write $\mathbf{ask}\ (c_{i_1}) \rightarrow P_{i_1} + \ldots + \mathbf{ask}\ (c_{i_n}) \rightarrow P_{i_n}$ if $I = \{i_1, \ldots, i_n\}$. If no ambiguity arises, we shall omit the "$\mathbf{ask}(c) \rightarrow$" when $c = true$. The *blind-choice* process $\sum_{i \in I} \mathbf{ask}\ (true) \rightarrow P_i$, for example, can be written $\sum_{i \in I} P_i$. We shall omit the "$\sum_{i \in I}$" when I is a singleton. We use **stop** as an abbreviation of the empty summation $\sum_{i \in \emptyset} P_i$.

$\exists_x$ is a *hiding operator*, namely it indicates that in $\exists_x P$ the variable x is *local* to P. The occurrences of x in $\exists_x P$ are said to be bound. The bound variables of P, $bv(P)$, are those with a bound occurrence in P, and its free variables, $fv(P)$, are those with an unbound occurrence[1].

[1] Notice that we also defined $fv(.)$ on constraints in the previous section.

Infinite processes. To specify infinite behavior, ccp provides parametric process definitions. A process $p(\mathbf{z})$ is said to be a *procedure call* with identifier p and actual parameters $\mathbf{z}$. We presuppose that for each procedure call $p(z_1 \ldots z_m)$ there exists a unique *procedure definition* possibly *recursive*, of the form $p(x_1 \ldots x_m) \stackrel{\text{def}}{=} P$ where $fv(P) \subseteq \{x_1, \ldots, x_m\}$. Furthermore we require recursion to be *guarded*: I.e., each procedure call within P must occur within an ask process. The behavior of $p(z_1 \ldots z_m)$ is that of $P[z_1 \ldots z_m / x_1 \ldots x_m]$, i.e., P with each x_i replaced with z_i (applying α-conversion to avoid clashes). We shall use $\mathcal{D}$ to denote the set of all process definitions.

Remark 1 (Choice-free fragment of ccp). Henceforth, we use ccp\+ to refer to the fragment of ccp without nondeterministic choice. More precisely ccp\+ processes are those in which every occurrence of $\sum_{i \in I} \mathbf{ask}\ (c_i) \rightarrow P_i$ has its index set I of cardinality 0 or 1.

2.3 Reduction Semantics

A configuration is a pair $\langle P, d \rangle$ representing a *state* of a system; d is a constraint representing the global store, and P is a process, i.e., a term of the syntax given above. We use *Conf* with typical elements $\gamma, \gamma', \ldots$ to denote the set of all configurations. We will use $Conf_{\,\mathtt{ccp}\backslash+}$ for the configurations whose processes are in the ccp\+ fragment.

The operational semantics of ccp is given by an *unlabeled* transition relation between configurations: a transition $\gamma \longrightarrow \gamma'$ intuitively means that the configuration γ can reduce to γ'. We call these kind of unlabeled transitions *reductions* and we use $\longrightarrow^*$ to denote the reflexive and transitive closure of $\longrightarrow$.

Formally, the reduction semantics of ccp is given by the relation $\longrightarrow$ defined in Table 1. Rules R1 and R2 are easily seen to realize the intuitions described in Section 2.2. Rule R3 states that $\sum_{i \in I} \mathbf{ask}\ (c_i) \rightarrow P_i$ can evolve to P_j whenever the global store d entails c_j and $j \in I$.

Rule R4 is somewhat more involved, first we extend the syntax by introducing a process $\exists_x^e P$ representing the evolution of a process of the form $\exists_x P$, where e is the local information (*local store*) produced during this evolution. The process $\exists_x P$ can be seen as a particular case of $\exists_x^e P$: it represents the situation in which the local store is empty. Namely, $\exists_x P = \exists_x^{true} P$.

Intuitively, $\exists_x^e P$ behaves like P, except that the variable x possibly present in P must be considered local, and that the information present in e has to be taken into account. It is convenient to distinguish between the *external* and the *internal* points of view. From the internal point of view, the variable x, possibly occurring in the global store d, is hidden. This corresponds to the usual scoping rules: the x in d is *global*, hence "covered" by the local x. Therefore, P has no access to the information on x in d, and this is achieved by filtering d with $\exists_x$. Furthermore, P can use the information (which may also concern the local x) that has been produced locally and accumulated in e. In conclusion, if the visible store at the external level is d, then the store that is visible internally by P is $e \sqcup \exists_x d$. Now, if P is able to make a step, thus reducing to P' and transforming the local store into e', what we see from the external point of view is

Table 1. Reduction semantics for ccp (symmetric rule for R2 is omitted). $\mathcal{D}$ is the set of process definitions.

$$\text{R1}\ \langle \mathbf{tell}(c), d\rangle \longrightarrow \langle \mathbf{stop}, d \sqcup c\rangle \qquad \text{R2}\ \frac{\langle P, d\rangle \longrightarrow \langle P', d'\rangle}{\langle P \parallel Q, d\rangle \longrightarrow \langle P' \parallel Q, d'\rangle}$$

$$\text{R3}\ \frac{j \in I \text{ and } c_j \sqsubseteq d}{\langle \sum_{i\in I} \mathbf{ask}\ (c_i) \rightarrow P_i, d\rangle \longrightarrow \langle P_j, d\rangle} \qquad \text{R4}\ \frac{\langle P, e \sqcup \exists_x d\rangle \longrightarrow \langle P', e' \sqcup \exists_x d\rangle}{\langle \exists_x^e P, d\rangle \longrightarrow \langle \exists_x^{e'} P', d \sqcup \exists_x e'\rangle}$$

$$\text{R5}\ \frac{\langle P[\boldsymbol{z}/\boldsymbol{x}], d\rangle \longrightarrow \gamma'}{\langle p(\boldsymbol{z}), d\rangle \longrightarrow \gamma'} \text{ where } p(\boldsymbol{x}) \stackrel{\text{def}}{=} P \text{ is a process definition in } \mathcal{D}$$

that the process is transformed into $\exists_x^{e'} P'$, and that the information $\exists_x e$ present in the global store is transformed into $\exists_x e'$.[2]

2.4 Barbed Semantics

In [2], the authors introduced a *barbed semantics* for ccp. Barbed equivalences have been introduced in [19] for CCS, and have become a classical way to define the semantics of formalisms equipped with unlabeled reduction semantics. Intuitively, *barbs* are basic observations (predicates) on the states of a system. In the case of ccp, barbs are taken from the underlying set Con_0 of the constraint system.

Definition 2 (Barbs). *A configuration $\gamma = \langle P, d\rangle$ is said to satisfy the barb c, written $\gamma \downarrow_c$, iff $c \in Con_0$ and $c \sqsubseteq d$. Similarly, γ satisfies a weak barb c, written $\gamma \Downarrow_c$, iff there exist γ' s.t. $\gamma \longrightarrow^* \gamma' \downarrow_c$.*

Example 2. Consider the constraint system from Example 1 and let $Vars = \{x\}$. Let $\gamma = \langle \mathbf{ask}\ (x > 10) \rightarrow \mathbf{tell}(x > 42), x > 10\rangle$. We have $\gamma \downarrow_{x>5}$ since $(x > 5) \sqsubseteq (x > 10)$ and $\gamma \Downarrow_{x>42}$ since $\gamma \longrightarrow \langle \mathbf{tell}(x > 42), x > 10\rangle \longrightarrow \langle \mathbf{stop}, (x > 42)\rangle \downarrow_{x>42}$. □

In this context, the equivalence proposed is the *saturated bisimilarity* [8,6]. Intuitively, in order for two states to be saturated bisimilar, then (i) they should expose the same barbs, (ii) whenever one of them moves then the other should reply and arrive at an equivalent state (i.e. follow the bisimulation game), (iii) they should be equivalent under all the possible contexts of the language.

Using this idea, in [2], the authors propose a saturated bisimilarity for ccp where condition (iii) requires the bisimulations to be *upward closed* instead of closing under any process context. A *process context* C is a term with a single hole • such that if we replace • with a process P, we obtain a process term $C[P]$. For example, for the parallel context $C = \bullet \parallel R$ we obtain $C[P] = P \parallel R$.

[2] For more details about the operational semantics we refer the reader to [2].

Definition 3 (Saturated Barbed Bisimilarity). *A saturated barbed bisimulation is a symmetric relation $\mathcal{R}$ on configurations s.t. whenever $(\gamma_1, \gamma_2) \in \mathcal{R}$ with $\gamma_1 = \langle P, c\rangle$ and $\gamma_2 = \langle Q, d\rangle$ implies that:*

(i) if $\gamma_1 \downarrow_e$ then $\gamma_2 \downarrow_e$,
(ii) if $\gamma_1 \longrightarrow \gamma_1'$ then there exists γ_2' s.t. $\gamma_2 \longrightarrow \gamma_2'$ and $(\gamma_1', \gamma_2') \in \mathcal{R}$,
(iii) for every $a \in Con_0$, $(\langle P, c \sqcup a\rangle, \langle Q, d \sqcup a\rangle) \in \mathcal{R}$.

We say that γ_1 and γ_2 are saturated barbed bisimilar ($\gamma_1 \dot{\sim}_{sb} \gamma_2$) if there is a saturated barbed bisimulation $\mathcal{R}$ s.t. $(\gamma_1, \gamma_2) \in \mathcal{R}$. We write $P \dot{\sim}_{sb} Q$ iff $\langle P, true\rangle \dot{\sim}_{sb} \langle Q, true\rangle$.

We shall prove that the closure condition (iii) is enough to make $\dot{\approx}_{sb}$ a congruence in ccp\+. This means that $P \dot{\approx}_{sb} Q$ implies $C[P] \dot{\approx}_{sb} C[Q]$ for every process context. However, this is not the case for ccp with nondeterministic choice as we shall demonstrate later on.

Weak saturated barbed bisimilarity ($\dot{\approx}_{sb}$) is obtained from Definition 3 by replacing the strong barbs in condition *(i)* for its weak version ($\Downarrow$) and the transitions in condition *(ii)* for the reflexive and transitive closure of the transition relation ($\longrightarrow^*$).

Definition 4 (Weak Saturated Barbed Bisimilarity). *A weak saturated barbed bisimulation is a symmetric relation $\mathcal{R}$ on configurations s.t. whenever $(\gamma_1, \gamma_2) \in \mathcal{R}$ with $\gamma_1 = \langle P, c\rangle$ and $\gamma_2 = \langle Q, d\rangle$ implies that:*

(i) if $\gamma_1 \Downarrow_e$ then $\gamma_2 \Downarrow_e$,
(ii) if $\gamma_1 \longrightarrow^* \gamma_1'$ then there exists γ_2' s.t. $\gamma_2 \longrightarrow^* \gamma_2'$ and $(\gamma_1', \gamma_2') \in \mathcal{R}$,
(iii) for every $a \in Con_0$, $(\langle P, c \sqcup a\rangle, \langle Q, d \sqcup a\rangle) \in \mathcal{R}$.

We say that γ_1 and γ_2 are weak saturated barbed bisimilar ($\gamma_1 \dot{\approx}_{sb} \gamma_2$) if there exists a weak saturated barbed bisimulation $\mathcal{R}$ s.t. $(\gamma_1, \gamma_2) \in \mathcal{R}$. We shall write $P \dot{\approx}_{sb} Q$ iff $\langle P, true\rangle \dot{\approx}_{sb} \langle Q, true\rangle$.

We now illustrate $\dot{\sim}_{sb}$ and $\dot{\approx}_{sb}$ with the following two examples.

Example 3. Consider the constraint system from Example 1 and let $Vars = \{x\}$. Take $P = \textbf{ask}\ (x > 5) \rightarrow \textbf{stop}$ and $Q = \textbf{ask}\ (x > 7) \rightarrow \textbf{stop}$. One can check that $P \dot{\not\sim}_{sb} Q$ since $\langle P, x > 5\rangle \longrightarrow$, while $\langle Q, x > 5\rangle \not\longrightarrow$. Then consider $\langle P{+}Q, true\rangle$ and observe that $\langle P + Q, true\rangle \dot{\sim}_{sb} \langle P, true\rangle$. Indeed, for all constraints e, s.t. $x > 5 \sqsubseteq e$, both the configurations evolve into $\langle \textbf{stop}, e\rangle$, while for all e s.t. $x > 5 \not\sqsubseteq e$, both configurations cannot proceed. Since $x > 5 \sqsubseteq x > 7$, the behavior of Q is somehow absorbed by the behavior of P. □

Example 4. Take P and Q as in Example 3. One can check that $P \dot{\approx}_{sb} Q$. First notice that $\langle P, true\rangle \not\longrightarrow$ and also $\langle Q, true\rangle \not\longrightarrow$. Now note that for all e it is the case that both configurations evolve to a γ where $\gamma \Downarrow_e$. Intuitively, none of the configurations adds information to the store and, since $\dot{\approx}_{sb}$ does not care about the silent transitions, then P and Q should be weakly bisimilar. □

Finally, notice that in ccp\+ configurations are *confluent* in the following sense.

Proposition 1 (Confluence [26]). *Let $\gamma \in Conf_{\text{ccp}\backslash+}$. If $\gamma \longrightarrow^* \gamma_1$ and $\gamma \longrightarrow^* \gamma_2$ then there exists γ' such that $\gamma_1 \longrightarrow^* \gamma'$ and $\gamma_2 \longrightarrow^* \gamma'$.*

The proposition above will be a cornerstone for the results we shall obtain in ccp\+.

3 Congruence Issues

A typical question in the realm of process calculi, and concurrency in general, is whether a given process equivalence is a *congruence*. In other words, whether the fact that P and Q are equivalent implies that they are still equivalent in any context. More precisely, a given equivalence $\bowtie$ is said to be a congruence if $P \bowtie Q$ implies $C[P] \bowtie C[Q]$ for every process context C[3]. The congruence issue is fundamental for algebraic as well as practical reasons; one may not be content with having $P \bowtie Q$ equivalent but $R \parallel P \not\bowtie R \parallel Q$. Nevertheless, some of the representative equivalences in concurrency are not congruences. For example, in CCS [17], trace equivalence and strong bisimilarity are congruences but weak bisimilarity is not because it is not preserved by summation contexts. So given a notion of equivalence one may wonder in what contexts the equivalence is preserved. For instance, the problem with weak bisimilarity can be avoided by using guarded-summation (see [18]).

We shall see that $\dot{\approx}_{sb}$ is a congruence for ccp\+. However, this is not the case in the presence of nondeterministic choice. Moreover, unlike CCS, the problem arises even in the presence of guarded summation/choice. In fact, our counterexample reveals that the problem is intrinsic to ccp.

3.1 Observational Equivalence

In this section we shall introduce the standard notion of observational equivalence ($\sim_o$) [26] for ccp as well as its relation with $\dot{\approx}_{sb}$.

The notion of *fairness* is central to the definition of observational equivalence for ccp. We introduce this notion following [12]. Any derivation of a transition involves an application of R1 or R3. We say that P is *active* in a transition $t = \gamma \longrightarrow \gamma'$ if there exists a derivation of t where rule R1 or R3 is used to produce a transition of the form $\langle P, d\rangle \longrightarrow \gamma''$. Moreover, we say that P is *enabled* in γ if there exists γ' such that P is active in $\gamma \longrightarrow \gamma'$. A computation $\gamma_0 \longrightarrow \gamma_1 \longrightarrow \gamma_2 \longrightarrow \ldots$ is said to be *fair* if for each process enabled in some γ_i there exists $j \geq i$ such that the process is active in $\gamma_j \longrightarrow \gamma_{j+1}$.

Note that a finite fair computation is guaranteed to be *maximal*, namely no outgoing transitions are possible from its last configuration.

The standard notion of observables for ccp are the *results* computed by a process for a given initial store. The result of a computation is defined as the least upper bound of all the stores occurring in the computation, which, due to the monotonic properties of ccp, form an increasing chain. More formally:

Definition 5 (Result). *Given a finite or infinite computation ξ of the form:*

$$\xi = \langle Q_0, d_0\rangle \longrightarrow \langle Q_1, d_1\rangle \longrightarrow \langle Q_2, d_2\rangle \longrightarrow \ldots$$

The result of ξ, denoted by Result(ξ), *is the constraint $\bigsqcup_i d_i$.*

[3] Recall that the expression $C[P]$ denotes the process that results from replacing in C, the hole $\bullet$ with P. For example $C = R \parallel \bullet$ then $C[P] = R \parallel P$.

Note that for a finite computation the result coincides with the store of the last configuration. Now since ccp\+ is confluent (Proposition 1), the following theorem from [26] states that all the fair computations of a configuration have the same result.

Proposition 2 ([26]). *Let γ be ccp\+ configuration and let ξ_1 and ξ_2 be two computations of γ. If ξ_1 and ξ_2 are fair, then* $\mathsf{Result}(\xi_1) = \mathsf{Result}(\xi_2)$.

Before introducing the notion of observational equivalence we need some notation. Below we define the set of possible computations of a given configuration.

Definition 6 (Set of Computations). *The set of computations starting from γ, denoted* $\mathsf{Comp}(\gamma)$*, is defined as:*

$$\mathsf{Comp}(\gamma) = \{\xi \mid \xi = \gamma \longrightarrow \gamma' \longrightarrow \gamma'' \longrightarrow \ldots\}$$

Now we introduce the notion of observables. Intuitively, the set of observables of γ is the set of results of the fair computations starting from γ.

Definition 7 (Observables). *Let $\mathcal{O} : Proc \to Con_0 \to 2^{Con}$ be given by:*

$$\mathcal{O}(P)(d) = \{e \mid \xi \in \mathsf{Comp}(\langle P, d\rangle), \xi \textit{ is fair and } \mathsf{Result}(\xi) = e\}.$$

Using these elements we define the notion of observational equivalence. Two configurations are deemed equivalent if they have the same set observables for any given store.

Definition 8 (Observational Equivalence). *We say that P and Q are observational equivalent, written $P \sim_o Q$, iff $\mathcal{O}(P) = \mathcal{O}(Q)$.*

Notice that in the case of ccp\+, as defined in [26], the set of observables is a singleton because of Proposition 2.

Remark 2. Let $\langle P, d\rangle \in Conf_{\mathsf{ccp}\backslash +}$. Note that $\mathcal{O} : Proc \to Con_0 \to Con$ because of Proposition 2 and it is defined as $\mathcal{O}(P)(d) = \mathsf{Result}(\xi)$ where ξ is any fair computation of $\langle P, d\rangle$.

In [2] it was shown that, in ccp\+, weak saturated barbed bisimilarity and observation equivalence coincide. Recall that $P \dot{\approx}_{sb} Q$ means $\langle P, true\rangle \dot{\approx}_{sb} \langle Q, true\rangle$.

Proposition 3 ([2]). *Let P and Q be ccp\+ processes. Then $P \sim_o Q$ iff $P \dot{\approx}_{sb} Q$.*

Nevertheless, the above theorem does not hold for ccpwith nondeterministic choice. We can show this by using a counter-example reminiscent from the standard one for CCS. Let $P = (\mathbf{ask}\ (b) \ \to\ \mathbf{tell}(c)) + (\mathbf{ask}\ (b) \ \to\ \mathbf{tell}(d))$ and $Q = \mathbf{ask}\ (b) \ \to ((\mathbf{ask}\ (true) \ \to\ \mathbf{tell}(c)) + (\mathbf{ask}\ (true) \ \to\ \mathbf{tell}(d)))$. One can verify that $P \sim_o Q$ but $P \not\dot{\approx}_{sb} Q$. However, the $(\Leftarrow)$ direction of the theorem does hold as we show next.

Theorem 1. *If $P \dot{\approx}_{sb} Q$ then $P \sim_o Q$.*

Table 2. Labeled semantics for ccp (symmetric rule for LR2 is omitted)

$$\text{LR1}\ \langle \mathbf{tell}(c), d\rangle \xrightarrow{true} \langle \mathbf{stop}, d \sqcup c\rangle \qquad \text{LR2}\ \frac{\langle P, d\rangle \xrightarrow{\alpha} \langle P', d'\rangle}{\langle P \parallel Q, d\rangle \xrightarrow{\alpha} \langle P' \parallel Q, d'\rangle}$$

$$\text{LR3}\ \frac{j \in I \text{ and } \alpha \in \min\{a \in Con_0 \mid c_j \sqsubseteq d \sqcup a\}}{\langle \sum_{i\in I} \mathbf{ask}\ (c_i) \rightarrow P_i, d\rangle \xrightarrow{\alpha} \langle P_j, d \sqcup \alpha\rangle}$$

$$\text{LR4}\ \frac{\langle P[z/x], e[z/x] \sqcup d\rangle \xrightarrow{\alpha} \langle P', e' \sqcup d \sqcup \alpha\rangle}{\langle \exists_x^e P, d\rangle \xrightarrow{\alpha} \langle \exists_x^{e'[x/z]} P'[x/z], \exists_x(e'[x/z]) \sqcup d \sqcup \alpha\rangle}$$

$$\text{with } x \notin fv(e'), z \notin fv(P) \cup fv(e \sqcup d \sqcup \alpha)$$

$$\text{LR5}\ \frac{\langle P[z/x], d\rangle \xrightarrow{\alpha} \gamma'}{\langle p(z), d\rangle \xrightarrow{\alpha} \gamma'} \text{ where } p(x) \stackrel{\text{def}}{=} P \text{ is a process definition in } \mathcal{D}$$

3.2 Congruence

We begin this section by showing that weak bisimilarity is a congruence in a restricted sense: It is preserved by all the contexts from the choice-free fragment. For this purpose it is convenient to recall the labeled semantics of ccp as well as the (labeled) weak bisimilarity introduced in [2].

Labeled Semantics. In a labeled transition of the form

$$\langle P, d\rangle \xrightarrow{\alpha} \langle P', d'\rangle$$

the label $\alpha \in Con_0$ represents a *minimal* information (from the environment) that needs to be added to the store d to reduce from $\langle P, d\rangle$ to $\langle P', d'\rangle$, i.e., $\langle P, d \sqcup \alpha\rangle \longrightarrow \langle P', d'\rangle$. As a consequence, the transitions labeled with the constraint *true* are in one to one correspondence with the reductions defined in the previous section. For this reason, hereafter we will sometimes write $\longrightarrow$ to mean $\xrightarrow{true}$.

The LTS $(Conf, Con_0, \longrightarrow)$ is defined by the rules in Table 2. The rule LR3, for example, says that $\langle \sum_{i\in I} \mathbf{ask}\ (c_i) \rightarrow P_i, d\rangle$ can evolve to $\langle P_j, d \sqcup \alpha\rangle$ if $j \in I$ and the environment provides a minimal constraint α that added to the store d entails the guard c_j, i.e., $\alpha \in \min\{a \in Con_0 \mid c_j \sqsubseteq d \sqcup a\}$. Notice that Assumption 1 guarantees the existence of α. The rule LR4 follows the same approach as R4, however it uses variable substitution instead of hiding with the existential operator.[4] The other rules are easily seen to realize the intuition given in Section 2.2.

We can now introduce the notion of weak bisimilarity ($\dot{\approx}$) from [2]. In [2] it is shown that $\dot{\approx}$ coincides with $\approx_{sb}$ and, by exploiting the labeled semantics, avoids the upward closure from condition (iii) in $\approx_{sb}$.

[4] See [2] for a detailed explanation of the rule LR4.

Definition 9 (Weak Bisimilarity). *A weak bisimulation is a symmetric relation $\mathcal{R}$ on configurations such that whenever $(\gamma_1, \gamma_2) \in \mathcal{R}$ with $\gamma_1 = \langle P, c\rangle$ and $\gamma_2 = \langle Q, d\rangle$:*

(i) if $\gamma_1 \downarrow_e$ then $\gamma_2 \Downarrow_e$,
(ii) if $\gamma_1 \xrightarrow{\alpha} \gamma_1'$ then $\exists \gamma_2'$ s.t. $\langle Q, d \sqcup \alpha\rangle \longrightarrow^* \gamma_2'$ and $(\gamma_1', \gamma_2') \in \mathcal{R}$.

We say that γ_1 and γ_2 are weakly bisimilar, written $\gamma_1 \approx \gamma_2$, if there exists a weak bisimulation $\mathcal{R}$ such that $(\gamma_1, \gamma_2) \in \mathcal{R}$. We write $P \approx Q$ iff $\langle P, true\rangle \approx \langle Q, true\rangle$.

To illustrate this definition consider the following example.

Example 5. Let $\gamma_1 = \langle \textbf{tell}(true), true\rangle$ and $\gamma_2 = \langle \textbf{ask}\ (c) \rightarrow \textbf{tell}(d), true\rangle$. We can show that $\gamma_1 \dot{\approx} \gamma_2$ when $d \sqsubseteq c$. Intuitively, this corresponds to the fact that the implication $c \Rightarrow d$ is equivalent to $true$ when c already entails d. The LTSs of γ_1 and γ_2 are the following: $\gamma_1 \longrightarrow \langle \textbf{stop}, true\rangle$ and $\gamma_2 \xrightarrow{c} \langle \textbf{tell}(d), c\rangle \longrightarrow \langle \textbf{stop}, c\rangle$. It is now easy to see that the symmetric closure of the relation

$$\mathcal{R} = \{(\gamma_2, \gamma_1),\ (\gamma_2, \langle \textbf{stop}, true\rangle),\ (\langle \textbf{tell}(d), c\rangle, \langle \textbf{stop}, c\rangle), (\langle \textbf{stop}, c\rangle, \langle \textbf{stop}, c\rangle)\}$$

is a weak bisimulation as in Definition 9. □

The following result from [2] states that weak bisimilarity coincides with weak saturated barbed bisimilarity (Definition 4).

Proposition 4 ([2]). $\dot{\approx}_{sb} = \dot{\approx}$.

We can now prove that $\dot{\approx}_{sb}$ is a congruence in ccp\+.

Theorem 2. *Let P and Q be ccp\+ processes and assume that $P \dot{\approx}_{sb} Q$. Then for every process context $C[\bullet]$ in ccp\+ we have $C[P] \dot{\approx}_{sb} C[Q]$.*

Notice that this result implies that observational equivalence ($\sim_o$) is a congruence. Unfortunately the theorem above does not hold for ccp with nondeterministic choice, as shown next.

Theorem 3. *There exists P', Q, R in ccp s.t. (a) $P' \dot{\approx}_{sb} Q$ but (b) $P' \parallel R \not\dot{\approx}_{sb} Q \parallel R$.*

Proof. *To prove this claim we let* $P = (\textbf{ask}\ (true) \rightarrow \textbf{tell}(c)) + (\textbf{ask}\ (true) \rightarrow \textbf{tell}(d))$, $P' = P \parallel \textbf{tell}(e)$ *and* $Q = (\textbf{ask}\ (true) \rightarrow \textbf{tell}(c \sqcup e)) + (\textbf{ask}\ (true) \rightarrow \textbf{tell}(d \sqcup e))$ *with* $c \not\sqsubseteq d, c \not\sqsubseteq e, d \not\sqsubseteq c, d \not\sqsubseteq e, e \not\sqsubseteq c, e \not\sqsubseteq d$.

For (a) we can show that $\langle P', true\rangle \dot{\approx}_{sb} \langle P, e\rangle \dot{\approx}_{sb} \langle Q, true\rangle$. The first equation is trivial. For the second we define a relation on configurations $\mathcal{R}$. The set of pairs in $\mathcal{R}$ are those linked in Figure 1. It can easily be verified that (the symmetric closure of) $\mathcal{R}$ is a weak bisimulation (see Definition 9). The point (a) then follows from Proposition 4.

For proving the part (b) of the above claim, we let $R = (\textbf{ask}\ (e) \rightarrow \textbf{tell}(\alpha)) + (\textbf{ask}\ (e) \rightarrow \textbf{tell}(\beta))$. *We shall prove that no weak bisimulation can contain the pair $(\langle P \parallel R, e\rangle, \langle Q \parallel R, true\rangle)$. The results then follows from Proposition 4 and the fact that $\langle P' \parallel R, true\rangle \dot{\approx}_{sb} \langle P \parallel R, e\rangle$ which can be easily verified.*

Consequently, let us assume that $\langle P \parallel R, e\rangle \longrightarrow \langle P \parallel \textbf{tell}(\alpha), e\rangle$ *by executing the left summand of R. By condition (ii) of weak bisimulation* $\langle Q \parallel R, true\rangle$ *must match the move. We have two cases:*

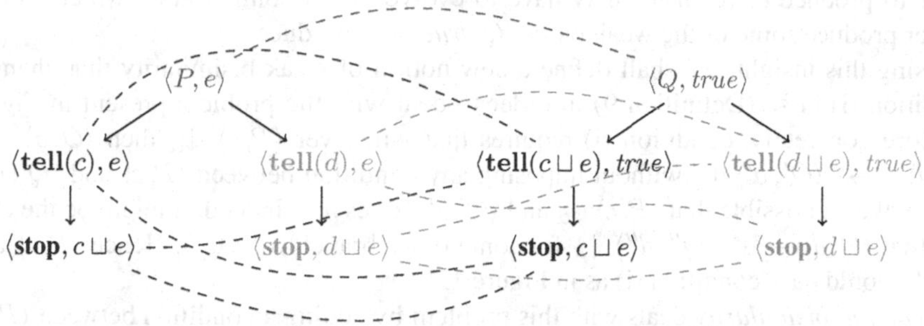

Fig. 1. Let $P = (\textbf{ask}\,(true) \rightarrow \textbf{tell}(c)) + (\textbf{ask}\,(true) \rightarrow \textbf{tell}(d))$ and $Q = (\textbf{ask}\,(true) \rightarrow \textbf{tell}(c \sqcup e)) + (\textbf{ask}\,(true) \rightarrow \textbf{tell}(d \sqcup e))$. The linked configurations are weakly bisimilar.

- *$\langle Q \parallel R, true\rangle$ does not make a transition. And now let us suppose that $\langle Q \parallel R, true\rangle \xrightarrow{e} \langle Q \parallel \textbf{tell}(\beta), true\rangle$. This means that $\langle P \parallel \textbf{tell}(\alpha), e\rangle$ now has to match this transition. However $\langle Q \parallel \textbf{tell}(\beta), true\rangle \longrightarrow \langle Q, \beta\rangle \Downarrow_\beta$ while $\langle P \parallel \textbf{tell}(\alpha), e\rangle \not\Downarrow_\beta$. Thus we cannot satisfy condition (i) of weak bisimulation.*
- *$\langle Q \parallel R, true\rangle$ makes a transition. To match the move it should also execute the left summand of R. However, since e is not the store of $\langle Q \parallel R, true\rangle$, Q must be executed first. and this means executing of one of summands in Q to be able to add e to the store. If the left summand of Q is executed, we get $\langle Q \parallel R, true\rangle \longrightarrow^* \langle \textbf{tell}(\alpha), c \sqcup e\rangle$. In this case we could then take the move $\langle P \parallel \textbf{tell}(\alpha), e\rangle \longrightarrow \langle \textbf{tell}(d) \parallel \textbf{tell}(\alpha), e\rangle$. But then $\langle \textbf{tell}(\alpha), c \sqcup e\rangle \Downarrow_c$ and notice that $\langle \textbf{tell}(d) \parallel \textbf{tell}(\alpha), e\rangle \not\Downarrow_c$, thus we cannot satisfy condition (i) of weak bisimulation. The case where the right summand of Q is executed is symmetric.*

4 Weak Full Bisimilarity

In the previous section we showed that $\approx$ (and $\approx_{sb}$) for the full ccp is not entirely satisfactory since it is not a congruence. By building on $\approx$, in this section we propose a new equivalence which we call *(weak) full bisimilarity*, written $\approx_f$. This new equivalence does not quantify over infinitely many process contexts in its definition yet we will show that is a congruence. Furthermore, we will also prove that adequacy of $\approx_f$ by showing that it is the largest congruence included in $\dot{\approx}_{sb}$.

4.1 More Than Weak Barbs

The key to figure out the element missing in the definition of $\dot{\approx}_{sb}$ (Definition 4) lies in Figure 1. If we look at the configurations in the figure we can see that while $\langle P, e\rangle$ is able to *produce* a barb e without choosing between c and d, $\langle Q, true\rangle$ is not. The definition of $\dot{\approx}_{sb}$ tries to capture this in the condition (i), namely by checking that $\langle P, e\rangle \Downarrow_e$ then requiring that $\langle Q, true\rangle \Downarrow_e$. However, this condition does not capture the fact that in order to produce e, $\langle Q, true\rangle$ may have to evolve into a configuration which can no longer produce some of the weak barbs $\langle Q, true\rangle$ can produce. [5]

Using this insight, we shall define a new notion of weak bisimilarity that changes condition (i) in $\dot{\approx}$ (Definition 9) in order to deal with the problem present in Figure 1. More concretely, condition (i) requires that whenever $\langle P, c\rangle \downarrow_\alpha$ then $\langle Q, d\rangle \Downarrow_\alpha$, $\langle Q, d\rangle \longrightarrow^* \langle Q', d'\rangle \downarrow_\alpha$ without imposing any condition between $\langle P, c\rangle$ and $\langle Q', d'\rangle$. This makes it possible that $\langle P, c\rangle \downarrow_\beta$ and $\langle Q', d'\rangle$ does *not*: indeed, it might be the case that that $\langle Q, d\rangle \longrightarrow^* \langle Q'', d''\rangle \downarrow_\beta$ for some other branch $\langle Q'', d''\rangle$. Hence $\langle P, c\rangle$ and $\langle Q, d\rangle$ would pass condition (i) as in Figure 1.

Weak full bisimilarity deals with this problem by adding a condition between $\langle P, c\rangle$ and $\langle Q', d'\rangle$, namely $\langle Q, d\rangle \Downarrow_c$ has to hold by reaching a bisimilar configuration: $\langle P, c\rangle$ has to be weakly bisimilar $\langle Q', d'\rangle$.

Definition 10 (Weak Full Bisimilarity). *A weak full bisimulation is a symmetric relation $\mathcal{R}$ on configurations s.t. whenever $(\gamma_1, \gamma_2) \in \mathcal{R}$ with $\gamma_1 = \langle P, c\rangle$ and $\gamma_2 = \langle Q, d\rangle$ implies that:*

(i) there is $\gamma_2' = \langle Q', d'\rangle$ such that $\langle Q, d\rangle \longrightarrow^* \gamma_2'$ where $c \sqsubseteq d'$ and $(\gamma_1, \gamma_2') \in \mathcal{R}$,
(ii) if $\gamma_1 \xrightarrow{\alpha} \gamma_1'$ then there exists $\gamma_2' = \langle Q', d'\rangle$ s.t. $\langle Q, d \sqcup \alpha\rangle \longrightarrow^* \gamma_2'$ where $c' \sqsubseteq d'$ and $(\gamma_1', \gamma_2') \in \mathcal{R}$.

We say that γ_1 and γ_2 are weak fully bisimilar ($\gamma_1 \approx_f \gamma_2$) if there exists a weak full bisimulation $\mathcal{R}$ s.t. $(\gamma_1, \gamma_2) \in \mathcal{R}$. We write $P \approx_f Q$ iff $\langle P, true\rangle \approx_f \langle Q, true\rangle$.

In the definition above, the fist condition states that $\langle Q, d\rangle$ has to produce c by reaching a (weakly) bisimilar configuration. The second condition is the bisimulation game from $\dot{\approx}$ (Definition 9) plus a condition requiring the store c' to be matched too.

To better explain this notion consider again the counterexample to $\dot{\approx}$ from Figure 1.

Example 6. Let $\langle P, e\rangle, \langle Q, true\rangle$ as in Figure 1. Let us build a relation $\mathcal{R}$ that is a weak full bisimulation where $(\langle P, e\rangle, \langle Q, true\rangle) \in \mathcal{R}$. By condition (i) in Definition 10 we need a $\gamma_2' = \langle Q', d'\rangle$ s.t. $\langle Q, d\rangle \longrightarrow^* \gamma_2'$ and $e \sqsubseteq d'$ and $(\gamma_1, \gamma_2') \in \mathcal{R}$. We have two options $Q' = \mathbf{stop}$ and $d' = c \sqcup e$ or $d' = d \sqcup e$.[6] However, if we take $(\langle P, e\rangle, \langle \mathbf{stop}, c \sqcup e\rangle) \in \mathcal{R}$ we have that $\langle P, e\rangle \Downarrow_d$ while $\langle \mathbf{stop}, c \sqcup e\rangle \not\Downarrow_d$. A similar argument works for $\langle \mathbf{stop}, d \sqcup e\rangle$. Therefore, no weak full bisimulation may contain $(\langle P, e\rangle, \langle Q, true\rangle)$. Hence $\langle P, e\rangle \not\approx_f \langle Q, true\rangle$. □

4.2 Congruence Issues

We shall now prove that full bisimilarity is a congruence w.r.t all possible contexts in ccp. Namely, whenever γ and γ' are in $\approx_f$ then they can be replaced for one another in any context.

Theorem 4. *Let P and Q be ccp processes and assume that $P \approx_f Q$. Then for every process context $C[\bullet]$ we have that $C[P] \approx_f C[Q]$.*

---

[5] In the case of ccp\+ this is not a concern given that in this fragment weak barbs are always preserved during evolution.

[6] The cases for $Q' = \mathbf{tell}(c \sqcup e)$ or $Q' = \mathbf{tell}(d \sqcup e)$ with $d' = true$ are equivalent.

Proof. Here we consider the parallel case; the other cases are trivial or easier to verify. We shall prove that $\mathcal{R} = \{(\langle P \parallel R, c\rangle, \langle Q \parallel R, d\rangle) \mid \langle P, c\rangle \approx_f \langle Q, d\rangle\}$ *is a weak full bisimulation as in Definition 10. To prove (i), since* $\langle P, c\rangle \approx_f \langle Q, d\rangle$ *we have that* $\langle Q, d\rangle \longrightarrow^* \langle Q', d'\rangle$ *where* $c \sqsubseteq d'$ *and* $\langle Q', d'\rangle \dot{\approx} \langle P, c\rangle$ **(1)**. *Therefore by* R2 *we get* $\langle Q \parallel R, d\rangle \longrightarrow^* \langle Q' \parallel R, d'\rangle$ *and by* **(1)** *we can conclude that* $(\langle Q' \parallel R, d'\rangle, \langle P \parallel R, c\rangle) \in \mathcal{R}$. *To prove (ii) let us assume that* $\langle P \parallel R, c\rangle \xrightarrow{\alpha} \langle P_1, c_1\rangle$. *We proceed by induction (on the depth) of the inference of* $\langle P \parallel R, c\rangle \xrightarrow{\alpha} \langle P_1, c'\rangle$. *Using* LR2 *(left), then* $P_1 = (P' \parallel R)$ *with* $\langle P, c\rangle \xrightarrow{\alpha} \langle P', c'\rangle$ *by a shorter inference. Since* $\langle P, c\rangle \approx_f \langle Q, d\rangle$ *then* $\langle Q, d \sqcup \alpha\rangle \longrightarrow^* \langle Q', d'\rangle$ *where* $\langle P', c'\rangle \approx_f \langle Q', d'\rangle$ *and* $c' \sqsubseteq d'$ **(3)**. *By* R2 *we have* $\langle Q \parallel R, d \sqcup \alpha\rangle \longrightarrow^* \langle Q' \parallel R, d'\rangle$ *and from* **(3)** *we can conclude that* $(\langle P' \parallel R, c\rangle, \langle Q' \parallel R, d'\rangle) \in \mathcal{R}$. *Using* LR2 *(right), then* $P_1 = (P \parallel R')$ *and* $c' = (c \sqcup \alpha \sqcup e)$ *with* $\langle R, c\rangle \xrightarrow{\alpha} \langle R', c'\rangle$ *by a shorter inference. From* **(1)** *we know that* $\langle Q, d\rangle \longrightarrow^* \langle Q', d'\rangle$ *where* $c \sqsubseteq d'$ *and* $\langle Q', d'\rangle \dot{\approx} \langle P, c\rangle$. *Hence* $\langle Q \parallel R, d \sqcup \alpha\rangle \longrightarrow^* \langle Q' \parallel R, d' \sqcup \alpha\rangle$. *Now since* $c \sqsubseteq d'$ *then by monotonicity* $\langle R, d' \sqcup \alpha\rangle \longrightarrow \langle R, d''\rangle$ *where* $d'' = d' \sqcup \alpha \sqcup e$. *Therefore by* R2 *we get* $\langle Q \parallel R, d \sqcup \alpha\rangle \longrightarrow^* \langle Q' \parallel R', d''\rangle$ *and from* **(1)** *and monotonicity* $\langle P, c'\rangle = \langle P, c \sqcup \alpha \sqcup e\rangle \dot{\approx} \langle Q', d' \sqcup \alpha \sqcup e\rangle = \langle Q', d''\rangle$. *Using this we can conclude that* $(\langle P \parallel R', c'\rangle, \langle Q' \parallel R', d''\rangle) \in \mathcal{R}$.

Note that $\approx_f$ is more distinguishing than $\dot{\approx}$ and the result above shows that this level of granularity is needed to obtain a weak bisimilarity that is a congruence for ccp.

4.3 Relation with Observational Equivalence

In section 3.1 we described the relation between weak (saturated) bisimilarity ($\dot{\approx}_{sb}$, Definition 4) and the standard observational equivalence ($\sim_o$, Definition 8) for ccp. Concretely, we know that, in ccp\+, $\dot{\approx}_{sb}$ coincides with $\sim_o$, while for the full ccp $\dot{\approx}_{sb}$ implies $\sim_o$ but the converse does not hold. In this section we shall see the relation between weak full bisimilarity ($\approx_f$, Definition 10) and $\sim_o$. We shall prove that $\approx_f$ coincides with $\sim_o$ in ccp\+ by proving that $\approx_f$ corresponds to $\dot{\approx}_{sb}$ in the choice-free fragment of ccp. Furthermore, for the full language of ccp, we shall prove that $\approx_f$ implies $\sim_o$ again by showing that $\approx_f$ implies $\dot{\approx}_{sb}$ in ccp.

Let us start by showing that $\approx_f$ and $\dot{\approx}$ coincide in ccp\+. This theorem strongly relies on the confluent nature of ccp\+ (Proposition 1).

Theorem 5. *Let* $\gamma, \gamma' \in Conf_{\text{ccp}\backslash+}$, $\gamma \approx_f \gamma'$ *iff* $\gamma \dot{\approx} \gamma'$.

The corollary below follows from Proposition 3 and 4, and Theorem 5.

Corollary 1. *Let P and Q be ccp\+ processes. Then* $P \approx_f Q$ *iff* $P \sim_o Q$.

We shall now prove that $\approx_f$ implies $\sim_o$ for the full ccp. In order to do this we first prove that $\approx_f$ implies $\dot{\approx}_{sb}$.

Theorem 6. *If* $\gamma \approx_f \gamma'$ *then* $\gamma \dot{\approx} \gamma'$.

The corollary below follows from Theorem 1 and 6, and Proposition 4.

Corollary 2. *If* $P \approx_f Q$ *then* $P \sim_o Q$.

The above statement allows us to use the co-inductive techniques of full bisimulation to prove observational equivalence.

Table 3. Summary of the contributions. Recall that $\dot{\approx}_{sb}$ stands for the weak saturated barbed bisimilarity (Definition 4), $\sim_o$ is the standard observational equivalence (Definition 8), $\dot{\approx}$ represents weak bisimilarity (Definition 9), $\approx_f$ is the notion of weak full bisimilarity proposed in this paper (Definition 10) and $\dot{\cong}$ stands for the behavioral congruence (Definition 11). $C[\bullet]\backslash+$ stands for the contexts where the summation operator does not occur, while $C[\bullet]$ represents any possible context, hence the summation operator may occur in $C[\bullet]$. For this reason we put N/A (Not Applicable) in the row corresponding to ccp\+. Notice that the correspondence $\dot{\approx} = \dot{\approx}_{sb} = \sim_o$ comes from [2].

Language	Relation among equivalences	Congruence w.r.t. $C[\bullet]$	Congruence w.r.t. $C[\bullet]\backslash+$
ccp\+	$\dot{\cong} = \approx_f = \dot{\approx} = \dot{\approx}_{sb} = \sim_o$	N/A	$\dot{\cong}, \approx_f, \dot{\approx}, \dot{\approx}_{sb}, \sim_o$
ccp	$\dot{\cong} = \approx_f \subseteq \dot{\approx} = \dot{\approx}_{sb} \subseteq \sim_o$	$\dot{\cong}, \approx_f$	$\dot{\cong}, \approx_f, \dot{\approx}, \dot{\approx}_{sb}$

4.4 Behavioral Congruence

Finally, we prove that $\approx_f$ is the largest congruence included in $\dot{\approx}$ by showing that it coincides with the congruence $\dot{\cong}$ defined next.

Definition 11 (Behavioral Congruence). *We say that P is behaviorally congruent to Q, denoted $P\dot{\cong}Q$, iff for every process context $C[\bullet]$ we have $C[P] \dot{\approx} C[Q]$. We use $\langle P, e\rangle\dot{\cong}\langle Q, d\rangle$ to denote $(P \parallel \mathbf{tell}(e))\dot{\cong}(Q \parallel \mathbf{tell}(d))$.*

We now state that $\approx_f$ coincides with $\dot{\cong}$ for ccp with nondeterministic choice.

Theorem 7. *$\langle P, e\rangle \approx_f \langle Q, d\rangle$ iff $\langle P, e\rangle\dot{\cong}\langle Q, d\rangle$.*

5 Conclusions and Related Work

In this paper we showed that the weak saturated barbed bisimilarity ($\dot{\approx}_{sb}$) proposed in [2] is not a congruence for ccp. Nevertheless, we also showed that the upward closure, i.e. condition (iii), is enough to make $\dot{\approx}_{sb}$ a congruence in the choice-free fragment (ccp\+). We then proposed a new notion of bisimilarity, called weak full bisimilarity ($\approx_f$), and we proved that it is a congruence for the full ccp despite the fact that $\approx_f$ does not require any quantification over a (potentially) infinite number of contexts in its definition. Furthermore, we showed that $\approx_f$ implies the standard observational equivalence ($\sim_o$) for ccp from [26]. Finally we demonstrated that $\approx_f$ is not too restrictive by showing that it is the largest congruence included in $\dot{\approx}_{sb}$. See Table 3 for a summary of the contributions of this paper. This is the first weak behavioral ccp congruence for ccp with nondeterministic choice that does not require implicit quantification over all contexts.

Most of the related work has already been discussed in the introduction (Section 1). There has been other attempts for finding a good notion of bisimilarity for ccp such as [25] and [16]. In [25] the authors propose a ccp bisimilarity that requires processes to match the exact label in the bisimulation game, a condition which is standard in process calculi realm, however this notion is known to be too distinguishing for ccp as shown in [2]. As for [16], their notion of (strong) bisimilarity resembles to the saturated barbed bisimilarity from [2] and, although they do not give a notion of weak bisimilarity, the results in this paper can be related directly.

We plan to adapt the algorithms from [3,23] to verify $\approx_f$. We conjecture that the decision procedure for $\approx_{sb}$ can be exploited to check $\approx_f$ by modifying the way the (weak) barbs are considered. Furthermore, in this paper we obtained a notion of weak bisimilarity that is a congruence even if we do not consider a label for observing the tell actions. Since ccp is an asynchronous language, not observing the tell follows the philosophy of considering as labels the minimal information needed to proceed, namely a tell process does not need a stimulus from the environment to post its information in the store. Following the same reasoning, we plan to investigate whether it is possible to define a labeled semantics for the asynchronous π-calculus ($A\pi$) [18,24] with a τ label for the output transitions, instead of a co-action, and we shall check if a notion of bisimilarity similar to ours would also be a congruence.

References

1. Arbab, F., Rutten, J.J.M.M.: A coinductive calculus of component connectors. In: WADT, pp. 34–55 (2002)
2. Aristizábal, A., Bonchi, F., Palamidessi, C., Pino, L., Valencia, F.: Deriving labels and bisimilarity for concurrent constraint programming. In: Hofmann, M. (ed.) FOSSACS 2011. LNCS, vol. 6604, pp. 138–152. Springer, Heidelberg (2011)
3. Aristizábal, A., Bonchi, F., Pino, L., Valencia, F.D.: Partition refinement for bisimilarity in CCP. In: Ossowski, S., Lecca, P. (eds.) 27th Annual ACM Symposium on Applied Computing (SAC 2012), pp. 88–93. ACM (2012)
4. Bartoletti, M., Zunino, R.: A calculus of contracting processes. In: 25th Annual IEEE Symposium on Logic in Computer Science (LICS 2010), pp. 332–341. IEEE Computer Society (2010)
5. Bengtson, J., Johansson, M., Parrow, J., Victor, B.: Psi-calculi: Mobile processes, nominal data, and logic. In: 24th Annual IEEE Symposium on Logic in Computer Science (LICS 2009), pp. 39–48. IEEE Computer Society (2009)
6. Bonchi, F., Gadducci, F., Monreale, G.V.: Reactive systems, barbed semantics, and the mobile ambients. In: de Alfaro, L. (ed.) FOSSACS 2009. LNCS, vol. 5504, pp. 272–287. Springer, Heidelberg (2009)
7. Bonchi, F., Gadducci, F., Monreale, G.V.: Towards a general theory of barbs, contexts and labels. In: Yang, H. (ed.) APLAS 2011. LNCS, vol. 7078, pp. 289–304. Springer, Heidelberg (2011)
8. Bonchi, F., König, B., Montanari, U.: Saturated semantics for reactive systems. In: 21th IEEE Symposium on Logic in Computer Science (LICS 2006), pp. 69–80. IEEE Computer Society (2006)
9. Buscemi, M.G., Montanari, U.: Open bisimulation for the concurrent constraint pi-calculus. In: Drossopoulou, S. (ed.) ESOP 2008. LNCS, vol. 4960, pp. 254–268. Springer, Heidelberg (2008)
10. de Boer, F.S., Pierro, A.D., Palamidessi, C.: Nondeterminism and infinite computations in constraint programming. Theoretical Computer Science 151(1), 37–78 (1995)

11. De Nicola, R.: Behavioral equivalences. In: Encyclopedia of Parallel Computing, pp. 120–127 (2011)
12. Falaschi, M., Gabbrielli, M., Marriott, K., Palamidessi, C.: Confluence in concurrent constraint programming. Theoretical Computer Science 183(2), 281–315 (1997)
13. Fokkink, W., Pang, J., Wijs, A.: Is timed branching bisimilarity a congruence indeed? Fundam. Inform. 87(3-4), 287–311 (2008)
14. Knight, S., Palamidessi, C., Panangaden, P., Valencia, F.D.: Spatial and epistemic modalities in constraint-based process calculi. In: Koutny, M., Ulidowski, I. (eds.) CONCUR 2012. LNCS, vol. 7454, pp. 317–332. Springer, Heidelberg (2012)
15. Leifer, J.J., Milner, R.: Deriving bisimulation congruences for reactive systems. In: Palamidessi, C. (ed.) CONCUR 2000. LNCS, vol. 1877, pp. 243–258. Springer, Heidelberg (2000)
16. Mendler, N.P., Panangaden, P., Scott, P.J., Seely, R.A.G.: A logical view of concurrent constraint programming. Nordic Journal of Computing 2(2), 181–220 (1995)
17. Milner, R.: A Calculus of Communication Systems. LNCS, vol. 92. Springer, Heidelberg (1980)
18. Milner, R.: Communicating and mobile systems: the π-calculus. Cambridge University Press (1999)
19. Milner, R., Sangiorgi, D.: Barbed bisimulation. In: Kuich, W. (ed.) ICALP 1992. LNCS, vol. 623, pp. 685–695. Springer, Heidelberg (1992)
20. Monk, J., Henkin, L., Tarski, A.: Cylindric Algebras (Part I). North-Holland (1971)
21. Palamidessi, C., Saraswat, V.A., Valencia, F.D., Victor, B.: On the expressiveness of linearity vs persistence in the asychronous pi-calculus. In: 21th IEEE Symposium on Logic in Computer Science (LICS 2006), pp. 59–68. IEEE Computer Society (2006)
22. Pino, L.F., Bonchi, F., Valencia, F.D.: A behavioral congruence for concurrent constraint programming with nondeterministic choice (extended version). Technical report, INRIA/DGA and LIX, École Polytechnique, France (2013), `http://www.lix.polytechnique.fr/~luis.pino/files/ictac14-extended.pdf`
23. Pino, L.F., Bonchi, F., Valencia, F.D.: Efficient computation of program equivalence for confluent concurrent constraint programming. In: Peña, R., Schrijvers, T. (eds.) 15th International Symposium on Principles and Practice of Declarative Programming (PPDP 2013), pp. 263–274. ACM (2013)
24. Sangiorgi, D., Walker, D.: The π-Calculus - a theory of mobile processes. Cambridge University Press (2001)
25. Saraswat, V.A., Rinard, M.C.: Concurrent constraint programming. In: Allen, F.E. (ed.) 17th Annual ACM Symposium on Principles of Programming Languages (POPL 1991), pp. 232–245. ACM Press (1990)
26. Saraswat, V.A., Rinard, M.C., Panangaden, P.: Semantic foundations of concurrent constraint programming. In: Wise, D.S. (ed.) 18th Annual ACM Symposium on Principles of Programming Languages (POPL 1991), pp. 333–352. ACM Press (1991)

Distributed Testing of Concurrent Systems: Vector Clocks to the Rescue *

Hernán Ponce-de-León[1], Stefan Haar[1], and Delphine Longuet[2]

[1] INRIA and LSV, École Normale Supérieure de Cachan and CNRS, France
ponce@lsv.ens-cachan.fr , stefan.haar@inria.fr
[2] Univ Paris-Sud, LRI UMR8623, Orsay, F-91405
longuet@lri.fr

Abstract. The **ioco** relation has become a standard in model-based conformance testing. The **co-ioco** conformance relation is an extension of this relation to concurrent systems specified with true-concurrency models. This relation assumes a global control and observation of the system under test, which is not usually realistic in the case of physically distributed systems. Such systems can be partially observed at each of their points of control and observation by the sequences of inputs and outputs exchanged with their environment. Unfortunately, in general, global observation cannot be reconstructed from local ones, so global conformance cannot be decided with local tests. We propose to append time stamps to the observable actions of the system under test in order to regain global conformance from local testing.

1 Introduction

The aim of testing is to execute a software system, the *implementation*, on a set of input data selected so as to find discrepancies between actual behavior and intended behavior described by the *specification*. Model-based testing requires a behavioral description of the system under test. One of the most popular formalisms studied in conformance testing is that of *input output labeled transition systems* (IOLTS). In this framework, the correctness (or conformance) relation the system under test (SUT) and its specification must verify is formalized by the **ioco** relation [1]. This relation has become a standard, and it is used as a basis in several testing theories for extended state-based models: restrictive transition systems [2, 3], symbolic transition systems [4, 5], timed automata [6, 7], multi-port finite state machines [8].

Model-based testing of concurrent systems has been studied in the past [9–11], but mostly in the context of interleaving, or trace, semantics, which is known to suffer from the state space explosion problem. Concurrent systems are naturally modeled as a *network of finite automata*, a formal class of models that can be captured equivalently by *safe Petri nets*. Partial order semantics of

* This work was funded by the DIGITEO / DIM-LSC project TECSTES, convention DIGITEO Number 2011-052D - TECSTES.

G. Ciobanu and D. Méry (Eds.): ICTAC 2014, LNCS 8687, pp. 369–387, 2014.

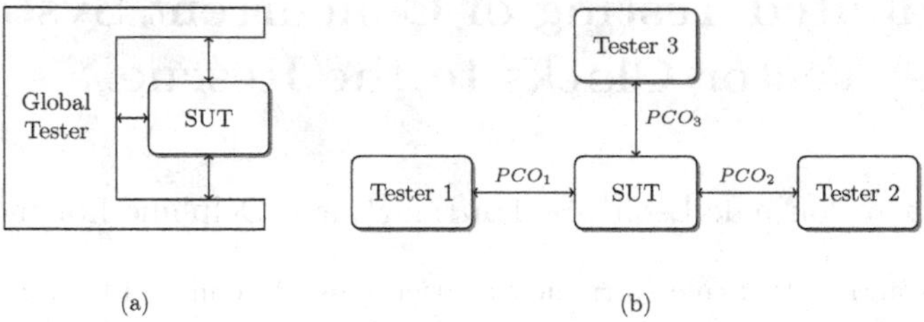

Fig. 1. The global and distributed testing architectures

a Petri net is given by its *unfolding* [12, 13]. Test case generation for concurrent systems based on unfoldings has been studied in [14, 15]. In the same direction, we proposed an extension of the **ioco** conformance relation to concurrent systems, called **co-ioco**, using both interleaving and partial order semantics [16]. We developped a full testing framework for **co-ioco** [17], but in this work, concurrency is only interpreted as independence between actions: actions specified as independent cannot be implemented by interleavings. We introduced a new semantics for unfoldings [18], allowing some concurrency to be implemented by interleavings, while forcing other concurrency to be preserved. The kind of concurrency we consider in this article arises from the distribution of the system, for this reason we restrict to partial order semantics only.

Our previous work [16–18] assume a global tester which controls and observes the whole system (see Fig. 1.a). If the system is distributed, the tester interacts with every component, but the observation of such interaction is global. When global observation of the system cannot be achieved, the testing activity needs to be distributed. In a distributed testing environment (see Fig. 1.b), the testers stimulate the implementation by sending messages on points of control and observation (PCOs) and partially observe the reactions of the implementation on these same PCOs. It is known that, in general, global traces cannot be reconstructed from local observations (see for example [19]). This reduces the ability to distinguish different systems. There are three mainly investigated solutions to overcome this problem: *(i)* the conformance relation needs to be weaken considering partial observation [8, 20]; *(ii)* testers are allowed to communicate to coordinate the testing activity [21]; *(iii)* stronger assumptions about the implementations are needed. In this paper, we follow the third approach and assume that each component has a local clock.

Related Work. According to these three directions, the following solutions have been proposed for testing global conformance in distributed testing architectures.

(i) Hierons et al. [8] argue that when the SUT is to be used in a context where the separate testers at the PCOs do not directly communicate with one another, the requirements placed on the SUT do not correspond to traditional implementation relations. In fact, testing the SUT using a method based on a standard implementation relation, such as **ioco**, may return an incorrect verdict. The authors of [8] consider different scenarios, and a dedicated implementation

relation for each of them. In the first scenario, there is a tester at each PCO, and these testers are pairwise independent. In this scenario, it is sufficient that the local behavior observed by a tester is consistent with some global behavior in the specification: this is captured by the **p-dioco** conformance relation. In the second scenario, a tester may receive information from other testers, and the local behaviors observed at different PCOs could be combined. Consequently, a stronger implementation relation, called **dioco**, is proposed. They show that **ioco** and **dioco** coincide when the system is composed of a single component, but that **dioco** is weaker than **ioco** when there are several components. Similar to this, Longuet [20] studies different ways of globally and locally testing a distributed system specified with Message Sequence Charts, by defining global and local conformance relations. Moreover, conditions under which local testing is equivalent to global testing are established under trace semantics.

(ii) Jard et al. [21] propose a method for constructing, given a global tester, a set of testers (one for each PCO) such that global conformance can be achieved by these testers. However, they assume that testers can communicate with each other in order to coordinate the testing activity. In addition, they consider the interaction between testers and the SUT as asynchronous.

(iii) Bhateja and Mukund [22] propose an approach where they assume each component has a local clock and they append tags to the messages generated by the SUT. These enriched behaviors are then compared against a tagged version of the specification. Hierons et al. [23] make the same assumption about local clocks. If the clocks agree exactly then the sequence of observations can be reconstructed. In practice the local clocks will not agree exactly, but some assumptions regarding how they can differ can be made. They explore several such assumptions and derive corresponding implementation relations. In this article, we also assume local clocks, but we use partial order semantics.

Contribution. The aim of this paper is to propose a formal framework for the distributed testing of concurrent systems from network of automata specifications, without relying on communications between testers. We show that some, but not all, situations leading to non global conformance w.r.t **co-ioco** can be detected by local testers without any further information of the other components; moreover we prove that, when vector clocks [24, 25] are used, the information held by each component suffices to reconstruct the global trace of an execution from the partial observations of it at each PCO, and that global conformance can thus be decided by distributed testers.

The paper is organized as follows. Section 2 recalls basic notions about network of automata and Petri nets, while Section 3 introduces their partial order semantics. Section 4 introduces the testing hypotheses and our **co-ioco** conformance relation. Finally, in Section 5, we distribute the testing architecture and show how global conformance can be achieved locally using time stamped traces.

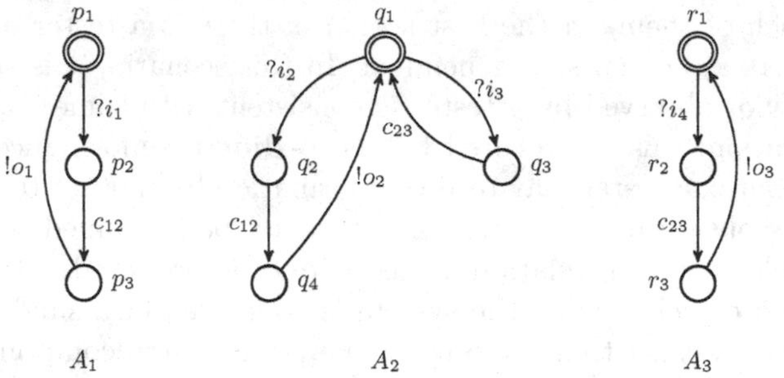

Fig. 2. Network of automata composed of 3 components

2 Model of the System

A sound software engineering rule for building complex systems is to divide the whole system in smaller and simpler components, each solving a specific task. This means that, in general, complex systems are actually collections of simpler components running in parallel. We use automata to model local behaviors, while global behaviors are modeled by networks of automata. We show that networks of automata are captured equivalently by Petri nets where explicit representation of concurrency avoids the state space explosion produced by interleavings.

Network of Automata. We consider a distributed system composed of n components that communicate with each other synchronizing on communication actions. The local model of a component is defined as a deterministic finite automaton (Q, Σ, Δ, q_0), where Q is a finite set of states, Σ is a finite set of actions, $\Delta : Q \times \Sigma \to Q$ is the transition function and $q_0 \in Q$ is the initial state. We distinguish between the controllable actions $\Sigma_{\mathcal{I}n}$ (inputs proposed by the environment), the observable ones $\Sigma_{\mathcal{O}ut}$ (outputs produced by the system), and communication actions $\Sigma_{\mathcal{C}}$ (invisible for the environment), i.e. $\Sigma = \Sigma_{\mathcal{I}n} \uplus \Sigma_{\mathcal{O}ut} \uplus \Sigma_{\mathcal{C}}$. Several components can communicate over the same communication action, but we assume that observable actions from different components are disjoint,[1] i.e. components only share communication actions.

Example 1. Fig. 2 shows a network of automata with three components A_1, A_2 and A_3. Input and output actions are denoted by ? and ! respectively. Components A_1 and A_2 communicate by synchronizing over c_{12} while A_2, A_3 do it over c_{23}. Components A_1 and A_3 do not communicate.

I/O Petri Nets. A *net* is a tuple $N = (P, T, F)$ where *(i)* $P \neq \emptyset$ is a set of *places*, *(ii)* $T \neq \emptyset$ is a set of *transitions* such that $P \cap T = \emptyset$, *(iii)* $F \subseteq (P \times T) \cup (T \times P)$ is a set of *flow arcs*. A *marking*[2] is a set M of places which represents the current "state" of the system. Let $\mathcal{I}n$ and $\mathcal{O}ut$ be two disjoint non-empty sets

---
[1] Action a from component A_i is labeled by a_i if necessary.
[2] We restrict to 1-safe nets.

of *input* and *output* labels respectively. A *labeled Petri net* is a tuple $\mathcal{N} = (P, T, F, \lambda, M_0)$, where *(i)* (P, T, F) is a finite net; *(ii)* $\lambda : T \to (\mathcal{I}n \uplus \mathcal{O}ut)$ labels transitions by input/output actions; and *(iii)* $M_0 \subseteq P$ is an *initial marking.* Denote by $T^{\mathcal{I}n}$ and $T^{\mathcal{O}ut}$ the input and output transition sets, respectively; that is, $T^{\mathcal{I}n} \triangleq \lambda^{-1}(\mathcal{I}n)$ and $T^{\mathcal{O}ut} \triangleq \lambda^{-1}(\mathcal{O}ut)$. Elements of $P \cup T$ are called the *nodes* of $\mathcal{N}$. For a transition $t \in T$, we call ${}^\bullet t = \{p \mid (p,t) \in F\}$ the *preset* of t, and $t^\bullet = \{p \mid (t,p) \in F\}$ the *postset* of t. These definitions can be extended to sets of transitions. In figures, we represent as usual places by empty circles, transitions by squares, F by arrows, and the marking of a place p by black tokens in p. A transition t is *enabled* in marking M, written $M \xrightarrow{t}$, if $\forall p \in {}^\bullet t,\ M(p) = 1$. This enabled transition can *fire*, resulting in a new marking $M' = (M \backslash {}^\bullet t) \cup t^\bullet$. This firing relation is denoted by $M \xrightarrow{t} M'$. A marking M is *reachable* from M_0 if there exists a *firing sequence*, i.e. transitions $t_0 \ldots t_n$ such that $M_0 \xrightarrow{t_0} M_1 \xrightarrow{t_1} \ldots \xrightarrow{t_n} M$. The set of markings reachable from M_0 is denoted $\mathbf{R}(M_0)$.

$\mathcal{N}$ is called *deterministically labeled* iff no two transitions with the same label are simultaneously enabled, i.e. for all $t_1, t_2 \in T$ and $M \in \mathbf{R}(M_0)$ we have $(M \xrightarrow{t_1} \wedge M \xrightarrow{t_2} \wedge \lambda(t_1) = \lambda(t_2)) \Rightarrow t_1 = t_2$. Deterministic labeling ensures that the system behavior is locally discernible through labels, either through distinct inputs or through observation of different outputs.

When testing reactive systems, we need to differentiate situations where the system can still produce some outputs and those where the system cannot evolve without an input from the environment. Such situations are captured by the notion of *quiescence* [26]. A marking is said quiescent if it only enables input transitions, i.e. $M \xrightarrow{t}$ implies $t \in T^{\mathcal{I}n}$.

From Automata to Nets. The translation from an automaton $A = (Q, \Sigma, \Delta, q_0)$ to a labeled Petri net $\mathcal{N}_A = (P, T, F, \lambda, M_0)$ is immediate: *(i)* places are the states of the automaton, i.e. $P = Q$; *(ii)* for every transition $(s_i, a, s_i') \in \Delta$ we add t to T and set ${}^\bullet t = \{s_i\}, t^\bullet = \{s_i'\}$ and $\lambda(t) = a$; *(iii)* the initial state is the only place marked initially, i.e. $M_0 = \{q_0\}$.

The joint behavior of a system composed of automata $A_1, \ldots, A_n$ is modeled by $\mathcal{N}_{A_1} \times \cdots \times \mathcal{N}_{A_n}$ where $\times$ represents the product of labeled nets [27] and we only synchronize on communication transitions (which are invisible for the environment and thus labeled by τ). As different components are deterministic and they only share communication actions, the net obtained by this product is deterministically labeled. Product of nets prevents the state space explosion problem, as the number of places in the final net is linear w.r.t the number of components while product of automata produces an exponential number of states. Product of nets naturally allows to distinguish its components by means of a distribution [28]. A distribution $D : P \cup T \to \mathcal{P}(\{1, \ldots, n\})$ is a function that relates each place/transition with its corresponding automata. In the case of communication actions, the distribution relates the synchronized transition with the automata that communicate over it.

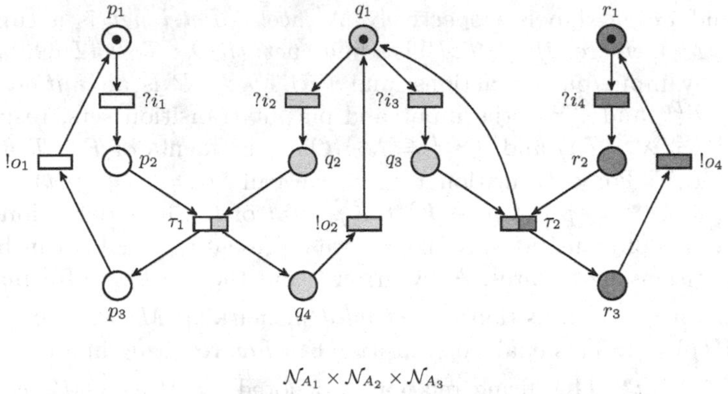

Fig. 3. Distributed Petri net

Example 2. Fig. 3 shows the net obtained from the automata in Fig. 2 and its distribution D represented by colors. The transition corresponding to action $?i_1$ corresponds to the first component, i.e. $D(?i_1) = \{1\}$, while communication between A_1 and A_2 is converted into a single transition τ_1 with $D(\tau_1) = \{1, 2\}$.

Proposition 1. *For every component A and its corresponding net $\mathcal{N}_A$ we have $\forall t \in T : |^\bullet t| = 1$. In the net obtained by the product between components, this property is only violated by communication transitions. Therefore, any input or output event is enabled by exactly one place.*

Proof. Immediate from construction. □

3 Partial Order Semantics

The partial order semantics associated to a Petri net is given by its unfolding where execution traces are not sequences but *partial orders*, in which concurrency is represented by absence of precedence. We recall here the basic notions.

3.1 Unfoldings of Petri Nets

The unfolding of a net [13] is an acyclic (and usually infinite) structure that represents the behavior of a system by explicit representation of its branching. Unfolding can be expressed by event structures in the sense of Winskel et al [29]. An *input/output labeled event structure (IOLES)* over an alphabet $L = \mathcal{I}n \uplus \mathcal{O}ut$ is a 4-tuple $\mathcal{E} = (E, \leq, \#, \lambda)$ where *(i)* E is a set of events, *(ii)* $\leq \subseteq E \times E$ is a partial order (called *causality*) satisfying the property of *finite causes*, i.e. $\forall e \in E : |\{e' \in E \mid e' \leq e\}| < \infty$, *(iii)* $\# \subseteq E \times E$ is an irreflexive symmetric relation (called *conflict*) satisfying the property of *conflict heredity*, i.e. $\forall e, e', e'' \in E : e \# e' \wedge e' \leq e'' \Rightarrow e \# e''$, *(iv)* $\lambda : E \rightarrow (\mathcal{I}n \uplus \mathcal{O}ut)$ is a labeling mapping. In addition, we assume every IOLES $\mathcal{E}$ has a unique minimal event $\perp_{\mathcal{E}}$.

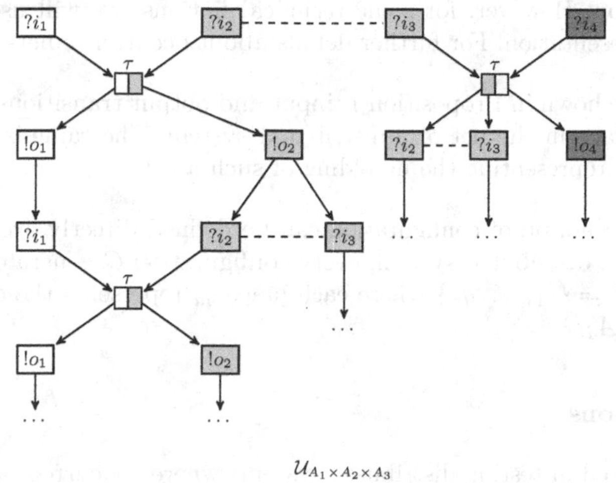

Fig. 4. Unfolding as an event structure

An event structure together with a distribution form a distributed IOLES. In such structures, we can distinguish events of different components, i.e. $E_d \triangleq \{e \in E \mid d \in D(e)\}$ for $d \in \{1, \ldots, n\}$. Such a distinction allows to project the unfolding onto a single component by just considering the events in E_d, and the restrictions of $\leq, \#$ and λ to E_d. The projection of an event structure to component $d \in \{1, \ldots, n\}$ is denoted by $\mathcal{E}_d$. We denote the class of all distributed input/output labeled event structures over L by $\mathcal{IOLES}(L)$.

Example 3. Fig. 4 shows the initial part of the unfolding of the net $\mathcal{N}_{A_1} \times \mathcal{N}_{A_2} \times \mathcal{N}_{A_3}$ given as a distributed event structure with its distribution represented by colors. As usual, we represent events by rectangles, causality by arrows and direct conflict with dashed lines. As in the case of the net in Fig. 3, communication events belong to more than one component.

The *local configuration* of an event e in $\mathcal{E}$ is defined as $[e]_{\mathcal{E}} \triangleq \{e' \in E \mid e' \leq e\}$, and its set of *causal predecessors* is $\langle e \rangle_{\mathcal{E}} \triangleq [e] \backslash \{e\}$. Two events $e, e' \in E$ are said to be concurrent (e **co** e') iff neither $e \leq e'$ nor $e' \leq e$ nor $e \# e'$ hold; $e, e' \in E$ are in *immediate conflict* ($e_1 \#^{\mu} e_2$) iff $[e_1] \times [e_2] \cap \# = \{(e_1, e_2)\}$. A *configuration* of an IOLES is a non-empty set $C \subseteq E$ that is *(i) causally closed*, i.e. $e \in C$ implies $[e] \subseteq C$, and *(ii) conflict-free*, i.e. $e \in C$ and $e \# e'$ imply $e' \notin C$. Note that we define, for technical convenience, all configurations to be non-empty; the initial configuration of $\mathcal{E}$, containing only $\bot_{\mathcal{E}}$ and denoted by $\bot_{\mathcal{E}}$, is contained in every configuration of $\mathcal{E}$. We denote the set of all configurations of $\mathcal{E}$ by $\mathcal{C}(\mathcal{E})$.

Unfoldings are usually represented by a subclass of Petri nets called occurrence nets. Occurrence nets are isomorphic to event structures [29]: one can easily forget about places of the net by adding conflict whenever two transitions compete for a resource, i.e. their presets intersect. Most of the notions presented in this paper are explained in terms of event structures since they facilitate the presentation. However, for some technical notions, we will use the occurrence net representation. For further details about occurrence nets, see [13].

Remark 1. As shown in Proposition 1, input and output transitions are enabled by only one place in the net of a distributed system. The same is true for the occurrence net representing the unfolding of such a net.

Remark 2. The notion of configuration can be defined directly over occurrence nets. Thus, in a distributed system, every configuration C generates a marking of the form $C^\bullet = \{q_1, \ldots, q_n\}$ where each place q_d represents the current state of component A_d.

3.2 Executions

We are interested in testing distributed systems where concurrent actions occur in different components of the system. That is, the specifications we consider do not impose any order of execution between concurrent events. Labeled partial orders can then be used to represent executions of such systems.

Labeled Partial Orders. A *labeled partial order* (lpo) is a tuple $lpo = (E, \leq, \lambda)$ where E is a set of events, $\leq$ is a reflexive, antisymmetric, and transitive relation, and $\lambda : E \to L$ is a labeling mapping to a fix alphabet L. We denote the class of all labeled partial orders over L by $\mathcal{LPO}(L)$. Consider $lpo_1 = (E_1, \leq_1, \lambda_1)$ and $lpo_2 = (E_2, \leq_2, \lambda_2) \in \mathcal{LPO}(L)$. A bijective function $f : E_1 \to E_2$ is an isomorphism between lpo_1 and lpo_2 iff *(i)* $\forall e, e' \in E_1 : e \leq_1 e' \Leftrightarrow f(e) \leq_2 f(e')$ and *(ii)* $\forall e \in E_1 : \lambda_1(e) = \lambda_2(f(e))$. Two labeled partial orders lpo_1 and lpo_2 are isomorphic if there exists an isomorphism between them. A *partially ordered multiset* (pomset) is an isomorphism class of lpos. We will represent such a class by one of its objects. Denote the class of all non empty pomsets over L by $\mathcal{POMSET}(L)$.

The observable behavior of a system can be captured by abstracting the internal actions from the executions of the system. A pomset ω is the τ-abstraction of another pomset μ, denoted by $abs(\mu) = \omega$, iff there exist $lpo_\mu = (E_\mu, \leq_\mu, \lambda_\mu) \in \mu$ and $lpo_\omega = (E_\omega, \leq_\omega, \lambda_\omega) \in \omega$ such that $E_\omega = \{e \in E_\mu \mid \lambda_\mu(e) \neq \tau\}$ and $\leq_\omega$ and λ_ω are the restrictions of $\leq_\mu$ and λ_μ to this set. Pomsets are observations; the observable evolution of the system is captured by the following definition:

Definition 1. *For $\mathcal{E} = (E, \leq, \#, \lambda) \in \mathcal{IOLES}(L)$, $\omega \in \mathcal{POMSET}(L)$ and $C, C' \in \mathcal{C}(\mathcal{E})$, define*[3]

$$
\begin{aligned}
C \overset{\omega}{\Longrightarrow} C' &\triangleq \exists lpo = (E_\mu, \leq_\mu, \lambda_\mu) \in \mu : E_\mu \subseteq E \backslash C, C' = C \cup E_\mu, \\
&\quad \leq \cap\, (E_\mu \times E_\mu) = \leq_\mu \text{ and } \lambda_{|E_\mu} = \lambda_\mu \text{ and } abs(\mu) = \omega \\
C \overset{\omega}{\Longrightarrow} \quad &\triangleq \exists C' : C \overset{\omega}{\Longrightarrow} C'
\end{aligned}
$$

[3] The notation $\lambda_{|E}$ denotes the restriction of λ to the set E.

We can now define the notions of traces and of configurations reachable from a given configuration by an observation. Our notion of traces is similar to that of Ulrich and König [15].

Definition 2. *For $\mathcal{E} \in \mathcal{IOLES}(L)$, $\omega \in \mathcal{POMSET}(L), C, C' \in \mathcal{C}(\mathcal{E})$, define*

$$traces(\mathcal{E}) \triangleq \{\omega \in \mathcal{POMSET}(L) \mid \bot_{\mathcal{E}} \stackrel{\omega}{\Longrightarrow}\}$$
$$C \textbf{ after } \omega \triangleq \{C' \mid C \stackrel{\omega}{\Longrightarrow} C'\}$$

Note that for deterministically labeled I/O Petri nets, the corresponding IOLES is deterministic and the set of reachable configurations is a singleton.

In a distributed system, global observation of the whole system is not available in general, i.e. the system is partially observed. This partial observation is captured by the projection of a global execution onto one of its component. As in the case of event structures, the projection of an execution only considers events belonging to a single component and restricts $\leq$ and λ to it. The projection of an execution ω onto component A_d is denoted by ω_d.

4 Testing Framework for I/O Petri Nets

4.1 Testing Hypotheses

We assume that the specification of the system under test is given as a network of deterministic automata $A_1, \ldots, A_n$ over alphabet $L = \mathcal{I}n \uplus \mathcal{O}ut$, whose global behavior is given by the distributed I/O Petri net $\mathcal{N} = \mathcal{N}_{A_1} \times \cdots \times \mathcal{N}_{A_n}$. To be able to test an implementation against such a specification, we make a set of testing assumptions, the first one being usual in testing. See [17, 18] for more details on these assumptions.

Assumption 1. *The behavior of the SUT can be modeled by a distributed I/O Petri net over alphabet L.*

In order to detect outputs depending on extra inputs, we also assume that the specification does not contain cycles of outputs actions, so that the number of expected outputs after a given trace is finite.

Assumption 2. *The net $\mathcal{N}$ has no cycle containing only output transitions.*

Third, in order to allow the observation of both the outputs produced by the system and the inputs it can accept, markings where conflicting inputs and outputs are enabled should not be reachable.[4] Such markings prevent from observing the inputs enabled in a given configuration, which we will see is one of the key points of our conformance relation.

Assumption 3. *The unfolding of the net $\mathcal{N}$ has no immediate conflict between input and output events, i.e. $\forall e_1 \in E^{\mathcal{I}n}, e_2 \in E^{\mathcal{O}ut} : \neg(e_1 \#^{\mu} e_2)$.*

[4] Gaudel et al [3] assume a similar property called *IO-exclusiveness*.

4.2 Conformance Relation

A formal testing framework relies on the definition of a conformance relation to be satisfied by the SUT and its specification. Our conformance relation is defined in terms of the inputs refused and the outputs produced by the system.

In partial order semantics, we need any set of outputs to be entirely produced by the SUT before we send a new input; this is necessary to detect outputs depending on extra inputs. For this reason we define the expected outputs from a configuration C as the pomset of outputs leading to a quiescent configuration. Such a configuration always exists, and must be finite by Assumption 2.

The notion of quiescence is inherited from nets, i.e. a configuration C is quiescent iff $C \stackrel{\omega}{\Longrightarrow}$ implies $\omega \in \mathcal{POMSET}(\mathcal{I}n)$. By abuse of notation we denote by δ the pomset reduced to only one event labeled by δ, and assume as usual that quiescence is observable by this pomset, i.e. C is quiescent iff $C \stackrel{\delta}{\Longrightarrow}$.

Definition 3. *For $\mathcal{E} \in \mathcal{IOLES}(L)$, $C \in \mathcal{C}(\mathcal{E})$, the outputs produced by C are*

$$out_{\mathcal{E}}(C) \triangleq \{!\omega \in \mathcal{POMSET}(\mathcal{O}ut) \mid C \stackrel{!\omega}{\Longrightarrow} C' \ \wedge\ C' \stackrel{\delta}{\Longrightarrow}\} \cup \{\delta \mid C \stackrel{\delta}{\Longrightarrow}\}.$$

The **ioco** theory assumes input enabledness of the implementation [1], i.e. in any state of the implementation, every input action is enabled. By constrast, we do not make this assumption, which is not always realistic [2, 3], and extend our conformance relation to consider refusals of inputs. For further discussion about the consequences of dropping the input-enabledness assumption, see [18].

Definition 4. *For $\mathcal{E} \in \mathcal{IOLES}(L)$ and $C \in \mathcal{C}(\mathcal{E})$, the possible inputs in C are*

$$poss_{\mathcal{E}}(\emptyset) \triangleq \mathcal{POMSET}(\mathcal{I}n)$$
$$poss_{\mathcal{E}}(C) \triangleq \{?\omega \in \mathcal{POMSET}(\mathcal{I}n) \mid C \stackrel{?\omega}{\Longrightarrow}\}$$

Remark 3. We intend our conformance relation to be conservative w.r.t **ioco** for systems with just one component. In order to compare the possible inputs of the specification with those of the SUT after a trace that cannot be executed in the SUT, we define $\text{poss}_{\mathcal{E}}(\emptyset)$ as the set of all possible inputs. To overcome the same problem, Gaudel et al [3] consider only traces of the specification that can also be executed in the implementation.

Consider a given marking $C^\bullet = \{q_1, \ldots, q_n\}$ and a configuration C_d of a component A_d with $d \in \{1, \ldots, n\}$ such that $C_d^\bullet = \{q_d\}$. An event which is not enabled in C_d cannot be enabled in C. The following result is central and will help proving that global conformance can be achieved by local testers.

Proposition 2. *Let C (C_d) be a configuration of a distributed system (of the system component A_d) with the corresponding cut $C^\bullet = \{q_1, \ldots, q_n\}$ ($C_d^\bullet = \{q_d\}$). Then:*

1. *if $?i \notin poss_{\mathcal{E}_d}(C_d)$, then $?i \notin poss_{\mathcal{E}}(C)$,*
2. *if $!o \notin out_{\mathcal{E}_d}(C_d)$, then for all $!\omega \in out_{\mathcal{E}}(C)$ we have $!\omega_d \neq !o$.*

Proof. If an input or output event is not enabled in configuration C_d, then by Remark 1, there is no token in condition q_d. This absence prohibits such an event to be part of an execution of any larger configuration (w.r.t set inclusion). □

Notice the distinction between possible inputs and produced outputs. Whenever the system reaches a configuration C that enables input actions in every component, i.e. $?i_d \in \text{poss}_{\mathcal{E}_d}(C_d)$ for all $d \in \{1, \ldots, n\}$, from the global point of view, not only $?i_1$ **co** ... **co** $?i_n$ is possible for the system, but also every single input, i.e. $?i_d \in \text{poss}_{\mathcal{E}}(C)$. The same is not true for produced outputs. Consider a system that enables output $!o_d$ in component A_d, leading to a quiescent configuration in A_d, i.e. $!o_d \in \text{out}_{\mathcal{E}_d}(C_d)$. If other components also enable outputs actions, $!o_d \notin \text{out}_{\mathcal{E}}(C)$ as the global configuration after $!o_d$ is not quiescent.

Our **co-ioco** conformance relation for labeled event structures can be informally described as follows. The behavior of a correct **co-ioco** implementation after some observations (obtained from the specification) should respect the following restrictions: (1) the outputs produced by the implementation should be specified; (2) if a quiescent configuration is reached, this should also be the case in the specification; (3) any time an input is possible in the specification, this should also be the case in the implementation. These restrictions are formalized by the following conformance relation.

Definition 5 ([17]). *Let $\mathcal{S}$ and $\mathcal{I}$ be respectively the specification and implementation of a distributed system; then*

$$\begin{aligned} \mathcal{I}\ \textbf{co-ioco}\ \mathcal{S} \Leftrightarrow\ & \forall \omega \in \textit{traces}(\mathcal{S}) : \\ & \textit{poss}_{\mathcal{S}}(\bot\ \textbf{after}\ \omega) \subseteq \textit{poss}_{\mathcal{I}}(\bot\ \textbf{after}\ \omega) \\ & \textit{out}_{\mathcal{I}}(\bot\ \textbf{after}\ \omega)\ \subseteq\ \textit{out}_{\mathcal{S}}(\bot\ \textbf{after}\ \omega) \end{aligned}$$

Non conformance of the implementation is given by the absence of a given input or an unspecified output or quiescence in a configuration of the implementation. In a distributed system, a configuration defines the local state of each components as shown in Remark 2. Thus, non conformance of a distributed system is due to one of the following reasons:

(NC1) An input which is possible in a state of a component in the specification is not possible in its corresponding state in the implementation,

(NC2) A state of a component in the implementation produces an output or is quiescent while the corresponding state of the specification does not,

(NC3) The input (resp. output) actions that the configuration is ready to accept (resp. produce) are the same in both implementation and specification, but they do not form the same partial order, i.e. concurrency is added or removed.

The next section shows how we can detect these situations in a distributed testing environment.

5 Global Conformance by Distributed Testers

The **co-ioco** framework assumes a global view of the distributed system. In practice this assumption may not be satisfied and we can only observe the system partially, i.e. only the behavior of a local component in its PCO is observed. In a distributed testing environment, we place a local tester at each PCO. In a pure distributed testing setting, these testers cannot communicate with each other during testing, and there is no global clock. We propose here a method allowing to decide global conformance by the conformance of every single component.

5.1 Local Testing

The last section described the three possible reasons for which a system may not conform to its specification. Non-conformance resulting from (NC1) and (NC2) can be locally tested under **co-ioco** by transforming each component into a net.

Theorem 1. *If $\mathcal{S}$ and $\mathcal{I}$ are, respectively, the specification and implementation of a distributed system, then $\mathcal{I}$ **co-ioco** $\mathcal{S}$ implies that for every $d \in \{1, \ldots, n\}$, $\mathcal{I}_d$ **co-ioco** $\mathcal{S}_d$.*

Proof. Assume there exists $d \in \{1, \ldots, n\}$ for which $\neg(\mathcal{I}_d$ **co-ioco** $\mathcal{S}_d)$, then there exists $\sigma \in \text{traces}(\mathcal{S}_d)$ such that one of the following holds:

- There exists $?i \in \text{poss}_{\mathcal{S}_d}(\bot$ **after** $\sigma)$, but $?i \notin \text{poss}_{\mathcal{I}_d}(\bot$ **after** $\sigma)$. Consider the global trace $\omega = \langle ?i \rangle_{\mathcal{S}}$ which enables $?i$ in $\mathcal{S}$, i.e. $?i \in \text{poss}_{\mathcal{S}}(\bot$ **after** $\omega)$. As $?i$ is not possible in $\mathcal{I}_d$, by Proposition 2 we have $?i \notin \text{poss}_{\mathcal{I}}(\bot$ **after** $\omega)$, and therefore $\neg(\mathcal{I}$ **co-ioco** $\mathcal{S})$.
- There exists $!o \in \text{out}_{\mathcal{I}_d}(\bot$ **after** $\sigma)$ such that $!o \notin \text{out}_{\mathcal{S}_d}(\bot$ **after** $\sigma)$. Consider the global trace $\omega = \langle !o \rangle_{\mathcal{I}}$ which enables $!o$ in $\mathcal{I}$, i.e. there exists $!\omega \in \text{out}_{\mathcal{I}}(\bot$ **after** $\omega)$ such that $!\omega_d = !o$. As $!o$ is not enabled in $\mathcal{S}_d$, by Proposition 2 we know that $!o$ cannot be enabled after ω in $\mathcal{S}$. Therefore, $!\omega \notin \text{out}_{\mathcal{S}}(\bot$ **after** $\omega)$ and $\neg(\mathcal{I}$ **co-ioco** $\mathcal{S})$.
- $\delta \in \text{out}_{\mathcal{I}_d}(\bot$ **after** $\sigma)$, while $\delta \notin \text{out}_{\mathcal{S}_d}(\bot$ **after** $\sigma)$. Let C_d be the configuration reached by component A_d of the implementation after σ and denote $C_d^\bullet = \{q_d\}$. Consider ω such that it leads the implementation to a quiescent configuration C with $q_d \in C^\bullet$ (such configuration always exists by Assumption 2); we have $\delta \in \text{out}_{\mathcal{I}}(\bot$ **after** $\omega)$. As the reached configuration in $\mathcal{S}_d$ is not quiescent, it enables some output and $\delta \notin \text{out}_{\mathcal{S}}(\bot$ **after** $\omega)$, therefore $\neg(\mathcal{I}$ **co-ioco** $\mathcal{S})$. □

The simplest kind of conformance relations that we can obtain in a distributed architecture are those that only consider the observation of the system executions at each PCO without any further information. Such kind of relations include **p-dioco** [8], where the local behavior need to be consistent only with *some* global behavior. Stronger relations can be obtained if consistency between local observations is considered, as in the case of **dioco** [8] where local behaviors must be projections of the *same* global behavior. However, even this kind of relations do not test the dependencies between actions occurring on different components. Relations that assume global observation are usually stronger, as it is shown by the implication **ioco** $\Rightarrow$ **p-dioco** or by the theorem above.

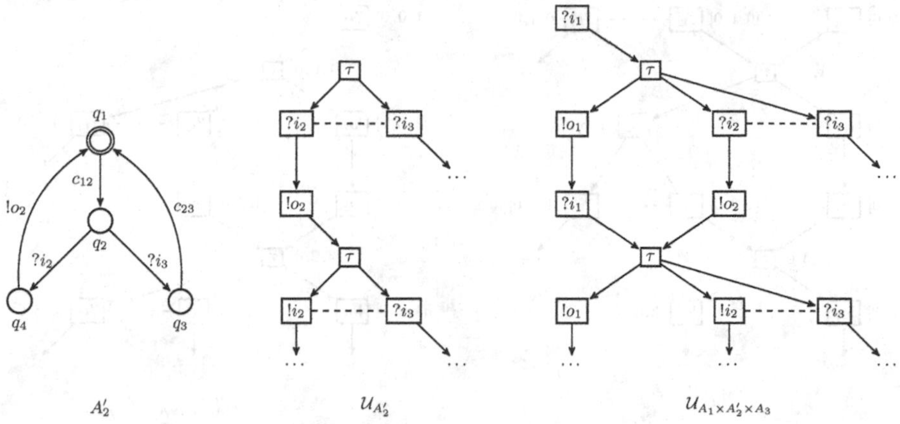

Fig. 5. Non conformant implementation

Example 4. Consider a component A_2' where inputs $?i_2$ and $?i_3$ are not possible before this component synchronizes with A_1 and therefore they cannot occur before $?i_1$ occurs in A_1. Let $\mathcal{I} = A_1 \times A_2' \times A_3$ and $\mathcal{S} = A_1 \times A_2 \times A_3$. Component A_2', its unfolding and the unfolding of $\mathcal{I}$ are shown in Fig. 5. Component A_2' conforms to A_2 as every possible input and produced outputs are implemented (only the order of synchronization events change, but those are not observable). Therefore, $\mathcal{I}_2$ **co-ioco** $\mathcal{S}_2$ and clearly, as **co-ioco** is reflexive, $\mathcal{I}_1$ **co-ioco** $\mathcal{S}_1$ and $\mathcal{I}_3$ **co-ioco** $\mathcal{S}_3$. In addition, the local behaviors $?i_1!o_1?i_1!o_1$ and $?i_2!o_2?i_2$ from components A_1 and A_2', respectively, are projections of the same global behavior in $\mathcal{S}$, even if the causalities between components are not preserved. The **co-ioco** relation preserves causalities between actions in different components: the possible input of the specification $?i_1$ **co** $?i_2 \in \text{poss}_{\mathcal{S}}(\bot)$ is not possible in the implementation, the actions are the same, but there is extra causality between them, i.e. $?i_1?i_2 \in \text{poss}_{\mathcal{I}}(\bot)$. We can conclude that $\neg(\mathcal{I}$ **co-ioco** $\mathcal{S})$ even if every component of the implementation conforms to the specification w.r.t **co-ioco** and local behaviors are projections of the same global behavior.

5.2 Adding Time Stamps

The example above shows that global conformance cannot always be achieved by local testers that do not communicate between themselves. This is exactly what happens in situation (NC3). However, as components of the implementation need to synchronize, we propose to use such synchronization to interchange some information that allows the testers to recompute the partial order between actions in different components using *vector clocks* [24, 25].

We assume each component A_d has a local clock that counts the number of interactions between itself and the environment, together with a local table of the

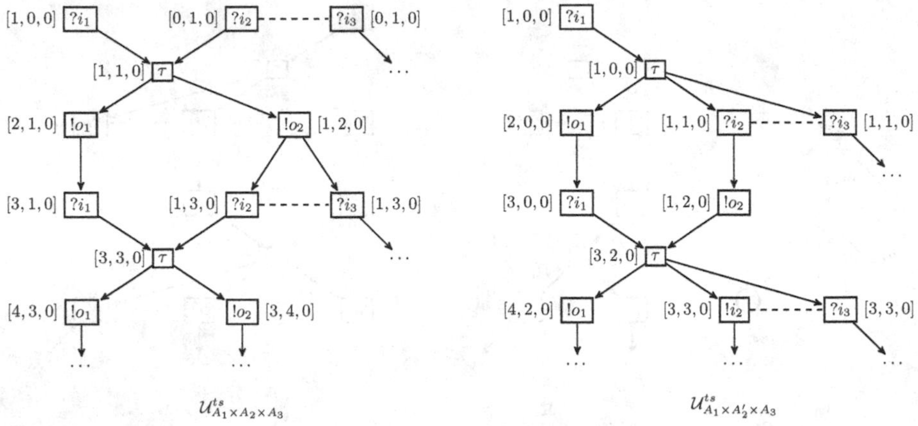

Fig. 6. Part of the time stamped unfolding $\mathcal{S}$ and $\mathcal{I}'$

form $[t_1^d, \ldots, t_n^d]$ with information about the clocks of every component (information about other components may not be updated). Each time two components communicate via synchronization, their local tables are updated.

We add the information about the tables to the model, i.e. events of the unfolding are tuples representing both the actions and the current values of the table. The unfolding structure allows to compute such tables very efficiently: when event e occurs in A_d, the value of the clock j in the table of A_d is equal to the number of input and outputs events from component j in the past of e, i.e. $t_j^d = \mid [e] \cap (E_j^{\mathcal{I}n} \uplus E_j^{\mathcal{O}ut}) \mid$. The unfolding algorithm [13] can be modified to consider time stamps as it is shown in Algorithm 1.[5] The behavior of system $\mathcal{E}$ where time stamps are considered is denoted by $\mathcal{E}^{ts}$.

Example 5. Consider the time stamped unfolding $\mathcal{U}^{ts}_{A_1 \times A_2 \times A_3}$ on Fig. 6. The first occurrence of action $!o_1$ is stamped by $[2,1,0]$, meaning that it is the second interaction with the environment in component A_1, and at least there was one interaction between the environment and component A_2 before the occurrence of $!o_1$. The information of component A_2 is propagated to the table of component A_1 after their synchronize over the first occurrence of τ.

The global trace of an execution of the system can be reconstructed from the local traces of this execution observed in PCOs using the information provided by time stamps.

Example 6. Consider Fig. 7 and the time stamped local traces σ_1 and σ_2 of the first and second components. From event $(!o_2, 1, 2, 0)$, we know that at least one event from the first component precedes $!o_2$, and as $?i_1$ is the first action in this component, we can add the causality $(?i_1, 1, 0, 0) \leq (!o_1, 1, 2, 0)$ as shown in the partial order ω.

[5] $PE(B)$ are the events that can be added to the unfolding based on the current prefix, see [13] for more details.

Algorithm 1. Time stamped unfolding algorithm

Require: A I/O Petri net $\mathcal{N} = (T, P, F, M_0, \lambda)$ where $M_0 = \{s_1, \ldots, s_n\}$ and a distribution $D : T \cup P \rightarrow \{1, \ldots, n\}$.
Ensure: The time stamped unfolding of $\mathcal{N}$

$B := (s_0, \emptyset), \ldots, (s_k, \emptyset)$
$E = \emptyset$
$pe := PE(B)$
while $pe \neq \emptyset$ **do**
  choose an event $e = (t, {}^{\bullet}t)$ in pe
  for $d \in \{1, \ldots, n\}$ **do**
    $t_d(e) := |\{(t', {}^{\bullet}t') \in E \mid D(t') = d \wedge \lambda(t) \neq \tau\}| + 1$
  end for
  append to E the event $e \times t_1(e) \times \cdots \times t_n(e)$
  for every place s in $t^{\bullet}$ add to B a condition (s, e)
  $pe := PE(B)$;
end while
return (B, E)

Given two time stamped LPOs $\omega_i = (E_i, \leq_i, \lambda_1)$ and $\omega_j = (E_j, \leq_j, \lambda_2)$, their joint causality is given by the LPO $\omega_i + \omega_j = (E_i \uplus E_j, \leq_{ij}, \lambda_1 \uplus \lambda_2)$ where for each pair of events $e_1 = (a, t_1^i, \ldots, t_n^i) \in E_i$ and $e_2 \in E_i \uplus E_j$, we have

$$e_2 \leq_{ij} e_1 \Leftrightarrow e_2 \leq_i e_1 \vee |[e_2]_j| \leq t_j^i$$

In other words, e_2 globally precedes e_1 either if they belong to the same component and e_2 locally precedes e_1 or if e_2 is the kth event in component j and e_1 is preceded by at least k events in component j according to time stamps.

When communication between components is asynchronous, a configuration C is called *consistent* if for every sending message in C, its corresponding receive message is also in C. Mattern [24] shows that consistent configurations have unambiguous time stamps; hence global causality can be reconstructed from local observations in an unique way. Under synchronous communication, the send and receive actions are represented by the same event, and therefore every configuration is consistent.

Proposition 3. *When the communication between components is synchronous, the partial order obtained by + is unique.*

Proof. Since every configuration is consistent, the result is immediate [24]. □

Non conformance coming from (NC3) can be detected by testing the time stamped system in a distributed way.

Theorem 2. *If $\mathcal{S}$ and $\mathcal{I}$ are, respectively, the specification and implementation of a distributed system, then $\forall d \in \{1, \ldots, n\}$: $\mathcal{I}_d^{ts}$* **co-ioco** *$\mathcal{S}_d^{ts}$ implies $\mathcal{I}$* ***co-ioco*** *$\mathcal{S}$.*

Proof. Assume $\mathcal{I}_d^{ts}$ **co-ioco** $\mathcal{S}_d^{ts}$ for every $d \in \{1, \ldots n\}$. Let $\omega \in$ traces$(\mathcal{S})$ and consider the following situations:

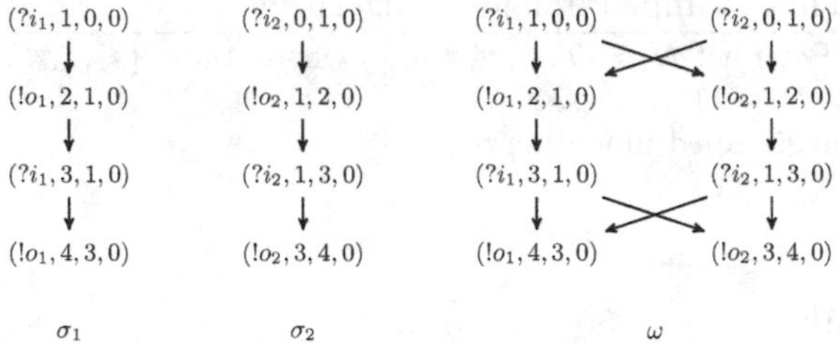

Fig. 7. From local traces to partial orders using time stamps

- If $?\omega \in \text{poss}_{\mathcal{S}}(\perp \textbf{ after } \omega)$, then for every d there exists a time stamped input $(?i_d, t_1^d, \ldots, t_n^d) \in \text{poss}_{\mathcal{S}_d^{ts}}(\perp \textbf{ after } \omega_d)$ such that $?\omega_d = ?i_d$ and $?\omega = ?i_1 + \cdots + ?i_n$. As for all d, we have $\mathcal{I}_d^{ts}$ **co-ioco** $\mathcal{S}_d^{ts}$, then $(?i_d, t_1^d, \ldots, t_n^d) \in \text{poss}_{\mathcal{I}_d^{ts}}(\perp \textbf{ after } \omega_d)$. By Proposition 3, $?\omega \in \text{poss}_{\mathcal{I}}(\perp \textbf{ after } \omega)$.
- If $!\omega \in \text{out}_{\mathcal{I}}(\perp \textbf{ after } \omega)$, then for every d there exists a time stamped output $(!o_d, t_1^d, \ldots, t_n^d) \in \text{out}_{\mathcal{I}_d^{ts}}(\perp \textbf{ after } \omega_d)$ such that $!\omega_d = !o_d$ and $!\omega = !o_1 + \cdots + !o_n$. As every component of the implementation conforms to its specification, we have $(!o_d, t_1^d, \ldots, t_n^d) \in \text{out}_{\mathcal{S}_d^{ts}}(\perp \textbf{ after } \omega_d)$. By Proposition 3, we have $!\omega \in \text{out}_{\mathcal{S}}(\perp \textbf{ after } \omega)$.
- If $\delta \in \text{out}_{\mathcal{I}}(\perp \textbf{ after } \omega)$, let C be the configuration reached by the implementation after ω and denote $C^\bullet = \{q_1, \ldots, q_n\}$. Configuration C is quiescent, so is each configuration C_d such that $C_d^\bullet = \{q_d\}$ and $\delta \in \text{out}_{\mathcal{I}_d^{ts}}(\perp \textbf{ after } \omega_d)$. Since $\mathcal{I}_d^{ts}$ **co-ioco** $\mathcal{S}_d^{ts}$ for each d, we have $\delta \in \text{out}_{\mathcal{S}_s^{ts}}(\perp \textbf{ after } \omega_d)$. This implies that the local configurations of the specification do not enable any output; by Remark 1, there is no output enabled in the global configuration and $\delta \in \text{out}_{\mathcal{S}}(\perp \textbf{ after } \omega)$.

These three cases allow us to conclude that $\mathcal{I}$ **co-ioco** $\mathcal{S}$. □

From a global test case to local test cases. We have shown that global conformance can be achieved by distributed testers. However testers need to consider time stamp information which cannot be computed locally. The test case generation algorithm that we proposed for concurrent systems [17] can easily be adapted to consider the time stamped unfolding presented in this article. The global test case obtained is a distributed IOLES and can therefore be projected to each component to obtain a distributed test case, i.e. a set of local test cases.

Example 7. Fig. 8 shows the initial part of global time stamped test case (restricted to components A_1 and A_2) and its projections $\mathcal{T}_1, \mathcal{T}_2$ over those components. These local test cases are supposed to be executed in parallel.

Consider the incorrect implementation $\mathcal{U}^{ts}_{A_1 \times A_2' \times A_3}$. If $?i_2$ is sent to A_2 before $?i_1$ in A_1, the implementation refuses the input and we detect the non conformance. However, as there is no interaction between $\mathcal{T}_1$ and $\mathcal{T}_2$, it can be the case that $?i_2$ is always sent after $?i_1$ and this refusal is never detected. If this is the case, after sending $?i_1$, the implementation produces $!o_1$ with time stamp

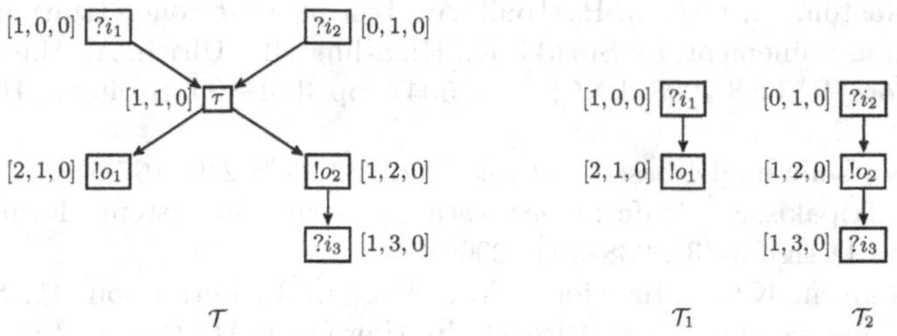

Fig. 8. Initial part of a global test case and its projections

$[2, 0, 0]$ which is not the time stamp expected by $\mathcal{T}_1$. Thus, non conformance is also detected.

We gave a sound and exhaustive test set for **co-ioco** [17]. The set of distributed test cases obtained by projecting global test cases of this complete test set naturally is also complete for **co-ioco** by Theorem 2. Therefore, it allows to decide global conformance w.r.t. **co-ioco** by distributed testing, with independent local testers.

6 Conclusion

We presented a distributed testing framework for concurrent systems specified as networks of automata or, equivalently, as 1-safe Petri nets. The **co-ioco** conformance relation introduced in our previous work is put into a distributed testing architecture, where nets are distributed, observation of the system is partial, and global configurations are represented by the collection of local states of components. When the implementation is equipped with local clocks, a global test case can be constructed adapting the test generation algorithm for **co-ioco** for handling time stamps. Global test cases can be projected into local test cases, which allow to achieve global conformance via conformance of local components.

This approach considers synchronous communication. Future work includes the extension to asynchronous communication not only between components, but also between the testers and the SUT.

References

1. Tretmans, J.: Test generation with inputs, outputs and repetitive quiescence. Software - Concepts and Tools 17(3), 103–120 (1996)
2. Heerink, L., Tretmans, J.: Refusal testing for classes of transition systems with inputs and outputs. In: FORTE. IFIP Conference Proceedings, vol. 107, pp. 23–38. Chapman & Hall (1997)
3. Lestiennes, G., Gaudel, M.C.: Test de systèmes réactifs non réceptifs. Journal Européen des Systèmes Automatisés 39(1-2-3), 255–270 (2005)

4. Faivre, A., Gaston, C., Le Gall, P., Touil, A.: Test purpose concretization through symbolic action refinement. In: Suzuki, K., Higashino, T., Ulrich, A., Hasegawa, T. (eds.) TestCom/FATES 2008. LNCS, vol. 5047, pp. 184–199. Springer, Heidelberg (2008)
5. Jéron, T.: Symbolic model-based test selection. ENTCS 240, 167–184 (2009)
6. Krichen, M., Tripakis, S.: Conformance testing for real-time systems. Formal Methods in System Design 34(3), 238–304 (2009)
7. Hessel, A., Larsen, K.G., Mikucionis, M., Nielsen, B., Pettersson, P., Skou, A.: Testing real-time systems using UPPAAL. In: Hierons, R.M., Bowen, J.P., Harman, M. (eds.) FORTEST. LNCS, vol. 4949, pp. 77–117. Springer, Heidelberg (2008)
8. Hierons, R.M., Merayo, M.G., Núñez, M.: Implementation relations for the distributed test architecture. In: Suzuki, K., Higashino, T., Ulrich, A., Hasegawa, T. (eds.) TestCom/FATES 2008. LNCS, vol. 5047, pp. 200–215. Springer, Heidelberg (2008)
9. Hennessy, M.: Algebraic Theory of Processes. MIT Press (1988)
10. Peleska, J., Siegel, M.: From testing theory to test driver implementation. In: Gaudel, M.-C., Wing, J.M. (eds.) FME 1996. LNCS, vol. 1051, pp. 538–556. Springer, Heidelberg (1996)
11. Schneider, S.: Concurrent and Real Time Systems: The CSP Approach, 1st edn. John Wiley & Sons, Inc., New York (1999)
12. McMillan, K.L.: A technique of state space search based on unfolding. Formal Methods in System Design 6(1), 45–65 (1995)
13. Esparza, J., Römer, S., Vogler, W.: An improvement of McMillan's unfolding algorithm. In: Margaria, T., Steffen, B. (eds.) TACAS 1996. LNCS, vol. 1055, pp. 87–106. Springer, Heidelberg (1996)
14. Jard, C.: Synthesis of distributed testers from true-concurrency models of reactive systems. Information & Software Technology 45(12), 805–814 (2003)
15. Ulrich, A., König, H.: Specification-based testing of concurrent systems. In: FORTE. IFIP Conf. Proc., vol. 107, pp. 7–22. Chapman & Hall (1998)
16. Ponce de León, H., Haar, S., Longuet, D.: Conformance relations for labeled event structures. In: Brucker, A.D., Julliand, J. (eds.) TAP 2012. LNCS, vol. 7305, pp. 83–98. Springer, Heidelberg (2012)
17. Ponce de León, H., Haar, S., Longuet, D.: Unfolding-based test selection for concurrent conformance. In: Yenigün, H., Yilmaz, C., Ulrich, A. (eds.) ICTSS 2013. LNCS, vol. 8254, pp. 98–113. Springer, Heidelberg (2013)
18. Ponce de León, H., Haar, S., Longuet, D.: Model based testing for concurrent systems with labeled event structures (2012), http://hal.inria.fr/hal-00796006
19. Bhateja, P., Gastin, P., Mukund, M., Kumar, K.N.: Local testing of message sequence charts is difficult. In: Csuhaj-Varjú, E., Ésik, Z. (eds.) FCT 2007. LNCS, vol. 4639, pp. 76–87. Springer, Heidelberg (2007)
20. Longuet, D.: Global and local testing from message sequence charts. In: SAC. Software Verification and Testing track, pp. 1332–1338. ACM (2012)
21. Jard, C., Jéron, T., Kahlouche, H., Viho, C.: Towards automatic distribution of testers for distributed conformance testing. In: FORTE. IFIP Conference Proceedings, vol. 135, pp. 353–368. Kluwer (1998)
22. Bhateja, P., Mukund, M.: Tagging make local testing of message-passing systems feasible. In: SEFM, pp. 171–180. IEEE Computer Society (2008)
23. Hierons, R.M., Merayo, M.G., Núñez, M.: Using time to add order to distributed testing. In: Giannakopoulou, D., Méry, D. (eds.) FM 2012. LNCS, vol. 7436, pp. 232–246. Springer, Heidelberg (2012)

24. Mattern, F.: Virtual time and global states of distributed systems. In: Parallel and Distributed Algorithms, pp. 215–226. North-Holland (1989)
25. Fidge, C.J.: Timestamps in Message-Passing Systems that Preserve the Partial Ordering. In: 11th Australian Computer Science Conference, University of Queensland, Australia, pp. 55–66 (1988)
26. Segala, R.: Quiescence, fairness, testing, and the notion of implementation. Information and Computation 138(2), 194–210 (1997)
27. Winskel, G.: Petri nets, morphisms and compositionality. In: Rozenberg, G. (ed.) Applications and Theory in Petri Nets. LNCS, vol. 222, pp. 453–477. Springer, Heidelbeg (1985)
28. van Glabbeek, R., Goltz, U., Schicke-Uffmann, J.-W.: On distributability of Petri nets - (extended abstract). In: Birkedal, L. (ed.) FOSSACS 2012. LNCS, vol. 7213, pp. 331–345. Springer, Heidelberg (2012)
29. Nielsen, M., Plotkin, G.D., Winskel, G.: Petri nets, event structures and domains, part I. Theoretical Computer Science 13, 85–108 (1981)

UTP Designs for Binary Multirelations

Pedro Ribeiro and Ana Cavalcanti

Department of Computer Science, University of York, UK
{pfr,alcc}@cs.york.ac.uk

Abstract. The total correctness of sequential computations can be established through different isomorphic models, such as monotonic predicate transformers and binary multirelations, where both angelic and demonic nondeterminism are captured. Assertional models can also be used to characterise process algebras: in Hoare and He's Unifying Theories of Programming, CSP processes can be specified as the range of a healthiness condition over designs, which are pre and postcondition pairs. In this context, we have previously developed a theory of angelic designs that is a stepping stone for the natural extension of the concept of angelic nondeterminism to the theory of CSP. In this paper we present an extended model of upward-closed binary multirelations that is isomorphic to angelic designs. This is a richer model than that of standard binary multirelations, in that we admit preconditions that rely on later or final observations as required for a treatment of processes.

Keywords: semantics, refinement, binary multirelations, UTP.

1 Introduction

In the context of sequential programs, their total correctness can be characterised through well-established models such as monotonic predicate transformers [1]. This model forms a complete lattice, where demonic choice corresponds to the greatest lower bound, while angelic choice is the least upper bound.

In [2] Rewitzky introduces the concept of binary multirelations, where the initial state of a computation is related to a set of final states. Amongst the different models studied [2,3], the theory of upward-closed binary multirelations is the most important as it has a lattice-theoretic structure. In this case, the set of final states corresponds to choices available to the angel, while those over the value of the set itself correspond to demonic choices.

The UTP of Hoare and He [4] is a predicative theory of relations suitable for the combination of refinement languages catering for different programming paradigms. In this context, the total correctness of sequential computations is characterised through the theory of designs, which are pre and postcondition pairs. Since the concept of angelic nondeterminism cannot be captured directly, binary multirelational encodings have been proposed [5,6,7].

While sequential computations can be given semantics using a relation between their initial and final state, reactive systems require a richer model that accounts

G. Ciobanu and D. Méry (Eds.): ICTAC 2014, LNCS 8687, pp. 388–405, 2014.

for the interactions with their environment. This is achieved in the UTP through the theory of reactive processes [4,8]. The combination of this theory and that of designs enables the specification of CSP processes in terms of designs that characterise the pre and postcondition of processes. We observe, however, that the theory of designs encompasses programs whose preconditions may also depend on the final or later observations of a computation. As a consequence, the general theory of designs allows these observations to be ascertained irrespective of termination. For instance, the precondition of the CSP process $a \to \mathit{Chaos}$ requires that no after observation of the trace of events is prefixed by event a.

In order to extend the concept of angelic nondeterminism to CSP, we have previously developed a theory of angelic designs. The most challenging aspect tackled pertains to the treatment of sequential composition, where it departs from the norm for UTP theories: instead of sequential composition being relational composition we have a different treatment [5] inspired on the definition of sequential composition for binary multirelations.

The main contribution of this work is a new theory of binary multirelations that caters for sets of final states where termination may not be necessarily enforced. Thus is in line with the general notion of UTP designs, with the added benefit that binary multirelations can handle both angelic and demonic nondeterminism. Our contribution is not only an extended model of upward-closed binary multirelations isomorphic to angelic designs, but also a solid basis for understanding the treatment of sequential composition in such models. To facilitate this analysis here, we also present links, Galois connections and isomorphisms, between the theories of interest. The links validate our new theory, and identify its potential role in a treatment of CSP processes.

Our long term aim is the development of a model of CSP where the angelic choice operator is a counterpart to that of the refinement calculus, that is, it avoids divergence [9]. For example, if we consider the angelic choice $a \to \mathit{Chaos} \sqcup a \to \mathit{Skip}$, then this would ideally be resolved in favour of $a \to \mathit{Skip}$. An application of this notion is found, for instance, in the context of a modelling approach for the verification of implementations of control systems [10].

The structure of this paper is as follows. In section 2 we introduce the UTP and the theories of interest. In section 3 the main contribution of this paper is discussed. In section 4 we establish the relationship between the new model and the theory of angelic designs. Finally in section 5 we present our conclusions.

2 Preliminaries

As mentioned before, the UTP is an alphabetized, predicative theory of relations suitable for modelling different programming paradigms [4]. UTP theories are characterised by three components: an alphabet, a set of healthiness conditions and a set of operators. The alphabet $\alpha(P)$ of a relation P can be split into $in\alpha(P)$, which contains undashed variables corresponding to the initial observations of a computation, and $out\alpha(P)$ containing dashed counterparts for the after or final observations. Refinement is defined as universal reverse implication.

2.1 Designs

In the UTP theory of designs [4,11] the alphabet consists of program variables and two auxiliary Boolean variables *ok* and *ok′* that record when a program starts, and when it terminates. A design is specified as follows.

Definition 1 (Design). $(P \vdash Q) \mathrel{\widehat=} (ok \wedge P) \Rightarrow (Q \wedge ok')$

P and Q are relations that together form a pre and postcondition pair, such that if the program is started, that is *ok* is *true*, and P is satisfied, then it establishes Q and terminates successfuly, with ok' being *true*.

A design can be expressed in this form if, and only if, it is a fixed point of the healthiness conditions **H1** and **H2** [4], whose functional composition is reproduced below, where $P^o = [o/ok']$, that is o is substituted for ok', with t corresponding to *true* and f to *false*.

Theorem 1. $\mathbf{H1} \circ \mathbf{H2}(P) = (\neg\, P^f \vdash P^t)$

The healthiness condition **H1** states that any observations can be made before a program is started, while **H2** requires that if a program may not terminate, then it must also be possible for it to terminate. In other words, it is not possible to require nontermination explicitly. The healthiness conditions of the theory are monotonic and idempotent, and so the model is a complete lattice [4].

When designs are used to model sequential computations, the precondition $\neg\, P^f$ of a design P is in fact not a relation, but rather a condition that only refers to undashed variables. Designs that observe this property are fixed points of the healthiness condition **H3**, whose definition is reproduced below [4].

Definition 2. $\mathbf{H3}(P) = P \,;\, \mathit{I\!I}_{\mathcal{D}}$

This is a healthiness condition that requires the skip of the theory, defined below as $\mathit{I\!I}_{\mathcal{D}}$ [4,11], to be a right-unit for sequential composition.

Definition 3. $\mathit{I\!I}_{\mathcal{D}} \mathrel{\widehat=} (true \vdash x' = x)$

The design $\mathit{I\!I}_{\mathcal{D}}$ once started keeps the value of every program variable x unchanged and terminates successfuly. In order to discuss the consequences of designs that do not satisfy **H3**, we consider the following example.

Example 1. $(x' \neq 2 \vdash x' = 1) = ok \Rightarrow ((x' = 1 \wedge ok') \vee x' = 2)$

This is a design that once started can either establish the final value of the program variable x as 1 and terminate, or alternatively can establish the final value of x as 2 but then termination is not necessarily required. This is unexpected behaviour in the context of a theory for sequential programs. However, in the theory of CSP [4,8], processes are expressed as the image of non-**H3** designs through the function **R** that characterises reactive programs.

2.2 Binary Multirelations

As mentioned before, the theory of binary multirelations as introduced by Rewitzky [2] is a theory of relations between an initial state and a set of final states.

We define these relations through the following type *BM*, where *State* is a type of records with a component for each program variable.

Definition 4 (Binary Multirelation). $BM \mathrel{\widehat=} State \leftrightarrow \mathbb{P}\, State$

For instance, the program that assigns the value 1 to the only program variable x when started from any initial state is defined as follows.

Example 2. $x :=_{BM} 1 = \{s : State, ss : \mathbb{P}\, State \mid (x \mapsto 1) \in ss\}$

Following [5], $(x \mapsto 1)$ denotes a record whose only component is x and its respective value is 1. For conciseness, in the definitions that follow, the types of s and ss may be omitted but are exactly the same as in example 2.

The target set of a binary multirelation can be interpreted as either encoding angelic or demonic choices [2,5]. We choose to present a model where the set of final states encodes angelic choices. This choice is justified in [12,5] as maintaining the refinement order of the UTP theories.

Demonic choices are encoded by the different ways in which the set of final states can be chosen. For example, the program that angelically assigns the value 1 or 2 to the only program variable x is specified by the following relation, where $\sqcup_{BM}$ is the angelic choice operator for binary multirelations.

Example 3. $x :=_{BM} 1 \sqcup_{BM} x :=_{BM} 2 = \{s, ss \mid (x \mapsto 1) \in ss \wedge (x \mapsto 2) \in ss\}$

This definition allows any superset of the set $\{(x \mapsto 1), (x \mapsto 2)\}$ to be chosen. The choice of values 1 and 2 for the program variable x are available in every set of final states ss, and so are available in every demonic choice.

The subset of *BM* of interest is that of upward-closed multirelations [2,3]. The following predicate [5] characterises this subset for a relation B.

Definition 5. $\mathbf{BMH} \mathrel{\widehat=} \forall s, ss_0, ss_1 \bullet ((s, ss_0) \in B \wedge ss_0 \subseteq ss_1) \Rightarrow (s, ss_1) \in B$

If a particular initial state s is related to a set of final states ss_0, then it is also related to any superset of ss_0. This means that if it is possible to terminate in some final state that is in ss_0, then the addition of any other final states to that same set does not change the final states available for angelic choice, which correspond to those in the distributed intersection of all sets of final states available for demonic choice. Alternatively, the set of healthy binary multirelations can be characterised by the fixed points of the following function.

Definition 6. $\mathbf{bmh_{up}}(B) \mathrel{\widehat=} \{s, ss \mid \exists\, ss_0 : \mathbb{P}\, State \bullet (s, ss_0) \in B \wedge ss_0 \subseteq ss\}$

This equivalence is established by the following lemma 1.

Lemma 1. $\mathbf{BMH} \Leftrightarrow \mathbf{bmh_{up}}(B) = B$

Proof of these and other results can be found in [7].

The refinement order for healthy binary multirelations B_0 and B_1 is given by subset inclusion [5], as reproduced below.

Definition 7 $(\sqsubseteq_{BM})$**.** $B_0 \sqsubseteq_{BM} B_1 \mathrel{\widehat=} B_0 \supseteq B_1$

This partial order over *BM* forms a complete lattice. It allows an increase in the degree of angelic nondeterminism and a decrease in demonic nondeterminism, with angelic choice as set intersection and demonic choice as set union.

For binary multirelations that are upward-closed, that is, which satisfy **BMH**, the definition of sequential composition is as follows.

Lemma 2. *Provided B_0 satisfies* **BMH**.

$$B_0 \mathbin{;_{BM}} B_1 = \{s_0 : State, ss : \mathbb{P}\, State \mid (s_0, \{s_1 : State \mid (s_1, ss) \in B_1\}) \in B_0\}$$

It considers every initial state s_0 in B_0 and set of final states ss of B_1, such that ss is a set that could be reached through some initial state s_1 of B_1 that is available to B_0 as a set of final states.

2.3 Angelic Designs

As discussed earlier, both angelic and demonic nondeterminism can be modelled in the UTP through a suitable encoding of multirelations. The first of these has been proposed in [5], where the alphabet consists of input program variables and a sole output variable ac', a set of final states. Those states in ac' correspond to angelic choices, while the choice over the value of ac' itself corresponds to demonic choices. Upward closure is enforced by the following healthiness condition, where v and v' refer to every variable other than ac and ac'.

Definition 8. $\mathbf{PBMH}(P) \mathrel{\widehat{=}} P \mathbin{;} ac \subseteq ac' \wedge v' = v$

PBMH requires that if it is possible for P to establish a set of final states ac', then any superset can also be established. (In the theory of [5], there are no other variables v', while here we consider a more general theory.)

Following the approach in [5] we have previously developed a theory of angelic designs [6]. The alphabet includes the variables ok and ok' from the theory of designs, a single input state s and a set of final states ac'. The healthiness conditions are **H1** and **H2** and **A**, whose definition is the functional composition of **A0** and **A1** as reproduced below [6].

Definition 9.

$$\mathbf{A0}(P) \mathrel{\widehat{=}} P \wedge ((ok \wedge \neg\, P^f) \Rightarrow (ok' \Rightarrow ac' \neq \emptyset))$$
$$\mathbf{A1}(P) \mathrel{\widehat{=}} (\neg\, \mathbf{PBMH}(P^f) \vdash \mathbf{PBMH}(P^t))$$
$$\mathbf{A}(P) \mathrel{\widehat{=}} \mathbf{A0} \circ \mathbf{A1}(P)$$

The healthiness condition **A0** requires that when a design terminates successfully, then there must be some final state in ac' available for angelic choice. **A1** requires that the final set of states in both the postcondition and the negation of the precondition are upward closed. We observe that **A1** can also be expressed as the application of **PBMH** to the whole of the design P.

Since all of the healthiness conditions of the theory commute, and they are all idempotent and monotonic [6], so is their functional composition. Furthermore, because the theory of designs is a complete lattice and **A** is both idempotent and monotonic, so is the theory of angelic designs.

The theory of angelic designs is based on non-homogeneous relations. As a consequence the definition of sequential composition departs from the norm for other UTP theories, where usually sequential composition is relational composition. Instead, the definition is layered upon the sequential composition operator $;_{\mathcal{A}}$ of [5], whose definition in the context of this theory, we reproduce below.

Definition 10. $P\ ;_{\mathcal{A}}\ Q \mathrel{\widehat{=}} P[\{s \mid Q\}/ac']$

The resulting set of angelic choices is that of Q, such that they can be reached from an initial state of Q that is available for P as a set ac' of angelic choices. This is a result that closely resembles that for binary multirelations, except for the fact that it is expressed using substitution. In the next section, we present a set-theoretic model of binary multirelations, like that in section 2.2, but extended to cater for angelic designs.

3 Extending Binary Multirelations

Based on the theory of binary multirelations, we introduce a new type of relations $BM_{\perp}$ by considering a different type $State_{\perp}$ for the target set of states.

Definition 11. $State_{\perp} == State \cup \{\perp\}$, $BM_{\perp} == State \leftrightarrow \mathbb{P}\, State_{\perp}$

Each initial state is related to a set of final states of type $State_{\perp}$, a set that may include the special state $\perp$, which denotes that termination is not guaranteed. If a set of final states does not contain $\perp$, then the program must terminate.

For example, consider the program that assigns the value 1 to the variable x, but may or may not terminate. This is specified by the following relation, where $:=_{BM_{\perp}}$ is the assignment operator that does not require termination.

Example 4. $x :=_{BM_{\perp}} 1 = \{s : State, ss : \mathbb{P}\, State_{\perp} \mid s \oplus (x \mapsto 1) \in ss\}$

Every initial state s is related to a set of final states ss where the state obtained from s by overriding the value of the component x with 1 is included. Since ss is of type $State_{\perp}$, all sets of final states in ss include those with and without $\perp$.

In the following section 3.1 we define the healthiness conditions of the new theory of binary multirelations of type $BM_{\perp}$. In section 3.2 we explore important properties of the new model. Finally in section 3.3 we explore the relationship between the new model and the original theory of binary multirelations.

3.1 Healthiness Conditions

Having defined a new type of relations, in what follows we introduce the healthiness conditions that characterise the relations in the theory.

BMH0. The first healthiness condition of interest enforces upward closure [2] for sets of final states that are necessarily terminating, and in addition enforces the same property for sets of final states that are not required to terminate.

Definition 12 (BMH0).

$$\forall s : State, ss_0, ss_1 : \mathbb{P}\, State_\perp \bullet$$
$$((s, ss_0) \in B \land ss_0 \subseteq ss_1 \land (\perp \in ss_0 \Leftrightarrow \perp \in ss_1)) \Rightarrow (s, ss_1) \in B$$

It states that for every initial state s, and for every set of final states ss_0 in a relation B, any superset ss_1 of that final set of states is also associated with s as long as $\perp$ is in ss_0 if, and only if, it is in ss_1. That is, **BMH0** requires upward closure for sets of final states that terminate, and for those that may or may not terminate, but separately. The definition of **BMH0** can be split into two conjunctions as shown in the following lemma 3.

Lemma 3

$$\mathbf{BMH0} \Leftrightarrow \left(\begin{array}{l} \left(\begin{array}{l} \forall s : State, ss_0, ss_1 : \mathbb{P}\, State_\perp \bullet \\ ((s, ss_0) \in B \land ss_0 \subseteq ss_1 \land \perp \in ss_0 \land \perp \in ss_1) \Rightarrow (s, ss_1) \in B \end{array} \right) \\ \land \\ \mathbf{BMH} \end{array} \right)$$

This confirms that for sets of final states that terminate this healthiness condition enforces **BMH** exactly as in the original theory of binary multirelations [2].

BMH1. The second healthiness condition **BMH1** requires that if it is possible to choose a set of final states where termination is not guaranteed, then it must also be possible to choose an equivalent set of states where termination is guaranteed. This healthiness condition is similar in nature to **H2** of the theory of designs.

Definition 13 (BMH1)

$$\forall s : State, ss : \mathbb{P}\, State_\perp \bullet (s, ss \cup \{\perp\}) \in B \Rightarrow (s, ss) \in B$$

This healthiness condition excludes relations that only offer sets of final states that may not terminate. Consider the following example.

Example 5. $\{s : State, ss : \mathbb{P}\, State_\perp \mid (x \mapsto 1) \in ss \land \perp \in ss\}$

This relation describes an assignment to the only program variable x where termination is not guaranteed. However, it discards the inclusive situation where termination may indeed occur. The inclusion of a corresponding final set of states that requires termination does not change the choices available to the angel as it is still impossible to guarantee termination.

BMH2. The third healthiness condition reflects a redundancy in the model, namely, that both the empty set and $\{\perp\}$ characterise abortion.

Definition 14 (BMH2). $\forall s : State \bullet (s, \emptyset) \in B \Leftrightarrow (s, \{\perp\}) \in B$

Therefore we require that every initial state s is related to the empty set of final states if, and only if, it is also related to the set of final states $\{\perp\}$.

If we consider **BMH1** in isolation, it covers the reverse implication of **BMH2** because if $(s, \{\perp\})$ is in the relation, so is $(s, \emptyset)$. However, **BMH2** is stronger than **BMH1** by requiring $(s, \{\perp\})$ to be in the relation if $(s, \emptyset)$ is also in the relation. The reason for this redundancy is to facilitate the linking between theories, in particular with the original theory. We come back to this point in section 3.3.

The new model of binary multirelations is characterised by the conjunction of the healthiness conditions **BMH0**, **BMH1** and **BMH2**, which we refer to as $\mathbf{BMH_{0,1,2}}$. An alternative characterisation in terms of fixed points is available in [7]. That characterisation has enabled us, for instance, to establish that all healthiness conditions are monotonic.

BMH3. The fourth healthiness condition characterises a subset of the model that corresponds to the original theory of binary multirelations.

Definition 15 (BMH3)

$$\forall s : State \bullet (s, \emptyset) \notin B \Rightarrow (\forall ss : \mathbb{P}\, State_\perp \bullet (s, ss) \in B \Rightarrow \perp \notin ss)$$

If an initial state s is not related to the empty set, then it must be the case that for all sets of final states ss related to s, $\perp$ is not included in the set ss.

This healthiness condition excludes relations that do not guarantee termination for particular initial states, yet establish some set of final states. example 4 is an example of such a relation. This is also the case for the original theory of binary multirelations. If it is possible for a program not to terminate when started from some initial state, then execution from that state must lead to arbitrary behaviour. This is the same intuition for **H3** of the theory of designs [4].

This concludes the discussion of the healthiness conditions. The relationship with the original model of binary multirelations is discussed in section 3.3.

3.2 Operators

Having defined the healthiness conditions, in this section we introduce the most important operators of the theory. These enable the discussion of interesting properties observed in the new model.

Assignment. In this model there is in fact the possibility to define two distinct assignment operators. The first one behaves exactly as in the original theory of binary multirelations $x :=_{BM} e$. This operator does not need to be redefined, since $BM \subseteq BM_\perp$. The new operator that we define below, however, behaves rather differently, in that it may or may not terminate.

Definition 16. $x :=_{BM_\perp} e \mathrel{\widehat=} \{s : State, ss : \mathbb{P}\, State_\perp \mid s \oplus (x \mapsto e) \in ss\}$

This assignment guarantees that for every initial state s, there is some set of final states available for angelic choice where x has the value of expression e. However, termination is not guaranteed. While the angel can choose the final value of x it cannot possibly guarantee termination in this case.

Angelic Choice. Angelic choice is defined as set intersection just like in the original theory of binary multirelations.

Definition 17. $B_0 \sqcup_{BM_\perp} B_1 \mathrel{\widehat=} B_0 \cap B_1$

For every set of final states available for demonic choice in B_0 and B_1, only those that can be chosen both in B_0 and B_1 are available. As the refinement ordering in the new model is exactly the same as in the theory of binary multirelations, the angelic choice operator, being the least upper bound, has the same properties with respect to the extreme points of the lattice.

An interesting property of angelic choice that is observed in this model is illustrated by the following lemma 4. It considers the angelic choice between two assignments of the same expression, yet only one is guaranteed to terminate.

Lemma 4. $(x :=_{BM_\perp} e) \sqcup_{BM_\perp} (x :=_{BM} e) = (x :=_{BM} e)$

This result can be interpreted as follows: given an assignment which is guaranteed to terminate, adding a corresponding angelic choice that is potentially non-terminating does not in fact introduce any new choices.

Demonic Choice. The demonic choice operator is defined by set union, exactly as in the original theory of binary multirelations.

Definition 18. $B_0 \sqcap_{BM_\perp} B_1 \mathrel{\widehat=} B_0 \cup B_1$

For every initial state, a corresponding set of final states available for demonic choice in either, or both, of B_0 and B_1, is included in the result.

Similarly to the angelic choice operator, there is a general result regarding the demonic choice over the two assignment operators, terminating and not necessarily terminating. This is shown in the following lemma 5.

Lemma 5. $(x :=_{BM} e) \sqcap_{BM_\perp} (x :=_{BM_\perp} e) = (x :=_{BM_\perp} e)$

If there is an assignment for which termination is not guaranteed, then the demonic choice over this assignment and a corresponding one that is guaranteed to terminate is the same as the assignment that does not require termination. In other words, if it is possible for the demon to choose between two similar sets of final states, one that is possibly non-terminating and one that terminates, then the one for which termination is not guaranteed dominates the choice.

Sequential Composition. The definition of sequential composition in this new model is not immediately obvious. In fact, one of the reasons for developing this theory is that it provides a more intuitive approach to the definition of sequential composition in the theory of angelic designs. To illustrate the issue, we consider the following example from the theory of designs, where a non-**H3**-design is sequentially composed with $\mathbf{I\!I}_{\mathcal{D}}$, the Skip of the theory.

Example 6.

$$
\begin{aligned}
&(x' = 1 \vdash true) \ ; \ \mathbf{I\!I}_{\mathcal{D}} && \{\text{Definition of } \mathbf{I\!I}_{\mathcal{D}}\} \\
&= (x' = 1 \vdash true) \ ; \ (true \vdash x' = x) && \{\text{Sequential composition for designs}\} \\
&= (\neg\, (x' \neq 1 \ ; \ true) \wedge \neg\, (true \ ; \ false) \vdash true \ ; \ x' = x) && \{\text{Sequential composition}\} \\
&= (\neg\, (\exists\, x_0 \bullet x_0 \neq 1 \wedge true) \wedge \neg\, (\exists\, x_0 \bullet true \wedge false) \vdash \exists\, x_0 \bullet true \wedge x' = x_0) \\
& && \{\text{Predicate calculus and one-point rule}\} \\
&= (\neg\, true \wedge \neg\, false \vdash true) && \{\text{Predicate calculus and property of designs}\} \\
&= true
\end{aligned}
$$

The result is *true*, the bottom of designs [4], whose behaviour is arbitrary. This result can be generalised for the sequential composition of any non-**H3**-design.

The behaviour just described provides the motivation for the definition of sequential composition in the new binary multirelational model.

Definition 19

$$
B_0 \ ;_{BM_\perp} \ B_1 \mathrel{\widehat=} \left\{ \begin{array}{l|l} s_0 : State, ss_0 : \mathbb{P}\, State_\perp & \exists\, ss : \mathbb{P}\, State_\perp \bullet (s_0, ss) \in B_0 \wedge \\ & (\perp \in ss \vee ss \subseteq \{s_1 : State \mid (s_1, ss_0) \in B_1\}) \end{array} \right\}
$$

For sets of final states where termination is guaranteed, that is, $\perp$ is not in the set of intermediate states ss, this definition matches that of the original theory. If $\perp$ is in ss, and hence termination is not guaranteed, then the result of the sequential composition is arbitrary as it can include any set of final states.

If we assume that B_0 is **BMH0**-healthy, then the definition of sequential composition can be split into the set union of two sets as shown in theorem 2.

Theorem 2. *Provided* B_0 *is* **BMH0***-healthy.*

$$
B_0 \ ;_{BM_\perp} \ B_1 = \left(\begin{array}{l} \{s_0, ss_0 \mid (s_0, State_\perp) \in B_0\} \\ \cup \\ \{s_0, ss_0 \mid (s_0, \{s_1 \mid (s_1, ss_0) \in B_1\}) \in B_0\} \end{array} \right)
$$

The first set considers the case when B_0 leads to sets of final states where termination is not required ($State_\perp$). The second set considers the case where termination is required and matches the result of lemma 2. This concludes our discussion of the main results regarding the operators of the theory.

3.3 Relationship with Binary Multirelations

Having presented the most important operators of the theory, in this section we focus our attention on the relationship between the new model and the original theory of binary multirelations. The first step consists in the definition of a pair of linking functions, *bmb2bm* that maps from the new model into the original theory of binary multirelations, and *bm2bm*, a mapping in the opposite direction.

The relationship between the theories of interest is illustrated in fig. 1 where each theory is labelled according to its healthiness conditions. In addition to the

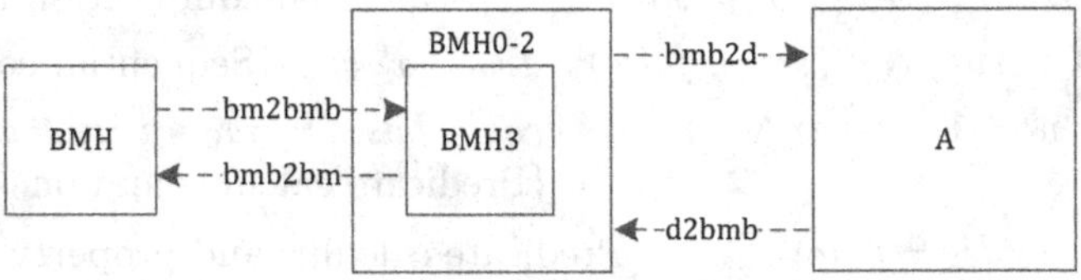

Fig. 1. Theories and links

relationship between both models of binary multirelations, fig. 1 also shows the relationship between the new model of binary multirelations and the theory of angelic designs characterised by **A**. The latter is the focus of section 4.

From $BM_\perp$ to BM. The function *bmb2bm*, defined below, maps binary multirelations in the new model, of type $BM_\perp$, to those in the original model.

Definition 20 (*bmb2bm*)

$$bmb2bm : BM_\perp \nrightarrow BM$$
$$bmb2bm(B) \mathrel{\widehat=} \{s : State, ss : \mathbb{P}\, State_\perp \mid ((s, ss) \in B \land \perp \notin ss)\}$$

It is defined by considering every pair (s, ss) in B such that $\perp$ is not in ss. We consider the following example, where *bmb2bm* is applied to the potentially non-terminating assignment of e to the program variable x.

Example 7. $bmb2bm(x :=_{BM_\perp} e) = (x :=_{BM} e)$

The results corresponds to assignment in the original theory. theorem 3 shows that the application of *bmb2bm* to an $\mathbf{BMH_{0,1,2,3}}$-healthy relation yields a **BMH**-healthy relation.

Theorem 3. *Provided B is* $\mathbf{BMH_{0,1,2,3}}$*-healthy.*

$$\mathbf{bmh_{up}} \circ bmb2bm(B) = bmb2bm(B)$$

This result confirms that *bmb2bm* yields relations that are in the original theory. The proof of this theorem and other proofs omitted below are found in [7].

From BM to $BM_\perp$. The function *bm2bmb* maps from relations in the original model, of type BM, into the new theory. Its definition is presented below.

Definition 21 (*bm2bmb*)

$$bm2bmb : BM \nrightarrow BM_\perp$$

$$bm2bmb(B) \mathrel{\widehat=} \{\, s : State, ss : \mathbb{P}\, State_\perp \mid ((s, ss) \in B \land \perp \notin ss) \lor (s, \emptyset) \in B \,\}$$

It considers every pair (s, ss) in a relation B where $\perp$ is not in the set of final states ss, or if B is aborting for a particular initial state s, then the result is the universal relation of type $BM_\perp$. A similar result to theorem 3 exists for the application of *bm2bmb* [7], where it yields $\mathbf{BMH_{0,1,2,3}}$-healthy relations.

Based on these results we can establish that *bm2bmb* and *bmb2bm* form a bijection for healthy relations as ascertained the following theorems 4 and 5.

Theorem 4. *Provided B is $\mathbf{BMH_{0,1,2,3}}$-healthy.* $bm2bmb \circ bmb2bm(B) = B$

Theorem 5. *Provided B is $\mathbf{BMH}$-healthy.* $bmb2bm \circ bm2bmb(B) = B$

These results show that the subset of the theory that is **BMH0-BMH3**-healthy is isomorphic to the original theory of binary multirelations [2]. This confirms that while our model is more expressive, it is still possible to express every program that could be specified using the original model. This concludes the discussion of the new theory. In the following section we discuss the relationship with the theory of angelic designs.

4 Relationship with UTP Designs

In this section we establish that the predicative model of **A**-healthy designs is isomorphic to the new theory of binary multirelations. We begin our discussion by defining a pair of linking functions: *d2bmb*, that maps from **A**-healthy designs into the new model of binary multirelations, and *bmb2d*, mapping in the opposite direction. The relationship between the theories is illustrated in fig. 1.

4.1 From Designs to Binary Multirelations

The first function of interest is *d2bmb*, whose definition is presented below.

Definition 22 (d2bmb)

$$d2bmb : \mathbf{A} \nrightarrow BM_\perp$$

$$d2bmb(P) \mathrel{\widehat=} \left\{ s : State, ss : \mathbb{P}\, State_\perp \;\middle|\; \begin{array}{l} ((\neg\, P^f \Rightarrow P^t)[ss/ac'] \land \perp \notin ss) \\ \lor \\ (P^f[ss \setminus \{\perp\}/ac'] \land \perp \in ss) \end{array} \right\}$$

For a given design $P = (\neg P^f \vdash P^t)$, the set construction of $d2bmb(P)$ is split into two disjuncts. In the first disjunct, we consider the case where P is started and terminates successfully, with ok and ok' both being substituted with *true*. The resulting set of final states ss, for which termination is required ($\bot \notin ss$) is obtained by substituting ss for ac' in P. The second disjunct considers the case where ok is also *true*, but ok' is *false*. This corresponds to the situation where P does not terminate. In this case, the set of final states is obtained by substituting $ss \setminus \{\bot\}$ for ac' and requiring $\bot$ to be in the set of final states ss.

As a consequence of P satisfying **H2**, we ensure that if there is some set of final states captured by the second disjunct with $\bot$, then there is also a corresponding set of final states without $\bot$ that is captured by the first disjunct.

In order to illustrate the result of applying $d2bmb$, we consider the following example 8. It specifies a program that either assigns the value 1 to the sole program variable x and successfully terminates, or assigns the value 2 to x, in which case termination is not required.

Example 8

$$d2bmb((x \mapsto 2) \notin ac' \vdash (x \mapsto 1) \in ac') \qquad \{\text{Definition of } d2bmb \text{ and designs}\}$$

$$= \left\{ \begin{array}{l} s : State, ss : \mathbb{P}\, State_\bot \\ \mid ((x \mapsto 2) \notin ac' \Rightarrow (x \mapsto 1) \in ac')[ss/ac'] \land \bot \notin ss) \\ \mid \lor \\ \mid (((x \mapsto 2) \in ac')[ss \setminus \{\bot\}/ac'] \land \bot \in ss) \end{array} \right\}$$

$$\{\text{Predicate calculus and substitution}\}$$

$$= \left\{ \begin{array}{l} s : State, ss : \mathbb{P}\, State_\bot \\ \mid ((x \mapsto 2) \in ss \land \bot \notin ss) \lor ((x \mapsto 1) \in ss \land \bot \notin ss) \\ \mid \lor \\ \mid ((x \mapsto 2) \in (ss \setminus \{\bot\}) \land \bot \in ss) \end{array} \right\}$$

$$\{\text{Property of sets and predicate calculus}\}$$

$$= \{s : State, ss : \mathbb{P}\, State_\bot \mid (x \mapsto 2) \in ss \lor ((x \mapsto 1) \in ss \land \bot \notin ss)\}$$

$$\{\text{Definition of } \sqcap_{BM_\bot} \text{ and } :=_{BM_\bot} \text{ and } :=_{BM}\}$$

$$= (x :=_{BM_\bot} 2) \sqcap_{BM_\bot} (x :=_{BM} 1)$$

The function $d2bmb$ yields a program with the same behaviour, but specified using the binary multirelational model. It is the demonic choice over two assignments, one requires termination while the other does not.

The following theorem 6 establishes that the application of $d2bmb$ to **A**-healthy designs yields relations that are **BMH0-BMH2**-healthy.

Theorem 6. $\mathbf{bmh_{0,1,2}} \circ d2bmb(\mathbf{A}(P)) = d2bmb(\mathbf{A}(P))$

This concludes our discussion regarding the linking function $d2bmb$.

4.2 From Binary Multirelations to Designs

The second linking function of interest is $bmb2d$, which maps binary multirelations to **A**-healthy predicates. Its definition is presented below.

Definition 23 ($bmb2d$)

$$bmb2d : BM_\perp \rightarrow\!\!\!\!\!\rightarrow \mathbf{A} \qquad bmb2d(B) \mathrel{\widehat=} ((s, ac' \cup \{\perp\}) \notin B \vdash (s, ac') \in B)$$

It is defined as a design, such that for a particular initial state s, the precondition requires $(s, ac' \cup \{\perp\})$ not to be in B, while the postcondition establishes that (s, ac') is in B. This definition can be expanded into a more intuitive representation, by expanding the design, according to the following lemma 6.

Lemma 6. $bmb2d(B) = ok \Rightarrow \begin{pmatrix} ((s, ac') \in B \land \perp \notin ac' \land ok') \\ \lor \\ (s, ac' \cup \{\perp\}) \in B \end{pmatrix}$

The behaviour of $bmb2d$ is defined by two disjuncts. The first one considers the case where B requires termination, and hence $\perp$ is not part of the set of final states of the pair in B. The second disjunct considers sets of final states that do not require termination, in which case ok' can be either *true* or *false*.

The following theorem 7 establishes that $bmb2d(B)$ yields **A**-healthy designs provided that B is **BMH0-BMH2**-healthy.

Theorem 7. *Provided B is $\mathbf{BMH_{0,1,2}}$-healthy.* $\mathbf{A} \circ bmb2d(B) = bmb2d(B)$

This result confirms that $bmb2d$ is closed with respect to **A** when applied to relations that are **BMH0-BMH2**-healthy. This concludes our discussion of $bmb2d$. In the following section 4.3 we focus our attention on the isomorphism.

4.3 Isomorphism

In this section we show that $d2bmb$ and $bmb2d$ form a bijection. The following theorem 8 establishes that $d2bmb$ is the inverse function of $bmb2d$ for relations that are **BMH0-BMH2**-healthy. While theorem 9 establishes that $bmb2d$ is the inverse function of $d2bmb$ for designs that are **A**-healthy. Together these results establish that the models are isomorphic.

Theorem 8. *Provided B is* **BMH0-BMH2**-*healthy.* $d2bmb \circ bmb2d(B) = B$

Theorem 9. *Provided P is* **A**-*healthy.* $bmb2d \circ d2bmb(P) = P$

These results establish that the same programs can be characterised using two different approaches. The binary multirelational model provides a set-theoretic approach, while the predicative theory proposed can easily be linked with other UTP theories of interest. This dual approach enables us to justify the definitions of certain aspects of the theory. This includes the healthiness conditions, and the operators, which we discuss in the following section 4.4. The most intuitive and appropriate model can be used in each case. The results obtained in either model can then be related using the linking functions.

4.4 Linking Results

In this section we discuss the most important results obtained from linking both the theory of angelic designs and the new model of binary multirelations.

Refinement. As discussed earlier, the theory of angelic designs [6] is a complete lattice under the refinement ordering, here denoted by $\sqsubseteq_{\mathcal{D}}$, which is universal reverse implication. In the theory of binary multirelations, refinement is subset inclusion, as denoted by $\sqsubseteq_{BM_\perp}$. theorem 10 establishes their correspondence.

Theorem 10. *Provided B_0 and B_1 are* **BMH0-BMH2**-*healthy.*

$$bmb2d(B_0) \sqsubseteq_{\mathcal{D}} bmb2d(B_1) \Leftrightarrow B_0 \sqsubseteq_{BM_\perp} B_1$$

It is reassuring to find that the refinement ordering of the theory of angelic designs corresponds to the subset ordering in the binary multirelational model.

Sequential Composition. Amongst the operators discussed in the context of the theories of interest, sequential composition is, perhaps, the most challenging. In the new model of binary multirelations, this is due to the addition of potential non-termination, while in the theory of angelic designs, the difficulty pertains to the use of non-homogenenous relations and the definition of $;_{\mathcal{A}}$.

In the theory of angelic designs, sequential composition is defined as follows.

Definition 24. $P ;_{\mathbf{\mathcal{D}ac}} Q \mathrel{\widehat=} \exists ok_0 \bullet P[ok_0/ok'] ;_{\mathcal{A}} Q[ok_0/ok]$

As discussed earlier, this is a definition that is layered upon $;_{\mathcal{A}}$ [6]. It resembles relational composition, with the notable difference that instead of conjunction we use the operator $;_{\mathcal{A}}$. When considering **A**-healthy designs, sequential composition can be expressed as an **A**-healthy design as established by theorem 11.

Theorem 11. *Provided P and Q are* **A**-*healthy designs.*

$$P ;_{\mathbf{\mathcal{D}ac}} Q = (\neg (P^f ;_{\mathcal{A}} true) \land \neg (P^t ;_{\mathcal{A}} Q^f) \vdash P^t ;_{\mathcal{A}} (\neg Q^f \Rightarrow Q^t))$$

This is a result similar to the one for designs [4,11], except for the use of the operator $;_{\mathcal{A}}$ and the postcondition, which is different. The implication in the postcondition acts as a filter that eliminates final states of P that fail to satisfy the precondition of Q. We consider the following example, where there is an angelic choice between assigning 1 and 2 to the only program variable b, followed by the program that maintains the state unchanged provided b is 1.

Example 9.

$$\begin{pmatrix} (true \vdash \{b \mapsto 1\} \in ac') \\ \sqcup \\ (true \vdash \{b \mapsto 2\} \in ac') \end{pmatrix} ;_{\mathbf{\mathcal{D}ac}} (s.b = 1 \vdash s \in ac') = (true \vdash \{b \mapsto 1\} \in ac')$$

The angelic choice is resolved as the assignment of 1 to b, which avoids aborting.

Finally, we have established through theorem 12 that the sequential composition operators of our theories are in correspondence.

Theorem 12. *Provided P and Q are* **A**-*healthy designs.*

$$bmb2d(d2bmb(P) ;_{BM_\perp} d2bmb(Q)) = P ;_{\mathbf{\mathcal{D}ac}} Q$$

This is a reassuring result that provides a dual characterisation for the sequential composition of angelic designs, both in a predicative model and in terms of sets.

Demonic Choice. The demonic choice operator of angelic designs ($\sqcap_{\mathcal{D}ac}$) defined as disjunction, corresponds exactly to the demonic choice operator ($\sqcap_{BM_\perp}$) of the binary multirelational model, defined as set union.

Theorem 13. $bmb2p(B_0 \sqcap_{BM_\perp} B_1) = bmb2p(B_0) \sqcap_{\mathcal{D}ac} bmb2p(B_1)$

This result confirms the correspondence of demonic choice in both models.

Angelic Choice. Similarly, the angelic choice operator ($\sqcup_{\mathcal{D}ac}$), defined as conjunction, is in correspondence with that of binary multirelations, ($\sqcup_{BM_\perp}$) which is defined as set intersection.

Theorem 14. $bmb2p(B_0 \sqcup_{BM_\perp} B_1) = bmb2p(B_0) \sqcup_{\mathcal{D}ac} bmb2p(B_1)$

Proof.

$bmb2p(B_0) \sqcup_{\mathcal{D}ac} bmb2p(B_1)$ {Definition of *bmb2p*}

$$= \left(\begin{array}{l} ((s, ac' \cup \{\perp\}) \notin B_0 \vee \perp \in ac' \vdash (s, ac') \in B_0 \wedge \perp \notin ac') \\ \sqcup_{\mathcal{D}ac} \\ ((s, ac' \cup \{\perp\}) \notin B_1 \vee \perp \in ac' \vdash (s, ac') \in B_1 \wedge \perp \notin ac') \end{array}\right)$$

{Definition of $\sqcup_{\mathcal{D}ac}$}

$$= \left(\begin{array}{l} ((s, ac' \cup \{\perp\}) \notin B_0 \vee \perp \in ac' \vee (s, ac' \cup \{\perp\}) \notin B_1 \vee \perp \in ac') \\ \vdash \\ \left(\begin{array}{l} ((s, ac' \cup \{\perp\}) \notin B_0 \vee \perp \in ac') \Rightarrow ((s, ac') \in B_0 \wedge \perp \notin ac') \\ \wedge \\ ((s, ac' \cup \{\perp\}) \notin B_1 \vee \perp \in ac') \Rightarrow ((s, ac') \in B_1 \wedge \perp \notin ac') \end{array}\right) \end{array}\right)$$

{Predicate calculus}

$$= \left(\begin{array}{l} ((s, ac' \cup \{\perp\}) \notin B_0 \vee \perp \in ac' \vee (s, ac' \cup \{\perp\}) \notin B_1) \\ \vdash \\ \left(\begin{array}{l} ((s, ac' \cup \{\perp\}) \in B_0 \vee (s, ac') \in B_0) \\ \wedge \\ ((s, ac' \cup \{\perp\}) \in B_1 \vee (s, ac') \in B_1) \end{array}\right) \wedge \perp \notin ac' \end{array}\right)$$

{Assumption: B_0 and B_1 are **BMH1**-healthy}

$$= \left(\begin{array}{l} ((s, ac' \cup \{\perp\}) \notin B_0 \vee \perp \in ac' \vee (s, ac' \cup \{\perp\}) \notin B_1) \\ \vdash \\ \left(\begin{array}{l} (((s, ac' \cup \{\perp\}) \in B_0 \wedge (s, ac') \in B_0) \vee (s, ac') \in B_0) \\ \wedge \\ (((s, ac' \cup \{\perp\}) \in B_1 \wedge (s, ac') \in B_1) \vee (s, ac') \in B_1) \end{array}\right) \wedge \perp \notin ac' \end{array}\right)$$

{Predicate calculus: absorption law}

$$= \left(\begin{array}{l} ((s, ac' \cup \{\perp\}) \notin B_0 \vee \perp \in ac' \vee (s, ac' \cup \{\perp\}) \notin B_1) \\ \vdash \\ (s, ac') \in B_0 \wedge (s, ac') \in B_1 \wedge \perp \notin ac' \end{array}\right)$$

{Predicate calculus and property of sets}

$$= ((s, ac' \cup \{\perp\}) \notin (B_0 \cap B_1) \vee \perp \in ac' \vdash (s, ac') \in (B_0 \cap B_1) \wedge \perp \notin ac')$$

{Definition of *bmb2p* and $\sqcup_{BM_\perp}$}

$$= bmb2p(B_0 \sqcup_{BM_\perp} B_1)$$ □

In [7] we have established a number of other properties regarding the angelic choice operator and sequential composition, namely that sequential composition does not, in general, distribute through angelic choice from neither the left nor the right, and that angelic and demonic choice distribute through one another. The latter follows directly from the properties of sets and the characterisation of angelic and demonic choice in the binary multirelational model.

5 Conclusion

Angelic nondeterminism has traditionally been considered in the context of theories of total correctness for sequential computations. Amongst these, isomorphic models include the universal monotonic predicate transformers of the refinement calculus [1,13,14], and binary multirelations [2], where both angelic and demonic nondeterminism are captured. The corresponding characterisation in a relational setting, such as the UTP, has been achieved via multirelational encodings [5,6].

Morris and Tyrrel [15,16], and Hesselink [17], have considered angelic nondeterminism in the context of functional languages, by characterising it at the expression or term level. A generalised algebraic structure has been proposed by Guttmann [18], where both the monotonic predicate transformers and multirelations are characterised as instances.

Tyrrell et al. [19] have proposed an axiomatized algebra of processes resembling CSP where external choice is angelic choice, however, in their model deadlock is indistinguishable from divergence. Roscoe [20] has proposed an angelic choice operator in the context of an operational combinator semantics for CSP. However, its semantics is far from being a counterpart to the angelic choice operator of the refinement calculus, where, if possible, abortion can be avoided.

The theory that we have introduced here presents itself as a natural extension of Rewitzky's [2] binary multirelations, by including information pertaining to the possibility for non-termination. This is a concept found in the general theory of UTP designs, where preconditions can refer to the value of later or final states, an essential property for the characterisation of CSP processes.

The development of links between the new theory and angelic designs provides two complementary views of the same computations. This dual approach has enabled us to characterise certain aspects more easily by choosing the most appropriate model. It is reassuring that the healthiness conditions and operators of both models are in correspondence. Our long term aim is the definition of a UTP theory of CSP that includes all standard CSP operators, and, additionally, an angelic choice operator that avoids divergence.

References

1. Back, R., Wright, J.: Refinement calculus: a systematic introduction. Graduate texts in computer science. Springer (1998)
2. Rewitzky, I.: Binary Multirelations. In: de Swart, H., Orłowska, E., Schmidt, G., Roubens, M. (eds.) Theory and Applications of Relational Structures as Knowledge Instruments. LNCS, vol. 2929, pp. 256–271. Springer, Heidelberg (2003)

3. Martin, C.E., Curtis, S.A., Rewitzky, I.: Modelling Nondeterminism. In: Kozen, D. (ed.) MPC 2004. LNCS, vol. 3125, pp. 228–251. Springer, Heidelberg (2004)
4. Hoare, C.A.R., Jifeng, H.: Unifying Theories of Programming. Prentice Hall International Series in Computer Science (1998)
5. Cavalcanti, A., Woodcock, J., Dunne, S.: Angelic nondeterminism in the unifying theories of programming. Formal Aspects of Computing 18, 288–307 (2006)
6. Ribeiro, P., Cavalcanti, A.: Designs with Angelic Nondeterminism. In: 2013 International Symposium on Theoretical Aspects of Software Engineering (TASE), pp. 71–78 (2013)
7. Ribeiro, P.: Designs with Angelic Nondeterminism. Technical report, University of York (February 2013), `http://www-users.cs.york.ac.uk/pfr/reports/dac.pdf`
8. Cavalcanti, A., Woodcock, J.: A Tutorial Introduction to CSP in *Unifying Theories of Programming*. In: Cavalcanti, A., Sampaio, A., Woodcock, J. (eds.) PSSE 2004. LNCS, vol. 3167, pp. 220–268. Springer, Heidelberg (2006)
9. Ribeiro, P., Cavalcanti, A.: Angelicism in the Theory of Reactive Processes. In: Unifying Theories of Programming (to appear, 2014)
10. Cavalcanti, A., Mota, A., Woodcock, J.: Simulink Timed Models for Program Verification. In: Liu, Z., Woodcock, J., Zhu, H. (eds.) Theories of Programming and Formal Methods. LNCS, vol. 8051, pp. 82–99. Springer, Heidelberg (2013)
11. Woodcock, J., Cavalcanti, A.: A Tutorial Introduction to Designs in Unifying Theories of Programming. In: Boiten, E.A., Derrick, J., Smith, G.P. (eds.) IFM 2004. LNCS, vol. 2999, pp. 40–66. Springer, Heidelberg (2004)
12. Cavalcanti, A., Woodcock, J.: Angelic Nondeterminism and Unifying Theories of Programming. Technical report, University of Kent (2004)
13. Morgan, C.: Programming from specifications. Prentice Hall (1994)
14. Morris, J.M.: A theoretical basis for stepwise refinement and the programming calculus. Sci. Comput. Program. 9, 287–306 (1987)
15. Morris, J.M.: Augmenting Types with Unbounded Demonic and Angelic Nondeterminacy. In: Kozen, D. (ed.) MPC 2004. LNCS, vol. 3125, pp. 274–288. Springer, Heidelberg (2004)
16. Morris, J.M., Tyrrell, M.: Terms with unbounded demonic and angelic nondeterminacy. Science of Computer Programming 65(2), 159–172 (2007)
17. Hesselink, W.H.: Alternating states for dual nondeterminism in imperative programming. Theoretical Computer Science 411(22-24), 2317–2330 (2010)
18. Guttmann, W.: Algebras for correctness of sequential computations. Science of Computer Programming 85(Pt. B), 224–240 (2014); Special Issue on Mathematics of Program Construction (2012)
19. Tyrrell, M., Morris, J.M., Butterfield, A., Hughes, A.: A Lattice-Theoretic Model for an Algebra of Communicating Sequential Processes. In: Barkaoui, K., Cavalcanti, A., Cerone, A. (eds.) ICTAC 2006. LNCS, vol. 4281, pp. 123–137. Springer, Heidelberg (2006)
20. Roscoe, A.W.: Understanding concurrent systems. Springer (2010)

The Arithmetic of Recursively Run-Length Compressed Natural Numbers

Paul Tarau

Department of Computer Science and Engineering
University of North Texas
tarau@cs.unt.edu

Abstract. We study arithmetic properties of a new tree-based canonical number representation, *recursively run-length compressed* natural numbers, defined by applying recursively a run-length encoding of their binary digits.

We design arithmetic operations with recursively run-length compressed natural numbers that work a block of digits at a time and are limited only by the representation complexity of their operands, rather than their bitsizes.

As a result, operations on very large numbers exhibiting a regular structure become tractable.

In addition, we ensure that the average complexity of our operations is still within constant factors of the usual arithmetic operations on binary numbers.

Keywords: run-length compressed numbers, hereditary numbering systems, arithmetic algorithms for giant numbers, representation complexity of natural numbers.

1 Introduction

Notations like Knuth's "up-arrow" [1] have been shown to be useful in describing very large numbers. However, they do not provide the ability to actually *compute* with them, as, for instance, addition or multiplication with a natural number results in a number that cannot be expressed with the notation anymore.

The main focus of this paper is a new tree-based numbering system that *allows* computations with numbers comparable in size with Knuth's "up-arrow" notation. Moreover, these computations have *worst and average case complexity that is comparable with the traditional binary numbers*, while their *best case complexity outperforms binary numbers by an arbitrary tower of exponents factor.*

For the curious reader, it is basically a *hereditary number system* [2], based on recursively applied *run-length* compression of the usual binary digit notation. It favors giant numbers in neighborhoods of towers of exponents of two, with super-exponential gains on their arithmetic operations. Moreover, the proposed notation is *canonical* i.e., each number has a unique representation (contrary to the traditional one where any number of leading zeros can be added).

G. Ciobanu and D. Méry (Eds.): ICTAC 2014, LNCS 8687, pp. 406–423, 2014.

We adopt a *literate programming* style, i.e. the code described in the paper forms a self-contained Haskell module (tested with ghc 7.6.3), also available as a separate file at `http://www.cse.unt.edu/~tarau/research/2014/RRL.hs`. We hope that this will encourage the reader to experiment interactively and validate the technical correctness of our claims.

The code in this paper can be seen as a compact and mathematically obvious specification rather than an implementation fine-tuned for performance. Faster but more verbose equivalent code can be derived in procedural or object oriented languages by replacing lists with (dynamic) arrays and some instances of recursion with iteration.

We mention, for the benefit of the reader unfamiliar with Haskell, that a notation like `f x y` stands for $f(x, y)$, `[t]` represents sequences of type `t` and a type declaration like `f :: s -> t -> u` stands for a function $f : s \times t \rightarrow u$.

Our Haskell functions are always represented as sequences of recursive equations guided by pattern matching, conditional to constraints (simple relations following `|` and before the `=` symbol). Locally scoped helper functions are defined in Haskell after the `where` keyword, using the same equational style.

The composition of functions `f` and `g` is denoted `f . g`. Note also that the result of the last evaluation is stored in the special Haskell variable `it`.

The paper is organized as follows. Section 2 discusses related work. Section 3 introduces our tree represented recursively run-length compressed natural numbers. Section 4 describes constant time successor and predecessor operations on tree-represented numbers. Section 5 describes novel algorithms for arithmetic operations taking advantage of our number representation. Section 6 defines a concept of representation complexity and studies best and worst cases. Section 7 describes an example of computation with very large numbers using recursively run-length compressed numbers. Section 8 concludes the paper.

2 Related Work

The first instance of a hereditary number system, at our best knowledge, occurs in the proof of Goodstein's theorem [2], where replacement of finite numbers on a tree's branches by the ordinal ω allows him to prove that a "hailstone sequence" visiting arbitrarily large numbers eventually turns around and terminates.

Conway's surreal numbers [3] can also be seen as inductively constructed trees. While our focus will be on efficient large natural number arithmetic, surreal numbers model games, transfinite ordinals and generalizations of real numbers.

Several notations for very large numbers have been invented in the past. Examples include Knuth's *up-arrow* notation [1], covering operations like the *tetration* (a notation for towers of exponents). In contrast to the tree-based natural numbers we propose in this paper, such notations are not closed under addition and multiplication, and consequently they cannot be used as a replacement for ordinary binary or decimal numbers.

This paper is similar in purpose with [4] which describes a more complex hereditary number system (based on run-length encoded "bijective base 2" numbers, introduced in [5] pp. 90-92 as "m-adic" numbers). In contrast to [4], we are using here the familiar binary number system, and we represent our numbers as the *free algebra* of ordered rooted multiway trees, rather than the more complex data structure used in [4].

The tree representation that we will use is a representative of the Catalan family of combinatorial objects [6], on which, in [7], arithmetic operations are seen as operating on balanced parenthesis languages.

An emulation of Peano and conventional binary arithmetic operations in Prolog, is described in [8]. Their approach is similar as far as a symbolic representation is used. The key difference with our work is that our operations work on tree structures, and as such, they are not based on previously known algorithms.

In [9] a binary tree representation enables arithmetic operations which are simpler but limited in efficiency to a small set of "sparse" numbers.

In [10] integer decision diagrams are introduced providing a compressed representation for sparse integers, sets and various other data types. However likewise [9] and [11], and in contrast to those proposed in this paper, they only compress "sparse" numbers, consisting of relatively few 1 bits in their binary representation.

3 The Data Type of Recursively Run-length Compressed Natural Numbers

First, we define a data type for our tree represented natural numbers, that we call *recursively run-length compressed numbers* to emphasize that *binary* rather than *unary* encoding is recursively used in their representation.

Definition 1. *The data type* T *of the set of recursively run-length compressed numbers is defined by the Haskell declaration:*

```haskell
data T = F [T] deriving (Eq,Show,Read)
```

that automatically derives the equality relation "==", as well as reading and string representation. The data type T *corresponds precisely to ordered rooted multiway trees with empty leaves, but for shortness, we will call the objects of type* T terms. *The "arithmetic intuition" behind the type* T *is the following:*

- *the term* `F []` *(empty leaf) corresponds to zero*
- *in the term* `F xs`*, each* `x` *on the list* `xs` *counts the number* `x+1` *of* $b \in \{0, 1\}$ *digits, followed by* alternating *counts of* `1-b` *and* `b` *digits, with the convention that the most significant digit is* 1
- *the same principle is applied recursively for the counters, until the empty sequence is reached.*

One can see this process as run-length compressed base-2 numbers, unfolded as trees with empty leaves, after applying the encoding recursively. Note that we count `x+1` as we start at `0`. By convention, as the last (and most significant) digit is `1`, the last count on the list `xs` is for `1` digits. For instance, the first level of the encoding of `123` as the (big-endian) binary number `1101111` is `[1,0,3]`.

The following simple fact allows inferring parity from the number of subtrees of a tree.

Proposition 1. *If the length of* `xs` *in* `F xs` *is odd, then* `F xs` *encodes an odd number, otherwise it encodes an even number.*

Proof. Observe that as the highest order digit is always a `1`, the lowest order digit is also `1` when length of the list of counters is odd, as counters for `0` and `1` digits alternate.

This ensures the correctness of the Haskell definitions of the predicates `odd_` and `even_`.

```
odd_ :: T → Bool
odd_ (F []) = False
odd_ (F (_:xs)) = even_ (F xs)

even_ :: T → Bool
even_ (F []) = True
even_ (F (_:xs)) = odd_ (F xs)
```

Note that while these predicates work in time proportional to the length of the list `xs` in `F xs`, *with a (dynamic) array-based list representation that keeps track of the length or keeps track of the parity bit explicitly, one can assume that they are constant time*, as we will do in the rest of the paper.

Definition 2. *The function* $n : \mathtt{T} \to \mathbb{N}$ *shown in equation* (1) *defines the unique natural number associated to a term of type* `T`.

$$n(a) = \begin{cases} 0 & \text{if } a = \texttt{F []}, \\ 2^{n(\texttt{x})+1} n(\texttt{F xs}) & \text{if } a = \texttt{F (x:xs)} \text{ is } \texttt{even\_}, \\ 2^{n(\texttt{x})+1} n(\texttt{F xs}) - 1 & \text{if } a = \texttt{F (x:xs)} \text{ is } \texttt{odd\_}. \end{cases} \tag{1}$$

For instance, the computation of `n(F [F [],F [F [],F []]])` using equation (1) expands to $(2^{0+1}(2^{(2^{0+1}(2^{0+1}-1))+1} - 1)) = 14$, which, in binary, is `[0,1,1,1]` where the first level expansion `[0,2]`, corresponds to `F []` → `0` and `F [F [],F []]` → 2. The Haskell equivalent[1] of equation (1) is:

```
n (F []) = 0
n a@(F (x:xs)) | even_ a = 2^(n x + 1)*(n (F xs))
n a@(F (x:xs)) | odd_ a = 2^(n x + 1)*(n (F xs)+1)-1
```

The following example illustrates the values associated with the first few natural numbers.

```
0: F []
1: F [F []])
2: F [F [],F []]
3: F [F [F []]]
```

One might notice that our trees are in bijection with objects of the *Catalan family*, e.g., strings of balanced parentheses, for instance 0 → `F []` → (), 1 → `F [F []]` → (()), 14 → `F [F [],F [F [],F []]]` → (()(()())).

[1] As a Haskell note, the pattern `a@p` indicates that the parameter `a` has the same value as its expanded version matching the patten `p`.

Definition 3. *The function* t : $\mathbb{N} \rightarrow$ T *defines the unique tree of type* T *associated to a natural number as follows:*

```
t 0 = F []
t k | k>0 = F zs where
  (x,y) = split_on (parity k) k
  F ys = t y
  zs = if x==0 then ys else t (x-1) : ys

  parity x = x `mod` 2

  split_on b z | z>0 && parity z == b = (1+x,y) where
    (x,y) = split_on b  ((z-b) `div` 2)
  split_on _ z = (0,z)
```

It uses the helper function split_on, which, depending on parity b, extracts a block of contiguous 0 or 1 digits from the lower end of a binary number. It returns a pair (x,y) consisting of a count x of the number of digits in the block, and the natural number y representing the digits left over after extracting the block. Note that div, occurring in both functions, is integer division.

The following holds:

Proposition 2. *Let* id *denote* $\lambda x.x$ *and* $\circ$ *function composition. Then, on their respective domains:*

$$t \circ n = id, \quad n \circ t = id \ . \tag{2}$$

Proof. By induction, using the arithmetic formulas defining the two functions.

4 Successor (s) and Predecessor (s')

We will now specify successor and predecessor on data type T through two mutually recursive functions, s and s'.

```
s :: T → T
s (F []) = F [F []] -- 1
s (F [x]) = F [x,F []] -- 2

s a@(F (F []:x:xs)) | even_ a = F (s x:xs) -- 3
s a@(F (x:xs)) | even_ a = F (F []:s' x:xs) -- 4

s a@(F (x:F []:y:xs)) | odd_ a = F (x:s y:xs) -- 5
s a@(F (x:y:xs)) | odd_ a = F (x:F []:(s' y):xs) -- 6
```

```
s':: T → T
s' (F [F []]) = F [] -- 1
s' (F [x,F []]) = F [x] -- 2

s' b@(F (x:F []:y:xs)) | even_ b = F (x:s y:xs) -- 6
s' b@(F (x:y:xs)) | even_ b = F (x:F []:s' y:xs) -- 5
```

```
s' b@(F (F []:x:xs)) | odd_ b = F (s x:xs) -- 4
s' b@(F (x:xs)) | odd_ b = F (F []:s' x:xs) -- 3
```

Note that the two functions work *on a block of* 0 *or* 1 *digits at a time.* They are based on simple arithmetic observations about the behavior of these blocks when incrementing or decrementing a binary number by 1. The following holds:

Proposition 3. *Denote* $\mathtt{T}^+ = \mathtt{T} - \{\mathtt{F\ []}\}$. *The functions* $s : \mathtt{T} \rightarrow \mathtt{T}^+$ *and* $s' : \mathtt{T}^+ \rightarrow \mathtt{T}$ *are inverses.*

Proof. It follows by structural induction after observing that patterns for rules marked with the number `-- k` in `s` correspond one by one to patterns marked by `-- k` in `s'` and vice versa.

More generally, it can be shown that Peano's axioms hold and as a result $< \mathtt{T}, \mathtt{F}[], \mathtt{s} >$ is a *Peano algebra.*

Note also that if parity information is kept explicitly, the calls to `odd_` *and* `even_` *are constant time, as we will assume in the rest of the paper.*

Proposition 4. *The worst case time complexity of the* `s` *and* `s'` *operations on* n *is given by the* iterated logarithm $O(log_2^*(n))$, *where* log_2^* *counts the number of times* log_2 *can be applied before reaching* 0.

Proof. Note that calls to `s` and `s'` in `s` or `s'` happen on terms at most logarithmic in the bitsize of their operands. The recurrence relation counting the worst case number of calls to `s` or `s'` is: $T(n) = T(log_2(n)) + O(1)$, which solves to $T(n) = O(log_2^*(n))$.

Note that this is much better than the logarithmic worst case for binary umbers (when computing, for instance, binary 111...111+1=1000...000).

Proposition 5. `s` *and* `s'` *are constant time, on the average.*

Proof. Observe that the average size of a contiguous block of 0s or 1s in a number of bitsize n has the upper bound 2 as $\sum_{k=0}^{n} \frac{1}{2^k} = 2 - \frac{1}{2^n} < 2$. As on 2-bit numbers we have an average of 0.25 more calls, we can conclude that the total average number of calls is constant, with upper bound $2 + 0.25 = 2.25$.

A quick empirical evaluation confirms this. When computing the successor on the first $2^{30} = 1073741824$ natural numbers, there are in total 2381889348 calls to `s` and `s'`, averaging to 2.2183 per computation. The same average for 100 successor computations on 5000 bit random numbers oscillates around 2.22.

5 Arithmetic Operations

We will now describe algorithms for basic arithmetic operations that take advantage of our number representation.

5.1 A Few Other Average Constant Time Operations

Doubling a number db and reversing the db operation (hf) are quite simple. For instance, db proceeds by adding a new counter for odd numbers and incrementing the first counter for even ones.

```
db (F []) = F []
db a@(F xs) | odd_ a = F (F []:xs)
db a@(F (x:xs)) | even_ a = F (s x:xs)
```

```
hf (F []) = F []
hf (F (F []:xs)) = F xs
hf (F (x:xs)) = F (s' x:xs)
```

Note that such efficient implementations follow directly from simple number theoretic observations.

For instance, exp2, computing an exponent of 2 , has the following definition in terms of s'.

```
exp2 (F []) = F [F []]
exp2 x = F [s' x,F []]
```

As log2 shows, exp2 is also easy to invert with a similar amount of work:

```
log2 (F [F []]) = F []
log2 (F [y,F []]) = s y
```

Note that this definition works on powers of 2, see subsection 5.4 for a general version.

Proposition 6. *The operations* db, hf, exp2 *and* log2 *are average constant-time and iterated logarithm in the worst case.*

Proof. At most 1 call to s or s' is made in each definition therefore these operations have the same worst and average complexity as s and s'.

5.2 Optimizing Addition and Subtraction for Numbers with few Large Blocks of 0s and 1s

We derive efficient addition and subtraction that *work on one run-length compressed block at a time*, rather than by individual 0 and 1 digit steps. The functions leftshiftBy, leftshiftBy' and respectively leftshiftBy'' correspond to $2^n k$, $(\lambda x.2x+1)^n(k)$ and $(\lambda x.2x+2)^n(k)$.

```
leftshiftBy :: T → T → T
leftshiftBy (F []) k = k
leftshiftBy _ (F []) = F []
leftshiftBy x k@(F xs) | odd_ k = F ((s' x):xs)
leftshiftBy x k@(F (y:xs)) | even_ k = F (add x y:xs)
```

```
leftshiftBy' x k = s' (leftshiftBy x (s k))
```

```
leftshiftBy'' x k = s' (s' (leftshiftBy x (s (s k))))
```

The last two are derived from the identities:

$$(\lambda x.2x + 1)^n(k) = 2^n(k + 1) - 1 \tag{3}$$

$$(\lambda x.2x + 2)^n(k) = 2^n(k + 2) - 2 \tag{4}$$

They are part of a *chain of mutually recursive functions* as they are already referring to the add function, to be implemented later. Note also that instead of naively iterating, they implement a more efficient algorithm, working "one block at a time". For instance, when detecting that its argument counts a number of 1s, leftshiftBy' just increments that count. As a result, the algorithm favors numbers with relatively few large blocks of 0 and 1 digits.

We are now ready for defining addition. The base cases are

```
add :: T → T → T
add (F []) y = y
add x (F []) = x
```

In the case when both terms represent even numbers, the two blocks add up to an even block of the same size.

```
add x@(F (a:as)) y@(F (b:bs)) |even_ x && even_ y = f (cmp a b) where
  f EQ = leftshiftBy (s a) (add (F as) (F bs))
  f GT = leftshiftBy (s b) (add (leftshiftBy (sub a b) (F as)) (F bs))
  f LT = leftshiftBy (s a) (add (F as) (leftshiftBy (sub b a) (F bs)))
```

In the case when the first term is even and the second odd, the two blocks add up to an odd block of the same size.

```
add x@(F (a:as)) y@(F (b:bs)) |even_ x && odd_ y = f (cmp a b) where
  f EQ = leftshiftBy' (s a) (add (F as) (F bs))
  f GT = leftshiftBy' (s b) (add (leftshiftBy (sub a b) (F as)) (F bs))
  f LT = leftshiftBy' (s a) (add (F as) (leftshiftBy' (sub b a) (F bs)))
```

In the case when the second term is even and the first odd the two blocks also add up to an odd block of the same size.

```
add x y |odd_ x && even_ y = add y x
```

In the case when both terms represent odd numbers, we use the identity (5):

$$(\lambda x.2x + 1)^k(x) + (\lambda x.2x + 1)^k(y) = (\lambda x.2x + 2)^k(x + y) \tag{5}$$

```
add x@(F (a:as)) y@(F (b:bs)) | odd_ x && odd_ y = f (cmp a b) where
  f EQ = leftshiftBy'' (s a) (add (F as) (F bs))
  f GT = leftshiftBy'' (s b) (add (leftshiftBy' (sub a b) (F as)) (F bs))
  f LT = leftshiftBy'' (s a) (add (F as) (leftshiftBy' (sub b a) (F bs)))
```

Note the presence of the comparison operation cmp, to be defined later, also part of our chain of mutually recursive operations. Note also the local function f that in each case ensures that a block of the same size is extracted, depending on which of the two operands a or b is larger. The code for the subtraction function sub is similar:

```
sub :: T → T → T
sub x (F []) = x
sub x@(F (a:as)) y@(F (b:bs)) | even_ x && even_ y = f (cmp a b) where
  f EQ = leftshiftBy (s a) (sub (F as) (F bs))
  f GT = leftshiftBy (s b) (sub (leftshiftBy (sub a b) (F as)) (F bs))
  f LT = leftshiftBy (s a) (sub (F as) (leftshiftBy (sub b a) (F bs)))
```

The case when both terms represent 1 blocks the result is a 0 block

```
sub x@(F (a:as)) y@(F (b:bs)) | odd_ x && odd_ y = f (cmp a b) where
  f EQ = leftshiftBy (s a) (sub (F as) (F bs))
  f GT = leftshiftBy (s b) (sub (leftshiftBy' (sub a b) (F as)) (F bs))
  f LT = leftshiftBy (s a) (sub (F as) (leftshiftBy' (sub b a) (F bs)))
```

The case when the first block is 1 and the second is a 0 block:

```
sub x@(F (a:as)) y@(F (b:bs)) | odd_ x && even_ y = f (cmp a b) where
  f EQ = leftshiftBy' (s a) (sub (F as) (F bs))
  f GT = leftshiftBy' (s b) (sub (leftshiftBy' (sub a b) (F as)) (F bs))
  f LT = leftshiftBy' (s a) (sub (F as) (leftshiftBy (sub b a) (F bs)))
```

Finally, when the first block is 0 and the second is 1 an identity dual to (5) is used:

```
sub x@(F (a:as)) y@(F (b:bs)) | even_ x && odd_ y = f (cmp a b) where
  f EQ = s (leftshiftBy (s a) (sub1 (F as) (F bs)))
  f GT =
    s (leftshiftBy (s b) (sub1 (leftshiftBy (sub a b) (F as)) (F bs)))
  f LT =
    s (leftshiftBy (s a) (sub1 (F as) (leftshiftBy' (sub b a) (F bs))))

sub1 x y = s' (sub x y)
```

Note that these algorithms collapse to the ordinary binary addition and subtraction most of the time, given that the average size of a block of contiguous 0s or 1s is 2 bits (as shown in Prop. 4), so their average performance is within a constant factor of their ordinary counterparts. On the other hand, the algorithms favor deeper trees made of large blocks, representing giant "towers of exponents"-like numbers by working (recursively) one block at a time rather than 1 bit at a time, resulting in possibly super-exponential gains.

5.3 Comparison

The comparison operation `cmp` provides a total order (isomorphic to that on $\mathbb{N}$) on our type `T`. It relies on `bitsize` computing the number of binary digits constructing a term in `T`. It is part of our mutually recursive functions, to be defined later.

We first observe that only terms of the same bitsize need detailed comparison, otherwise the relation between their bitsizes is enough, *recursively.* More precisely, the following holds:

Proposition 7. *Let* `bitsize` *count the number of digits of a base-2 number, with the convention that it is* 0 *for* 0. *Then* bitsize(x) <bitsize(y) $\Rightarrow x < y$.

Proof. Observe that their lexicographic enumeration ensures that the bitsize of base-2 numbers is a non-decreasing function.

The comparison operation also proceeds one block at a time, and it also takes some inferential shortcuts, when possible.

```
cmp :: T → T → Ordering
cmp (F []) (F []) = EQ
cmp (F []) _ = LT
cmp _ (F []) = GT
cmp (F [F []]) (F [F [],F []]) = LT
cmp  (F [F [],F []]) (F [F []]) = GT
cmp x y | x' /= y'  = cmp x' y' where
  x' = bitsize x
  y' = bitsize y
cmp (F xs) (F ys) =
  compBigFirst True True (F (reverse xs)) (F (reverse ys))
```

The function `compBigFirst` compares two terms known to have the same `bitsize`. It works on reversed (highest order digit first) variants, computed by `reverse` and it takes advantage of the block structure using the following proposition:

Proposition 8. *Assuming two terms of the same bitsizes, the one with its first before its highest order digit* 1 *is larger than the one with its first before its highest order digit* 0.

Proof. Observe that "highest order digit first" numbers are lexicographically ordered with $0 < 1$.

As a consequence, `cmp` only recurses when *identical* blocks lead the sequence of blocks, otherwise it infers the `LT` or `GT` relation.

```
compBigFirst _ _ (F []) (F []) = EQ
compBigFirst False False (F (a:as)) (F (b:bs)) = f (cmp a b) where
   f EQ = compBigFirst True True (F as) (F bs)
   f LT = GT
   f GT = LT
compBigFirst True True (F (a:as)) (F (b:bs)) = f (cmp a b) where
   f EQ = compBigFirst False False (F as) (F bs)
   f LT = LT
   f GT = GT
compBigFirst False True x y = LT
compBigFirst True False x y = GT
```

5.4 Bitsize

The function `bitsize` computes the number of digits, except that we define it as `F []` for `F []`, corresponding to `0`. It concludes the chain of mutually recursive functions starting with the addition operation add. It works by summing up (using Haskell's foldr) the counts of 0 and 1 digit blocks composing a tree-represented natural number.

```
bitsize :: T → T
bitsize (F []) = (F [])
bitsize (F (x:xs)) = s (foldr add1 x xs)

add1 x y = s (add x y)
```

It follows that the base-2 integer logarithm is then computed as

```
ilog2 = s' . bitsize
```

The iterated logarithm log_2^* can be also defined as

```
ilog2star :: T → T
ilog2star (F []) = F []
ilog2star x = s (ilog2star (ilog2 x))
```

5.5 Multiplication, Optimized for Large Blocks of 0s and 1s

Devising a similar optimization as for add and sub for multiplication is actually easier.

When the first term represents an even number we apply the leftshiftBy operation and we reduce the other case to this one.

```
mul :: T → T → T
mul x y = f (cmp x y) where
  f GT = mul1 y x
  f _ = mul1 x y

  mul1 (F []) _ = F []
  mul1 a@(F (x:xs)) y | even_ a =  leftshiftBy (s x) (mul1 (F xs) y)
  mul1 a y | odd_ a = add y (mul1 (s' a) y)
```

Note that when the operands are composed of large blocks of alternating 0 and 1 digits, the algorithm is quite efficient as it works (roughly) in time depending on the the number of blocks in its first argument rather than the the number of digits. The following example illustrates a blend of arithmetic operations benefiting from complexity reductions on giant tree-represented numbers:

```
*RRL> let term1 = sub (exp2 (exp2 (t 12345))) (exp2 (t 6789))
*RRL> let term2 = add (exp2 (exp2 (t 123))) (exp2 (t 456789))
*RRL> bitsize (bitsize (mul term1 term2))
F [F [],F [],F [],F [F [],F []],F [F [],F [],F []],F [F []]]
*RRL> n it
12346
```

This hints toward a possibly new computational paradigm where arithmetic operations are not limited by the size of their operands, but only by their representation complexity. We will make this concept more precise in section 6.

5.6 Power

After specializing our multiplication for a squaring operation,

```
square x = mul x x
```

we can implement a simple but efficient "power by squaring" operation for x^y, as follows:

```
pow :: T → T → T
pow _ (F []) = F [F []]
pow a@(F (x:xs)) b | even_ a = F (s' (mul (s x) b):ys) where
  F ys = pow (F xs) b
pow a b@(F (y:ys)) | even_ b = pow (superSquare y a) (F ys) where
  superSquare (F []) x = square x
  superSquare k x = square (superSquare (s' k) x)
pow x y = mul x (pow x (s' y))
```

It works well with fairly large numbers, by also benefiting from efficiency of multiplication on terms with large blocks of 0 and 1 digits:

```
*RRL> n (bitsize (pow (t 10) (t 100)))
333
*RRL> pow (t 32) (t 10000000)
F [F [F [F [],F [F []]],F [F [F []],F []],F [F [F []]],
  F [],F [],F [],F [F [F []],F []],F [],F []],F []]
```

5.7 Division Operations

An integer division algorithm is given here, but it does not provide the same complexity gains as, for instance, multiplication, addition or subtraction.

```
div_and_rem :: T → T → (T, T)
div_and_rem x y | LT == cmp x y = (F [],x)
div_and_rem x y | y /= F [] = (q,r) where
  (qt,rm) = divstep x y
  (z,r) = div_and_rem rm y
  q = add (exp2 qt) z
```

The function divstep implements a step of the division operation.

```
divstep n m = (q, sub n p) where
  q = try_to_double n m (F [])
  p = leftshiftBy q m
```

The function try_to_double doubles its second argument while smaller than its first argument and returns the number of steps it took. This value will be used by divstep when applying the leftshiftBy operation.

```
try_to_double x y k =
  if (LT==cmp x y) then s' k
  else try_to_double x (db y) (s k)
```

Division and remainder are obtained by specializing div_and_rem.

```
divide n m = fst (div_and_rem n m)
remainder n m = snd (div_and_rem n m)
```

6 Representation Complexity

While a detailed analysis of our algorithms is beyond the scope of this paper, arguments similar to those about the average behavior of s and s' can be carried out to prove that *their average complexity matches their traditional counterparts*, using the fact, shown in the proof of Prop. 4, that the average size of a block of contiguous 0 or 1 bits is at most 2.

6.1 Complexity as Representation Size

To evaluate the best and worst case space requirements of our number representation, we introduce here a measure of *representation complexity*, defined by the function tsize that counts the nodes of a tree of type T (except the root).

```
tsize :: T → T
tsize (F xs) = foldr add1 (F []) (map tsize xs)
```

It corresponds to the function $c : \mathtt{T} \to \mathbb{N}$ defined as follows:

$$c(t) = \begin{cases} 0 & \text{if } \mathtt{t} = \mathtt{F}\ [\,], \\ \sum_{x \in \mathtt{xs}} (1 + c(x)) & \text{if } \mathtt{t} = \mathtt{F\ xs}. \end{cases} \tag{6}$$

The following holds:

Proposition 9. *For all terms* $t \in \mathtt{T}$, tsize t $\leq$ bitsize t.

Proof. By induction on the structure of t, observing that the two functions have similar definitions and corresponding calls to tsize return terms inductively assumed smaller than those of bitsize.

The following example illustrates their use:

```
*RRL> map (n.tsize.t) [0,100,1000,10000]
[0,7,9,13]
*RRL> map (n.tsize.t) [2^16,2^32,2^64,2^256]
[5,6,6,6]
*RRL> map (n.bitsize.t) [2^16,2^32,2^64,2^256]
[17,33,65,257]
```

6.2 Best and Worst Cases

Next we define the higher order function iterated that applies k times the function f.

```
iterated f (F []) x = x
iterated f k x = f (iterated f (s' k) x)
```

We can exhibit, for a given bitsize, a best case

```
bestCase k = iterated wTree k (F []) where wTree x = F [x]
```

and a worst case

```
worstCase k = iterated f k (F []) where f (F xs) = F (F []:xs)
```

The function `bestCase` computes the iterated exponent of 2 and then applies the predecessor to it. For $k = 4$ it corresponds to

$$(2^{(2^{(2^{(2^{0+1}-1)+1}-1)+1}-1)+1} - 1) = 2^{2^{2^2}} - 1 = 65535.$$

The following examples illustrate these functions:

```
*RRL> bestCase (t 4)
F [F [F [F [F []]]]]
*RRL> n it
65535
*RRL> n (bitsize (bestCase (t 4)))
16
*RRL> n (tsize (bestCase (t 4)))
4

*RRL> worstCase (t 4)
F [F [],F [],F [],F []]
*RRL> n it
10
*RRL> n (bitsize (worstCase (t 4)))
4
*RRL> n (tsize (worstCase (t 4)))
4
```

Our concept of representation complexity is only a weak approximation of Kolmogorov complexity [12]. For instance, the reader might notice that our worst case example is computable by a program of relatively small size. However, as `bitsize` is an upper limit to `tsize`, we can be sure that we are within constant factors from the corresponding bitstring computations, even on random data of high Kolmogorov complexity. Note also that an alternative concept of representation complexity can be defined by considering the (vertices+edges) size of the DAG obtained by folding together identical subtrees.

6.3 A Concept of Duality

As our multiway trees with empty leaves are members of the Catalan family of combinatorial objects, they can be seen as binary trees with empty leaves, as defined by the bijection `toBinView` and its inverse `fromBinView`.

```
toBinView :: T → (T, T)
toBinView (F (x:xs)) = (x,F xs)

fromBinView :: (T, T) → T
fromBinView (x,F xs) = F (x:xs)
```

Therefore, we can transform the tree-representation of a natural number by swapping left and right branches under a binary tree view, recursively. The corresponding Haskell code is:

```
dual (F []) = F []
dual x = fromBinView (dual b, dual a) where (a,b) = toBinView x
```

As clearly `dual` is an *involution* (i.e., `dual` ∘ `dual` is the identity of `Cat`), the corresponding permutation of ℕ will bring together huge and small natural numbers sharing representations of the same size, as illustrated by the following example.

```
*RRL> map (n.dual.t) [0..20]
[0,1,3,2,4,15,7,6,12,31,65535,16,8,255,127,5,11,8191,4294967295,32,65536]
```

For instance, 18 and 4294967295 have dual representations of the same size, except that the wide tree associated to 18 maps to the tall tree associated to 4294967295, as illustrated by Fig. 1, with trees folded to DAGs by merging together shared subtrees. As a result, significantly different bitsizes can result for a term and its dual.

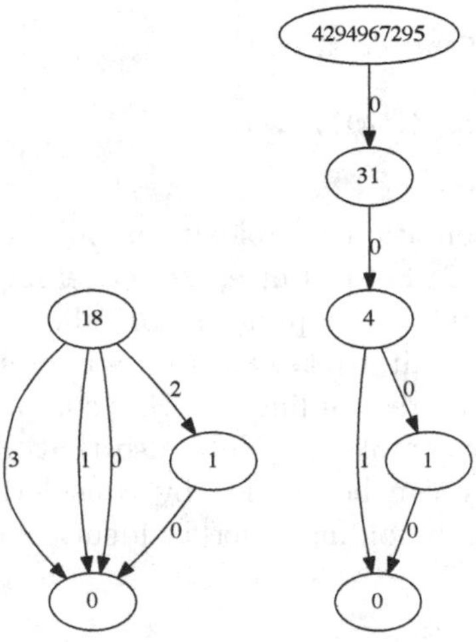

(a) t 18 (b) dual (t 18)

Fig. 1. Duals, with trees folded to DAGs

```
*RRL> t 18
F [F [],F [],F [F []],F []]
*RRL> dual (t 18)
F [F [F [F [F []],F []]]]
*RRL> n (bitsize (t 18))
5
*RRL> n (bitsize (dual (t 18)))
32
```

It follows immediately from the definitions of the respective functions, that as an extreme case, the following holds:

Proposition 10. $\forall$ `x dual (bestCase x)` $=$ `worstCase x`.

The following example illustrates this equality, with a tower of exponents `1000` tall, reached by `bestCase`.

```
*Cats> dual (bestCase (t 10000)) == worstCase (t 10000)
True
```

Note that these computations run easily on objects of type `T`, while their equivalents would dramatically overflow memory on bitstring-represented numbers.

7 A Case Study: Computing the Collatz/Syracuse Sequence for Huge Numbers

An application that achieves something one cannot do with traditional arbitrary bitsize integers is to explore the behavior of interesting conjectures in the "new world" of numbers limited not by their sizes but by their representation complexity. The Collatz conjecture states that the function

$$collatz(x) = \begin{cases} \frac{x}{2} & \text{if } x \text{ is even,} \\ 3x+1 & \text{if } x \text{ is odd.} \end{cases} \tag{7}$$

reaches 1 after a finite number of iterations. An equivalent formulation, by grouping together all the division by 2 steps, is the function:

$$collatz'(x) = \begin{cases} \frac{x}{2^{\nu_2(x)}} & \text{if } x \text{ is even,} \\ 3x+1 & \text{if } x \text{ is odd.} \end{cases} \tag{8}$$

where $\nu_2(x)$ denotes the *dyadic valuation of* x, i.e., the largest exponent of 2 that divides x. One step further, the *syracuse function* is defined as the odd integer k' such that $n = 3k+1 = 2^{\nu_2(n)}k'$. One more step further, by writing $k' = 2m+1$ we get a function that associates $k \in \mathbb{N}$ to $m \in \mathbb{N}$.

The function `tl` computes efficiently the equivalent of

$$tl(k) = \frac{\frac{k}{2^{\nu_2(k)}} - 1}{2} \tag{9}$$

Together with its hd counterpart, it is defined as

```
hd = fst . decons

tl = snd . decons

decons a@(F (x:xs)) | even_ a = (s x,hf (s' (F xs)))
decons a = (F [],hf (s' a))
```

where the function decons is the inverse of

```
cons (x,y) = leftshiftBy x (s (db y))
```

corresponding to $2^x\ (2y+1)$. Then our variant of the *syracuse function* corresponds to

$$syracuse(n) = tl(3n+2) \tag{10}$$

which is defined from $\mathbb{N}$ to $\mathbb{N}$ and can be implemented as

```
syracuse n = tl (add n (db (s n)))
```

The function nsyr computes the iterates of this function, until (possibly) stopping:

```
nsyr (F []) = [F []]
nsyr n = n : nsyr (syracuse n)
```

It is easy to see that the Collatz conjecture is true if and only if nsyr terminates for all n, as illustrated by the following example:

```
*RRL> map n (nsyr (t 2014))
[2014,755,1133,1700,1275,1913,2870,1076,807,1211,1817,2726,1022,383,
 575,863,1295,1943,2915,4373,6560,4920,3690,86,32,24,18,3,5,8,6,2,0]
```

The next examples will show that computations for nsyr can be efficiently carried out for giant numbers, that, with the traditional bitstring-representation, would easily overflow the memory of a computer with as many transistors as the atoms in the known universe.

And finally something we are quite sure has never been computed before, we can also start with a *tower of exponents 100 levels tall*:

```
*RRL> take 100 (map(n.tsize) (nsyr (bestCase (t 100))))
[100,199,297,298,300,...,440,436,429,434,445,439]
```

Note that we have only computed the decimal equivalents of the representation complexity tsize of these numbers, that obviously would not fit themselves in a decimal representation.

8 Conclusion

We have provided in the form of a literate Haskell program a specification of a tree-based number system where trees are built by recursively applying run-length encoding on the usual binary representation until the empty leaves corresponding to 0 are reached.

We have shown that arithmetic *computations* like addition, subtraction, multiplication, bitsize, exponent of 2, that favor giant numbers with *low representation complexity*, are performed in constant time, or time proportional to their representation complexity. We have also studied the best and worst case representation complexity of our representations and shown that, as representation complexity is bounded by bitsize, computations and data representations are within constant factors of conventional arithmetic even in the worst case.

References

1. Knuth, D.E.: Mathematics and Computer Science: Coping with Finiteness. Science 194(4271), 1235–1242 (1976)
2. Goodstein, R.: On the restricted ordinal theorem. Journal of Symbolic Logic (9), 33–41 (1944)
3. Conway, J.H.: On Numbers and Games, 2nd edn. AK Peters, Ltd. (2000)
4. Tarau, P., Buckles, B.: Arithmetic Algorithms for Hereditarily Binary Natural Numbers. In: Proceedings of SAC 2014, ACM Symposium on Applied Computing, PL track, Gyeongju, Korea. ACM (March 2014)
5. Salomaa, A.: Formal Languages. Academic Press, New York (1973)
6. Stanley, R.P.: Enumerative Combinatorics. Wadsworth Publ. Co., Belmont (1986)
7. Tarau, P.: Computing with Catalan Families. In: Dediu, A.-H., Martín-Vide, C., Sierra-Rodríguez, J.-L., Truthe, B. (eds.) LATA 2014. LNCS, vol. 8370, pp. 565–575. Springer, Heidelberg (2014)
8. Kiselyov, O., Byrd, W.E., Shan, C.-C.: Pure, declarative, and constructive arithmetic relations (declarative pearl). In: Garrigue, J., Hermenegildo, M.V. (eds.) FLOPS 2008. LNCS, vol. 4989, pp. 64–80. Springer, Heidelberg (2008)
9. Tarau, P., Haraburda, D.: On Computing with Types. In: Proceedings of SAC 2012, ACM Symposium on Applied Computing, PL track, Riva del Garda (Trento), Italy, pp. 1889–1896 (March 2012)
10. Vuillemin, J.: Efficient Data Structure and Algorithms for Sparse Integers, Sets and Predicates. In: 19th IEEE Symposium on Computer Arithmetic, ARITH 2009, pp. 7–14 (June 2009)
11. Tarau, P.: Declarative Modeling of Finite Mathematics. In: PPDP 2010: Proceedings of the 12th International ACM SIGPLAN Symposium on Principles and Practice of Declarative Programming, pp. 131–142. ACM, New York (2010)
12. Li, M., Vitányi, P.: An introduction to Kolmogorov complexity and its applications. Springer-Verlag New York, Inc., New York (1993)

Synchronous Parallel Composition in a Process Calculus for Ecological Models

Mauricio Toro[1,*], Anna Philippou[1], Christina Kassara[2], and Spyros Sfenthourakis[3]

[1] Department of Computer Science, University of Cyprus
{mtoro,annap}@cs.ucy.ac.cy
[2] Department of Biology, University of Patras, Greece
cristina.kassara@gmail.com
[3] Department of Biology, University of Cyprus
sfendourakis.spyros@ucy.ac.cy

Abstract. In this paper we extend PALPS, a process calculus proposed for the spatially-explicit, individual-based modeling of ecological systems, with a synchronous parallel operator. The semantics of the resulting calculus, S-PALPS, is defined at the level of populations as opposed to the level of individuals as was the case with PALPS, thus, allowing a considerable reduction in a system's state space. Furthermore, we provide a translation of the calculus into the model checker PRISM for simulation and analysis. We apply our framework to model and study the population dynamics of the Eleonora's falcon in the Mediterranean sea.

1 Introduction

Population ecology is a sub-field of ecology that studies changes in the size and age composition of populations, and the biological and environmental processes influencing those changes. Its main aim is to gain a better understanding of population dynamics and make predictions about how populations will evolve and how they will respond to specific management schemes. To achieve this goal, scientists have been constructing models of ecosystems. These models are abstract representations of the systems in question which are studied to gain understanding of the real systems (see, e.g. [4]).

Recently, we have witnessed an increasing trend towards the use of formal frameworks for reasoning about biological and ecological systems [20,15,7]. In our work, we are interested in the application of process algebras for studying the population dynamics of ecological systems. Process algebras provide a number of features that make them suitable for capturing these systems. In contrast to the traditional approach to modeling ecological systems using ordinary differential equations which describe a system in terms of changes in the population as a whole, process algebras are suited towards the so-called individual-based approach of modeling populations, as they enable one to describe the evolution of each individual of the population as a process and, subsequently, to compose a set of individuals (as well as their environment) into a complete

* This work was carried out during the tenure by the first author of an ERCIM "Alain Bensoussan" Fellowship Programme. The research leading to these results has received funding from the European Union Seventh Framework Programme (FP7/2007-2013) under grant agreement no 246016.

G. Ciobanu and D. Méry (Eds.): ICTAC 2014, LNCS 8687, pp. 424–441, 2014.

ecological system. Features such as time, probability and stochastic behavior, which have been extensively studied within the context of process algebras, can be exploited to provide more accurate models. Furthermore, following a model construction formal frameworks such as process algebras can be used in association with model-checking tools for automatically analyzing properties of models as opposed to just simulating trajectories as is typically carried out in ecological studies.

In our previous work we presented PALPS, a process algebra developed for modeling and reasoning about spatially-explicit individual-based systems [14]. In PALPS, individuals are modeled as processes consisting of a species and a location that may change dynamically. Individuals may engage in any of the basic processes of reproduction, dispersal, predation and death and they may communicate with other individuals residing at the same location. We have also associated PALPS with a translation to the probabilistic model checker PRISM with the prospect of making more advanced analysis of ecological models as opposed to just simulations. Our initial experiments of [14,13] delivered promising results via the use of statistical model checking provided by PRISM. However, it also revealed two limitations of our approach.

The first limitation regards reproduction and the dynamic evolution of the size of a population. In particular, given a system consisting of a set of individuals, our PRISM translation associated one PRISM module to each individual. Given this, when an individual produces an offspring, one would require that a new module would be created dynamically. However, PRISM does not support the dynamic creation of modules. Thus, we resorted to placing a limit max on the maximum number of individuals that could be active at any point in time and defining max modules which oscillated between the *active* and the *inactive* state as individuals are born and die, respectively.

The second weakness of PALPS relates to the semantics of the parallel composition operator. To begin with, the interleaving nature of parallel composition comes in contrast to the usual approach of modeling adopted by ecologists where it is considered that execution evolves in phases during which all individuals of a population engage in some process, such as birth, dispersal, reproduction, etc. For instance, given a species which first migrates and then reproduces, it would be expected that all individuals of the population migrate before reproduction takes place. However, in PALPS, it would be possible to interleave actions so that some individuals reproduce before the whole population engages in the migration process. In addition, the high level of nondeterminism of the semantics leads to a state-space explosion. For instance, consider individuals P_i, $1 \leq i \leq 5$, of species $\mathbf{s}$ at some location ℓ, each executing an action a_i and then becoming Q_i. In PALPS we would write this system as $S \stackrel{\text{def}}{=} P_1{:}\langle \mathbf{s}, \ell\rangle | \ldots | P_5{:}\langle \mathbf{s}, \ell\rangle$, where $P_i \stackrel{\text{def}}{=} a_i.Q_i$. Then, according to the operational semantics of PALPS, S may execute the 5 actions $a_1, \ldots, a_5$ in any order. As a result, there exist 5! possible executions of these actions eventually leading to state $S' \stackrel{\text{def}}{=} Q_1{:}\langle \mathbf{s}, \ell\rangle | \ldots | Q_5{:}\langle \mathbf{s}, \ell\rangle$. This phenomenon leads to a very quick explosion of the state space. To alleviate this problem, in [13], we proposed the use of policies within the PALPS framework. Policies were defined as an entity that may place a priority on the order of execution between actions. On the one hand, they enable the modeling of *process ordering* often used in ecological models as described above and, on the other hand they reduce the state space. In the example, if we consider a policy that assigns increasing priorities to actions a_1 to a_5, then, there is only one possible execution to reach state S'. Note, however, that this method will give reduced or no benefits in the case where some or all of the a_i's coincide.

In this work, our goal has been to address the above-mentioned issues by proposing a new semantics of PALPS and an associated PRISM translation that disassociates the number of modules from the maximum number of individuals, while capturing more faithfully the synchronous evolution of populations and removing as much unnecessary nondeterminism as possible. Our proposal, consists of a synchronous extension S-PALPS which features the following two key design decisions.

1. To address the first problem relating to the dynamic evolution of population sizes, we structure our calculus at the level of local populations, grouping together identical individuals located at the same location. This is achieved by the introduction of the new construct $P{:}\langle \mathrm{s}, \ell, q\rangle$ which refers to q individuals of species s at location ℓ.
2. To address the second problem, we provide a synchronous semantics of PALPS which implements the concept of *maximum parallelism*: at any given time all individuals that may execute an action will do so simultaneously. For example, system S considered above, will evolve to state S' in just one step, during which the actions $a_1, \ldots, a_5$ will be executed concurrently. As a result, the new parallel composition construct achieves a reduction in the state space while enabling the process-based execution adopted by ecologists.

We provide S-PALPS with an encoding to the PRISM language and we prove its correctness. In this translation, it is natural to define one module for each component of the form $P{:}\langle \mathrm{s}, \ell, q\rangle$. As a result, the number of modules of which the model is comprised is independent of the number of existing individuals.

As an example, we study the Eleonora's Falcon (falco eleonorae) [22] in S-PALPS. Eleonora's falcon is a migrant species that breeds on Mediterranean islands and winters on islands of the Indian Ocean and along the eastern African coast. A large part of the world population concentrates on a small number of islands in the Aegean Sea. In Europe, the species is considered as rare and hence of local conservation importance. because its survival in Europe is highly dependent on the breeding conditions on the islands on which it concentrates. We employ our methodology to investigate the population dynamics of the species by statistical model checking in PRISM.

Various formal frameworks have been proposed in the literature for modeling ecosystems. One strand is based, like PALPS, on process calculi such as WSCCS [20]. WSCCS is a probabilistic extension of CCS [11] with synchronous communication that has been employed in various ecological studies by the author and others [19,9]. Like PALPS, it follows the discrete-time approach to modeling but does not include the notion of space. A different approach is that of *P systems* [15], conceived as a class of distributed and parallel computing inspired by the compartmental structure and the functioning of living cells. P-systems fall in the category of rewriting systems, where a structure may evolve by the application of rewriting rules. The semantics of P-systems is closely related to S-PALPS: rules are usually applied with maximal parallelism while several proposals have been considered on resolving the nondeterminism that may arise when more than one combination of rules is possible, e.g. [12,6]. Probabilistic P systems have been applied to model the population dynamics of various ecosystems [5,3,6] as well as to study evolution problems [2]. Finally, we mention that Stochastic P systems have been translated into PRISM in [18]. However, as far as we know, there has been no work on the use of model checking for probabilistic P-Systems.

The structure of the remainder of the paper is as follows. In Section 2 we present the syntax and the semantics of S-PALPS. In Section 3 we present a translation of S-PALPS into the Markov-decision-process component of the PRISM language and we state the correctness of the translation. We then apply our techniques to study the population dynamics of the Eleonora's falcon in Sections 4 and 5. Section 6 concludes the paper.

2 Synchronous PALPS

In this section we introduce Synchronous PALPS, S-PALPS. S-PALPS extends PALPS in two ways. Firstly, S-PALPS differs with PALPS in the treatment of the parallel composition: in the semantics of S-PALPS this is treated synchronously, in the sense that in any composition $P|Q$ the actions of P and Q are taken simultaneously. Secondly, S-PALPS offers a new construct for modeling multiplicity of individuals: we write $P{:}\langle \mathbf{s}, \ell, q\rangle$ for q copies of individual P of species $\mathbf{s}$ and location ℓ. Other changes implemented to S-PALPS are the removal of the nondeterministic choice at the individual level, which is replaced by a conditional choice, and the inclusion of the parallel composition at the individual level which allows an explicit modeling of reproduction.

2.1 Syntax

Similarly to PALPS, in S-PALPS we consider a system as a set of individuals operating in space, each belonging to a certain species and inhabiting a location. Individuals who reside at the same location may communicate with each other upon channels, e.g. for preying, or they may migrate to a new location. S-PALPS models probabilistic events with the aid of a probabilistic operator and uses a discrete treatment of time.

The syntax of S-PALPS is based on the following basic entities: (1) $\mathbf{S}$ is a set of species ranged over by $\mathbf{s}, \mathbf{s}'$. (2) $\mathbf{Loc}$ is a set of locations ranged over by ℓ, ℓ'. The habitat is then implemented via a relation $\mathbf{Nb}$, where $(\ell, \ell') \in \mathbf{Nb}$ exactly when ℓ and ℓ' are neighbors. (3) $\mathbf{Ch}$ is a set of channels ranged over by lower-case strings.

The syntax of S-PALPS is given at two levels, the individual level ranged over by P and the system level ranged over by S. Their syntax is defined via the following BNFs.

$$P ::= \mathbf{0} \mid \eta.P \mid \sum_{i\in I} p_i{:}P_i \mid \gamma?\,(P_1, P_2) \mid P_1|P_2 \mid C$$

$$S ::= \mathbf{0} \mid P{:}\langle \mathbf{s}, \ell, q\rangle \mid S_1 \,\|\, S_2 \mid S\backslash L$$

where $L \subseteq \mathbf{Ch}$, I is an index set, $p_i \in (0,1]$ with $\sum_{i\in I} p_i = 1$, C ranges over a set of process constants $\mathcal{C}$, each with an associated definition of the form $C \stackrel{\mathrm{def}}{=} P$, and

$$\eta ::= a \mid \overline{a} \mid go\,\ell \mid \surd \qquad \gamma ::= a \mid \overline{a}$$

Beginning with the individual level, P can be one of the following: Process $\mathbf{0}$ represents the inactive individual, that is, an individual who has ceased to exist.

Process $\eta.P$ describes the action-prefixed process which executes action η before proceeding as P. In turn, an activity η can be an input action on a channel a, written simply as a, a complementary output action on a channel a, written as $\overline{a}$, a movement action with destination ℓ, $go\,\ell$, or the time-passing action, $\surd$. Actions of the form a, and $\overline{a}$, $a \in \mathbf{Ch}$, are used to model arbitrary activities performed by an individual; for instance, eating, preying and reproduction. The tick action $\surd$ measures a tick on a global clock.

Process $\sum_{i \in I} p_i{:}P_i$represents the probabilistic choice between processes P_i, $i \in I$. The process randomly selects an index $i \in I$ with probability p_i, and then evolves to process P_i. We write $p_1{:}P_1 \oplus p_2{:}P_2$ for the binary form of this operator.

Operator $\gamma?\,(P_1, P_2)$, is an operator new to S-PALPS. Its behavior depends on the availability of a communication on a certain channel as described by γ. Specifically, if a communication is available according to γ then the flow of control proceeds according to P, if not, the process proceeds as Q. This operator is a deterministic operator as, in any scenario, the process $\gamma?\,(P, Q)$ proceeds as either P or Q but not both, depending on the availability of the complementary action of γ in the environment in which the process is running. This construct has in fact replaced the nondeterministic choice of PALPS with the intention of yielding more tractable models. We believe this construct to be sufficient and appropriate for modeling ecosystems where choices are typically resolved either probabilistically or based on some precedence relation.

Moving on to the population level, population systems are built by composing in parallel sets of located individuals. A set of q individuals of species $\mathbf{s}$ located at location ℓ is defined as $P{:}\langle \mathbf{s}, \ell, q\rangle$. In a composition $S_1 \| S_2$ the components may proceed while synchronizing on their actions following a set of restrictions. These restrictions enforce that probabilistic transitions take precedence over the execution of other actions and that time proceeds synchronously in all components of a system. That is, for $S_1 \| S_2$ to execute a $\surd$ action, both S_1 and S_2 must be willing to execute $\surd$. Action $\surd$ measures a tick on a global clock. These time steps are abstract in the sense that they do not necessarily have a defined length and, in practice, $\surd$ is used to separate the rounds of an individual's behavior. In the case of multi-species systems these actions must be carefully placed in order to synchronize species with possibly different time scales.

Finally, $S \backslash L$ models the restriction of channels in L within S. This construct plays an important role in defining valid systems: We define a *valid* system to be any process of the form $S \backslash L$ where, for all of S's subprocesses of the form $a?(P, Q)$ and $\overline{a}?(P, Q)$ we have that $a \in L$. Hereafter, we consider only processes that are valid systems.

Example 1. Let us consider a species $\mathbf{s}$ where individuals cycle through a dispersal phase followed by a reproduction phase. In S-PALPS, we may model $\mathbf{s}$ by P_0, where

$$P_0 \stackrel{\text{def}}{=} \sum_{\ell \in \mathbf{Nb}(\mathsf{myloc})} \frac{1}{4} : go\,\ell.\surd.P_1 \qquad P_1 \stackrel{\text{def}}{=} p{:}\surd.(P_0|P_0) \;\oplus\; (1-p){:}\surd.(P_0|P_0|P_0)$$

According to the definition, during the dispersal phase, an individual moves to a neighboring location which is chosen probabilistically among the neighboring locations of the current location (myloc) of the individual. Subsequently, the flow of control proceeds according to P_1 which models the probabilistic production of one (case of $P_0|P_0$)

or two offspring (case of $P_0|P_0|P_0$). A system that contains two individuals at a location ℓ and one at location ℓ' can be modeled as

$$System \stackrel{\text{def}}{=} (P_0{:}\langle \mathbf{s}, \ell, 2\rangle | P_0{:}\langle \mathbf{s}, \ell', 1\rangle).$$

Let us now extend the example into a two-species system. In particular, consider a competing species which preys on s defined as:

$$Q_0 \stackrel{\text{def}}{=} \overline{prey}?(\surd.Q_1, \surd.Q_2) \qquad Q_1 \stackrel{\text{def}}{=} Q_0|Q_0 \qquad Q_2 \stackrel{\text{def}}{=} \overline{prey}?(\surd.Q_1, \mathbf{0})$$

An individual of species s' looks for a prey. This is implemented by the conditional process $\overline{prey}?(\surd.Q_1, \surd.Q_2)$. If it succeeds in communicating on channel *prey*, which implies that a prey is available, the individual will produces an offspring. If it fails for two consecutive time units it dies.

To implement the possibility of preying on the side of species s, the definition must be extended by introducing the complementary input actions on channel *prey* at the appropriate places:

$$P_0 \stackrel{\text{def}}{=} prey?\,(\mathbf{0}, \sum_{\ell \in \mathbf{Nb}(\mathsf{myloc})} \frac{1}{4} : go\,\ell.\surd.P_1)$$

$$P_1 \stackrel{\text{def}}{=} prey?\,(\mathbf{0},\ p{:}(\surd.(P_0|P_0)) \ \oplus\ (1-p) : \surd.(P_0|P_0|P_0))$$

2.2 Semantics

We may now define the semantics of S-PALPS. This is defined in terms of a structural congruence, $\equiv$, presented in Table 1 and a structural operational semantics presented in Tables 2 and 3. Beginning with Table 1, of greatest interest are the following con-

Table 1. Structural congruence relation

(I1) $P \equiv P|0$	(S1) $S \equiv S\|0$
(I2) $P_1|P_2 \equiv P_2|P_1$	(S2) $S_1\|S_2 \equiv S_2\|S_1$
(I3) $(P_1|P_2)|P_3 \equiv P_1|(P_2|P_3)$	(S3) $(S_1\|S_2)\|S_3 \equiv S_1\|(S_2\|S_3)$
(S4) $(P_1|P_2){:}\langle \mathbf{s}, \ell, q\rangle \equiv P_1{:}\langle \mathbf{s}, \ell, q\rangle \parallel P_2{:}\langle \mathbf{s}, \ell, q\rangle$	(S5) $P{:}\langle \mathbf{s}, \ell, 0\rangle \equiv 0$
(S6) $P{:}\langle \mathbf{s}, \ell, q\rangle \parallel P{:}\langle \mathbf{s}, \ell, r\rangle \equiv P{:}\langle \mathbf{s}, \ell, q+r\rangle$	(S7) $0{:}\langle \mathbf{s}, \ell, q\rangle \equiv 0$

gruences: Equivalence (S4) states that operator "$:\langle\ldots\rangle$" distributes over the parallel composition construct and equivalence (S6) states that the parallel composition of q individuals of type P of species s at location ℓ and r of the same individuals is equivalent to a system with $q+r$ individuals.

Moving on to the structural operational semantics of S-PALPS, this is given in terms of two transition relations: the non-deterministic relation $\longrightarrow_n$ and the probabilistic relation $\longrightarrow_p$. A transition of the form $S \xrightarrow{\mu}_n S'$ means that a system S may execute action μ and become S'. A transition of the form $S \xrightarrow{w}_p S'$ means that a configuration S may evolve into configuration S' with probability w. Whenever the type of the transition is irrelevant to the context, we write $S \xrightarrow{\alpha} S'$ to denote either $S \xrightarrow{\mu}_n S'$ or $S \xrightarrow{w}_p S'$. We write μ to range over system non-probabilistic activities, which we call actions. Actions are built based on activities of individuals which we call events and denote by β. Events β may have one of the following forms:

- $a_{\ell,\mathbf{s}}$ and $\overline{a}_{\ell,\mathbf{s}}$ denote the execution of a and $\overline{a}$ respectively at location ℓ by individuals of species s.
- $a?_{\ell,\mathbf{s}}$ and $\overline{a}?_{\ell,\mathbf{s}}$ denote the conditional execution of a and $\overline{a}$ respectively at location ℓ by individuals of species s. (This arises in processes of the form $\gamma?(P, Q)$.)
- $\tau_{a,\ell,\mathbf{s}}$ denotes an internal action that has taken place on channel a at location ℓ where the output was carried out by an individual of species s. This may arise when two complementary actions take place at the same location ℓ or when a move action takes place at location ℓ by an individual of species s.
- $\surd$ denotes the time passing action.

In turn μ can have one of the following forms:

- $\beta_1^{k_1}\#\ldots\#\beta_n^{k_n}$ where for all $1 \leq i \leq n$, $\beta_i \neq \surd$, $k_i \geq 1$ and $n \geq 1$, denotes the simultaneous execution of k_i actions of type β_i for $1 \leq i \leq n$, and
- $\surd$ denotes the time passing action.

We may now move on to the semantics of S-PALPS. We begin with the semantics of processes of the form $P{:}\langle \mathbf{s}, \ell, q\rangle$. We discuss these rules separately below:

- Rule (Act) states that a system composed of q individuals, where each can perform an action η, can perform simultaneously q times the action η.
- Rule (Tick) states that a system of q individuals, where each can perform action $\surd$, can also perform action $\surd$.
- Rule (Go) states that a system of q individuals, where each individual can perform a movement action, can perform simultaneously q times the action $\tau_{go,\ell,\mathbf{s}}$.
- Rule (Choice) states that a system of q individuals, executing the conditional choice $\gamma?(P, Q)$ may have any number $m \leq q$ of its components executing the action $(\gamma?_{\ell,\mathbf{s}})^m$ and proceedings to state P whereas the remaining $q - m$ of its components will proceed to Q. Note that the nondeterminism apparent in this rule will be resolved once this process is placed in a wider system context. Recall that a valid system including this process would have the form $(\gamma?(P, Q)\|S)\backslash L$, where the channel of action γ belongs to L. As a result, the semantics of the hiding operator $\backslash L$ will resolve the nondeterminism by selecting the value m where m is the number of times action $\overline{\gamma}$ is available in S.
- Rule (PSum) says that a system of q individuals each consisting of the probabilistic choice $\sum_{1\leq i\leq n} p_i{:}P_i$, can evolve into a parallel composition of q_i instances

Table 2. Transition rules for single populations

(Act)	$(\eta.P){:}\langle \mathbf{s}, \ell, q\rangle \xrightarrow{(\eta_{\ell,\mathbf{s}})^q}_n P{:}\langle \mathbf{s}, \ell, q\rangle$	$\eta \neq go\,\ell', \surd$
(Tick)	$(\surd.P){:}\langle \mathbf{s}, \ell, q\rangle \xrightarrow{\surd}_n P{:}\langle \mathbf{s}, \ell, q\rangle$	
(Go)	$(go\,\ell'.P){:}\langle \mathbf{s}, \ell, q\rangle \xrightarrow{(\tau_{go,\ell,\mathbf{s}})^q}_n P{:}\langle \mathbf{s}, \ell', q\rangle$	$(\ell, \ell') \in \mathbf{Nb}$
(Choice)	$(\gamma?(P,Q)){:}\langle \mathbf{s}, \ell, q\rangle \xrightarrow{(\gamma?_{\ell,\mathbf{s}})^m}_n P{:}\langle \mathbf{s}, \ell, m\rangle \| Q{:}\langle \mathbf{s}, \ell, q-m\rangle,$	$0 \leq m \leq q$
(PSum)	$(\sum_{1\leq i\leq n} p_i{:}P_i){:}\langle \mathbf{s}, \ell, q\rangle \xrightarrow{w\langle p_1:q_1,\ldots,p_n:q_n\rangle}_p P_1{:}\langle \mathbf{s}, \ell, q_1\rangle \| \ldots \| P_n{:}\langle \mathbf{s}, \ell, q_n\rangle,\ \sum q_i = q$	
(RConst)	$\dfrac{P{:}\langle \mathbf{s}, \ell, q\rangle \xrightarrow{\alpha} P'{:}\langle \mathbf{s}, \ell, q\rangle}{C{:}\langle \mathbf{s}, \ell, q\rangle \xrightarrow{\alpha} P'{:}\langle \mathbf{s}, \ell, q\rangle}\quad C \stackrel{\text{def}}{=} P{:}\langle \mathbf{s}, \ell, q\rangle$	

of process P_i for each $1 \leq i \leq n$, for all combinations of the $q_i \geq 0$, where $\sum_{1\leq i\leq n} q_i = q$, with probability w given as

$$w_{\langle p_1:q_1,\ldots p_n:q_n\rangle} = \prod_{1\leq i\leq n} p_i^{q_i} \cdot \binom{q - \sum_{1\leq j\leq i-1} q_j}{q_i}$$

- Rule (RConst) expresses the semantics of process constants in the expected way. Recall that $\xrightarrow{\alpha}$ ranges over both $\xrightarrow{\mu}_n$ and $\xrightarrow{w}_p$.

We point that we have not included a rule for the process $(P_1|P_2){:}\langle \mathbf{s}, \ell, q\rangle$ as its semantics is given using structural congruence via equivalence (S4) and rule (Struct) presented below. We may now define the semantics for general systems in Table 3:

- Rule (Time) specifies that if two systems may execute a timed action then their parallel composition may also execute a timed action.
- Rule (Par1) considers the case where one of the components in a parallel composition may execute a timed action and the other a non-timed action. According to the rule the non-timed action takes precedence over the timed action. The latter is postponed until both processes may execute the timed action. Note that this as well as the previous rule ensure that time evolves according to a global clock.
- Rules (Par2) and (Par4) consider probabilistic actions of a parallel composition. The first one specifies that if both components of the composition may execute a probabilistic transition then the composition executes a probabilistic transition with probability the product of the two probabilities. The second rule states that if exactly one of the processes may execute a probabilistic transition then the parallel composition may also execute the transition.
- Rule (Par3) says that if two systems can perform non-deterministic actions β^k and μ, respectively, then their parallel composition can perform the combination of the

two actions assuming that neither of them is the $\surd$ action. The combination of these actions is defined according to Definition 1 below.
- Rule (Res) states that a restricted process may only execute actions involving channels that do not belong to the restriction set L.
- Rule (Struct) specifies that congruent processes have the same transitions.

Table 3. Transition rules for systems

$$\text{(Time)}\ \frac{S_1 \xrightarrow{\surd}_n S_1', S_2 \xrightarrow{\surd}_n S_2'}{S_1\|S_2 \xrightarrow{\surd}_n S_1'\|S_2'} \qquad \text{(Par1)}\ \frac{S_1 \xrightarrow{\mu_1}_n S_1', S_2 \xrightarrow{\surd}_n S_2', \mu_1 \neq \surd}{S_1\|S_2 \xrightarrow{\mu_1}_n S_1'\|S_2}$$

$$\text{(Par2)}\ \frac{S_1 \xrightarrow{w_1}_p S_1', S_2 \xrightarrow{w_2}_p S_2'}{S_1\|S_2 \xrightarrow{w_1\cdot w_2}_p S_1'\|S_2'} \qquad \text{(Par3)}\ \frac{S_1 \xrightarrow{\beta^k}_n S_1', S_2 \xrightarrow{\mu}_n S_2', \mu \neq \surd}{S_1\|S_2 \xrightarrow{\beta^k \diamond \mu}_n S_1'\|S_2'}$$

$$\text{(Par4)}\ \frac{S_1 \xrightarrow{w}_p S_1', S_2 \not\rightarrow_p}{S_1\|S_2 \xrightarrow{w}_p S_1'\|S_2} \qquad \text{(Res)}\ \frac{S \xrightarrow{\alpha} S', \{a | a_{\mathbf{s},\ell}, \overline{a}_{\mathbf{s},\ell} \in \alpha\} \cap L = \emptyset}{S\backslash L \xrightarrow{\alpha} S'\backslash L}$$

$$\text{(Struct)}\ \frac{S \equiv S', S' \xrightarrow{\alpha} S''}{S \xrightarrow{\alpha} S''}$$

We conclude the semantics with the definition of operator $\diamond$. This operation combines a species action β^k and an action μ by grouping together all actions that are the same and turning complementary transitions into τ actions. Formally:

Definition 1. Consider actions β^k and $\mu \neq \surd$ then

$$\beta^k \diamond \mu = \begin{cases} (\tau_{a,\ell,\mathbf{s}})^k \# \mu' & \text{if } \beta = a_{\mathbf{s},\ell}, \mu = (\overline{a}_{\mathbf{s},\ell})^k \# \mu' \\ (\tau_{a,\ell,\mathbf{s}})^k \# (\overline{a}_{\mathbf{s},\ell})^{k'-k} \# \mu' & \text{if } \beta = a_{\mathbf{s},\ell}, \mu = (\overline{a}_{\mathbf{s},\ell})^{k'} \# \mu', k < k' \\ (\tau_{a,\ell,\mathbf{s}})^{k'} \# (a_{\mathbf{s},\ell})^{k-k'} \# \mu' & \text{if } \beta = a_{\mathbf{s},\ell}, \mu = (\overline{a}_{\mathbf{s},\ell})^{k'} \# \mu', k > k' \\ (a_{\mathbf{s},\ell})^{k+k'} \# \mu' & \text{if } \beta = a_{\mathbf{s},\ell}, \mu = (a_{\mathbf{s},\ell})^{k'} \# \mu', k' \geq 0 \\ (\overline{a}_{\mathbf{s},\ell})^{k+k'} \# \mu' & \text{if } \beta = \overline{a}_{\mathbf{s},\ell}, \mu = (\overline{a}_{\mathbf{s},\ell})^{k'} \# \mu', k' \geq 0 \\ (\tau_{a,\ell,\mathbf{s}})^{k+k'} \# \mu' & \text{if } \beta = \tau_{a,\ell,\mathbf{s}}, \mu = (\tau_{a,\ell,\mathbf{s}})^{k'} \# \mu', k' \geq 0 \\ (\tau_{a,\ell,\mathbf{s}})^k \# \mu' & \text{if } \beta = a?_{\mathbf{s},\ell}, \mu = (\overline{a}_{\mathbf{s},\ell})^{k'} \# \mu', k' \geq k \\ (\tau_{a,\ell,\mathbf{s}})^k \# \mu' & \text{if } \beta = \overline{a}?_{\mathbf{s},\ell}, \mu = (a_{\mathbf{s},\ell})^{k'} \# \mu', k' \geq k \\ \bot & \text{otherwise} \end{cases}$$

Example 2. Consider $P_1 \stackrel{\text{def}}{=} a?(P_2, P_3)$, $Q_1 \stackrel{\text{def}}{=} a.Q_2$ and $R_1 \stackrel{\text{def}}{=} \overline{a}.R_2$. Further, suppose that $S \stackrel{\text{def}}{=} (P_1{:}\langle \mathbf{s}_1, \ell, 3\rangle \| Q_1{:}\langle \mathbf{s}_2, \ell, 4\rangle \| R_1{:}\langle \mathbf{s}_3, \ell, 5\rangle)\backslash\{a\}$. Then we have:

$$P_1{:}\langle \mathbf{s}_1, \ell, 3\rangle \xrightarrow{(a?_{\ell,\mathbf{s}_1})^i}_n P_2{:}\langle \mathbf{s}_1, \ell, i\rangle \| P_3{:}\langle \mathbf{s}_1, \ell, 3-i\rangle, \quad 0 \leq i \leq 3$$

$$Q_1{:}\langle \mathbf{s}_2, \ell, 4\rangle \xrightarrow{(a_{\ell,\mathbf{s}_2})^4}_n Q_2{:}\langle \mathbf{s}_2, \ell, 4\rangle \qquad R_1{:}\langle \mathbf{s}_3, \ell, 5\rangle \xrightarrow{(\overline{a}_{\ell,\mathbf{s}_3})^5}_n R_2{:}\langle \mathbf{s}_3, \ell, 5\rangle$$

Additionally,

$$Q_1{:}\langle \mathsf{s}_2, \ell, 4\rangle \| R_1{:}\langle \mathsf{s}_3, \ell, 5\rangle \xrightarrow{(\tau_{\alpha,\ell,\mathsf{s}_3})^4 \# (\overline{a}_{\ell,\mathsf{s}_3})^1}_n Q_2{:}\langle \mathsf{s}_2, \ell, 4\rangle \| R_2{:}\langle \mathsf{s}_3, \ell, 5\rangle$$

and now $P_1{:}\langle \mathsf{s}, \ell, 3\rangle$, by the definition of $\diamond$, may only communicate with the system above via its action $(a?_{\ell,\mathsf{s}_1})^1$, thus yielding:

$$S \xrightarrow{(\tau_{\alpha,\ell,\mathsf{s}_3})^5}_n (P_2{:}\langle \mathsf{s}_1, \ell, 1\rangle \| P_3{:}\langle \mathsf{s}_1, \ell, 2\rangle \| Q_2{:}\langle \mathsf{s}_2, \ell, 4\rangle \| R_2{:}\langle \mathsf{s}_3, \ell, 5\rangle) \backslash \{a\}$$

3 Translating S-PALPS into PRISM

In this section we turn to the problem of model checking S-PALPS. As is the case of PALPS, the operational semantics of S-PALPS gives rise to transition systems that can be easily translated to Markov decision processes (MDPs). As such, model checking approaches that have been developed for MDPs can also be applied to S-PALPS models. PRISM is one such tool developed for the analysis of probabilistic systems. Specifically, it is a probabilistic model checker for Markov decision processes, discrete time Markov chains, and continuous time Markov chains. For our study we are interested in the MDP support of the tool which offers model checking and simulation capabilities.

In [14] we defined a translation of PALPS into the MDP subset of the PRISM language and we explored the possibility of employing PRISM to perform analysis of the semantic models derived from PALPS processes. In this paper, we redefine a translation which implements the synchronous parallel operator of S-PALPS. In the remainder of this section, we give a brief presentation of the PRISM language, sketch an encoding of S-PALPS into PRISM and state its correctness. The full details can be found in [21].

3.1 The PRISM Language

The PRISM language is a simple, state-based language, based on guarded commands. A PRISM model consists of a set of *modules* which can interact with each other on shared actions following the CSP-style of communication [1]. Each module possesses a set of *local variables* which can be written by the module and read by all modules. In addition, there are *global variables* which can be read and written by all modules. The behavior of a module is described by a set of *guarded commands*. When modeling Markov decision processes, these commands take the form:

```
[act] guard -> p1 : u1 + ... + pm :um;
```

where `act` is an optional action label, `guard` is a predicate over the set of variables, $p_i \in (0, 1]$ and u_i are updates of the form: $(x'_1 = u_{i,1})$ & ... & $(x'_k = u_{i,k})$ where $u_{i,j}$ is a function over the variables. Intuitively, such an action is enabled in global state s if s satisfies `guard`. If a command is enabled then it may be executed in which case, with probability p_i, the update u_i is performed by setting the value of each variable x_j to $u_{i,j}(s)$ (where x'_j denotes the new value of variable x_j). We refer the reader to [1] for the full description and the semantics of the PRISM language.

3.2 Encoding S-PALPS into the PRISM Language

Consider an S-PALPS model. This consists of a set of locations, the neighborhood relation **Nb** and a process *System*. In turn, the process *System* satisfies $System \equiv (P_1{:}\langle \mathbf{s}_1, \ell_1, q_1\rangle \| \ldots \| P_n{:}\langle \mathbf{s}_n, \ell_n, q_n\rangle)\backslash L$, where each P_i is a process that may evolve to a set of states, say P_i^j, $1 \leq j \leq m_i$. This allows us to conclude that in any state $System'$ reachable from $System$, we have $System' \equiv (\prod_{i \in I, j \in J, \ell \in \mathbf{Loc}} P_i^j{:}\langle \mathbf{s}_i, \ell, q_{i,j,\ell}\rangle)\backslash L$, that is, at any point in time, there may be an arbitrary number of individuals in each location and of each of the reachable states of the populations.

Based on this observation, our translation of *System* in PRISM consists of a set of $(m_1 + \ldots + m_n) \cdot K$ modules, where K is the total number of locations. Each module captures the behavior of the individuals in the specific state and location. Note that the total number of modules is stable and independent of the precise number of individuals existing in the model. This comes in contrast with our PALPS translation of [14] where the translation of a model contained one module for each individual a fact that resulted in restrictions in space and expressiveness.

In addition to these module definitions, a system translation in PRISM should contain the following global information relating to the system.

- For each species $\mathbf{s}_i$ and each state j in the process description of $\mathbf{s}_i$, the model contains the K global variables $s_{i,j,k}$ $1 \leq k \leq K$ capturing the number of individuals of species $\mathbf{s}_i$ in state j for location k. The variables should be appropriately initialized based on the definition of *System*.
- For each channel a on which synchronization may take place we introduce a variable a_y which counts the number of available inputs on a at location y and a variable $\overline{a_y}$ which counts the number of available outputs on a at location y.
- There exists a global variable *pact* which may take values from $\{0, 1\}$ and expresses if there is a probabilistic action enabled. It is used to give precedence to probabilistic actions over nondeterministic actions. Initially, $pact = 0$. Furthermore, all non-probabilistic actions have $pact = 0$ as a precondition.
- There exists a global variable *tact* which may take values from $\{0, 1\}$ and expresses whether a timed action may take place. For such an action to take place it must be that $tact = 1$. If any process is unable to execute the $\surd$ action then it sets *tact* to 0.

As an example, consider processes P_1, Q_1, R_1 and *System* of Example 2 and suppose that our the system is located on a habitat consisting of 2 patches $\{1, 2\}$. Then the system's PRISM translation should contain the following global variables:

global $s_{1,1,1} : [0, max]$ init 3;
global $s_{2,1,1} : [0, max]$ init 4;
global $s_{3,1,1} : [0, max]$ init 5;
global $s_{1,1,2}, s_{2,1,2}, s_{3,1,2}, s_{1,2,1}, s_{2,2,1}, s_{3,2,1}, s_{1,2,2}, s_{2,2,2}, s_{3,2,2} : [0, max]$ init 0;
global $a_1, \overline{a}_1, a_2, \overline{a}_2 : [0, max]$ init 0;
global $pact, tact : [0, 1]$ init 0;

We now continue to describe how a specific module is described by considering the above example. Specifically, consider process $Q_1 \stackrel{\text{def}}{=} a.Q_2$ and an initial population

$Q_1{:}\langle \mathbf{s}_2, \ell, 4\rangle$. Then, according to the semantics of S-PALPS, these 4 individuals should synchronize on channel a and become individuals in state Q_2. To model this in PRISM these 4 individuals should *flow* from their current state to their next state. To achieve this we need to make the necessary updates on the global variables $\mathbf{s}_{2,1,1}$ and $\mathbf{s}_{2,2,1}$: $\mathbf{s}'_{2,1,1} = \mathbf{s}_{2,1,1} - 4$ and $\mathbf{s}'_{2,2,1} = \mathbf{s}_{2,2,1} + 4$. Furthermore, if state Q_2 is a probabilistic state, the module should set $pact' = 1$.

Now to implement the synchronization of the module with all other modules executing an action we need to execute a sequence of actions as illustrated below:

module $S_{2,1,1}$
$st_{2,1,1} : [0..5]\ init\ 1;$
$n_{2,1,1} : [0..max]$

$[\,]$	$(st_{2,1,1} = 1)\&(s_{2,1,1} > 0) \longrightarrow (st'_{2,1,1} = 2)\&(tact' = 0)\&(n'_{2,1,1} = s_{2,1,1});$
$[\,]$	$(st_{2,1,1} = 1)\&(s_{2,1,1} = 0) \longrightarrow (st'_{2,1,1} = 2)\&(n'_{2,1,1} = s_{2,1,1});$
$[synch]$	$(st_{2,1,1} = 2) \longrightarrow (st'_{2,1,1} = 3);$
$[\,]$	$(st_{2,1,1} = 3)\&(pact = 0) \longrightarrow (a'_\ell = a_\ell + n_{2,1,1})\&(st'_{2,1,1} = 4);$
$[a_\ell]$	$(st_{2,1,1} = 4) \longrightarrow (st'_{2,1,1} = 5);$
$[\,]$	$(st_{2,1,1} = 5) \longrightarrow (a'_\ell = 0)\&\mathrm{updates}(Q_1, Q_2)\&(st'_{2,1,1} = 1);$
$[prob]$	$(st_{2,1,1} = 3)\&(pact = 1) \longrightarrow (st'_{2,1,1} = 1)$

endmodule

Variable $st_{2,1,1}$ in module $S_{2,1,1}$ (initially set to 1) will guide the flow of execution through a sequence of actions. It begins by testing whether there are active individuals of this module and then proceeds to synchronize with the other modules. This synchronization will take place on action $synch$. Subsequently, if there are 1 or more modules in a probabilistic state the module will synchronize with them via action $prob$, otherwise, the module will proceed to make its necessary updates: $s'_{2,1,1} = s_{2,1,1} - n_{2,1,1} \& s'_{2,2,1} = s_{2,2,1} + n_{2,1,1}$. Furthermore, if $Q_2 = \surd.Q_3$ then the update $tact = 1$ is included, whereas if $Q_2 = p_1 : T_1 \oplus \ldots p_n : T_n$, then the update $pact = 1$ is included.

Let us now discuss some characteristics of the above translation which are also relevant to the translations of process constructs other than $a.P$. To begin with, the module begins by setting $tact = 0$, assuming that there are active individuals in this state. Thus, it is ensured that nondeterministic actions take precedence over timed actions. In addition, variable $n_{2,1,1}$ is used to store the initial population of the module. This is necessary because other processes may 'flow' into this module and the value of $s_{2,1,1}$ may not correctly reflect the initial size of the population. Furthermore, we point out that if a probabilistic action is available ($pact = 1$) then the process will synchronize on this action and return to its initial state. We also note that it is not possible to collapse e.g. states 2 and 3 of the module because PRISM does not allow to execute updates on global variables within synchronization actions. Finally, we observe that in the case of channel communication, the module records the number of available inputs and outputs on a channel at a certain location (update $a'_\ell = a_\ell + n$) and continues to synchronize on action a_ℓ. This is required for translating the restriction construct where we must check that the number of inputs and outputs performed on the channel are equal.

In a similar manner we may translate all constructs of the S-PALPS syntax by allowing processes to flow from one module to the next. The only source of additional complexity concerns the probabilistic operator. This is because, given a set of q individuals executing a probabilistic action, the set of possible resulting locations is dependent on q (see rule (Prob) in the semantics). To address this point, in a translation of a probabilistic choice, we enumerate the possible outcomes for all possible values of q.

Regarding the correctness of the proposed translation, we have proved the correspondence between an S-PALPS model S and its PRISM translation, denoted by $[\![S]\!]$, by showing that a move of S can be simulated by its translation $[\![S]\!]$ in a finite number of steps and vice versa. In proving this we employ the following notation: Given a PRISM model M, we write $M \stackrel{\alpha,p_i}{\longrightarrow} M'$ if M contains an action `[`α`] guard -> `$p_1 : u_1 + \ldots + p_m : u_m$ where `guard` is satisfied in model M and execution of u_i gives rise to model M'. If a command has no label α and $p_i = 1$ then we write $M \longrightarrow M'$. Finally, we write $M \stackrel{\alpha_1 \# \ldots \# \alpha_n}{\Longrightarrow} M'$ if $M (\longrightarrow)^* \stackrel{\alpha_1,1}{\longrightarrow} (\longrightarrow)^* \ldots (\longrightarrow)^* \stackrel{\alpha_m,1}{\longrightarrow} (\longrightarrow)^* M'$.

Theorem 1. For any PALPS process S and its PRISM translation $[\![S]\!]$ we have:

1. If $S \stackrel{\surd}{\longrightarrow}_n S'$ then $[\![S]\!] \stackrel{synch}{\Longrightarrow} \stackrel{tick}{\Longrightarrow} [\![S']\!]$.
2. If $S \stackrel{\mu}{\longrightarrow}_n S'$ then $[\![S]\!] \stackrel{synch}{\Longrightarrow} \stackrel{\mu}{\Longrightarrow} [\![S']\!]$.
3. If $S \stackrel{w}{\longrightarrow}_p S'$ then $[\![S]\!] \stackrel{synch}{\Longrightarrow} \stackrel{prob,w}{\longrightarrow} [\![S']\!]$.

Theorem 1 establishes that each transition of S can be mimicked by its translation in a sequence of steps. A similar result holds in the opposite direction. The details and proofs of these results can be found in [21].

4 Case Study: Eleonora's Falcon Population Dynamics

In this section we study the Eleonora's Falcon [22] using S-PALPS. Eleonora's falcon is a migrant species that breeds on Mediterranean islands and winters on islands of the Indian Ocean and along the eastern African coast. A large part of the world population concentrates on a small number of islands in the Aegean Sea [8]. In Europe, the species is considered as rare and hence of local conservation importance because, although not globally threatened, its world population is below 10,000 breeding pairs and its survival in Europe is highly dependent on the breeding conditions on the islands on which it concentrates. In particular, the breeding calendar of the Eleonora's falcon overlaps with the summer months when tourism peaks in most Mediterranean islands while the climatic changes may also have consequences on the reproduction of the species.

The life cycle of the Eleonora's Falcon is defined as follows. The juveniles disperse from the island during their first year of life. It takes them approximately four years to achieve sexual maturity and they only come back to the island once they reach this age. When they return, they choose a nest. For the sake of model simplicity we consider two types of nests in terms of provision of shelter to the breeding pairs and their young: exposed nests (e.g., to predators, sun, humans and wind) and less-exposed nests. The choice of the nest determines the survival probability of the offspring. According to studies, first-year breeders usually do not choose less-exposed nests. This choice is reserved for mature adults, who are not guaranteed to acquire a less-exposed nest due to the limited number of such nests [17]. In what follows we construct a model of the Eleonora's falcon ecosystem in S-PALPS.

Spatial domains. We consider two spatial domains which we model as two S-PALPS locations: The island where the colony lives, ℓ_1, and the territory outside the island, ℓ_2. The spatial location of the nests on the island is not crucial for our model, hence the use of a single breeding location.

Species. To enable the modeling of the system we define two S-PALPS species in our model: the *Eleonora's falcon* (f) and the *less-exposed nests* (le). We then model the selection of less-exposed nests as a predator-prey problem.

Processes. We associate each of the above species with an S-PALPS description. To model nests, we create a group of n less-exposed nests as $LeNest{:}\langle le, l_1, n\rangle$ such that

$$LeNest \stackrel{\text{def}}{=} prey.LeNest' + \surd.LeNest \qquad LeNest' \stackrel{\text{def}}{=} \overline{release}.LeNest + \surd.LeNest'$$

The life cycle of a falcon begins in the newborn/juvenile state (process J_0 below). In this state an individual disperses to location ℓ_2 and waits for 4 years which, in our model, consists of 4 occurrences of action $\surd$, before becoming a first-year breeder adult (process $A_1^{\bigcirc}$ below). Note that not all juveniles will mature to adults. In fact, a juvenile may die with a mortality rate of 78% [16].

$$J_0 \stackrel{\text{def}}{=} go\,\ell_2.\surd.\surd.\surd.\surd.J_4 \qquad J_4 \stackrel{\text{def}}{=} (0.78 : 0 \oplus 0.22 : A_1^{\bigcirc})$$

Moving on to the adult population, we observe that while male adults are responsible for choosing the nest and the female, and to hunt, in our model, for the sake of simplicity, we have opted to abstract away from a falcon's gender. We believe that this simplification does not affect the faithfulness of the model as there is no indication that the percentages of males and females differ significantly, nor that the probability to die during dispersal depends on the gender, and also because adult males and females live in pairs and are considered monogamous.

Thus we model by A the notion of an adult pair. There are two types of such adult pairs: first-year breeders who have no experience in choosing less-exposed nests and second-year or older adult pairs whose experience allows them to select less-exposed nests, if such nests are available [17]. Depending on the nest that a pair chooses, there are different probabilities to have an offspring of size 0,1,2 or 3 during the breeding season. We adopt the reproduction rates from [22] appropriately weighted so that only half of the offspring is produced (to account for pairs). In the model below we write ε_i for the probability that an offspring of size i is produced in an exposed nest and λ_i for the probability that an offspring of size i is produced in a less-exposed nest. Furthermore, we write A_1, A and M for a first-year breeder pair, a mature pair in the phase of reproduction and a mature pair in the phase of possible mortality, respectively. Finally, we use the superscripts $^{\bigcirc}$, $^{◖}$ and $^{\bullet}$ to denote a state of no nest, an exposed nest and a less-exposed nest, respectively.

The behavior of a pair proceeds as follows. A first-year breeder pair, returns to the island. It chooses an exposed nest and proceeds as a mature adult pair in an exposed nest. A mature adult pair selects a less-exposed nest, if one is available (i.e. there is an input available on channel *prey*) and an exposed nest, otherwise. It then produces offspring, leaves the island and goes through a mortality phase. If it survives it executes action $\surd$ and returns to its initial phase. The mortality rate of an adult pair is equal to 13%. Note that, in the mortality phase, a pair in a less-exposed nest releases its nest.

$$
\begin{aligned}
A_1^{\bigcirc} &\stackrel{\text{def}}{=} go\,\ell_1.A^{\CIRCLE} \\
A^{\bigcirc} &\stackrel{\text{def}}{=} \overline{prey}?(A^{\LEFTcircle}, A^{\CIRCLE}) \\
A^{\CIRCLE} &\stackrel{\text{def}}{=} \varepsilon_0 : M^{\CIRCLE} \oplus \varepsilon_1 : J_0|M^{\CIRCLE} \oplus \varepsilon_2 : J_0|J_0|M^{\CIRCLE} \oplus \varepsilon_3 : J_0|J_0|J_0|M^{\CIRCLE} \\
A^{\LEFTcircle} &\stackrel{\text{def}}{=} \lambda_0 : M^{\LEFTcircle} \oplus \lambda_1 : J_0|M^{\LEFTcircle} \oplus \lambda_2 : J_0|J_0|M^{\LEFTcircle} \oplus \lambda_3 : J_0|J_0|J_0|M^{\LEFTcircle} \\
M^{\CIRCLE} &\stackrel{\text{def}}{=} go\,\ell_2.(0.13 : 0 \oplus 0.87 : \surd.A^{\bigcirc}) \\
M^{\LEFTcircle} &\stackrel{\text{def}}{=} \overline{release}.go\,\ell_2.\,(0.13 : 0 \oplus 0.87 : \surd.A^{\bigcirc})
\end{aligned}
$$

Our system is defined below. It consists of n nests and m adult pairs with no nest.

$$System \stackrel{\text{def}}{=} (LeNest{:}\langle le, l_1, n\rangle | A^{\bigcirc}{:}\langle f, l_1, m\rangle) \backslash \{prey\}$$

5 Analysis in PRISM

In this section, we report on some of the results we obtained by applying our methodology for studying the population dynamics of the Eleonora's falcon. To begin with we translated our PALPS model into PRISM by following the encoding presented in Section 3. For our experiments, we took advantage of the model checking capabilities of PRISM and we checked properties by using the *model-checking by simulation* option, referred to as *confidence interval (CI)* simulation method. The property we experimented with is `R =?[I = k]`. This property is a reward-based property that computes the average state instant reward at time *k*. We were interested to study the expected size of the population. For this, we associate to each state a reward representing this size.

We were interested in studying various properties of this model. One of these properties involved assessing the stability of the model for different sizes of the initial population. To achieve this, we considered initial populations of 20, 40 and 60 adult pairs and we studied the growth of the population for a duration of approximately 10 years. These results are reported in Figure 1. Subsequently, we were interested to study the composition of the population in terms of juvenile and adult pairs. In Figure 2 we present the results obtained for an initial population of 40 adult pairs.

Another property we were interested to study is the sensitivity of the population to changes in the local conditions. These conditions may affect the probabilities associated with reproduction and, in particular, the survival rate of the offspring of a falcon pair. To study this property we analyzed the impact of changing the reproduction rates in both exposed and less-exposed nests. Specifically, we increased (decreased) the probabilities of 0 fledglings surviving by 3% and 6% while appropriately decreasing (increasing) the probabilities of 1, 2 and 3 fledglings surviving. These results are presented in Figure 3. We observe that the results reveal a fair degree of stability in the evolution of the species and a relative insensitivity to small changes in the local conditions. For a complete presentation of the results obtained we refer the reader to [21].

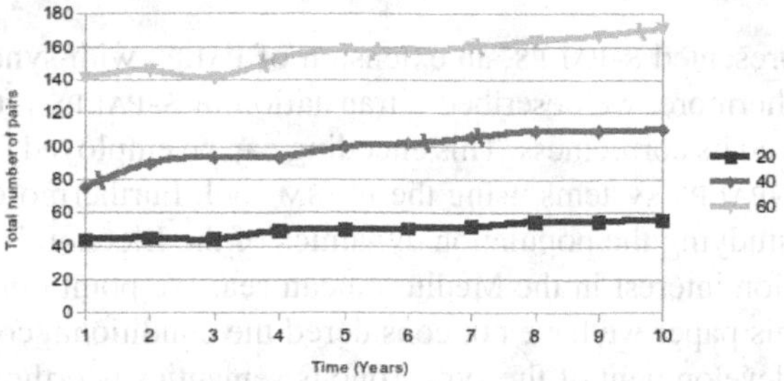

Fig. 1. Expected number of total pairs (juveniles and adults) vs time for an initial population of 20, 40 and 60 pairs of adults

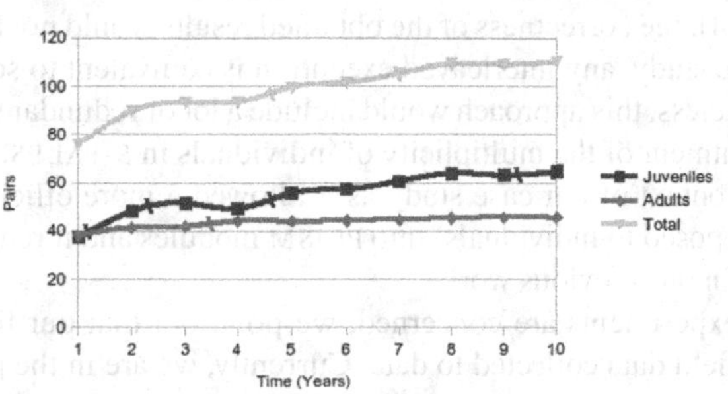

Fig. 2. Expected number of total pairs, juveniles pairs and adult pairs vs time for an initial population of 40 pairs of adults

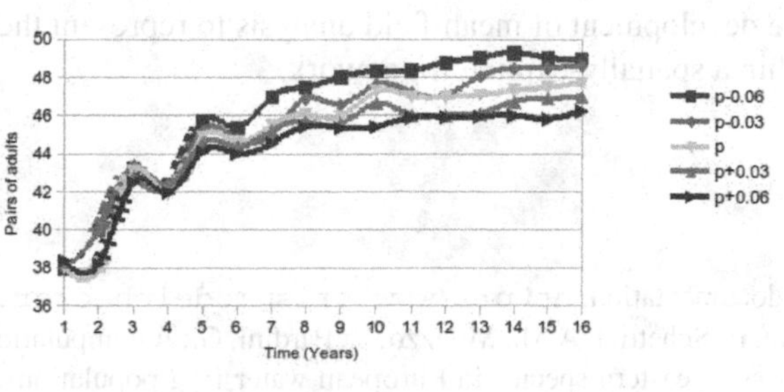

Fig. 3. Expected number of pairs of adults vs time with an initial population of 40 pairs of adults, for different values of the probability p that zero fledglings survive from an offspring of a pair

6 Conclusions

In this paper we presented S-PALPS, an extension of PALPS with synchronous parallel composition. Furthermore, we described a translation of S-PALPS into the PRISM language and we proved its correctness. This encoding can be employed for simulating and model checking S-PALPS systems using the PRISM tool. Furthermore, we applied our methodology for studying the population dynamics of the Eleonora's falcon, a species of local conservation interest in the Mediterranean sea. We point out that for the sake of simplicity, in this paper we have not considered the conditional construct of PALPS. We note that the development of the synchronous semantics is orthogonal to this construct. This can be observed in [21] where we present the complete framework where synchronous parallelism and conditional behaviors are combined.

We have observed that the adoption of a synchronous parallel composition in S-PALPS enables a more faithful model of our case study: The use of the synchronous parallel composition was especially useful as it separated the phases of migration (action $go\,\ell_1$) from the reproduction-related actions. Note that if these actions were allowed to interleave, as in [14], the correctness of the obtained results would not be affected: in the context of the case study, any interleaved execution is equivalent to some synchronous execution. Nonetheless, this approach would include a lot of redundant nondeterminism. Regarding the treatment of the multiplicity of individuals in S-PALPS, this proved to be beneficial in the context of our case study as it allowed a more efficient translation of populations (as opposed to individuals) into PRISM modules and it removed restrictions that were present in our previous work.

As far as our experiments are concerned, we point out that our findings appear to be compatible to field data collected to date. Currently, we are in the process of further calibrating our model based on a wider range of field data and extending our analysis on the population dynamics of the species. Furthermore, we are implementing a tool to automatically translate PALPS systems into PRISM models. Our intention is to provide a complete methodology that allows the user to model check classes of properties using PRISM without the need of the user manually manipulating the PRISM code.

As future work, we are interested in applying our methodology to other case studies from the local habitat and, in particular, to employ model checking for studying their behavior. Finally, an interesting future research direction would be extend the work of [10] towards the development of mean-field analysis to represent the average behavior of systems within a spatially-explicit framework.

References

1. Online PRISM documentation, `http://www.prismmodelchecker.org/doc/`
2. Barbuti, R., Bove, P., Schettini, A.M., Milazzo, P., Pardini, G.: A computational formal model of the invasiveness of eastern species in European water frog populations. In: Counsell, S., Núñez, M. (eds.) SEFM 2013. LNCS, vol. 8368, pp. 329–344. Springer, Heidelberg (2014)
3. Basuki, T.A., Cerone, A., Barbuti, R., Maggiolo-Schettini, A., Milazzo, P., Rossi, E.: Modelling the dynamics of an aedes albopictus population. In: Proceedings of AMCA-POP 2010. EPTCS, vol. 33, pp. 18–36 (2010)

4. Berec, L.: Techniques of spatially explicit individual-based models: construction, simulation, and mean-field analysis. Ecological Modeling 150, 55–81 (2002)
5. Besozzi, D., Cazzaniga, P., Pescini, D., Mauri, G.: Modelling metapopulations with stochastic membrane systems. BioSystems 91(3), 499–514 (2008)
6. Cardona, M., Colomer, M.A., Margalida, A., Palau, A., Pérez-Hurtado, I., Pérez-Jiménez, M.J., Sanuy, D.: A computational modeling for real ecosystems based on P systems. Natural Computing 10(1), 39–53 (2011)
7. Fu, S.C., Milne, G.: A Flexible Automata Model for Disease Simulation. In: Sloot, P.M.A., Chopard, B., Hoekstra, A.G. (eds.) ACRI 2004. LNCS, vol. 3305, pp. 642–649. Springer, Heidelberg (2004)
8. Kassara, C., Fric, J., Gschweng, M., Sftenthourakis, S.: Complementing the puzzle of Eleonora's falcon (Falco eleonorae) migration: new evidence from an eastern colony in the Aegean Sea. Journal of Ornithology 153, 839–848 (2012)
9. McCaig, C., Norman, R., Shankland, C.: Process Algebra Models of Population Dynamics. In: Horimoto, K., Regensburger, G., Rosenkranz, M., Yoshida, H. (eds.) AB 2008. LNCS, vol. 5147, pp. 139–155. Springer, Heidelberg (2008)
10. McCaig, C., Norman, R., Shankland, C.: From individuals to populations: A mean field semantics for process algebra. Theoretical Computer Science 412(17), 1557–1580 (2011)
11. Milner, R.: A Calculus of Communicating Systems. Springer (1980)
12. Pescini, D., Besozzi, D., Mauri, G., Zandron, C.: Dynamical probabilistic P- systems. Journal of Foundations of Computer Science 17(1), 183–204 (2006)
13. Philippou, A., Toro, M.: Process Ordering in a Process Calculus for Spatially-Explicit Ecological Models. In: Counsell, S., Núñez, M. (eds.) SEFM 2013. LNCS, vol. 8368, pp. 345–361. Springer, Heidelberg (2014)
14. Philippou, A., Toro, M., Antonaki, M.: Simulation and Verification for a Process Calculus for Spatially-Explicit Ecological Models. Scientific Annals of Computer Science 23(1), 119–167 (2013)
15. Păun, G.: Membrane Computing: An Introduction. Springer (2002)
16. Ristow, D., Scharlau, W., Wink, M.: Population structure and mortality of Eleonora's falcon Falco eleonorae. In: Raptors in the Modern World Working Group on Birds of Prey, pp. 321–326 (1989)
17. Ristow, D., Wink, M., Wink, C., Friemann, H.: Biologie des Eleonorenfalken (Falco eleonorae). Das Brutreifealter der Weibchen 124, 291–293 (1983)
18. Romero-Campero, F.J., Gheorghe, M., Bianco, L., Pescini, D., Jesús Pérez-Jímenez, M., Ceterchi, R.: Towards Probabilistic Model Checking on P Systems Using PRISM. In: Hoogeboom, H.J., Păun, G., Rozenberg, G., Salomaa, A. (eds.) WMC 2006. LNCS, vol. 4361, pp. 477–495. Springer, Heidelberg (2006)
19. Sumpter, D.J.T., Blanchard, G.B., Broomhear, D.S.: Ants and Agents: a Process Algebra Approach to Modelling Ant Colony Behaviour. Bulletin of Mathematical Biology 63, 951–980 (2001)
20. Tofts, C.: Processes with probabilities, priority and time. Formal Aspects of Computing 6, 536–564 (1994)
21. Toro, M., Philippou, A., Kassara, C., Sfendourakis, S.: Synchronous parallel composition in a process calculus for ecological models. Technical report, Department of Computer Science, University of Cyprus (2014), `http://www.cs.ucy.ac.cy/~annap/spalps-tr.pdf`
22. Xirouchakis, S., Fric, J., Kassara, C., Portolou, D., Dimalexis, A., Karris, G., Barboutis, C., Latsoudis, P., Bourdakis, S., Kakalis, E., Sftenthourakis, S.: Variation in breeding parameters of Eleonora's falcon (Falco eleonorae) and factors affecting its reproductive performance. Ecological Research 27, 407–416 (2012)

Finite Vector Spaces as Model of Simply-Typed Lambda-Calculi

Benoît Valiron[1,⋆] and Steve Zdancewic[2,⋆⋆]

[1] PPS, UMR 7126, Univ Paris Diderot, Sorbonne Paris Cité, F-75205 Paris, France
[2] University of Pennsylvania, Philadelphia, US

Abstract. In this paper we use finite vector spaces (finite dimension, over finite fields) as a non-standard computational model of linear logic. We first define a simple, finite PCF-like lambda-calculus with booleans, and then we discuss two finite models, one based on finite sets and the other on finite vector spaces. The first model is shown to be fully complete with respect to the operational semantics of the language. The second model is not complete, but we develop an algebraic extension of the finite lambda calculus that recovers completeness. The relationship between the two semantics is described, and several examples based on Church numerals are presented.

1 Introduction

A standard way to study properties of functional programming languages is via denotational semantics. A denotational semantics (or model) for a language is a mathematical representation of its programs [32], and the typical representation of a term is a function whose domain and codomain are the data-types of input and output. This paper is concerned with a non-standard class of models based on finite vector spaces.

The two languages we will consider are based on PCF [27] – the laboratory mouse of functional programming languages. PCF comes as an extension of simply-typed lambda-calculus with a call-by-name reduction strategy, basic types and term constructs, and can be easily extended to handle specific effects. Here, we define $\mathbf{PCF}_f$ as a simple lambda-calculus with pairs and booleans, and $\mathbf{PCF}_f^{alg}$, its extension to linear combinations of terms.

There has been much work and progress on various denotational models of PCF, often with the emphasis on trying to achieve full abstraction. The seminal works are using term models [21], cpos [22] or game semantics [1], while more recent works use quantitative semantics of linear logic [12] and discuss probabilistic extensions [10] or non-determinism [6].

As a category, a model for a PCF language is at least required to be cartesian closed to model internal morphisms and pairing. An expressive class of cartesian closed categories can be made of models of linear logic, by considering the (co)Kleisli category stemming from the modality "!". Although the models that are usually considered are

⋆ Partially supported by the ANR project ANR-2010-BLAN-021301 LOGOI.
⋆⋆ This work has been funded in part by NFS grant CCF-1421193.

G. Ciobanu and D. Méry (Eds.): ICTAC 2014, LNCS 8687, pp. 442–459, 2014.

rich and expressive [6, 9, 10], "degenerate" models nevertheless exist [15, 24]. The consequences of the existence of such models of PCF have not been explored thoroughly.

In this paper, we consider two related finitary categories: the category of finite sets and functions **FinSet** and the category of finite vector spaces and linear functions **FinVec**, i.e. finite-dimensional vector spaces over a finite field. The adjunction between these two categories is known in the folklore to give a model of linear logic [23], but the computational behavior of the corresponding coKleisli category $\mathbf{FinVec}_!$ as a model of PCF has not been studied until now.

The primary motivation for this work is simple curiosity: What do the vectors interpreting lambda calculus terms look like? Though not the focus of this paper, one could imagine that the ability to encode programming language constructs in the category of vector spaces might yield interesting applications. For instance, a Matlab-like programming language that natively supports rich datatypes and first-class functions, all with the same semantic status as "vectors" and "matrices." A benefit of this design would be the possibility of "typed" matrix programming, or perhaps sparse matrix representations based on lambda terms and their semantics. The algebraic lambda calculus sketched in this paper is a (rudimentary) first step in this direction. Conversely, one could imagine applying techniques from linear algebra to lambda calculus terms. For instance, finite fields play a crucial role in cryptography, which, when combined with programming language semantics, might lead to new algorithms for homomorphic encryption.

The goal here is more modest, however. The objective of the paper is to study how the two models **FinSet** and $\mathbf{FinVec}_!$ fit with respect to the language $\mathbf{PCF}_f$ and its algebraic extension $\mathbf{PCF}_f^{alg}$. In particular, we consider the usual three gradually more constraining properties: *adequacy*, *full abstraction* and *full completeness*. A semantics is *adequate* if whenever terms of some observable type (`Bool` for example) are operationally equivalent then their denotations match. An adequate semantics is "reasonable" in the sense that programs and their representations match at ground type. The semantics is *fully abstract* if operational equivalence and equality of denotation are the same thing for all types. In this situation, programs and their denotations are in correspondence at all types, but the model can contain non-representable elements. Finally, the semantics is *fully complete* if moreover, every element in the image of a type A is representable by a term in the language. With such a semantics, the set of terms and its mathematical representation are fully correlated. If a semantics is fully complete, then it is fully abstract and if it is fully abstract, then it is adequate.

Results. This paper presents the first account of the interpretation of two PCF-like languages in finite vector spaces. More specifically, we show that the category of finite sets **FinSet** forms a fully complete model for the language $\mathbf{PCF}_f$, and that the coKleisli category $\mathbf{FinVec}_!$ is adequate but not fully-abstract: this model has too many points compared to what one can express in the language. We present several examples of the encoding of Church numerals to illustrate the model. We then present an algebraic extension $\mathbf{PCF}_f^{alg}$ of $\mathbf{PCF}_f$ and show that $\mathbf{FinVec}_!$ forms a fully complete model for this extension. We discuss the relationship between the two languages and show how to encode the extension within $\mathbf{PCF}_f$.

Related Works. In the literature, finite models for lambda-calculi are commonly used. For example, Hillebrand analyzes databases as finite models of the simply-typed lambda calculus [14]. Salvati presents a model based on finite sets [25], while Selinger presents models based on finite posets [28]. Finally, Solovev [29] relate the equational theory of cartesian closed categories with the category of finite sets.

More general than vector spaces, various categories of modules over semirings, as standard models of linear logic have been studied as computational models: sets and relations [6], finiteness spaces [9], probabilistic coherent spaces [10], *etc.*

As models of linear logic, finite vector spaces are folklore [23] and appear as side examples of more general constructions such as Chu spaces [24] or glueing [15]. Computationally, Chu spaces (and then to some extent finite vector spaces) have been used in connection with automata [24]. Finally, recently finite vector spaces have also been used as a toy model for quantum computation (see e.g. [16, 26]).

Algebraic lambda-calculi, that is, lambda-calculi with a vectorial structure have been first defined in connection with finiteness spaces [11,31]. Another approach [2,3] comes to a similar type of language from quantum computation. The former approach is call-by-name while the latter is call-by-value. A general categorical semantics has been developed [30] but no other concrete models have been considered.

Plan of the paper. The paper is shaped as follows. Section 2 presents a finite PCF-style language $\mathbf{PCF}_f$ with pairs and booleans, together with its operational semantics. Section 3 presents the category **FinSet** of finite sets and functions, and discusses its properties as a model of the language $\mathbf{PCF}_f$. Section 4 describes finite vector spaces and shows how to build a model of linear logic from the adjunction with finite sets. Section 4.4 discusses the corresponding coKleisli category as a model of $\mathbf{PCF}_f$ and presents some examples based on Church numerals. As $\mathbf{PCF}_f$ is not fully-abstract, Section 5 explains how to extend the language to better match the model. Finally, Section 6 discusses various related aspects: the relationship between $\mathbf{PCF}_f$ and its extension, other categories in play, and potential generalization of fields.

2 A Finite PCF-Style Lambda Calculus

We pick a minimal finite PCF-style language with pairs and booleans. We call it $\mathbf{PCF}_f$: it is intrinsically typed (i.e. Church-style: all subterms are defined with their type) and defined as follows.

$$\begin{array}{lcl} M, N, P & ::= & x \mid \lambda x.M \mid MN \mid \pi_l(M) \mid \pi_r(M) \mid \langle M, N \rangle \mid \star \mid \\ & & \texttt{tt} \mid \texttt{ff} \mid \texttt{if}\, M \,\texttt{then}\, N \,\texttt{else}\, P \mid \texttt{let}\, \star = M \,\texttt{in}\, N \\ A, B & ::= & \texttt{Bool} \mid A \to B \mid A \times B \mid \mathbf{1}. \end{array}$$

Values, including "lazy" pairs (that is, pairs of arbitrary terms, as opposed to pairs of values), are inductively defined by $U, V ::= x \mid \lambda x.M \mid \langle M, N \rangle \mid \star \mid \texttt{tt} \mid \texttt{ff}$. The terms consist of the regular lambda-terms, plus specific term constructs. The terms $\texttt{tt}$ and $\texttt{ff}$ respectively stand for the booleans True and False, while $\texttt{if} - \texttt{then} - \texttt{else} -$ is the boolean test operator. The type $\texttt{Bool}$ is the type of the booleans. The term $\star$ is the unique value of type $\mathbf{1}$, and $\texttt{let}\,\star = - \,\texttt{in}\, -$ is the evaluation of a "command", that is, of a term evaluating to $\star$. The term $\langle -, - \rangle$ is the pairing operation, and π_l and π_r stand for the left and right projections. The type operator $(\times)$ is used to type pairs, while $(\to)$ is used to type lambda-abstractions and functions.

Table 1. Typing rules for the language $\mathbf{PCF}_f$

$$\frac{}{\Delta, x : A \vdash x : A} \quad \frac{}{\Delta \vdash \star : \mathbf{1}} \quad \frac{}{\Delta \vdash \mathtt{tt}, \mathtt{ff} : \mathtt{Bool}} \quad \frac{\Delta, x : A \vdash M : B}{\Delta \vdash \lambda x.M : A \to B} \quad \frac{\Delta \vdash M : A_l \times A_r}{\Delta \vdash \pi_i(M) : A_i}$$

$$\frac{\Delta \vdash M : A \to B \quad \Delta \vdash N : A}{\Delta \vdash MN : B} \quad \frac{\Delta \vdash M : A \quad \Delta \vdash N : B}{\Delta \vdash \langle M, N \rangle : A \times B} \quad \frac{\Delta \vdash M : \mathtt{Bool} \quad \Delta \vdash N_1, N_2 : A}{\Delta \vdash \mathtt{if}\, M\, \mathtt{then}\, N_1\, \mathtt{else}\, N_2 : A} \quad \frac{\Delta \vdash M : \mathbf{1} \quad \Delta \vdash N : A}{\Delta \vdash \mathtt{let}\, \star = M\, \mathtt{in}\, N : A}$$

A typing judgment is a sequent of the form $\Delta \vdash M : A$, where Δ is a typing context: a collection of typed variables $x : A$. A typing judgment is said to be *valid* when there exists a valid typing derivation built out of the rules in Table 1.

Note that since terms are intrinsically typed, for any valid typing judgment there is only one typing derivation. Again because the terms are intrinsically typed, by abuse of notation when the context is clear we use $M : A$ instead of $\Delta \vdash M : A$.

Notation 1. When considering typing judgments such as $x : A \vdash M : B$ and $y : B \vdash N : C$, we use categorical notation to denote the composition: $M; N$ stands for the (typed) term $x : A \vdash (\lambda y.N)M : C$, also written as $A \xrightarrow{M} B \xrightarrow{N} C$. We also extend pairs to finite products as follows: $\langle M_1, M_2, \dots \rangle$ is the term $\langle M_1, \langle M_2, \langle \dots \rangle \rangle \rangle$. Projections are generalized to finite products with the notation π_i projecting the i-th coordinate of the product. Types are extended similarly: $A \times \cdots \times A$, also written as $A^{\times n}$, is defined as $A \times (A \times (\cdots))$.

2.1 Small Step Semantics

The language is equipped with a call-by-name reduction strategy: a term M reduces to a term M', denoted with $M \to M'$, when the reduction can be derived from the rules of Table 2. We use the notation $\to^*$ to refer to the reflexive transitive closure of $\to$.

Lemma 2. *(1) For any well-typed term $M : A$, either M is a value or M reduces to some term $N : A$. (2) The only closed value of type $\mathbf{1}$ is $\star$ and the only closed values of type* `Bool` *are* `tt` *and* `ff`*. (3) The language $\mathbf{PCF}_f$ is strongly normalizing.*

Proof. The fact that the language $\mathbf{PCF}_f$ is strongly normalizing comes from the fact that it can be easily encoded in the strongly normalizing language system F [13]. □

Table 2. Small-step semantics for the language $\mathbf{PCF}_f$

$$(\lambda x.M)N \to M[N/x] \qquad \pi_l\langle M, N \rangle \to M \qquad \mathtt{if\ tt\ then}\, M\, \mathtt{else}\, N \to M$$

$$\mathtt{let}\, \star = \star\, \mathtt{in}\, M \to M \qquad \pi_r\langle M, N \rangle \to N \qquad \mathtt{if\ ff\ then}\, M\, \mathtt{else}\, N \to N$$

$$\frac{M \to M'}{MN \to M'N} \quad \frac{M \to M'}{\pi_l(M) \to \pi_l(M')} \quad \frac{M \to M'}{\pi_r(M) \to \pi_r(M')}$$

$$\frac{M \to M'}{\mathtt{if}\, M\, \mathtt{then}\, N_1\, \mathtt{else}\, N_2 \to \mathtt{if}\, M'\, \mathtt{then}\, N_1\, \mathtt{else}\, N_2} \quad \frac{M \to M'}{\mathtt{let}\, \star = M\, \mathtt{in}\, N \to \mathtt{let}\, \star = M'\, \mathtt{in}\, N}$$

Table 3. Denotational semantics for the language $\mathbf{PCF}_f$

$$[\![\Delta, x : A \vdash x : A]\!]^{\mathrm{set}} : (d,a) \longmapsto a \qquad [\![\Delta \vdash \langle M, N \rangle : A \times B]\!]^{\mathrm{set}} : d \longmapsto \langle [\![M]\!]^{\mathrm{set}}(d), [\![N]\!]^{\mathrm{set}}(d) \rangle$$

$$[\![\Delta \vdash \mathtt{tt} : \mathtt{Bool}]\!]^{\mathrm{set}} : d \longmapsto \mathtt{tt} \qquad [\![\Delta \vdash MN : B]\!]^{\mathrm{set}} : d \longmapsto [\![M]\!]^{\mathrm{set}}(d)([\![N]\!]^{\mathrm{set}}(d))$$

$$[\![\Delta \vdash \mathtt{ff} : \mathtt{Bool}]\!]^{\mathrm{set}} : d \longmapsto \mathtt{ff} \qquad [\![\Delta \vdash \pi_l(M) : A]\!]^{\mathrm{set}} = [\![M]\!]^{\mathrm{set}}; \pi_l$$

$$[\![\Delta \vdash \star : \mathbf{1}]\!]^{\mathrm{set}} : d \longmapsto \star \qquad [\![\Delta \vdash \pi_r(M) : B]\!]^{\mathrm{set}} = [\![M]\!]^{\mathrm{set}}; \pi_r$$

$$[\![\Delta \vdash \lambda x.M : A \to B]\!]^{\mathrm{set}} = d \longmapsto (a \mapsto [\![M]\!]^{\mathrm{set}}(d,a)) \quad [\![\Delta \vdash \mathtt{let}\,\star = M \,\mathtt{in}\, N : A]\!]^{\mathrm{set}} = [\![N]\!]^{\mathrm{set}}$$

$$[\![\Delta \vdash \mathtt{if}\, M \,\mathtt{then}\, N \,\mathtt{else}\, P : A]\!]^{\mathrm{set}} = d \longmapsto \begin{cases} [\![N]\!]^{\mathrm{set}}(d) \text{ if } [\![M]\!]^{\mathrm{set}}(d) = \mathtt{tt}, \\ [\![P]\!]^{\mathrm{set}}(d) \text{ if } [\![M]\!]^{\mathrm{set}}(d) = \mathtt{ff}. \end{cases}$$

2.2 Operational Equivalence

We define the operational equivalence on terms in a standard way. A *context* $C[-]$ is a "term with a hole", that is, a term consisting of the following grammar:

$$\begin{aligned} C[-] ::= {} & x \mid [-] \mid \lambda x.C[-] \mid C[-]N \mid MC[-] \mid \pi_l(C[-]) \mid \pi_r(C[-]) \mid \langle C[-], N \rangle \mid \\ & \langle M, C[-] \rangle \mid \star \mid \mathtt{tt} \mid \mathtt{ff} \mid \mathtt{if}\, C[-] \,\mathtt{then}\, N \,\mathtt{else}\, P \mid \mathtt{if} M \mathtt{then}\, C[-] \,\mathtt{else}\, P \mid \\ & \mathtt{if}\, M \,\mathtt{then}\, N \mathtt{else}\, C[-] \mid \mathtt{let}\,\star = C[-] \,\mathtt{in}\, M \mid \mathtt{let}\,\star = M \,\mathtt{in}\, C[-]. \end{aligned}$$

The hole can bind term variables, and a well-typed context is defined as for terms. A closed context is a context with no free variables.

We say that $\Delta \vdash M : A$ and $\Delta \vdash N : A$ are operationally equivalent, written $M \simeq_{\mathrm{op}} N$, if for all closed contexts $C[-]$ of type `Bool` where the hole binds Δ, for all b ranging over $\mathtt{tt}$ and $\mathtt{ff}$, $C[M] \to^* b$ if and only if $C[N] \to^* b$.

2.3 Axiomatic Equivalence

We also define an equational theory for the language, called *axiomatic equivalence* and denoted with $\simeq_{\mathrm{ax}}$, and mainly used as a technical apparatus. The relation $\simeq_{\mathrm{ax}}$ is defined as the smallest reflexive, symmetric, transitive and fully-congruent relation verifying the rules of Table 2, together with the rules $\lambda x.Mx \simeq_{\mathrm{ax}} M$ and $\langle \pi_l(M), \pi_r(M) \rangle \simeq_{\mathrm{ax}} M$. A relation $\sim$ is said to be *fully-congruent* on $\mathbf{PCF}_f$ if whenever $M \sim M'$, for all contexts $C[-]$ we also have $C[M] \sim C[N]$. The two additional rules are standard equational rules for a lambda-calculus [17].

Lemma 3. *If $M : A$ and $M \to N$ then $M \simeq_{\mathrm{ax}} N$.* □

3 Finite Sets as a Concrete Model

Finite sets generate the full sub-category **FinSet** of the category **Set**: objects are finite sets and morphisms are set-functions between finite sets. The category is cartesian closed [29]: the product is the set-product and the internal hom between two sets X and Y is the set of all set-functions from X to Y. Both sets are finite: so is the hom-set.

We can use the category **FinSet** as a model for our PCF language $\mathbf{PCF}_f$. The denotation of types corresponds to the implicit meaning of the types: $[\![\mathbf{1}]\!]^{\text{set}} := \{\star\}$, $[\![\texttt{Bool}]\!]^{\text{set}} := \{\,\texttt{tt}, \texttt{ff}\,\}$, the product is the set-product $[\![A \times B]\!]^{\text{set}} := [\![A]\!]^{\text{set}} \times [\![B]\!]^{\text{set}}$, while the arrow is the set of morphisms: $[\![A \to B]\!]^{\text{set}} := \mathbf{FinSet}([\![A]\!]^{\text{set}}, [\![B]\!]^{\text{set}})$. The set $\{\texttt{tt}, \texttt{ff}\}$ is also written Bool. Similarly, the set $\{\star\}$ is also written $\mathbf{1}$. The denotation of a typing judgment $x_1 : A_1, \dots x_n : A_n \vdash M : B$ is a morphism $[\![A_1]\!]^{\text{set}} \times \cdots \times [\![A_n]\!]^{\text{set}} \to [\![B]\!]^{\text{set}}$, and is inductively defined as in Table 3. The variable d is assumed to be an element of $[\![\Delta]\!]^{\text{set}}$, while a and b are elements of $[\![A]\!]^{\text{set}}$ and $[\![B]\!]^{\text{set}}$ respectively.

This denotation is sound with respect to the operational equivalence.

Lemma 4. *If $M \simeq_{\text{ax}} N : A$ then $[\![M]\!]^{\text{set}} = [\![N]\!]^{\text{set}}$.* □

Theorem 5. *The model is sound with respect to the operational equivalence: Suppose that $\Delta \vdash M, N : A$. If $[\![M]\!]^{\text{set}} = [\![N]\!]^{\text{set}}$ then $M \simeq_{\text{op}} N$.*

Proof. Suppose that $M \not\simeq_{\text{op}} N$ and let Δ be $\{x_i : A_i\}_i$. Then, because of Lemma 2, there exists a context $C[-]$ such that $C[M] \to^* \texttt{tt}$ and $C[N] \to^* \texttt{ff}$. It follows that $(\lambda z.C[z\,x_1 \dots x_n])(\lambda x_1 \dots x_n.M) \simeq_{\text{ax}} \texttt{tt}$ and $(\lambda z.C[z\,x_1 \dots x_n])(\lambda x_1 \dots x_n.N) \simeq_{\text{ax}} \texttt{ff}$. If the denotations of M and N were equal, so would be the denotations of the terms $(\lambda x_1 \dots x_n.M)$ and $(\lambda x_1 \dots x_n.N)$. Lemmas 3 and 4 yield a contradiction. □

FinSet and the language $\mathbf{PCF}_f$ are somehow two sides of the same coin. Theorems 6 and 7 formalize this correspondence.

Theorem 6 (Full Completeness). *For every morphism $f : [\![A]\!]^{\text{set}} \to [\![B]\!]^{\text{set}}$ there exists a valid judgment $x : A \vdash M : B$ such that $f = [\![M]\!]^{\text{set}}$.*

Proof. We start by defining inductively on A two families of terms $M_a : A$ and $\delta_a : A \to \texttt{Bool}$ indexed by $a \in [\![A]\!]^{\text{set}}$, such that $[\![M_a]\!]^{\text{set}} = a$ and $[\![\delta_a]\!]^{\text{set}}$ sends a to tt and all other elements to ff. For the types $\mathbf{1}$ and Bool, the terms $M_\star$, $M_{\texttt{tt}}$ and $M_{\texttt{ff}}$ are the corresponding constants. The term $\delta_\star$ is $\lambda x.\star$, $\delta_{\texttt{tt}}$ is $\lambda x.x$ while $\delta_{\texttt{ff}}$ is the negation. For the type $A \times B$, one trivially calls the induction step. The type $A \to B$ is handled by remembering that the set $[\![A]\!]^{\text{set}}$ is finite: if $g \in [\![A \to B]\!]^{\text{set}}$, the term M_g is the lambda-term with argument x containing a list of `if-then-else` testing with δ_a whether x is equal to a, and returning $M_{g(a)}$ if it is. The term δ_g is built similarly. The judgement $x : A \vdash M : B$ asked for in the theorem is obtained by setting M to $(M_f)x$. □

Theorem 7 (Equivalence). *Suppose that $\Delta \vdash M, N : A$. Then $[\![M]\!]^{\text{set}} = [\![N]\!]^{\text{set}}$ if and only if $M \simeq_{\text{op}} N$.*

Proof. The left-to-right implication is Theorem 5. We prove the right-to-left implication by contrapositive. Assume that $[\![M]\!]^{\text{set}} \neq [\![N]\!]^{\text{set}}$. Then there exists a function $f : \mathbf{1} \to [\![A]\!]^{\text{set}}$ and a function $g : [\![B]\!]^{\text{set}} \to [\![\texttt{Bool}]\!]^{\text{set}}$ such that the boolean $f; [\![M]\!]^{\text{set}}; g$ is different from $f; [\![N]\!]^{\text{set}}; g$. By Theorem 6, the functions f and g are representable by two terms N_f and N_g. They generate a context that distinguishes M and N: this proves that $M \not\simeq_{\text{op}} N$. □

Corollary 8. *Since it is fully complete, the semantics* **FinSet** *is also adequate and fully abstract with respect to* $\mathbf{PCF}_f$. □

Example 9. Consider the Church numerals based over $\mathbf{1}$: they are of type $(\mathbf{1} \to \mathbf{1}) \to (\mathbf{1} \to \mathbf{1})$. In **FinSet**, there is only one element since there is only one map from $\mathbf{1}$ to $\mathbf{1}$. As a consequence of Theorem 7, one can conclude that all Church numerals $\lambda fx.f(f(\cdots(fx)\cdots))$ of type $(\mathbf{1} \to \mathbf{1}) \to (\mathbf{1} \to \mathbf{1})$ are operationally equivalent. Note that this is not true in general as soon as the type is inhabited by more elements.

Example 10. How many operationally distinct Church numerals based over `Bool` are there ? From Theorem 7, it is enough to count how many distinct denotations of Church numerals there are in $[\![(\texttt{Bool} \to \texttt{Bool}) \to (\texttt{Bool} \to \texttt{Bool})]\!]^{\text{set}}$. There are exactly 4 distinct maps $\texttt{Bool} \to \texttt{Bool}$. Written as pairs (x, y) when $f(\texttt{tt}) = x$ and $f(\texttt{ff}) = y$, the maps *tt*, *tf*, *ft* and *ff* are respectively $(\texttt{tt}, \texttt{tt})$, $(\texttt{tt}, \texttt{ff})$, $(\texttt{ff}, \texttt{tt})$ and $(\texttt{ff}, \texttt{ff})$.

Then, if the Church numeral $\bar{n}$ is written as a tuple $(\bar{n}(tt), \bar{n}(tf), \bar{n}(ft), \bar{n}(ff))$, we have the following equalities: $\bar{0} = (tf, tf, tf, tf)$, $\bar{1} = (tt, tf, ft, ff)$, $\bar{2} = (tt, tf, tf, ff)$, $\bar{3} = (tt, tf, ft, ff)$, and one can show that for all $n \geq 1$, $[\![\bar{n}]\!]^{\text{set}} = [\![\overline{n+2}]\!]^{\text{set}}$. There are therefore only 3 operationally distinct Church numerals based on the type `Bool`: the number $\bar{0}$, then all even non-null numbers, and finally all odd numbers.

4 Finite Vector Spaces

We now turn to the second finitary model that we want to use for the language $\mathbf{PCF}_f$: finite vector spaces. We first start by reminding the reader about this algebraic structure.

4.1 Background Definitions

A *field* [19] K is a commutative ring such that the unit 0 of the addition is distinct from the unit 1 of the multiplication and such all non-zero elements of K admit an inverse with respect to the multiplication. A *finite field* is a field of finite size. The *characteristic* q of a field K is the minimum (non-zero) number such that $1 + \cdots + 1 = 0$ (q instances of 1). If there is none, we say that the characteristic is 0. For example, the field of real numbers has characteristic 0, while the field $\mathbb{F}_2$ consisting of 0 and 1 has characteristic 2. The *order* of a finite field is the order of its multiplicative group.

A *vector space* [18] V over a field K is an algebraic structure consisting of a set $|V|$, a binary addition $+$ and a scalar multiplication $(\cdot) : K \times V \to V$, satisfying the equations of Table 6 (taken unordered). The *dimension* of a vector space is the size of the largest set of independent vectors. A particular vector space is the vector space *freely generated from a space* X, denoted with $\langle X \rangle$: it consists of all the formal finite linear combinations $\sum_i \alpha_i \cdot x_i$, where x_i belongs to X and α_i belongs to K. To define a linear map f on $\langle X \rangle$, it is enough to give its behavior on each of the vector $x \in X$: the image of $\sum_i \alpha_i \cdot x_i$ is then by linearity imposed to be $\sum_i \alpha_i \cdot f(x_i)$.

In this paper, the vector spaces we shall concentrate on are *finite vector spaces*, that is, vector spaces of finite dimensions over a finite field. For example, the 2-dimensional space $\mathbb{F}_2 \times \mathbb{F}_2$ consists of the four vectors $\binom{0}{0}, \binom{0}{1}, \binom{1}{0}, \binom{1}{1}$ and is a finite vector space. It is also the vector space freely generated from the 2-elements set $\{\texttt{tt}, \texttt{ff}\}$: each vectors respectively corresponds to 0, $\texttt{tt}$, $\texttt{ff}$, and $\texttt{tt} + \texttt{ff}$.

Once a given finite field K has been fixed, the category **FinVec** has for objects finite vector spaces over K and for morphisms linear maps between these spaces. The category is symmetric monoidal closed: the tensor product is the algebraic tensor product, the unit of the tensor is $I = K = \langle\star\rangle$ and the internal hom between two spaces U and V is the vector space of all linear functions $U \multimap V$ between U and V. The addition and the scalar multiplication over functions are pointwise.

4.2 A Linear-Non-Linear Model

It is well-known [20] that the category of finite sets and functions and the category of finite vector spaces and linear maps form an adjunction

$$\mathbf{FinSet} \underset{G}{\overset{F}{\rightleftarrows}} \mathbf{FinVec}. \tag{1}$$

The functor F sends the set X to the vector space $\langle X\rangle$ freely generated from X and the set-map $f : X \to Y$ to the linear map sending a basis element $x \in X$ to the base element $f(x)$. The functor G sends a vector space U to the same space seen as a set, and consider any linear function as a set-map from the corresponding sets.

This adjunction makes **FinVec** into a model of linear logic [23]. Indeed, the adjunction is symmetric monoidal with the following two natural transformations:

$$\begin{array}{llll} m_{X,Y} : \langle X \times Y\rangle \to \langle X\rangle \otimes \langle Y\rangle & & m_1 : \langle 1\rangle \to I & \\ (x, y) & \mapsto x \otimes y, & \star & \mapsto 1. \end{array}$$

This makes a *linear-non-linear category* [4], equivalent to a linear category, and is a model of intuitionistic linear logic [5].

4.3 Model of Linear Logic

The adjunction in Eq. (1) generates a linear comonad on **FinVec**. If A is a finite vector space, we define the finite vector space $!A$ as the vector space freely generated from the

Table 4. Modeling the language $\mathbf{PCF}_f$ in **FinVec**

$$\llbracket \Delta, x : A \vdash x : A\rrbracket^{\text{vec}} : d \otimes b_a \longmapsto a \qquad \llbracket \Delta \vdash \langle M, N\rangle : A \times B\rrbracket^{\text{vec}} : d \longmapsto \llbracket M\rrbracket^{\text{vec}}(d) \otimes \llbracket N\rrbracket^{\text{vec}}(d)$$

$$\llbracket \Delta \vdash \texttt{tt} : \texttt{Bool}\rrbracket^{\text{vec}} : d \longmapsto \texttt{tt} \qquad \llbracket \Delta \vdash MN : B\rrbracket^{\text{vec}} : d \longmapsto \llbracket M\rrbracket^{\text{vec}}(d)(\llbracket N\rrbracket^{\text{vec}}(d))$$

$$\llbracket \Delta \vdash \texttt{ff} : \texttt{Bool}\rrbracket^{\text{vec}} : d \longmapsto \texttt{ff} \qquad \llbracket \Delta \vdash \pi_l(M) : A\rrbracket^{\text{vec}} = \llbracket M\rrbracket^{\text{vec}}; \pi_l$$

$$\llbracket \Delta \vdash \star : \mathbf{1}\rrbracket^{\text{vec}} : d \longmapsto \star \qquad \llbracket \Delta \vdash \pi_r(M) : B\rrbracket^{\text{vec}} = \llbracket M\rrbracket^{\text{vec}}; \pi_r$$

$$\llbracket \Delta \vdash \lambda x.M : A \to B\rrbracket^{\text{vec}} = d \longmapsto (b_a \mapsto \llbracket M\rrbracket^{\text{vec}}(d \otimes b_a))$$

$$\llbracket \Delta \vdash \texttt{let}\,\star = M\,\texttt{in}\,N : A\rrbracket^{\text{vec}} = d \longmapsto \alpha \cdot \llbracket N\rrbracket^{\text{vec}}(d) \qquad \text{where } \llbracket M\rrbracket^{\text{vec}}(d) = \alpha \cdot \star.$$

$$\llbracket \Delta \vdash \texttt{if}\,M\,\texttt{then}\,N\,\texttt{else}\,P : A\rrbracket^{\text{vec}} = d \longmapsto \alpha \cdot \llbracket N\rrbracket^{\text{vec}}(d) + \beta \cdot \llbracket P\rrbracket^{\text{set}}(d)$$

$$\text{where } \llbracket \Delta \vdash M : \texttt{Bool}\rrbracket^{\text{vec}}(d) = \alpha \cdot \texttt{tt} + \beta \cdot \texttt{ff}.$$

set $\{b_v\}_{v \in A}$: it consists of the space $\langle b_v \mid v \in A \rangle$. If $f : A \to B$ is a linear map, the map $!f : !A \to !B$ is defined as $b_v \mapsto b_{f(v)}$. The comultiplication and the counit of the comonad are respectively $\delta_A : !A \to !!A$ and $\epsilon_A : !A \to A$ where $\delta_A(b_v) = b_{b_v}$ and $\epsilon_A(b_v) = v$. Every element $!A$ is a commutative comonoid when equipped with the natural transformations $\Delta_A : !A \to !A \otimes !A$ and $\Diamond_A : !A \to I$ where $\Delta_A(b_v) = b_v \otimes b_v$ and $\Diamond(b_v) = 1$. This makes the category **FinVec** into a linear category.

In particular, the coKleisli category $\mathbf{FinVec}_!$ coming from the comonad is cartesian closed: the product of A and B is $A \times B$, the usual product of vector spaces, and the terminal object is the vector space $\langle 0 \rangle$. This coKleisli category is the usual one: the objects are the objects of **FinVec**, and the morphisms $\mathbf{FinVec}_!(A, B)$ are the morphisms $\mathbf{FinVec}(!A, B)$. The identity $!A \to A$ is the counit and the composition of $f : !A \to B$ and $!B \to C$ is $f; g := !A \xrightarrow{\delta_A} !!A \xrightarrow{!f} !B \xrightarrow{g} C$.

There is a canonical full embedding E of categories sending $\mathbf{FinVec}_!$ on **FinSet**. It sends an object U to the set of vectors of U (i.e. it acts as the forgetful functor on objects) and sends the linear map $f : !U \to V$ to the map $v \mapsto f(b_v)$.

This functor preserves the cartesian closed structure: the terminal object $\langle 0 \rangle$ of $\mathbf{FinVec}_!$ is sent to the set containing only 0, that is, the singleton-set **1**. The product space $U \times V$ is sent to the set of vectors $\{\langle u, v \rangle \mid u \in U, v \in V\}$, which is exactly the set-product of U and V. Finally, the function space $!U \to V$ is in exact correspondence with the set of set-functions $U \to V$.

Remark 11. The construction proposed as side example by Hyland and Schalk [15] considers finite vector spaces with a field of characteristic 2. There, the modality is built using the exterior product algebra, and it turns out to be identical to the functor we use in the present paper. Note though, that their construction does not work with fields of other characteristics.

Remark 12. Quantitative models of linear logic such as finiteness spaces [9] are also based on vector spaces; however, in these cases the procedure to build a comonad does not play well with the finite dimension the vector spaces considered in this paper: the definition of the comultiplication assumes that the space $!A$ is infinitely dimensional.

4.4 Finite Vector Spaces as a Model

Since $\mathbf{FinVec}_!$ is a cartesian closed category, one can model terms of $\mathbf{PCF}_f$ as linear maps. Types are interpreted as follows. The unit type is $[\![\mathbf{1}]\!]^{\text{vec}} := \{\alpha \cdot \star \mid \alpha \in K\}$. The boolean type is $[\![\texttt{Bool}]\!]^{\text{vec}} := \{\sum_i \alpha_i \cdot \texttt{tt} + \beta_i \cdot \texttt{ff} \mid \alpha_i, \beta_i \in K\}$. The product is the usual product space: $[\![A \times B]\!]^{\text{vec}} := [\![A]\!]^{\text{vec}} \times [\![B]\!]^{\text{vec}}$, whereas the arrow type is $[\![A \to B]\!]^{\text{vec}} := \mathbf{FinVec}(![\![A]\!]^{\text{vec}}, [\![B]\!]^{\text{vec}})$. A typing judgment $x_1 : A_1, \ldots, x_n : A_n \vdash M : B$ is represented by a morphism of **FinVec** of type

$$![\![A_1]\!]^{\text{vec}} \otimes \cdots \otimes ![\![A_n]\!]^{\text{vec}} \longrightarrow [\![B]\!]^{\text{vec}}, \tag{2}$$

inductively defined as in Table 4. The variable d stands for a base element $b_{u_1} \otimes \ldots \otimes b_{u_n}$ of $[\![\Delta]\!]^{\text{vec}}$, and b_a is a base element of $[\![A]\!]^{\text{vec}}$. The functions π_l and π_r are the left and right projections of the product.

Note that because of the equivalence between $!(A \times B)$ and $!A \otimes !B$, the map in Eq. (2) is a morphism of $\mathbf{FinVec}_!$, as desired.

Example 13. In **FinSet**, there was only one Church numeral based on type $\mathbf{1}$. In $\mathbf{FinVec}_!$, there are more elements in the corresponding space $!(!\mathbf{1} \multimap \mathbf{1}) \multimap (!\mathbf{1} \multimap \mathbf{1})$ and we get more distinct Church numerals.

Assume that the finite field under consideration is the 2-elements field $\mathbb{F}_2 = \{0, 1\}$. Then $[\![\mathbf{1}]\!]^{\text{vec}} = \mathbf{1} = \{0 \cdot \star, 1 \cdot \star\} = \{0, \star\}$. The space $!\mathbf{1}$ is freely generated from the vectors of $\mathbf{1}$: it therefore consists of just the four vectors $\{0, b_0, b_\star, b_0 + b_\star\}$. The space of morphisms $[\![\mathbf{1} \to \mathbf{1}]\!]^{\text{vec}}$ is the space $!\mathbf{1} \multimap \mathbf{1}$. It is generated by two functions: f_0 sending b_0 to $\star$ and $b_\star$ to 0, and $f_\star$ sending b_v to v. The space therefore also contains 4 vectors: 0, f_0, $f_\star$ and $f_0 + f_\star$. Finally, the vector space $!(!\mathbf{1} \multimap \mathbf{1})$ is freely generated from the 4 base elements $b_0, b_{f_0}, b_{f_\star}$ and $b_{f_0+f_\star}$, therefore containing 16 vectors. Morphisms $!(!\mathbf{1} \multimap \mathbf{1}) \multimap (!\mathbf{1} \multimap \mathbf{1})$ can be represented by 2×4 matrices with coefficients in $\mathbb{F}_2$. The basis elements b_v are ordered as above, as are the basis elements f_w, as shown on the right. The Church numeral $\bar{0}$ sends all of its arguments to the identity function, that is, $f_\star$. The Church numeral $\bar{1}$ is the identity. So their respective matrices are $\left(\begin{smallmatrix} 0&0&0&0 \\ 1&1&1&1 \end{smallmatrix}\right)$ and $\left(\begin{smallmatrix} 0&1&0&1 \\ 0&0&1&1 \end{smallmatrix}\right)$. The next two Church numerals are $\bar{2} = \left(\begin{smallmatrix} 0&0&0&1 \\ 0&1&1&1 \end{smallmatrix}\right)$ and $\bar{3} = \left(\begin{smallmatrix} 0&1&0&1 \\ 0&0&1&1 \end{smallmatrix}\right)$, which is also $\bar{1}$. So $\mathbf{FinVec}_!$ with the field of characteristic 2 distinguishes null, even and odds numerals over the type $!\mathbf{1}$.

$$\begin{array}{cccc} & & & \\ \end{array}\begin{pmatrix} \cdot & \cdot & \cdot & \cdot \\ \cdot & \cdot & \cdot & \cdot \end{pmatrix}\begin{matrix} f_0 \\ f_\star \end{matrix} \qquad b_0 \; b_{f_0} b_{f_\star} b_{f_0+f_\star}$$

Note that this characterization is similar to the **FinSet** Example 10, except that there, the type over which the Church numerals were built was `Bool`. Over $\mathbf{1}$, Example 9 stated that all Church numerals collapse.

Example 14. The fact that $\mathbf{FinVec}_!$ with the field of characteristic 2 can be put in parallel with **FinSet** when considering Church numerals is an artifact of the fact that the field has only two elements. If instead one chooses another field $K = \mathbb{F}_p = \{0, 1, \ldots, p-1\}$ of characteristic p, with p prime, then this is in general not true anymore. In this case, $[\![\mathbf{1}]\!]^{\text{vec}} = \{0, \star, 2 \cdot \star, \ldots, (p-1) \cdot \star\}$, and $!\mathbf{1} \multimap \mathbf{1}$ has dimension p with basis elements f_i sending $b_{i \cdot \star} \mapsto \star$ and $b_{j \cdot \star} \mapsto 0$ when $i \neq j$. It therefore consists of p^p vectors. Let us represent a function $f : !\mathbf{1} \multimap \mathbf{1}$ with $\mathtt{x}_0 \ldots \mathtt{x}_{\mathtt{p}-1}$ where $f(b_{i \cdot \star}) = \mathtt{x}_i \cdot \star$. A morphism $!(!\mathbf{1} \multimap \mathbf{1}) \multimap (!\mathbf{1} \multimap \mathbf{1})$ can be represented with a $p^p \times p$ matrix. The basis elements $b_{\mathtt{x}_0 \ldots \mathtt{x}_{\mathtt{p}-1}}$ of $!(!\mathbf{1} \multimap \mathbf{1})$ are ordered lexicographically: $b_{0 \ldots 00}, b_{0 \ldots 01}, b_{0 \ldots 02}, \ldots, b_{0 \ldots 0(p-1)}, \ldots, b_{(p-1) \ldots (p-1)}$, as are the basis elements $f_0, f_1, \ldots, f_{p-1}$.

The Church numeral $\bar{0}$ is again the constant function returning the identity, that is, $\sum_i i \cdot f_i$. The numeral $\bar{1}$ sends $\mathtt{x}_0 \cdots \mathtt{x}_{\mathtt{p}-1}$ onto the function sending $b_{i \cdot \star}$ onto $\mathtt{x}_i \cdot \star$. The numeral $\bar{2}$ sends $\mathtt{x}_0 \cdots \mathtt{x}_{\mathtt{p}-1}$ onto the function sending $b_{i \cdot \star}$ onto $\mathtt{x}_{\mathtt{x}_i} \cdot \star$. The numeral $\bar{3}$ sends $\mathtt{x}_0 \cdots \mathtt{x}_{\mathtt{p}-1}$ onto the function sending $b_{i \cdot \star}$ onto $\mathtt{x}_{\mathtt{x}_{\mathtt{x}_i}} \cdot \star$. And so on.

In particular, each combination $\mathtt{x}_0 \cdots \mathtt{x}_{p-1}$ can be considered as a function $\mathtt{x} : \{0, \ldots p-1\} \to \{0, \ldots p-1\}$. The sequence $(\mathtt{x}^0, \mathtt{x}^1, \mathtt{x}^2, \ldots)$ eventually loops. The order of the loop is $lcm(p)$, the least common multiple of all integers $1, \ldots, p$, and for all $n \geq p-1$ we have $\mathtt{x}^n = \mathtt{x}^{n+lcm(p)}$: there are $lcm(p) + p - 1$ distinct Church numerals in the model $\mathbf{FinVec}_!$ with a field of characteristic p prime.

For $p = 2$ we recover the 3 distinct Church numerals. But for $p = 3$, we deduce that there are 8 distinct Church numerals (the 8 corresponding matrices are reproduced in

Table 5. The 8 Church numerals over type **1** in **FinVec**$_!$ with $K = \mathbb{F}_3$

$$\overline{0} = \begin{pmatrix} 0&0&0&0&0&0&0&0&0&0&0&0&0&0&0&0&0&0&0&0&0&0&0&0&0&0&0 \\ 1&1&1&1&1&1&1&1&1&1&1&1&1&1&1&1&1&1&1&1&1&1&1&1&1&1&1 \\ 2&2&2&2&2&2&2&2&2&2&2&2&2&2&2&2&2&2&2&2&2&2&2&2&2&2&2 \end{pmatrix}$$

$$\overline{1} = \begin{pmatrix} 0&0&0&0&0&0&0&0&0&1&1&1&1&1&1&1&1&1&2&2&2&2&2&2&2&2&2 \\ 0&0&0&1&1&1&2&2&2&0&0&0&1&1&1&2&2&2&0&0&0&1&1&2&2&2&2 \\ 0&1&2&0&1&2&0&1&2&0&1&2&0&1&2&0&1&2&0&1&2&0&1&2&0&1&2 \end{pmatrix}$$

$$\overline{2+6n} = \begin{pmatrix} 0&0&0&0&0&0&0&0&0&0&0&0&1&1&1&2&2&2&0&1&2&0&1&2&0&1&2 \\ 0&0&0&1&1&1&0&1&2&1&1&1&1&1&1&0&1&2&2&2&2&1&1&1&0&1&2 \\ 0&0&2&0&1&2&0&2&2&1&0&2&1&1&2&1&2&2&2&0&2&2&1&2&2&2&2 \end{pmatrix}$$

$$\overline{3+6n} = \begin{pmatrix} 0&0&0&0&0&0&0&0&0&1&1&1&1&1&1&0&1&2&2&0&2&2&1&2&2&2&2 \\ 0&0&0&1&1&1&0&2&2&0&0&0&1&1&1&1&2&2&0&1&2&1&1&1&2&2&2 \\ 0&0&2&0&1&2&0&1&2&0&1&2&1&1&2&2&1&2&0&2&2&0&1&2&0&1&2 \end{pmatrix}$$

$$\overline{4+6n} = \begin{pmatrix} 0&0&0&0&0&0&0&0&0&0&0&0&1&1&1&1&2&2&0&2&2&0&1&2&0&1&2 \\ 0&0&0&1&1&1&0&1&2&1&1&1&1&1&1&2&1&2&2&0&2&1&1&1&0&1&2 \\ 0&0&2&0&1&2&0&2&2&1&0&2&1&1&2&0&2&2&2&1&2&2&1&2&2&2&2 \end{pmatrix}$$

$$\overline{5+6n} = \begin{pmatrix} 0&0&0&0&0&0&0&0&0&1&1&1&1&1&1&2&1&2&2&1&2&2&1&2&2&2&2 \\ 0&0&0&1&1&1&0&2&2&0&0&0&1&1&1&0&2&2&0&2&2&1&1&1&2&2&2 \\ 0&0&2&0&1&2&0&1&2&0&1&2&1&1&2&1&1&2&0&0&2&0&1&2&0&1&2 \end{pmatrix}$$

$$\overline{6+6n} = \begin{pmatrix} 0&0&0&0&0&0&0&0&0&0&0&0&1&1&1&0&2&2&0&0&2&0&1&2&0&1&2 \\ 0&0&0&1&1&1&0&1&2&1&1&1&1&1&1&1&1&2&2&1&2&1&1&1&0&1&2 \\ 0&0&2&0&1&2&0&2&2&1&0&2&1&1&2&2&2&2&2&2&2&2&1&2&2&2&2 \end{pmatrix}$$

$$\overline{7+6n} = \begin{pmatrix} 0&0&0&0&0&0&0&0&0&1&1&1&1&1&1&1&1&2&2&2&2&2&1&2&2&2&2 \\ 0&0&0&1&1&1&0&2&2&0&0&0&1&1&1&2&2&2&0&0&2&1&1&1&2&2&2 \\ 0&0&2&0&1&2&0&1&2&0&1&2&1&1&2&0&1&2&0&1&2&0&1&2&0&1&2 \end{pmatrix}$$

Table 5). As there is almost a factorial function, the number of distinct Church numerals grows fast as p grows: With $\mathbb{F}_5$, there are 64 distinct numerals, and with $\mathbb{F}_7$ there are 426 distinct numerals.

Example 15. Let us briefly reprise Example 10 in the context of **FinVec**$_!$. Even with a field of characteristic 2, the vector space $[\![(\texttt{Bool} \to \texttt{Bool}) \to (\texttt{Bool} \to \texttt{Bool})]\!]^{\text{vec}}$ is relatively large: `Bool` has dimension 2 and consists of 4 vectors, !`Bool` then has dimension 4 and consists of 16 vectors. The dimension of the homset !`Bool` $\multimap$ `Bool` is 8, and it contains $2^8 = 256$ vectors. Using the representation of the two previous examples, a Church numeral is then a matrix of size 256×8.

Let us represent a function !`Bool` $\multimap$ `Bool` as a tuple $(\mathtt{x}_{ij}^k)_{i,j,k}$ lexicographically ordered $\mathtt{x}_{00}^0, \mathtt{x}_{00}^1, \mathtt{x}_{01}^0, \mathtt{x}_{01}^1, \mathtt{x}_{10}^0, \mathtt{x}_{10}^1, \mathtt{x}_{11}^0, \mathtt{x}_{11}^1$, representing the map sending $b_{i\cdot\mathtt{tt}+j\cdot\mathtt{ff}}$ to $\mathtt{x}_{ij}^0 \cdot \mathtt{tt} + \mathtt{x}_{ij}^1 \cdot \mathtt{ff}$. These form the basis elements of the range of the matrix. The domain of the matrix consists of all the 256 combinations of 0/1 values that these can take. Ordered lexicographically, they form the basis of the domain of the matrix.

As before, the Church numeral $\bar{0}$ is constant while $\bar{1}$ is the identity. The numeral $\bar{2}$ sends each of the 8-tuples $(\mathtt{x}_{ij}^k)_{i,j,k}$ to the 8-tuple $(x^0_{\mathtt{x}^0_{a,b},\mathtt{x}^1_{a,b}}, \mathtt{x}^1_{\mathtt{x}^0_{a,b},\mathtt{x}^1_{a,b}})_{a,b\in\{0,1\}}$, and so forth. So for example, the negation sending $b_{a\cdot\mathtt{tt}+b\cdot\mathtt{ff}}$ to $a \cdot \mathtt{ff} + b \cdot \mathtt{tt}$ is the 8-tuple $(0,0,1,0,0,1,1,1)$ and is sent by $\bar{2}$ to the tuple $(0,0,0,1,1,0,1,1)$ which is indeed the identity.

If one performs the calculation, one finds out that in **FinVec**$_!$, over the type `Bool`, there are exactly 15 distinct Church numerals. The numerals $\bar{0}$, $\bar{1}$ and $\bar{2}$ are uniquely determined, and then the semantics distinguishes the equivalence classes $\{i + 12n \mid n \in \mathbb{N}\}$, for $i = 3, 4, \ldots 14$.

4.5 Properties of the FinVec Model

As shown in the next results, this semantics is both sound and adequate with respect to the operational equivalence. Usually adequacy uses non-terminating terms. Because the language is strongly normalizing, we adapt the notion. However, because there are usually more maps between $[\![A]\!]^{\mathrm{vec}}$ and $[\![B]\!]^{\mathrm{vec}}$ than between $[\![A]\!]^{\mathrm{set}}$ and $[\![B]\!]^{\mathrm{set}}$ (as shown in Examples 13, 14 and 15), the model fails to be fully abstract.

Lemma 16. *If* $M \simeq_{\mathrm{ax}} N : A$ *then* $[\![M]\!]^{\mathrm{vec}} = [\![N]\!]^{\mathrm{vec}}$. □

Theorem 17. *If* $\Delta \vdash M, N : A$ *and* $[\![M]\!]^{\mathrm{vec}} = [\![N]\!]^{\mathrm{vec}}$ *then* $M \simeq_{\mathrm{op}} N$.

Proof. The proof is similar to the proof of Theorem 5 and proceeds by contrapositive, using Lemmas 2, 2, 3 and 16. □

Theorem 18 (Adequacy). *Given two closed terms* M *and* N *of type* `Bool`, $[\![M]\!]^{\mathrm{vec}} = [\![N]\!]^{\mathrm{vec}}$ *if and only if* $M \simeq_{\mathrm{op}} N$.

Proof. The left-to-right direction is Theorem 17. For the right-to-left direction, since the terms M and N are closed of type `Bool`, one can choose the context $C[-]$ to be $[-]$, and we have $M \to^* b$ if and only if $N \to^* b$. From Lemma 2, there exists such a boolean b: we deduce from Lemma 3 that $M \simeq_{\mathrm{ax}} N$. We conclude with Lemma 16. □

Remark 19. The model **FinVec**$_!$ is not fully abstract. Indeed, consider the two valid typing judgments x : `Bool` ⊢ `tt` : `Bool` and x : `Bool` ⊢ `if` x `then tt else tt` : `Bool`. The denotations of both of these judgments are linear maps $![\![\mathtt{Bool}]\!]^{\mathrm{vec}} \to [\![\mathtt{Bool}]\!]^{\mathrm{vec}}$. According to the rules of Table 4, the denotation of the first term is the constant function sending all non-zero vectors b_- to `tt`.

For the second term, suppose that $v \in ![\![\mathtt{Bool}]\!]^{\mathrm{vec}}$ is equal to $\sum_i \gamma_i \cdot b_{\alpha_i \cdot \mathtt{tt} + \beta_i \cdot \mathtt{ff}}$. Let $\nu = \sum_i \gamma_i(\alpha_i + \beta_i)$. Then since $[\![x : \mathtt{Bool} \vdash x : \mathtt{Bool}]\!]^{\mathrm{vec}}(v) = \nu$, the denotation of the second term is the function sending v to $\nu \cdot [\![x : \mathtt{Bool} \vdash \mathtt{tt} : \mathtt{Bool}]\!]^{\mathrm{vec}}(v)$, equal to $\nu \cdot \mathtt{tt}$ from what we just discussed. We conclude that if $v = b_0$, then $\nu = 0$: the denotation of x : `Bool` ⊢ `if` x `then tt else tt` : `Bool` sends b_0 to 0.

Nonetheless, they are clearly operationally equivalent in **PCF**$_f$ since their denotation in **FinSet** is the same. The language is not expressive enough to distinguish between these two functions. Note that there exists operational settings where these would actually be different, for example if we were to allow divergence.

Remark 20. Given a term A, another question one could ask is whether the set of terms $M : A$ in **PCF**$_f$ generates a free family of vectors in the vector space $[\![A]\!]^{\mathrm{vec}}$. It turns out not: The field structure brought into the model introduces interferences, and algebraic sums coming from operationally distinct terms may collapse to a representable element. For example, supposing for simplicity that the characteristic of the field is $q = 2$, consider the terms $T_{\mathtt{tt},\mathtt{tt}}$, $T_{\mathtt{ff},\mathtt{ff}}$, $T_{\mathtt{tt},\mathtt{ff}}$ and $T_{\mathtt{ff},\mathtt{tt}}$ defined as $T_{\mathtt{y},\mathtt{z}} = \lambda x.$`if` x `then y else z`, all of types `Bool` → `Bool`. They are clearly operationally distinct, and their denotations live in `!Bool` ⊸ `Bool`. They can be written as a 2 × 4 matrices along the bases $(b_0, b_{\mathtt{tt}}, b_{\mathtt{ff}}, b_{\mathtt{tt}+\mathtt{ff}})$ for the domain and $(\mathtt{tt}, \mathtt{ff})$ for the range. The respective images of the 4 terms are $\left(\begin{smallmatrix}0&1&1&0\\0&0&0&0\end{smallmatrix}\right)$, $\left(\begin{smallmatrix}0&0&0&0\\0&1&1&0\end{smallmatrix}\right)$, $\left(\begin{smallmatrix}0&1&0&1\\0&0&1&1\end{smallmatrix}\right)$, $\left(\begin{smallmatrix}0&0&1&1\\0&1&0&1\end{smallmatrix}\right)$ and clearly, $[\![T_{\mathtt{tt},\mathtt{tt}}]\!]^{\mathrm{vec}} = [\![T_{\mathtt{ff},\mathtt{ff}}]\!]^{\mathrm{vec}} + [\![T_{\mathtt{tt},\mathtt{ff}}]\!]^{\mathrm{vec}} + [\![T_{\mathtt{ff},\mathtt{tt}}]\!]^{\mathrm{vec}}$.

So if the model we are interested in is **FinVec**$_!$, the language is missing some structure to correctly handle the algebraicity.

Table 6. Rewrite system for the algebraic fragment of $\mathbf{PCF}_f^{alg}$

$\alpha \cdot M + \beta \cdot M \to (\alpha+\beta) \cdot M$	$M+N \to N+M$		$(M+N)+P \to M+(N+P)$
$\alpha \cdot M + M \to (\alpha+1) \cdot M$	$0 \cdot M \to 0$	$1 \cdot M \to M$	$\alpha \cdot (M+N) \to \alpha \cdot M + \alpha \cdot N$
$M+M \to (1+1) \cdot M$	$\alpha \cdot 0 \to 0$	$0+M \to M$	$\alpha \cdot (\beta \cdot M) \to (\alpha\beta) \cdot M$

5 An Algebraic Lambda Calculus

To solve the problem, we extend the language $\mathbf{PCF}_f$ by adding an algebraic structure to mimic the notion of linear distribution existing in $\mathbf{FinVec}_!$. The extended language $\mathbf{PCF}_f^{alg}$ is a call-by-name variation of [2,3] and reads as follows:

$$\begin{aligned} M,N,P ::= {} & x \mid \lambda x.M \mid MN \mid \pi_l(M) \mid \pi_r(M) \mid \langle M,N \rangle \mid \star \mid \mathtt{tt} \mid \mathtt{ff} \mid \\ & \mathtt{if}\, M\, \mathtt{then}\, N\, \mathtt{else}\, P \mid \mathtt{let}\, \star = M\, \mathtt{in}\, N \mid 0 \mid M+N \mid \alpha \cdot M, \\ A,B ::= {} & 1 \mid \mathtt{Bool} \mid A \to B \mid A \times B. \end{aligned}$$

The scalar α ranges over the field. The values are now $U,V ::= x \mid \lambda x.M \mid \langle M,N \rangle \mid \star \mid \mathtt{tt} \mid \mathtt{ff} \mid 0 \mid U+V \mid \alpha \cdot U$.. The typing rules are the same for the regular constructs. The new constructs are typed as follows: for all A, $\Delta \vdash 0 : A$, and provided that $\Delta \vdash M,N : A$, then $\Delta \vdash M+N : A$ and $\Delta \vdash \alpha \cdot M : A$. The rewrite rules are extended as follows.

1) A set of algebraic rewrite rules shown in Table 6. We shall explicitly talk about *algebraic rewrite rules* when referring to these extended rules. The top row consists of the associativity and commutativity (AC) rules. We shall use the term *modulo AC* when referring to a rule or property that is true when not regarding AC rules. For example, modulo AC the term $\star$ is in normal form and $\alpha \cdot M + (N + \alpha \cdot P)$ reduces to $\alpha \cdot (M + P) + N$. The reduction rules from Γ will be called *non-algebraic*.

2) The relation between the algebraic structure and the other constructs: one says that a construct $c(-)$ is *distributive* when for all M,N, $c(M+M) \to c(M)+c(N)$, $c(\alpha \cdot M) \to \alpha \cdot c(M)$ and $c(0) \to 0$. The following constructs are distributive: $(-)P$, $\mathtt{if}\,(-)\,\mathtt{then}\,P_1\,\mathtt{else}\,P_2$, $\pi_i(-)$, $\mathtt{let}\,\star = (-)\,\mathtt{in}\,N$, and the pairing construct factors: $\langle M,N \rangle + \langle M',N' \rangle \to \langle M+M', N+N' \rangle$, $\alpha \cdot \langle M,N \rangle \to \langle \alpha \cdot M, \alpha \cdot N \rangle$ and $0^{A \times B} \to \langle 0^A, 0^B \rangle$.

3) Two congruence rules. If $M \to M'$, then $M+N \to M'+N$ and $\alpha \cdot M \to \alpha \cdot M'$.

Remark 21. Note that if $(M_1+M_2)(N_1+N_2)$ reduces to $M_1(N_1+N_2)+M_2(N_1+N_2)$, it does *not* reduce to $(M_1+M_2)N_1+(M_1+M_2)N_2$. If it did, one would get an inconsistent calculus [3]. For example, the term $(\lambda x.\langle x,x \rangle)(\mathtt{tt}+\mathtt{ff})$ would reduce both to $\langle \mathtt{tt},\mathtt{tt} \rangle + \langle \mathtt{ff},\mathtt{ff} \rangle$ and to $\langle \mathtt{tt},\mathtt{tt} \rangle + \langle \mathtt{ff},\mathtt{ff} \rangle + \langle \mathtt{tt},\mathtt{ff} \rangle + \langle \mathtt{ff},\mathtt{tt} \rangle$. We'll come back to this distinction in Section 6.3.

The algebraic extension preserves the safety properties, the characterization of values and the strong normalization. Associativity and commutativity induce a subtlety.

Lemma 22. *The algebraic fragment of* $\mathbf{PCF}_f^{alg}$ *is strongly normalizing modulo AC.*

Proof. The proof can be done as in [3], using the same measure on terms that decreases with algebraic rewrites. The measure, written a, is defined by $a(x) = 1$, $a(M + N) = 2 + a(M) + a(N)$, $a(\alpha \cdot M) = 1 + 2a(M)$, $a(0) = 0$. □

Lemma 23 (Safety properties mod AC). *A well-typed term $M : A$ is a value or, if not, reduces to some $N : A$ via a sequence of steps among which one is* not *algebraic.* □

Lemma 24. *Any value of type $\mathbf{1}$ has AC-normal form 0, $\star$ or $\alpha \cdot \star$, with $\alpha \neq 0, 1$.* □

Lemma 25. *Modulo AC, $\mathbf{PCF}_f^{alg}$ is strongly normalizing.*

Proof. The proof is done by defining an intermediate language $\mathbf{PCF}_{f\,int}$ where scalars are omitted. Modulo AC, this language is essentially the language $\lambda_{-\text{wLK}^{\rightarrow}}$ of [7], and is therefore SN. Any term of $\mathbf{PCF}_f^{alg}$ can be re-written as a term of $\mathbf{PCF}_{f\,int}$. With Lemma 23, by eliminating some algebraic steps a sequence of reductions in $\mathbf{PCF}_f^{alg}$ can be rewritten as a sequence of reductions in $\mathbf{PCF}_{f\,int}$. We conclude with Lemma 22, saying there is always a finite number of these eliminated algebraic rewrites. □

5.1 Operational Equivalence

As for $\mathbf{PCF}_f$, we define an operational equivalence on terms of the language $\mathbf{PCF}_f^{alg}$. A *context* $C[-]$ for this language has the same grammar as for $\mathbf{PCF}_f$, augmented with algebraic structure: $C[-] ::= \alpha \cdot C[-] \mid C[-] + N \mid M + C[-] \mid 0$.

For $\mathbf{PCF}_f^{alg}$, instead of using closed contexts of type `Bool`, we shall use contexts of type $\mathbf{1}$: thanks to Lemma 24, there are distinct normal forms for values of type $\mathbf{1}$, making this type a good (and slightly simpler) candidate.

We therefore say that $\Delta \vdash M : A$ and $\Delta \vdash N : A$ are operationally equivalent, written $M \simeq_{\text{op}} N$, if for all closed contexts $C[-]$ of type $\mathbf{1}$ where the hole binds Δ, for all b normal forms of type $\mathbf{1}$, $C[M] \rightarrow^* b$ if and only if $C[N] \rightarrow^* b$.

5.2 Axiomatic Equivalence

The axiomatic equivalence on $\mathbf{PCF}_f^{alg}$ consists of the one of $\mathbf{PCF}_f$, augmented with the added reduction rules.

Lemma 26. *If $M : A$ and $M \rightarrow N$ then $M \simeq_{\text{ax}} N$.* □

5.3 Finite Vector Spaces as a Model

The category $\mathbf{FinVec}_!$ is a denotational model of the language $\mathbf{PCF}_f^{alg}$. Types are interpreted as for the language $\mathbf{PCF}_f$ in Section 4.4. Typing judgments are also interpreted in the same way, with the following additional rules. First, $[\![\Delta \vdash 0 : A]\!]^{\text{vec}} = 0$. Then $[\![\Delta \vdash \alpha \cdot M : A]\!]^{\text{vec}} = \alpha \cdot [\![\Delta \vdash M : A]\!]^{\text{vec}}$. Finally, we have $[\![\Delta \vdash M + N : A]\!]^{\text{vec}} = [\![\Delta \vdash M : A]\!]^{\text{vec}} + [\![\Delta \vdash N : A]\!]^{\text{vec}}$.

Remark 27. With the extended term constructs, the language $\mathbf{PCF}_f^{alg}$ does not share the drawbacks of $\mathbf{PCF}_f$ emphasized in Remark 19. In particular, the two valid typing judgments $x : \texttt{Bool} \vdash \texttt{tt} : \texttt{Bool}$ and $x : \texttt{Bool} \vdash \texttt{if}\, x\, \texttt{then}\, \texttt{tt}\, \texttt{else}\, \texttt{tt} : \texttt{Bool}$ are now operationally distinct. For example, if one chooses the context $C[-] = (\lambda x.[-])0$, the term $C[\texttt{tt}]$ reduces to $\texttt{tt}$ whereas the term $C[\texttt{if}\, x\, \texttt{then}\, \texttt{tt}\, \texttt{else}\, \texttt{tt}]$ reduces to 0.

Lemma 28. *If $M \simeq_{\mathrm{ax}} N : A$ in $\mathbf{PCF}_f^{alg}$ then $[\![M]\!]^{\mathrm{vec}} = [\![N]\!]^{\mathrm{vec}}$.* □

Theorem 29. *Let $\Delta \vdash M, N : A$ be two valid typing judgments in $\mathbf{PCF}_f^{alg}$. If $[\![M]\!]^{\mathrm{vec}} = [\![N]\!]^{\mathrm{vec}}$ then we also have $M \simeq_{\mathrm{op}} N$.*

Proof. The proof is similar to the proof of Theorem 5: Assume $M \not\simeq_{\mathrm{op}} N$. Then there exists a context $C[-]$ that distinguishes them. The call-by-name reduction preserves the type from Lemma 23, and $C[M]$ and $C[N]$ can be rewritten as the terms $(\lambda y.C[y\, x_1 \ldots x_n])\lambda x_1 \ldots x_n.M$ and $(\lambda y.C[y\, x_1 \ldots x_n])\lambda x_1 \ldots x_n.N$, and these are axiomatically equivalent to distinct normal forms, from Lemmas 25 and 26. We conclude from Lemmas 26 and 28 that the denotations of M and N are distinct. □

5.4 Two Auxiliary Constructs

Full completeness requires some machinery. It is obtained by showing that for every type A, for every vector v in $[\![A]\!]^{\mathrm{vec}}$, there are two terms $M_v^A : A$ and $\delta_v^A : A \to \mathbf{1}$ such that $[\![M_v^A]\!]^{\mathrm{vec}} = v$ and $[\![\delta_v^A]\!]^{\mathrm{vec}}$ sends b_v to $\star$ and all other $b_{\_}$'s to 0.

We first define a family of terms $\mathtt{exp}^i : \mathbf{1} \to \mathbf{1}$ inductively on i: $\mathtt{exp}^0 = \lambda x.\star$ and $\mathtt{exp}^{i+1} = \lambda x.\texttt{let}\, \star = x\, \texttt{in}\, \mathtt{exp}^i(x)$. One can show that $[\![\mathtt{exp}^i(\alpha \cdot \star)]\!]^{\mathrm{vec}} = \alpha^i \cdot \star$. Then assume that o is the order of the field. Let $\texttt{iszero} : \mathbf{1} \to \mathbf{1}$ be the term $\mathtt{exp}^o$. The denotation of $\texttt{iszero}$ is such that $[\![\texttt{izero}(\alpha \cdot \star)]\!]^{\mathrm{vec}} = 0$ if $\alpha = 0$ and $\star$ otherwise.

The mutually recursive definitions of δ_v^A and M_v^A read as follows.

At type $A = \mathbf{1}$. The term $M_{\alpha\cdot\star}^{\mathbf{1}}$ is simply $\alpha \cdot \star$. The term $\delta_{\alpha\cdot\star}^{\mathbf{1}}$ is $\lambda x.\texttt{iszero}(x - \alpha \cdot \star)$.

At type $A = \texttt{Bool}$. As for the type $\mathbf{1}$, the term $M_{\alpha\cdot\texttt{tt}+\beta\cdot\texttt{ff}}^{\texttt{Bool}}$ is simply $\alpha \cdot \texttt{tt} + \beta \cdot \texttt{ff}$. The term $\delta_{\alpha\cdot\texttt{tt}+\beta\cdot\texttt{ff}}^{\texttt{Bool}}$ is reusing the definition of $\delta^{\mathbf{1}}$: it is the term $\lambda x.\texttt{let}\, \star = \delta_{\alpha\cdot\star}^{\mathbf{1}}(\texttt{if}\, x\, \texttt{then}\, \star\, \texttt{else}\, 0)\, \texttt{in}\, \delta_{\beta\cdot\star}^{\mathbf{1}}(\texttt{if}\, x\, \texttt{then}\, 0\, \texttt{else}\, \star)$.

At type $A = B \times C$. If $v \in [\![A]\!]^{\mathrm{vec}} = [\![B]\!]^{\mathrm{vec}} \times [\![C]\!]^{\mathrm{vec}}$, then $v = \langle u, w \rangle$, with $u \in [\![B]\!]^{\mathrm{vec}}$ and $w \in [\![C]\!]^{\mathrm{vec}}$. By induction, one can construct M_u^B and M_w^C: the term $M_v^{B\times C}$ is $\langle M_u^B, M_w^C \rangle$. Similarly, one can construct the terms δ_u^B and δ_w^C: the term $\delta_v^{B\to C}$ is $\lambda x.\texttt{let}\, \star = \delta_u^B\, \pi_l(x)\, \texttt{in}\, \delta_w^C\, \pi_r(x)$.

At type $A = B \to C$. Consider $f \in [\![A]\!]^{\mathrm{vec}} = ![\![B]\!]^{\mathrm{vec}} \multimap [\![C]\!]^{\mathrm{vec}}$. The domain of f is finite-dimensional: let $\{b_{u_i}\}_{i=1\ldots n}$ be its basis, and let w_i be the value $f(b_{u_i})$. Then, using the terms $\delta_{u_i}^B$ and $M_{w_i}^C$, one can define $M_v^{B\to C}$ as the term $\sum_i \lambda x.\texttt{let}\, \star = \delta_{u_i}^B\, x\, \texttt{in}\, M_{w_i}^C$. Similarly, one can construct $\delta_{w_i}^C$ and $M_{u_i}^B$, and from the construction in the previous paragraph we can also generate $\delta_{\langle w_1,\ldots w_n \rangle}^{C^{\times n}} : C^{\times n} \to \texttt{Bool}$. The term $\delta_v^{B\to C}$ is then defined as $\lambda f.\delta_{\langle w_1,\ldots w_n \rangle}^{C^{\times n}} \langle f\, M_{u_1}^B, \ldots, f\, M_{u_1}^B \rangle$.

5.5 Full Completeness

We are now ready to state completeness, whose proof is simply by observing that any $v \in [\![A]\!]^{\mathrm{vec}}$ can be realized by the term $M_v^A : A$.

Theorem 30 (Full Completeness). *For any type A, any vector v of $[\![A]\!]^{\mathrm{vec}}$ in* $\mathbf{FinVec}_!$ *is representable in the language* $\mathbf{PCF}_f^{alg}$. □

Theorem 31. *For all M and N, $M \simeq_{\mathrm{op}} N$ if and only if $[\![M]\!]^{\mathrm{vec}} = [\![N]\!]^{\mathrm{vec}}$.* □

A corollary of the full completeness is that the semantics **FinVec** is also adequate and fully abstract with respect to $\mathbf{PCF}_f^{alg}$.

6 Discussion

6.1 Simulating the Vectorial Structure

As we already saw, there is a full embedding of category $E : \mathbf{FinVec}_! \hookrightarrow \mathbf{FinSet}$. This embedding can be understood as "mostly" saying that the vectorial structure "does not count" in $\mathbf{FinVec}_!$, as one can simulate it with finite sets. Because of Theorems 7 and 31, on the syntactic side algebraic terms can also be simulated by the regular $\mathbf{PCF}_f$.

In this section, for simplicity, we assume that the field is $\mathbb{F}_2$. In general, it can be any finite size provided that the regular lambda-calculus $\mathbf{PCF}_f$ is augmented with q-bits, i.e. base types with q elements (where q is the characteristic of the field).

Definition 32. The *vec-to-set* encoding of a type A, written $\mathrm{VtoS}A$, is defined inductively as follows: $\mathrm{VtoS}(\mathtt{1}) = \mathtt{Bool}$, $\mathrm{VtoS}(\mathtt{Bool}) = \mathtt{Bool} \times \mathtt{Bool}$, $\mathrm{VtoS}(A \times B) = \mathrm{VtoS}(A) \times \mathrm{VtoS}(B)$, and $\mathrm{VtoS}(A \to B) = \mathrm{VtoS}(A) \to \mathrm{VtoS}(B)$.

Theorem 33. *There are two typing judgments $x : A \vdash \phi_A^{\mathrm{vec}} : \mathrm{VtoS}(A)$ and $x : \mathrm{VtoS}(A) \vdash \bar{\phi}_A^{\mathrm{vec}} : A$, inverse of each other, in* $\mathbf{PCF}_f^{alg}$ *such that any typing judgment $x : A \vdash M : B$ can be factored into $A \xrightarrow{\phi_A^{\mathrm{vec}}} \mathrm{VtoS}(A) \xrightarrow{\tilde{M}} \mathrm{VtoS}(B) \xrightarrow{\bar{\phi}_B^{\mathrm{vec}}} B$, where $\tilde{M}$ is a regular lambda-term of* $\mathbf{PCF}_f$.

Proof. The two terms ϕ_A^{vec} and $\bar{\phi}_A^{\mathrm{vec}}$ are defined inductively on A. For the definition of $\phi_{\mathtt{Bool}}^{\mathrm{vec}}$ we are reusing the term δ_v of Section 5.4. The definition is in Table 7 □

6.2 Categorical Structures of the Syntactic Categories

Out of the language $\mathbf{PCF}_f$ one can define a syntactic category: objects are types and morphisms $A \to B$ are valid typing judgments $x : A \vdash M : B$ modulo operational

Table 7. Relation between $\mathbf{PCF}_f$ and $\mathbf{PCF}_f^{alg}$

$$\phi_{\mathtt{1}}^{\mathrm{vec}} = \mathtt{let}\,\star = \delta_0 x\,\mathtt{in}\,\mathtt{tt} + \mathtt{let}\,\star = \delta_\star x\,\mathtt{in}\,\mathtt{ff} \qquad \bar{\phi}_{\mathtt{1}}^{\mathrm{vec}} = \mathtt{if}\,x\,\mathtt{then}\,0\,\mathtt{else}\,\star$$

$$\begin{aligned}\phi_{\mathtt{Bool}}^{\mathrm{vec}} = {} & \mathtt{let}\,\star = \delta_0 x\,\mathtt{in}\,\langle\,\mathtt{tt},\mathtt{tt}\,\rangle + \mathtt{let}\,\star = \delta_{\mathtt{tt}} x\,\mathtt{in}\,\langle\,\mathtt{tt},\mathtt{ff}\,\rangle \\ & + \mathtt{let}\,\star = \delta_{\mathtt{ff}} x\,\mathtt{in}\,\langle\,\mathtt{ff},\mathtt{tt}\,\rangle + \mathtt{let}\,\star = \delta_{\mathtt{tt}+\mathtt{ff}} x\,\mathtt{in}\,\langle\,\mathtt{ff},\mathtt{ff}\,\rangle\end{aligned}$$

$$\bar{\phi}_{\mathtt{Bool}}^{\mathrm{vec}} = \mathtt{if}\,(\pi_l x)\,\mathtt{then}\,(\mathtt{if}\,(\pi_r x)\,\mathtt{then}\,0\,\mathtt{else}\,\mathtt{tt})\,\mathtt{else}\,(\mathtt{if}\,(\pi_r x)\,\mathtt{then}\,\mathtt{ff}\,\mathtt{else}\,\mathtt{tt}+\mathtt{ff})$$

$$\phi_{B\times C}^{\mathrm{vec}} = \langle\, x;\pi_l;\phi_B^{\mathrm{vec}}, x;\pi_r;\phi_C^{\mathrm{vec}}\,\rangle, \qquad \phi_{B\to C}^{\mathrm{vec}} = \lambda y.x(y;\bar{\phi}_B^{\mathrm{vec}});\phi_C^{\mathrm{vec}},$$

$$\bar{\phi}_{B\times C}^{\mathrm{vec}} = \langle\, x;\pi_l;\bar{\phi}_B^{\mathrm{vec}}, x;\pi_r;\bar{\phi}_C^{\mathrm{vec}}\,\rangle, \qquad \bar{\phi}_{B\to C}^{\mathrm{vec}} = \lambda y.x(y;\phi_B^{\mathrm{vec}});\bar{\phi}_C^{\mathrm{vec}}.$$

equivalence. Because of Theorem 7, this category is cartesian closed, and one can easily see that the product of $x : A \vdash M : B$ and $x : A \vdash N : C$ is $\langle M, N \rangle : B \times C$, that the terminal object is $\star : \mathbf{1}$, that projections are defined with π_l and π_r, and that the lambda-abstraction plays the role of the internal morphism.

The language $\mathbf{PCF}_f^{alg}$ almost defines a cartesian closed category: by Theorem 31, it is clear that pairing and lambda-abstraction form a product and an internal hom. However, it is missing a terminal object (the type $\mathbf{1}$ doesn't make one as $x : A \vdash 0 : \mathbf{1}$ and $x : A \vdash \star : \mathbf{1}$ are operationally distinct). There is no type corresponding to the vector space $\langle 0 \rangle$. It is not difficult, though, to extend the language to support it: it is enough to only add a type $\mathbf{0}$. Its only inhabitant will then be the term 0: it make a terminal object for the syntactic category.

Finally, Theorem 33 is essentially giving us a functor $\mathbf{PCF}_f^{alg} \to \mathbf{PCF}_f$ corresponding to the full embedding E. This makes a full correspondence between the two models **FinSet** and $\mathbf{FinVec}_!$, and $\mathbf{PCF}_f$ and $\mathbf{PCF}_f^{alg}$, showing that computationally the algebraic structure is virtually irrelevant.

6.3 (Co)Eilenberg-Moore Category and Call-by-value

From a linear category with modality ! there are two canonical cartesian closed categories: the coKleisli category, but also the (co)Eilenberg-Moore category: here, objects are still those of **FinVec**, but morphisms are now $!A \to !B$.

According to [30], such a model would correspond to the call-by-value (or, as coined by [8] *call-by-base*) strategy for the algebraic structure discussed in Remark 21.

6.4 Generalizing to Modules

To conclude this discussion, let us consider a generalization of finite vector spaces to finite modules over finite semi-rings.

Indeed, the model of linear logic this paper uses would work in the context of a finite semi-ring instead of a finite field, as long as addition and multiplication have distinct units. For example, by using the semiring $\{0, 1\}$ where $1 + 1 = 1$ one recover sets and relations. However, we heavily rely on the fact that we have a finite field K in the construction of Section 5.4, yielding the completeness result in Theorem 30.

This particular construction works because one can construct any function between any two finite vector spaces as polynomial, for the same reason as any function $K \to K$ can be realized as a polynomial.

References

1. Abramsky, S., Jagadeesan, R., Malacaria, P.: Full abstraction for PCF. Inf. and Comp. 163, 409–470 (2000)
2. Arrighi, P., Díaz-Caro, A., Valiron, B.: A type system for the vectorial aspects of the linear-algebraic lambda-calculus. In: Proc. of DCM (2011)
3. Arrighi, P., Dowek, G.: Linear-algebraic λ-calculus: higher-order, encodings, and confluence. In: Voronkov, A. (ed.) RTA 2008. LNCS, vol. 5117, pp. 17–31. Springer, Heidelberg (2008)

4. Benton, N.: A mixed linear and non-linear logic: Proofs, terms and models. Technical report, Cambridge U (1994)
5. Bierman, G.: On Intuitionistic Linear Logic. PhD thesis, Cambridge U (1993)
6. Bucciarelli, A., Ehrhard, T., Manzonetto, G.: A relational semantics for parallelism and non-determinism in a functional setting. A. of Pure and App. Logic 163, 918–934 (2012)
7. de Groote, P.: Strong normalization in a non-deterministic typed lambda-calculus. In: Nerode, A., Matiyasevich, Y.V. (eds.) Logical Foundations of Computer Science. LNCS, vol. 813, pp. 142–152. Springer, Heidelberg (1994)
8. Díaz-Caro, A.: Du Typage Vectoriel. PhD thesis, U. de Grenoble (2011)
9. Ehrhard, T.: Finiteness spaces. Math. Str. Comp. Sc. 15, 615–646 (2005)
10. Ehrhard, T., Pagani, M., Tasson, C.: The computational meaning of probabilistic coherence spaces. In: Proc. of LICS (2011)
11. Ehrhard, T., Regnier, L.: The differential lambda-calculus. Th. Comp. Sc. 309, 1–41 (2003)
12. Girard, J.-Y.: Linear logic. Th. Comp. Sc. 50, 1–101 (1987)
13. Girard, J.-Y., Lafont, Y., Taylor, P.: Proof and Types. CUP (1990)
14. Hillebrand, G.G.: Finite Model Theory in the Simply Typed Lambda Calculus. PhD thesis, Brown University (1991)
15. Hyland, M., Schalk, A.: Glueing and orthogonality for models of linear logic. Th. Comp. Sc. 294, 183–231 (2003)
16. James, R.P., Ortiz, G., Sabry, A.: Quantum computing over finite fields. Draft (2011)
17. Lambek, J., Scott, P.J.: Introduction to Higher-Order Categorical Logic. CUP (1994)
18. Lang, S.: Algebra. Springer (2005)
19. Lidl, R.: Finite fields, vol. 20. CUP (1997)
20. Mac Lane, S.: Categories for the Working Mathematician. Springer (1998)
21. Milner, R.: Fully abstract models of typed lambda-calculi. Th. Comp. Sc. 4, 1–22 (1977)
22. Plotkin, G.: LCF considered as a programming language. Th. Comp. Sc. 5, 223–255 (1977)
23. Pratt, V.R.: Re: Linear logic semantics (barwise). On the TYPES list (February 1992), `http://www.seas.upenn.edu/~sweirich/types/archive/1992/msg00047.html`
24. Pratt, V.R.: Chu spaces: Complementarity and uncertainty in rational mechanics. Technical report, Stanford U (1994)
25. Salvati, S.: Recognizability in the simply typed lambda-calculus. In: Ono, H., Kanazawa, M., de Queiroz, R. (eds.) WoLLIC 2009. LNCS, vol. 5514, pp. 48–60. Springer, Heidelberg (2009)
26. Schumacher, B., Westmoreland, M.D.: Modal quantum theory. In: Proc. of QPL (2010)
27. Scott, D.S.: A type-theoretic alternative to CUCH, ISWIM, OWHY. Th. Comp. Sc. 121, 411–440 (1993)
28. Selinger, P.: Order-incompleteness and finite lambda reduction models. Th. Comp. Sc. 309, 43–63 (2003)
29. Soloviev, S.: Category of finite sets and cartesian closed categories. J. of Soviet Math. 22(3) (1983)
30. Valiron, B.: A typed, algebraic, computational lambda-calculus. Math. Str. Comp. Sc. 23, 504–554 (2013)
31. Vaux, L.: The algebraic lambda-calculus. Math. Str. Comp. Sc. 19, 1029–1059 (2009)
32. Winskel, G.: The Formal Semantics of Programming Languages. MIT Press (1993)

A Decidable Recursive Logic for Weighted Transition Systems

Kim Guldstrand Larsen, Radu Mardare, and Bingtian Xue

Aalborg University, Denmark

Abstract. In this paper we develop and study the Recursive Weighted Logic (RWL), a multi-modal logic that expresses qualitative and quantitative properties of labelled weighted transition systems (LWSs). LWSs are transition systems labelled with actions and real-valued quantities representing the costs of transitions with respect to various resources. RWL uses first-order variables to measure local costs. The main syntactic operators are similar to the ones of timed logics for real-time systems. In addition, our logic is endowed, with simultaneous recursive equations, which specify the weakest properties satisfied by the recursive variables. We prove that unlike in the case of the timed logics, the satisfiability problem for RWL is decidable. The proof uses a variant of the region construction technique used in literature with timed automata, which we adapt to the specific settings of RWL. This paper extends previous results that we have demonstrated for a similar but much more restrictive logic that can only use one variable for each type of resource to encode logical properties.

Keywords: labelled weighted transition system, multi-modal logic, maximal fixed point computation.

1 Introduction

For industrial practice, especially in the area of embedded systems, an essential problem is how to deal with the high complexity of the systems, while still meeting the requirements of correctness, predictability, performance and also resource constraints. In this respect, for embedded systems, verification should not only consider functional properties but also non-functional properties such as those related to resource constraints. Within the area of model checking a number of state-machine based modelling formalisms have emerged, which allow for such quantitative properties to be expressed. For instance, the timed automata [AD90] and its extensions to weighted timed automata [BFH$^+$01, ATP01] allow time-constraints to be modelled and efficiently analysed.

In order to specify and reason about not only the qualitative behaviours of (embedded) systems but also about their quantitative consumptions of resources, we consider a multi-modal logic – Recursive Weighted Logic (RWL) – defined for a semantics based on labelled weighted transition systems (LWS). Our notion of weighted transition systems is more than a simple instance of *weighted automata* [DKV09], since an LWS can also be infinite or/and infinitely branching. The transitions of LWSs are labelled with both actions and real numbers. The numbers represent the costs of the corresponding transitions in terms of resources. In order to use a variant of the region construction

G. Ciobanu and D. Méry (Eds.): ICTAC 2014, LNCS 8687, pp. 460–476, 2014.

technique developed for timed automata in [AD90, ACD90], we only consider non-negative labels in this paper.

RWL is an extension of the *weighted modal logic* [LM13] with only maximal fixed points. The maximal fixed points which are defined by simultaneous recursive equations [Lar90, CKS92, CS93], allows us to encode properties of infinite behaviours including safety and cost-bounded liveness. They specify the weakest properties satisfied by the recursive variables. RWL is also endowed with modal operators that predicate about the values of resource-variables, which allow us to specify and reason about the quantitative properties related to resources, e.g., energy and time. While in an LWS we can have real-valued labels, the modalities of the logic only encode rational values. This will not restrict too much the expressive power of RWL, since we can characterize a transition using an infinite convergent sequences of rationals that approximate the real-valued resource.

To encode various resource-constrains in RWL, we use resource-variables, similar to the clock-variables used in the timed logics [ACD93a, HNSY92, AILS07]. We use *resource valuation* to assign non-negative real values to resource-variables. Previously, we restricted our attention to only one resource-variable for each type of resources [LMX14]. This guaranteed the decidability of the logic and the finite model property. However, this restriction bounds the expressiveness of the logic. For example, consider a system where one cannot consume one or more than one unit of energy. On the other hand, the system is required to do some action infinitely often which costs non-zero amount of energy. This system cannot be specified in the logic with only one resource-variable for each type of resources, because we need two resource-variables to measure the same resource – energy in this example. Here, we allow multiple resource-variables for each type of resource, which measure the resource in different ways. In this paper we only discuss the event related resource-variables. More precisely, for each type of resource and each action, we associate one resource-variable. Every time the system performs this action, all the resource-variables associated to this action are reset after the corresponding transition, meaning that the resource valuation will map those resource-variables to zero. This is useful for encoding various interesting scenarios.

Even though RWL does not enjoy the finite model property, we may apply a variant of the region construction technique developed for timed automata [AD90, ACD90, LLW95], to obtain symbolic LWSs of the satisfiable formulas. These symbolic LWSs provide an abstract semantics for LWSs, allowing us to reason about satisfiability by investigating these symbolic models that are finite. We propose a model construction algorithm, which constructs a symbolic LWS for a given satisfiable (consistent) RWL formula. The symbolic model can be eventually used to decide the existence of the concrete LWSs and generate them – possibly infinite – which are models of the given formula.

Our decidability result is important and, in a sense surprising, being that the satisfiability problem is known to be undecidable for logics very similar to ours, such as TCTL [ACD93b], T_μ [HNSY92], L_ν [LLW95] and timed modal logic (TML) [LMX].

The paper is organized as follows: in the following section we present the notion of labelled weighted transition system; in Section 3 we introduce the recursive weighted logic with its syntax and semantics; Section 4 is dedicated to the region construction technique and the symbolic models of LWSs; in Section 5 we prove the decidability of the satisfiability problem for RWL. The paper also includes a conclusive section where we summarize the results and describe future research directions.

2 Labelled Weighted Transition Systems

A *labelled weighted transition system* (LWS) is a transition system that has several types of resources and the transitions labelled both with actions and (non-negative) real numbers. In Figure 1 is represented such a system in which there are three types of resource; each number is used to represent the costs of the corresponding transitions in terms of one type of resource, e.g., energy or time.

Definition 1 (Labelled Weighted Transition System). *A LWS is a tuple*

$$\mathcal{W} = (M, \mathcal{K}, \Sigma, \theta)$$

where M is a non-empty set of states, *$\mathcal{K} = \{e_1, \ldots, e_k\}$ is the finite set of (k types of) resources, Σ a non-empty set of* actions *and $\theta : M \times (\Sigma \times (\mathcal{K} \to \mathbb{R}_{\geq 0})) \to 2^M$ is the* labelled transition function.

For simplicity, hereafter we use a vector of real numbers instead of the function from the set of the resources $\mathcal{K}$ to real numbers, i.e., for $f : \mathcal{K} \to \mathbb{R}_{\geq 0}$ defined as $f(e_i) = r_i$ for all $i = 1, \ldots, k$, we write $\overline{u} = (r_1, \ldots, r_k) \in \mathbb{R}^k_{\geq 0}$ instead. On the other hand, for a vector of real numbers $\overline{u} \in \mathbb{R}^k_{\geq 0}$, $\overline{u}(e_i)$ denotes the i-th number of the vector $\overline{u}$, which represents the cost of the resource e_i during the transition.

Instead of $m' \in \theta(m, a, \overline{u})$ we write $m \xrightarrow{\overline{u}}_a m'$.

To clarify the role of the aforementioned concepts consider the following example.

Example 1. In Figure 1, we show the LWS

$$\mathcal{W} = (M, \Sigma, \mathcal{K}, \theta),$$

where $M = \{m_0, m_1, m_2\}$, $\mathcal{K} = \{e_1, e_2, e_3\}$, $\Sigma = \{a, b\}$, and θ defined as follows: $m_0 \xrightarrow{(3,4,5)}_a m_1$, $m_0 \xrightarrow{(\pi,\pi,0)}_b m_2$ and $m_1 \xrightarrow{(\sqrt{2},1.9,7)}_a m_2$.

$\mathcal{W}$ has three states m_0, m_1, m_2, three kinds of resource e_1, e_2, e_3 and two actions a, b. The state m_0 has two transitions: one a-transition – which costs 3 units of e_1, 4 units of e_2 and 5 units of e_3 – to m_1 and one b-transition – which costs π units of e_1 and e_2 respectively (and does not cost any e_3) – to m_2. If the system does an a-transition from m_0 to m_1, the amounts of the resource e_1, e_2 and e_3 increase with 3, 4 and 5 units respectively – that the system gains by doing the a-transition. ■

In the rest of this paper, we fix a set Σ of actions, and for simplicity we omit it in the description of LWSs and the logic defined in the next section.

The concept of *weighted bisimulation* is a relation between the states of a given LWS that equates states with identical (action- and weighted-) behaviors.

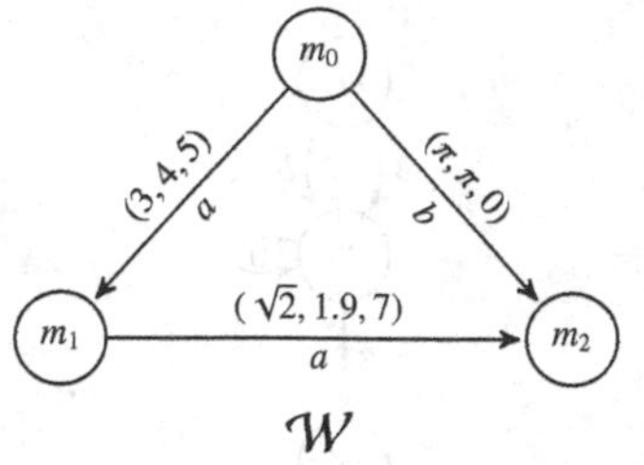

Fig. 1. Labelled Weighted Transition System

Definition 2 (Weighted Bisimulation). *Given a LWS $\mathcal{W} = (M, \mathcal{K}, \theta)$, a* weighted bisimulation *is an equivalence relation $R \subseteq M \times M$ such that whenever $(m, m') \in R$,*

- if $m \xrightarrow{\overline{u}}_a m_1$, then there exists $m'_1 \in M$ s.t. $m' \xrightarrow{\overline{u}}_a m'_1$ and $(m_1, m'_1) \in R$;
- if $m' \xrightarrow{\overline{u}}_a m'_1$, then there exists $m_1 \in M$ s.t. $m \xrightarrow{\overline{u}}_a m_1$ and $(m_1, m'_1) \in R$.

If there exists a weighted bisimulation relation R such that $(m, m') \in R$, we say that m and m' are bisimilar, *denoted by $m \sim m'$.*

As for the other types of bisimulation, the previous definition can be extended to define the weighted bisimulation between distinct LWSs by considering bisimulation relations on their disjoint union. *Weighted bisimilarity* is the largest weighted bisimulation relation; if $\mathcal{W}_i = (M_i, \mathcal{K}_i, \theta_i)$, $m_i \in M_i$ for $i = 1, 2$ and m_1 and m_2 are bisimilar, we write $(m_1, \mathcal{W}_1) \sim (m_2, \mathcal{W}_2)$.

The next examples shows the role of the weighted bisimilarity.

Example 2. In Figure 2, $\mathcal{W}_1 = (M_1, \mathcal{K}_1, \theta_1)$ is a LWS with five states and one type of resources, where $M_1 = \{m_0, m_1, m_2, m_3, m_4\}$, $\mathcal{K}_1 = \{e\}$ and θ_1 is defined as: $m_0 \xrightarrow{3}_a m_1$, $m_0 \xrightarrow{2}_b m_2$, $m_1 \xrightarrow{0}_d m_2$, $m_1 \xrightarrow{3}_c m_3$, $m_2 \xrightarrow{0}_d m_1$ and $m_2 \xrightarrow{3}_c m_4$.

It is easy to see that $m_3 \sim m_4$ because both of them can not do any transition. Besides, $m_1 \sim m_2$ because both of them can do a c-transition with cost 3 to some states which are bisimilar (m_3 and m_4), and a d-action transition with cost 0 to each other. m_0 is not bisimilar to any states in $\mathcal{W}_1$.

$\mathcal{W}_2 = (M_2, \mathcal{K}_2, \theta_2)$ is a LWS with three states, where $M_2 = \{m'_0, m'_1, m'_2\}$, $\mathcal{K}_2 = \mathcal{K}_1$ and θ_2 is defined as: $m'_0 \xrightarrow{3}_a m'_1$, $m'_0 \xrightarrow{2}_b m'_1$, $m'_1 \xrightarrow{0}_d m'_1$ and $m'_1 \xrightarrow{3}_c m'_2$.

We can see that: $(m_0, \mathcal{W}_1) \sim (m'_0, \mathcal{W}_2)$, $(m_1, \mathcal{W}_1) \sim (m'_1, \mathcal{W}_2)$, $(m_2, \mathcal{W}_1) \sim (m'_1, \mathcal{W}_2)$,
$(m_3, \mathcal{W}_1) \sim (m'_2, \mathcal{W}_2)$, $(m_4, \mathcal{W}_1) \sim (m'_2, \mathcal{W}_2)$.

Notice that $(m''_0, \mathcal{W}_3) \nsim (m'_0, \mathcal{W}_2)$, because $(m''_1, \mathcal{W}_3) \nsim (m'_1, \mathcal{W}_2)$. Besides, $m''_1 \nsim m''_2$, because m''_1 can do a d-action with weight 2 while m''_2 cannot and m''_2 can do a d-action with weight 1 while m''_1 cannot. ∎

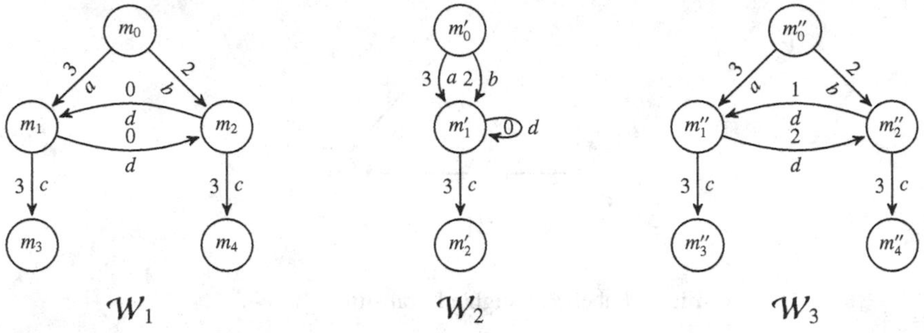

Fig. 2. Weighted Bisimulation

3 Recursive Weighted Logic

In this section we introduce a multi-modal logic that encodes properties of LWSs called *Recursive Weighted Logic* (RWL).

To encode various resource-constrains in RWL, we use resource-variables, similar to the clock-variables used in timed logics [ACD93a, HNSY92, AILS07]. In this paper, we introduce event related resource-variables to measure the resources in different ways corresponding to different actions, i.e., for each action $a \in \Sigma$, we associate resource-variables $x_a^1, \ldots, x_a^k$ for each types of resource $e_1, \ldots, e_k$ respectively. In the following, we use $\mathcal{V}_i = \{x_a^i \mid a \in \Sigma\}$ to denote the set of the resource-variables associated for the type of resource e_i, $\mathcal{V}_a = \{x_a^i \mid i = 1, \ldots, k\}$ to denote the set of the resource-variables associated with the action a and $\mathcal{V} = \bigcup_{i=1,\ldots,k} \mathcal{V}_i = \bigcup_{a \in \Sigma} \mathcal{V}_a$ to denote the set of all the resource-variables. Note that for any i, j such that $i \neq j$, $\mathcal{V}_i \cap \mathcal{V}_j = \emptyset$, and for any a, b such that $a \neq b$, $\mathcal{V}_a \cap \mathcal{V}_b = \emptyset$.

In addition to the classic boolean operators (except negation), our logic is firstly endowed with a class of recursive (formula) variables $X_1, \ldots, X_n$, which specify properties of infinite behaviours. We denote by $\mathcal{X}$ the set of recursive formula variables.

Secondly, RWL is endowed with a class of modalities of arity 1, named *transition modalities*, of type $[a]$ or $\langle a \rangle$, for $a \in \Sigma$, which are defined as the classical transition modalities with reset operation of all the resource-variables associated with the corresponding action followed. More precisely, every time the system does an a-action, all the resource-variables $x \in \mathcal{V}_a$ will be reset after this transition, i.e., x is interpreted to zero after every a-action, for all $x \in \mathcal{V}_a$.

Besides, our logic is also endowed with a class of modalities of arity 0 called *state modalities* of type $x \bowtie r$, for $\bowtie \in \{\leq, \geq, <, >\}$, $r \in \mathbb{Q}_{\geq 0}$ and $x \in \mathcal{V}$, which predicates about the value of the resource-variable x at the current state.

Before proceeding with the maximal fixed points, we define the basic formulas of RWL and their semantics firstly. Based on them, we will eventually introduce the recursive definitions - the maximal equation blocks - which extend the semantics of the basic formulas.

Definition 3 (Syntax of Basic Formulas). *For arbitrary* $r \in \mathbb{Q}_{\geq 0}$, $a \in \Sigma$, $x \in \mathcal{V}$, $\bowtie \in \{\leq, \geq, <, >\}$ *and* $X \in \mathcal{X}$, *let*

$$\mathcal{L}: \quad \phi := \top \mid \bot \mid x \bowtie r \mid \phi \wedge \phi \mid \phi \vee \phi \mid [a]\phi \mid \langle a \rangle \phi \mid X \,.$$

Before looking at the semantics for the basic formulas, we define the notion of *resource valuation* and *extended states*.

Definition 4 (Resource Valuation). *A resource valuation is a function* $l : \mathcal{V} \to \mathbb{R}_{\geq 0}$ *that assigns (non-negative) real numbers to the resource-variables in* $\mathcal{V}$.

A resource valuation assigns non-negative real values to all the resource-variables and the assignment is interpreted as the amount of resources measured by the corresponding resource-variable in a given state of the system. We denote by L the class of resource valuations.

We write l_i to denote the valuation for all resource-variables $x \in \mathcal{V}_i$ under the resource valuation l, i.e., for any $x \in \mathcal{V}$,

$$l_i(x) = \begin{cases} l(x), & x \in \mathcal{V}_i \\ \text{undefined}, & \text{otherwise} \end{cases}$$

Similarly, we write l_a to denote the valuation for all resource-variables $x \in \mathcal{V}_a$ under the resource valuation l, i.e., for any $x \in \mathcal{V}$,

$$l_a(x) = \begin{cases} l(x), & x \in \mathcal{V}_a \\ \text{undefined}, & \text{otherwise} \end{cases}$$

If l is a resource valuation and $x \in \mathcal{V}$, $s \in \mathbb{R}_{\geq 0}$ we denote by $l[x \mapsto s]$ the resource valuation that associates the same values as l to all variables except x, to which it associates the value s, i.e., for any $y \in \mathcal{V}$,

$$l[x \mapsto s](y) = \begin{cases} s, & y = x \\ l(y), & \text{otherwise} \end{cases}$$

Moreover, for $\mathcal{V}' \subseteq \mathcal{V}$ and $s \in \mathbb{R}_{\geq 0}$, we denote by $l[\mathcal{V}' \mapsto s]$ the resource valuation that associates the same values as l to all variables except those in $\mathcal{V}'$, to which it associates the value s, i.e., for any $y \in \mathcal{V}$,

$$l[\mathcal{V}' \mapsto s](y) = \begin{cases} s, & y \in \mathcal{V}' \\ l(y), & \text{otherwise} \end{cases}$$

For $\overline{u} \in \mathbb{R}^k_{\geq 0}$, $l + \overline{u}$ is defined as: for any $i \in \{1, \ldots, k\}$, for any $x \in \mathcal{V}_i$,

$$(l + \overline{u})(x) = l(x) + \overline{u}(e_i).$$

A pair (m, l) is called *extended state* of a given LWS $\mathcal{W} = (M, \mathcal{K}, \theta)$, where $m \in M$ and $l \in L$. Transitions between extended states are defined by:

$$(m, l) \longrightarrow_a (m', l') \text{ iff } m \xrightarrow{\overline{u}}_a m' \text{ and } l' = l + \overline{u}.$$

Given a LWS $\mathcal{W} = (M, \mathcal{K}, \theta)$, we interpret the RWL basic formulas over an extended state (m, l) and an environment ρ which maps each recursive formula variables to subsets of $M \times L$. The *LWS-semantics* of RWL basic formulas is defined inductively as follows.

$\mathcal{W}, (m, l), \rho \models \top$ – always,

$\mathcal{W}, (m, l), \rho \models \bot$ – never,

$\mathcal{W}, (m, l), \rho \models x \bowtie r$ iff $l(x) \bowtie r$,

$\mathcal{W}, (m, l), \rho \models \phi \wedge \psi$ iff $\mathcal{W}, (m, l), \rho \models \phi$ and $\mathcal{W}, (m, l), \rho \models \psi$,

$\mathcal{W}, (m, l), \rho \models \phi \vee \psi$ iff $\mathcal{W}, (m, l), \rho \models \phi$ or $\mathcal{W}, (m, l), \rho \models \psi$,

$\mathcal{W}, (m, l), \rho \models [a]\phi$ iff for arbitrary $(m', l') \in M \times L$ such that $(m, l) \longrightarrow_a (m', l')$, $\mathcal{W}, (m', l'[\mathcal{V}_a \mapsto 0]), \rho \models \phi$,

$\mathcal{W}, (m, l), \rho \models \langle a \rangle\phi$ iff exists $(m', l') \in M \times L$ such that $(m, l) \longrightarrow_a (m', l')$ and $\mathcal{W}, (m', l'[\mathcal{V}_a \mapsto 0]), \rho \models \phi$,

$\mathcal{W}, (m, l), \rho \models X$ iff $(m, l) \in \rho(X)$.

Definition 5 (Syntax of Maximal Equation Blocks). *Let $\mathcal{X} = \{X_1, \ldots, X_n\}$ be a set of recursive formula variables. A maximal equation block B is a list of (mutually recursive) equations:*

$$X_1 = \phi_1$$
$$\vdots$$
$$X_n = \phi_n$$

in which X_i are pairwise-distinct over $\mathcal{X}$ and ϕ_i are basic formulas over $\mathcal{X}$, for all $i = 1, \ldots, n$.

Each maximal equation block B defines an *environment* for the recursive formula variables $X_1, \ldots, X_n$, which is the weakest property that the variables satisfy.

We say that an arbitrary formula ϕ is *closed with respect to a maximal equation block* B if all the recursive formula variables appearing in ϕ are defined in B by some of its equations. If all the formulas ϕ_i that appears in the right hand side of some equation in B is closed with respect to B, we say that B is *closed.*

Given an environment ρ and $\overline{\Upsilon} = \langle \Upsilon_1, \ldots, \Upsilon_n \rangle \in (2^{M \times L})^n$, let

$$\rho_{\overline{\Upsilon}} = \rho[X_1 \mapsto \Upsilon_1, \ldots, X_n \mapsto \Upsilon_n]$$

be the environment obtained from ρ by updating the binding of X_i to Υ_i.

Given a maximal equation block B and an environment ρ, consider the function

$$f_B^\rho : (2^{M \times L})^n \longrightarrow (2^{M \times L})^n$$

defined as follows:

$$f_B^\rho(\overline{\Upsilon}) = \langle [\![\phi_1]\!]\rho_{\overline{\Upsilon}}, \ldots, [\![\phi_n]\!]\rho_{\overline{\Upsilon}} \rangle,$$

where $[\![\phi]\!]\rho = \{(m, l) \in M \times L \mid \mathcal{W}, (m, l), \rho \models \phi\}$.

Observe that $(2^{M \times L})^n$ forms a complete lattice with the ordering, join and meet operations defined as the point-wise extensions of the set-theoretic inclusion, union and intersection, respectively. Moreover, for any maximal equation block B and environment ρ, f_B^ρ is monotonic with respect to the order of the lattice and therefore, according to the Tarski fixed point theorem [Tar55], it has a greatest fixed point that we denote by $\nu\overline{X}.f^{\rho}_{B}$. This fixed point can be characterized as follows:

$$\nu\overline{X}.f^{\rho}_{B} = \bigcup\{\overline{\Upsilon} \mid \overline{\Upsilon} \subseteq f^{\rho}_{B}(\overline{\Upsilon})\}.$$

Consequently, a maximal equation block defines an environment that satisfies all its equations, i.e.,

$$[\![B]\!]\rho = \nu\overline{X}.f^{\rho}_{B}.$$

When B is closed, i.e. there is no free recursive formula variable in B, it is not difficult to see that for any ρ and ρ', $[\![B]\!]\rho = [\![B]\!]\rho'$. So, we just take $\rho = 0$ and write $[\![B]\!]$ instead of $[\![B]\!]0$. In the rest of the paper we will only discuss this kind of equation blocks. (For those that are not closed, we only need to have the initial environment which maps the free recursive variables to subsets of $M \times L$.)

Now we are ready to define the general semantics of RWL: for an arbitrary LWS $\mathcal{W} = (M, \mathcal{K}, \theta)$ with $m \in M$, an arbitrary resource valuation $l \in L$ and arbitrary RWL-formula ϕ closed w.r.t. a maximal equation block B,

$$\mathcal{W}, (m, l) \models_B \phi \text{ iff } \mathcal{W}, (m, l), [\![B]\!] \models \phi.$$

The symbol $\models_B$ is interpreted as satisfiability for the block B. Whenever it is not the case that $\mathcal{W}, (m, l) \models_B \phi$, we write $\mathcal{W}, (m, l) \not\models_B \phi$. We say that a formula ϕ is *B-satisfiable* if there exists at least one LWS that satisfies it for the block B in one of its states under at least one resource valuation; ϕ is a *B-validity* if it is satisfied in all states of any LWS under any resource valuation - in this case we write $\models_B \phi$.

To exemplify the expressiveness of RWL, we propose the following example of system with recursively-defined properties.

Example 3. Consider a system which only has one type of resource, e.g., energy. It involves three actions: a, b and c, to which three resource-variables x_a, x_b and x_c are associated respectively. Those resource-variables are used to measure the amount of energy in the system. The specifications of the system are as follows:

- The system cannot cost one or more than one unit of energy;
- The system has the following (action) trace: *abcbcbc*..., i.e., it does an a-action followed by infinitely repeating the sequence bc of actions, during which both b and c will have some non-zero cost.

In our logic the above mentioned requirements can be encoded as follows:

$$\phi = \langle a \rangle X,$$
$$B = \left\{ \begin{array}{l} X = x_a < 1 \wedge \langle b \rangle (Y \wedge x_c > 0), \\ Y = x_a < 1 \wedge \langle c \rangle (X \wedge x_b > 0) \end{array} \right\}$$

■

4 Regions and Symbolic Models

Before proceeding with the definitions of regions and symbolic models, we take a further look at Example 3 in the above section. It is not difficult to see that there exists a model satisfying the formula ϕ under the maximal equation block B, but it must be infinite. This is because x_a is synchronised with x_b and x_c, which are constantly growing, while x_a is bounded by 1. This example proves that RWL does not enjoy the finite model property.

In this section we introduce the region technique for LWS, which is inspired by that for timed automata of Alur and Dill [AD90, ACD90]. It provides an abstract semantics of LWSs in the form of finite labelled transition systems with the truth value of RWL formulas being maintained.

Here we introduce the regions defined for a given maximal constant $N \in \mathbb{N}$. For the case where the maximal constant is a rational number $\frac{p}{q}$, where $p, q \in \mathbb{N}$, we only need to get the regions for the maximal constant p_i first and divide all the regions by q. In fact for this case we could, alternatively, assume that all the constraints involve natural numbers, since the constraints that occur in one formula are finitely many (for instance, we can multiply all the rationals with the same well-chosen integer; by this operation the truth values of the correspondingly modified formulas are preserved).

For $r \in \mathbb{R}_{\geq 0}$, let $\lfloor r \rfloor \stackrel{\text{def}}{=} \max\{z \in \mathbb{Z} \mid z \leq r\}$ denote the integral part of r, and let $\{r\} = r - \lfloor r \rfloor$ denote its fractional part. Moreover, we have $\lceil r \rceil \stackrel{\text{def}}{=} \min\{z \in \mathbb{Z} \mid z \geq r\}$.

Definition 6. *Let $N \in \mathbb{N}$ be a given maximal constant and let $\mathcal{V}_i$ be a set of resource-variables for resource e_i. Then $l_i, l'_i : \mathcal{V}_i \to \mathbb{R}_{\geq 0}$ are equivalent with respect to N, denoted by $l_i \doteq l'_i$ iff:*

1. *$\forall x \in \mathcal{V}_i$, $l_i(x) > N$ iff $l'_i(x) > N$;*
2. *$\forall x \in \mathcal{V}_i$ s.t. $0 \leq l_i(x) \leq N$, $\lfloor l_i(x) \rfloor = \lfloor l'_i(x) \rfloor$ and $\{l_i(x)\} = 0 \Leftrightarrow \{l'_i(x)\} = 0$;*
3. *$\forall x, y \in \mathcal{V}_i$ s.t. $0 \leq l_i(x), l_i(y) \leq N$, $\{(l_i(x)\} \leq \{l_i(y)\} \Leftrightarrow \{(l'_i(x)\} \leq \{l'_i(y)\}$.*

The equivalence classes under $\doteq$ are called *regions*. $[l_i]$ denotes the region which contains the labelling l_i for resource-variables $x \in \mathcal{V}_i$ and $R_{N_i}^{\mathcal{V}_i}$ denotes the set of all regions for the set $\mathcal{V}_i$ of resource-variables for resource e_i and the constant N_i. Notice that for a given $N_i \in \mathbb{N}$, $R_{N_i}^{\mathcal{V}_i}$ is finite.

For a region $\delta \in R_{N_i}^{\mathcal{V}_i}$, we define the *successor* region as the region δ' – denoted by $\delta \rightsquigarrow \delta'$ – iff:

$$\text{for any } l_i \in \delta, \text{ there exists } d \in \mathbb{R}_{\geq 0} \text{ s.t. } l_i + d \in \delta'.$$

As we mentioned before, for the case where the maximal constant is a rational number $\frac{p_i}{q_i}$ where $p_i, q_i \in \mathbb{N}$, we only need to get the regions for the maximal constant p first and divide all the regions by q_i to achieve the set of all regions for the set $\mathcal{V}_i$ of resource-variables for resource e_i and the constant $\frac{p_i}{q_i}$ – denoted by $R_{p_i/q_i}^{\mathcal{V}_i}$. Let $\mathcal{R}^{\mathcal{V}} = \{\overline{[l]} = ([l_1], \ldots, [l_k]) \mid [l_i] \in R_{p_i/q_i}^{\mathcal{V}_i}, \frac{p_i}{q_i} \in \mathbb{Q}_{\geq 0} \text{ for any } i \in \{1, \ldots, k\}\}$.

We will now define the fundamental concept of a *symbolic model* of LWS. Every extended state (m, l) is replaced by a so-called *extended symbolic state* $(m, \overline{[l]})$. Whenever we have transition between two extended states, there should also be a transition between the corresponding symbolic states, i.e.:

$$(m, \overline{[l]}) \longrightarrow_a (m', \overline{[l']}) \text{ iff } (m, l) \longrightarrow_a (m', l').$$

Definition 7. *Given $\mathcal{R}^{\mathcal{V}}$ and a non-empty set of states M^s, a* symbolic LWS *is a tuple*

$$\mathcal{W}^s = (\Pi^s, \Sigma^s, \theta^s)$$

where $\Pi^s \subseteq M^s \times \mathcal{R}^\mathcal{V}$ *is a non-empty set of* symbolic states $\pi^s = (m, \overline{\delta})$*,* Σ^s *a non-empty set of actions and* $\theta^s : \Pi^s \times (\Sigma^s) \rightarrow 2^{\Pi^s}$ *is the* symbolic labelled transition function*, which satisfies the following:*

$$\textit{if } (m', \overline{\delta'}) \in \theta((m, \overline{\delta}), a), \textit{ then } \overline{\delta} \rightsquigarrow \overline{\delta'}.$$

Given a symbolic LWS, we can define the symbolic satisfiability relation $\models^s$ with $\pi = (m, \overline{\delta}) \in \Pi^s$ as follows:

$\mathcal{W}^s, \pi, \rho^s \models^s \top$ – always,
$\mathcal{W}^s, \pi, \rho^s \models^s \bot$ – never,
$\mathcal{W}^s, \pi, \rho^s \models^s x \bowtie r$ iff for any $w \in \overline{\delta}(x)$, $w \bowtie r$,
$\mathcal{W}^s, \pi, \rho^s \models^s \phi \wedge \psi$ iff $\mathcal{W}^s, \pi, \rho^s \models^s \phi$ and $\mathcal{W}^s, \pi, \rho^s \models^s \psi$,
$\mathcal{W}^s, \pi, \rho \models^s \phi \vee \psi$ iff $\mathcal{W}^s, \pi, \rho^s \models^s \phi$ or $\mathcal{W}^s, \pi, \rho^s \models^s \psi$,
$\mathcal{W}^s, \pi, \rho^s \models^s [a]\phi$ iff for any $\pi' = (m', \overline{\delta'}) \in \Pi^s$ s.t. $\pi \rightarrow_a \pi'$, $\mathcal{W}^s, (m', \overline{\delta'}[\mathcal{V}_a \mapsto 0]), \rho^s \models^s \phi$,
$\mathcal{W}^s, \pi, \rho^s \models^s \langle a \rangle\phi$ iff there exists $\pi' = (m', \overline{\delta'}) \in \Pi^s$ s.t. $\pi \rightarrow_a \pi'$ and $\mathcal{W}^s, (m', \overline{\delta'}[\mathcal{V}_a \mapsto 0]), \rho^s \models^s \phi$,
$\mathcal{W}^s, \pi, \rho^s \models^s X$ iff $m \in \rho^s(X)$,
where $\overline{\delta}[\mathcal{V}_a \mapsto 0]$ is defined as $\overline{\delta}[\mathcal{V}_a \mapsto 0](x) = 0$ for any $x \in \mathcal{V}_a$ and $\overline{\delta}[\mathcal{V}_a \mapsto 0](y) = \overline{\delta}(y)$ for any $y \in \mathcal{V}/\mathcal{V}_a$.

Similarly we can define the symbolic B-satisfiability relation $\models^s_B$ as in Section 3:

$$\mathcal{W}^s, \pi \models^s_B \phi \;\text{ iff }\; \mathcal{W}^s, \pi, [\![B]\!] \models^s \phi.$$

Lemma 1. *Let* ϕ *be a RWL formula closed w.r.t a maximal equation block* B*. If it is satisfied by a symbolic LWS* $\mathcal{W}^s = (\Pi^s, \Sigma^s, \theta^s)$ *i.e.* $\mathcal{W}^s, \pi \models^s_B \phi$ *with* $\pi = (m, \overline{\delta}) \in \Pi^s$*, then there exists a LWS* $\mathcal{W} = (M, \Sigma, \mathcal{K}, \theta)$ *and a resource valuation* $l \in L$ *such that* $\mathcal{W}, (m, l) \models_B \phi$ *with* $m \in M$*.*

Proof. Let $\Sigma = \Sigma^s$, $\mathcal{K}$ be the set of the resources appearing in $\mathcal{R}^\mathcal{V}$ and $l \in \overline{\delta}$. The transition function is defined as: $(m_1, \overline{\delta_1}, l_1) \xrightarrow{\overline{u}}_a (m_2, \overline{\delta_2}, l_2)$ iff,

$$(m_1, \overline{\delta_1}) \longrightarrow_a (m_2, \overline{\delta_2}), \text{ for } i = 1, 2, l_i \in \overline{\delta_i} \text{ and } l_2 = (l_1 + \overline{u})[\mathcal{V}_a \mapsto 0].$$

We define the transition relation starting from $(m, \overline{\delta}, l)$. Let M be the set of all the states in the form of $(m', \overline{\delta'}, l')$ defined for the transitions as the above. Note that it might be infinite. It is easy to verify that $\mathcal{W}, ((m, \overline{\delta}, l), l) \models_B \phi$. ∎

5 Satisfiability of Recursive Weighted Logic

In this section, we prove that it is decidable whether a given formula ϕ which is closed w.r.t. a maximal equation block B of RWL is satisfiable. We also present a decision procedure for the satisfiability problem of RWL. The results rely on a syntactic characterization of satisfiability that involves a notion of *mutually-consistent sets* that we define later.

Consider an arbitrary formula $\phi \in \mathcal{L}$ which is closed w.r.t. a maximal equation block B. In this context we define the following notions:

- Let $\Sigma[\phi, B]$ be the set of all actions $a \in \Sigma$ such that a appears in some transition modality of type $\langle a \rangle$ or $[a]$ in ϕ or B.

- For any $e_i \in \mathcal{K}$ and $x \in \mathcal{V}_i$, let $Q_i[\phi, B] \subseteq \mathbb{Q}_{\geq 0}$ be the set of all $r \in \mathbb{Q}_{\geq 0}$ such that r is in the label of some state or transition modality of type $x \bowtie r$ that appears in the syntax of ϕ or B.

- We denoted by g_i the *granularity of e_i in ϕ*, defined as the least common denominator of the elements of $Q_i[\phi, B]$. Let $R^{\mathcal{V}_i}_{p_i/g_i}[\phi, B]$ be the set of all regions for resource e_i, where $\frac{p_i}{g_i} = \max Q_i[\phi, B]$. Let

$$\mathcal{R}^{\mathcal{V}}[\phi, B] = \{\overline{\delta} = (\delta_1, \ldots, \delta_k) \mid \delta_i \in R^{\mathcal{V}_i}_{p_i/g_i}[\phi, B] \text{ for any } i \in \{1, \ldots, k\}\}.$$

For $r \in \mathbb{R}_{\geq 0}$, we use $r \in \overline{\delta}(x)$ to denote $r \in \delta_i(x)$, for any $i \in \{1, \ldots, k\}$ and $x \in \mathcal{V}_i$.

Observe that $\Sigma[\phi, B]$, $Q_i[\phi, B]$, $R^{\mathcal{V}_i}_{p_i/g_i}[\phi, B]$ and $\mathcal{R}^{\mathcal{V}}[\phi, B]$ are all finite (or empty).

At this point we can start our model construction. We fix a formula $\phi_0 \in \mathcal{L}$ that is closed w.r.t. a given maximal equation block B and, supposing that the formula admits a model, we construct a model for it. Let

$$\mathcal{L}[\phi_0, B] = \{\phi \in \mathcal{L} \mid \Sigma[\phi, B] \subseteq \Sigma[\phi_0, B], Q_i[\phi, B] \subseteq Q_i[\phi_0, B]\}.$$

Here we are going to construct a *symbolic model* first. To construct the symbolic model we will use as symbolic states tuples of type $(\Gamma, \overline{\delta}) \in 2^{\mathcal{L}[\phi_0, B]} \times \mathcal{R}^{\mathcal{V}}[\phi_0, B]$, which are required to be maximal in a precise way. The intuition is that a state $(\Gamma, \overline{\delta}) \subseteq 2^{\mathcal{L}[\phi_0, B]} \times \mathcal{R}^{\mathcal{V}}[\phi_0, B]$ shall symbolically satisfy the formula ϕ in our model, whenever $\phi \in \Gamma$. From this symbolic model we can generalize a LWS - might be infinite - which is a model of the given RWL formula. Our construction is inspired from the region construction proposed in [LLW95] for timed automata, which adapts of the classical filtration-based model construction used in modal logics [HC96, HKT01, Wal00].

Let $\Omega[\phi_0, B] \subseteq 2^{\mathcal{L}[\phi_0, B]} \times \mathcal{R}^{\mathcal{V}}[\phi_0, B]$. Since $\mathcal{L}[\phi_0, B]$ and $\mathcal{R}^{\mathcal{V}}[\phi_0, B]$ are both finite, $\Omega[\phi_0, B]$ is finite.

Definition 8. *For any $(\Gamma, \overline{\delta}) \subseteq \Omega[\phi_0, B]$, $(\Gamma, \overline{\delta})$ is said to be* maximal *iff:*
1. $\top \in \Gamma$, $\bot \notin \Gamma$;
2. $x \bowtie r \in \Gamma$ *iff for any* $w \in \mathbb{R}_{\geq 0}$ *s.t.* $w \in \overline{\delta}(x)$, $w \bowtie r$;
3. $\phi \wedge \psi \in \Gamma$ *implies* $\phi \in \Gamma$ *and* $\psi \in \Gamma$;
   $\phi \vee \psi \in \Gamma$ *implies* $\phi \in \Gamma$ *or* $\psi \in \Gamma$;
4. $X \in \Gamma$ *implies* $\phi \in \Gamma$, *for* $X = \phi \in B$.

The following definition establishes the framework on which we will define our model.

Definition 9. *Let $C \subseteq 2^{\Omega[\phi_0, B]}$. C is said to be mutually-consistent if for any $(\Gamma, \overline{\delta}) \in C$, whenever $\langle a \rangle \psi \in \Gamma$, then there exists $(\Gamma', \overline{\delta'}) \in C$ s.t.:*
1. *there exists* $\overline{\delta''}$ *s.t.* $\overline{\delta} \rightsquigarrow \overline{\delta''}$ *and* $\overline{\delta'} = \overline{\delta''}[\mathcal{V}_a \mapsto 0]$;
2. $\psi \in \Gamma'$;
3. *for any* $[a]\psi' \in \Gamma$, $\psi' \in \Gamma'$.

We say that $(\Gamma, \overline{\delta})$ is *consistent* if it belongs to some mutually-consistent set.

The following lemma proves a necessary precondition for the model construction.

Lemma 2. *Let $\phi \in \mathcal{L}$ be a formula closed w.r.t. a maximal equation block B. Then ϕ is satisfiable iff there exist $\Gamma \subseteq \mathcal{L}[\phi_0, B]$ and $\overline{\delta} \in \mathcal{R}^{\mathcal{V}}[\phi_0, B]$ s.t. $(\Gamma, \overline{\delta})$ is consistent and $\phi \in \Gamma$.*

Proof. ($\Longrightarrow$): Suppose ϕ is satisfied in the LWS $\mathcal{W} = (M, \Sigma, \mathcal{K}, \theta)$ under the resource valuation $l \in L$, i.e., there exists $m \in M$ s.t. $\mathcal{W}, (m, l) \models_B \phi$. We construct

$$C = \{(\Gamma, \overline{\delta}) \in \Omega[\phi_0, B] \mid \exists m \in M \text{ s.t. for any } \psi \in \Gamma, \exists l \in \overline{\delta} \text{ s.t. } \mathcal{W}, (m, l) \models_B \psi\}.$$

It is not difficult to verify that C is a mutually-consistent set.

($\Longleftarrow$): Let C be a mutually-consistent set.

We construct a symbolic LWS $\mathcal{W}^s = (\Pi^s, \Sigma^s, \theta^s)$, where $\Pi^s = C$, $\Sigma^s = \Sigma[\phi_0, B]$ and for $(\Gamma, \overline{\delta}), (\Gamma', \overline{\delta'}) \in C$, the transition relation $(\Gamma, \overline{\delta}) \longrightarrow_a (\Gamma', \overline{\delta'})$ is defined iff

1. there exists $\overline{\delta''} \in \mathcal{R}^{\mathcal{V}}[\phi_0, B]$ s.t. $\overline{\delta} \rightsquigarrow \overline{\delta''}$ and $\overline{\delta'} = \overline{\delta''}[\mathcal{V}_a \mapsto 0]$;
2. whenever $[a]\psi \in \Gamma$ then $\psi' \in \Gamma'$.

Let $\rho^s(X) = \{(\Gamma, \overline{\delta}) \mid X \in \Gamma\}$ for $X \in \mathcal{X}$. With this construction we can prove the following implication by a simple induction on the structure of ϕ, where $(\Gamma, \overline{\delta}) \in \Pi^s$:

$$\phi \in \Gamma \text{ implies } \mathcal{W}^s, (\Gamma, \overline{\delta}), \rho^s \models^s \phi.$$

We prove that ρ^s is a fixed point of B under the assumption that $X = \phi_X \in B$:

$\Gamma \in \rho^s(X)$ implies $(X, \overline{\delta}) \in \Gamma$ by the construction of ρ^s, which implies $(\phi_X, \overline{\delta}) \in \Gamma$ by the definition of $\Omega[\phi_0, B]$. Then, by the implication we just proved, $\mathcal{W}^s, \Gamma, \rho^s \models^s \phi_X$.

Thus ρ^s is a fixed point of B. Since $[\![B]\!]$ is the maximal fixed point, $\rho^s \subseteq [\![B]\!]$.

Then for any $(\phi, \overline{\delta}) \in \Gamma \in C$, we have $\mathcal{W}^s, (\Gamma, \overline{\delta}), \rho^s \models^s \phi$, which further implies $\mathcal{W}^s, (\Gamma, \overline{\delta}), [\![B]\!] \models^s \phi$ because $\rho^s \subseteq [\![B]\!]$.

Hence, $\phi \in \Gamma$ and $(\Gamma, \overline{\delta}) \in C$ implies $\mathcal{W}^s, \Gamma \models^s_B \phi$.

By Lemma 1, there exists a LWS $\mathcal{W} = (M, \Sigma, \mathcal{K}, \theta)$ and a resource valuation $l \in L$ such that $\mathcal{W}, (m, l) \models_B \phi$ with $m \in M$. ∎

To summarize, the above lemmas allow us to conclude the model constructions.

Theorem 1. *For any satisfiable RWL formula ϕ closed w.r.t. a maximal equation block B, there exists a finite symbolic LWS $\mathcal{W}^s = (\Pi^s, \Sigma^s, \theta^s)$ such that $\mathcal{W}^s, \pi \models^s_B \phi$ for some $\pi \in \Pi^s$. Reversely, if a RWL formula ϕ is satisfied by a symbolic model, then it is satisfiable, i.e., there exists a LWS $\mathcal{W} = (M, \Sigma, \mathcal{K}, \theta)$ and resource valuation $l \in L$ such that $\mathcal{W}, (m, l) \models_B \phi$ for some $m \in M$.*

Lemma 2 and Theorem 1 provide a decision procedure for the satisfiability problem of RWL. Given a RWL formula ϕ_0 closed w.r.t. a maximal equation block B, the algorithm constructs the model

$$\mathcal{W} = (M, \Sigma, \mathcal{K}, \theta).$$

To do this, we first construct the symbolic LWS

$$\mathcal{W}^s = (\Pi^s, \Sigma^s, \theta^s)$$

where $\Sigma^s = \Sigma[\phi_0, B]$, with ϕ_0 is satisfied in some state $\pi \in \Pi^s$, i.e. $\mathcal{W}^s, \pi \models^s_B \phi_0$.

If ϕ_0 is satisfiable, then it is contained in some maximal set Γ, where $(\Gamma,\overline{\delta})$ is consistent together with some $\overline{\delta} \in \mathcal{R}^{\mathcal{V}}[\phi_0, B]$. Hence, ϕ_0 will be satisfied at some state π of $\mathcal{W}^s$. If ϕ_0 is not satisfiable, then the attempt to construct a model will fail; in this case the algorithm will halt and report the failure.

We start with a superset of the set of states of $\mathcal{W}$, then repeatedly delete states when we discover some inconsistency. This will give a sequence of approximations

$$\mathcal{W}_0^s \supseteq \mathcal{W}_1^s \supseteq \mathcal{W}_2^s \supseteq \ldots$$

converging to $\mathcal{W}^s$.

The domains Π_i^s, $i = 0, 1, 2, \ldots$, of these structures are defined below and they are s.t.

$$\Pi_0^s \supseteq \Pi_1^s \supseteq \Pi_2^s \supseteq \ldots.$$

The transition relation for $\mathcal{W}_i^s$ are defined as follows: for any $(\Gamma,\overline{\delta}), (\Gamma',\overline{\delta'}) \in \Pi_i^s$, $(\Gamma,\overline{\delta}) \longrightarrow_a (\Gamma',\overline{\delta'})$ iff

1. there exists $\overline{\delta''} \in \mathcal{R}^{\mathcal{V}}[\phi_0, B]$ s.t. $\overline{\delta} \rightsquigarrow \overline{\delta''}$ and $\overline{\delta'} = \overline{\delta''}[\mathcal{V}_a \mapsto 0]$;
2. whenever $[a]\psi \in \Gamma$ then $\psi \in \Gamma'$.

Here is the algorithm for constructing the domains Π_i^s of $\mathcal{W}_i^s$.

Algorithm

- **Step 1:** Construct $\Pi_0^s = \Omega[\phi_0, B]$.
- **Step 2:** Repeat the following for $i = 0, 1, 2, \ldots$ until no more states are deleted. Find a formula $[a]\phi \in \mathcal{L}[\phi_0, B]$ and a state $(\Gamma,\overline{\delta}) \in \Pi_i^s$ violating the following property

$$[\forall(\Gamma',\overline{\delta'}) \in \Pi_i^s, (\Gamma,\overline{\delta}) \longrightarrow_a (\Gamma',\overline{\delta'}) \Rightarrow \phi \in \Gamma'] \text{ implies } [a]\phi \in \Gamma.$$

that is, such that $\langle a\rangle\neg\phi \in \Gamma$, but for no Γ' and $\overline{\delta'}$ such that $(\Gamma,\overline{\delta}) \longrightarrow_a (\Gamma',\overline{\delta'})$ is it the case that $\neg\phi \in \Gamma'$.

Pick such an $[a]\phi$ and $(\Gamma,\overline{\delta})$. Delete $(\Gamma,\overline{\delta})$ from Π_i^s to get Π_{i+1}^s.

Step 2 can be justified intuitively as follows. To say that $(\Gamma,\overline{\delta})$ violates the above mentioned condition, it means that $(\Gamma,\overline{\delta})$ requires an a-transition to some state that does not satisfy ϕ; however, the left-hand side of the condition above guarantees that all the outcomes of an a-transition satisfy ϕ. This demonstrates that $(\Gamma,\overline{\delta})$ can not be in Π^s, since every state $(\Gamma,\overline{\delta})$ in Π^s satisfies ψ, whenever $\psi \in \Gamma$.

The algorithm must terminate, since there are only finitely many states initially, and at least one state must be deleted during each iteration of step 2 in order to continue. Then, ϕ is satisfiable if and only if, upon termination there exists $(\Gamma,\overline{\delta}) \in \Pi^s$ such that $\phi \in \Gamma$. Obviously, Π^s is a mutually-consistent set upon termination.

The correctness of this algorithm follows from the proof of Lemma 2. The direction ($\Leftarrow$) of the proof guarantees that all formulas in any Γ with $(\Gamma,\overline{\delta}) \in \Pi^s$ are satisfiable. The direction ($\Rightarrow$) of the proof guarantees that all satisfiable Γ will not be deleted from Π^s.

After we get the symbolic LWS $\mathcal{W}^s$, we can use the technique in Lemma 1 to generalize a LWS $\mathcal{W} = (M, \Sigma, \mathcal{K}, \theta)$, which might be infinite.

Suppose ϕ is satisfied by $(\Gamma,\overline{\delta}) \in \Pi^s$, i.e., $\mathcal{W}^s, (\Gamma,\overline{\delta}) \models_B^s \phi_0$. Let $\Sigma = \Sigma^s$, $\mathcal{K}$ be the set of the resources appearing in $\mathcal{R}^{\mathcal{V}}[\phi_0, B]$ and $l \in \overline{\delta}$. The transition function is defined as: $(\Gamma_1, \overline{\delta_1}, l_1) \xrightarrow{\overline{u}}_a (\Gamma_2, \overline{\delta_2}, l_2)$ iff,

$$(\Gamma_1, \overline{\delta_1}) \longrightarrow_a (\Gamma_2, \overline{\delta_2}), \text{ for } l_1 \in \overline{\delta_1}, l_2 \in \overline{\delta_2} \text{ and } l_2 = (l_1 + \overline{u})[\mathcal{V}_a \mapsto 0].$$

We define the transition relation starting from $(\Gamma, \overline{\delta}, l)$. Let M be the set of all the states in the form of $(\Gamma', \overline{\delta'}, l')$ defined for the transitions as the above. Note that the model defined as above might be infinite. It is easy to verify that $\mathcal{W}, ((\Gamma, \overline{\delta}, l), l) \models_B \phi_0$.

Theorem 1, also supported by the above algorithm, demonstrates the decidability of the B-satisfiability problem for RWL.

Theorem 2 (Decidability of B-Satisfiability). *For an arbitrary maximal equation block B, the B-satisfiability problem for RWL is decidable.*

Example 4. Now we can discuss the satisfiability of the formula ϕ in Example 3.

$$\phi = \langle a \rangle X,$$
$$B = \begin{Bmatrix} X = x_a < 1 \land \langle b \rangle (Y \land x_c > 0), \\ Y = x_a < 1 \land \langle c \rangle (X \land x_b > 0) \end{Bmatrix}$$

In Figure3 is the symbolic LWS for the above formula ϕ w.r.t B by applying our algorithm. Here the details of using the algorithm to get the model are not presented, limited by the length of the paper, which is very technical.

$\Gamma_0 = \{\phi, \langle a \rangle X\}$
$\Gamma_1 = \{X, x_a < 1, \langle b \rangle (Y \land x_c > 0)\}$
$\Gamma_2 = \{Y, x_c > 0, x_a < 1, \langle c \rangle (X \land x_b > 0)\}$
$\Gamma_3 = \{X, x_b > 0, x_a < 1, \langle b \rangle (Y \land x_c > 0)\}$
$\delta_0 = [x_a = x_b = x_c = 0]$
$\delta_1 = [x_a = 0, 0 < x_b = x_c < 1]$
$\delta_2 = [x_b = 0, 0 < x_a = x_c < 1]$
$\delta_3 = [x_b = 0, 0 < x_a < x_c < 1]$
$\delta_4 = [x_c = 0, 0 < x_b < x_a < 1]$
$\delta_5 = [x_b = 0, 0 < x_c < x_a < 1]$

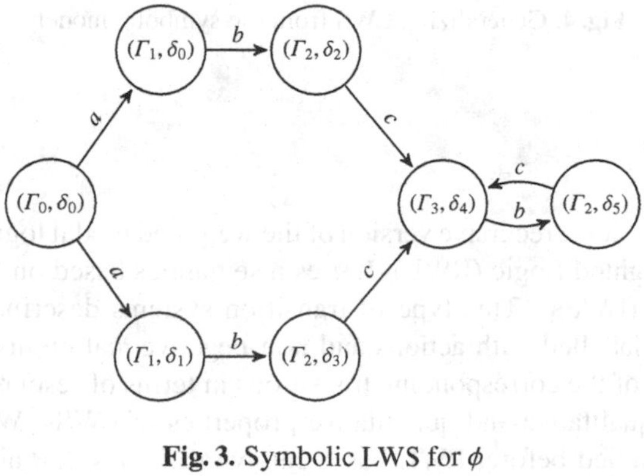

Fig. 3. Symbolic LWS for ϕ

From the symbolic model in Figure 3, one can generate a LWS, which in this case is infinite, where ϕ is satisfied in some state of it. In Figure 4, we show part of this infinite model.

$$
\begin{array}{ll}
l_0 = (0,0,0) & l_5 = (0.2,0,0.3) \\
l_1 = (0.1,0.1,0) & l_6 = (0.5,0.2,0) \\
l_2 = (\frac{\pi}{4},\frac{\pi}{4},0) & l_7 = (0.4,0.2,0) \\
l_3 = (0.3,0,0.3) & l_8 = (0.6,0,0.1) \\
l_4 = (\frac{\pi}{4},0,\frac{\pi}{4}) & l_9 = (0.5,0,0.1) \\
\ldots\ldots &
\end{array}
$$

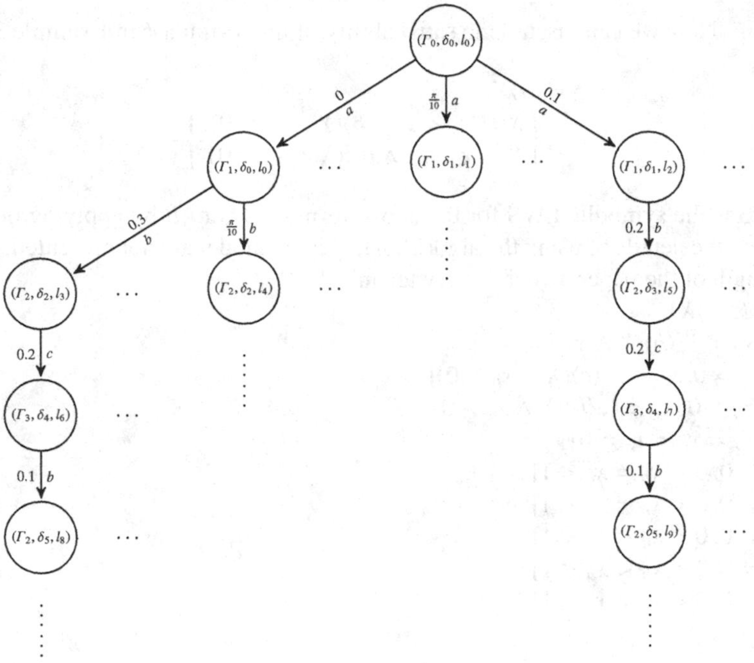

Fig. 4. Generalizing LWS from the symbolic model

6 Conclusion

In this paper we develop a recursive version of the weighted modal logic [LM13] that we call Recursive Weighted Logic (RWL). It uses a semantics based on labelled weighted transition systems (LWSs). This type of transition systems describes systems where the transitions are labelled with actions and non-negative real numbers. The numbers represent the costs of the corresponding transitions in terms of resources.

RWL encodes qualitative and quantitative properties of LWSs. With respect to the weighted logics studied before, RWL has recursive variables that allow us to encode circular properties of infinite behaviours including safety and cost-bounded liveness properties.

Even though RWL does not enjoy the finite model property, we proved that the satisfiability problem for our logic is still decidable. By adapting a variant of region construction technique for timed automata, we constructed a symbolic LWS for a given satisfiable RWL formula. This allows us to decide the existence of a concrete LWS and generate it. The system is possibly infinite but nevertheless it is a model for the given formula. Our decidability result is important and somehow surprising, being that the satisfiability problem is undecidable for the similar timed logics for real-time systems.

References

[ACD90] Alur, R., Courcoubetis, C., Dill, D.L.: Model-checking for real-time systems. In: LICS, pp. 414–425 (1990)

[ACD93a] Alur, R., Courcoubetis, C., Dill, D.L.: Model-checking in dense real-time. Inf. Comput. 104(1), 2–34 (1993)

[ACD93b] Alur, R., Courcoubetis, C., Dill, D.L.: Model-checking in dense real-time. Information and Computation 104(1), 2–34 (1993)

[AD90] Alur, R., Dill, D.L.: Automata for modeling real-time systems. In: Paterson, M. (ed.) ICALP 1990. LNCS, vol. 443, pp. 322–335. Springer, Heidelberg (1990)

[AILS07] Aceto, L., Ingólfsdóttir, A., Larsen, K.G., Srba, J.: Reactive Systems: modelling, specification and verification. Cambridge University Press (2007)

[ATP01] Alur, R., Torre, S.L., Pappas, G.J.: Optimal paths in weighted timed automata. In: Benedetto and Sangiovanni-Vincentelli [BSV01], pp. 49–62

[BFH+01] Behrmann, G., Fehnker, A., Hune, T., Larsen, K.G., Pettersson, P., Romijn, J., Vaandrager, F.W.: Minimum-cost reachability for priced timed automata. In: Benedetto and Sangiovanni-Vincentelli [BSV01], pp. 147–161

[BSV01] Di Benedetto, M.D., Sangiovanni-Vincentelli, A.L.: HSCC 2001. LNCS, vol. 2034. Springer, Heidelberg (2001)

[CKS92] Cleaveland, R., Klein, M., Steffen, B.: Faster model checking for the modal mu-calculus. In: Probst, D.K., von Bochmann, G. (eds.) CAV 1992. LNCS, vol. 663, pp. 410–422. Springer, Heidelberg (1993)

[CS93] Cleaveland, R., Steffen, B.: A linear-time model-checking algorithm for the alternation-free modal mu-calculus. Formal Methods in System Design 2(2), 121–147 (1993)

[DKV09] Droste, M., Kuich, W., Vogler, H. (eds.): Handbook of Weighted Automata. Springer (2009)

[HC96] Hughes, G.E., Cresswell, M.J.: A New Introduction to Modal Logic. Routhledge, London (1996)

[HKT01] Harel, D., Kozen, D., Tiuryn, J.: Dynamic Logic. The MIT Press (2001)

[HNSY92] Henzinger, T.A., Nicollin, X., Sifakis, J., Yovine, S.: Symbolic model checking for real-time systems. In: LICS, pp. 394–406 (1992)

[Lar90] Larsen, K.G.: Proof systems for satisfiability in Hennessy-Milner logic with recursion. Theor. Comput. Sci. 72(2&3), 265–288 (1990)

[LLW95] Laroussinie, F., Larsen, K.G., Weise, C.: From timed automata to logic - and back. In: Hájek, P., Wiedermann, J. (eds.) MFCS 1995. LNCS, vol. 969, pp. 529–539. Springer, Heidelberg (1995)

[LM13] Larsen, K.G., Mardare, R.: Complete proof system for weighted modal logic. In: Theoretical Computer Science (2013) (in press)

[LMX] Larsen, K.G., Mardare, R., Xue, B.: Adequacy and strongly-complete axiomatization for timed modal logic (under review)

[LMX14] Larsen, K.G., Mardare, R., Xue, B.: Decidability and expressiveness of recursive weighted logic. In: Ershov Informatics Conference (PSI) (to appear, 2014)

[Tar55] Tarski, A.: A lattice-theoretical fixpoint theorem and its applications. Pacific Journal of Mathematics 5(2), 285–309 (1955)

[Wal00] Walukiewicz, I.: Completeness of Kozen's axiomatisation of the propositional μ-calculus. Inf. Comput. 157(1-2), 142–182 (2000)

Author Index